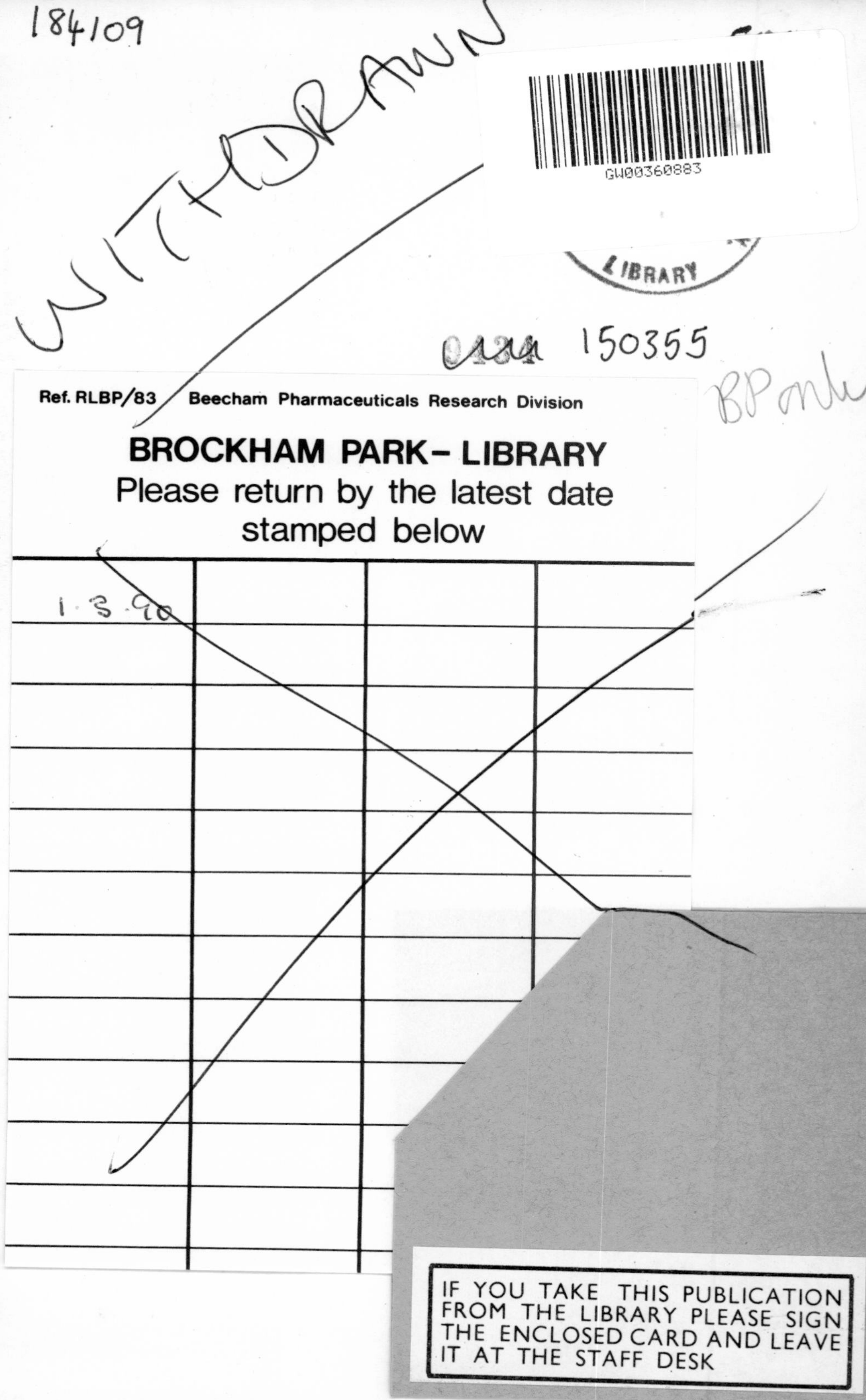

Advances in Natural Products Chemistry

Extraction and Isola tion of Biologically Active Compounds

Advances in Natural Products Chemistry

Extraction and Isolation of Biologically Active Compounds

Edited by
Sinsaku NATORI, Nobuo IKEKAWA and Makoto SUZUKI

A HALSTED PRESS BOOK

KODANSHA LTD.
Tokyo

JOHN WILEY & SONS
New York-Chichester-Brisbane-Toronto

Library of Congress Cataloging in Publication Data
Main entry under title:

Advances in natural products chemistry.

(Kodansha scientific books)
"Revised version of a book originally published in Japanese in 1977"—Pref.
"A Halsted Press book"
Includes indexes.
1. Biological products. 2. Natural products.
I. Natori, Shinsaku, 1923- . II. Ikekawa, Nobuo, 1926- . III. Suzuki, Makoto, 1929- IV. Title: Biologically active compounds. V. Series.
QH345.A29 1981 574.19'2 81-6637
ISBN 0-470-27245-7 (Halsted) AACR2

Published in Japan by

KODANSHA LTD.
12–21 Otowa 2-chome, Bunkyo-ku, Tokyo 112, Japan

Published by

HALSTED PRESS
a Division of John Wiley & Sons, Inc.
605 Third Avenue, New York, N.Y. 10016, U.S.A.

PRINTED IN JAPAN

List of Contributors

Numbers are the chapter numbers to which the author(s) contributed.
*: Editors; (c) refers to co-author(s)

Norio AIMI	Faculty of Pharmaceutical Sciences, Chiba University, *Chiba 260, Japan*	24 (c)
Takaaki AOYAGI	Institute of Microbial Chemistry, *Tokyo 141, Japan*	4
Shin-ichi AYABE	School of Pharmaceutical Sciences, Kitasato University, *Tokyo 108, Japan*	26 (c)
Hiroshi FUKAMI	Pesticide Research Institute, College of Agriculture, Kyoto University, *Kyoto 606, Japan*	31 (c)
Hiroshi FUKUI	Dept. of Pharmacognosy, Faculty of Pharmaceutical Sciences, Kyoto University, *Kyoto 606, Japan*	15 (c)
Tsutomu FURUYA	School of Pharmaceutical Sciences, Kitasato University, *Tokyo 108, Japan*	26 (c)
Nobuhiro FUSETANI	Laboratory of Marine Biochemistry, Faculty of Agriculture, the University of Tokyo, *Tokyo 113, Japan*	34
Minoru GOTO	Central Research Division, Kyoto Herbal Garden, Takeda Chemical Industries Co., Ltd., *Kyoto 606, Japan*	29 (c)
Toshio GOTO	Dept. of Food Science and Technology, Faculty of Agriculture, Nagoya University, *Nagoya 464, Japan*	37 (c)
Hiroshi HIKINO	Pharmaceutical Institute, Tohoku University, *Sendai 980, Japan*	32
Yoshimasa HIRATA	Faculty of Pharmacy, Meijo University, *Nagoya 468, Japan*	23
Nobuo IKEKAWA*	Laboratory of Chemistry for Natural Products, Tokyo Institute of Technology, *Yokohama 227, Japan*	33
Shoji INOUE	Faculty of Pharmacy, Meijo University, *Nagoya 468, Japan*	37 (c)
Hiroyuki INOUYE	Faculty of Pharmaceutical Sciences, Kyoto University, *Kyoto 606, Japan*	19
Kiyoshi ISONO	The Institute of Physical and Chemical Research, *Wako-shi 351, Japan*	2

Tadahiro KATO	Dept. of Chemistry, Faculty of Science, Tohoku University, *Sendai 980, Japan*	14 (c)
Toshio KAWASAKI	Laboratory of Plant Chemistry, Faculty of Pharmaceutical Sciences, Kyushu University, *Fukuoka 812, Japan*	21
Kazuyoshi KAWAZU	Dept. of Agricultural Chemistry, Faculty of Agriculture, Okayama University, *Okayama 700, Japan*	18
Isao KITAGAWA	Faculty of Pharmaceutical Sciences, Osaka University, *Suita 565, Japan*	22
Mineo KOJIMA	Institute for Biochemical Regulation, Faculty of Agriculture, Nagoya University, *Nagoya 464, Japan*	13 (c)
Koichi KOSHIMIZU	Dept. of Food Science and Technology, Faculty of Agriculture, Kyoto University, *Kyoto 606, Japan*	15 (c)
Takashi KUBOTA	School of Medicine, Kinki University, *Osaka 589, Japan*	30 (c)
Esturo KUROSAWA	Dept. of Chemistry, Faculty of Science, Hokkaido University, *Sapporo 060, Japan*	12
Akio MIYAKE	Instituto Zoolgia, Universitá di Pisa	30 (c)
Katsura MUNAKATA	Dept. of Agricultural Chemistry, Faculty of Agriculture, Nagoya University, *Nagoya 464, Japan*	17 (c)
Tadashi NAKAJIMA	Research Laboratories, Nippon Shoji Kaisha, Ltd., *Ibaraki 567, Japan*	10 (c)
Koji NAKANISHI	Dept. of Chemistry, Columbia University, *New York 10027, U.S.A.*	39
Toshio NAMBARA	Pharmaceutical Institute, Tohoku University, *Sendai 980, Japan*	38 (c)
Shinsaku NATORI*	National Institute of Hygienic Sciences, *Tokyo 158, Japan*	8 (c)
Ritsuo NISHIDA	Pesticide Research Institute, College of Agriculture, Kyoto University, *Kyoto 606, Japan*	31 (c)
Itsuo NISHIOKA	Faculty of Pharmaceutical Sciences, Kyushu University, *Fukuoka 812, Japan*	27
Haruji OSHIO	Central Research Division, Pharmacognostic Research Laboratories, Takeda Chemical Industries Co., Ltd., *Osaka 532, Japan*	29 (c)
Kiyoshi SAKAI	Central Research Laboratories, Sankyo Co., Ltd., *Tokyo 140, Japan*	36 (c)
Shin-ichiro SAKAI	Faculty of Pharmaceutical Sciences, Chiba University, *Chiba 260, Japan*	24 (c)
Sadao SAKAMURA	Dept. of Agricultural Chemistry, Faculty of Agriculture, Hokkaido University, *Sapporo 060, Japan*	7
Shoji SHIBATA	Meiji College of Pharmacy, *Tokyo 154, Japan*	28
Kazutake SHIMADA	Pharmaceutical Institute, Tohoku University, *Sendai 980, Japan*	38 (c)
Yuzuru SHIMIZU	Dept. of Pharmacognosy and Environmental Health Sciences, College of Pharmacy, University of Rhode Island, *Rhode Island 02881, U.S.A.*	11

Junzo SHOJI	School of Pharmaceutical Sciences, Showa University, *Tokyo 142, Japan*	20
Akinori SUZUKI	Dept. of Agricultural Chemistry, Faculty of Agriculture, The University of Tokyo, *Tokyo 113, Japan*	5 (c)
Kazuo T. SUZUKI	National Institute for Environmental Studies, *Tsukuba-gun 300–21, Japan*	9
Makoto SUZUKI*	Instrumental Analytical Chemistry, Faculty of Pharmacy, Meijo University, *Nagoya 468, Japan*	1
Ryuji TACHIKAWA	Central Research Laboratories, Sankyo Co., Ltd., *Tokyo 140, Japan*	36 (c)
Nobutaka TAKAHASHI	Dept. of Agricultural Chemistry, Faculty of Agriculture, the University of Tokyo, *Tokyo 113, Japan*	16
Norindo TAKAHASHI	Agricultural Laboratories, Tohoku University, *Sendai 980, Japan*	14 (c)
Yoshio TAKEDA	Faculty of Pharmaceutical Sciences, Tokushima University, *Tokushima 770, Japan*	
Tsunematsu TAKEMOTO	Parmaceutical Faculty, Tokushima Bunri University, *Tokushima 770, Japan*	10 (c)
Tomohisa TAKITA	Institute of Microbial Chemistry, *Tokyo 141, Japan*	3
Takashi TOKOROYAMA	Dept. of Chemistry, Faculty of Science, Osaka City University, *Osaka 558, Japan*	30 (c)
Saburo TAMURA	Emeritus Professor of the University of Tokyo, *Tokyo 113, Japan*	5 (c)
Tamio UENO	Pesticide Research Institute, College of Agriculture, Kyoto University, *Kyoto 606, Japan*	6
Makoto UMEDA	Yokohama City University School of Medical, *Yokohama 232, Japan*	8 (c)
Ikuzo URITANI	Lab. of Biochemistry, Faculty of Agriculture, Nagoya University, *Nagoya 464, Japan*	13 (c)
Kojiro WADA	Dept. of Agricultural Chemistry, Faculty of Agriculture, Nagoya University, *Nagoya 464, Japan*	17 (c)
Masami YOKOTA	Shizuoka College of Pharmacy, *Shizuoka 422, Japan*	25
Hiroshi ZENDA	Dept. of Pharmacy, Shinshu University Hospital, *Matsumoto 390, Japan*	35

Preface

The present book is a revised version of a book originally published in Japanese in 1977. It was conceived as a guide book to strategy in natural products research in the light of the remarkable progress in techniques of separation, physical methods, and procedures for structure elucidation during the past two decades. Our knowledge of the structures and reactions of secondary metabolites, and of the molecular basis of many biological phenomena has increased enormously during this time, and although many monographs and reviews of individual methods have been published, we considered that a broader guide to overall strategy might be useful. We therefore planned to collect typical examples of natural products research carried out in Japan, asking the authors to describe why a particular project was selected, how the materials were gathered and treated, and which points were most important in the extraction and detection procedures.

Studies on biologically active substances carried out during the past ten years were preferred. Some structural studies are included, but no synthetic work. We have also attempted to minimize overlapping of subject matter and techniques among the various authors.

In spite of limitations of space, many interesting reports on the contributors' approaches and techniques for the selection of materials, separation, detection and identification are presented, which we hope will stimulate further work in related fields. As an aid to the use of this book as a guide for experimental work, an index of experimental methods is included in addition to the subject index.

The editors are grateful to the contributors for their cooperation, and wish to record with regret the deaths of Professors Y. Kitahara and Y. Hashimoto during the preparation of this book. Thanks are also due to Mr. W.R.S. Steele and the staff of Kodansha for their linguistic and editorial assistance in preparing the final manuscript for publication.

March 1981

S. Natori
N. Ikegawa
M. Suzuki

Contents

CHAPTER **1**

Basic Macrolide Antibiotics

The term macrolide was coined by Woodward to describe a group of natural antibiotics, all of which possess a medium-sized lactone ring and one to three sugar moieties. Among these compounds, basic macrolides are widely used as chemotherapeutic agents active against gram-positive bacteria and mycoplasma (PPLO) strains.[1]

Basic macrolides can be classified into three major groups in terms of the ring size of the lactone moiety (12-, 14- and 16-membered rings); the basicity arises from the aminosugar moiety. Medically important basic macrolides include erythromycin and oleandomycin (14-membered group), and the leucomycin, josamycin and spiramycin series (16-membered group). Midecamycin, platenomycin, espinomycin and maridomycin have recently been discovered as members of the latter group, and studies on their structure-activity relationships and biosynthetic pathways have been made clear.[2]

The author has been engaged in chemical, biochemical and biosynthetic studies of platenomycin group compounds at the Microbial Chemistry Research Laboratory of Tanabe Seiyaku Co. Ltd., and this chapter describes some topics in the separation, purification and structure characterization of the basic macrolide, platenomycin.

Most of the 16-membered basic macrolides are produced by *Streptomyces* species. Platenomycin (PLM) is a product of *Streptomyces platensis* subsp. *malvinus* MCRL 0388, and two major products (A_1 and B_1) were isolated from cultured broth together with many minor components.

A general property of macrolide-producing strains seems to be the production of multiple components with rather closely related structures. Thus, effective separation is most important in the study of macrolide biochemistry.

In the case of PLM, silica gel chromatography was employed as the first separation step and thin-layer chromatography (tlc) with ultraviolet (uv) monitoring is effective for detection of the separated components.

The chemical structures of PLM components are summarized in Table 1.1. Three kinds of chromophores (I, II and III) are contained in the 16-membered lactone ring, which forms the skeleton of the macrolides. Chromophores I and II show strong uv absorption maxima (log $\varepsilon > 4.0$)

TABLE 1.1 Structures and diagnostic fragment ions of platenomycin components.

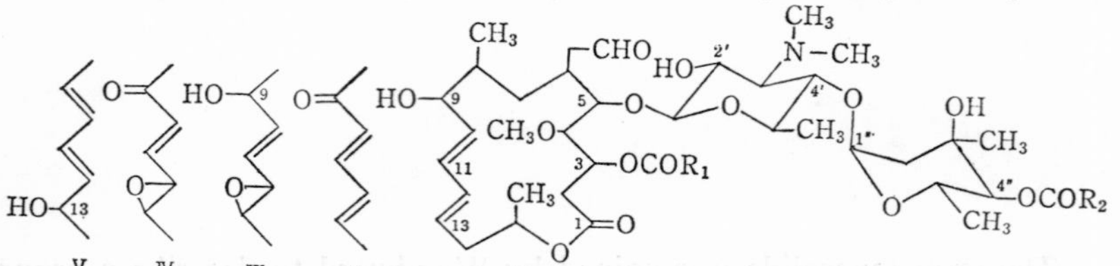

Platenomycin (PLM)	Chromophore	R_1	R_2	Diagnostic ions of acetyl derivatives				
				M^+	AGL^+	ADS^+	AMA^+	AM^+
PLM A_0	I	$-CH_2CH_2CH_3$	$-CH_2CH(CH_3)_2$	939	479	444	216	229
PLM A_1	I	$-CH_2CH_3$	$-CH_2CH(CH_3)_2$	925	465	444	216	229
PLM B_1	I	$-CH_2CH_3$	$-CH_2CH_3$	897	465	416	216	201
PLM C_2	I	$-CH_2CH_3$	$-CH_3$	883	465	402	216	187
PLM A_3	I	$-CH_3$	$-CH_2CH(CH_3)_2$	911	451	444	216	229
PLM B_3	I	$-CH_3$	$-CH_2CH_3$	883	451	416	216	201
Leucomycin A_3†[1]	I	$-CH_3$	$-CH_2CH(CH_3)_2$	911	451	444	216	229
PLM W_1	II	$-CH_2CH_3$	$-CH_2CH(CH_3)_2$	881	421	444	216	229
PLM W_2	II	$-CH_2CH_2CH_3$	$-CH_2CH(CH_3)_2$	895	435	444	216	229
Carbomycin B†[1]	II	$-CH_3$	$-CH_2CH(CH_3)_2$	867	407	444	216	229
PLM C_1	III	$-CH_2CH_3$	$-CH_2CH_3$	913	481	416	216	201
PLM C_3	III	$-CH_2CH_3$	$-CH_2CH(CH_3)_2$	941	481	444	216	229
PLM C_4	III	$-CH_3$	$-CH_2CH(CH_3)_2$	927	467	444	216	229
9-Dehydro-PLM C_1	IV	$-CH_2CH_3$	$-CH_2CH_3$	869	437	416	216	201
Carbomycin A†[1]	IV	$-CH_3$	$-CH_2CH(CH_3)_2$	883	423	444	216	229
PLM A_2	V	$-CH_2CH_3$	$-CH_2CH(CH_3)_2$	925	465	444	216	229
PLM B_2	V	$-CH_2CH_3$	$-CH_2CH_3$	897	465	416	216	201

†[1] Standard macrolide antibiotics.

†[2] Chromophore I, λ_{max}^{EtOH} 232 nm (log ε 4.35 — 4.45);
Chromophore II, λ_{max}^{EtOH} 280 nm (log ε 4.35 — 4.37);
Chromophore III, λ_{max}^{EtOH} end absorption;
Chromophore IV, λ_{max}^{EtOH} 239 nm (log ε 4.13 — 4.30);
Chromophore V, λ_{max}^{EtOH} 235 nm (log ε 4.22 — 4.36).

at 232 nm and 280 nm, respectively, but chromophore III shows only end absorption. However, active MnO_2 oxidation converts chromophore III into a new chromophore IV, which shows a strong maximum at 239 nm. Chromophore V (235 nm) is observed as an artifact of allyl rearrangement from chromophore I-containing components. These properties can be used to detect and identify the components on tlc plates.

The lactone-ring structure and the sequence of sugar moieties can readily be determined from the mass spectra, which also provide information regarding minor differences of structure. The sample size required for ms measurement is quite small (1–10 μg), so even extracts from tlc spots can be sufficient for ms analysis.

The study of PLM was started by separating thirteen components by silica gel/alumina chromatography and continued by a combination of tlc, uv and ms analysis, as mentioned above. The results are summarized in Table 1.1.

1.1 Separation and Identification of Platenomycins[3)]

Column Chromatography

The cultured broth was treated as summarized in Table 1.2, and the resulting crude PLM powder was then chromatographed on a silica gel column. Components separated with a benzene-acetone solvent system were clasified into four major groups (W, A, B and C) by silica gel tlc, and finally, nine fractions (I–IX) were obtained.

It was found that these fractions contained one to three components by alumina-kieselguhr tlc, and alumina column chromatography was employed for complete separation of the components except in the case of fraction II. For fraction II, final separation was carried out after acetylation. We finally obtained thirteen components belonging to the PLM series, as summarized in Tables 1.2 and 1.3.

PLM components thus separated, their final purities are tested by microbial assay (*B. subtilis* PCI 219) and UV absorbance at λ_{max} (nm).

[Experimental procedure 1] Silica gel column chromatography

Silica gel (Mallinckrodt; silicic acid), 1 kg, was charged in a glass column (100 cm × 8 cm) with benzene, and crude PLM white powder (14.5 g; cf. Table 1.2) in 300 ml of benzene was applied to the top of the column. After complete absorption of the sample, benzene (500 ml) was applied, and the column was developed with benzene-acetone (9:1,

TABLE 1.2 Isolation and separation of platenomycin components.

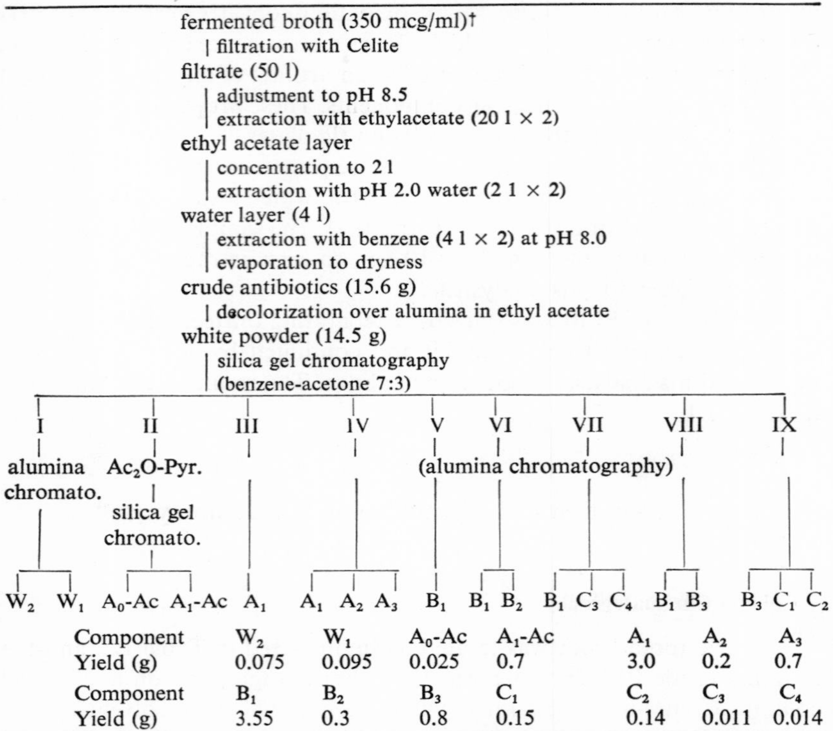

Component	W_2	W_1	A_0-Ac	A_1-Ac	A_1	A_2	A_3
Yield (g)	0.075	0.095	0.025	0.7	3.0	0.2	0.7
Component	B_1	B_2	B_3	C_1	C_2	C_3	C_4
Yield (g)	3.55	0.3	0.8	0.15	0.14	0.011	0.014

† Anti-microbial activity towards *Bacillus subtilis* PCI 219.

5 l) and benzene-acetone (7:3). The eluate was collected continuously in 50 ml fractions using an automatic fraction collecter. The results are shown in the left column of Table 1.3.

[Experimental procedure 2] Alumina column chromatography

For the separation of Fraction IV, 200 g of neutral alumina (Woelm; activity I) was charged in a 300 mm × 25 mm column with benzene, and Fraction IV (1.7 g) in 20 ml of benzene solution was loaded on the column. After benzene elution (200 ml), a benzene-ethyl acetate (7:3) system was used for separation. The eluate (5 ml fractions) was collected automatically and the separation proceeded as shown in the right column of Table 1.3.

For the separation of Fraction VIII, 80 g of neutral alumina (Woelm; activity III) was charged in the column with benzene, and Fraction VIII (1.0 g) in 15 ml of benzene solution was applied. After benzene elution

TABLE 1.3 Column chromatography for the purification of platenomycin.

Silica gel chromatography			Alumina chromatography				
Fraction number	Fraction	Yield (g)	Alumina (g)	Activity	Solvent system	Fraction number	Component
7–13	I	0.2	20	I	▲(7:3)	15–25 29–45	W_2 W_1
15–21	II	0.8	80		▽(85:15)	31–40 45–120	A_0-Ac A_1-Ac
22–61	III	2.5					A_1
62–84	IV	1.7	200	I	▲ (7:3)	61–85 111–125 134–160	A_1 A_2 A_3
86–130	V	2.8					B_1
131–150	VI	1.5	150	III	▽(8:2)	15–55 57–64	B_1 B_2
151–160	VII	0.2	40	III	▽(8:2)	16–45 47–50 52–56	B_1 C_3 C_4
161–182	VIII	1.0	80	III	▽(8:2)	15–35 47–71	B_1 B_3
183–200	IX	0.4	80	IV	▽(6:4)	21–35 36–45 46–55	B_3 C_1 C_2

▲ benzene-ethyl acetate ▽ benzene-acetone

(150 ml), benzene-acetone (8:2) was used for the separation. The eluate (5 ml fractions) was collected as before (Table 1.3).

The total yields of the thirteen PLM components thus obtained are summarized in Table 1.2.

Thin Layer Chromatography

Separation of platenomycin components: As shown in Table 1.1, the structures of PLM components differ as regards the individual chromophore, C-3 and C-4″ substituents. The C-4″ acyl homologs are easily separated on usual silica gel plates using benzene-acetone (2:1) or chloroform-methanol-acetic acid-water (79:11:8:2) as developing solvents. Components which have the same C-4″ but different C-3 acyl groups (e.g., (A_1/A_3 and B_1/B_3), and components in which both acyl groups are the same, but which contain slightly different chromophores (e.g., A_1/A_2 and B_1/B_2) cannot be completely separated on silica gel plates even if active alumina plates are used.

However, tlc plates made from alumina and kieselguhr (6:1) showed

TABLE 1.4 Thin-layer chromatography of platenomycin components (*Rf* values).

PLM component	*Rf*†1	*Rf*†2	*Rf*†3
W_2	0.56	0.92	0.72
W_1	0.37	0.88	0.68
A_0	0.22	0.75	0.51
A_1	0.20	0.72	0.50
A_2		0.60	0.48
A_3		0.53	0.43
B_1		0.58	0.36
B_2		0.45	0.34
B_3		0.38	0.33
C_1		0.30	0.29
C_2		0.43	0.30
C_3		0.41	0.39
C_4		0.29	0.35

†1 Alumina-kieselguhr: benzene-acetone (4:1).
†2 Alumina-kieselguhr: benzene-acetone (7:3).
†3 Silica gel: benzene-acetone (2:1).

good separation using a benzene-acetone solvent system. These plates are used without heat activation, and so contain a lot of water (9.2%; differential thermal analysis); this means that a kind of partition tlc occurs. Using these plates, we have obtained complete separation of PLM $A_1/A_2/A_3$ and $B_1/B_2/B_3$ (Table 1.4). These results made it possible to identify PLM minor components in the presence of other similar basic macrolides, e.g., many components of leucomycin and maridomycin.

Such success obtained by alumina-kieselguhr tlc has developed to the preparative separation of each PLM components using deactivated alumina column chromatography.

Detection and identification of the spots on tlc: All PLM components can be visualized as dark violet (silica gel) or purple (alumina) colored spots by spraying 40% sulfuric acid, followed by heating at 120° C. These color reactions are specific to the basic macrolide group, but cannot discriminate each component. A microbial technique using agar plates of *B. subtilis* is also available, but the identification of PLM components is rather difficult.

On the other hand, the characteristic uv absorption of the chromophore can be used to detect many kinds of PLM components.

Fig. 1.1 shows uv-monitoring traces on an alumina-kieselguhr tlc chromatogram (benzene-acetone, 7:3) for a mixture of PLM A_1, B_1, C_1 and W_1. When the uv spectrodensitometer (Hitachi MPF-2A with tlc scanning accessory) is set at 232 nm (λ_{max} of chromophore I; cf. Table 1.1), signals due to PLM A_1 and B_1 are observed at the expected positions. When it is set at 280 nm (chromophore II), a clear signal of PLM W_1 is seen. How-

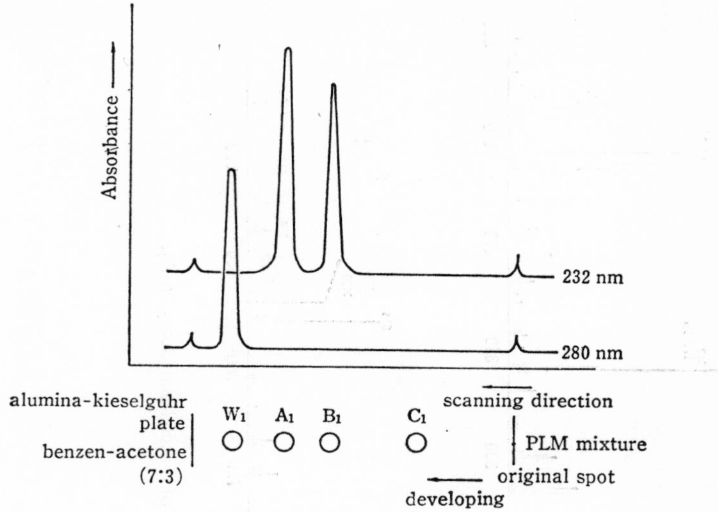

Fig. 1.1 Thin-layer chromatography of platenomycin components (automatic uv monitoring; Hitachi MPF-2A machine).

ever, PLM C_1 shows only end absorption, so no signal appears at the position of this spot. Therefore, combination analysis in conjunction with the sulfuric acid spraying method is necessary for the detection of PLM C series components in the mixture. Since major components of PLM show strong uv absorption, the uv monitoring method is very useful for purity determination and rough quantitation of some components.

Direct preliminary instrumental analysis of such tlc spots, e.g., by measurements of uv (double-beam spectrophotometer), ir and cd (micro KBr-tablet method)[4] and ms (direct-inlet system), may also provide valuable information on small amounts of samples.

[Experimental procedure 3] Preparation of alumina-kieselguhr plates

Neutral alumina (Woelm-TLC-or-tlc, 42 g) and kieselguhr (Merck, 7 g) are mixed well in a mortar, slowly adding 60 ml of water. After checking that there are no air bubbles in the slurry, it is charged on glass plates (20 cm × 20 cm × 5 mm) using a chromato-charger (applicator), and stored for one night at room temperature before use.

Mass Spectrometry[5]

Mass spectrometry is extremely effective for the identification and structure characterization of basic 16-membered macrolides. It is possible to determine the exact molecular weight and many structural features with

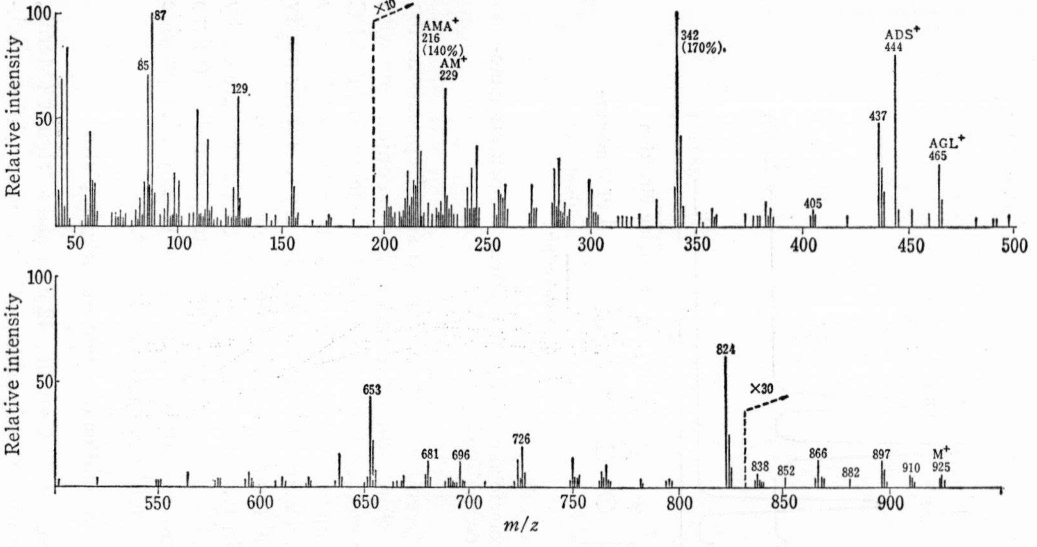

Fig. 1.2 Electron impact mass spectrum of diacetyl-platenomycin A_1.

μg samples. Sometimes, exact molecular weight and correct molecular formula determination of macrolide antibiotics is quite difficult, because they combine with some solvent molecule very tightly. Accordingly, we can say now that such problems mentioned above were completely solved by the introduction of mass spectrometry to this area.

In general, electron impact with the free bases of basic 16-membered macrolides results in extensive water elimination, giving poorly reproducible mass spectra. However, the acetyl derivatives show a clear molecular ion (M^+) which is an accurate indicator of molecular weight, and give regular fragment ions (AGL^+, ADS^+, AMA^+, and AM^+: see below). The latter ions reflect the specific structural features of the original macrolides.

The mass spectrum of diacetyl-PLM A_1 is shown in Fig. 1.2, with a clear molecular ion (*m/z* 925) and specific fragment ions. Such fragment ions have been assigned by measurement of the mass spectra of selected standard macrolides (acetyl derivatives of leucomycin A_3, carbomycin A and B) and by the shift method using the d_3-acetyl derivatives. High-resolution studies have provided the precise compositions of key fragment ions. The relationships of fragment ions and structure characteristics can be summarized by a "diagnostic fragmentation diagram (DFD)," as shown in Fig. 1.3. The shift method using acetyl/ d_3-acetyl derivatives has also been useful in the identification and structure elucidation of other groups of 16-membered macrolide antibiotics.

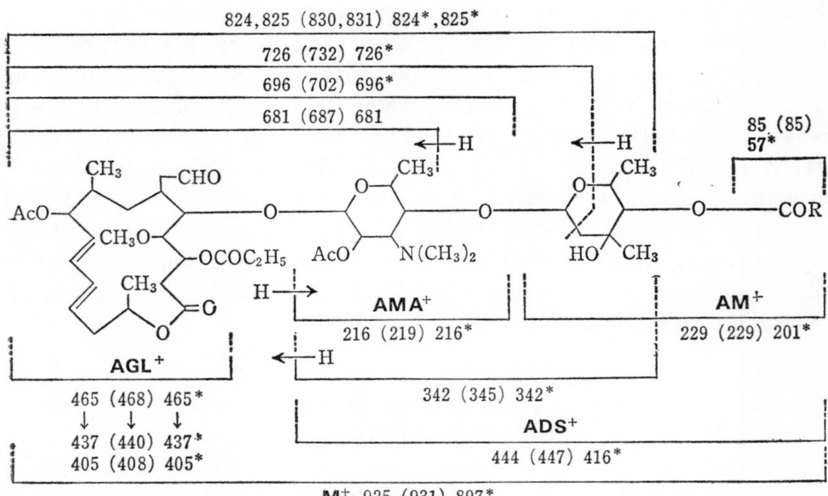

Fig. 1.3 Diagnostic fragmentation diagram of diacetyl-platenomycin A_1 and B_1*. — cf. (d_3-diacetyl-platenomycin A_1) —

The DFD in Fig. 1.3 clearly shows key fragment ions of diacetyl-PLM A_1 and B_1 and shifting ions in the former. Among these, the aglycone ion (AGL^+), acyl-disaccharide ion (ADS^+), acetyl-mycaminose ion (AMA^+) and acyl-mycarose ion (AM^+) are important ions which reflect the specific structure of each PLM component. They are listed in the right-hand column of Table 1.1, and illustrate the effectiveness of mass spectrometry for the discrimination and structure characterization of the thirteen kinds of PLM components.

It was found that some PLM components were identical with already known macrolides. Thus, PLM B_1 (a major component) was identical with midecamycin and espinomycin A_1. Further, the minor components PLM A_3, B_3 were identical with leucomycin A_3, A_6, respectively, and PLM C_1, C_3, C_4, with maridomycin III, I, II, respectively.

Recently, the chemical ionization technique has been developed and applied to more polar, heat-instable natural products. Some papers have already appeared concerning the basic macrolide antibiotics.[6] Using this novel, mild ionization technique, the sample molecule is ionized by ion-molecule reaction in an ion-plasma of the reagent gases, e.g. methane, isobutane and ammonia. Stable quasi-molecular ion ($QM^+ = MH^+$) formation is the first process in the case of sample molecules containing hetero-atoms, and almost no carbon-carbon bond cleavages are observed in general CI mass spectra.

The chemical ionization mass spectrum of intact PLM A_1 using isobutane as a reagent gas is shown in Fig. 1.4. Compared with the electron impact mass spectrum (Fig. 1.2), simplified patterns are observed: namely, a clear MH^+ peak (m/z 842) and peaks corresponding to the elimination of small neutral molecules, such as water (18 u), propionic acid (74 u), isovaleric acid (102 u), appear in the high-mass region. Specific ions such as $(AGL \cdot MA)H^+$ at m/z 614 and $(ADS \cdot OH)H^+$ at m/z 420 are seen, together with sugar ions (MA^+ at m/z 174, AM^+ at m/z 229), ADS^+ (m/z 402) and AGL^+ (m/z 423). These features are quite different from those of the electron impact mass spectrum of the diacetyl derivative, and the extension of the chemical ionization technique should be very helpful in the field of macrolide antibiotics.[9]

1.2 Separation and Identification in Biosynthetic Studies of Macrolide Antibiotics[7]

Production, Isolation and Purification of the Intermediates Produced by Blocked Mutants

In a search for blocked mutants which might accumulate a biosyn-

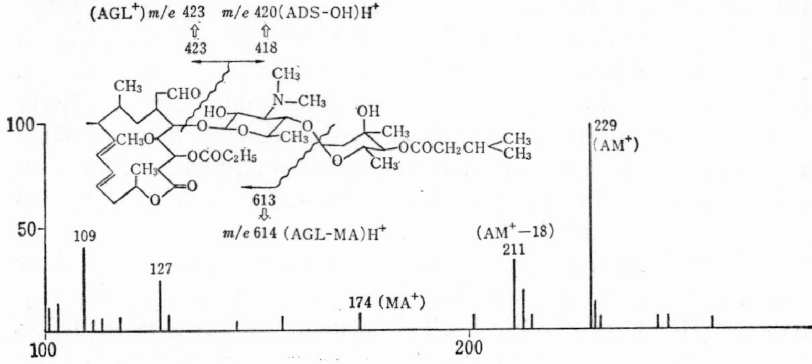

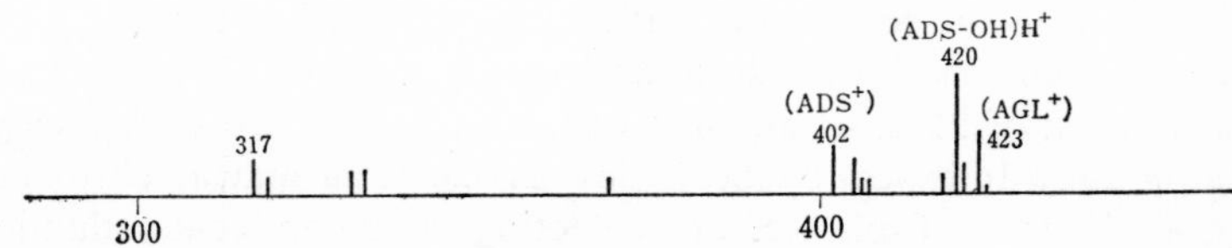

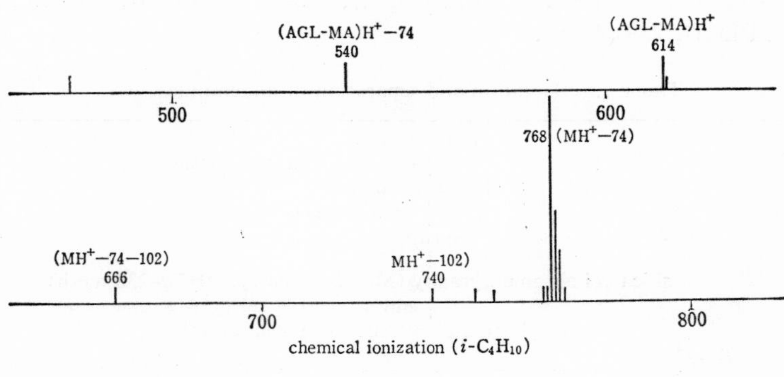

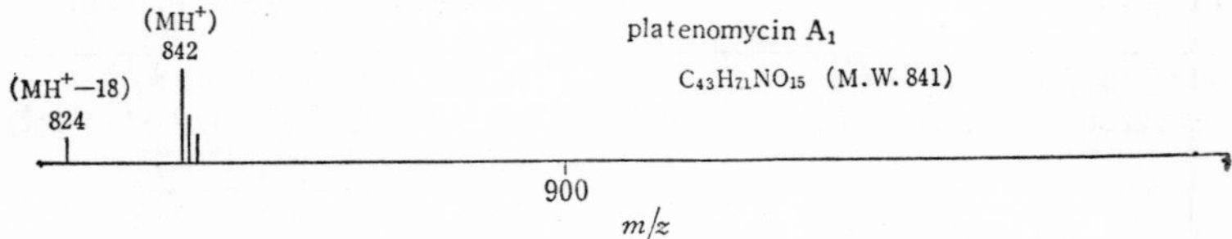

Fig. 1.4 Chemical ionization mass spectrum of platenomycin A_1.

thetic intermediate, *N*-methyl-*N'*-nitro-*N*-nitrosoguanidine (NTG) treatment and/or uv irradiation of PLM-producing *Streptomyces platensis* subsp. *malvinus* MCRL 0388 were carried out to cause mutagenesis. Twenty-four non-PLM-producing stable mutants thus obtained were tested for cosynthesis ability. Antibiotic cosynthesis by pairs of these mutants made it possible to detect producers of intermediates. Platenolides (PL), precursors in the PLM biosynthesis, were produced by two groups of blocked mutants (U-92, N-22).

PL-I and -II were major platenolides and contained PL-III as a minor component. The most striking structural characteristic is the 18-CH_3 group instead of 18-CHO in the PLM series. This ultimately led to the conclusion that the real origin of the C-5/C-6/C-17/C-18 moiety must be *n*-butyric acid.

The isolation and separation of PL is difficult, because they are rather polar-neutral substances with no antimicrobial activities. The separation and purification of PL-I, -II and -III were achieved by two-step silica gel chromatography followed by Sephadex LH-20 (Pharmacia) chromatography, as shown in Table 1.5. For detecting the components, alumina-kieselguhr tlc with 40% sulfuric acid spray was used (PL-I, yellow spot; PL-II, brown spot; PL-III, light blue spot), and uv monitoring was also applicable in this case [λ_{max}^{EtOH} nm (log ε); PL-I, 280 (4.38); PL-II, 232 (4.48); PL-III, 239 (4.05)].

TABLE 1.5 Isolation and separation of platenolides.

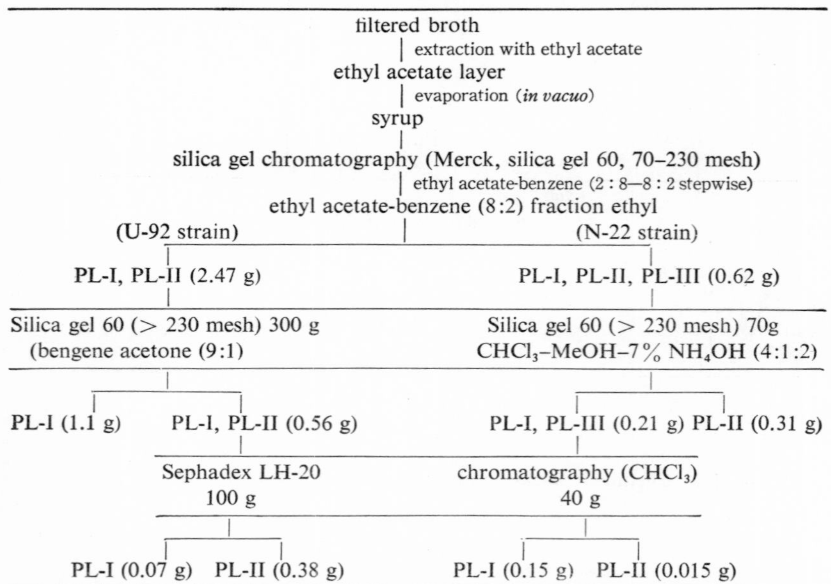

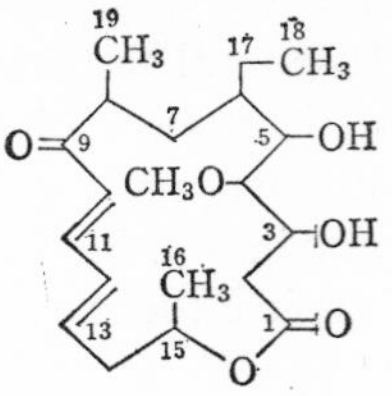

PL-I 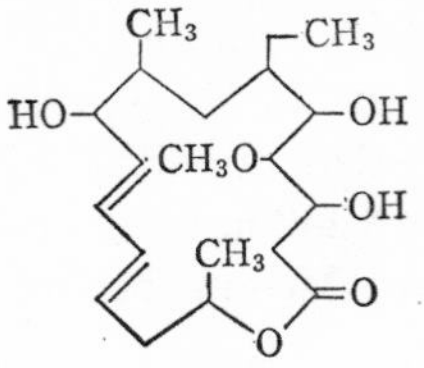 PL-II 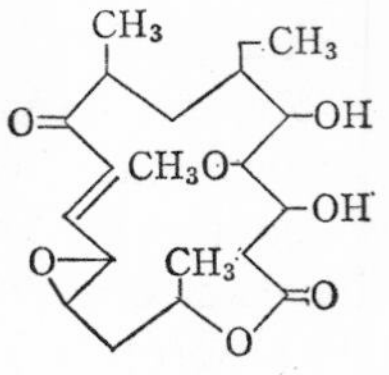 PL-III

Characterization and Identification of Biosynthetic Intermediates

Table 1.6 shows the tlc behavior of PL-I and -II, which have rather similar structures. The separation was quite good on alumina-kieselguhr

TABLE 1.6 Thin-layer chromatography of platenolides I and II (*Rf* values).†

		PL-I	PL-II
Silica gel GF sheet (Woelm)	ii	0.33	0.26
	iii	0.54	0.52
Alumina-kieselguhr (6:1)	i	0.58	0.40
	ii	0.32	0.20
	iii	0.56	0.34
	iv	0.48	0.32

† Solvent system i, benzene-acetone (1:1); ii, benzene-acetone (7:3); iii, ethyl acetate-acetone (8:2); iv, benzene-acetone-water (70:29:1).

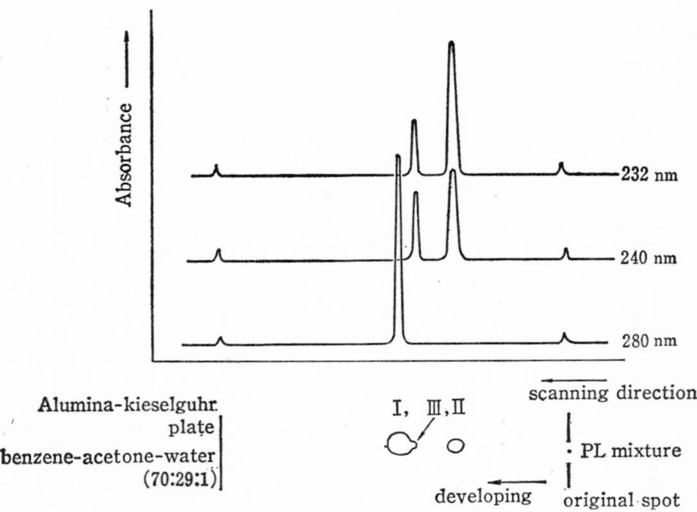

Fig. 1.5 Thin-layer chromatography of platenolide components (automatic uv monitoring; Hitachi MPF-2A machine). Alumina-kieselguhr plate; benzene-acetone-water (79:29:1).

plates and the *Rf* values themselves are lower than those of the PLM series. In the case of the minor component, PL-III, complete separation was not observed even on an alumina-kieselguhr plate with solvent iv. However, uv monitoring revealed a small spot incompletely separated from PL-I as a component having a 240 nm absorption maximum. PL-I and -II show signals at both 232 and 240 nm because of their close uv maxima and wide absorption curves.

Mass spectrometry is most effective for the characterization of the intermediates thus obtained, both by electron impact (Fig. 1.6, *a*) and chemical ionization (Fig. 1.6, *b* and *c*). For instance, the ions at m/z 369 ($QM^+ = MH^+$) and m/z 351 ($MH^+ - H_2O$) characterize PL-I. Interesting features can be observed in Fig. 1.6 *c*. This sample is PL-I containing a small amount of PL-III as a contaminant. In the high-mass region of the spectrum, ions can be seen at m/z 385 and 367, and these are clearly

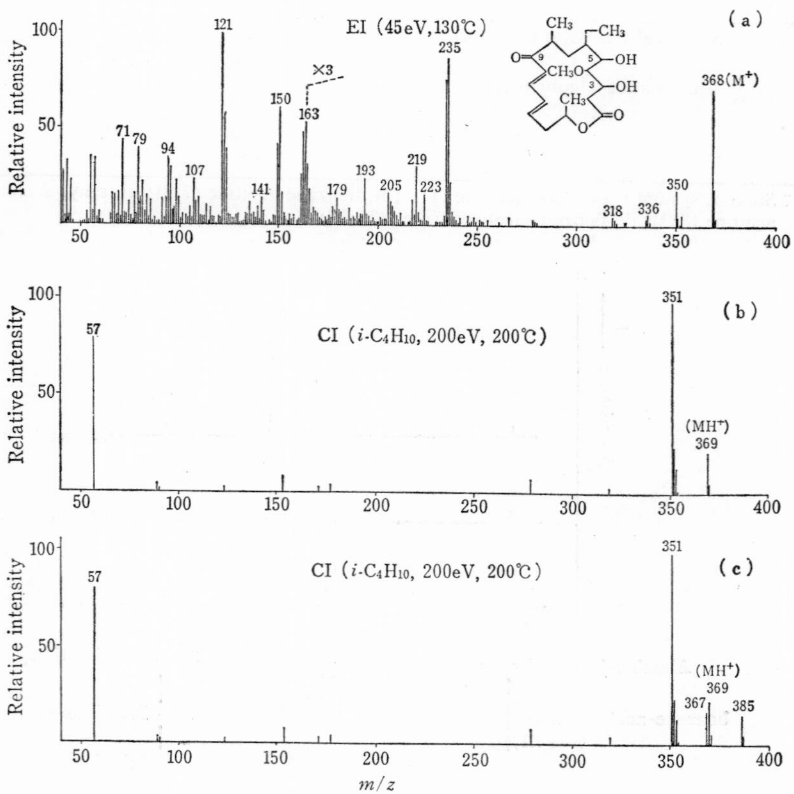

Fig. 1.6 Mass spectra of platenolide I.

QM^+ and ($QM^+ - H_2O$) for PL-III. The spectrum of Fig. 1.6 *c* demonstrates the wide applicability of chemical ionization mass spectrometry to intact natural products only available in small quantities.

The complete structure elucidation of platenolides was achieved by detailed analysis of ^{1}H-nmr and mass spectra, including those of chemically modified derivatives.[8]

1.3 Concluding Remarks

The separation, purification and structural analysis of platenomycins and their biosynthetic intermediates have been described as an example of the experimental approach to basic 16-membered macrolides.

The utilization of deactivated alumina chromatography with LH-20 chromatography and a new tlc system (alumina-kieselguhr plate) with uv-monitoring shows that scope exists for novel separation and detection methods.

For structural analysis effective application of mass spectrometry is a major topic and its usefulness will be developed widely in the field of macrolide biochemistry. Further, the application of chemical ionization and other mild ionization techniques to macrolides and other more polar natural products clearly has great potential.

REFERENCES

1) W. Keller-Schierlein, *Progress in the Chemistry of Organic Natural Products*, **30**, 313 (1973); S. Inoue, *Sci. Reports of Meiji Seika Kaisha*, **13**, 100 (1973); *ibid*(, **14**, 28, (1974).
2) S. Omura, A. Nakagawa, *J. Antibiot.* 28, 401 (1975).
3) A. Kinumaki, I. Takamori, Y. Sugawara, M. Suzuki, T. Okuda, *ibid.*, **27**, 102, 107, 117 (1974) (cf. *Tetr. Lett.*, 1971, 435).
4) T. Takakuwa, F. Kaneuchi, *Jasco Report*, **8**, 178 (1971).
5) M. Suzuki, *J. Synth. Org. Chem., Japan*, **30**, 784 (1972); A. Kinumaki, M. Suzuki, *J. Antibiot.*, **25**, 480 (1972).
6) L.A. Mitscher, H.D.H. Showalter, *Chem. Commun.*, **1972**, 796; L.A. Mitscher, H.D.H. Showalter, *J. Antibiot.*, **26**, 55 (1973).
7) T. Furumai, M. Suzuki, *ibid.*, **28**, 770, 775, 783, 789 (1975).
8) A. Kinumaki, K.I. Harada, T. Furumai, M. Suzuki, *ibid.*, **29**, 1209 (1976).
9) M. Suzuki, K.I. Harada, N. Takeda, A. Tatematsu, *Heterocycles*, **15**,1213 (1981).

CHAPTER 2

Nucleoside Peptide Antibiotics: Isolation of the Polyoxins

As early as the 1950's, a search for new antibiotics for agricultural use had started in Japan. Rice blast disease, which is caused by a pathogenic fungus, *Piricularia oryzae,* was a primary target. The nucleoside antibiotic blasticidin S and the aminoglycoside antibiotic kasugamycin were discovered and proved to be useful as agricultural fungicides. In 1963, we started a search for antibiotics which could prevent sheath-blight disease of rice plants; this is another serious disease of rice plants, caused by a pathogenic fungus, *Pellicularia sasakii.* Screening was carried out by means of the *in vivo* pot test using growing rice plants inoculated with the pathogen. We discovered a polyoxin-producing streptomycete, which was later designated as *Streptomyces cacaoi* var. *asoensis.*[1]

Since the antimicrobial spectrum of this antibiotic was very selective, being restricted to a few species of phytopathogenic fungi, it was presumed to be a new type of antibiotic at the early stage of the research. Upon purification, it was found that the antibiotic is composed of many analogous components.[1–3] Separation of the components was performed in parallel with structure studies. The structure of polyoxin A, main component of the polyoxin complex, was established first. Isolation of polyoxins B-M was achieved subsequently. In all, 13 polyoxins (polyoxins A through M) have been isolated and characterized. The structures are summarized in Fig. 2.1. [4–5]

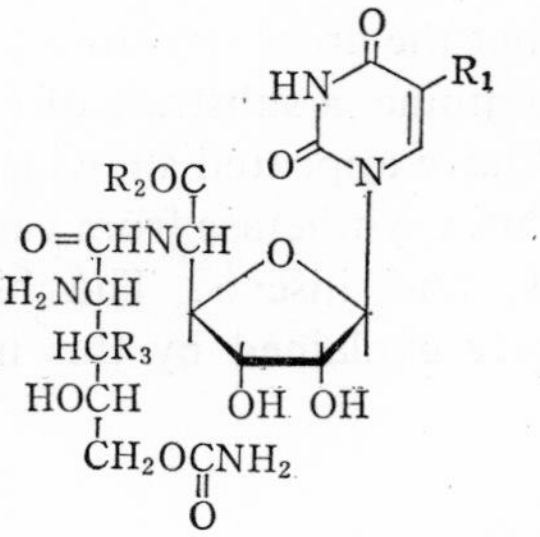

Polyoxin	R_1	R_2	R_3
A	$-CH_2OH$	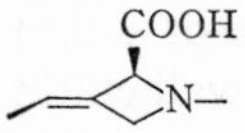 COOH N–	$-OH$
B	$-CH_2OH$	HO–	$-OH$
D	$-COOH$	HO–	$-OH$
E	$-COOH$	HO–	$-H$
F	$-COOH$	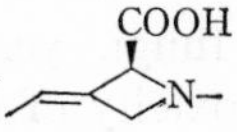 COOH N–	$-OH$
G	$-CH_2OH$	HO–	$-H$
H	$-CH_3$	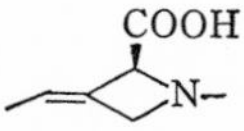 COOH N–	$-OH$
J	$-CH_3$	HO–	$-OH$
K	$-H$	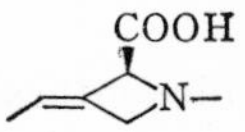 COOH N–	$-OH$
L	$-H$	HO–	$-OH$
M	$-H$	H–	$-H$

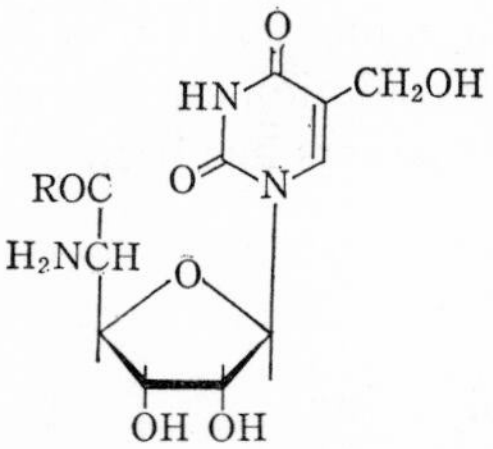

	R
C	HO–
I	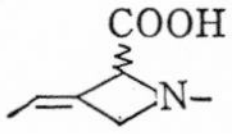 COOH N–

Fig. 2.1 Structures of polyoxins.

2.1 Structures and Modes of Action of Polyoxins

A hybrid nucleoside and peptide structure is characteristic of the polyoxins. The basic skeleton common to all of the polyoxins is a 5-substituted uracil nucleoside of 5-amino-5-deoxy-D-allofuranosyl uronic acid. This is uridine carrying an L-α-amino acid structure on its 5′-carbon. Additional L-α-amino acids, i.e., 5-*O*-carbamoyl-2-amino-2-deoxy-L-xylonic acid (or its 3-deoxy compound) and 3-ethylidene-L-azetidine-2-carboxylic acid are amide-linked either through the 5′-amino group or the carboxyl group on C-5′.

The mode of action of the polyoxins has been shown to be inhibition of fungal cell wall chitin biosynthesis. This was predicted when the structure was elucidated. We have pointed out[4] that the gross structure of the polyoxins mimics that of UDP-*N*-acetylglucosamine, a substrate of chitin synthetase. Subsequently, various papers[6–12] have appeared showing that the polyoxins are competitive inhibitors for chitin synthetase from various sources, i.e., fungi, yeasts, Basidiomycetes, and insects. The highly selective cytotoxicity of the polyoxins is largely explained by this mode of action.

2.2 Biosynthesis

Polyoxin biosynthesis is schematically shown in outline in Fig. 2.2. The biosynthetic studies were done mostly by feeding experiments utilizing ^{14}C-, ^{13}C, and ^{3}H-labeled compounds with whole cells of *Streptomyces cacaoi*. In this study, isolation of the polyoxin complex from the culture filtrate, isolation of the constituent amino acids after hydrolysis, and analysis of the label distribution by degradation are described for a number of candidate precursors. A series of papers has been published on this subject.[13–19]

2.3 Isolation of the Polyoxins

Bioassay

A simple assay method is an absolute requisite for the isolation of bioactive compounds. Although the polyoxins were first discovered using the *in vivo* pot test, a conventional paper disc-agar plate method[20] would be more convenient for use during isolation. In fact, *Piricularia oryzae* or *Alternaria kikuchiana* was used as a test organism. In most cases, semiquantitative determination is sufficient for the purpose of isolation. Usually, a simple standard plot of the logarithmic concentration against the diameter of the inhibition zone using the crude compound is sufficiently helpful for further purification. Fig. 2.3 shows a standard curve for polyoxin A using *Piricularia oryzae* as a test organism. Because bioassay is time-consuming (2 days in this case), it is advisable to utilize uv absorpt-

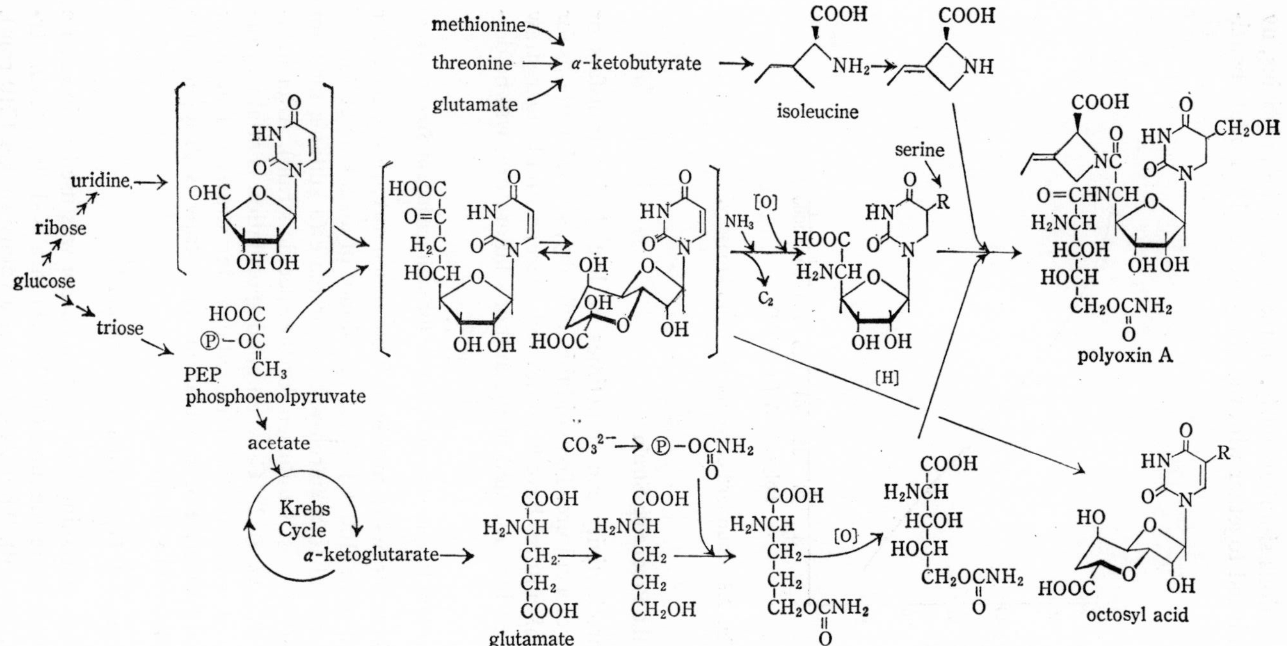

Fig. 2.2 Biosynthesis of the polyoxins.

tion and color reactions simultaneously. In the case of the polyoxins, uv absorption was utilized together with the ninhydrin and periodate-benzidine reactions.

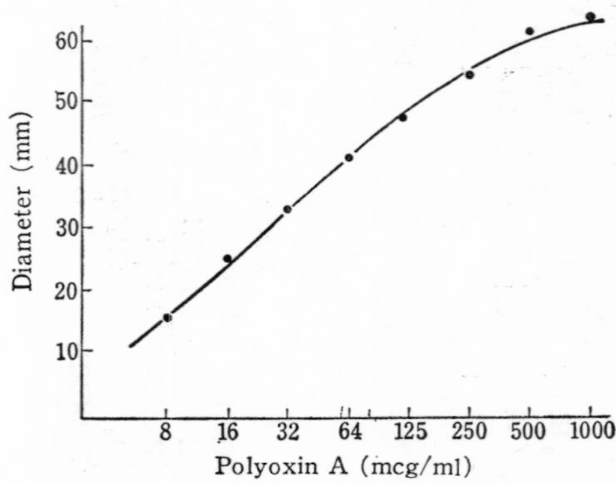

Fig. 2.3 Standard curve for polyoxin bioassay.

Maintenance of the Producing Strain

Antibiotic productivity can somet'mes be lost during transfers or simply on storage of a culture. To avoid this difficulty, several methods for the preservation of cultures have been devised. Here, a simple agar-straw method[21] with freezing, adopted in the author's laboratory, will be described.

[Experimental procedure 1] Agar straw method for the maintenance of *Streptomyces*

Plastic drinking straws (6 mm by 6–7 cm in length) are sterilized for several days in an oven (100° C). A 0.5 ml aliquot of a shaking culture of *S. cacaoi* is transferred to a starch-yeast agar plate (a suitable medium for good sporulation must be chosen for the specific strain). After incubation for several days at 28° C, when good sporulation is attained, the spores on agar are forced into the sterilized drinking straws and stored at −80° C in a deep-freezer. For inoculation, one straw is dropped into fermentation medium in a flask.

[Experimental procedure 2] Fermentation of *S. cacaoi*[15]

An agar straw containing spores of *S. cacaoi* was used to inoculate 60 ml of the seed culture medium (Table 2.1) in a K-1 flask (cylinder-bottle type flask, 500 ml). After 24–27 hr on a rotary shaker (210 rpm), the cells reached the logarithmic phase, and 1 ml aliquots were transferred to the main culture medium (Table 2.1). Fermentation was then carried out for 96–120 hr under the same conditions. A typical fermentation profile is shown in Fig. 2.4.

TABLE 2.1 Composition of culture medium for polyoxin fermentation.

	Seed culture (%)	Main culture (%)
Soluble Starch	1	3–9
Glucose	1	1
Soybean flour	2	2
Dry yeast	1	4
Na NO_3	0.2	0.2
K_2HPO_4	0.2	0.2

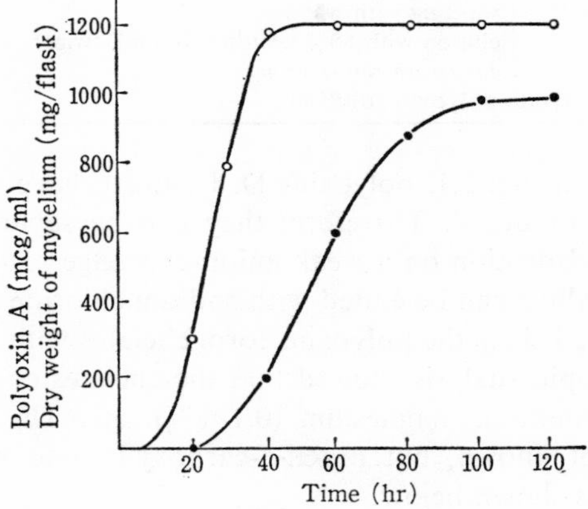

Fig. 2.4 Profile of polyoxin fermentation.
●, Polyoxin A (mcg/ml); ○, dry weight of mycelium (mg/flask).

Isolation of the Polyoxin Complex

A preliminary extraction test using a small amount of culture filtrate showed polyoxin complex to have the following properties. (i) It is stable in neutral and acidic solution. (ii) It cannot be extracted by organic solvents from the aqueous solution. (iii) It can be adsorbed on activated carbon over a wide pH range and eluted with aqueous acetone. (iv) Paper electrophoresis showed that it is an amphoteric compound having an isoelectric point near pH 4–5. Ion-exchange is often effective for the isolation of such a compound. Since polyoxin is unstable to alkali, anion-exchange resin cannot be used. However, it can be adsorbed on sulfonic acid type cation-exchange resin. Elution was carried out with sodium chloride solution or more effectively with dilute ammonium hydroxide solution. A typical extraction procedure for the preparation of crude polyoxin is described in Table 2.2 and in "Experimental procedure 3."

TABLE 2.2 Preparation of crude polyoxin.

culture broth 2 l (9% starch medium, 120 hr)
↓ pH 2 with 10% HCl, 2% Celite
↓ Filtration or centrifugation
filtrate
↓
Dowex 50 W X-8 (H-type, 100–200 mesh) (400 ml)
↓ 0.6 N NH_4OH (1.5 l)
eluate 1.5 l ($A_{262} = 65{,}000$)
↓ neutralization with 10% HCl
↓ activated carbon (65 g), filtration
carbon
↓ washing with water
↓ elution with 50% acetone (700 ml, twice)
↓ concentration *in vacuo*
crude polyoxin solution

As shown in Fig. 2.1, polyoxins D, E, and F have an extra carboxyl group on C-5 of uracil. Therefore, they can be separated from other polyoxins by adsorption on a weak anion-exchange resin (Amberlite IR-4B, Cl form), which can be eluted with sodium chloride solution. Cations must be removed since the polyoxins form chelates with metal salts.

Spectroscopic analysis showed that the chelates contain calcium (> 1%), zinc, aluminum, magnesium (0.1–1%), etc. The structure of the products is not known, but uracil-5-carboxylic acid may play an important role, as shown below.

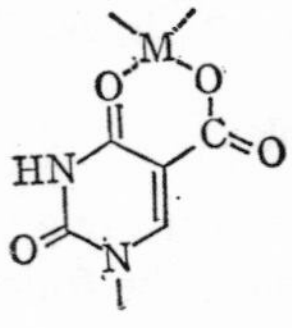

The purification procedure for polyoxins is shown schematically in Table 2.3 and described in detail below. A carbon adsorption procedure (adsorption on carbon and elution therefrom) was used in order to remove inorganic salts[22] and aliphatic amino acids and peptides, yielding a crude polyoxin solution. The batch-wise procedure shown in Table 2.3 may be replaced by granulated carbon column adsorption, as described in "Experimental procedure 6."

TABLE 2.3 Purification of the polyoxin complex.

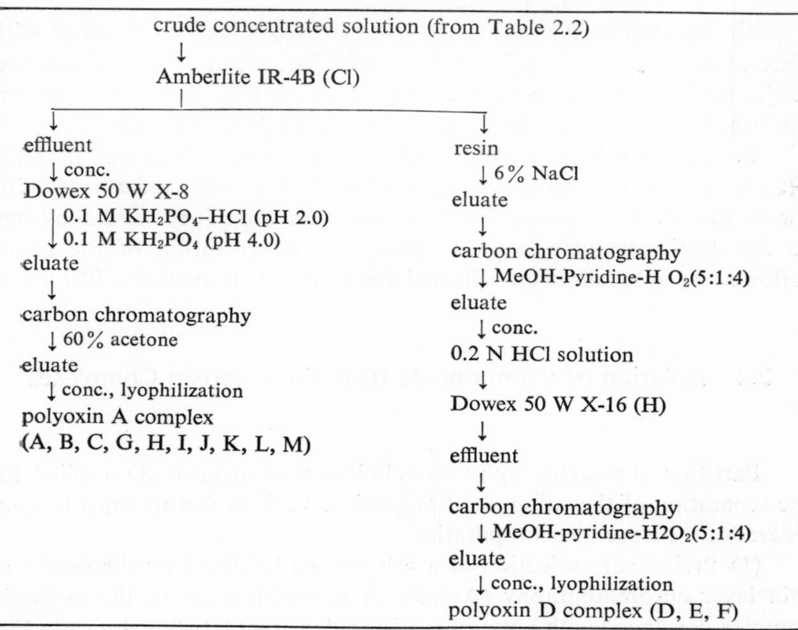

crude concentrated solution (from Table 2.2)
↓
Amberlite IR-4B (Cl)

Left branch:
effluent
↓ conc.
Dowex 50 W X-8
| 0.1 M KH_2PO_4–HCl (pH 2.0)
↓ 0.1 M KH_2PO_4 (pH 4.0)
eluate
↓
carbon chromatography
↓ 60% acetone
eluate
↓ conc., lyophilization
polyoxin A complex
(A, B, C, G, H, I, J, K, L, M)

Right branch:
resin
↓ 6% NaCl
eluate
↓
carbon chromatography
↓ MeOH-Pyridine-H O_2(5:1:4)
eluate
↓ conc.
0.2 N HCl solution
↓
Dowex 50 W X-16 (H)
↓
effluent
↓
carbon chromatography
↓ MeOH-pyridine-H2O_2(5:1:4)
eluate
↓ conc., lyophilization
polyoxin D complex (D, E, F)

[Experimental procedure 3] Purification of the polyoxin A complex

The crude polyoxin solution was passed through a column of 30 ml of Amberlite IR-4B (Cl form, 100–200 mesh). The effluent and washings were combined ($A_{262} = 30{,}000$), concentrated *in vacuo*, and lyophilized, affording 2 g of crude powder. Next, buffered Dowex 50W chromatography was carried out. To increase dissociation of the amino group and to depress that of the carboxyl group, buffer having acidic pH was selected. A column of Dowex 50W (150 ml) was prepared with 0.5 M KH_2PO_4–HCl (pH 2.0), finally replacing the solution with 0.1 M of the same buffer. The crude powder (2 g) was dissolved in a small amount of the same buffer and applied to the top of the column. After development with 500 ml of the same buffer, elution was continued with 3 l of M/20 KH_2PO_4 (pH 4.3). The uv-absorbing fractions were combined and adsorbed on carbon (1 g of carbon per 1000 O.D. units), which was eluted twice with 50% acetone. After concentration followed by lyophilization, 1.5 g of polyoxin A complex was obtained as a white powder. It contains all the components except polyoxins D, E, and F.

[Experimental procedure 4] Purification of the polyoxin D complex

The column of Amberlite IR-4B described above was washed with water, then eluted with 6% NaCl solution. The eluate ($A_{272} = 7800$) was subjected to the usual batch-wise carbon adsorption process (8 g of carbon was used) with methanol-pyridine-water (5:1:4) as a solvent.

To remove inorganic cations, the antibiotics were dissolved in 0.2 N HCl and passed through a column of Dowex 50W X-16 (H form, 100–200 mesh, 10 ml). The effluent and washings were combined, and subjected to the usual carbon adsorption process. Concentration of the eluate followed by lyophilization afforded the polyoxin D complex (0.7 g).

2.4 Isolation of Components from the Polyoxin Complexes

Partition chromatography on cellulose was successfully applied for the separation of the component polyoxins. Various factors must be considered to ensure good separation.

(1) Preliminary selection of a solvent can be based on the results of thin layer chromatography (Avicel). A chromatogram of the polyoxin complex developed with butanol-acetic acid–water (4:1:2) is shown in Fig. 2.5. For column chromatography, a solvent system should be selected to give an *Rf* value in the range of 0.1–0.15 for the target compound. For example, Fig.2.5 shows that butanol-acetic acid-water (4:1:2) is suitable for the separation of polyoxins J, G, L, and B. For the separation of polyoxins H, K, and A, butanol-acetic acid-water (4:1:1.5) is preferred. A column was prepared using butanol-acetic acid-water (4:1:1).

(2) Avicel (microcrystalline, for column chromatography) gave good separation. To remove air bubbles, a solvent slurry of cellulose powder was evacuated for several hours with an aspirator. A 25 mm × 1000 mm column was used for separation of 1 g of the polyoxin complex (a long column gives better separation).

(3) For the application of a sample to a column, the following procedure usually gives good results. An aqueous solution of a sample is mixed with an adequate amount of cellulose powder to give a slurry, which is then lyophilized. The dried powder is triturated with a small amount of solvent to give a slurry, which is then applied to the top of the column, using the least possible amount of solvent.

(4) The column was developed with butanol-acetic acid-water (4:1:1.5) and (4:1:2) either by stepwise or gradient elution. A suitable flow speed is 0.5 ml/min for this column size. Each fraction was monitored

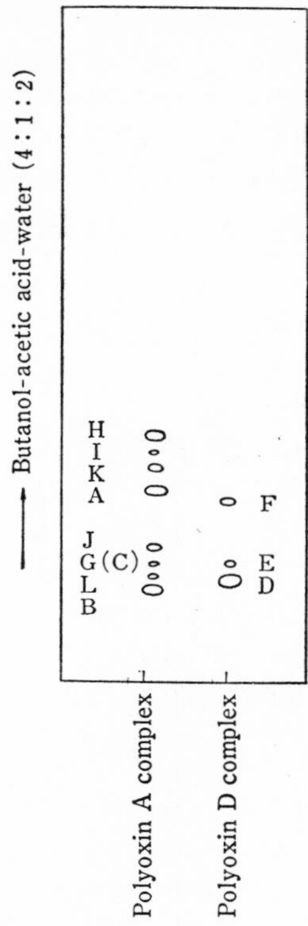

Fig. 2.5 Thin layer chromatogram of the polyoxin complex (cellulose plate F, Merck).

by thin layer chromatography and fractions giving a single spot were combined appropriately. The main components, polyoxins A and B, could be separated by a single operation. However, for the separation of other minor components, repeated chromatography may be necessary. The same column may be used several times for the same purpose. Separation of the polyoxin D complex is easier, since it contains only three components (D, E, and F).

Crystallization and Elementary Analysis

Aminoacyl nucleosides generally crystallize poorly. Only polyoxins A, C, and H have been crystallized (from aqueous ethanol). Therefore, the molecular formulae obtained from elementary analysis and electrometric titration require careful confirmation. Hydration is often very strong in this class of compounds and special care must be taken in the drying of analytical samples. A small sample for analysis should be finely powdered and spread uniformly over the bottom of a platinum boat as a thin layer. Dried air or nitrogen can be introduced in a gentle stream into an Abderhalden apparatus through an adjustable stopper for efficient drying.[23)] Constant weight can be achieved in 4 to 5 hr at 100° C.

Isolation of Constituent Amino Acids and Bases

The hydrolysis of polyoxin A is shown schematically in Fig. 2.6. Acid or alkaline hydrolysis gave the constituent amino acids.

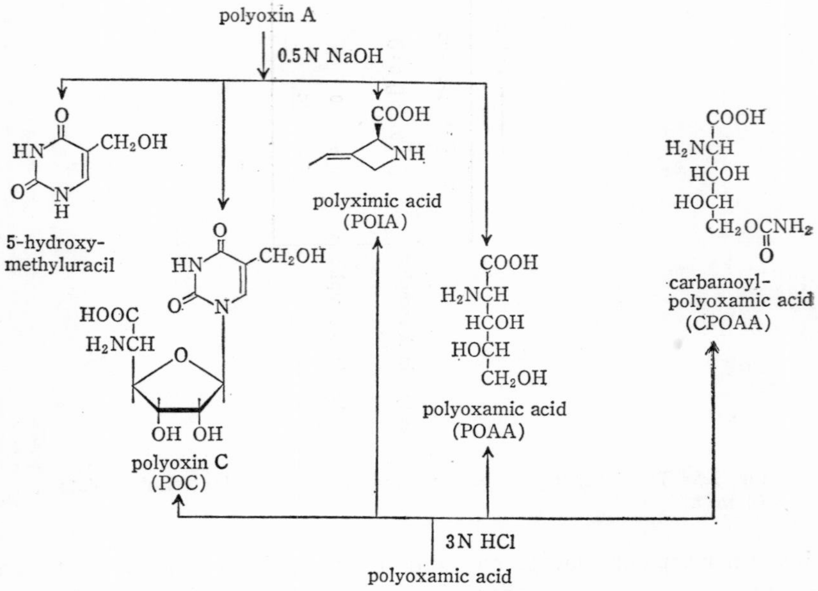

Fig. 2.6 Hydrolysis of polyoxin A.

[Experimental procedure 5] Alkaline hydrolysis of polyoxin A and isolation of the constituents

Polyoxin A complex (800 mg) was dissolved in 20 ml of 0.5 N sodium hydroxide and the solution was heated for 2 hr on a steam bath. The hydrolysate was passed through a column of Dowex 50W X-8 (H form, 100–200 mesh, 20 ml) and the column was washed with water. The effluent and washings were combined, and concentrated *in vacuo* to a small volume. Preparative chromatography on Whatman 3MM paper (46 × 57 cm) was used to separate pyrimidine bases. Approximately 50 mg of sample can be applied to one paper. The descending method was preferred because the development speed is faster and development can be extended after the solvent front reaches the bottom of the paper to obtain better separation. Complete separation of uracil, thymine, and hydroxymethyluracil can be achieved with butanol-acetic acid-water (4:1:2) or water-saturated butanol.[24] Each base was extracted from the paper with water using the centrifuge method: a strip of paper to be eluted was moistened with solvent (usually water or dilute ammonium hydroxide), packed in aluminum foil, and folded into a long-cylindrical shape. One end of this is fixed to the top of a small centrifuge tube with a plastic stopper. Centrifugation at 2000 rpm for several minutes resulted in over 90% recovery.* After rechromatography with water, the bases were crystallized from water.

Constituent amino acids, polyoxin C (POC), polyoximic acid (POIA), polyoxamic acid (POAA), etc., were eluted from the resin with 0.6 N ammonium hydroxide. POC has an acidic uracil group (p*Ka'* 9.6) and can be separated from other amino acids by DEAE-cellulose chromatography (Whatman DE–52). Triethylamine-carbonate elution buffer (0.2M, pH 7.2) is convenient, since it can be removed by evaporation. POIA and POAA were eluted first, followed by POC. Separation of POIA and POAA was achieved by cellulose column chromatography or by paper chromatography. Each amino acid was finally crystallized from aqueous alcohol (approximately 30 mg each).

Carbon chromatography is also useful for the separation of POIA, POAA, and POC, as described below.

As shown in Fig. 2.6, carbamoylpolyoxamic acid (CPOAA) can be obtained by acid hydrolysis. POAA and CPOAA have similar properties and are difficult to separate by ion-exchange chromatography or partition chromatography. However, adsorption chromatography on carbon was very effective.

[Experimental procedure 6] Isolation of carbamoylpolyoxamic acid by acid hydrolysis of the polyoxin A complex

Polyoxin A complex (1 g) was dissolved in 16 ml of 3N hydrochloric acid and the solution was heated for 2 hr on a steam bath. The hydrolysate was passed through a column of Amberlite IR-45 (OH form, 30 ml) to remove chloride. The effluent was passed through a column of Dowex 50W X-8 (H form, 60 ml), which was then eluted with 0.6 N ammonium hy-

---

* If the number of papers to be eluted is large, descending elution can be carried out by stapling a leader paper (Whatman, 3MM, 10–15 cm long, doubled over) to the end of the papers to be eluted and immersing it in a solvent trough for up to 12 hr in a chromatographic chamber.

droxide. POIA was removed by cellulose chromatography (see section 2.4). An aqueous solution of the residual sample was passed through a column containing 20 g of granulated carbon* (Shirasagi, from Takeda Chemical Ind. Co.). After washing with 100 ml of water, linear gradient elution (water, 200 ml ⟶ 20% acetone, 200 ml) was carried out. POAA CPOAA, and POC were eluted in this order.

References

1) K. Isono, J. Nagatsu, Y. Kawashima, S. Suzuki, *Agric. Biol. Chem.*, **29**, 848 (1965).
2) K. Isono, J. Nagatsu, K. Kobinata, K. Sasaki, S. Suzuki, *ibid.*, **31**, 190 (1967).
3) K. Isono, J. Kobinata, S. Suzuki, *ibid.*, **32**, 792 (1968).
4) K. Isono, K. Asahi, S. Suzuki, *J. Am. Chem. Soc.*, **91**, 7490 (1969).
5) K. Isono, S. Suzuki, M. Tanaka, T. Nanbata, K. Shibuya, *Tetr. Lett.*, 1970, 425.
6) A. Endo, T. Misato, *Biochem. Biophy. Res. Commun.*, **37**, 718 (1969).
7) A. Endo, K. Kakiki, T. Misato, *J. Bacteriol*, **104**, 189 (1970); *ibid.*, **103**, 588 (1970).
8) F. A. Keller, E. Cabib, *J. Biol. Chem.*, **246**, 160 (1971).
9) M. Hori, K. Kakiki, S. Suzuki, T. Misato, *Agric. Biol. Chem.*, **35**, 1280 (1971).
10) M. Hori, J. Eguchi, K. Kakiki, T. Misato, *J. Antibiot.*, **27**, 260 (1974).
11) Y. N. Yan, *J. Biol. Chem.*, **249**, 1973 (1974).
12) S. Bartnicki-Garcia, E. Lippman, *J. Gen. Microbiol.*, **71**, 301 (1972).
13) K. Isono, P. F. Crain, T. J. Odiorne, J. A. McCloskey, R. J. Suhadolnik, *J. Am. Chem. Soc.*, **95**, 5788 (1973).
14) K. Isono, S. Funayama, R. J. Suhadolnik, *Biochemistry,* **14**, 2992 (1975).
15) S. Funayama, K. Isono, *ibid.*, **14**, 5568 (1975).
16) K. Isono, R. J. Suhadolnik, *Arch. Biochem. Biophys.*, **173**, 141 (1976).
17) S. Funayama, K. Isono, *Biochemistry*, **16**, 3121 (1977).
18) K. Isono, T. Sato, K. Hirasawa, S. Funayama, S. Suzuki, *J. Am. Chem. Soc.*, **100**, 3937 (1978).
19) T. Sato, K. Hirasawa, J. Uzawa, T. Inaba, K. Isono, *Tetr. Lett.* **1979**, in press.
20) F. Kavanagh, *Methods in Enzymology* (ed. J. H. Hash), vol. 43, *Antibiotics,* p. 55, Academic Press (1975).
21) T. Uematsu, R. J. Suhadolnik, *Biochemistry*, **11**, 4669 (1972).
22) E. Heftmann, *Chromatography*, p. 656, Reinhold (1967).
23) H. Honma, K. Tada, O. Kamimori, Y. Aoshima, *Bunseki Kagaku* (in Japanese), **19**, 277 (1970).
24) G. R. Wyatt, The Nucleic Acids (ed. E. Chargaff, J. N. Davidson), vol. 1, p. 243, Academic Press (1955).

Before use, the granulated carbon was boiled in water for several minutes to prevent the formation of air bubbles.

CHAPTER 3

Bleomycin-Phleomycin Group Antibiotics

Tomohisa TAKITA

Bleomycin (BLM) is a group of antitumor glycopeptide antibiotics produced by *Streptomyces verticillus*. It is used clinically for the treatment of squamous cell carcinoma, malignant lymphoma, etc. Bleomycin was discovered in 1963[1] after a search for phleomycin-like antibiotics with no renal toxicity. Phleomycin itself was discovered in 1956 as an antimicrobial antibiotic.[2]

A mixture of BLMs produced by *S. verticillus* was isolated by cation-exchange resin absorption, followed by carbon, alumina and Sephadex column chromatographies. Each component was separated by CM-Sephadex C-25 column chromatography, developing with a linear gradient of ammonium formate. BLMs thus obtained were blue-colored amorphous powers containing one atom of copper per molecule. The copper can be removed by treatment with hydrogen sulfide to give colorless metal-free BLM, and the metal complex can be regenerated by addition of cupric ions to the metal-free form. Both the copper complex and metal-free BLMs show antibacterial and antitumor activity.

Recently, the total structure of BLM was elucidated (Fig. 3.1).[3] The structural relation between BLM and phleomycin (PHM) was also clarified (see Fig. 3.1).[4] The BLM-PHM family includes YA-56X (zorbamycin) and Y, zorbonomycins B and C, victomycin, platomycins A and B, and tallysomycins A and B.[5] In this chapter, some of the most important experimental procedures in the chemical studies on BLM will be described.

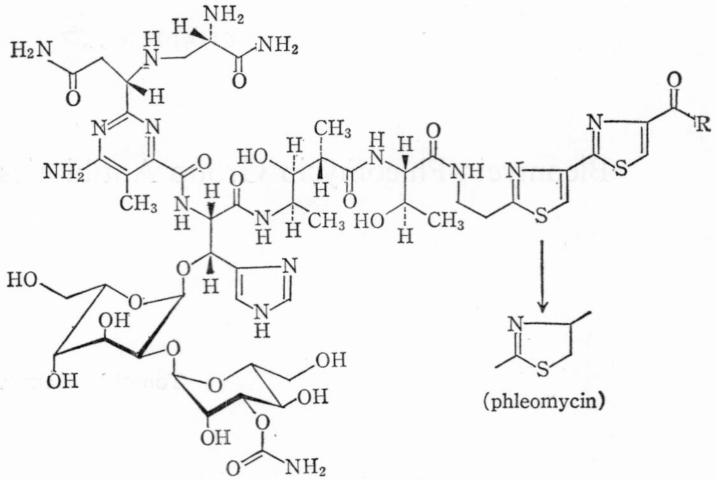

(phleomycin)

Bleomycin	R=
A_1	$NH-(CH_2)_3-\underset{\downarrow \atop O}{S}-CH_3$
Dimethyl-A_2	$NH-(CH_2)_3-S-CH_3$
A_2	$NH-(CH_2)_3-\overset{+}{\underset{\vert \atop CH_3}{S}}-CH_3X^-$
A_2'-a	$NH-(CH_2)_4-NH_2$
A_2'-b	$NH-(CH_2)_3-NH_2$
A_2'-c	$NH-(CH_2)_2-$(imidazolyl, N / NH) 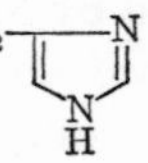
A_5	$NH-(CH_2)_3-NH-(CH_2)_4-NH_2$
A_6	$NH-(CH_2)_3-NH-(CH_2)_4-NH-(CH_2)_3-NH_2$
B_1'	NH_2
B_2	$NH-(CH_2)_4-NH-\underset{\Vert \atop NH}{C}-NH_2$
B_4	$NH-(CH_2)_4-NH-\underset{\Vert \atop NH}{C}-NH-(CH_2)_4-NH-\underset{\Vert \atop NH}{C}-NH_2$

Fig. 3.1 The structures of bleomycin and phleomycin (R is the terminal amine of BLMs).

3.1 Isolation of Phleomycin and Bleomycin

PHM was first isolated as a crude material in 1956. Later, it was obtained in a purified form by the following procedure.[6)] The culture filtrate of PHM was passed through a column containing Amberlite IRC-50, a carboxylic acid resin, which had previously been neutralized with one-fourth equivalent of NaOH. Formerly, the carboxylic acid form of the resin was used as the adsorbent, but it caused decomposition of the acid-labile PHM during adsorption. PHM was adsorbed more effectively on the partially neutralized resin than on the free carboxylic acid form. Washing of the adsorbed resin with 50% acetone before elution produced a further improvement. PHM was eluted with 50% acetone containing 0.5 N HCl. The eluate containing PHM appeared just before the acidic eluate, but it contained a high concentration of sodium chloride. Desalting was difficult by methanol extraction or by alumina chromatography, but precipitation with Reinecke's salt followed by washing with acetone containing 1% NH_4OH gave a blue-colored material of reasonable quality.

Introduction of a resin such as Amberlite XAD or Diaion HP made the purification procedure much easier. Adsorption and elution proceeded under neutral conditions, avoiding decomposition. These resins have a macroreticular matrix composed of styrene-divinylbenzene copolymer, and have no ion-exchange functional groups. The adsorption is due to hydrophobic forces. In the event of incomplete adsorption, the addition of 1–5% sodium chloride to increase the polarity of the solvent strengthened the adsorption. In this case, water-washing before elution must be minimized to avoid losses of the adsorbed material. Methanol or aqueous acetone are generally used for elution, but a solvent containing 0.01–0.001 N HCl is recommended for the elution of basic substances.

The applicability of Amberlite XAD-2, an adsorption resin, compared with that of Amberlite IRC-50, a carboxylic acid resin, is illustrated in Table 3.1 in the case of isolation of vitamin B_{12}.

TABLE 3.1 Isolation of vitamin B_{12} on Amberlites IRC-50 and XAD-2.

Adsorbent	Adsorption capacity† (mg/ml)		Elution (methanol)	
	leakage	saturation	max. concentration (ppm)	eluent required (1/1 resin)
Amberlite IRC-50	0.03	0.14	150	5
Amberlite XAD-2	3.5	5.2	7200	2

† Conditions: concentration of solution, 15 ppm; SV = 8.

Isolation of BLM is much easier than that of PHM due to its stability to acid. (The thiazoline ring of PHM is unstable under both acidic and alkaline conditions.) BLM was also isolated from its culture filtrate by adsorption on Amberlite IRC-50 followed by carbon, alumina and Sephadex column chromatographies. The mixture of BLMs thus obtained can be separated by CM-Sephadex C-25 column chromatography, as first applied for the isolation of PHM by Ikekawa *et al.*[7]

CM-Sephadex has a carboxyl function, like Amberlite IRC-50. The skeleton of IRC-50 is a styrene-divinylbenzene copolymer. It shows hydrophobic binding and π-electron binding forces in addition to ion-exchange properties. On the other hand, the skeleton of CM-Sephadex is dextran, which shows only ion-exchange adsorption. The advantage of CM-Sephadex is that the retention time of each component of BLM is expected to be dependent on the structure of the terminal amine.

[Experimental procedure 1] Isolation of bleomycin

The culture filtrate (4.8 l) of *Streptomyces verticillus,* which contained 440 mg potency units of BLM (BLM A2: 1 mg unit/mg), was passed through a column of Amberlite XAD-2 (500 ml, 4.0 × 40 cm) (space velocity: SV 1.0). The column was washed with 800 ml of water (SV 1.0), and then eluted with 0.05 N HCl-MeOH (1 :4 v/v) (SV 0.5). The bioactive eluate (1.5 l) contained 348 mg units of BLM (yield 79%). It was concentrated to 30 ml under reduced pressure after neutralization with Dowex 44 (OH form), then 120 ml of methanol was added. The aqueous methanol was passed through a column of neutral alumina (100 ml, 3.0 × 14 cm), and the adsorbed BLM was eluted with 80% methanol to give 400 ml of a pale blue bioactive fraction (yield 69%). The active eluate was dried, and then dissolved in a small amount of methanol. It was passed through a column of Sephadex LH-20 (100 ml, 2.2 × 26 cm), developing with methanol. The active eluate was dried to give 135 mg of BLM mixture (1.76 mg unit/mg, yield 54%).

BLM mixture (1.00 g, 1.49 mg unit/mg), obtained by the experimental procedure described above, was deparated by CM-Sephadex C-25 column chromatography (600 ml, 2.5 × 120 cm), developing with a linear gradient formed from 3 l each of 0.05 and 1.0 M ammonium formate. The eluate

TABLE 3.2 Fraction numbers and yields of BLM (see the text).

Component	Fraction No.	Yield (mg)
A_1, demethyl A_2	19– 24	53
B_1'	25– 28	14
A_2	43– 48	307
A_2'-a, b, c	51– 54	48
B_2	56– 59	135
A_5	81– 83	33
B_4	86– 88	15
A_6	107–109	4
B_6	112–119	5

was fractionated (50 ml each) to give 120 fractions. Table 3.2 lists the fraction numbers and yield of each component of BLM.

3.2 High-Performance Liquid Chromatography of BLM

For the kinetic study of epimerization, isomerization and other minor modifications of BLM, it was necessary to establish a rapid and quantitative small-scale analysis procedure for BLM. This was achieved by high-performance liquid chromatography using Lichrosorb SI60, eluting with a mixture of solvents A and B (described later), or with a linear gradient of solvents A and B.

When metal-free BLM is kept in aqueous alcohol containing triethylamine, the carbamoyl group attached to the 3-*O*-position of the mannose migrates to the 2-*O*-position, and an equilibrium is established between them. The latter is called iso-bleomycin.[8] Under the same conditions, the copper complex of BLM gives epi-bleomycin,[9] not iso-bleomycin. The epimerization occurs at the α-methine carbon of the pyrimidine ring 2-substituent (Fig. 3.2). The epimerization proceeds irreversibly as a first-order reaction. The reason why epimerization occurs but isomerization does not in the case of the copper complex of BLM is that metal coordination by the pyrimidine ring nitrogen and the carbamoyl group

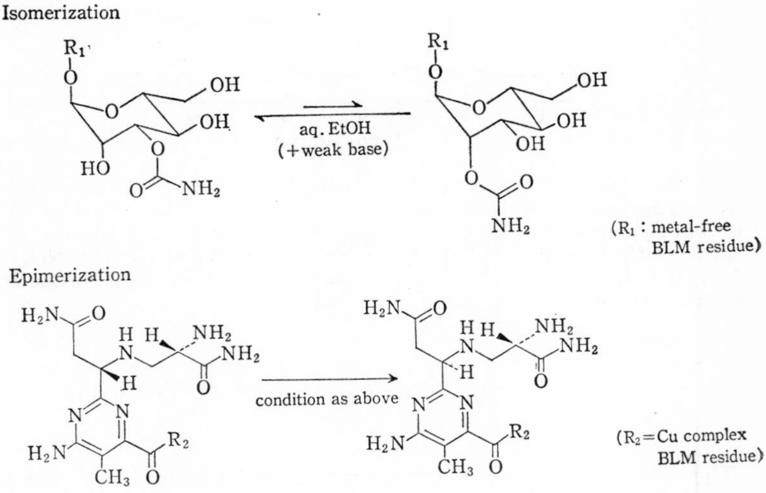

Isomerization

aq. EtOH (+weak base)

(R_1 : metal-free BLM residue)

Epimerization

condition as above

(R_2=Cu complex BLM residue)

Fig. 3.2 Isomerization and epimerization of bleomycin.

occurs.[10] Fixation of the carbamoyl group at the sixth coordination site prevents the migration of the carbamoyl group, and the coordination of the pyrimidine ring nitrogen increases the acidity of the α-methine hydrogen at the pyrimidine ring 2-substituent.

[Experimental procedure 2] High-performance liquid chromatography of BLM[11]

A 10% aqueous solution of BLM B2 was diluted with two volumes of 2% triethylamine ethanol solution. The solution was kept at 70° C for 1 hr, then the reaction mixture was dried under reduced pressure. The dried material (Cu complex) was used for chromatography. The metal-free product was obtained by treatment with hydrogen sulfide. As shown in Fig. 3.3, separation of the Cu complexes was much better than that of the metal-free products.

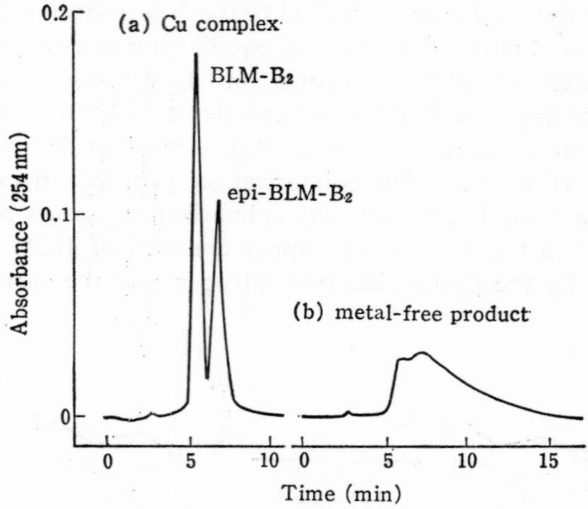

Fig. 3.3 High-performance liquid chromatography of BLM. The sample (33 μg) was applied to a Perkin-Elmer 1220 instrument with a Lichrosorb SI 60 (5 μm) column (150 × 2.6 mm), eluting with an A:B (= 2:8) mixture of A (10% CH_3COONH_4–CH_3OH–H_2O (1:20:19)) and B (10% CH_3COONH_4–CH_3OH–CH_3CH_2OH (1:10:29)) at 0.5 ml/min. Temperature, 60°C; pressure 600 psi; detection, uv (254 nm).

3.3 Isolation of the Sugar Component of Bleomycin

The existence of a sugar component in the BLM molecule was suggested by the strong IR absorption centered at 1050 cm^{-1}. The sugar component was not liberated in an appreciable amount when BLM was hydrolyzed in 6 N HCl at 105° for 20 hr. However, the authors have established a new method of methanolysis using Amberlyst 15 as an acid catalyst to obtain the sugar component in good yield.[12] Amberlyst 15 is a sulfonic acid resin with a macro-reticular structure for non-aqueous acid catalysis. This method should be applicable for general methanolysis, especially for glycopeptides.

BLM was dissolved in methanol, then Amberlyst 15 was added to the solution and the mixture was refluxed for 20 hr. The liberated amine component was adsorbed on the resin, and the sugar component could thus be isolated easily. The isolated methylglycoside was a mixture, which was acetylated and then separated by silica gel chromatography to give methyl 2, 4, 6-tri-*O*-acetyl-3-*O*-carbamoyl-α-D-mannopyranoside (S-I), and the α (S-II-α) and β (S-II-β) anomers of methyl tetra-*O*-acetyl-L-gulopyranoside.

[Experimental procedure 3] Isolation of the sugar component of BLM

Metal-free BLM A2 (6.4 g) was dissolved in 300 ml of methanol, then 90 ml of Amberlyst 15, previously washed twice with hot methanol, was added and the mixture was refluxed for 20 hr (it boiled very suddenly). After reaction, the resin was filtered with the aid of Celite. The clear filtrate was neutralized with Dowex 3 (OH form), and concentration gave 1.8 g of syrup. This was dissolved in ethanol and benzene, and the water contained was removed by azeotropic distillation. After removal of the solvent, the solid material was dissolved in 12 ml of pyridine, then 5.5 ml of acetic anhydride was added under ice-cooling, and the reaction mixture was kept overnight at room temperature. The acetyl derivative (3.0 g) was obtained by usual work-up. It was separated by silica gel chromatography using $CHCl_3$: methanol (100:1). S-I was thus separated from the mixture of S-II-α and -β. The *Rf* values on tlc under the above conditions were: S-I (0.17), S-II (0.43–0.47). S-I was obtained as needles after removal of the solvent (0.86 g). The mixture of S-II-α and -β was obtained as a syrup (1.58 g), which was separated by silica gel chromatography using toluene: methylethylketone (4:1). The *Rf* values on tlc were: S-II-α, 0.27; S-II-β, 0.36. The yields were: S-II-α, 0.20 g; S-II-β, 1.08 g. Both gave crystals after prolonged storage at room temperature.

3.4 Transformation of BLM A2 into Bleomycinic Acid, the Starting Material for Semi-Synthetic BLM

Natural BLMs differ in their terminal amine moiety (see Fig. 3.1). The common part of BLMs is named bleomycinic acid (BLM acid), which was obtained by enzymatic hydrolysis of BLM B2 using acylagmatine amidohydrolase isolated from *Fusarium anguiodies.*[13] BLM acid is the starting material for semi-synthetic BLMs, which may provide a better tumor therapeutic agent than the present mixture of natural BLM. BLM acid can also be obtained by chemical transformation from BLM A2, the main component of natural BLMs, via BLM demethyl-A2 (Fig. 3.4).[14] BLM demethyl-A2 is present in natural BLMs as a minor component. However, it is not a biosynthetic product, but is formed from BLM A2 by spontaneous degradation during storage.

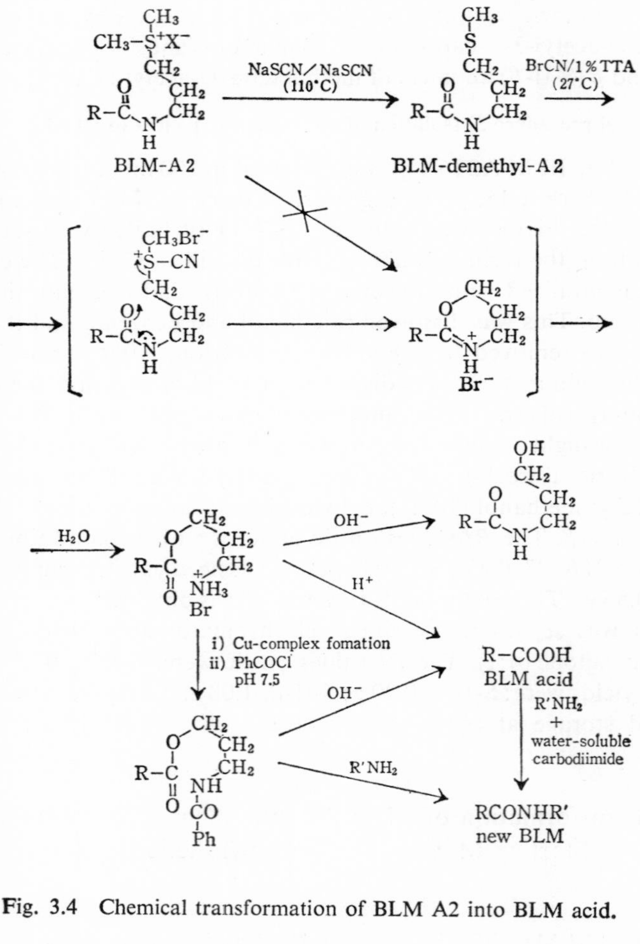

Fig. 3.4 Chemical transformation of BLM A2 into BLM acid.

[Experimental procedure 4] Transformation of BLM A2 into BLM acid

BLM A2 Cu complex (22 g) was heated at 100° C under reduced pressure for 16 hr. The product was dissolved in 200 ml of 0.05 M ammonium chloride and then applied to a column of CM-Sephadex (800 ml) pretreated with the same solvent. BLM demethyl-A2 was obtained by elution with 0.1 M ammonium chloride. The inorganic salt was removed by Amberlite XAD-2 treatment to give 8.91 g of BLM demethyl-A2

BLM demethyl-A2 (1 g) was dissolved in 20 ml of 1% aqueous trifluoroacetic acid, and 2.0 g of cyanogen bromide was added. The reaction mixture was stirred for 18 hr at room temperature. After removal of excess cyanogen bromide under reduced pressure, the reaction mixture was neutralized with NaOH and then applied to a column of Sephadex C-25 (100 ml) pre-equilibrated with 0.05 M pyridine-acetate buffer, pH 4.5. The 3-aminopropyl ester of BLM acid was eluted with 0.5 M pyridine-acetate buffer. After desalting, 495 mg of the product was obtained.

The product was dissolved in 5 ml of water and the solution was adjusted to pH 4.0 by addition of HCl. The solution was heated at 105° C for 6 hr in a sealed tube, and the hydrolysis product was isolated in the usual manner to give BLM acid (265 mg) as a blue-colored amorphous powder.

REFERENCES

1) H. Umezawa, K. Maeda, T. Takeuchi, Y. Okami, *J. Antibiot.*, **19A**, 200 (1966).
2) K. Maeda, H. Kosaka, K. Yagishita, H. Umezawa, *ibid.*, **9A**, 82 (1956).
3) T. Takita, Y. Muraoka, T. Nakatani, A. Fujii, Y. Umezawa, H. Naganawa, H. Umezawa, *ibid.*, **31**, 801 (1978).
4) T. Takita, Y. Muraoka, T. Yoshioka, A. Fujii, K. Maeda, H. Umezawa, *ibid.*, **12**, 755 (1972).
5) H. Kawaguchi, H. Tsukiura, K. Tomita, M. Konishi, K. Saito. S. Kobaru, K. Numata, K. Fujisawa, T. Miyaki, M. Hatori, H. Koshiyama, *ibid.*, **30**, 779 (1977).
6) T. Takita, *ibid.*, **12A**, 285 (1959).
7) M. Ikekawa, F. Iwami, H. Hiranaka, H. Umezawa, *ibid.*, **17A**, 194 (1964).
8) Y. Nakayama, M. Kunishima, S. Omoto, T. Takita, H. Umezawa, *ibid.*, **26**, 400 (1973).
9) Y. Muraoka, H. Kobayashi, A. Fujii, M. Kunishima, T. Fujii, Y. Nakayama, T., Takita, H. Umezawa, *ibid.*, **29**, 853 (1976).
10) T. Takita, Y. Muraoka, T. Nakatani, A. Fujii, Y. Iitaka, H. Umezawa, *ibid.*, **31**, 1073 (1978).
11) Y. Muraoka, *Bleomycin: Chemical, Biochemical and Biological Aspects*, p. 90, Springer-Verlag (1979).
12) T. Takita, K. Maeda, H. Umezawa, S. Omoto, S. Umezawa, *J. Antibiot.*, **22**, 237 (1969)
13) H. Umezawa, Y. Takahashi, A. Fujii, T. Saino, T. Shirai, T. Takita, *ibid.*, **26**, 117 (1973).

14) T. Takita, A. Fujii, T. Fukuoka, H. Umezawa, *J. Antibiot.*, **26**, 252 (1973).

General References

i) J.H. Hash (ed.), *Methods in Enzymology*, vol. 43, *Antibiotics*, Academic Press (1975).
ii) S.M. Hecht (ed.), *Bleomycins: Chemical, Biochemical and Biological Aspects*, Springer-Verlag (1979).
iii) H. Umezawa, T. Takita, *Structure and Bonding*, 40 37 (1980).

CHAPTER 4

Protease Inhibitors

Enzyme inhibitors are useful tools in analysis of the mechanisms of biological functions and even of disease processes. Some are known to have clinical potential. Since proteolytic enzymes are involved in the processes of blood coagulation, fibrinolysis, kinin formation and the complement reaction, cell fusion, cell proliferation, immunity, tumorigenesis, metastasis, and so forth, the effects of enzyme inhibitors on these processes are now attracting considerable attention.

Studies on enzyme inhibitors, which were initiated by Umezawa and his group, have proved very fruitful, especially as regards inhibitors of proteolytic enzymes. All inhibitors previously isolated from animals and plants are macromolecular proteins, but those discovered in our laboratory are all peptides of low molecular weight. The pathway of biosynthesis of these inhibitors is very similar to that of peptide antibiotics, and is controlled by genes located on plasmids. The present chapter describes the protease inhibitors produced in actinomycete culture filtrates and discusses their structure-activity relationships and biological activities. Readers are referred to our reviews[1–6] of recent progress.

4.1 Screening of Enzyme Inhibitors and their Isolation and Identification

The establishment of a rapid, simple, inexpensive and quantitative assay procedure is essential for the screening of inhibitors. Culture filtrates of microbial origin usually contain various hydrolytic enzymes, and it is preferable to carry out a preliminary test for inhibitors using samples heated at 100° C for 5 min. Enzyme reactions are usually followed colorimetrically or by the determination of radioactivity after incubating labeled substrates with the enzymes.

Estimation of Inhibitory Activity

The inhibition of enzyme reactions was calculated as follows:

$$\text{Percent inhibition} = \frac{A - B}{A} \times 100$$

where A is the product formed in the reaction mixture without inhibitor, and B is the product formed in the reaction mixture with inhibitor. The probit of percent inhibition was plotted on the ordinate and the logarithm of the inhibitor concentration or a corresponding value was plotted on the abscissa. By using standard curves obtained in this way, relating inhibitor concentration to the percentage inhibition, the concentration of inhibitor in culture filtrates or inhibitor solutions was determined. The 50% inhibition concentration (IC_{50}) was obtained from Fig. 4.1. and this value was employed to express the activity.

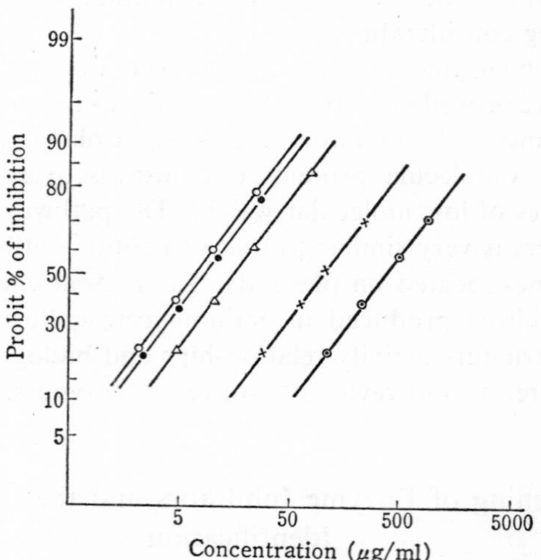

Fig. 4.1 Inhibitory activities of leupeptin during the purification process. ◎——◎, Eluate from carbon; X—X, carbon-chromatography fraction; △——△, alumina chromatography fraction; ○—○, leupeptin-Pr (pure); ●——●, leupeptin-Ac (pure).

Culture of Actinomycetes

Media for actinomycetes are usually composed of glucose, glycerol, lactose, sucrose, starch, etc., as carbon sources and of meat extract, peptone, N-Z amine, yeast extract, soybean meal, etc., as nitrogen sources. Time courses of the production of inhibitors, pH and sugar content are shown below. Optimal medium composition and culture conditions were established on the basis of these results. The pH and sugar content are useful markers for monitoring the progress of fermentation because of the ease with which they can be determined. Fig. 4.2 shows the time course of pepstatin production.

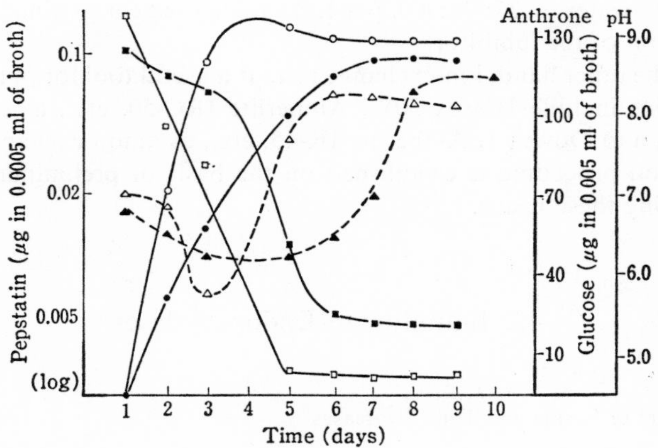

Fig. 4.2 Time course of pepstatin production by *S. testaceus*. Open symbols, medium A, closed symbols, medium B. Circles, activity; triangles, pH; squares, anthrone.

Medium A: starch 1.0%, glucose 1.0%, meat ext. 0.75%, polypeptone 0.75%, NaCl 0.3%, $MgSO_4 \cdot 7H_2O$ 0.1%, $CuSO_4 \cdot 5H_2O$ 0.0007%, $FeSO_4 \cdot 7H_2O$ 0.0001 %, $MnCl_2 \cdot 4H_2O$ 0.0008%, $ZnSO_4 \cdot 7H_2O$ 0.0002%. pH 7.0
Medium B: starch 1.0%, glucose 1.0%, soybean meal 1.5%, K_2HPO_4 0.1%, $MgSO_4 \cdot 7H_2O$ 0.1%, NaCl 0.3%, $CuSO_4 \cdot 5H_2O$ 0.0007%, $FeSO_4 \cdot 7H_2O$ 0.0001 %, $MnCl_2 \cdot 4H_2O$ 0.0008%, $ZnSO_4 \cdot 7H_2O$ 0.0002%. pH 7.0

Preliminary Examination of Inhibitory Activity

When activities were detected in the culture media, the mycelia were separated by filtration. The stability of the activities in the filtrates was examined by heating the filtrates at 60° C for 30 min at pH 2, 7 and 9. Extraction of the filtrates with equal volumes of butanol and butyl acetate was carried out after adjusting the pH of the filtrates to 2 and 8, respectively.

Adsorption of the activities on active carbon (2%, Shirasagi, Takeda Chem. Co., Japan) was also checked at pH 2 and 8. One-fifth portions were resuspended in 80% methanol (1/2.5 volume of the filtrate) and in 50% acetone-water and adjusted to pH 2 and 8, respectively, for extraction. Another portion was extracted with 50% propanol-water. The mycelia were extracted with methanol (1/4 volume of the culture media). The activities in the filtrate, extracts and water-phase were determined. If activities are detected in the solvent phase after pH 2 extraction or eluted from carbon at pH 8, the inhibitors are acidic substances. If activities are detected in the solvent phase after pH 8 extraction or eluted from carbon at pH 2, they are basic substances. Furthermore, the mobility (*Rm*) on high-voltage electrophoresis on filter paper (3,500 V, 15 min, formic acetic acid-water = 25: 75:900, pH 1.8) can be used for preliminary characterization of the inhibitor.

On the other hand, ion-exchange resin is a useful tool for purification. Such resins include Dowex 50W, Amberlite IRC-50, etc., as cation exchangers, and Dowex 1, Amberlite IR-45, etc., as anion exchangers. The purification procedure is established on the basis of preliminary experiments using these resins.

4.2 Inhibitors of Endopeptidases

Inhibitors of Serine and Thiol Proteases

Leupeptin: Leupeptin is an inhibitor of plasmin, trypsin, papain and cathepsin B.

$$\text{R–L–Leu–L–Leu–NH–}\underset{|}{\overset{\displaystyle CH(OH)_2}{}}\text{CH–CH}_2\text{–CH}_2\text{–CH}_2\text{–NH–C}\begin{smallmatrix}\diagup NH_2\\ \diagdown\!\diagdown NH\end{smallmatrix}$$

$$\Updownarrow$$

$$\text{R–L–Leu–L–Leu–NH–}\overset{\displaystyle CHO}{\overset{|}{\text{CH}}}\text{–CH}_2\text{–CH}_2\text{–CH}_2\text{–NH–C}\begin{smallmatrix}\diagup NH_2\\ \diagdown\!\diagdown NH\end{smallmatrix}$$

$$\Updownarrow$$

$$\text{R–L–Leu–L–Leu–NH–}C_6H_{10}(\text{HO})(\text{C}(=\text{NH})\text{NH}_2)$$

Leupeptin (R = CH_3CO or CH_3CH_2CO)

The screening of actinomycete culture filtrates for activity to inhibit trypsin or plasmin led to the discovery of leupeptin.[7–9] The pmr signal of the proton on the aldehyde carbon in aqueous solution indicates that leupeptin occurs mainly in the hydrate and hydroxypiperidine forms.[10] Trypsin or plasmin cleaves peptide bonds on the carboxyl side of arginine or lysine in peptides, and the *C*-terminal argininal residue is an absolute requirement for the inhibition of trypsin or plasmin by leupeptin.[11] The purification procedure, composition of the medium and chemical properties of leupeptin are shown in Tables 4.1, 4.2 and 4.3. The di-*n*-butyl acetal of leupeptin is useful in its purification. Care should be exercised in the use of ion exchange resin, since it may cause racemization of leupeptin.

TABLE 4.1 Isolation and purification of leupeptin.

S. roseus
| cultured at 27°C for 71 hr

culture filtrate (9 l, pH 7.3, IC_{50} = 0.015 ml) — mycelia
| 135 g activated carbon, filtration
| 80% MeOH (pH 2)
crude powder I
| 22 g, IC_{50} = 95 μg
carbon chromatography
| 110 g (diameter 5 cm)
| 80% MeOH (pH 2)
crude powder II
| 11g, IC_{50} = 53 μg
alumina chromatography
| 750 g (diameter 7 cm)
| MeOH
crude powder III
| 2.3 g, IC_{50} = 12 μg
| dissolved in 30 ml of BuOH and refluxed for 3 hr
silica gel chromatography
| 110 g (diameter 5 cm)
| BuOH–BuOAC–AcOH–H_2O (4:8:1:1)
leupeptin di-*n*-butyl acetal
↓ 1.3 g, heating at 60°C in 0.01 N HCl for 3 hr
leupeptin
colorless powder 1 g, IC_{50} = 8 μg

[Experimental procedure 1] Determination of plasmin-inhibiting activity by leupeptin[11]

0.5 ml of plasminogen solution (see below), 0.3 ml of 0.05 M phosphate buffer-saline (PBS, pH 7.2), and 0.1 ml of the buffer with or without an inhibitor were placed in a series of 15 × 100 mm test tubes at 37°C. To each test tube, 0.1 ml of streptokinase (200 μ, Varidase, Lederle Lab.,

TABLE 4.2 Activities, compositions of media and producing strains of protease inhibitors.

Inhibitors	Composition of media	Inhibited enzymes	Producing strains
Leupeptin	glucose 1%, starch 1%, peptone 2.0%, NaCl 0.5%, $MgSO_4 \cdot 7H_2O$ 0.1%, $MnCl_2 \cdot 4H_2O$ 0.0008%, $CuSO_4 \cdot 5H_2O$ 0.0007%, $ZnSo_4 \cdot 7H_20$ 0.0002%, $FeSO_4 \cdot 7H_2O$ 0.0001%	trypsin, plasmin, papain, cathepsin B	*Streptomyces roseus, S. roseochromogenes, S. lavendulae, S. albireticuli, S. thioluteus, S. chartleusis, S. noboritoensis* and more than 11 other species
Antipain	glucose 2%, N-Z amine 1%, yeast extract 0.2%, NaCl 0.3%, $MgSO_4 \cdot 7H_2O$ 0.1%, K_2HPO_4 0.1%, $CuSO_4 \cdot 5H_2O$ 0.0007%, $ZnSo_4 \cdot 7H_2O$ 0.0002%, $FeSO_4 \cdot 7H_2O$ 0.0008%, $MnCl_2 \cdot 4H_2O$ 0.0002%	trypsin, papain, cathepsins A and B	*S. michiganensis, S. violascens, S. mauvecolor, S. yokosukanensis*
Chymostatin	glycerol 2.5%, meat extract 0.5%, peptone 0.5%, yeast extract 1.0%, NaCl 0.2%, $MgSO_4 \cdot 7H_2O$ 0.05%, K_2HPO_4 0.05%, $CaCO_3$ 0.32%	chymotrypsins, papain, cathepsins A, B and D	*S. hygroscopicus, S. lavendulae*
Elastatinal	glucose 3.0%, soybean meal 2.0%, NaCl 0.3%, NH_4Cl 0.25%, $CaCO_3$ 0.6%	Elastase	*S. griseoruber* and more than 2 other species
Pepstatin	glucose 1.0%, starch 1.0%, peptone 0.75%, meat extract 0.75%, NaCl 0.3%, $MgSO \cdot 7H_2O$ 0.1%, K_2HPO_4 0.0007%, $CuSO_4 \cdot 5H_2O$ 0.0001%, $FeSO_4 \cdot 7H_2O$ 0.0008%, $MnCl_2 \cdot 4H_2O$ 0.0002%, $ZnSO_4 \cdot 7H_2O$ 0.0002%	pepsin, gastricsin rennin, cathepsin D, renin	*S. testaceus, S. argenteolus* var. *toyonakensis, S. naniwaensis, S. parvisporogenes* and more than 5 other species
Phosphoramidon	glycerol 2.5%, meat extract 0.5%, polypeptone 0.5%, yeast extract 1.0%, NaCl 0.2%, $MgSO_4 \cdot 7H_2O$ 0.05%, K_2HPO_4 0.05%, $CaCO_3$ 0.32%	thermolysins, collagenase	*S. tanashiensis* and more than 5 other species
Bestatin	glucose 2.0%, starch 2.0%, soybean meal 2.0%, yeast extract 0.5%, NaCl 0.25%, $CaCO_3$ 0.32%, $CuSO_4 \cdot 5H_2O$ 0.0005%, $MnCl_2 \cdot 4H_2O$ 0.0005%, $ZnSO_4 \cdot 7H_2O$ 0.005%	aminopeptidase B, leucine aminopeptidase	*S. olivoreticuli* and more than 5 other species

TABLE 4.3 Chemical properties of protease inhibitors.

Inhibitors	Molecular formula MW	mp (°C)	$[\alpha]_D$	hvpe Rm (Ala=1)	tlc (silica gel) (Rf)	Color test	Solubility
Leupeptin•HCl	$C_{20}H_{38}N_6O_4 \cdot HCl$ 462.0	143–146 (dec.)	$[\alpha]_D^{23}$ −75° (c = 1, H_2O)	0.75	0.6 (PrOH–H_2O 7:3) 0.2 (($(CH_3)_2CO$–H_2O 1:1)	Rydon-Smith 2,4-DNP, TTC, Sakaguchi	H_2O, MeOH, EtOH, BuOH, AcOH, DMF, DMSO
Antipain•HCl	$C_{27}H_{44}N_{10}O_6 \cdot HCl$ 641.2	170–177 (dec.)	$[\alpha]_D^{20}$ −10° (c = 1, H_2O)	0.9	0.4 (BuOH–BuOAc–AcOH –H_2O 4:2:1:1)	Rydon-Smith, 2,4-DNP, TTC Sakaguchi	H_2O, MeOH, DMF, DMSO
Chymostatin A	$C_{31}H_{41}N_7O_6$ 607.7	205–207 (dec.)	$[\alpha]_D^{22}$ +9° (c = 0.25, AcOH)	0.4	0.45(BuOH–MeOH–H_2O 4:1:2)	Rydon-Smith, 2,4-DNP, TTC FCNP	AcOH, DMSO
Elastatinal	$C_{21}H_{36}N_8O_7$ 512.6	196–204 (dec.)	$[\alpha]_D^{25}$ +2° (c = 1, H_2O)	0.58	0.31(BuOH–AcOH–H_2O 4:1:1)	Rydon-Smith, 2,4-DNP, TTC FCNP	H_2O, MeOH, pyridine, DMSO
Pepstatin A	$C_{34}H_{63}N_5O_9$ 685.9	228–229 (dec.)	$[\alpha]_D^{27}$ −90° (c = 0.3, MeOH)	0	0.76(BuOH–BuOAc–AcOH–H_2O 4:4:1:1)	Rydon-Smith	MeOH, EtOH, AcOH, pyridine, DMSO
Phosphoramidon	$C_{23}H_{34}N_3O_{10}P \cdot 2Na$ 589.5	173–178 (dec.)	$[\alpha]_D^{20}$ −33.6° (c = 1, H_2O)	−0.32	0.32 (BuOH–AcOH–H_2O 4:1:1)	Ehrlich, Ammonium molybdate-perchloric acid, Rydon Smith	H_2O, MeOH, DMSO
Bestatin	$C_{16}H_{24}N_2O_4$ 308.4	233–236 (dec.)	$[\alpha]_D^{20}$ −30.6° (c = 1, 1 N HCl)	0.68	0.3 (BuOAc–BuOH–AcOH–H_2O 4:4:1:1)	Ninhydrin, Rydon-Smith	AcOH, DMSO, MeOH, H_2O

Abbreviations: 2,4-DNP, 2,4-dinitrophenyl hydrazine; TTC, triphenyltetrazolium chloride; FCNP, nitroprusside-ferricyanide; DMF, dimethylformamide; DMSO, dimethylsulfoxide; hvpe, high-voltage paper electrophoresis (3,500 V, 15 min, formic acid-acetic acid-water = 25:75:900, pH 1.8).

U.S.A.) was added and the mixture was incubated for 5 min. The reaction was initiated by rapidly mixing in 2.0 ml of fibrinogen (2%, Armour Pharm. Co., U.S.A.). Exactly 20 min later, the reaction was stopped by adding 1.5 ml of 1.7 M perchloric acid. After 1 hr at room temperature, the mixture was centrifuged (3,000 rpm, 5 min), and the absorbance of the supernatant was measured at 280 nm. Plasminogen solution was prepared as follows: 100 ml of human serum was diluted with 2 l of water, and the pH was adjusted to 5.2 with 2 N acetic acid. After 30 min, the sediment was collected by centrifugation at 3,000 rpm for 10 min and suspended in 100 ml of water. It was again collected by centrifugation and dissolved in 100 ml of buffer. This solution was stable for several weeks at 4° C.

Antipain: Antipain is an inhibitor of papain, trypsin and cathepsins A and B.

```
                    H2N-C=NH                      H2N-C=NH
  (C6H5)                |                              |
                       NH      H3C  CH3               NH
    |                   |         \/                   |
   CH2               (CH2)3       CH               (CH2)3
    |                   |          |                   |
HOOC-CH-NH-CO-NH-CH-CO-NH-CH-CO-NH-CH-CHO
    (S)               (S)        (S)
```

antipain

While testing the activity of actinomycete culture filtrates to inhibit a thiol protease such as papain, an active agent which was named antipain was found in *S. michiganensis* and *S. yokosukanensis,* as well as in more than 10 other species.[12] The same compound was also isolated by screening for Sakaguchi-positive products.[13] The *C*-terminal aldehyde structure which was first found in leupeptin is also present in antipain, and the *C*-terminal argininal residue is an absolute requirement for the inhibition of papain by antipain.

The purification, composition of the medium and chemical properties of antipain are shown in Tables 4.2 through 4.4.

[Experimental procedure 2] Determination of papain-inhibiting activity by antipain[12]

1.0 ml of casein solution (2%, pH 7.4), 0.7 ml of 0.05 M borate buffer (containing 0.05 M NaCl, pH 7.4) and 0.1 ml of the buffer with or without an inhibitor were placed in a series of test tubes at 37° C. After 3 min, 0.2 ml of papain solution (2 mg/ml containing 4 mg/ml L-cysteine HCl (E. Merck, Germany)) was added and mixed well. Exactly 20 min later, the

TABLE 4.4 Isolation and purification of antipain.

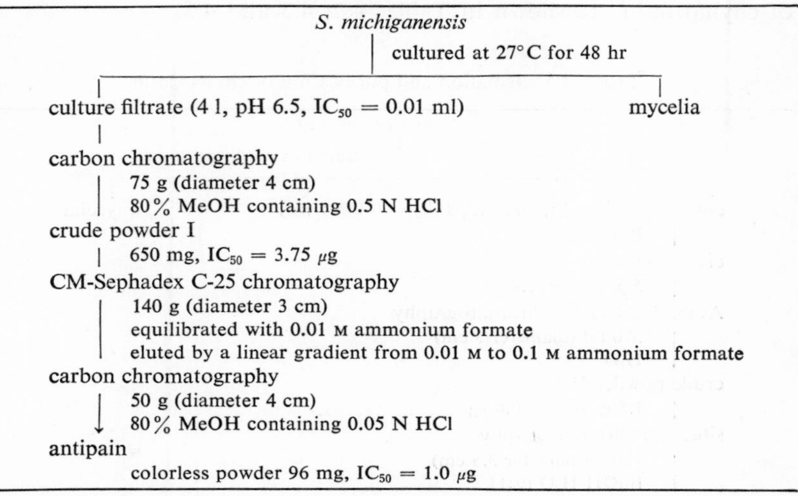

reaction was stopped by adding 2.0 ml of 1.7 M perchloric acid and the absorbance of the supernatant was read at 280 nm.

S. michiganensis
cultured at 27° C for 48 hr
culture filtrate (4 l, pH 6.5, IC50 = 0.01 ml)
mycelia
carbon chromatography
75 g (diameter 4 cm)
80% MeOH containing 0.5 N HCl
crude powder I
650 mg, IC50 = 3.75 μg
CM-Sephadex C-25 chromatography
140 g (diameter 3 cm)
equilibrated with 0.01 M ammonium formate
eluted by a linear gradient from 0.01 M to 0.1 M ammonium formate
carbon chromatography
50 g (diameter 4 cm)
80% MeOH containing 0.05 N HCl
antipain
colorless powder 96 mg, IC50 = 1.0 μg

Chymostatin: Chymostatin is an inhibitor of chymotrypsins.

HOOC–CH(CH₂Ph)(S)–NH–CO–NH–CH(S)[2-iminohexahydro-(4S)-pyrimidyl]–CO–NH–CH(S)[CH₂–CH(CH₃)₂]–CO–NH–CH(CH₂Ph)–CHO

chymostatin

Leu may be replaced by Val or Ile in other chymostatins

Our screening studies, utilizing the determination of antichymotrypsin activity, resulted in the isolation of an inhibitor, named chymostatin, from more than 10 strains of *Streptomyces*.[14,15] Purified chymostatin was a mixture of chymostatins A, B and C. Chymostatin A, which was the main component, contained leucine, while B contained valine, and C contained isoleucine. All the chymostatins contained an unusual amino acid, 2-(2-iminohexahydro-(4S)-pyrimidyl)-(S)-glycine. Chymotrypsin cleaves the carboxyl side of phenylalanine in peptides, and chymostatin (containing *C*-terminal phenylalaninal) inhibits this enzyme.

The purification, composition of the medium and chemical properties of chymostatin are shown in Tables 4.2, 4.3 and 4.5.

TABLE 4.5 Isolation and purification of chymostatin.

S. hygroscopicus
| cultured at 27° C for 96 hr

culture filtrate (3 l, pH 8.3, IC_{50} = 0.0005 ml) — mycelia
| BuOH
crude powder I
| 5 g, IC_{50} = 1.8 μg
Amberlite CG–50 chromatography
| 370 ml (diameter 3 cm)
| H_2O
crude powder II
| 1.5 g, IC_{50} = 0.6 μg
silica gel chromatography
| 130 g (diameter 3.5 cm)
| BuOH-H_2O (9:1)
crude powder III
| 0.8 g, IC_{50} = 0.4 μg
↓ MeOH recrystallization
chymostatin
colorless powder 0.3 g, IC_{50} = 0.3 μg

[Experimental procedure 3] Determination of chymotrypsin-inhibiting activity by chymostatin[14)]

1.0 ml of casein solution (2%, pH 7.4), 0.7 ml of 0.05 M borate buffer (containing 0.05 M NaCl, pH 7.4) and 0.1 ml of the buffer with or without an inhibitor were placed in a series of test tubes at 37° C. After 3 min, 0.2 ml of chymotrypsin solution (4 μg, Sigma Chem. Co., U.S.A) was added and mixed well. The subsequent procedure was as described for measuring antipain activity.

Elastatinal: Elastatinal is a specific inhibitor of elastase.

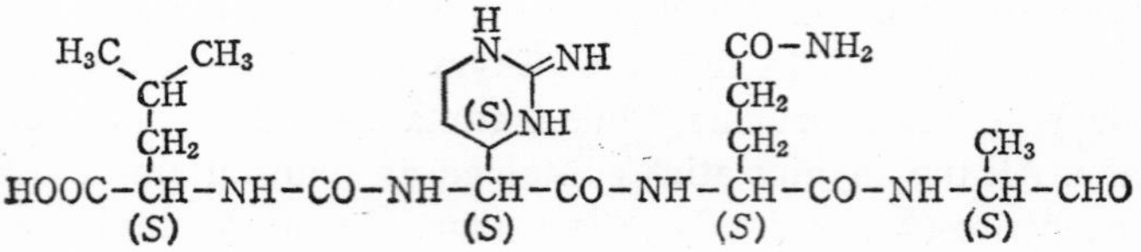

```
H3C   CH3                 H               CO-NH2
   \ /                    N    NH         |
   CH                   /  \ //           CH2
   |                   | (S) NH           |
   CH2                  \  /              CH2            CH3
   |                     |                |              |
HOOC-CH-NH-CO-NH-CH-CO-NH-CH-CO-NH-CH-CHO
    (S)              (S)         (S)       (S)
```

elastatinal

As noted above, leupeptin, antipain and chymostatin, which are serine protease inhibitors, do not show inhibitory activity towards elastase. By testing for antielastase activity in *Streptomyces* culture filtrates, elastatinal was discovered in *S. griseoruber*.[16,17] Elastatinal contains an unusual amino acid which also occurs in chymostatin. Elastase cleaves the carboxyl side of alanine in peptides, and elastatinal (containing *C*-terminal alaninal) inhibits this enzyme.

The purification, composition of the medium and chemical properties of elastatinal are shown in Tables 4.2, 4.3 and 4.6.

TABLE 4.6 Isolation and purification of elastatinal.

S. griseoruber
| cultured at 27°C for 48 hr

culture filtrate (7.5 l, pH 6.6, IC_{50} = 0.01 ml) — mycelia

Amberlite XAD–4 chromatography
| (diameter 4.5 cm)
| acetone-H_2O (1:1)

crude powder I
| 4.3 g, IC_{50} = 7.0 μg

Dowex 1–X2 chromatography
| 350 ml (acetate form, diameter 2.5 cm)
| H_2O

crude powder II
| 3.7 g, IC_{50} = 7.1 μg

SP-Sephadex C-25 chromatography
| 140 ml (diameter 2 cm)
| equilibrated with 50 mM pyridine-formic acid buffer (pH 3.1)
| eluted with 75 mM pyridine-formic acid buffer (pH 4.1)

crude powder III
| 0.85 g, IC_{50} = 2.6 μg

carbon chromatography
| 70 ml (diameter 2 cm)
↓ acetone-H_2O (8:1)

elastatinal
colorless powder 0.3 g, IC_{50} = 1.8 μg

[Experimental procedure 4] Determination of the elastase-inhibiting activity of elastatinal[16]

1.0 ml of elastin-Congo red (2 mg, Boehringer, Mannheim, Germany), 0.7 ml of 0.2 M Tris-HCl buffer (pH 8.8) and 0.1 ml of the buffer with or without an inhibitor were placed in a series of test tubes at 37° C. After 3 min, 0.2 ml of the elastase solution (10 μg, Boehringer, Mannheim, Germany) was added and mixed well. Exactly 30 min later, the reaction was

Table 4.7 Inhibitory activities of leupeptin, antipain, chymostatin, elastatinal, pepstatin, phosphoramidon and bestatin on various proteases.

Enzyme	Substrate	IC_{50} (μg/ml)						
		Leupeptin	Antipain	Chymostatin	Elastatinal	Pepstatin	Phosphoramidon	Bestatin
Trypsin	casein	2.0	0.26	>250.0	>250.0	>250.0	>250.0	>250.0
Plasmin	fibrinogen	8.0	93.0	>250.0	>250.0	>250.0	>250.0	>250.0
Papain	casein	2.5	2.0	2.0	>250.0	>250.0	>250.0	>250.0
Chymotrypsin	casein	>500.0	>250.0	0.15	>250.0	>250.0	>250.0	>250.0
Elastase	elastin-congo red	>250.0	>250.0	>250.0	1.8	>250.0	>250.0	<250.0
Pepsin	casein	>500.0	>250.0	>250.0	>250.0	0.01	>250.0	>250.0
Thermolysin	casein	>250.0	>250.0	>250.0	>250.0	>250.0	0.4	>250.0
Cathepsin A	Z-Glu-Tyr†[1]	>500.0	1.2	62.5	>250.0	>125.0	>250.0	>250.0
Cathepsin B	BAA†[2]	0.44	0.6	2.6	—	>125.0	>250.0	>250.0
Cathepsin C	Ser-Tyr-NA†[3]	>250.0	>250.0	>250.0	>250.0	>250.0	>250.0	>250.0
Cathepsin D	hemoglobin	109.0	>250.0	49.0	>250.0	0.01	>250.0	>250.0
Renin	peptide†[4]	>250.0	>250.0	>250.0	>250.0	4.5	>250.0	>250.0
Aminopeptidase B	Arg-NA†[5]	>250.0	>250.0	>250.0	>250.0	>250.0	>250.0	0.05
Leucine aminopeptidase	Leu-NA†[6]	>250.0	>250.0	>250.0	>250.0	>250.0	>250.0	0.01

†[1] Carbobenzoxy–L-glutamyl–L-tyrosine, †[2] N^{α}-Benzoyl-L-arginine amide HCl.
†[3] L-Seryl–L-tyrosine 2-naphthylamide, †[4] His–Pro–Phe–His–Leu–Leu–(^{3}H-Val)–Tyr–Ser.
†[5] L-Arginine 2-naphthylamide. †[6] L–Leucine 2–naphthylamide.

terminated by adding 2.0 ml of 0.5 M phosphate buffer (pH 6.0), and the absorbance of the supernatant was measured at 492 nm.

As mentioned above, leupeptin, antipain, chymostatin and elastatinal contain *C*-terminal aldehyde groups, and the results indicate that the *α*-acylamino aldehydes are comparable to the acylamino portion of susceptible substrates, causing the strong inhibitory actions (Table 4.7). These inhibitors inhibit carrageenin edema. Leupeptin ointment (1 %) applied to a burn immediately suppresses pain and blister formation. Oral administration of leupeptin and antipain was found to inhibit chemical tumorigenesis in mouse skin[18] and the colon,[19] and to inhibit vascular metastasis of hepatoma to the lung in rats.

Inhibitors of carboxyl proteases including renin

Pepstatins, pepstanones and hydroxypepstatins are inhibitors of pepsin, gastricsin, rennin, cathepsin D and renin.

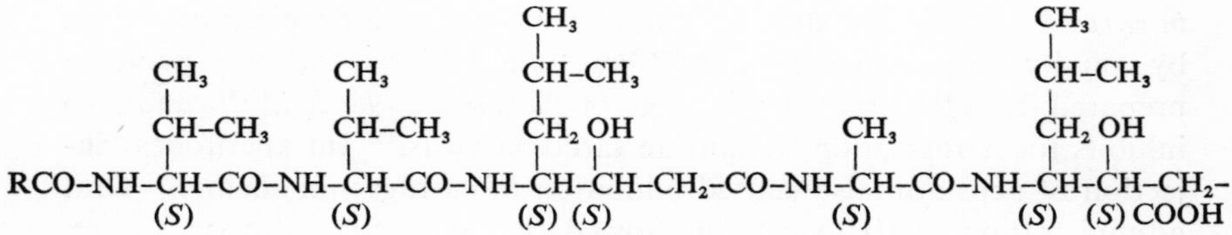

```
                                 CH3                                   CH3
                                  |                                     |
       CH3          CH3          CH-CH3                                CH-CH3
        |            |            |                                     |
       CH-CH3       CH-CH3       CH2 OH                 CH3            CH2 OH
        |            |            |   |                  |              |   |
RCO-NH-CH-CO-NH-CH-CO-NH-CH-CH-CH2-CO-NH-CH-CO-NH-CH-CH-CH2-
       (S)          (S)          (S) (S)                (S)            (S) (S) COOH
                                 pepstatin

                                 CH3                                   CH3
                                  |                                     |
       CH3          CH3          CH-CH3                                CH-CH3
        |            |            |                                     |
       CH-CH3       CH-CH3       CH2 OH                 CH3            CH2 O
        |            |            |   |                  |              |  ||
RCO-NH-CH-CO-NH-CH-CO-NH-CH-CH-CH2-CO-NH-CH-CO-NH-CH-C-CH3
                                 pepstanone

                                 CH3                                   CH3
                                  |                                     |
       CH3          CH3          CH-CH3                                CH-CH3
        |            |            |                                     |
       CH-CH3       CH-CH3       CH2 OH                 CH2OH          CH2 OH
        |            |            |   |                  |              |   |
RCO-NH-CH-CO-NH-CH-CO-NH-CH-CH-CH2-CO-NH-CH-CO-NH-CH-CH-CH2-
                                                                            COOH
                              hydroxypepstatin
```

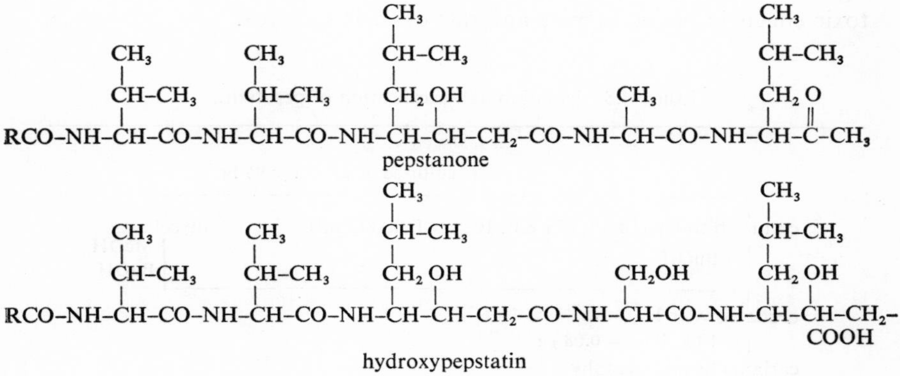

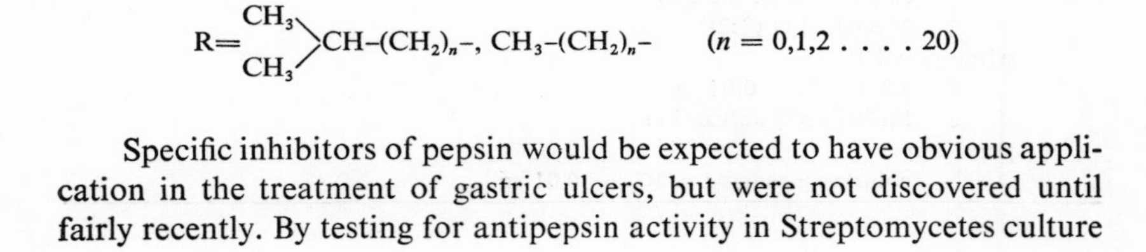

$$R = \begin{matrix} CH_3 \\ CH_3 \end{matrix}\!\!>\! CH\text{-}(CH_2)_n\text{-},\ CH_3\text{-}(CH_2)_n\text{-} \qquad (n = 0,1,2 \ldots . 20)$$

Specific inhibitors of pepsin would be expected to have obvious application in the treatment of gastric ulcers, but were not discovered until fairly recently. By testing for antipepsin activity in Streptomycetes culture filtrates, pepstatins, hydroxypepstatins and pepstanones were discovered in *S. testaceus, S. argenteolus* var. *toyonakensis* and various other *Streptomyces* species.[20–24]

Pepstatins, pepstanones and hydroxypepstatins were found to show almost equal activity against pepsin and cathepsin D[25,26] (Table 4.7). The production of pepstatin by *S. testaceus* is shown in Fig. 4.2, and the purification, composition of medium and chemical properties of pepstatin are shown in Tables 4.2, 4.3 and 4.8.

Pepstatin shows strong inhibition of cathepsin D prepared from swine liver,[26] rabbit liver and human liver,[27] beef lung, rabbit lung, rabbit alveolar macrophages and oil-induced rabbit peritoneal macrophages,[28] glycogen-induced rat peritoneal macrophages,[29] pig brain[30] and mouse leukemia L1210 cells.[31] Pepstatin is very effective against Shay rat ulcer, which is caused by ligation of the pylorus, indicating that excreted pepsin may be a major factor in causing this ulcer.[25] Pepstatin is known to inhibit leukokinin formation and ascitic fluid accumulation.[32] It inhibits renin *in vitro*[26,33–35] and also *in vivo*,[36] and has been used for renin purification by affinity chromatography.[37–40] Pepstatin inhibits carboxyl protease prepared from the erythrocytic stage of *Plasmodium beghei*,[41,42] and also inhibits focus formation by murine sarcoma virus.[43] Intraperitoneal injection of pepstatin also caused inhibition of carrageenin edema.[25] Oral administration of [^{3}H]pepstatin showed that it is not absorbed to any significant extent. The oral administration of pepstatin did not induce toxic effects in pregnant mice and rats or their embryos.

TABLE 4.8 Isolation and purification of pepstatin.

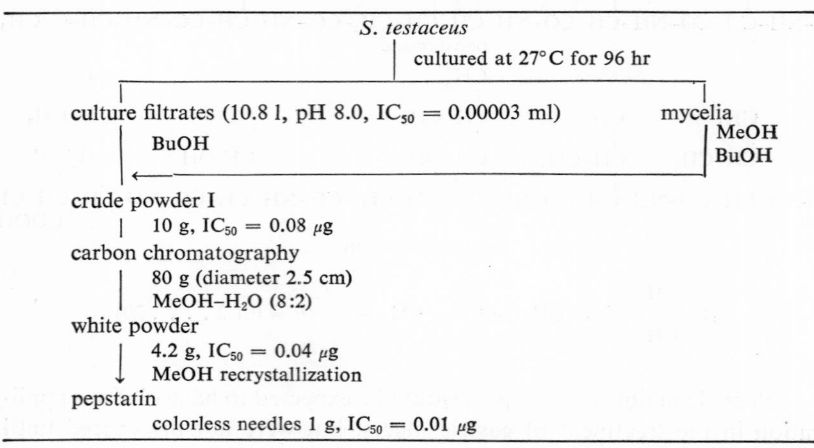

[Experimental procedure 5] Determination of pepsin and renin-inhibiting activity by pepstatin[25,26]

Antipepsin activity: 1.0 ml of casein solution (600 mg of purified casein dissolved with warming in 100 ml of 0.75% lactic acid, pH 2.0), 0.8 ml of 0.02 M KCl–HCl buffer (pH 2.0) and 0.1 ml of the buffer with or without an inhibitor were placed in a series of test tubes at 37° C. After 3 min, 0.1 ml of pepsin solution (4 μg, Sigma Chem. Co., U.S.A.) was added and mixed well. The reaction was terminated 30 min later by adding 2.0 ml of 1.7 M perchloric acid. The absorbance of the protein-free supernatant was measured at 280 nm.

Antirenin activity: 0.5 ml of His–Pro–Phe–His–Leu–Leu–(^{3}H–Val)–Tyr–Ser (300 μg, 8×10^5 cpm), 0.3 ml of 0.05 M phosphate buffer (containing 0.05 M poly-vinylpyrrolidone, pH 7.5) and 0.1 ml of the buffer with or without an inhibitor were placed in a series of test tubes at 37° C. After 3 min, 0.1 ml of renin solution (0.1 dog unit, hog kidney, General Biochem., U.S.A.) was added and mixed well. Exactly 60 min later, the reaction mixture was put in a boiling water bath for 1 min, then it was passed through a column of Dowex 50 × 8 (NH_4^+ form, 200–400 mesh, 0.5 × 5.0 cm) at room temperature, followed by 1 ml of distilled water. The radioactivity of Leu-(^{3}H-Val)-Tyr-Ser in the eluate and the wash was determined in a Beckman LS-250 liquid scintillation system using 8 ml of Bray's scintillation solution.

Inhibitors of Metalloendopeptidases

Phosphoramidon is an inhibitor of metalloendopeptidases.

```
                                              HN—(indole)
            CH3   CH3
               \ /
               CH
                |
    OH         CH2                   CH2
    |           |                     |
O←P—NH—CH—CO—NH—CH—COOH
    |
    O
    |
  (sugar: H3C, HO, OH, OH)
```

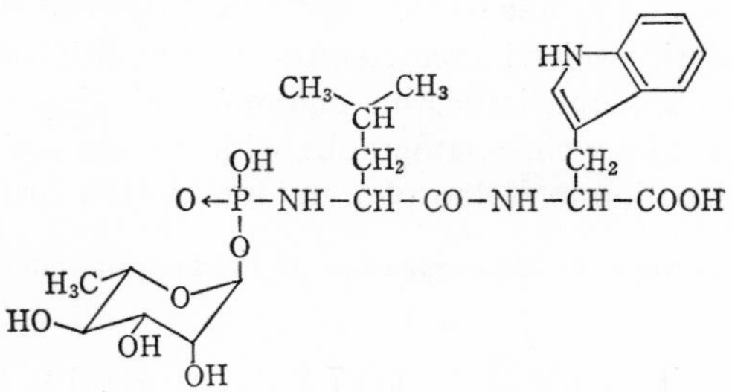

phosphoramidon

We obtained the thermolysin inhibitor, phosphoramidon, from *S. tanashiensis* in our screening studies for inhibitors of metalloendopeptidase.[44] The same compound was also isolated by screening for Ehrlich-positive products.[45] The purification, composition of medium and chemical properties of phosphoramidon are shown in Tables 4.2, 4.3 and 4.9.

TABLE 4.9 Isolation and purification of phosphoramidon.

S. tanashiensis
| cultured at 27° C for 72 hr

culture filtrate (7.7 l, pH 6.0, IC_{50} = 0.0028 ml) — mycelia
| 150 g active carbon
| MeOH (pH 8.0)
crude powder I
| 11.4 g, IC_{50} = 9.0 μg
DEAE-Sephadex A-25 chromatography
| 100 ml (diameter 3 cm)
| equilibrated with 1 M acetic acid
| eluted with 0 to 1 M NaCl (in 1 M acetic acid)
crude powder II
| 0.6 g, IC_{50} = 0.85 μg
Sephadex LH-20 chromatography
| 450 ml (diameter 2.5 cm)
↓ MeOH
phosphoramidon
colorless powder 0.2 g, IC_{50} = 0.4 μg

Phosphoramidon inhibits various metalloendopeptidases such as thermolysin and collagenase (Table 4.7). Several analogs have been prepared and their activities have been compared with those of phosphoramidon (Rha–P–Leu–Trp) as follows: P–Leu–Trp > Rha–P–Leu–Trp > Rha–P–Leu–His > Leu–Trp > Rha–P–Leu·OCH_3 = Rha–P. These data indicate that the P–Leu–Trp moiety is the active group in phosphoramidon. The inhibition types of P–Leu–Trp, Rha–P–Leu–Trp and Rha–P–Leu–His are competitive with respect to carbobenzoxy-glycyl-L-leucineamide. Phosphoramidon and its analogs showed a uv difference maximum at 295 nm with thermolysin, which increased on further addition of inhibitors up to an equimolar amount, indicating equimolar binding of inhibitors to thermolysin. An affinity chromatography with phosphramidon is effective for enzyme purification.[46,47] Phosphoramidon has low toxicity.

[Experimental procedure 6] Determination of thermolysin-inhibiting activity by phosphoramidon[44]

1.0 ml of casein solution (2%, pH 7.5), 0.7 ml of 0.1 M Tris–HCl buffer (containing 0.02 M $CaCl_2$ and 0.06 M NaCl, pH 7.5) and 0.1 ml of the buffer with or without an inhibitor were placed in a series of test tubes at 37° C. After 3 min, 0.1 ml of thermolysin solution (0.75 μg, Nakarai Chem. Co., Japan) was added and mixed well. Exactly 30 min later, the reaction was stopped by adding 2.0 ml of 1.7 M perchloric acid and the absorbance of the supernatant was read at 280 nm.

4.3 Inhibitors of Exopeptidases

Bestatin is an inhibitor of aminopeptidase B.

```
                              CH3
                              |
                              CH-CH3
                              |
          NH2 OH              CH2
          |   |               |
(C6H5)-CH2-CH--CH-CO-NH-CH-COOH
          (R) (S)             (S)
                 bestatin
```

The authors have investigated various functions of mammalian cells in terms of the reactions occurring on the cell surface. We found that the activities of aminopeptidase, alkaline phosphatase and esterase were located on the mammalian cell surface.[48-50] Considering the biological roles of these enzymes located on the cell surface, we undertook a search for inhibitors of aminopeptidase B and obtained bestatin from a culture filtrate of strain MD976-C7, which was identified as *S. olivoreticuli*.[51-53] The purification, composition of the medium and chemical properties of bestatin are shown in Tables 4.2, 4.3 and 4.10.

Table 4.10 Isolation and purification of bestatin.

S. olivoreticuli
| cultured at 27°C for 96 hr

culture filtrate (15 l, pH 6.5, IC_{50} = 0.02 ml — mycelia
| BuOH (pH 2.0)
crude powder I
| 8 g, IC_{50} = 12.0 μg
Dowex 50W-X8 Chromatography
| equilibrated with 0.2 M pyridine-acetic acid buffer (pH 3.0)
| eluted with 0.2 to 1.0 M pyridine-acetic acid buffer (pH 3.0–4.75)
crude powder II
| 0.88 g, IC_{50} = 1.8 μg
Sephadex LH-20 chromatography
| 500 ml (diameter 2.5 cm)
crude powder III
| 0.17 g, IC_{50} = 0.4 μg
silica gel chromatography
| equilibrated with BuOAc–BuOH–AcOH–H_2O (6:4:1:1)
↓ eluted with same solvent
bestatin
colorless needles 0.024 g, IC_{50} = 0.1 μg

Bestatin strongly inhibited the activities of aminopeptidase B and leucine aminopeptidase (Table 4.7) on the surface of mammalian cells including polymorphonuclear leukocytes, macrophages, lymphocytes and tumor cells, indicating that bestatin binds to these enzymes on the cell surface.

Bestatin has 3 asymmetric centers, and 8 stereoisomers have been synthesized. The S configuration of C-2 of the (2S, 3R)-3-amino-2-hydroxy-4-phenylbutanoyl moiety is essential for its activity.[55)] We found that bestatin and its analogs with the 2S configuration could enhance delayed-type hypersensitivity.[56)] Bestatin was shown to produce immune resistance and to enhance the therapeutic effect of bleomycin on tumors. The dosage of bestatin which enhanced the delayed-type hypersensitivity also enhanced the therapeutic effect of adriamycin.[57,58)]

[Experimental procedure 7] Determination of aminopeptidase B-inhibiting activity by bestatin[51)]

0.25 ml of L-arginine β-naphthylamide (2 mM, Peptide Institute, Japan), 0.5 ml of 0.1 M Tris-HCl buffer (pH 7.0) and 0.1 ml of the buffer with or without an inhibitor were placed in a series of test tubes at 37° C. After 3 min, 0.15 ml of aminopeptidase B solution (prepared by Hopsu[54)]) was added and mixed well. Exactly 30 min later, the reaction was stopped by adding 1.0 ml of Fast Garnet reagent (diazonium salt Garnet GBC (1 mg/ml, Sigma Chem. Co., U.S.A.) in 1 M acetic acid buffer, pH 4.2, containing 10% Tween 20) and left to stand for 15 min at room temperature. The absorbance of the supernatant was read at 525 nm.

4.4 Concluding Remarks

Enzyme inhibitors are powerful tools for analyzing various aspects of the homeostasis in living organisms and even for elucidating disease processes. We undertook a search for inhibitors against proteases related to various diseases and found several new inhibitors of small molecular size in culture filtrates of actinomycetes. These inhibitors can be utilized for many purposes: for the identification of the types of enzymes in various tissues or cells, as active ligands in affinity chromatography and as specific tools for analyzing the biological roles of enzymes in various processes and functions of the living organisms. Some may also become therapeutically useful drugs. There is clearly considerable scope for further research.

References

1) H. Umezawa, *Enzyme Inhibitors of Microbial Origin*, University of Tokyo Press (1972).
2) T. Aoyagi, H. Umezawa, *Proteases and Biological Control* (ed. E. Reich *et al.*), p. 429, Cold Spring Harbor Symposium(1975).
3) H. Umezawa, *Methods in Enzymology* (ed. L. Lorand), vol. 45, p. 678, Academic Press (1976).
4) H. Umezawa, T. Aoyagi, *Proteinases in Mammalian Cells and Tissues* (ed. J.T. Dingle *et al.*), vol. 2, p. 637, ASP Biological and Medical Press (1977).
5) T. Aoyagi, *Enzyme Inhibitors* (in Japanese), Kyoritsu zensho 224, Kyoritsu Press (1978).
6) T. Aoyagi, *Bioactive Peptides Produced by Microorganisms* (ed. H. Umezawa *et al.*), p. 129, Kodansha Scientific (1978).
7) T. Aoyagi, T. Takeuchi, A. Matsuzaki, K. Kawamura, S. Kondon, M. Hamada, H. Umezawa, *J. Antibiot.*, **22**, 283 (1969).
8) S. Kondo, K. Kawamura, J. Iwanaga, M. Hamada, T. Aoyagi, K. Maeda, T. Takeuchi, H. Umezawa, *Chem. Pharm. Bull.*, **17**, 1896 (1969).
9) K. Kawamura, S. Kondo, K. Maeda, H. Umezawa, *ibid.*, **17**, 1902 (1969).
10) K. Maeda, K. Kawamura, S. Kondo, T. Aoyagi, T. Takeuchi, H. Umezawa, *J. Antibiot.*, **24**, 402 (1971).
11) T. Aoyagi, S. Miyata, M. Nanbo, F. Kojima, M. Ishizuka, T. Takeuchi, H. Umezawa, *ibid.*, **22**, 558 (1969).
12) H. Suda, T. Aoyagi, M. Hamada, T. Takeuchi, H. Umzawa, *ibid.*, **25**, 263 (1972)
13) S. Umezawa, K. Tatsuta, K. Fujimoto, T. Tsuchiya, H. Umezawa, H. Naganawa, *ibid.*, **25**, 267 (1972).
14) H. Umezawa, T. Aoyagi, H. Morishima, S. Kunimoto, M. Matsuzaki, M. Hamada, T. Takeuchi, *ibid.*, **23**, 425 (1970).
15) K. Tastuta, N. Mikami, K. Fujimoto, S. Umezawa, H. Umezawa, T. Aoyagi, *ibid.*, **26**, 625 (1973).
16) H. Umezawa, T. Aoyagi, A. Okura, H. Morishima, T. Takeuchi, Y. Okami, *ibid.*, **26**, 787 (1973).
17) A. Okura, H. Morishima, T. Takita, T. Aoyagi, T. Takeuchi, H. Umezawa, *ibid.*, **28**, 337 (1975).
18) M. Hozumi, M. Ogawa, T. Sugimura, T. Takeuchi, H. Umezawa, *Cancer Res.*, **32**, 1725 (1972).
19) T. Matsushima, R. S. Yamamoto, K. Hara, T. Sugimura, T. Takeuchi, H. Umezawa, *Igaku no Ayumi* (in Japanese), **88**, 710 (1974).
20) H. Umezawa, T. Aoyagi, H. Morishima, M. Matsuzaki, M. Hamada, T. Takeuchi, *J. Antibiot.*, **23**, 259 (1970).
21) H. Morishima, T. Takita, T. Aoyagi, T. Takeuchi, H. Umezawa, *ibid.*, **23**, 263 (1970).
22) T. Miyano, M. Tomiyasu, H. Iizuka, S. Tomisaka, T. Takita, T. Aoyagi, H. Umezawa, *ibid.*, **25**, 489 (1972).
23) H. Umezawa, T. Miyano, T. Murakami, T. Takita, T. Aoyagi, T. Takeuchi, H. Naganawa, H. Morishima, *ibid.*, **27**, 615 (1973).
24) T. Aoyagi, Y. Yagisawa, M. Kumagai, M. Hamada, H. Morishima, T. Takeuchi, H. Umezawa, *ibid.*, **26**, 539 (1973).
25) T. Aoyagi, S. Kunimoto, H. Morishima, T. Takeuchi, H. Umezawa, *ibid.*, **24**, 687 (1971).
26) T. Aoyagi, H. Morishima, R. Nishizawa, S. Kunimoto, T. Takeuchi, H. Umezawa, *Ibid.*, **25**, 689 (1972).
27) A.J. Barrett, J.T. Dingle, *Biochem. J.*, **127**, 439 (1972).

28) M.H. McAdoo, A.M. Dannenberg, C.J. Hayes, S.P. James, J.H. Sanner, *Infect. Immun.*, **7**, 655 (1973).
29) T. Kato, K. Kojima, T. Murachi, *Biochim. Biophys. Acta,* **289**, 187 (1972).
30) N. Marks, *Science,* **181**, 649 (1973).
31) W. Bowers, C.F. Beyer, N. Yago, *Biochim. Biophys. Acta,* **497**, 272 (1977).
32) L.M. Greenbaum, P. Grebow, M. Johnston, A. Prekash, G. Semente, *Cancer Res.*, **35**, 706 (1975).
33) F.G.J. Lazar, H. Orth, *Science,* **175**, 656 (1972).
34) M. Overturf, M. Lenard, W.M. Kirkendall, *Biochem. Pharm.*, **23**, 671 (1974).
35) R.L. Johnson, A.M. Poisner, *ibid.*, **26**, 639 (1977).
36) R.P. Miller, C.J. Poper, C.W. Wilson, E. Devito, *ibid.*, **21**, 2941 (1972).
37) P. Corvol, C. Devaux, J. Menard, *FEBS Lett.*, **34**, 189 (1973).
38) K. Murakami, T. Inagami, A.M. Michelakis, S. Cohen, *Biochem. Biophys. Res. Commun.*, **54**, 482 (1973).
39) C. Devaux, P. Corvol, J. Menard, *Biochim. Biophys. Acta,* **359**, 421 (1974).
40) K. Murakami, T. Inagami, *Biochem. Biophys. Res. Commun.*, **62**, 757 (1975).
41) M.R. Levy, S.C. Chow, *Biochim. Biophys. Acta,* **334**, 423 (1974).
42) M.R. Levy, S.C. Chow, *Experientia,* **31**, 52 (1975).
43) Y. Yuasa, H. Shimojo, T. Aoyagi, H. Umezawa, *J. Natl. Cancer Inst.*, **54**, 1255 (1975).
44) H. Suda, T. Aoyagi, T. Takeuchi, H. Umezawa, *J. Antibiot.*, **26**, 621 (1973).
45) S. Umezawa, K. Tatsuta, O. Izawa, T. Tsuchiya, H. Umezawa, *Tetr. Lett.*, **1972**, 97.
46) T. Komiyama, T. Aoyagi, T. Takeuchi, H. Umezawa, *Biochem. Biophys. Res. Commun.*, **65**, 352 (1975).
47) T. Komiyama, H. Suda, T. Aoyagi, T. Takeuchi, H. Umezawa, K. Fujimoto, S. Umezawa, *Arch. Biochem. Biophys.*, **171**, 727 (1975).
48) T. Aoyagi, H. Suda, M. Nagai, K. Ogawa, J. Suzuki, T. Takeuchi, H. Umezawa, *Biochim. Biophys. Acta*, **452**, 131 (1976).
49) T. Aoyagi, H. Suda, M. Nagai, H. Tobe, J. Suzuki, T. Takeuchi, H. Umezawa, *Biochem. Biophys. Res. Commun.*, **80**, 435 (1978).
50) T. Aoyagi, M. Nagai, M. Iwabuchi, W.S. Liaw, T. Andoh, H. Umezawa, *Cancer Res.*, **38**, 3505 (1978).
51) H. Umezawa, T. Aoyagi, H. Suda, M. Hamada, T. Takeuchi, *J. Antibiot.*, **29**, 97 (1976).
52) H. Suda, T. Takita, T. Aoyagi, H. Umezawa, *ibid.*, **29**, 100 (1976).
53) H. Nakamura, H. Suda, T. Takita, T. Aoyagi, H. Umzawa, *ibid.*, **29**, 102 (1976).
54) V.K. Hopsu, K.K. Mäkinen, G.G. Glenner, *Arch. Biochem. Biophys.*, **114**, 557 (1966).
55) R. Nishizawa, T. Saino, T. Takita, H. Suda, T. Aoyagi, H. Umezawa, *J. Med. Chem.*, **20**, 510 (1977).
56) H. Umezawa, M. Ishizuka, T. Aoyagi, T. Takeuchi, *J. Antibiot.*, **29**, 857 (1976).
57) T. Aoyagi, M. Ishizuka, T. Takeuchi, H. Umezawa, *Japan. J. Antibiot.*, **30** (Suppl.), 121 (1977).
58) H. Umezawa, *ibid.*, **30** (Suppl.), 138 (1977).

CHAPTER 5

Toxins Produced by Insect or Plant Pathogenic Fungi

A vast number of biologically active substances have been isolated as metabolites of insect or plant pathogenic fungi, and many of their structures have been determined.[1,2] Some of them are considered to play a role in the interaction between the host and parasite. When a toxic substance produced by the pathogenic fungus is harmful to the host and its formation in the infected host is evident, the substance is considered to be a toxin participating in the mycosis. Among the biologically active substances isolated from cultures of pathogenic fungi, however, few have been proved to play a role in the disease caused by the fungi, but this may be mainly because the amount of toxin formed in the diseased host is too small to be detected by chemical methods.

We have isolated a number of biologically active compounds from pathogenic fungi and, among them, destruxins (isolated from *Metarrhizium anisopliae*) and tenuazonic acid (from *Alternaria longipes*) have been shown to be formed in the diseased hosts. This chapter outlines our experimental results on these compounds.

5.1 Destruxins: insecticidal toxins produced by *Metarrhizium anisopliae*

Insect diseases caused by fungi are generally called muscardines, and a number of the pathogenic fungi involved have been isolated and characterized. Among these pathogenic fungi, *M. anisopliae* was known to kill insects after only limited invasion without severe destruction of insect organs, and it was therefore expected to produce some toxic substance(s) participating in the death of the insects.

In 1961, Kodaira showed that *M. anisopliae* produced a toxic substance(s) in the culture filtrate and mycelia which could kill insects on injection, and succeeded in isolating two highly insecticidal principles termed destruxins A and B.[3] The structure of destruxin B was determined as **3** by Tamura *et al.* and later destruxin A was assigned as **6** by Suzuki *et al.*[4] Kuyama *et al.* confirmed the proposed structure of destruxin B by total synthesis.[5]

Destruxin B **3** is different from destruxin A **6** in the hydroxy acid moiety, and **6** is considered to be derived from **3** by the elimination of elements corresponding to a methane molecule. We thus attempted to isolate the intermediary metabolites in the pathway from destruxin B to destruxin A through careful fractionation of the culture filtrate of *M. anisopliae*.

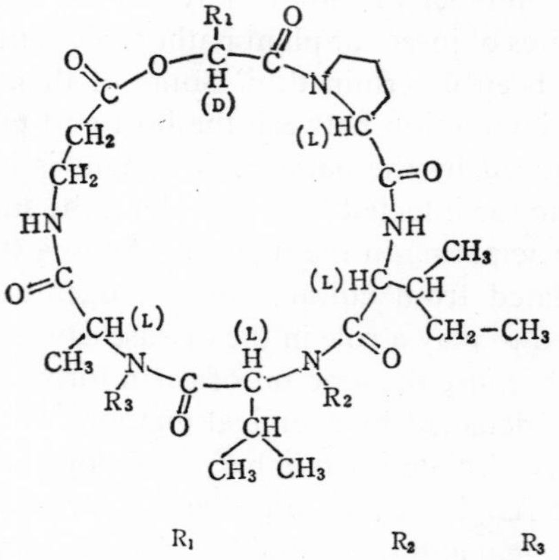

		R_1	R_2	R_3
Protodestruxin	1	$-CH_2-CH<{}^{CH_3}_{CH_3}$	$-H$	$-H$
Desmethyldestruxin B	2	$-CH_2-CH<{}^{CH_3}_{CH_3}$	$-H$	$-CH_3$
Destruxin B	3	$-CH_2-CH<{}^{CH_3}_{CH_3}$	$-CH_3$	$-CH_3$
Destruxin C	4	$-CH_2-CH<{}^{CH_3}_{CH_2OH}$	$-CH_3$	$-CH_3$
Destruxin D	5	$-CH_2-CH<{}^{CH_3}_{COOH}$	$-CH_3$	$-CH_3$
Destruxin A	6	$-CH_2-CH=CH_2$	$-CH_3$	$-CH_3$

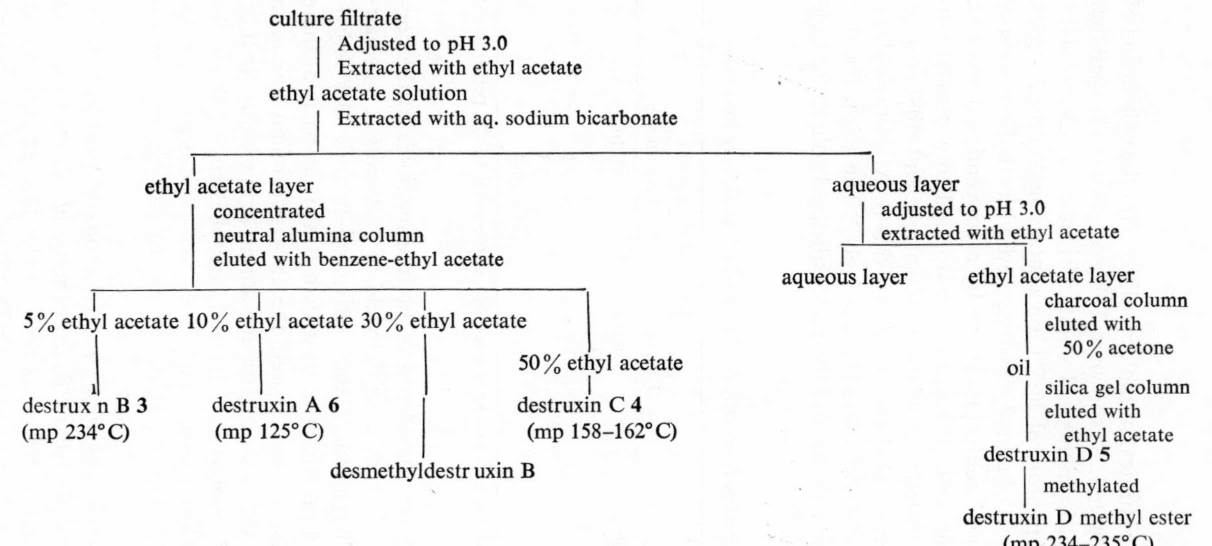

Fig. 5.1 Isolation of destruxins.

Isolation of Destruxins

Prior to fractionation studies, suitable conditions for the cultivation of *M. anisopliae* were investigated. As shown in Table 5.1, it was clarified that destruxins were produced in high yield upon cultivation in Czapek-Dox medium containing 0.5% peptone at 26.5° C on a rotary shaker for 3 weeks.

The culture filtrate thus obtained was extracted with ethyl acetate and fractionated into acidic and neutral fractions by a conventional method. As shown in Fig. 5.1, the neutral fraction was applied to a neutral alumina column, which was eluted with mixtures of benzene and ethyl acetate. On increasing the fraction of ethyl acetate, destruxins B, A, desmethyldestruxin B **2** and destruxin C **4** were eluted successively.[6] The acidic fraction was purified on a charcoal column and then on a silica gel column to yield destruxin D **5**.[6]

TABLE 5.1 Yields of destruxins from culture fitrate of *Metarrhizium anisopliae.*

Destruxin	Cultivation period†	
	1 week	3 weeks
Desmethyldestruxin B	1 mg/l	20 mg/l
Destruxin B	20	60
Destruxin C	3	21
Destruxin D	2	13
Destruxin A	2	65

† The fungus was cultivated on Czapek-Dox medium containing 0.5% peptone on a rotary shaker at 26.5° C.

Destruxins C and D are considered to be intermediates in the conversion of destruxin B to destruxin A; the yield of destruxin A was markedly increased by prolongation of the cultivation period. Desmethyldestruxin B seems to be a precursor of destruxin B. On the basis of these biogenetic considerations, we anticipated the existence of an unknown destruxin lacking the *N*-methyl moiety on any amino acid residue. In fact, we isolated the desired compound, termed protodestruxin, from the culture filtrate of a methionine-requiring mutant of the fungus and elucidated its structure as **1**.[7] Total synthesis of protodestruxin was accomplished by Lee *et al.*[8]

In order to examine the solution conformations of protodestruxin, desmethyldestruxin B and destruxin B, Naganawa *et al.* measured the proton-deuterium exchange reaction as well as the temperature dependence of the peptide proton resonances in DMSO-d_6 by PMR spectroscopy. As shown in Figs. 5.2 and 5.3 in the molecules of destruxin B and desmethyldestruxin B, the peptide protons of β-alanine and isoleucine residues showed little temperature dependence of the chemical shifts,

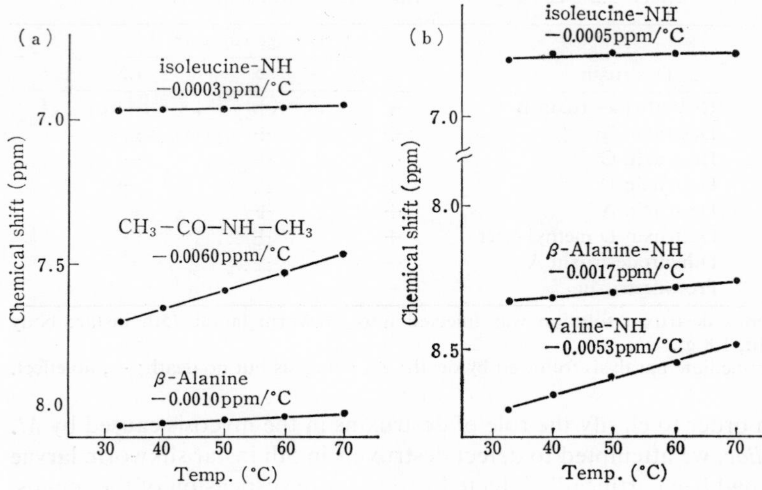

Fig. 5.2 Temperature dependence of the peptide proton resonance. (a) destruxin B; (b) desmethyl destruxin B.

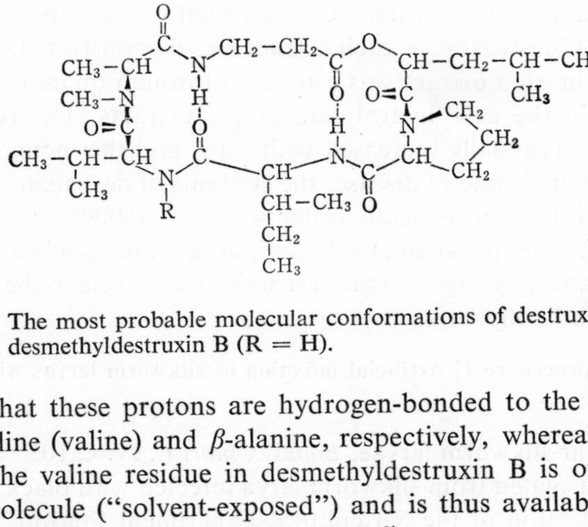

Fig. 5.3 The most probable molecular conformations of destruxin B (R = CH_3) and desmethyldestruxin B (R = H).

indicating that these protons are hydrogen-bonded to the carbonyls of *N*-methylvaline (valine) and *β*-alanine, respectively, whereas the peptide proton of the valine residue in desmethyldestruxin B is oriented away from the molecule ("solvent-exposed") and is thus available to be biologically methylated to afford destruxin B.[9)]

Role of Destruxins in the Insect Disease

The toxicities of destruxins to silkworm larvae on injection are summarized in Table 5.2. Protodestruxin showed no toxicity, indicating the participation of *N*-methyl group(s) in the biological activity.

TABLE 5.2 Toxicity of destruxins to silkworm larvae.†1

Destruxin	Dosage (μg/g)†2		
	15	5	1.5
Demethyldestruxin B	+	+	—
Destruxin B	+	+	±
Destruxin C	+	+	—
Destruxin D	+	±	—
Destruxin A	+	+	±
Destruxin D methyl ester	+	±	—
Dihydrodestruxin A	+	±	—
Protodestruxin	—	—	—

†1 Aqueous destruxin solution was injected into silkworm larvae (5th instar; body weight, 1.8 g).
†2 +, immediate paralysis followed by death; ±, paralysis but no death; —, no effect.

In order to clarify the role of destruxins in the mycosis caused by *M. anisopliae*, we attempted to detect destruxins in 5th instar silkworm larvae which had been artificailly infected with a spore suspension of the fungus. The 5th instar larvae infected with the spore suspension died after 5 or 6 days. The diseased larvae were collected after 4 days (24 hr before death), 5 days (just before death) and immediately after death, and extracted. After partial purification, the extracts were subjected to mass spectrometry and the presence of destruxins in each extract was demonstrated. The content of destruxins in each extract was estimated through quantitative analysis of β-alanine in the acid hydrolysate of the extracts. Destruxins in the diseased insect gradually increased with time, and the increase was most marked at the final state of disease; the contents of destruxins were 0.0004 μmol/larva 24 hr before death (after 4 days), 0.0006 μmol/larva just before death (after 5 days) and 0.02 μmol/larva immediately after death.[10)] These results strongly suggest that destruxins play a role in the mechanism of insect death by mycosis.

[Experimental procedure 1] Artificial infection of silkworm larvas with *M. anisopliae*

Fifth instar silkworm larvae, *Bombyx mori* L. cv. C 105 × J 124, and *M. anisopliae* isolated from silkworm larva infected with black muscardine (from the collection of the Sericultural Experiment Station, Ministry of Agriculture, Forestry and Fisheries, No. 205) were used for the experiments.

For spore formation, the fungus was grown at 25° C for 20 days on agar containing a decoction of silkworm pupae and 2% sucrose. The harvested spores were suspended at a density of 10^6 spores/μl in water containing 0.05% Tween 40. The spore suspension was applied with a small brush to the body surfaces of 5th instar larvae just after exuviation at a dose of approximately 5×10^7 spores/larva. In order to promote infection, the silkworm larvae were reared under rather high humidity for 2 days. Four days after spore inoculation, brown spots characteristic of the fungus infection appeared on the cuticules of the insects (Fig. 5.4), and the existence of hyphal bodies of the fungus in the blood was confirmed by microscopy (Fig. 5.5).

Fig. 5.4 Silkworm larvae infected with *Metarrhizium anisopliae.* White arrows indicate the brown spots characteristic of infection which appeared on the fourth day after spore inoculation.

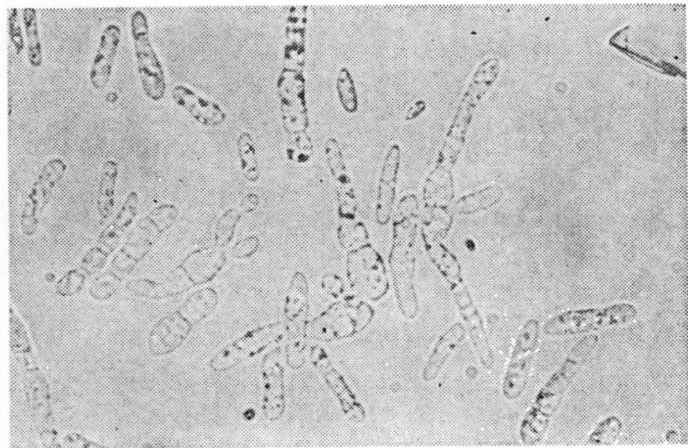

Fig. 5.5 Hyphal bodies of *Metarrhizium anisopliae* observed in the blood of infected silkworm larvae under a microscope (x400).

The living larvae infected with the fungus were collected on the 4th and 5th days after inoculation and immediately frozen to death at –20° C. Larvae that had died of the disease were collected and immediately frozen at –20° C. The corpses were kept at –20° C until used for analysis.

[Experimental procedure 2] Detection of destruxins in the diseased silkworm larvae

Three hundred diseased larvae collected on the 4th day after inoculation were homogenized with 3 l of methanol-acetone mixture (2:1, v/v) and filtered. The filtrate was concentrated *in vacuo* and the aqueous residue was extracted with ethyl acetate at pH 3. The extract was separated into netural, acidic and basic fractions by a conventional method. After drying over anhydrous sodium sulfate followed by concentration *in vacuo*, the neutral fraction yielded 5 g of oil

The neutral oil (5 g) was dissolved in 500 ml of 90% aqueous methanol and the solution was extracted with 1 l of *n*-hexane. The methanol layer was concentrated to give 1.2 g of oil. The oil thus obtained was dissolved in a minimal volume of acetone and mixed with 5 g of refined Celite. The impregnated Celite was dried *in vacuo* and applied to a charcoal column (Wako, charcoal for chromatography, 30 g), which had been packed with water. The column was eluted with 1 l of acetone-water (1: 1, v/v) and then 3 l of acetone. The combined eluate was concentrated to afford 900 mg of oil, which was dissolved in 250 ml of 90% aqueous methanol. The methanol solution was extracted twice with 250 ml of *n*-hexane, and the methanol layer gave 350 mg of oil.

Alumina for chromatography (Merck) was pretreated with ethyl acetate to give neutral alumina, 30 g of which was packed into a column with *n*-hexane. The extracted oil (350 mg) was dissolved in a mixture of ben-

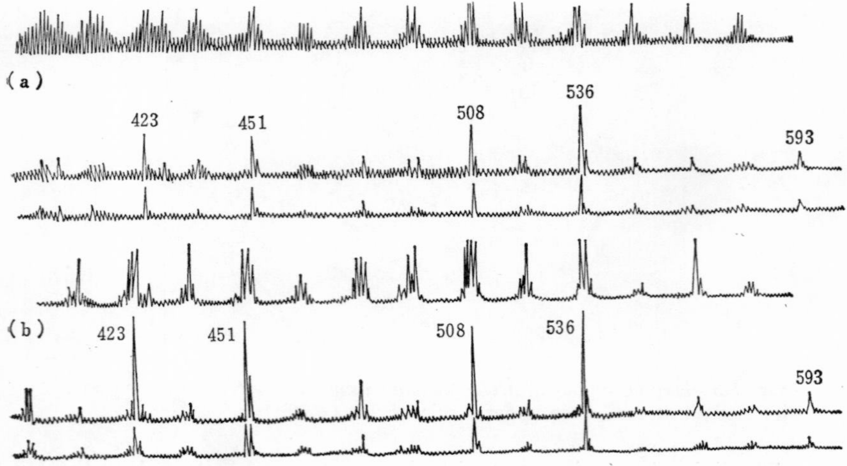

Fig. 5.6 Mass spectra of an extract of diseased silkworm larvae (a) and of authentic destruxin B (b) taken with a Hitachi RMU 6 mass spectrometer (direct inlet, 180° inlet temperature, 70 eV).

zene (6 ml) and *n*-hexane (1 ml), and charged onto the netural alumina column. After elution with 300 ml of benzene and 300 ml of a mixture of benzene-ethyl acetate (2:1, v/v), the column was eluted with 1.5 l of ethyl acetate, affording ca. 8 mg of oil.

The fractions from the column was analyzed by tlc and mass spectrometry to detect destruxins.[11] As shown in Fig. 5.6, the presence of destruxin B was confirmed in the eluate with ethyl acetate.

5.2 Tenuazonic acid: a halo-inducing toxin of *Alternaria longipes*[12]

Various fungi belonging to *Alternaria* species are known to be phytopathogenic and to produce brown spots on the host leaves. *Alternaria longipes* is a pathogenic fungus causing brown-spot disease of the tobacco plant; the typical symptom is the appearance of chlorosis regions called "halos" around brown necrotic spots.

Several years ago, we noticed that the culture filtrate of *A. longipes* contained a toxin which could produce the characteristic brown spots on tobacco leaves, and attempted to isolate the toxin. During the course of investigations, the toxin was found to inhibit the growth of lettuce seedlings. Since bioassay with tobacco leaves is rather tedious, the assay using lettuce seedings was often applied to detect the toxin, assuming that the halo induction and growth inhibition were caused by the same principle. Through the procedure illustrated in Fig. 5.7, tenuazonic acid **7** was identified as a halo-inducing toxin contained in the culture filtrate of the fungus.

When an aqueous solution of tenuazonic acid (250–1000 ppm) was applied to a tobacco leaf at a dose of 20 μl/spot and the treated leaves were kept at 28°C under high humidity for 60 hr, characteristic brown spots surrounded by halos were induced.

As shown in Fig. 5.8, tenuazonic acid gives characteristic uv spectra due to the presence of a β-triketone system in the molecule. The features of the spectra in neutral and basic methanol solutions are similar, while the absorption maxima in acid solution show blue shifts. This phenomenon enabled us to detect tenuazonic acid in the diseased tissue of tobacco plants in the field.

Tobacco leaves infected with *A. longipes* were collected from fields and the brown spots regions were cut from the leaves. The brown spots regions were classified into three groups; early, middle and final stages of disease. The tissues at early stages of infection were extracted with 80% aqueous acetone and partially purified through the procedure illustrated in Fig. 5.9. The partially purified extract was subjected to uv measurement

acidic extract (10 g)
| silica gel column
| eluted with benzene-acetone

eluate with benzene-acetone (95:5, v/v) — eluate with benzene
| silica gel column
| eluted with *n*-hexane-ethyl acetate (9:1, v/v)
oil
| Silic AR-CC 4 column
| eluted with benzene
oil
| dissolved in aqueous methanol
| treated with copper acetate
| extracted with chloroform
crude copper salt
| recrystallized from aqueous methanol
copper salt
| dissolved in chloroform
| washed with 2 N HCl and water
tenuazonic acid (120 mg)
$[\alpha]_D^{20}$ −121°

Fig. 5.7 Isolation of tenuazonic acid from the culture filtrate of *Alternari longipes.*

diseased leaves
|
early stage region
|
acidic fraction
| silica gel column
| eluted with benzene-acetone
eluate with benzene-acetone (95:5, v/v)
| Silic AR-CC 4 column
| eluted with *n*-hexane-ethyl acetate
eluate with *n*-hexane-ethyl acetate (9:1, v/v)
| Avicel tlc, *n*-butanol satd. with 0.5 N ammonia
Rf 0.45–0.55 fraction
| extracted with acetone
fraction containing tenuazonic acid
↓
uv measurement

Fig. 5.8 Extraction and partial purification of tenuazonic acid from diseased tissues.

to detect tenuazonic acid. As shown in Fig. 5.10, the spectra in neutral and basic solutions showed absorption maxima at 280 nm, whereas the spectrum in acid solution gave an absorption maximum at 277 nm; the features of the spectra were similar to those of the spectra of authentic

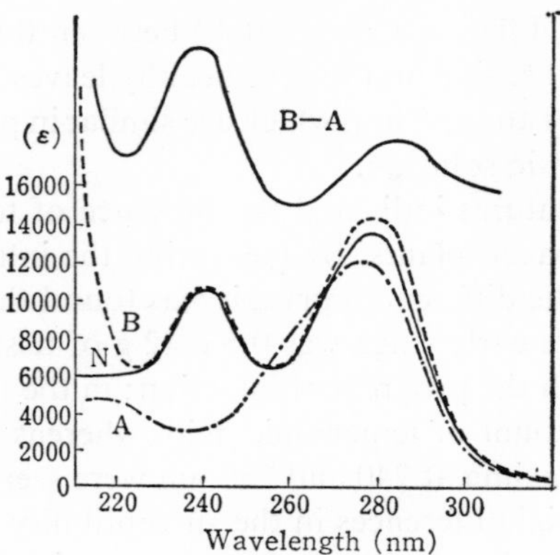

Fig. 5.9 Ultraviolet spectra of tenuazonic acid. A, in 0.09 N HCl–MeOH; B, in 0.09 N NaOH-MeOH; C, in H_2O–MeOH (1:9); B-A, difference spectrum.

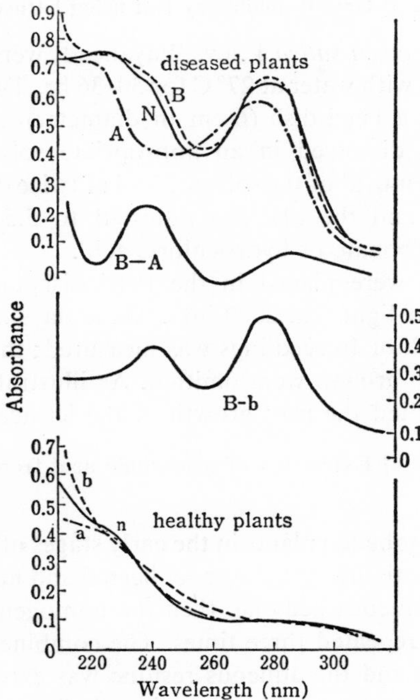

Fig. 5.10 Detection of tenuazonic acid in diseased plants by uv measurements. A, a, in 0.09 N HCl–MeOH; B, b, in 0.09 N NaOH–MeOH; N, n, in in H_2O-MeOH (1:9).

tenuazonic acid. The difference curve (B–b) between the spectrum of the extract of the diseased region and that of healthy leaves in basic methanol showed maxima at 240 and 280 nm, which are similar in position to those of tenuazonic acid in basic solution.

These spectral features indicated the presence of tenuazonic acid in diseased tissue of tobacco plants. By measuring the intensities of the absorption maxima in the difference curve, it was found that the quantity of tenuazonic acid in the early stage was 0.5 g/42 g of tissue and that it decreased markedly with the progress of infection; in the middle stage there was only a minor amount of tenuazonic acid, whereas in the case of the final stage no clear maxima at 240 and 280 nm were seen.

Further, intraracial differences in the susceptibility of tobacco plants to *A. longipes* were correlated with the degree of sensitivity of the plants to tenuazonic acid. It thus appears that tenuazonic acid is a toxin produced by the fungus in tobacco plants *in vivo*.

[Experimental procedure 3] Growth inhibitory test using lettuce seedlings

Lettuce seeds, *Lactuca sativa* L. cv. Wayahead, were germinated on filter paper moistened with water at 27° C for 30–36 hr. Two sheets of filter paper were placed in a Petri dish (6 cm in diameter) and impregnated with sample solution dissolved in an appropriate solvent (acetone or ethyl acetate). After removal of the solvent, 3 ml of twice diluted Hoagland medium was added and the pH was adjusted to 5.5–6.0 with dilute aqueous sodium bicarbonate or hydrochloric acid.

Twelve seedlings were placed in the Petri dish and incubated at 26° C for 3 days under light (ca. 2000 lux). Three days after the start of incubation, the growth of 10 seedlings was measured; those showing minimum and maximum growth were omitted. As illustrated in Fig. 5.11, tenuazonic acid inhibited the root growth of the lettuce seedlings.

[Experimental procedure 4] Extraction of tenuazonic acid from diseased regions of tobacco leaves

Leaves (68 kg) of tobacco plants in the early stages of the disease were collected. The diseased tissues (42 g) were collected and homogenized with 10 volumes of 80% aqueous acetone, then the homogenate was filtered. The procedures were repeated three times. The combined mixtures were concentrated *in vacuo* and the aqueous residue was extracted at pH 2.0 with ethyl acetate. The extract was conventionally fractionated to obtain the acidic fraction, which gave 2.28 g of an oil.

The oil (2.28 g) was dissolved in a minimal volume of benzene and the resulting solution was applied to a silica gel column (Mallinckrodt, 80 g), which was eluted successively with 800 ml of benzene and 800 ml

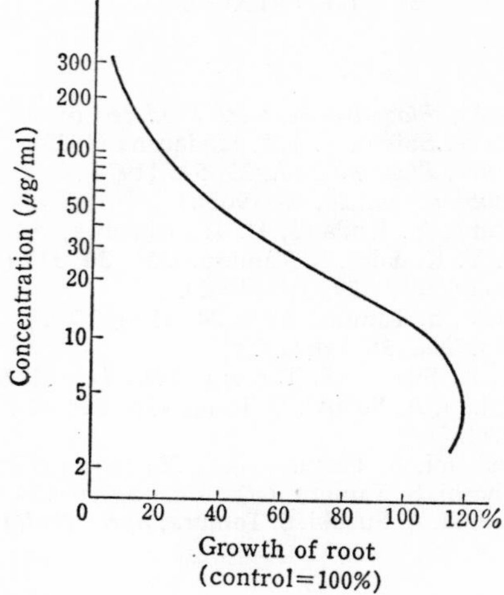

Fig. 5.11 Inhibitory activity of tenuazonic acid against root growth of lettuce seedlings.

of benzene-acetone (95:5, v/v). The eluate with benzene-acetone was concentrated to give 517 mg of an oil, which was further subjected to a Silic AR-CC 4 column chromatography (Mallinckrodt, 20 g). This was eluted successively with 100 ml of benzene-*n*-hexane (1:1, v/v) and 300 ml of *n*-hexane-ethyl acetate (1:1, v/v). The eluate with *n*-hexane-ethyl acetate was concentrated to afford 50 mg of a greenish oil, which was dissolved in a minimal volume of acetone and applied to Avicel plates for tlc. The plates were developed with *n*-butanol saturated with 1.5 *N* aqueous ammonia. The fraction corresponding to *Rf* 0.45–0.55 was scraped off and extracted with acetone. The extract gave 22 mg of a colorless oil.

The oil was shown to contain 0.5 mg of tenuazonic acid by uv measurements.

Acknowledgements

We wish to thank Dr. K. Kawakami, Sericultural Experiment Station, Ministry of Agriculture, Forestry and Fisheries, and Dr. Y. Mikami, Central Research Institute, the Japan Tobacco and Salt Public Corporation, for their contributions to the experiments described in this chapter.

REFERENCES

1) S. Tamura, A. Suzuki, *Bioactive Peptides Produced by Microorganisms* (ed. H. Umezawa, T. Takita T. Shiba), p. 105, Kodansha (1978).
2) G.A. Strobel, *Ann. Rev. Plant Physiol.*, **25**, 541 (1974).
3) Y. Kodaira, *Agric. Biol. Chem.*, **26**, 36 (1962).
4) S. Tamfira, S. Kuyama, Y. Kodaira, H. Higashikawa, *ibid.*, **28**, 137 (1964); A. Suzuki, S. Kuyama, Y. Kodaira, S. Tamura, *ibid.*, **30**, 517 (1966).
5) S. Kuyama, S. Tamura, *ibid.*, **29**, 168 (1965).
6) A. Suzuki, H. Taguchi, S. Tamura, *ibid.*, **34**, 813 (1970).
7) A. Suzuki, S. Tamura, *ibid.*, **36**, 896 (1972).
8) S. Lee, N. Izumiya, A. Suzuki, S. Tamura, *Tetr. Lett.* **1975**, 883.
9) H. Naganawa, T. Takita, A. Suzuki, S. Tamura, S. Lee., N. Izumiya, *Agric. Biol. Chem.*, **40**, 2223 (1976).
10) A. Suzuki, K. Kawakami, S. Tamura, *ibid.*, **35**, 1641 (1971).
11) A. Suzuki, N. Takahashi. S. Tamura, *Org. Mass Spectry.*, **4**, 175 (1970).
12) Y. Mikami, Y. Nishijima, A. Suzuki, S. Tamura, *Agric. Biol. Chem.*, **35**, 611 (1971).

CHAPTER 6

Host-specific Toxins Produced by Some Fungal Pathogens of Plants

The complex processes of infectious plant diseases caused by some fungal pathogens are summarized schematically in Fig. 6.1. Various types of biochemical responses between host and parasite have been postulated by plant pathologists in plant diseases. It is widely known that small organic molecules play important roles, not only in disease development but also in the infection process: that is, in some cases toxic substances arise from a positive offensive reaction by the fungus itself, and in other cases antifungal substances such as prohibitin and phytoalexin are released in the host tissue as defensive substances. However, at present, the most intriguing problem in plant pathology is "host specificity"; in general, pathogens do not attack plants at random but invade specific plants to cause diseases. In other words, in plant diseases fixed combinations are generally observed between pathogens and host plants, for instances, *Pyricularia oryzae* and rice plant in rice blast disease, *Alternaria mali* and susceptible apple cultivars in leaf spot disease of apple and *Helminthosporium maydis* race T and Texas male sterile cultivar of maize in southern leaf blight disease of maize. It is assumed that complex biological mechanisms are involved in the host specificity, but in some cases the causal fungi discriminate their host plants and invade their tissues by exuding highly potent chemicals which display specific toxicity only to the host plants at the initial stage of the infection process. Such toxic chemicals produced by the causal fungi are named "host-specific toxins" by plant pathologists. They have the following properties.[1)]

(1) Host-specific toxins are metabolic products of pathogenic microor-

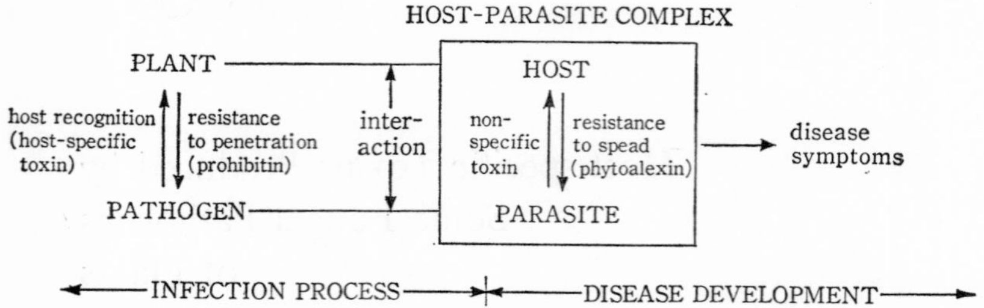

Fig. 6.1 Scheme for the host-parasite interaction in plant diseases.

ganisms which are toxic only to the host plants of these pathogens; that is, these substances have a high degree of toxicity to susceptible species of cultivars, but are essentially without toxicity for other nonhost plants.

(2) There is a quantitative parallelism between host-specific toxin producing ability and the pathogenicity of the pathogen; that is, the pathogenicity is proportional to the amount of the toxin(s) which is produced by the causal pathogen.

(3) A complete correspondence is observed between disease susceptibility and toxin sensitivity in the host plant. All susceptible cultivars are markedly affected by application of the toxin as well as by inoculation of the pathogenic strains, but all resistant cultivars or nonhost plants are tolerant to the toxin.

(4) The host-specific toxin reproduces all the symptoms and biochemical responses of the diseases caused by the corresponding pathogens.

(5) Spores of the pathogen contain the toxin(s), which is released on germination. This indicates that host-specific toxin(s) is involved in initial establishment of the pathogens in the host.

On this view, the host-specific toxin is a primary determinant of the disease and is differentiated from the nonspecific toxic substances which are produced in the host-parasite complex with development of the disease. At present, it is known that host-specific toxins participate in disease establishment in the following 12 instances: *Alternaria alternata* f. sp. *lycoperscici* [stem canker disease of tomato], *A. citri* (one pathotype) [brown spot disease of mandarin], *A. kikuchiana* [black spot disease of Japanese pear], *A. mali* [leaf spot disease of apple], *A. alternata* (one pathotype) [black spot disease of strawberry], *Corynespora cassiocola* [target leaf spot disease of tomato], *Helminthosporium carbonum* [northern leaf spot disease of maize], *H. maydis* race T [southern leaf blight disease of maize], *H. sacchari* [eye spot disease of sugarcane], *H. victoriae* [Victoria blight disease of oat], *Periconia circinata* [milo disease of sorghum] and

Phyllosticta maydis [yellow leaf blight of maize]. However, chemical studies of these host-specific toxins have encountered considerable difficulties, and consquently the chemical structures have been determined only in two cases, helminthospoloside from *H. sacchari*[2,3]* and AM-toxins I (alternariolide), II and III from *A. mali.*[4–7] In the case of *A. kikuchiana,* a toxic compound called altenine was reported and its structure was confirmed by synthesis,[8–10] but this compound is not definitely known to be a host-specific toxin owing to the lack of precise biological experiments. Otani *et al.* reported the isolation of AK-toxin against the same infectious disease with biological data indicating that the compound corresponds to a host-specific toxin,[11–14] so it is now necessary to determine whether or not altenine is the same substance as AK-toxin. The chemistry of AM-toxins has been extensively developed by two groups in Japan. Both of them proposed the same structure for AM-toxin I (alternariolide) at almost the same time,[4–6] and one of them reported the isolation and structure elucidation of AM-toxins II and III (Fig. 6.2).[4,7] These proposed structures were confirmed by total synthesis.[15,16] On the basis of plant pathological data,[17–19] these compounds meet the definition of a host-specific toxin described above. In the following section, based on the experiments on AM-toxins, several points of interest will be discussed on the basis that AM-toxins are considered at present to be representative host-specific toxins whose chemical structures are firmly established.

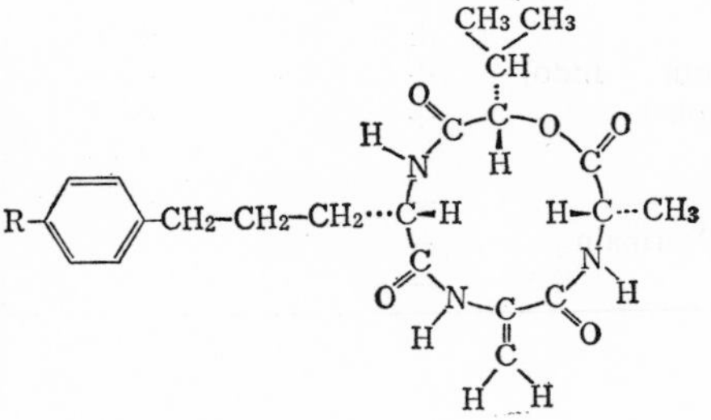

Fig. 6.2 Structures of AM-toxins.
AM-toxin I (alternariolide), $R = OCH_3$;
AM-toxin II, $R = H$;
AM-toxin III, $R = OH$

A. mali Roberts is a causal fungus of leaf spot disease of apple and produces necrotic blotches especially on the leaves, shoots and fruits of susceptible cultivars. When infection is severe, many of the leaves fall off by midsummer. The market value of infected fruits is considerably affected.

However, great differences in tolerance to this disease are observed between apple cultivars, as shown in Table 6.1. Indo, Starking Delicious and Mutsu belong to the very susceptible group, Ralls, Orei, Fuji and Golden Delicious show moderate susceptibility and McIntosh, American Summer Pearmain and Jonathan belong to the highly resistant group. As a result of a series of phytopathological experiments on this disease, Sawamura suggested the presence of a host-specific toxin in the culture filtrate of the causal fungus which induces the same necrotic symptoms as spore inoculation (Table 6.1).[21] Consequently, it seemed that this toxin was a key substance with important physiological significance in infection by the pathogen, *A. mali,* of the host plant. However, in order to permit useful chemical studies on this toxin, it was essential to establish a suitable biological assay system, to establish optimum culture conditions with a highly virulent strain of the pathogen, and to investigate whether the isolated substances met the definition mentioned above. These lines of study will be developed in detail in the sections that follow.

TABLE 6.1 Susceptibility of apple cultivars to spore suspension and culture filtrate of *A. mali.* (AKI-3)

Tested cultivar	Spore suspension inoculation	Culture filtrate dropping	Cutting in culture filtrate
Indo	+++	+++	+++
Starking Delicious	+++	+++	+++
Mutsu (Golden Delicious × Indo)	+++	+++	±
Ralls	±	++	+
Orei (Golden Delicious × Indo)	+	++	++
Fuji (Ralls × Delicious)	+	+	+
Golden Delicious	−	+	−
McIntosh	−	−	−
American Summer Pearmain	−	−	−
Jonathan	−	−	−

6.1 Biological Assay

In the study of biologically active substances, it is no exaggeration to say that the adoption of a suitable bioassay system has a major influence on the success of an investigation. The bioassay method adopted should meet the following criteria. The method should be suitable to examine the nature of the active principle and should be as simple as possible. The time required to determine the result of the assay should be short. High reproducibility is necessary with small amounts of the sample and materials. Quantitative results should be obtainable with good accuracy. As already described, Indo is the most susceptible cultivar to the disease and Jonathan is the most resistant to it. This relation should also hold in the bioassay of the host-specific toxins (AM-toxins I, II and III) in the culture filtrate of *A. mali*. However, many nonspecific toxins which induce wilting, necrosis and chlorosis have been isolated from culture broth of several phytopathogenic *Alternaria* species as follows: alternariol monomethyl ether, alternaric acid, zinnol, tenuazonic acid, tentoxin, altenine,[9,10] and 3,6,8-trihydroxy-3-methyl-3,4-dihydroisocoumarin.[22] If the active principle(s) in the culture filtrate of *A. mali* is investigated using only the leaves of Indo, nonspecific toxins might be detected, since some of them might induce necrosis on the assayed leaves. In fact, some nonspecific factors inducing necrosis, such as tenuazonic acid and tentoxin, were isolated from the culture broth of *A. mali*.[6] Therefore, culture conditions and isolation procedures suitable to obtain AM-toxins were examined by using leaves of both Indo (susceptible) and Jonathan (resistant), and the quantity of the toxins in the culture broth and organic solvent extracts was quantitatively monitored by the use of successively ten-fold diluted solutions in the bioassay system.[23] Since each cultivar of apple trees has been multiplied from its original mother tree by vegetative propagation such as grafting, all trees of the same cultivar should have the same genetic character and thus cultivars such as Indo and Jonathan are considered to be suitable materials for biological experiments, because they do not have any individual variation in principle. However, the fresh leaves of the cultivars sometimes show a small variation in susceptibility which depends on their positions on the tree, time of collection, temperature and assay method. For this reason, two bioassay systems were adopted after several detailed preliminary experiments.

[Experimental procedure 1] Dropping assay

Young fresh leaves of Indo (susceptible) and Jonathan (resistant) were cut from trees just before use and placed, upper side down, on moist cotton in a Petri dish.

To evaluate the titer of the toxin(s) contained in a filtrate cultured using Richards' medium, sample solutions were prepared by successive ten-fold dilutions with distilled water (each prepared sample solution is designated as 10^{-1} = 10-fold diluted solution, 10^{-2} = 100-fold diluted, and so on). Ten μl of each solution was placed as a droplet on the lower side of the leaves. As a control the same diluting procedures were applied to each blank solution of Richards' medium.

To evaluate the titer of the toxin(s) extracted with organic solvents, sample solutions were prepared by successive ten-fold dilutions with ethyl acetate. Ten μl of the sample solution was then spotted on a square silica gel thin layer plate (1×1 cm, Kiesel-gel G nach Stahl, 0.25 mm thick, E. Merck) and allowed to air-dry until the solvent was completely removed (some organic solvents also induce necrosis on the leaves). The silica gel containing the sample was then scraped from the glass plate and placed in a small circle on the lower side of a fresh apple leaf in a Petri dish. The silica gel on the leaf was wetted with 100 μl of distilled water and the leaf was incubated in a moist chamber at 28° C. As a control, silica gel treated only with ethyl acetate was applied using the same procedures. The leaf was examined for the induction of veinal necrosis after 18–20 hr.

[Experimental procedure 2] Cutting assay

Sample solutions were prepared by successive ten-fold dilutions with distilled water. Shoots with five fresh leaves were cut just before use, soaked in a flask containing the sample solution and incubated for 20 hr at 28° C to take up the toxin(s). The induced veinal necrosis was observed. The threshold value of the toxin concentration inducing necrosis is higher in this method as compared with the dropping assay.

6.2 Investigation of Optimal Culture Conditions

It is well known that the quantity and nature of fermentation products depend markedly on the selected strains and culture conditions. The yield of AM-toxins produced by *A. mali* also showed marked fluctuations depending on the pathogenicity of the strain, the kind of culture medium and other culture conditions.

Selection of Strains

An unknown disease of apple causing brown or blackish spots on the leaves and fruits, which was later named "leaf spot disease of apple" had been observed over the main apple-producing areas in Japan since 1956. At that time, outbreaks of the disease were restricted to such cultivars as Indo and Delicious and the isolated strains of the causal fungus showed pathogenicity only against such cultivars. However, the host range of the fungus has since increased to include the cultivars, Jonathan, American Summer Pearmain and McIntosh, which were formerly considered to be resistant. On the basis of detailed studies of the disease, Sekiguchi concluded that the disease of the resistant cultivars was caused by several strongly pathogenic strains which parasitized them, and that a quantitative correlation existed between the toxin-producing ability and the pathogenicity of the strains.[25] Thus, in order to isolate AM-toxins, it is desirable to use a highly potent and stable strain of *A. mali* which does not show any decrease of pathogenicity during a long period of successive culture. More than one thousand strains of *A. mali* were isolated from diseased leaves, shoots and fruits and their pathogenicity was examined by inoculation of spore suspensions and by dropping assay of culture filtrates. A strongly pathogenic strain, I-716, the culture filtrate of which induced necrosis on ten thousand-fold dilution, was thus selected for the isolation of AM-toxins.

Culture Conditions

The quantity of the toxins produced in liquid media was not affected by the agar slant medium from which the inoculum of *A. mali* was transferred. A potato-sucrose medium was selected as the slant medium because *A. mali* showed good mycelial growth in this medium. Using the bioassay system described in section 6.1, the toxin production was quantitatively investigated in several liquid media. Synthetic media such as Richards' and Czapek solution provided much larger amounts of the toxins in comparison with a natural medium such as potato-sucrose. In addition, filtrates cultured under stationary conditions in Roux bottles showed much

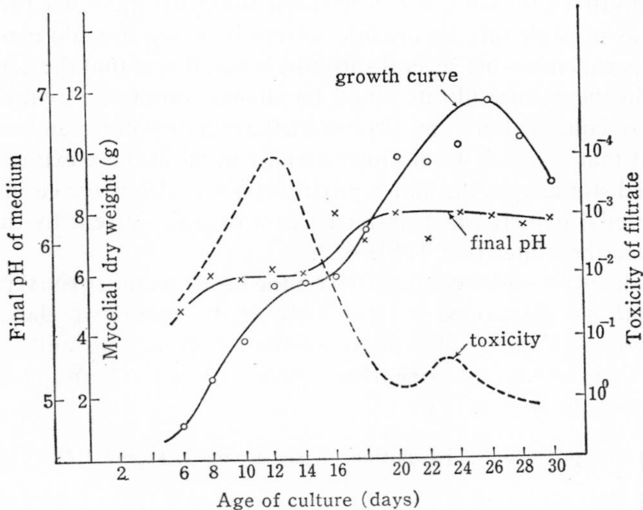

Fig. 6.3 Relation between the culture duration of *A. mali* strain No. I-716 on Richards' medium in the stationary state and the production of toxin in the culture filtrate.

higher activity than those cultured under shaking conditions. The quantity of toxins produced in liquid media was also affected by the temperature and the age of the culture. Fig. 6.3 shows the relation between the culture duration of *A. mali* in Richards' medium under stationary conditions and the quantity of the toxins in the culture filtrate. Based on these results, the best culture condition for the preparation of AM-toxins were concluded to be as follows: I–716 strain of *A. mali* was cultivated in Richards' medium under stationary conditions for 12 days at 28° C. This filtrate showed the maximal toxic activity at 10^4-fold dilution using the dropping assay.

6.3 Isolation of AM-Toxins

Before deciding on isolation procedures, it was necessary to examine some of the properties of the active principle(s), such as stability to pH and temperature change. According to the literature,[21] the principle(s) was at first thought to be stable to heating, but later it was concluded to be rather unstable on the basis of the preliminary finding that the culture filtrate lost its toxic activity on heating for 10 min at 100° C at pH 3.0 or 9.0, and even lost activity on storage overnight at room temperature under neutral conditions (pH 6.0). Thus, it was desirable to extract the active principle(s) as soon as possible into an organic solvent in which it could exist stably. After experiments using several solvents, it was found that the active principle(s) in the culture filtrate could be almost completely extracted by a mixture of equal volumes of dichloromethane, chloroform and ethyl acetate, and that the activity did not decrease at all in these extracts during prolonged storage. As the active principle(s) was also detected in the residual mycelial mat, it was transferred to a organic solvent by the procedures described later (see Table 6.2).

As already mentioned, research on the host-specific toxin(s) produced by *A. mali* was developed independently by two groups in Japan. Since the experimental procedures adopted for isolation were quite different, both of them will be described here. Alternariolide (Okuno *et al.*) is the same substance as AM-toxin I.

[Experimental procedure 3] Isolation of AM-toxins I, II and III[4,7,22]

Isolation procedures for AM-toxins I, II and III are summarized in Table 6.2. *A. mali* (strain I–716), maintained on agar slants, was inoculated into standard size Roux bottles, each containing 400 ml of Richards' medium, and cultured under stationary conditions at 28° C for 12 days to achieve maximal production of the toxins. The filtrate (64 l) of the culture

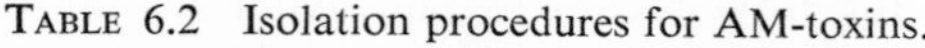

TABLE 6.2 Isolation procedures for AM-toxins.

- culture filtrate (pH 6.0) 64 l
  - —extracted with ethyl acetate
    - aqueous phase
    - ethyl acetate phase (combined with ethyl acetate phase from mycelial mat)
- mycelial mat
  - —soaked in acetone
  - —evaporation of acetone
  - aqueous solution
    - —extracted with ethyl acetate
      - ethyl acetate phase (combined)
      - aqueous phase

ethyl acetate phase
- —evaporation of ethyl acetate
- brown oily residue (11 g)
  - —dissolved in water containing a small amount of methanol
  - —extracted with cyclohexane
    - cyclohexane layer
    - aqueous layer
      - —extracted with benzene
        - benzene layer
        - aqueous layer
          - —extracted with dichloromethane
            - dichloromethane layer
              - —evaporation of dichloromethane
              - —silicic acid column chromatography
              - —eluted with chloroform-ethyl acetate mixture
                - —(7:3–6:4 v/v)
                  - mixture of AM-toxin I and II
                    - —rechromatographed under the same conditions
                      - —(7:3 v/v) AM-toxin II (50 mg)
                      - —(6:4 v/v) AM-toxin I (89 mg)
                - —(1:1 v/v) crude AM-toxin III
            - aqueous layer
              - —extracted with ethyl acetate
                - ethyl acetate layer
                  - —evaporation of ethyl acetate
                  - —silicic acid column chromatography
                  - —eluted with chloroform-acetone mixture (4:1 v/v)
                  - crude AM-toxin III
                    - —rechromatographed under the same conditions
                    - —preparative tlc with benzene-acetone mixture (1:1 v/v)
                    - AM-toxin III (a small quantity)
                - aqueous layer

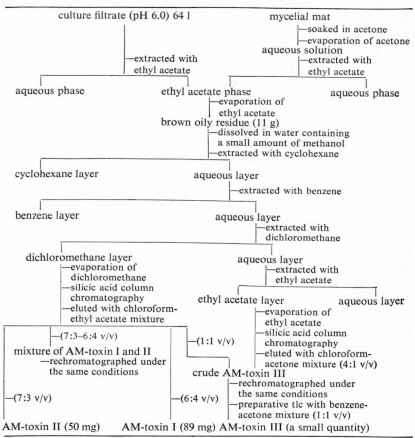

medium was extracted twice with ethyl acetate (30 l each) at pH 6.0. The residual mycelial mat was soaked in acetone overnight and then filtered. After removal of the acetone by evaporation, the residual brown liquid was extracted twice with ethyl acetate. The ethyl acetate extracts from the filtrate and the mycelial mat were combined and dried over anhydrous sodium sulfate. Removal of the solvent under reduced pressure gave a brown oily residue (11 g) containing various kinds of metabolites. The residue was dissolved in water containing a small amount of methanol and extracted successively with cyclohexane, benzene, dichloromethane and finally ethyl acetate. Of these extracts, the major activity inducing necrosis on apple leaves was observed in the dichloromethane extract and minor activity was observed in the ethyl acetate extract. After evaporation of the dichloromethane extract to dryness, the brown residue obtained (4 g) was subjected to silicic acid column chromatography (200 g, 100 mesh, Mallinckrodt), eluting with chloroform-ethyl acetate mixtures with stepwise increases in the ratio of the latter. The active principles were found in the eluate with chloroform-ethyl acetate mixture (7:3 ~ 6:4 by volume). Thin layer chromatography (Kieselgel GF_{254}, E. Merck) with cyclohexane-ethyl acetate-*n*-propanol (15:10:1 by volume) showed that AM-toxin II was eluted first, followed by a mixture of AM-toxins I and II and finally AM-toxin I. Rechromatography of the mixture of AM-toxins I and II with chloroform-ethyl acetate gave a small amount of each of them. Concentration of each eluate gave crude AM-toxins I and II, which were each recrystallized from ethyl acetate to yield pure crystals, AM-toxin I, mp 204–208° C (89 mg) and AM-toxin II, mp 212–214° (50 mg). After elution of the fractions containing AM-toxins I and II, elution of the same column was continued with the same solvent, increasing ratios of the latter component (1:1 by volume), to give a smaller amount of AM-toxin III, which was finally purified by preparative tlc (Kieselgel GF_{254}, 0.5 mm thick) with benzene-acetone mixture (1:1 by volume). At each step of the isolation AM-toxin III was monitored by bioassay and tlc (Kieselgel GF_{254}, E. Merck, 0.25 mm thick), its *Rf* values being 0.50 and 0.35 with benzene-acetone mixture (1:1 by volume) and with benzene-acetone-ether mixture (1:1:1 by volume), respectively (the *Rf* values of AM-toxin I were 0.63 and 0.50, respectively, under the same conditions). The active zone on the plate was scraped off and extracted with acetone. Concentration of the acetone extract gave colorless crystals of AM-toxin III, mp 228° C. A small amount of AM-toxin III was isolated from the dichloromethane extract as described above, but the bulk of it was isolated from the ethyl acetate extract. After evaporation of the ethyl acetate extract to dryness, the residue was subjected to the following procedures for purification: silicic acid column chromatography (Mallinckrodt) with chloroform-acetone mixture using increasing ratios of the latter component, rechromatography on a silicic acid column with benzene-acetone using increasing ratios of the latter component, and finally preparative tlc using the procedures described above.

Two points are noteworthy in connection with the isolation procedures. First, the chromatographic properties of AM-toxins I and II are very similar, as shown in Table 6.3. In order to separate them by column chromatography, the dichloromethane extract was studied by tlc (Kieselgel GF_{254}, E. Merck) with many kinds of solvent system. Stepwise gradient elution with chloroform-ethyl acetate mixture was adopted for the silicic acid column chromatography, because the spots of AM-toxins I and II

Table 6.3 Physicochemical properties of AM-toxins.

AM-toxin	mp (°C)	$[\alpha]_D$	Molecular formula	*Rf* value on tlc (Kieselgel GF, 0.25 mm)		
I	204–208	−79° ($c = 1.0$, $CHCl_3$, 19°)	$C_{23}H_{31}N_3O_6$	0.25†[1]	0.63†[2]	0.50†[3]
II	212–214	−1.4° ($c = 0.55$, DMSO, 27°)	$C_{22}H_{29}N_3O_5$	0.30†[1]	0.63†[2]	0.52†[3]
III	228		$C_{22}H_{29}N_3O_6$	0.15†[1]	0.50†[2]	0.35†[3]

†[1] Cyclohexane-ethyl acetate-*n*-propanol (15:10:1 v/v/v).
†[2] Benzene-acetone (1:1 v/v).
†[3] Benzene-acetone-ether (1:1:1 v/v/v).

AM-toxin	uv $\lambda_{max}^{methanol}$ nm (ε)	ir ν_{max}^{KBr} cm^{-1}					
		ν_{N-H}	ν_{C-H}	$\nu_{C=O\,(ester)}$	$\nu_{C=O\,(amide\ I)}$	$\nu_{C=O\,(amide\ II)}$	$\nu_{C-O-C\,(ester)}$
I	224 (20,600) 268 (2100), 277 (2200), 284 (1800)	3320, 3280	2940, 2860	1748	1658, 1635	1530	1048
II	224 (6000), 268 (800)	3320, 3280	2940, 2860	1742	1658, 1632	1510	1048

TABLE 6.4 Isolation procedures for alternariolide.

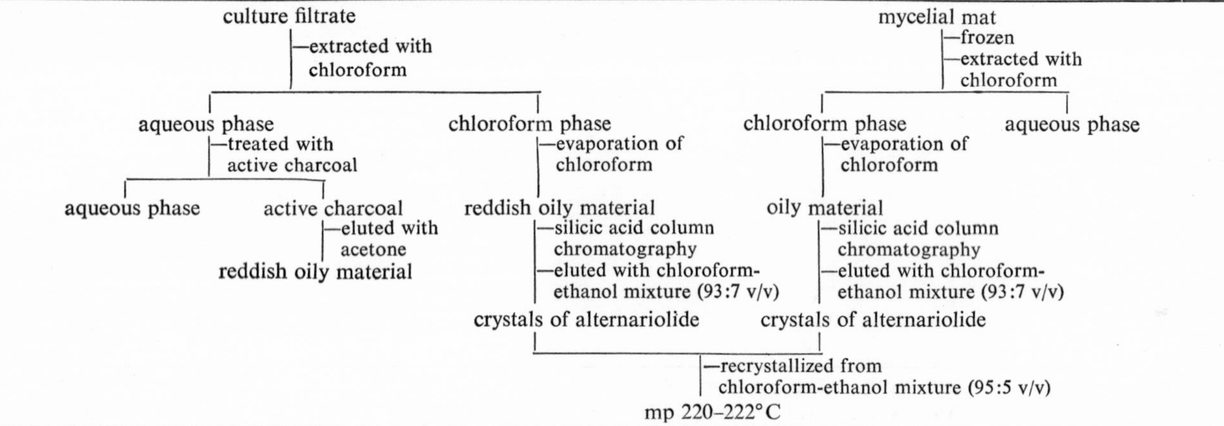

showed good separation in spite of the relatively low *Rf* values. Second, the contents of AM-toxin III in the extracts with dichloromethane and ethyl acetate were low: only 10 mg was isolated in total from 2 tons of the culture broth. In general, it is difficult to isolate such a small amount of active substance from large amounts of impurities. Fortunately, however, AM-toxin III showed quite strong biological activity (see Table 6.5) and had a quite different *Rf* value from other toxins on tlc (see Table 6.3).

[Experimental procedure 4] Isolation of alternariolide[24]

A. mali ROBERTS (AKI-3 strain) was grown in Richards' medium in stationary culture for 2–3 weeks at 37° C. The culture broth (pH 5.5–6.5) thus obtained was filtered and the filtrate was repeatedly extracted with chloroform (Table 6.4). The extract was concentrated *in vacuo* to give a reddish oily material. The aqueous solution was treated with active charcoal and the charcoal was separated. The dried charcoal was eluted with acetone and the acetone solution was concentrated *in vacuo* to give a reddish oil. The host-specific toxicity was examined by means of a biological test: whether or not necrotic brown spots formed when a few drops of the aqueous solution of the extracts were applied to green leaves of Indo. It was found that the chloroform extract showed much more potent toxicity than the acetone eluate. On the other hand, the chloroform extract of the frozen mycelia also exhibited strong toxicity and the chromatographic behavior of this extract was identical with that of the filtrate. The chloroform extract was subsequently chromatographed on silica gel, and elution with chloroform-ethanol (93:3 v/v) gave a crystalline material after evaporation to dryness. Recrystallization from chloroform-ethanol afforded pure crystals, mp 220–222° C, which showed a single spot on silica gel tlc (chloroform-ethanol, 95:5 v/v).

6.4 Biological Activity of AM-Toxins

The AM-toxins thus isolated in crystalline form not only showed extremely potent biological activity (Table 6.5), but solutions of these toxins also induced the same necrotic symptoms on apple leaves in the dropping assay as those observed on inoculation with spores of the causal fungus. In the cutting assay, the threshold concentrations of the toxins inducing necrosis of apple cultivars were found to be 10^{-5}–10^{-3} μg/ml for the extremely susceptible group, 10^{-3}–10^{1-} μg/ml for the moderately susceptible group and 10^{-1}–10^{1} μg/ml for the highly resistant group. These host-specific toxic activities show good correspondence with the results obtained

TABLE 6.5 Susceptibility of apple cultivars to AM-toxins (cutting assay).

Tested cultivar	AM-toxin I (μg/ml)	AM-toxin II (μg/ml)	AM-toxin III (μg/ml)
Indo	10^{-5}	5×10^{-3}	10^{-4}
Starking Delicious	10^{-3}	5×10^{-2}	
Fuji	5×10^{-3}	5×10^{-2}	
Ralls	5×10^{-3}	5×10^{-2}	
Golden Delicious	10^{-2}	5×10^{-2}	
McIntosh	10^{-1}	5×10^{-1}	
American Summer Pearmain	10^{0}	5×10^{-1}	
Jonathan	10^{0}	$> 10^{0}$	10^{1}

by spore inoculation and with empirical observations in orchards. *A. mali* has a wide potential host range, for instance, some cultivars of Japanese pear, in addition to apple cultivars, and AM-toxins were shown to induce veinal necrosis on all of the host plants of *A. mali*. This indicates that AM-toxins play an important role in host recognition during the early stage of infection. This view is supported by the following observation: on inoculation of an avirulent mutant of *A. mali* (nonpathogenic) with a droplet of the dilute toxin solution, the nonpathogenic mutant can invade the host tissue with the aid of the toxin just as the pathogenic strains do. Several of the cellular responses occur very quickly after exposure to the toxins. The mechanism of these effects is not clear yet. Electron microscope pictures have shown that AM-toxins cause changes in the plasma membrane and chloroplast of the host cell within 1 hr after treatment. Quite similar alterations of the host cell are observed in infected tissue of the host leaves upon inoculation of spores of *A. mali*.[20] Such morphologic responses to the toxin are in accord with the results of physiological experiments. Purified AM-toxin I is capable of inducing destruction of the microstructure of chloroplasts and electrolyte leakage from the host tissue almost immediately after treatment.[19] The relationship between the toxin concentration and electrolyte leakage in susceptible and resistant tissue is sigmoidal in form. However, a resistant cultivar tolerates the toxins at approximately 10,000 times higher concentration than a susceptible cultivar. These physiological and morphological observations suggest that the primary attack of the toxin may involve destruction of the membrane system.

6.5 Some Comments on Structure Determination of the Toxins

As already shown in Fig. 6.2, AM-toxins I, II and III are similar cyclic depsipeptides constructed from L-α-hydroxy-isovaleric acid, L-alanine, dehydroalanine and L-α-amino-δ-(p-methoxyphenyl)-valeric acid (AM-toxin I), L-α-amino-δ-phenyl-valeric acid (AM-toxin II) or L-α-amino-δ-p hydroxyphenyl)-valeric acid (AM-toxin III), which are abbreviated below as Hyv, Ala, Dha, Amv, Apv and Ahv, respectively. These structures were determined by chemical modification reactions such as acid hydrolysis, methanolysis and catalytic hydrogenation, and also by a consideration of the physicochemical properties listed in Table 6.3. Some interesting features of these procedures for dealing with cyclic depsipeptides have been selected from the original reports[4,7] and are discussed in the following sections.

Determination of Constituent Amino Acids

As a standard procedure, acid hydrolysis with 6 N hydrochloric acid in a sealed tube at 110° C is applied for determination of the constituents in a peptide. However, in some cases where unusual amino acids are present, such drastic conditions can induce undesirable side reactions. In the case of AM-toxin I, therefore, acid hydrolysates obtained under various conditions were analyzed quantitatively with an amino acid autoanalyser (Fig. 6.4).

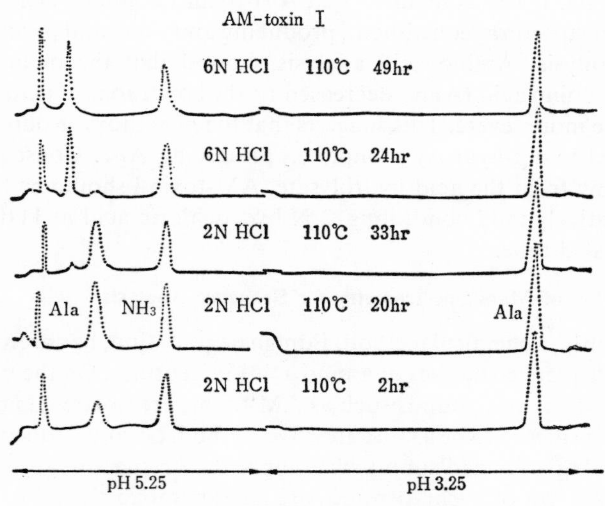

Fig. 6.4 Amino acid analysis of AM-toxin I.

[Experimental procedure 5] Acid hydrolysis of AM-toxin I for structure determination

AM-toxin I (1.0 mg) in 2 N or 6 N hydrochloric acid (3 ml) was heated in a sealed tube at 110° C for 2, 20 and 33 or for 24 and 46 hr, respectively. Each hydrolysate was quantitatively analyzed with an amino acid autoanalyzer, giving the chromatograms shown in Fig. 6.4. In addition each hydrolysate was concentrated under reduced pressure and the residue was chromatographed on a cellulose powder plate (Avicel, SF, Funakoshi), developing with *n*-butanol-acetic acid-water mixture (4:1:2 by volume) to give four spots, A, B, C and D. A and C were positive to iodine vapor, and B and D were positive to ninhydrin (*Rf* values: A, 0.24; B, 0.35; C, 0.76; D, 0.82). Spot A was also positive to 2,4-dinitrophenylhydrazine.

These results indicate that the hydrolysate contains ammonia (1 mol), Ala (B) (1 mol), an unknown unusual amino acid (D) (1 mol) and a carbonyl compound (A). On acid hydrolysis of a peptide, ammonia is usually released from the amide group of asparagine or glutamine or from the enamine group of an α,β-unsaturated-α-amino acid. Since the pmr spectrum of AM-toxin I indicates the presence of two vinyl protons ($\delta^{\mathrm{DMSO}}_{(\mathrm{CH_3})_4\mathrm{Si}}$: 1H 5.23 singlet, 1H 5.36 singlet) and the hydrolysate contains a carbonyl compound, it is likely that an α,β-unsaturated-α-amino acid is contained as a structural component of AM-toxin I. The 2,4-dinitrophenylhydrazone of A was identical with the 2,4-dinitrophenylhydrazone of authentic pyruvic acid. After hydrogenation of AM-toxin I, acid hydrolysis provided 2 mol of Ala. It was concluded that AM-toxin I contains dehydroalanine (Dha) as a structural constituent, producing ammonia and pyruvic acid on acid hydrolysis. Amino acid analysis showed that the quantity of the unknown amino acid (Amv) decreased as the conditions of acid hydrolysis were made more severe. This suggests that the *p*-methoxy group of Amv is hydrolyzed to a *p*-hydroxy group, i.e., producing Ahv. Consequently, to isolate Amv from the acid hydrolysate, AM-toxin I should be hydrolyzed under relatively mild conditions, 2 N hydrochloric acid at 1110° C for 20 hr in a sealed tube.

Application of Mass Spectrometry for Sequence Analysis[26]

Recently, a chemical method, Edman degradation, has been widely applied for sequence analyses of amino acids in peptides. On the other hand, when analogous compounds such as AM-toxins are isolated, mass spectrometry is considered to be a more effective method, because sequence analysis can be easily performed simply by comparing the spectra. However, since the mass spectra of cyclic peptides are more complex than those of linear peptides and, in some cases, recombination of amino acid moieties can take place to form larger fragment ions as a result of transannular effects, it is desirable that sequence determination should be performed with linear derivatives. In the case of AM-toxins, linear methyl ester derivatives were quantitatively obtained by methanolysis of the lactone linkage (these derivatives are abbreviated as AM-toxins I-, II- and III-CH_3OH in Fig. 6.5; the dihydroderivative produced by catalytic hydrogenation of AM-toxin I-CH_3OH is abbreviated as dihydro AM-toxin I-CH_3OH). As shown in Fig. 6.5, the sequence of the constituents in AM-toxins can be unequivocally determined by comparing the larger fragment ions of these linear derivatives. In addition to sequence analysis, the characteristic ions observed in the lower mass region of the spectra provide important infor-

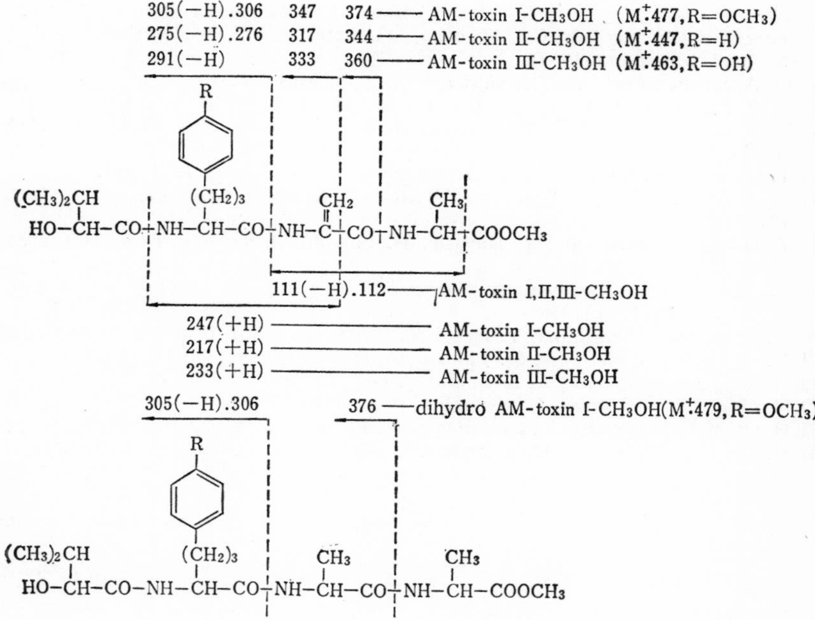

Fig. 6.5 Mass fragments of AM-toxin-CH_3OH's.

TABLE 6.6 Characteristic fragment ions of the components of AM-toxins.

Component	Characteristic ions (elemental composition)
Dha	*m/e* 42 (C_2H_4N), *m/e* 69 (C_3H_3NO)
Ala	*m/e* 44 (C_2H_6N), *m/e* 71 (C_3H_5NO)
Hyv	*m/e* 55 (C_4H_7), *m/e* 83 (C_5H_7O)
Amv	*m/e* 121 (C_8H_9O), *m/e* 134 ($C_9H_{10}O$), *m/e* 161 ($C_{11}H_{13}O$), *m/e* 178 ($C_{11}H_{16}NO$), *m/e* 188 ($C_{12}H_{12}O_2$), *m/e* 203 ($C_{12}H_{13}NO_2$)
Apv	*m/e* 91 (C_7H_7), *m/e* 104 (C_8H_8), *m/e* 131 ($C_{10}H_{11}$), *m/e* 148 ($C_{10}H_{14}N$), *m/e* 159 ($C_{11}H_{11}O$), *m/e* 175 ($C_{11}H_{13}NO$)
Ahv	*m/e* 107 (C_7H_7O), *m/e* 120 (C_8H_8O), *m/e* 147 ($C_{10}H_{11}O$), *m/e* 164 ($C_{10}H_{14}NO$), *m/e* 174 ($C_{11}H_{10}O$), *m/e* 191 ($C_{11}H_{13}NO_2$)

mation on the structure of each constituent, as shown in Table 6.6. Thus, mass spectrometry is a very important technique for the structure elucidation of related peptides.

References

1) R. P. Scheffer, R. B. Pringle, *Pathogen-produced determinants of disease and their effects on host plants.* In *The Dynamic Role of Molecular Constituents in Plant-parasite Interaction* (eds. C.J. Mirocha, I. Uritani), p. 217, Bruce Publishing Co., St. Paul, Minnesota (1967).
2) G. A. Strobel, *Proc. Natl. Acad. Sci. U.S.A.*, **70**, 1693 (1973).
3) G.W. Steiner, G.A. Strobel, *J. Biol. Chem.*, **246**, 4350 (1971).
4) T. Ueno, T. Nakashima, Y. Hayashi, H. Fukami, *Agric. Biol. Chem.*, **39**, 1115 (1975).
5) T. Okuno, Y. Ishita, K. Sawai, T. Matsumoto, *Chem. Lett.*, **1974**, 635.
6) T. Okuno, Y. Ishita, A. Sugawara, Y. Mori, K. Sawai, T. Matsumoto, *Tetr. Lett.* **1975**, 335.
7) T. Ueno, T. Nakashima, Y. Hayashi, H. Fukumi, *Agric. Biol. Chem.*, **39**, 2081 (1975).
8) N. Sugiyama, C. Kashima, M. Yamamoto, T. Sugaya, R. Mohri, *Bull. Chem. Soc. Japan*, **39**, 1573 (1966)
9) N. Sugiyama, C. Kashima, Y. Hosoi, T. Ikeda, R. Mohri, *ibid.*, **39**, 2470 (1966).
10) N. Sugiyama, C. Kashima, M. Yamamoto, R. Mohri, *ibid.*, **40**, 345 (1967).
11) H. Otani, S. Nishimura, K. Kohmoto, *J. Fac. Agr. Tottori Univ.*, **7**, 5 (1972).
12) H. Otani, S. Nishimura, K. Kohmoto, *ibid.*, **8**, 14 (1973).
13) H. Otani, S. Nishimura, K. Kohmoto, *Ann. Phytopathol. Soc. Japan*, **40**, 59 (1974).
14) H. Otani, S. Nishimura, K. Kohmoto, K. Yano, T. Seno. *ibid.*, **40**, 467 (1975).
15) S. Lee, H. Aoyagi, Y. Shimohigashi, N. Izumiya, T. Ueno, H. Fukami, *Tetr. Lett.* **1976**, 843.
16) Y. Shimohigashi, S. Lee, T. Kato, N. Izumiya, T. Ueno and H. Fukami, *Chem. Lett.* **1977**, 1411.
17) K. Kohmoto, T. Taniguchi, S. Nishimura, *Ann. Phytopathol. Soc. Japan*, **43**, 65 (1977).
18) I.D. Khan, K. Kohmoto, S. Nishimura, *ibid.*, **41**, 408 (1975).
19) K. Kohmoto, I. D. Khan, Y. Renbutsu, T. Taniguchi, S. Nishimura, *Physiol. Pl. Path.*, **8**, 141 (1976).
20) P. Park, M. Tsuda, Y. Hayashi, T. Ueno, *Can. J. Bot.*, **55**, 2383 (1977).
21) K. Sawamura, *Bull. Hort. Res. Stn. Japan, Series* **C-4**, 43 (1966).
22) K. Kameda, H. Aoki, H. Tanaka, M. Namiki, *Agric. Biol. Chem*, **37**, 2137 (1973).
23) T. Ueno, Y. Hayashi, T. Nakashima, H. Nishimura, K. Kohmoto, A. Sekiguchi, *Phytopathology*, **65**, 82 (1975).
24) T. Okuno, Y. Ishita, S. Nakayama, K. Sawai, T. Fujita, K. Sawamura, *Ann. Phytopathol. Soc. Japan*, **40**, 375 (1974).
25) A. Sekiguchi, *Bull. Hort. Res. Stn. Nagano Prefecture, Japan*, **12**, 1 (1976).
26) T. Ueno, T. Nakashima, M. Uemoto, H. Fukami, S. Lee, N. Izumiya, *Biomed. Mass Spectrom.*, **4**, 134 (1977).

CHAPTER 7

Phytotoxins Produced by Plant Pathogenic Fungi: Isolation and Bioassay

There have been a large number of reports on physiologically active compounds originating from plant pathogenic microorganisms. The observations that the culture liquids of two grass-invading plant pathogens, *Phyllosticta* sp. and *Cladosporium phlei*, when applied to the host plants, induced symptoms similar to those of diseased plants prompted us to investigate the causal agents. We describe here some representative studies on phytotoxins and related compounds.

7.1 Phytotoxins Produced by *Phyllosticta* sp.

In the course of plant pathogenic studies, Narita in 1966 isolated *Phyllosticta* sp. (a Fungi imperfecti) from the diseased leaf tissue of red clover, *Trifolium pratense* ("*kurohagare*" disease in Japanese). Later, one of us[1] found that the culture liquid of the fungus caused host foliage wilting and lesion with darkening when the leafstalk of the plant was inserted into the liquid. We succeeded in isolating an active compound having a phytotoxic effect and designated it phyllosinol **1**.[1] Its chemical structure, including stereochemistry, was established as identical with that of epoxydon,[2] which had been found by Closse *et al.* as an antitumor agent from the culture filtrates of *Phoma* S 1186.

A second substance, named phyllostine,[3] which is similar in phytotoxic activity to epoxydon, was isolated and its structure was elucidated as

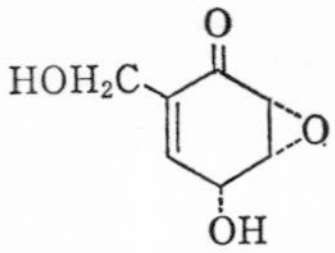

epoxydon (phyllosinol)
1

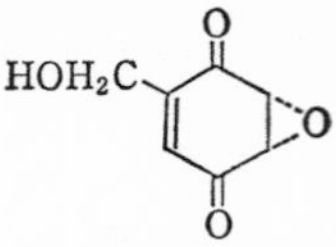

phyllostine
2

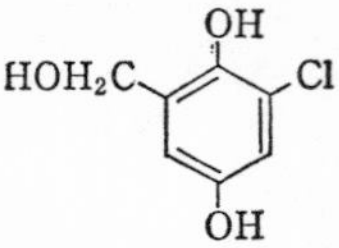

3-chlorogentisyl alcohol
3

2. In view of the structural similarity of compounds **1** and **2** to terremutin, terreic acid and other highly oxygenated derivatives,[4] we have extended our interest to studies on the biosynthesis of these compounds, and to biological activities other than phytotoxicity. A third substance, containing chlorine, appeared to be a new compound with reduced phytotoxicity, and the chemical structure was assigned as 3-chlorogentisyl alcohol **3**. Shortly afterward, 3-chlorogentisyl alcohol was also isolated from a *Phoma* species[5] and *Penicillium canadense*.[6]

Bioassay of Active Compounds

To monitor the separation and isolation of the active compounds, the following two bioassay techniques using red clover plants were chosen.

Leafy stem cutting test:[7,8] Red clover plants growing in the university pasture or in a greenhouse were obtained for tests. A healthy leafy stem 10 cm in length was trimmed off and the basal part inserted into 5 ml of an aqueous sample solution in a test tube (10 × 1 cm). This was kept under light (4000 lux) at 20° C for 72 hr, and the appearance of the plant was observed at 24, 48 and 72 hr, compared with a control kept in water instead of the sample solution.

Generally, darkening of leaflets occurs within 24 hr when an active compound is present at high concentration, while wilting and yellowing of the leaflets develops at 48–74 hr at low concentrations.

Leaf test: Red clover leaflets freed from leafstalk were placed on a water-moistened filter paper in a 9 cm diameter Petri dish, and a piece of absorbent cotton moistened with an aqueous sample solution was placed on the center of a single leaflet cut with a razor. After incubation under light at 20° C for 48 hr, the color changes of the leaflets were observed. The phytotoxic activity can be evaluated in terms of the extent of discoloration compared with the control using water.

Culture Conditions for Phytotoxin Production

Monitoring by means of bioassay, preliminary experiments showed that phytotoxic substances were produced in both potato-dextrose and red clover extract media. Therefore, a potato-dextrose medium was prepared as follows: 3% dextrose was added to the supernatant of the boiled extract of 900 g of potatoes in 3 l of water, and this basal medium was diluted with an equal volume of water prior to use.

The fungus from the stock culture was inoculated in 200 ml of the medium in 500 ml flasks, and incubated at 25° C for 35 days. A peak production of phytotoxic substances appeared at 3 weeks; the pH rose from 6.0 to 7.5, with a marked brown discoloration. Generally, the browner the medium, the more the phytotoxic activity. The use of Waksman flasks shortened the culture period to a third, although such flasks were not employed in this work.

Detection and Isolation of Phytotoxins

The chemical properties of the phytotoxic substances determined in preliminary experiments were as follows: (1) passed through dialysis tubing, (2) adsorbed on active carbon and extracted with acetone or ethyl acetate, (3) in acetone extract, easily colored red on standing at room temperature or on adding 1 N NaOH; more intense red coloration corresponded to stronger phytotoxicity, (4) characteristic uv absorption maximum at 238 nm.

The acetone extract was first subjected to paper chromatography using Toyo filter paper No. 50 (40 × 40 cm), irrigating with *n*-butanol-acetic acid-water (4:1:5 v/v) by the ascending technique. After drying, the paper was cut into 10 sheets of equal size and then extracted with water. On assay by the leafy stem cutting test, it was found that active substances were were located in the range of *Rf* 0.58 to 0.90 and could be visualized with the phenol reagent. For spraying, the phenol reagent was prepared by mixing of one part of Folin-Ciocalteu reagent, one part of water and two parts of 95% ethanol. Color development (dark blue) on paper chromatography and tlc was performed by spraying the reagent followed by exposure to ammonia vapor. In this case, the phytotoxic substances on chromatograms were readily detected by staining with the phenol reagent instead of by bioassay.

The tlc was carried out on Wako gel B-5 plates with benzene-95% methanol (8:1 v/v). As described above, an active substance was located at *Rf* 0.24; the spot was also stained dark blue with the phenol reagent and red with 20% aqueous sodium carbonate.

A large amount of the culture filtrate was treated as shown in Fig. 7.1. After several attempts, an active crystalline material **1**, mp 76–77° C, $[\alpha]_D^{23}$ + 102.5 was obtained and tentatively called phyllosinol (from *Phyllosticta* sp.).

Elementary analysis, molecular weight measurement by the vapor pressure method, and spectral data indicated the composition $C_7H_8O_4$ for **1**. On acetylation it afforded tetra-*O*-acetyl–2,3,5-trihydroxybenzyl alcohol.

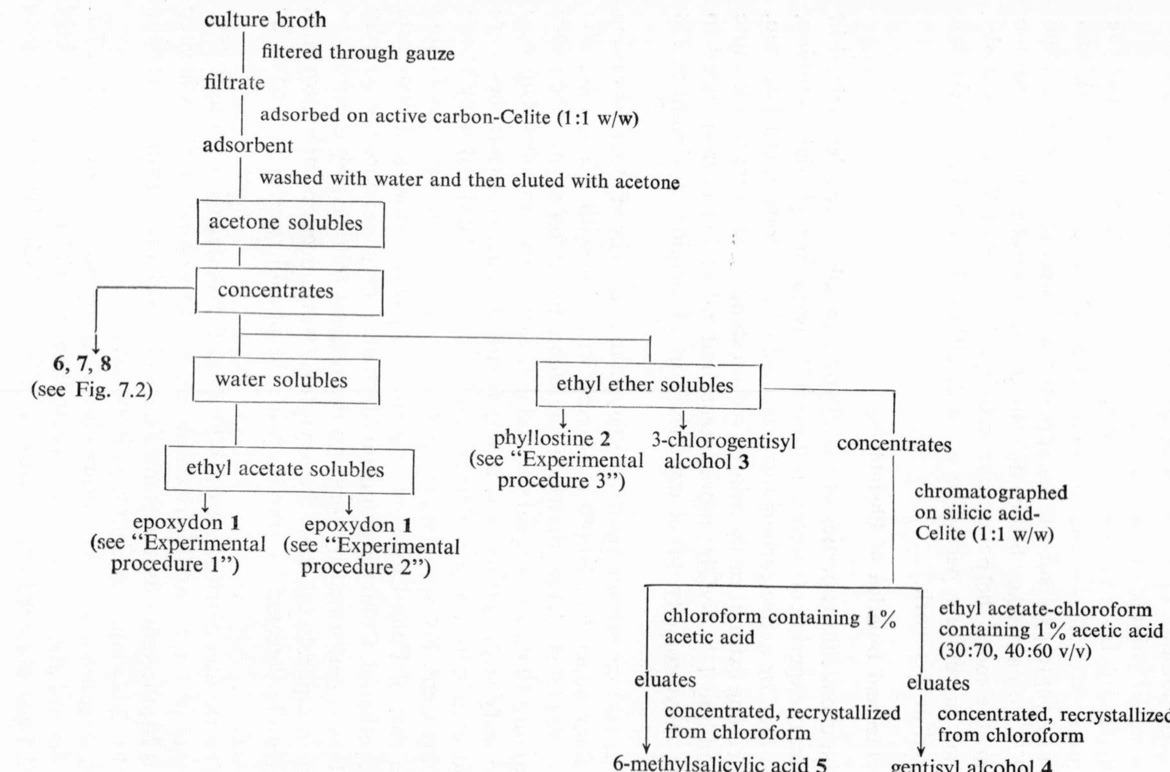

Fig. 7.1 Isolation of the metabolites produced by *Phyllosticta* sp.

Except that the melting point of **1** differed from that of epoxydon (mp 40–45° C), the properties are compatible with those of epoxydon or a diastereoisomer. Finally, it was established by cd measurement that phyllosinol was identical with epoxydon in all respects, including stereochemistry.

[Experimental procedure 1] Crystallization of epoxydon

Active carbon (Shirasagi brand)-Celite (1:1 w/w) (1.6 kg) was added to the culture filtrate (20 l), and the resulting slurry was packed in a column, then washed with water. The active materials were eluted with 20 l of acetone. The eluates were slightly acidified by adding acetic acid and concentrated under reduced pressure. The aqueous concentrate was shaken with ethyl ether several times and subsequently with ethyl acetate. Phytotoxic activity was detected in both solvent extracts. First, the ethyl acetate extract, after drying over sodium sulfate, was concentrated *in vacuo* and chromatographed on a column packed with 200 g of silicic acid (Mallinckrodt)-Celite (1:1 w/w). Elution was carried out with ethyl acetate in chloroform containing 1% acetic acid, the concentration of ethyl acetate being increased in 10% steps. On paper chromatography, the active substance showing *Rf* 0.64 in *n*-butanol-acetic acid-water (4:1:5 v/v) was eluted with ethyl acetate-chloroform (60:40 v/v) to yield a yellow oil after concentration.

For further purification, the partition chromatography of Bulen *et al.* was adopted. A column was packed with a slurry which consisted of silicic acid mixed with an equal quantity of 0.5 N sulfuric acid and dispersed in chloroform saturated with 0.5 N sulfuric acid. The active material was fractionated by elution with *n*-butanol in chloroform with 5% step increases of *n*-butanol concentration. The eluate with *n*-butanol—chloroform (15:85 v/v) contained a chromatographically pure component having phytotoxicity. After concentration, the residue was taken up in water and shaken with ethyl acetate, then the ethyl acetate layer was concentrated after drying over sodium sulfate.

The resulting syrupy residue was further concentrated in a vacuum desiccator to yield crystals which were used as a seed for crystallization in later experiments. Recrystallization from ethyl acetate gave colorless needles **1**, mp 76–77° C, $[\alpha]_D^{23} + 102.5$ ($c = 1$, ethanol). Measurement of the cd in dioxane showed positive Cotton effect at 341 nm ($\Delta_\varepsilon = +3.60$), identical with that of epoxydon.

[Experimental procedure 2] Facile crystallization of epoxydon 1

The ethyl acetate solubles in Fig. 7.1 were chromatographed on a silicic acid-Celite column as described above. The eluate with ethyl acetate-chloroform (70:30 v/v) was concentrated to a small volume, seeded with crystalline epoxydon and stored in a refrigerator to yield crystals **1** (approximately 1 g from 40 l of the culture filtrate). It was also possible to crystallize epoxydon directly from the ethyl acetate solubles by seeding if the culture filtrate contained sufficient epoxydon **1**. The yield was over twice that obtained by the previous method.

Phyllostine 2

Since the ether solubles in Fig. 7.1 contained another phytotoxic substance, chromatographic separations were carried out to obtain the pure material. The second phytotoxin was crystallized as pale yellowish plates, mp 56° C, $[\alpha]_D^{20}$ – 105.6 ($c = 1$, ethanol), and named phyllostine **2**. Elementary analysis and mass spectroscopy (M^+, m/e 154) indicated the molecular formula $C_7H_6O_4$. The following color reactions were positive: the sodium thiosulfate-phenolphtalein test for epoxide, the 2,4-dinitrophenylhydrazine test (red precipitate), and the 1,2-dinitrobenzene test for diketone. In addition, all the spectral data led to the formula **2** for phyllostine. The absolute structure was determined on the basis of chemical conversion of epoxydon into phyllostine. Recently, phyllostine was shown to be an important intermediate of the patulin biosynthetic pathway in *Penicillium uriticae.*[9)]

[Experimental procedure 3] Crystallization of phyllostine 2

The ether solubles in Fig. 7.1 were chromatographed on a silicic acid-Celite (1:1 w/w) column with methanol-chloroform (3:97 v/v) containing 1% acetic acid. The second phytotoxin was obtained as a crude material accompanied by 3-chlorogentisyl alcohol **3**. This was further chromatographed on a silicic acid–Celite (1:1 w/w) column with acetone-benzene (10:90 v/v), yielding pure phyllostine **2** (30 mg from 18 l of the culture filtrate).

[Experimental procedure 4] Chemical conversion of epoxydon 1 to phyllostine 2

Compound **1** (200 mg) was dissolved in 20 ml of ethyl acetate, and cooled in ice water, then 5 ml of chromic solution containing chromium trioxide (4 g) in water (3 ml) and acetic acid (20 ml) was added dropwise with stirring for 15 min. After standing in ice-water with stirring for a further 30 min, the reaction mixture was diluted with water and extracted with ethyl acetate. The extract was dried over sodium sulfate and then concentrated *in vacuo*. The resulting residues were subjected to chromatographic separation using acetone-benzene (10:90 v/v) as described in the previous example, giving phyllostine **2** in 21.6% yield, mp 56–57° C. The mixed melting point with natural phyllostine was not depressed.

3-Chlorogentisyl Alcohol 3

In the isolation of **2** described in "Experimental procedure 3," a third compound **3** with a lower phytotoxic activity was isolated from the chloroform-methanol (91:9 v/v) fraction as optically inactive crystals. Elementary analyses and mass spectrometry, *m/e* 174 (M^+, 100%), 175 ($M^+ + 1$, 15.4 %), 176 ($M^+ + 2$, 38.4%) indicated the molecular formula $C_7H_7O_3Cl$. On acetylation by the usual procedure it afforded a triacetate. The spectral data indicated it to be a new compound, 3-chlorogentisyl alcohol **3**. The structure **3** was confirmed by comparing the chemical and spectral data with those of a synthetic specimen prepared by us. Compound **3** was later isolated independently in two other laboratories from species of *Phoma* and *Penicillium*, and synthesised via different routes.

Gentisyl Alcohol 4 and 6-Methylsalicylic Acid 5[10]

Although the general properties of phytotoxins produced by the fungus were clarified by the isolation and identification of the compounds **1, 2** and **3** followed by the demonstration on their phytotoxicity, there were still unidentified compounds positive to the phenol reagent in the culture filtrates. These compounds appeared to be structurally related to the phytotoxic compounds. Therefore, in view of the possible biosynthetic implications, the ether solubles in Fig. 7.1 were analyzed by paper chromatography with *n*-butanol-acetic acid-water (4:1:5 v/v). Four spots showing *Rf* 0.96 (**5**, unidentified), 0.88 (**3**) 0.82 (**2**) and 0.75 (**4**, unidentified) were detected. On crystallization of the individual compounds, the unidentified

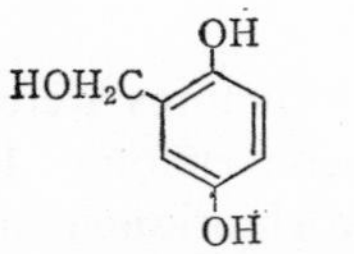

OH
HOH_2C
OH
gentisyl alcohol
4

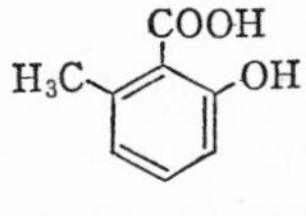

COOH
H_3C OH
6-methylsalicylic acid
5

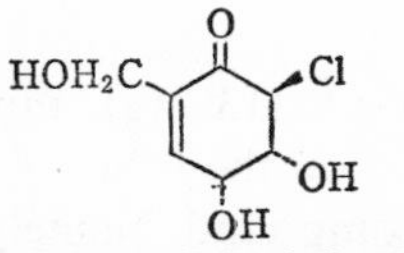

O
HOH_2C Cl
OH
OH
6(*S*)-chloro-4(*R*),5(*R*)-dihydroxy-2-hydroxymethylcyclohex-2-en-1-one
6

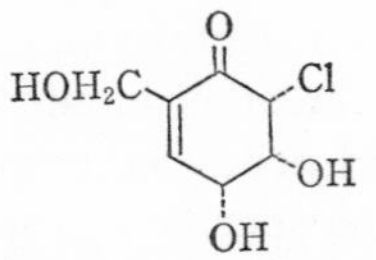

O
HOH_2C Cl
OH
OH
6(*R*)-chloro-4(*R*),5(*R*)-dihydroxy-2-hydroxymethylcyclohex-2-en-1-one
7

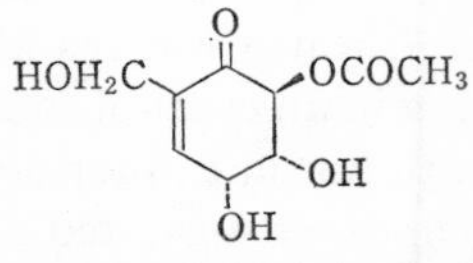

O
HOH_2C $OCOCH_3$
OH
OH
6(*S*)-acetoxy-4(*R*),5(*R*)-dihydroxy-2-hydroxymethylcyclohex-2-en-1-one
8

4, mp 104–105° C, and **5**, mp 168° C (sublimed at 115° C), were established to be gentisyl alcohol and 6-methylsalicylic acid, respectively, from the spectral and chemical data compared with those of authentic samples. An outline of the separation is also given in Fig. 7.1.

Minor Compounds[11] 6, 7, and 8 Related to Epoxydon 1

A chromatographic analysis of the acetone solubles in Fig. 7.1 using *n*-butanol-acetic acid-water (4:1:5 v/v) revealed the presence of some constituents in addition to compounds **1–5**, which were positive to the phenol reagent and had *Rf* values less than that of **1**. Following the procedure outlined in Fig. 7.2, 15 l of the culture filtrate provided the minor constituents, **6** (25 mg yield), mp 126–128° C, $[\alpha]^{20}_{D}-158°$; **7** (20 mg), mp 142–144° C, $[\alpha]^{20}_{D}-125°$; **8** (13 mg), mp 116–168° C, $[\alpha]^{20}_{D}-181$. Their structures were elucidated on the basis of spectral data and the chemical conversion of **1** to each compound. Namely, treatment of **1** with conc. hydrochloric acid in ethanol afforded the chlorine-containing compound **6** through stereospecific *trans* dioxal opening of the epoxide. Subsequently, **6** was worked up with 10% aqueous sodium bicarbonate in tetrahydrofuran to yield the isomer **7**, which was readily transformed to 3-chlorogentisyl alcohol **3** on standing in an ethanol solution for 4 weeks. On the other hand, treatment of **1** with boron trifluoride etherate in acetic acid produced compound **8**. These findings suggested a non-enzymic and stereospecific transformation of the metabolite **1** into compounds **6, 7, 8** and **3**, which might be formed successively.

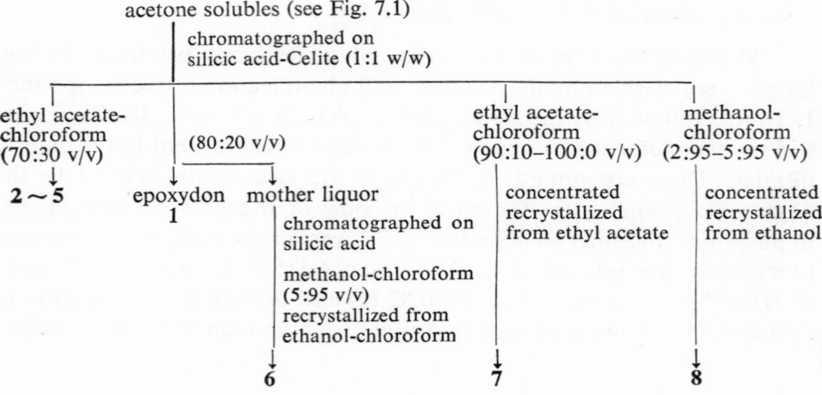

Fig. 7.2 Isolation of compounds **6, 7** and **8**.

7.2 Biosynthesis and Metabolism of Epoxydon 1[12]

A polyketide pathway for the biosynthesis of **1** was postulated based on the findings that compounds **1** and **2** coexist with compounds **4** and **5** in the culture medium, that the former two have carbon skeletons similar to that of gentisyl alcohol **4**, and that both **4** and **5** are known to be biosynthesised via the polyketide pathway in other organisms, although the possibility of the shikimic acid pathway cannot be ruled out. To confirm this pathway, the fungus was grown in the basal medium containing [U-^{14}C] CH_3COONa, which was shown to be incorporated into epoxydon. Next, optimum conditions for the incorporation of the labeled acetate were determined, and the formation of labeled **1** from [^{14}C]gentisyl alcohol was confirmed. As it was concluded that **1** is biosynthesised via the polyketide pathway, excluding the shikimic acid pathway, feeding experiments with [^{13}C]acetates were carried out in order to determine the labeled carbon positions in epoxydon by means of ^{13}C-nmr spectrometry. Additional experiments were attempted to examine whether or not a non-enzymic degradation of epoxydon takes place: sample solutions containing epoxydon in the basal medium were kept under sterile conditions at 25° C for 60 hr to 6 days, and the products were analyzed; **6**, **7** and **3** were identified by isolation techniques and tlc. These results support the view that compounds **6**, **7**, and **3** can be derived from **1** by a reaction mechanism similar to that mentioned on p. 100, where a non-enzymic conversion proceeds in the presence of chlorine ions (130 ppm) originating from potatoes. Thus the biosynthesis and the metabolic transformation of **1** by the fungus can be depicted as shown in Fig. 7.3.

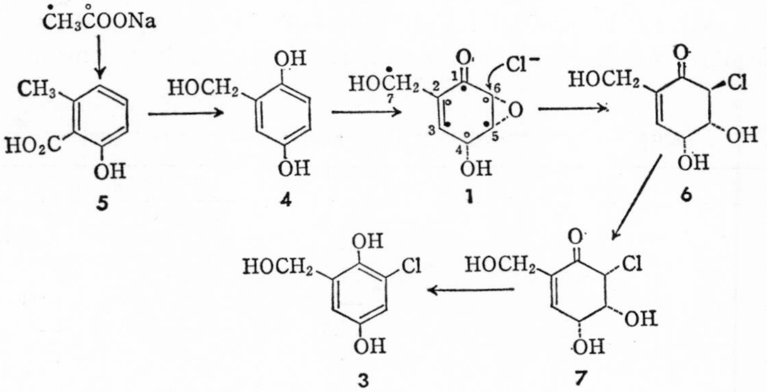

Fig. 7.3 Biosynthesis and metabolism of phytotoxins **1**, **2** and **3** in *Phyllosticta* sp.

[Experimental procedure 5] Isolation of [^{14}C]-epoxydon 1

Sodium [U-^{14}C]acetate was added to the basal medium at concentrations of 0.01, 0.03 and 0.05%, and the fungus was grown in a set of three flasks (500 ml), each containing 200 ml of the medium, for 20 days at 25° C. Following the procedure in "Experimental procedure 2," crystalline epoxydon **1**, mp 77° C, was obtained in the highest yield (1.3 mg, 3180 cpm/mg) in the case of 0.05% acetate. No crystals were obtained at higher concentration than 0.05%, at which decomposition of **1** might occur due to the higher pH of the culture medium.

[Experimental procedure 6] Preparation of [^{14}C]gentisyl alcohol 4

Using a method similar to that of Closse *et al.*, potassium iodide (200 mg) was added to [^{14}C]epoxydon (40 mg) in glacial acetic acid (30 ml), and the resulting solution was heated at 78° C for 40 min. After cooling in a ice-bath, the solution was poured into a cooled aqueous solution of sodium bisulfate (40 mg in 35 ml of water). The solution was extracted three times with ethyl acetate. The extract was treated by the method described in Fig. 7.1, yielding crystalline [^{14}C]gentisyl alcohol **4** (6.6 mg). The crystals were mixed with non-labeled gentisyl alcohol (6.7 mg) and recrystallised from chloroform. [^{14}C]Gentisyl alcohol thus obtained, showing a specific radioactivity of 810 cpm/mg, was used for feeding experiments.

[Experimental procedure 7] Isolation of [^{13}C]epoxydon

The fungus was grown under the same conditions as in "Experimental procedure 5" in four flasks containing the basal medium with either sodium [1-^{13}C]acetate (purity, 88%) or [2-^{13}C]acetate (86.2%) at 0.05% concentration. The culture filtrates were subjected to the isolation procedure to yield crystalline **1**, mp 78° C. The yield of ^{13}C-labeled epoxydons was 30 mg from [1-^{13}C]acetate and 80 mg from [2-^{13}C]acetate. Next, cmr measure-

TABLE 7.1 ^{13}C-nmr data for epoxydon (ppm from benzene).

Position†	Natural **1**	**1** from [2-^{13}C] acetate	**1** from [1-^{13}C]acetate
C-1	−66.5	−66.4	—
C-2	−6.7	—	−6.9
C-3	−13.4	−13.3	—
C-4	63.3	—	62.8
C-5	73.5	73.3	—
C-6	74.6	—	74.5
C-7	69.3	69.4	—

† The numbering is given in Fig. 7.3.

ments were carried out on the labeled epoxydons, and all the carbon positions enriched with [^{13}C]acetate were assigned by comparison with cmr data for natural epoxydon. The results are shown in Table 7.1.

7.3 Bioassay of Epoxydon **1** and Phyllostine **2**[8,9]

In the leafy stem cutting test, phytotoxicity is observed within 24–48 hr when **1** and **2** are employed at a concentration of 10 ppm. However, as mentioned later, root elongation is remarkably enhanced at very low concentrations of **1** and **2**, which have no phytotoxic effect.

Compounds **1** and **2** dissolved in N/50 phosphate buffer solutions at pH 5.6 were tested on seedlings of red clover, cress, pea and rice. It was found that these compounds were toxic to the plants, though the susceptibility varied, but generally promoted the growth of seedlings, particularly root elongation, in the concentration range from 10^{-5} to 10^{-8} M.

[Experimental procedure 8] Growth stimulation of rice plants

Seeds of *Oryzae sativa* L., (Norin 20) were soaked in tap water at 30° C for 2 days. Uniform germinated seeds with 1–2 mm shoot length were selected and planted on cloth nets on cylindrical bottles 5.7 cm in diameter and 7.5 cm in height.

The bottle was placed in a 200 ml beaker filled with test solution containing **1** or **2** in M/50 phosphate buffer at pH 5.6. They were then incubated for a definite number of days in a growth cabinet with 16 hr light exposure (23,000 lux) at 27° C and 8 hr in the dark at 20° C per day. The results showed that **1** at a concentration of 1×10^{-5} to 10^{-8} M promoted root elongation 1.4–1.5 times by the 10th day, whereas **2** at the same concentration promoted it 1.4–2 times compared to the control.

In addition, the stimulating action of compound **1** on root formation was evaluated using cuttings of *Azukia angularis* in comparison with the effects of known plant hormones such as IBA, IAA and NAA. The tests resulted in an increase of the number of adventitious roots as well as enlargement of the rooting region; the effects were similar to those of IAA. These activities were partly destroyed by simultaneous application of a sulfhydryl compound, L-cysteine. This observation led us to postulate that compound **1** formed an adduct with L-cysteine, because epoxide α,β-unsaturated carbonyl moieties in the molecule might be reactive. In fact, adduct formation of **1** with sulfyhydryl compounds was confirmed by model experiments using cysteamine and ethylmercaptan.

[Experimental procedure 9] Stimulating effects on adventitious root formation

Seeds of *Azukia angularis*, "*wase-tairyu*", were germinated in moist vermiculite in a growth cabinet. Plants 10 cm in length were trimmed 2 cm up the cotyledonal node and used as cuttings for tests. The basal parts of the cuttings were inserted into aqueous solutions of **1** or **2**, and kept at room temperature for 16 hr. After removal, the inserted section was washed with dist. water and again inserted into dist. water in a cabinet (4000 lux, 25° C) for 7 days. The numbers of roots and the region of rooting were determined, and the results are shown in Table 7.2. Compounds **1** and **2**, with greater activity of the latter, both apparently contribute to the increase in the number of adventitious roots and enlargement of the rooting region.

TABLE 7.2 Effects of various compounds on adventitious root formation from azukia cuttings†

Compounds	Concentration (M)	Root formation Number of roots	% of control	Region of root formation Length of rooting zone (mm)	% of control
Control (H_2O)		9.5 ± 2.0	100	3.8 ± 0.32	100
Epoxydon **1**	1×10^{-4}	14.2 ± 2.1	147	16.3 ± 1.27	420
Epoxydon **1**	2.5×10^{-4}	15.2 ± 2.3	160	30.2 ± 0.13	778
Phyllostine **2**	1×10^{-6}	10.2 ± 3.6	107	5.1 ± 1.82	132
Phyllostine **2**	1×10^{-5}	10.2 ± 3.6	107	4.7 ± 0.62	121
Phyllostine **2**	1×10^{-4}	16.5 ± 3.5	173	13.1 ± 1.18	337
Phyllostine **2**	2.5×10^{-4}	18.8 ± 5.1	197	25.3 ± 0.61	652

† Averages of 12 cuttings.

[Experimental procedure 10] Effects of the addition of L-cysteine on root formation

Solutions of compound **1** mixed with L-cysteine in M/50 phosphate buffer at pH 5.0 were allowed to stand at 5° C for 24 hr and then treated

TABLE 7.3 Effects of L-cysteine on epoxydon-induced root formation in azukia cuttings†

Compounds	Without epoxydon Number of roots	Without epoxydon Length of rooting zone (mm)	With epoxydon Number of roots	With epoxydon Length of rooting zone (mm)
Control (H_2O)	5.8 ± 0.1	3.0 ± 0	14.9 ± 0.7	31.7 ± 0.3
L-Cysteine (2×10^{-4} M)	5.5 ± 0.3	3.0 ± 0	10.9 ± 1.0	15.1 ± 0.1
L-Cysteine (4×10^{-4} M)	6.4 ± 0.1	3.0 ± 0.06	10.0 ± 0.7	16.9 ± 0.1

† Averages of 20 cuttings.

as in "Experimental procedure 9" for comparison with the control without L-cysteine. As shown in Table 7.3, the stimulating action on root formation is partly destroyed by the addition of L-cysteine.

7.4 Phytotoxin of *Cladosporium phlei*[13)]

A leaf spot disease of timothy, *Phleum pratense*, ("*hantenbyo*" in Japanese) often occurs in Hokkaido prefecture, where the climate is relatively cool and humid. The disease is caused by *Cladosporium phlei* (C.T. GREGORY) which forms a characteristic eye spot having a light greyish-fawn center with purplish margins on the plant leaves. Accordingly, an investigation was undertaken on phytotoxic substances in the culture filtrate of the fungus, and resulted in the isolation of a new active red pigment named phleichrome **9**. The spectral data (pmr, ir, uv) for **9** together with the green coloration in alkaline solution strongly suggested that the compound possessed a 4,9-dihydroxyperylene-3,10-quinone nucleus. In particular, the uv spectrum of **9** is comparable to that of cercosporin **10**. Furthermore, acetylation of **9** gives a tetraacetate whose mass spectrum showed a parent peak at *m*/*e* 718, 16 mass units larger than the parent peak of **10**. The pmr spectrum of the acetate was very similar to that

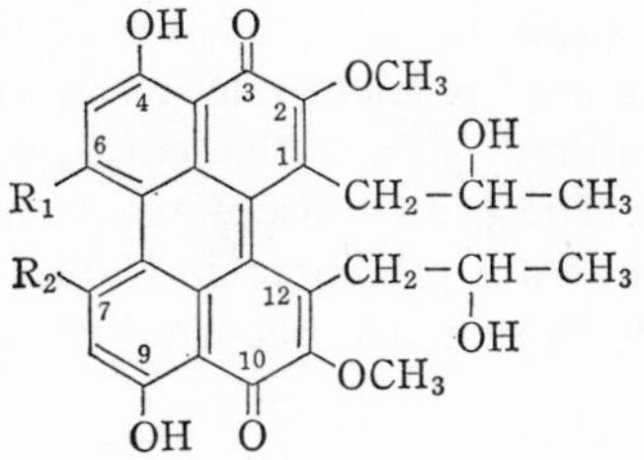

phleichrome **9** $R_1=R_2=OCH_3$

cercosporin **10** $\left.\begin{matrix}R_1\\R_2\end{matrix}\right\}=CH_2\begin{matrix}O-\\O-\end{matrix}$

of cercosporin tetraacetate, and a signal at δ 4.18 in the former replaced a methylenedioxy signal at δ 5.69 in the latter. Thus, the structure of phleichrome was assigned as 1,12-bis-(2-hydroxypropyl)-2,6,7,11-tetramethoxy-4,9-dihydroxyperylene-3,10-quinone **9**. It is known that compounds related to **9** occur in nature and have physiological activities. For example, hypericin and fagopyrin cause photo-dermatosis and cercosporin has antibiotic activity in the light.

[Experimental procedure 11] Isolation of phleichrome 9

Cladosporium phlei was seeded into the basal potato-dextrose medium and incubated at 20° C for 4 days. The culture filtrate (18 l) was concentrated *in vacuo* to an appropriate volume and then partitioned by successive extractions with ethyl ether and ethyl acetate. As mentioned later, the results of bioassay indicated strong phtotoxicity only in the ether extract. Thus, the ether extract was subjected to preparative tlc using plates made from silica gel containing 2% oxalic acid, irrigating with benzene-ethyl acetate (1:1 v/v). The active material was located near *Rf* 0.56 as a red band, and the pigment itself seemed to be an active principle. Therefore, the ether extract was chromatographed on a polyamide column using methanol-water (1:1 v/v) to yield a pure pigment fraction. The fraction, after concentration, was subjected to preparative tlc to yield phleichrome **9** as a red powder (20 mg).

[Experimental procedure 12] Bioassay of the phytotoxicity of phleichrome 9

Throughout the course of isolation of phytotoxic substances, bioassay was conducted by leaf tests with and without injury. The intensity of phytotoxic activity was estimated by comparison of the developed spots with the diseased spots produced by treatment with a spore suspension of the fungus. Phleichrome **9** produced large spots similar to the diseased spots when it was applied at concentrations of more than 10^{-4} M.

The method was as follows: healthy timothy leaves were injured with a punch, and test samples (0.02 ml) (aqueous solutions of fractions from the isolation stages, aqueous solutions of phleichrome, or aqueous spore suspension) were dropped on the injured parts of the leaves, which were kept in the dark at 20° C for 72 hr. The brown discoloration around injured parts was observed. The results are presented in Table 7.4. Clear diseased spots were produced by **9**. No change occurred in the control using water.

TABLE 7.4 Diseased spot formation with phleichrome **9** using timothy†

	Spore suspension	Ether solubles	Ethyl acetate solubles	Phleichrome (M)			
				0	1×10^{-5}	1×10^{-4}	1×10^{-3}
Diseased spot	+	+	±	−	−	+	+

† Symbols: ±, small spot; +, large spot.

REFERENCES

1) S. Sakamura, H. Niki, Y. Obata, R. Sakai, T. Matsumoto, *Agr. Biol. Chem.*, **33**, 698 (1969).
2) A. Closse, R. Mauli, H.P. Sigg, *Helv. Chim. Acta,* **49**, 204 (1966).
3) S. Sakamura, J. Ito, R. Sakai, *Agr. Biol. Chem.*, **34**, 153 (1970); *ibid.*, **35**, 105 (1971).
4) A. Ichihara, S. Sakamura, *CA*, **85**, 46018 v (1976), *Kagaku to Seibutsu* **14**, 78 (1976).
5) M.S. Frey, Ch. Tamm, *Helv. Chem. Acta,* **54**, 851 (1971).
6) N.J. McCorkinale, T.P. Roy, S.A. Hutchinson, *Tetrahedron,* **28**, 1107 (1972).
7) R. Sakai, R. Sato, S. Sakamura, *Plant Cell Physiol.*, **11**, 907 (1970).
8) R. Sakai, R. Sato, J. Ito, S. Sakamura, *Ann. Phytopathol. Soc. Japan,* **38**, 290 (1972).
9) J. Sekiguchi, G.M. Gaucher, *Biochemistry*, **17**, 1785 (1978).
10) S. Sakamura, T. Chida, J. Ito, R. Sakai, *Agr. Biol. Chem.*, **35**, 1810 (1971).
11) S. Sakamura, K. Nabeta, S. Yamada, A. Ichihara, *ibid.*, **35**, 1639 (1971); *ibid.*, **39**, 403 (1975).
12) K. Nabeta, A. Ichihara, S. Sakamura, *Chem. Commun.*, **1973**, 814; *Agr. Biol. Chem.*, 39, 409 (1975).
13) T. Yoshihara, T. Shimanuki, T. Araki, S. Sakamura, *ibid.*, **39**, 1683 (1975).
14) S. Yamazaki, A. Okubo, Y. Akiyama, K. Fuwa, *ibid.*, **39**, 287 (1975).
15) R.H. Thompson, *Naturally Occurring Quinones,* p. 576, Academic Press (1971).

CHAPTER 8

Surveys of Mycotoxins Monitored by Cytotoxicity Testing

The generic name, "mycotoxins," covers fungal secondary metabolites causing pathological and physiological abnormalities in man and warm-blooded animals. Mushroom poisons are generally excluded from this category by convention, and the fungi contaminating foods and feeds have been the principal subjects of study. Examples of mycotoxicosis and related problems caused by the contamination of foods and feeds by mycotoxins include ergotism caused by ergot alkaloids in European countries, *Fusarium* toxicosis in Siberia and the U.S.A., and the "yellow rice dispute" in 1954 in Japan. In 1960 "turkey X disease" in the U.K. was found to be due to metabolites of *Aspergillus flavus* in ground nut meal; metabolites such as aflatoxins A_1 **1** and B_1 **2** were not only identified as the causative agents of the acute toxicity but were also found to be strong carcinogens in experimental animals. This finding led to considerable interest in fungal secondary metabolites from a public health point of view.

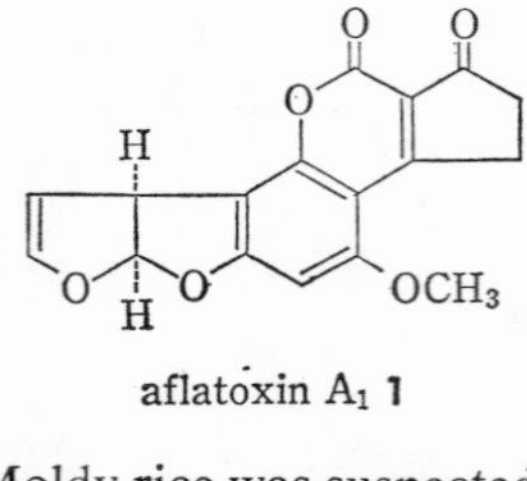

aflatoxin A_1 **1**

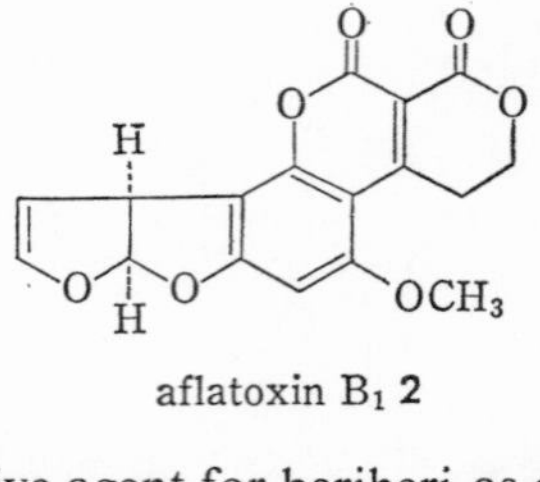

aflatoxin B_1 **2**

Moldy rice was suspected as a causative agent for beriberi as early as the end of the last century in Japan. Indeed, acute cardiac beriberi was differentiated from other types of the disease, which are now well known as a thiamine avitaminosis, and was proved to be a mycotoxicosis due to citreoviridin produced by *Penicillium citreoviride*. After World War II, when there was a shortage of food in Japan, pigmented rice was detected and the toxicity of the metabolites of *Penicillium islandicum* was discovered. *Fusarium* toxins have also been studied extensively in Japan in connection with food and feed poisonings.

Recent studies on mycotoxins have covered many fields of science such as mycology, epidemiology, pathology, toxicology, biochemistry and organic chemistry. Details of recent advances are well documented in several monographs and reviews.[1–8)]

From the viewpoint of food hygiene, naturally occurring toxicants should be considered in relation to both their toxicity and their distribution in foods and feeds. Many toxic substances of plant and animal origin are known,[9)] but in most cases there is little possibility of their intake into human bodies under normal conditions. In contrast, unintentional intake of mycotoxins via foodstuffs contaminated by molds occurs frequently, especially in areas where the climate is suitable for the growth of molds and where storage conditions for cereals are inadequate. Reflecting the large numbers of mold species known, the variety of secondary metabolites so far identified is great.[10)] Even a single species of a mold may produce many kinds of metabolites as the strain and the culture conditions vary. Thus chemical and biological studies of mold metabolites cover a very wide range. In view of public concern, the acute toxicity, chronic toxicity, teratogenicity, and carcinogenicity of many chemicals have been studied quite recently, but our knowledge of these properties of natural products is still quite limited. Nevertheless, carcinogenicity of the following mold metabolites, besides aflatoxins, has been experimentally confirmed: sterigmatocystin **3**, luteoskyrin **4**, rugulosin **5**, cyclochlorotin **6**, patulin **7**, penicillic acid **8**, griseofulvin **9**, and ochrotoxin A **10**. Clearly, general surveys of toxic metabolites of molds are urgently necessary.

Economical short-term bioassay methods are generally required for surveys on biologically active substances. For monitoring the separation and isolation of active principles from natural products, the bioassay methods should be sensitive (requiring only a small amount of sample), specific (not influenced by coexisting substances), simple, rapid, and applicable to large numbers of samples. In the case of mycotoxins, acute toxicity tests using experimental animals together with pathological examination to identify abnormalities related to chronic toxicity are an obvious choice. However, such tests using experimental animals are not suitable as a monitoring system for precise separation, e.g., by chro-

sterigmatocystin **3** 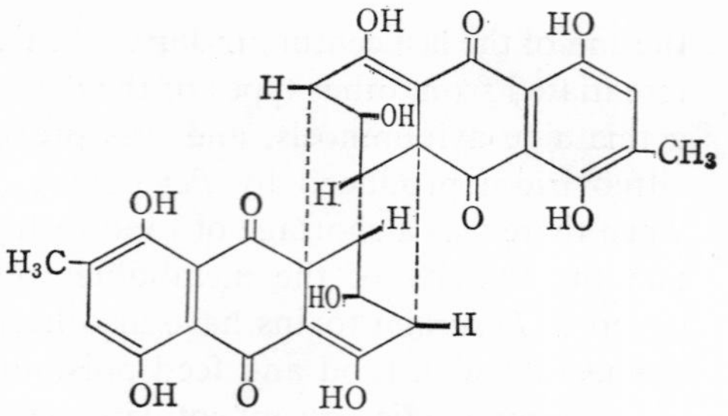 luteoskyrin **4**

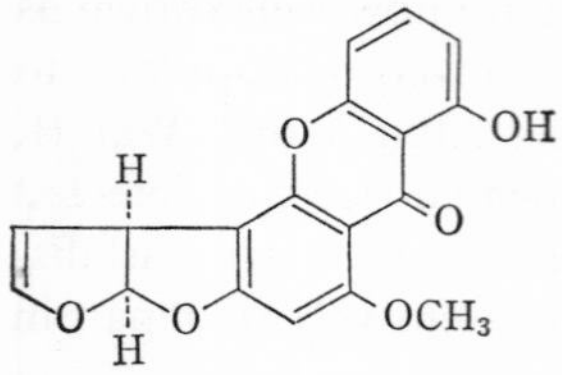

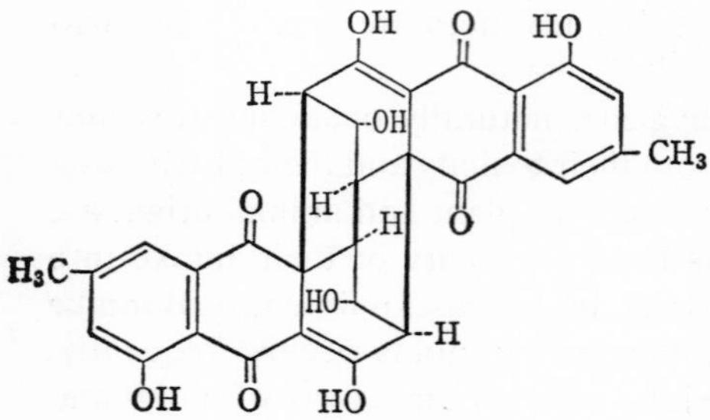

regulosin **5** 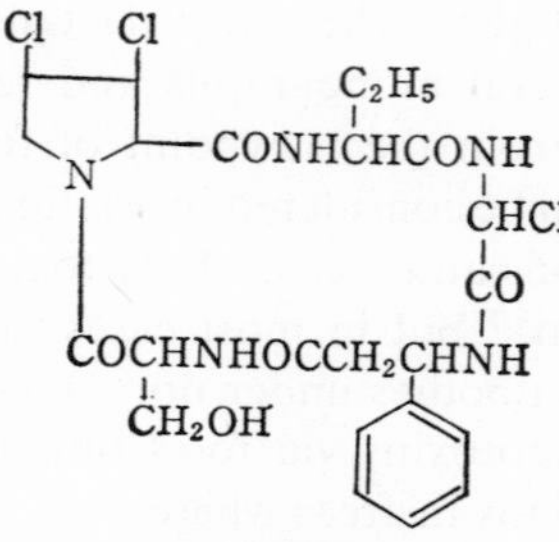 cyclochlorotin **6**

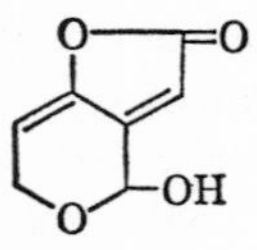

patulin **7** 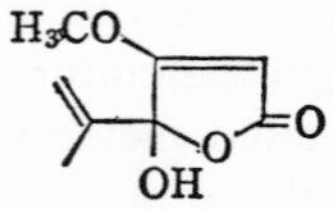 penicillic acid **8** 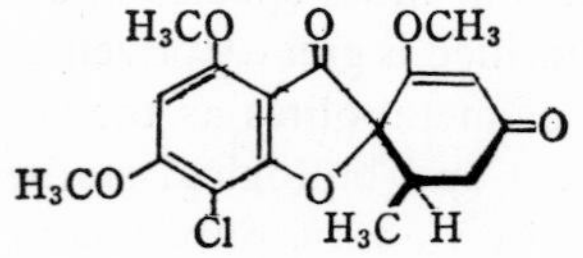 griseofulvin **9**

matography. Several toxicity tests at cellular and subcellular levels have been devised, and for the preliminary screening of potential carcinogens, mutagenicity tests using bacteria and mammalian cells are now widely used. In this chapter, surveys on mycotoxins carried out during the last ten years in our laboratories, employing cultured cells, will be described.

8.1 Outline of the Mycotoxin Surveys[11,12)]

A survey group composed of epidemiologists, pathologists, nutritionists, and mycologists was organized, and mold contamination was investigated in foodstuffs collected at six towns and villages in rural areas, where the mortality rate due to cerebrovascular and hepatic diseases is high. Among the fungi isolated, 405 strains belonging to 159 species, i.e., 124 strains of *Penicillium*, 101 of *Aspergillus*, and 180 of other genera, were selected for examination of their acute toxicity to mice and cytotoxicity to HeLa cells using the culture filtrates and chloroform extracts of the mycelia. Significant toxicities were observed with 102 strains.[13] The toxicities were evaluated on the basis of chemical and toxicological knowledge of these strains and 29 strains were selected for further surveys using both subacute toxicity tests with experimental animals and preliminary chemical examination combining thin-layer chromatography with the cytoxicity tests (see section 8.3). At this stage, several of the toxic metabolites were identified as known mycotoxins,[14] while further studies of some strains were abandoned due to loss of productivity of the toxic metabolites on storage of the strains. Later, 10 additional strains of 5 species were also examined.[15] Finally, seven strains were chosen for long-term feeding experiments with rats and mice using rice cultures, and precise pathological examinations were performed.[16]

In addition, detailed examination of the metabolites was conducted in the cases of *A. ochraceus*,[17] *P. charlesii*,[18] *P. purpurogenum*,[19] *Phoma* sp.,[20] *Chaetomium globosum*,[21-25] and *A. candidus*.[26-29] Production of rubratoxin B (**11**) by *P. purpurogenum*,[19] of mycophenolic acid (**12**) by *P. meleagrinum*,[14] and of penicillic acid (**8**) by several molds[14,17] was identified in the course of the study. As new mycotoxins, chaetoglobosins A-G and J (new cytochalasins[30]) from *C. globosum*,[21-25] terphenyllin and xanthoascin from *A. candidus*,[26-29] and chaetochromin from *C. thielavioideum*[31] were isolated and their structures were elucidated.

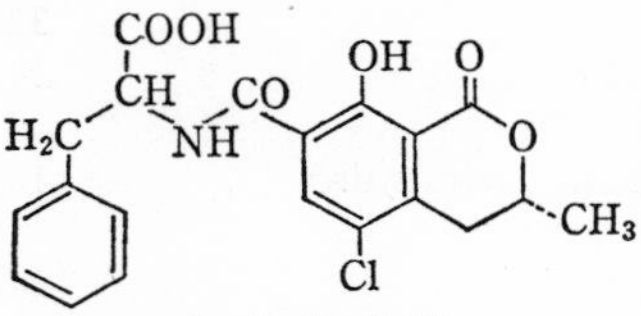

ochratoxin A **10**

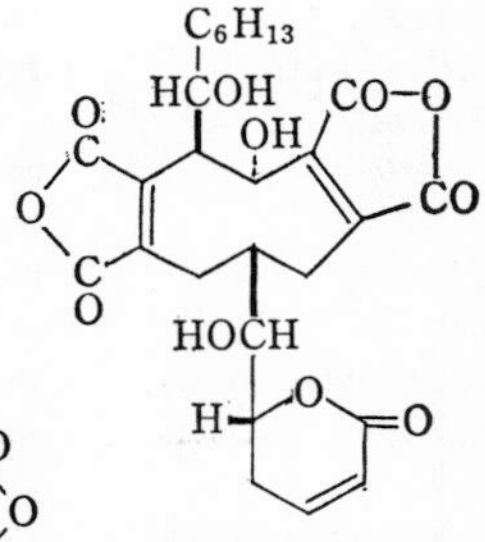

rubratoxin B **11**

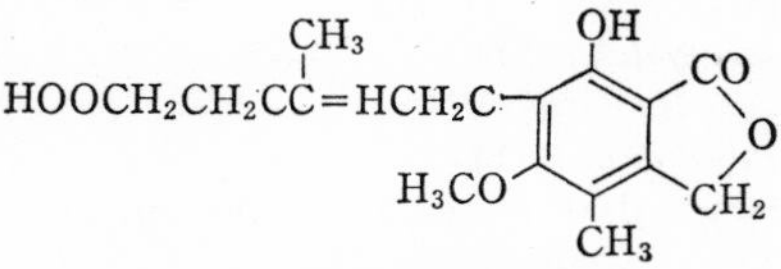

mycophenolic acid **12**

In the course of examination of the bioproduction of chaetoglobosins by *Chaetomium* spp. and allied fungi, *C. thielavioideum* and *Farrowia* sp. were unexpectedly found to produce the carcinogens sterigmatocystin (**3**) and *O*-methylsterigmatocystin.[32]

The metabolites identified in the course of these studies are summarized in Table 8.1.

TABLE 8.1 Mycotoxin-producing fungi and metabolites identified in surveys.

Species	Metabolites†[1]	References
Alternaria sp.	*alternariol, alternariol monomethyl ether*	15)
Aspergillus amstelodami	flavoglaucin, echinulin	14)
A. candidus	*terphenyllin* (**21**)†[2], *deoxyterphenyllin* (**23**)†[2], *xanthoascin* (**22**)†[2]	26–29)
A. clavatus	*patulin* (**7**)	14)
A. ochraceus	*penicillic acid* (**8**), *ochratoxin A* (**10**)	17)
A. ostianus	mellein†[3], *penicillic acid* (**8**)†[3]	14)
A. sulphureus	*penicillic acid* (**8**)	14)
Chaetomium amygdalisporum	*neocochliodinol*,†[2] mollicellin G†[3]	†[4]
C. caprinum	*chaetochromin*†[3]	†[4]
C. cochliodes	*chaetoglobosin C* (**16**)†[3], emodin†[3], chrysophanol	14)
C. elatum	*cochliodinol*†[3]	†[4]
C. globosum	*chaetoglobosins A-G, J* (**13–20**)†[2], flavipin†[3]	21–25)
C. mollipilium	*chaetoglobosins A-D* (**13–16**)†[3]	32)
C. murorum	*isocochliodinol*†[2]	†[4]
C. rectum	*chaetoglobosins A-E* (**13–17**)†[3]	32)
C. subaffine	*chaetoglobosins A-E* (**13–17**)†[3]	32)
C. subglobosum	*chetomin*†[3]	†[4]
C. tetrasporum	*chaetochromin*†[3]	†[4]
C. thielavioideum	*chaetocin*†[3], *sterigmatocystin* (**3**)†[3], *O-methylsterigmatocystin*†[3], eugenitin†[3], *chaetochromin*†[2]	31, 32)
C. udagawae	*sterigmatocystin* (**3**)†[3]	†[4]
Epicoccum nigrum	6-methylsalicylic acid†[3]	14)
Farrowia sp.	*sterigmatocystin* (**3**)†[3]	32)
Penicillium brevicompactum	*mycophenolic acid* (**12**)	14)
P. charlesii	parietin†[3], flavoglaucin†[3], tetrahydroauroglaucin†[2]	18)
P. cyclopium	puberulonic acid†[3], erythritol	14)
P. martensii	*penicillic acid* (**8**)	14)
P. meleagrinum	*mycophenolic acid* (**12**)†[3]	14)
P. oxalicum	*secalonic acid D*	†[4]
P. paraherquei	*penicillic acid* (**8**)†[3]	14)
P. purpurogenum	*rubratoxin B* (**11**)†[3]	19)
Phoma sp.	6-methylsalicylic acid, cynodontin, a chromanone	20)

†[1] Metabolites shown in italics are toxic.
†[2] New compounds.
†[3] Identified for the first time from this species.
†[4] Unpublished data of our laboratory.

8.2 Toxicity Tests with Cultured Cells

The toxicity bioassay of chemical substances using cultured cells has various advantages. This assay method is economical and rapid for the examination of many samples compared with animal experiments. It is useful for the detection and isolation of toxic principle(s) from raw materials by assaying various fractions during the separation process. The test method also requires relatively small amounts of samples. Cells from mammals, if necessary from man, can be used. In addition, tissue culture techniques have become quite straightforward with the commercial manufacture of media, sera, plastic vessels and so on, and the increasing availability of "clean" facilities. In this section we describe a cytotoxicity test using cultured human cells which we have employed for the isolation of toxic metabolite(s) from fungal cultures, together with further applications of the tissue culture technique.

As a toxicity assay, we have been employing a modified panel method,[13)] originally developed by Toplin.[33)] Briefly, HeLa cells grown on cover-slips fitting each cup of a panel or on the surface of each chamber of a Lab-Tek slide are treated with test chemicals for 3 days. By microscopic examination of the stained preparation, the approximate degree of toxicity and the induced morphological characteristics are examined. We always check morphological changes induced by test samples as well as the toxicity.

Morphological examination of cells can be useful on many occasions. One is during the purification of active chemicals. Thus, when a test raw material produces a characteristic morphological change, the toxic fraction can be identified not only in terms of the toxicity towards cells but also the induced morphology. In our experience, as described in section 8.3, we were able to detect two different toxic principles named terphenyllin and xanthoascin among the metabolites of *A. candidus*.[26,27)] They produced different morphological changes. Another advantage of morphological study is that it may provide a clue to the mechanism of action of test samples. When chemicals with a known mechanism of action are administered to cultured cells, the induced morphological changes are often very characteristic. The results obtained by the application of typical agents are shown in Table 8.2. The changes induced by test materials can then be compared with those caused by these materials.[13)] Morphological changes induced by different chemicals may vary widely, and another case where morphology was a useful indicator was in our work on the extraction of mycophenolic acid from the culture fluid of *P. brevi-compactum*[14)] (see section 8.3).

TABLE 8.2 Representative drug-induced morphological changes of HeLa cells (HE stain) classified according to mechanism of action.

Symbol	Class of drug	Cytoplasm	Nucleus	Nucleolus	Mitotic cells	Others
D	Inhibitor of DNA synthesis (FUDR, araC, hydroxyurea)	enlarged, polygonal	enlarged, fine granular chromatin	enlarged, irregular contour	scanty, abnormal	
A	Alkylating agent (mitomycin C, nitromine)	enlarged and small	enlarged and small, fine granular chromatin	enlarged, irregular contour	slightly increased, abnormal	polynuclear cells
R	Inhibitor of RNA synthesis (actinomycin D, proflavine)	enlarged, spindle-shaped, faintly stained	enlarged, spotty chromatin (fine nucleoplasm)	very small, round	not decreased	
S	Purine analog (8-azaguanine, 6-mercaptopurine)	small, spindle-shaped	small, clear nucleoplasm	small, round	not decreased, atrophic	polynuclear cell
P	Inhibitor of protein synthesis (cyclohexi-mide, fusarenon-X)	small, spindle-shaped, scanty	small, thick nuclear membrane	enlarged, irregular or round shape	decreased, atrophic	
C	Spindle-fiber poison (colcemid, vinblastine)	polymorphic	enlarged and small		increased, clumped chromosomes	polynuclear cells
M	Inhibitor of cytokinesis (cytochalasin B)		multiple nuclei of equal size		multipolar division	polynuclear cells
V	Cytoplasmic vacuolation	large vacuoles (not stained by fat staining)				

When toxicity is investigated using cultured cells such as HeLa cells, there is a tendency to detect substances which inhibit the growth of cells severely or induce mitotic abnormality. In our experience, it was difficult to examine organ-specific substances by the use of HeLa cells.[13] Such substances can be examined using primary cultured cells from appropriate organs, by the panel method.[34] For example, hepatotoxic aflatoxin B_1 or luteoskyrin strongly affects liver parenchymal cells when these toxins are applied to cultured rat liver cells. Cytochalasins are known to affect microfilaments, which consist of muscular components. When chaetoglobosin A, a kind of cytochalasins, was applied to rat muscle culture, the myotubes were shortened and formed spindle-shaped or round multinucleated giant cells.[34] This finding clearly demonstrated that chaetoglobosin A affected muscular components, suggesting that this effect might be the major mechanism of its action.

As carcinogenic mycotoxins are usually toxic to cells, toxicity assay can be utilized as the first step in their examination. However, test methods directly or indirectly related to carcinogenicity must be employed as the next step. Examinations of mutagenicity, ability to induce DNA breaks and chromosomal aberrations,[35] carcinogenicity *in vitro,* and so on are the usual methods. These approaches are very useful for defining the nature of damage caused by test compounds and are described in detail in the literature.

Recently, bacterial mutation tests have been developed as a simple, rapid and economical method to detect genetic damage. A good correlation between bacterial mutation and carcinogenicity was reported.[36–38] However, these tests are not suitable for antibacterial substances, or for testing compounds containing amino acids in the case of amino acids-requiring reverse mutation systems. In contrast, the cell culture method is able to identify not only mutagenicity but also carcinogenicity *in vitro*. Thus, the cell culture method, including the use of mammalian cells and even human cells, is very important.

Another problem is the examination of substances which require metabolic modification before exerting an effect on cells. In the case of carcinogenic or mutagenic substances, these are called procarcinogens or promutagens. In bacterial mutation assay these substances have no effect upon direct application, but become mutagenic when applied to the bacteria with rat liver microsomes and cofactors. This *in vitro* metabolic activation system has made the bacterial mutation assay more effective.[37,38] In cultured cell experiments, procarcinogens can also be examined using this metabolic activation system.[39]

8.3 Extraction and Isolation of Mycotoxins

Screening for Mycotoxins by tlc and Cytotoxicity Tests[14]

The first step in the separation and isolation of biologically active natural products is to extract the target substances selectively in good yield. In the case of compounds such as alkaloids and saponins, they can be separated by virtue of their specific physical and chemical properties. Mycotoxins have no such general characteristics, but mold metabolites in general are produced in a shorter period of time, contain smaller amounts of polyphenols such as lignins and tannins, and are formed individually in higher yields in comparison with those of higher plants. Furthermore, almost all mycotoxins so far known have molecular weights smaller than 500 and are extractable in nonpolar solvents, in contrast to bacterial toxins.

In the early stage of our work, the conditions for extraction of metabolites and for thin-layer chromatography, and the recovery from thin-layer plates using several solvents were quantitatively examined.[17] Simple procedures such as those described below were found to be applicable for the survey. Such simple procedures are not suitable for surveys of higher plant constituents. In combination with cytotoxicity tests using HeLa cells, preliminary examination of toxic metabolites was performed.

[Experimental procedure 1] Preliminary test for the isolation and identification of toxic metabolites[14]

A test strain was cultured in three 1 l Roux flasks containing 200 ml of modified Czapek-Dox medium, MY 20 medium, or potato-dextrose medium for 3 weeks at 25° C. The mycelia and the filtrate were separated by filtration. The mycelia were macerated with chloroform in a Waring blender and, after filtration, the solvent was evaporated off to give an extract. The mycelia were then extracted with methanol. The culture filtrate was extracted with ethyl acetate. The three extracts thus obtained were used for acute toxicity tests with mice and for the following experiments. The extracts were applied to thin-layer plates of silica gel HF_{254} and developed with chloroform-ethyl acetate (4–0.5:1) or benzene-ethyl acetate (4–1:1) according to the polarity of the metabolites in the extracts. Under uv lamps (254 and 365 nm) several spots were generally observed. The plate was separated into 3–10 zones, using the spots as markers, and these were extracted separately. The cytotoxicity of these fractions were tested as described below. If necessary, comparison with authentic mycotoxins and other known mold metabolites by thin-layer chromatography was carried out at this stage. In some cases, the eluates from the thin-layer zones crystallized. Such compounds were directly examined by ir and ^{1}H-nmr for identification. The chemical and toxicological data thus obtained were evaluated with reference to the information available in the literature.

[Experimental procedure 2] Cytotoxicity test

Water-soluble samples are directly diluted with the medium to 200 μg/ml. Lipophilic samples are dissolved in DMSO to 20 mg/ml and then diluted with the medium to 200 μg/ml. Each cup in a panel is fitted with a round cover-glass. Lab-Tek 4 chamber slides can replace the round cover-glasses in the panel. Usually 4 serial dilutions at a half log level are employed. The 2nd, 3rd and 4th cups or chambers receive 0.5 ml of the control medium, then the sample solution at 200 μg/ml is added at 0.5 ml to the first cup and at 0.23 ml to the 2nd cup. After thorough pipetting of the mixed solution in the 2nd cup or chamber, 0.23 ml of it is transferred into the 3rd one. After pipetting, 0.23ml from the 3rd cup or chamber is transferred into the 4th one. From the 4th, 0.23 ml is discarded after pipetting. A cell suspension at a density of 1×10^5 cells/ml prepared from logarithmically growing HeLa cells is then distributed into each cup or chamber and a DMSO control is prepared. After cultivation for 3 days, each cover-glass or LabTek chamber slide is washed with saline, fixed with Carnoy's solution and stained with hematoxylin and eosin. The prepared slides are examined microscopically.

In the above method, test samples are examined at concentrations of 100, 32, 10 and 3.2 μg/ml. These concentrations are varied depending on the toxicity of samples. The concentration of DMSO should not exceed 0.5% in the final experimental medium.

When primary cultured cells are used, it is necessary to examine the effect of test samples after obtaining suitable organ-specific cells. After preliminary cultivation, usually for 2 or 5 days, the cultures receive media containing test samples, and are cultivated for another 3 days. After fixation and staining, the preparations are examined as described in the case of HeLa cells.[34]

[Experimental procedure 3] Identification of mycophenolic acid (12)[14]

The filtrate of *P. brevi-compactum* exhibited R-type cytotoxicity, as shown in Table 8.2. The ethyl acetate extract of the filtrate was developed on a thin-layer plate with benzene-ethyl acetate (4:1). The chromatograms under uv irradiation are shown in Fig. 8.1. Fractions 1–8 were separately extracted and the fractions containing small amounts of materials, i.e., fractions 2–4, were combined to give 6 samples for cytotoxicity tests. The results are also shown in the figure. Fraction 6 was then recrystallized from benzene to give colorless needles of mp 138–139°. Yield, 57 mg from 1 l of medium. From the physical data (mp, ir, ^{1}H-nmr, and ms) the compound was identified as mycophenolic acid (**12**), previously isolated from other fungal sources.

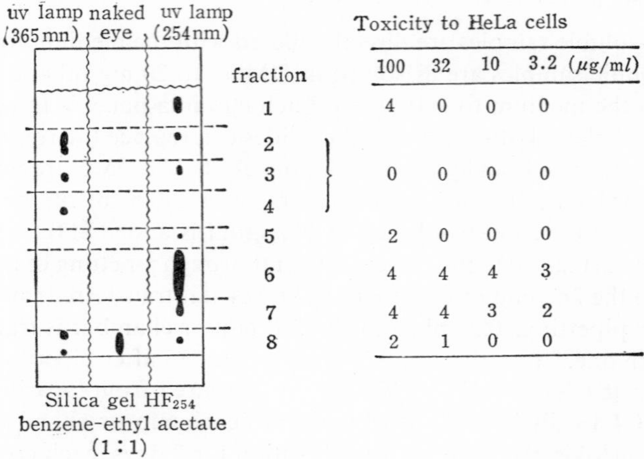

fraction	100	32	10	3.2 (μg/ml)
1	4	0	0	0
2–4	0	0	0	0
5	2	0	0	0
6	4	4	4	3
7	4	4	3	2
8	2	1	0	0

Fig. 8.1 Thin-layer chromatogram and cytotoxicity data for the extract of *P. brevi-compactum* (the degree of cytotoxicity is estimated on a scale ranging from 0 (no cellular damage) through 4 (complete cytolysis)).

Isolation of Chaetoglobosins A–G and J, Novel Cytotoxic Cytochalasins, from *Chaetomium globosum*[21–25,30]

In the course of our screening for mycotoxins, the culture filtrate and mycelial extract of *C. globosum* were found to show cytotoxicity to HeLa cells, with peculiar morphological changes, i.e., inhibition of cytoplasmic division without interfering with nuclear division, resulting in the formation of multinucleated cells.[14] Application of the methods described above for its separation met with little success, chiefly due to a decrease of the yield of the metabolites with changes of the metabolic activity of the strain. Guided by the HeLa cell assay, selection of strains and examinations of culture conditions and of separation methods were performed. Finally the causative agents, chaetoglobosins A, B and C, were isolated as crystalline compounds, and were shown to be indole derivatives giving a positive Ehrlich reaction. By further improvement of the conditions, the yields were increased and the minor congeners, chaetoglobosins D–G and J, were also isolated.

The cytotoxicity of chaetoglobosins is shown in Table 8.3. All the compounds other than chaetoglobosin J induced M-type abnormality of the cells (cf. Table 8.2).[22]

TABLE 8.3 Cytotoxicity† of chaetoglobosins to HeLa cells.

	Concentration (μg/ml)			
Chaetoglobosin	32	10	3.2	1.0
A	4	4	1	
B	4	3.5	0.5	
C	3.5	1	0.5	
D	4	4	2	
E	4	4	2	
F	4	2	1	
G	4	3	1	
J	4	4	3	1

† For details of the scoring system for cytotoxicity, see Fig. 8.1.

The presence of an indole skeleton in this series of compounds was indicated by uv and ms but it was difficult to obtain further information. However, the ^{1}H-nmr signals of the compounds were well separated, so decoupling experiments were performed. The results obtained suggested a similarity to the group of mold metabolites known as "cytochalasins."[30] Based on the data reported for cytochalasins, the plain structures of chaetoglobosins A and B were proposed.[21] Confirmation of the structures and the absolute configurations were obtained by X-ray analysis of chaetoglobosin A (**13**).[23,40] Since chaetoglobosins B and D were prepared from A by acid or base cleavage of the epoxy group, the structures **14** and **15**, respectively, were established.[23,24] Chaetoglobosins C and G are also isomers of A, B, and D. Since chaetoglobosins C and G were derived from A and B, respectively, by base-catalyzed isomerization of the 13-membered ring, the structures **16** and **19** were established with the aid of spectral data.[24,25] Chaetoglobosins E and F correspond to dihydro derivatives of A-D and G. By mutual correlation and bismuth trioxide oxidation of F to C, the structures **17** and **18** were confirmed for E and F, respectively.[24,25] Chaetoglobosin J was suggested to be the deoxy derivative (**20**) of A on the basis of the spectroscopic data, and this was confirmed by the deoxygenation of A to J upon WCl_6-BuLi treatment.[25]

Chaetoglobosins are a novel type of cytochalasins, in which the phenyl group in the compounds so far reported is replaced by an indole-3-yl group, i.e., the phenylalanine unit has been replaced by tryptophan.

Recently cytochalasins have attracted considerable interest among biologists because of their strong biological activities, especially with cultured mammalian cells.[30,41] Chaetoglobosins, novel indolyl-cytochalasins, exhibit the same effects as the previously known phenyl-cytochalasins against mammalian cells.[22,34] The acute toxicity of chaetoglobosin A towards mice and rats was precisely examined (LD_{50}(mice) ♂ 6.5 mg/kg, ♀ 17.5 mg/kg (sc)).[42] Mutagenicity was not observed,[35,43] but teratogenicity towards mice was observed with chaetoglobosin A.[44] The effect on sea urchin embryos[45] and the interaction with rabbit muscle actin[46] were also studied.

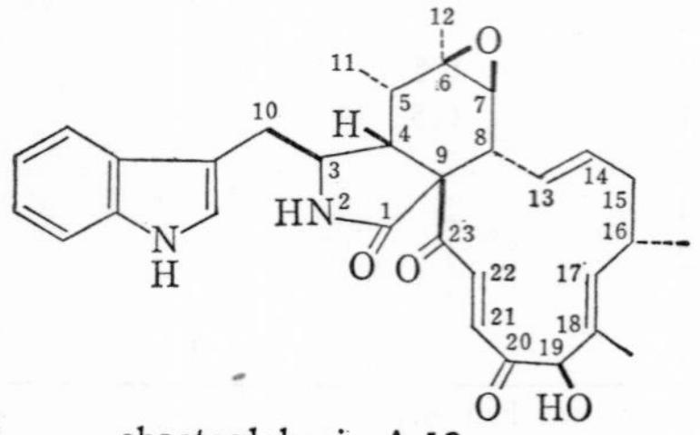

chaetoglobosin A **13**

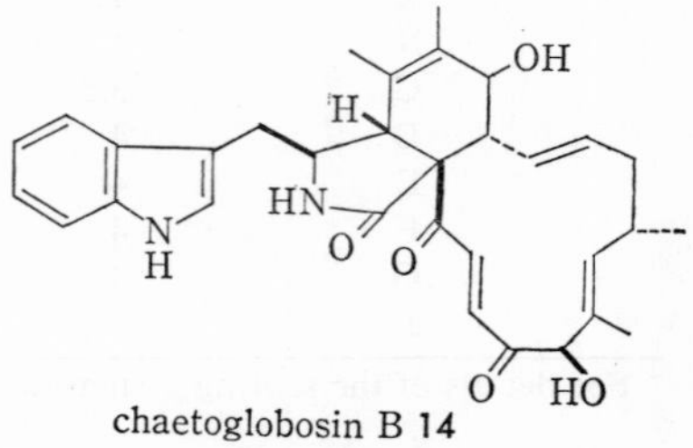

chaetoglobosin B **14**

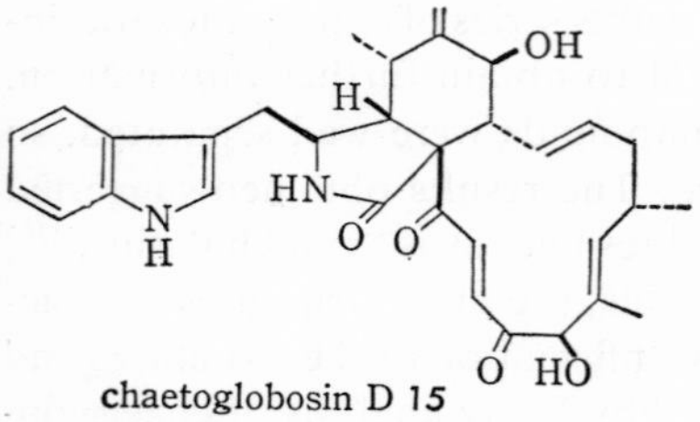

chaetoglobosin D **15**

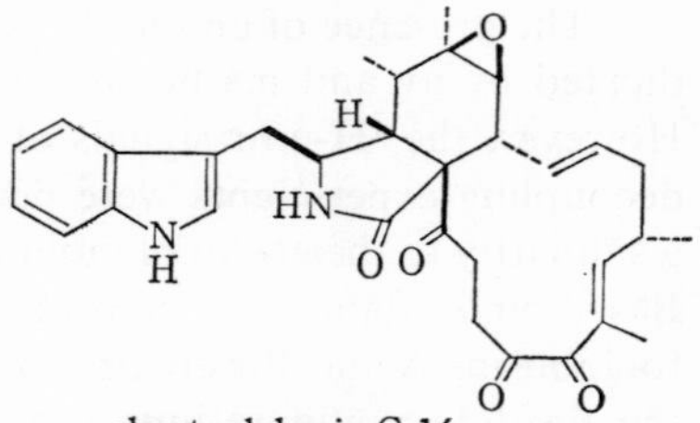

chaetoglobosin C **16**

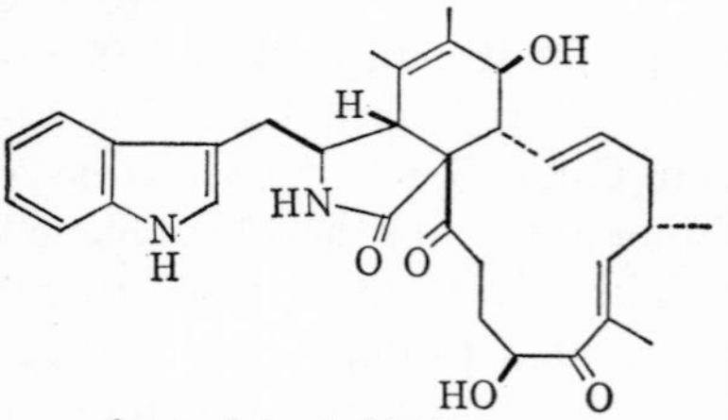

chaetoglobosin E **17**

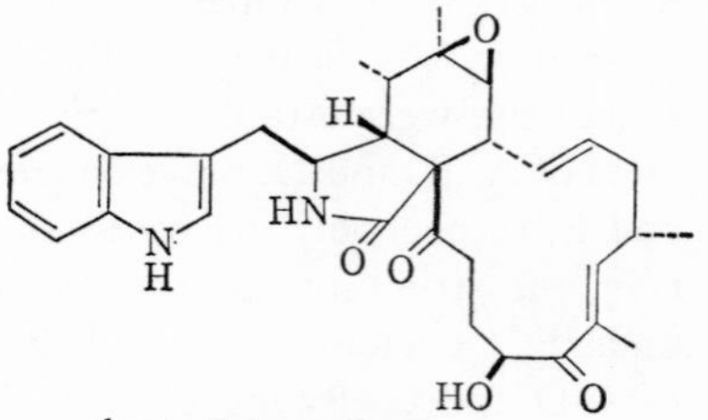

chaetoglobosin F **18**

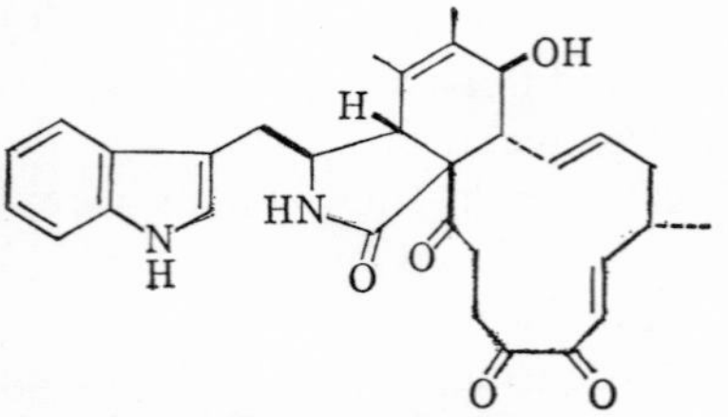

chaetoglobosin G **19**

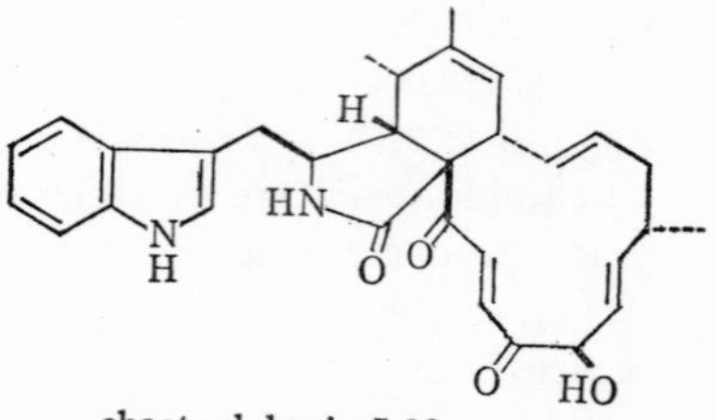

chaetoglobosin J **20**

[Experimental procedure 4] Isolation of chaetoglobosins

The strain was grown in 100 Roux flasks containing 200 g each of polished rice previously washed with water. The inoculated flasks were maintained at 25° C for 3 weeks with occasional shaking. The moldy rice was extracted twice with chloroform at room temperature. The dark brown extract (55 g) was chromatographed on a column of silica gel (2.2 kg) and eluted with a gradient system of chloroform-acetone. Fractions of 2 l were collected. After the elution of fatty materials and ergosterol, chaetoglobosin C (1.1 g) was obtained from fractions 21–26 (chloroform) by recrystallization from acetone. Recrystallization of fractions 31–42 (chloroform) from chloroform gave chaetoglobosin A (4.9 g). The mother liquor of chaetoglobosin A gave chaetoglobosin B (0.16 g). Fraction 44 (chloroform-acetone, 20:1) gave chaetoglobosin G (0.01 g). In a similar manner, fractions 46–50 (chloroform-acetone, 20:1) gave chaetoglobosin D (0.34 g) and fractions 51–55 (chloroform-acetone, 9:1) gave chaetoglobosin E (0.40 g). The mother liquor after separation of chaetoglobosins D and E was rechromatographed using the same system to obtain chaetoglobosins F (1.3 g) and D (0.49 g). Chaetoglobosin J was obtained by submerged culture, together with other compounds.

Isolation of Terphenyllin and Xanthoascin from *Aspergillus candidus*[26–29]

A. candidus is a mold that frequently contaminates cereals, but there was no previous report of mycotoxin production. In the course of our study, the culture filtrate of the fungus was found to exhibit R-type cytotoxicity.[13,26] In feeding experiments, rice infected with the strain induced noticeable pathological changes in the liver and heart.[16,26] The result of thin-layer chromatography-cytotoxicity testing of the extract of a culture on rice is shown in Fig. 8.2.

As indicated in the figure, fractions 6 and 2 showed cytotoxic activity (tentatively named toxins A and B, respectively), the former causing the same morphological change as that induced by the filtrate.

Toxins A and B were subsequently isolated[27] and named terphenyllin and xanthoascin, respectively. The skeleton of terphenyllin was established by measurements of physical data, by oxidations to *p*-hydroxybenzoic acid and to a *p*-terphenylquinone derivative, and by the synthesis of the trimethyl ether (**24**). The choice among the formulae **21**, **21′**, and **21″** for the relative positions of hydroxyl and methoxyl groups was made on the basis of ^{1}H-nmr and coloration reactions, and the structure (**21**) was assigned for terphenyllin.[27] By analogy, the structure (**23**) was proposed for deoxyterphenyllin.[27]

Spectral data (uv, ir, ^{1}H-nmr, and ^{13}C-nmr) for xanthoascin revealed

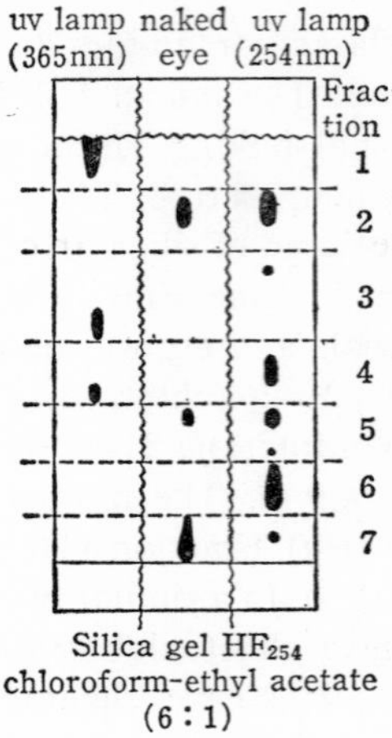

Fraction	Yield		Toxicity to HeLa cells 100	32	10	3.2(μg/ml)
1	4.9		4	0	0	0
2	1.3	(toxin B)	4	0.5	0	0
3	4.9		1	0	0	0
4	2.5		4	1	0	0
5	3.6		4	3	2	0
6	3.7	(toxin A)	4	3.5	2	0
7	3.3		3	1	0	0

Fig. 8.2 Thin-layer chromatogram and cytotoxicity data for the extract of *A. candidus* (details of the scoring system for cytotoxicity are given in Fig. 8.1).

its similarity to xanthocillin, a diphenylbutadiene derivative from *P. notatum*. Oxidation gave *p*-hydroxybenzoic acid and 2,2-dimethylchromane-6-carboxylic acid. The presence of isonitrile groups was confirmed by hydrolysis to the bis(formylamino) derivative. Thus, the structure (**22**) was established for xanthoascin.[28]

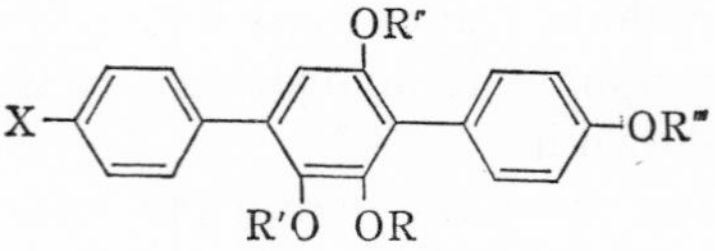

	X	R	R'	R"	R"'
21	OH	H	CH_3	CH_3	H
21'	OH	CH_3	H	CH_3	H
21"	OH	CH_3	CH_3	H	H
23	H	H	CH_3	CH_3	H
24	OCH_3	CH_3	CH_3	CH_3	CH_3

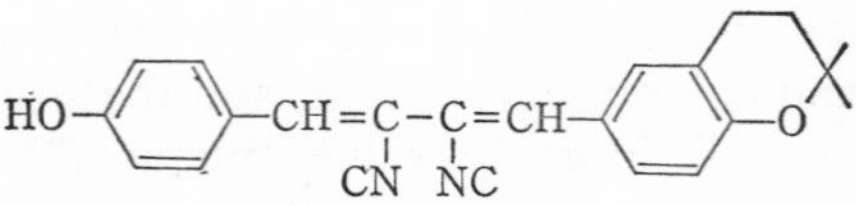

xanthoascin **22**

Pathological examination was carried out to investigate the hepato-

and cardiotoxicity of xanthoascin to experimental animals.[47–49] Xanthoascin was also shown to inhibit prostaglandin synthetase.[50]

[Experimental procedure 5] Isolation of terphenyllin and xanthoascin[27]

The moldy rice (28 kg) prepared as described in "Experimental procedure 4" at 25° C for 13 days was extracted three times with chloroform. Concentration of the first extract afforded a precipitate, which was removed by filtration (fraction 1). The mother liquor and the second and third extracts were combined and concentrated (fraction 2). The moldy rice was then extracted with methanol and the extract was shaken with ethyl acetate (fraction 3). Each extract were examined for acute toxicity towards rats and by thin-layer chromatography-cytotoxicity testing. Acute toxicity towards rats was observed in fraction 1, which contained xanthoascin as determined by thin-layer chromatography. The presence of terphenyllin in fractions 1, 2, and 3 was demonstrated by thin-layer chromatography-cytotoxicity testing. Silica gel chromatography of fraction 1 gave xanthoascin (**22**) (2.4 g), deoxyterphenyllin (**23**) (0.02 g), and terphenyllin (**21**) (1.4 g). Using the same procedures, ergosterol (0.2 g), **23** (0.4 g), and **21** (0.6 g) were obtained from fraction 2, and ergosterol (0.5 g) and **21** (1.1 g) from fraction 3. The cytotoxic effects of the metabolites are summarized in Table 8.4

TABLE 8.4 Cytotoxicity† of the metabolites of *A. candidus* to HeLa cells

Metabolite	Concentration (μg/ml)			
	100	32	10	3.1
Terphenyllin **21**	4	3	2	1
Xanthoascin **22**	4	4	4	1
Deoxyterphenyllin **23**	4	2	0	0

† For details of the scoring system for cytotoxicity, see Fig. 8.1.

8.4 CONCLUDING REMARKS

The recent development of powerful methods for separation and structure elucidation has greatly influenced the trend of natural products chemistry. Approaches to biologically active substances by chemical methods are now becoming a major feature of this field. However, studies on substances directly influencing human health, such as medicines and hazardous chemicals, face the great difficulty of finding a suitable bioassay system, and further collaboration between chemists and biologists will be necessary to develop versatile systems.

Acknowledgment

The work described above was the result of collaborative studies by mycologists, pathologists, and natural products chemists, and was conducted in collaboration with Prof. M. Saito, Drs. H. Kurata, K. Ohtsubo, T. Ishiko, S. Udagawa, F. Sakabe, and K. Yoshihira, Mrs. S. Sekita, and Miss C. Takahashi. Thanks are also due to those who carried out physical determinations and provided technical assistance.

References

1) G.N. Wogan (ed.), *Mycotoxins in Foodstuffs,* M.I.T. Press (1965).
2) A. Ciegler, S. Kadis, S.J. Ajl (eds.), *Microbial Toxins,* vols. 6, 7, Academic Press (1971).
3) I.F.H. Purchase (ed.), *Mycotoxins in Human Health,* MacMillan (1971).
4) I.F.H. Purchase (ed.), *Mycotoxins,* Elsevier (1973).
5) M. Enomoto, M. Saito, *Ann. Rev. Microbiol.*, **26**, 279 (1972).
6) J.V. Rodricks (ed.), *Mycotoxins and other Fungal Related Food Problems,* Am. Chem. Soc. (1976).
7) J.V. Rodricks, C.W. Hesseltine, M.A. Mehlman (eds.), *Mycotoxins in Human and Animal Health,* Pathotox Publishers (1977).
8) K. Uraguchi, M. Yamazaki (eds.), *Toxicology, Biochemistry and Pathology of Mycotoxins,* Kodansha-John Wiley & Sons (1978).
9) National Academy of Sciences (ed.), *Toxicants Occurring Naturally in Foods,* National Academy of Sciences (1973).
10) M.W. Miller, *The Pfizer Handbook of Microbial Metabolites,* McGraw-Hill (1961); S. Shibata, S. Natori, S. Udagawa, *List of Fungal Products,* University of Tokyo Press-Charles C. Thomas (1964); W.B. Turner, *Fungal Metabolites,* Academic Press (1971).
11) M. Saito, *Trans. Soc. Path. Japan,* **61**, 33 (1972).
12) S. Udagawa, O. Tsuruta, *Food and Moulds—an Introduction to Food Mycology* (in Japanese), p. 263, Ishiyaku Publishers (1975).
13) M. Saito, K. Ohtsubo, M. Umeda, M. Enomoto, H. Kurata, S. Udagawa, F. Sakabe, M. Ichinoe, *Japan. J. Exptl. Med.*, **41**, 1 (1971); M. Saito, T. Ishiko, M. Enomoto, K. Ohtsubo, M. Umeda, H. Kurata, S. Udagawa, S. Taniguchi, S. Sekita, *ibid.*, **44**, 63 (1974).
14) M. Umeda, T. Yamashita, M. Saito, S. Sekita, C. Takahashi, K. Yoshihira, S. Natori, H. Kurata, S. Udagawa, *ibid.*, **44**, 83 (1974).
15) K. Ohtsubo, M. Saito, T. Ishiko, M. Umeda, S. Sekita, K. Yoshihira, S. Natori, F Sakabe, S. Udagawa, H. Kurata, *ibid.*, **48**, 257 (1978).
16) K. Ohtsubo, M. Enomoto, T. Ishiko, M. Saito, F. Sakabe, S. Udagawa, H. Kurata, *ibid.*, **44**, 477 (1974).
17) S. Natori, S. Sakaki, H. Kurata, S. Udagawa, M. Ichinoe, M. Saito, M. Umeda, *Chem. Pharm. Bull.*, **18**, 2259 (1970).
18) K. Yoshihira, C. Takahashi, S. Sekita (née Sakaki), S. Natori, *ibid.*, **20**, 2727 (1972).
19) S. Natori, S. Sakaki, H. Kurata, S. Udagawa, M. Ichinoe, M. Saito, M. Umeda, K. Ohtsubo, *Appl. Microbiol.*, **19**, 613 (1970).
20) C. Takahashi, S. Sekita, K. Yoshihira, S. Natori, S. Udagawa, H. Kurata, M. Enomoto, K. Ohtsubo, M. Umeda, M. Saito, *Chem. Pharm. Bull.*, **21**, 2286 (1973).
21) S. Sekita, K. Yoshihira, S. Natori, H. Kuwano, *Tetr. Lett.* **1973**, 2109.

22) M. Umeda, K. Ohtsubo, M. Saito, S. Sekita, K. Yoshihira, S. Natori, S. Udagawa, F. Sakabe, H. Kurata, *Experientia*, **31**, 435 (1975).
23) J.V. Silverton, T. Akiyama, C. Kabuto, S. Sekita, K. Yoshihira, S. Natori, *Tetr. Lett.*, **1976**, 1349.
24) S. Sekita, K. Yoshihira, S. Natori, H. Kuwano, *ibid.*, **1976**, 1351.
25) S. Sekita, K. Yoshihira, S. Natori, H. Kuwano, *ibid.*, **1977**, 2771.
26) C. Takahashi, K. Yoshihira, S. Natori, M. Umeda, K. Ohtsubo, M. Saito, *Experientia*, **30**, 529 (1974).
27) C. Takahashi, K. Yoshihira, S. Natori, M. Umeda, *Chem. Pharm. Bull.*, **24**, 613 (1976)
28) C. Takahashi, S. Sekita, K. Yoshihira, S. Natori, *ibid.*, **24**, 2313 (1976).
29) C. Takahashi, S. Sekita, K. Yoshihira, S. Natori, S. Udagawa, F. Sakabe, H. Kurata, *Proc. Jap. Assoc. Mycotoxicol.* (in Japanese), No. 2, 26 (1976).
30) S. Natori, *Mycotoxins in Human and Animal Health* (ed. J.V. Rodricks, C.W. Hesseltine, M.A. Mehlman), p. 559, Pathotox Publishers (1977); M. Binder, Ch. Tamm, *Angew. Chem., Intern. Ed.*, **12**, 370 (1973).
31) S. Sekita, K. Yoshihira, S. Natori, *Chem. Pharm. Bull.*, **28**, 2428 (1980).
32) S. Udagawa, T. Muroi, H. Kurata, S. Sekita, K. Yoshihira, S. Natori, M. Umeda, *Canad. J. Microbiol.*, **25**, 170 (1979).
33) I. Toplin, *Cancer Res.*, **19**, 959 (1959).
34) M. Umeda, *Mycotoxins in Human and Animal Health* (eds. J.V. Rodricks, C.W. Hesseltine, M.A. Mehlman), p. 713, Pathotox Publishers (1977).
35) M. Umeda, T. Tsutsui, M. Saito, *Gann*, **68**, 619 (1977).
36) B.N. Ames, J. McCann, E. Yamasaki, *Mutation Res.*, **31**, 347 (1975).
37) J. McCann, E. Choi, E. Yamasaki, B. N. Ames., *Proc. Natl. Acad. Sci. U.S.A.*, **72**, 5135 (1975).
38) T. Sugimura, S. Sato, M. Nagao, T. Yahagi, T. Matsushima, Y. Seino, M. Takeuchi, T. Kawachi, *Fundamentals in Cancer Prevention* (eds. P.N. Magee *et al.*), p. 191, Univ. of Tokyo Press (1976).
39) M. Umeda, M. Saito, *Mutation Res.*, **30**, 249 (1975).
40) J.V. Silverton, C. Kabuto, T. Akiyama, *Acta Cryst.*, **B34**, 588 (1978).
41) S.B. Carter, *Nature,* **213**, 261 (1967); *idem, Endeavour*, **13**, 77 (1972); S.W. Tanenbaum (ed.), *Cytochalasins, Biochemical and Cell Biological Aspects*, North-Holland (1978).
42) K. Ohtsubo, M. Saito, S. Sekita, K. Yoshihira, S. Natori, *Japan. J. Exptl. Med.*, **48**, 105 (1978).
43) M. Nagao, M. Honda, S. Hamasaki, S. Natori, Y. Ueno, M. Yamasaki, Y. Seino, T. Yahagi, T. Sugimura, *Proc. Jap. Assoc. Mycotoxicol.* (in Japanese), No. 3/4, 41 (1976).
44) K. Ohtsubo, private communication.
45) J. Osame, T. Yoshizawa, Y. Kurokawa and N. Morooka, *Proc. Jap. Assoc. Mycotoxicol.* (in Japanese), No. 8, 22 (1978).
46) I. Löw, W. Jahn, Th. Wieland, S. Sekita, K. Yoshihira, S. Natori. *Anal. Biochem.*, **95**, 14 (1979).
47) K. Ohtsubo, T. Horiuchi, Y. Hatanaka, M. Saito, *Japan. J. Exptl. Med.,* **46**, 277 (1976).
48) K. Ohtsubo, M. Saito, *Ann. Nutr. Alim.*, **31**, 771 (1977).
49) Y. Ito, K. Ohtsubo, K. Yoshihira, S. Sekita, S. Natori, H. Tsunoda, *Japan. J. Exptl. Med.*, **48**, 187 (1978).
50) A. Endo, private communication.

CHAPTER 9

Biosynthetic Studies using Stable Isotopes

Most naturally occurring compounds studied so far have been isolated from plants and microorganisms. Biosynthetic studies of these compounds have generally involved one of two approaches. One deals with the biosynthetic routes from precursors and/or intermediates to the final metabolites, and the other deals with reaction mechanisms in the biosynthetic pathway. Such studies have been performed using tracers and there are three main kinds of labeling methods, i.e., (i) feeding with non-labeled natural or synthesized precursors and comparing the relative amounts of products obtained, (ii) feeding with radioisotopically labeled precursors and tracing the fate of the label, (iii) feeding with stable-isotopically labeled precursors and analyzing the fate of the label by mass spectrometry or nmr spectrometry. The present short review deals with the third approach, and in particular, with feeding experiments with stable-isotopically labeled, simple precursors and analysis of the results by nmr spectrometry.

There are no established generalized extraction and separation methods for nmr analysis of diverse compounds, so special adaptations of usual extraction and separation methods are also described for labeling and detecting naturally occurring compounds with stable isotopes.

9.1 Nuclear Magnetic Resonance Spectroscopy for Biosynthetic Studies[1–4,15]

All atomic nuclei with magnetic moment are active to nuclear magnetic resonance spectroscopy. Among those nuclei, the proton has been most widely studied due to its large magnetic moment and high natural abundance. Recent instrumental developments have permitted routine nmr analysis of nuclei with spectroscopically low sensitivity such as carbon, nitrogen, and deuterium. Such nmr spectroscopy is most informative for the elucidation of the chemical states of these nuclei in naturally occurring compounds, and these elements constitute the bulk of naturally occurring compounds.

Carbon-13 has the same half-spin as the proton, but there are some differences in measuring and interpreting carbon-13 nmr compared to proton nmr, e.g. in the homonuclear spin-spin couplings between protons or carbons and the heteronuclear spin-spin couplings between protons and carbons.

Homonuclear spin-spin couplings between adjacent or closely located protons such as $^{1}H–C–^{1}H$ and $^{1}H–C–C^{1}H$ are always observed in ^{1}H nmr spectra and give useful information. On the other hand, homonuclear spin-spin couplings between carbons are usually not observed due to the low natural abundance of ^{13}C nuclei, making ^{13}C nmr spectra rather simple. At the same time, the absence of homonuclear spin-spin coupled peaks between carbons in ^{13}C nmr spectra makes it difficult to assign carbon chemical shifts to particular carbon nuclei.

Heteronuclear spin-spin couplings are usually not observed in ^{1}H nmr spectra, but are always observed for carbons bearing protons in ^{13}C nmr spectra. Heteronuclear spin-spin couplings in ^{13}C nmr spectra provide information as to the numbers of protons directly attached to carbons; namely, carbons bearing 0, 1, 2, and 3 protons in the same magnetic state split into a singlet, doublet, triplet, and quartet, respectively. Small heteronuclear spin-spin couplings are also observed for carbons not directly attached to protons (long-range coupling) and this broadens the carbon signals. To overcome the complexity due to heteronuclear spin-spin couplings in ^{13}C nmr spectra, two kinds of decoupling techniques are usually applied for measuring ^{13}C nmr spectra. Those are the proton-noise decoupling and off-resonance decoupling techniques. The former technique decouples all couplings due to protons, including directly attached ones, so that all the carbon signals appear as singlets. The latter method decouples only couplings due to remote protons and coupled peaks are produced only by directly attached protons. Both decoupling methods give further information on carbons bearing protons due to the nuclear Overhauser effect.

Homonuclear and heteronuclear couplings are not restricted to the proton and/or carbon-13, but are also common among other nuclei, for example heteronuclear couplings between ^{13}C–D, ^{13}C–^{15}N, and so on. These homonuclear and heteronuclear couplings give information not obtainable by any other method, as shown later. At the same time, the study of heteronuclear couplings between carbon-13 and other nuclei permits direct measurement of nuclei attached directly to carbons, such as deuterium and nitrogen-15 by observing spin-spin coupled peaks in the ^{13}C nmr spectra.[2,5,6] Although deuterium, nitrogen-15 and other nuclei can

TABLE 9.1 Representative nuclei related to naturally occurring compounds and properties related to nmr spectroscopy.

Nuclei	Spin number	Relative sensitivity	Natural abundance
^{1}H	1/2	1.00	99.98
^{2}H	1	9.65×10^{-3}	0.015
^{3}H	1/2	1.21	—
^{13}C	1/2	1.59×10^{-2}	1.1
^{14}N	1	1.01×10^{-3}	99.6
^{15}N	1/2	1.04×10^{-3}	0.37
^{17}O	5/2	2.91×10^{-2}	0.04

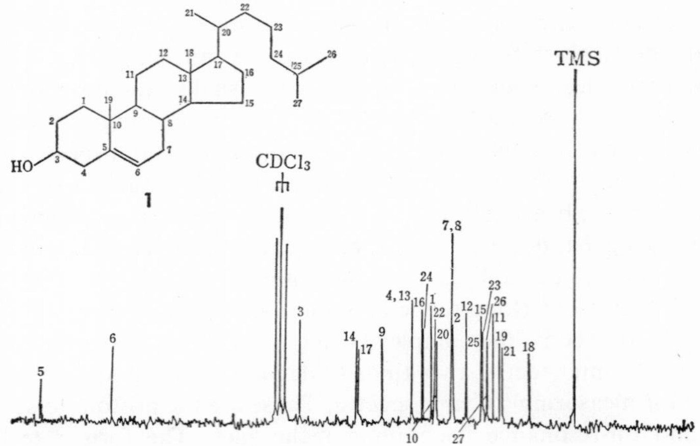

Fig. 9.1 ^{13}C FT-nmr spectrum of cholesterol **1**. Cholesterol (10 mg/ml $CdCl_3$) in a 10 mm nmr tube was measured with a JEOL FX-100 nmr spectrometer at 25.15 MHz; spectral width 5 KHz; data points 4K/4K; 3600 pulses (1.0 sec between pulses); pulse width 7 μsec (45°).

now be observed directly rather easily, the observation of heteronuclear spin-spin couplings simplifies the assignments of chemical shifts and sometimes makes it possible to observe nuclei with low nmr sensitivity by observing coupled peaks in more sensitive nmr spectra.

Relaxation phenomena for excited nuclei are also important for understanding FT-nmr spectra. Different measuring conditions give spectra with variable peak intensities due to relaxation phenomena; pulse-width and delay time should be carefully selected for each measurement. This is extremely important when peak intensities are compared between peaks due to natural abundance and enriched peaks.

Table 9.1 summarizes important properties of nmr spectroscopically active isotopes among elements in naturally occurring compounds. Fig. 9.1 shows a typical sample size required for routine nmr spectrometry. Recent developments in nmr spectrometers, of course, will make it possible to reduce the sample size still further and to get better resolution.

9.2 Labeling and Detection Methods

Biosynthetic studies using stable-isotopically labeled precursors with detection by nmr spectrometry give unequivocal and precise information unobtainable by other methods. There are two approaches, using singly labeled precursors or doubly labeled precursors.

Single labeling compares the relative peak intensities between labeled and natural abundance peaks in the nmr spectra.[7] Relative peak intensities in the same molecule are dependent on the measuring conditions; for example, applied pulse-width, repetition time, concentration of the sample solution, and so on. Therefore, the peak intensities must be carefully compared and a reasonably good incorporation of labeled precursors is necessary to permit a reliable conclusion; for example, peak intensities more than 50% higher than natural abundance peaks are required. As the single labeling method compares the relative peak intensities, the relative incorporation rates of precursors are more important than total incorporation rate, and dilution with natural metabolites must be avoided. When plants or microorganisms produce the target metabolite in high quantity, the metabolite should be depleted before feeding the labeled precursor in order to avoid undesirable dilution and reduction of the relative peak intensities.

Double labeling does not compare the relative peak intensities between labeled and natural abundance peaks, because the peaks from the doubly labeled precursor appear as new peaks due to spin-spin coupling.[8]

Spin-spin coupled peaks between low natural abundance nuclei such as ^{13}C–^{13}C, ^{13}C–^{15}N, or ^{13}C–D, are not usually observed in the nmr spectra of natural abundance samples. Therefore, only the peaks derived from doubly enriched precursors are accompanied by spin-spin coupled peaks. Although single labeling compares the relative peak intensities, the double labeling method involves determining whether there are any spin-spin coupled peaks or not, and this greatly reduces the detection limit. It is possible to dilute the sample with natural abundance material when the sample is labeled with doubly enriched precursors, as illustrated in Fig. 9.2.

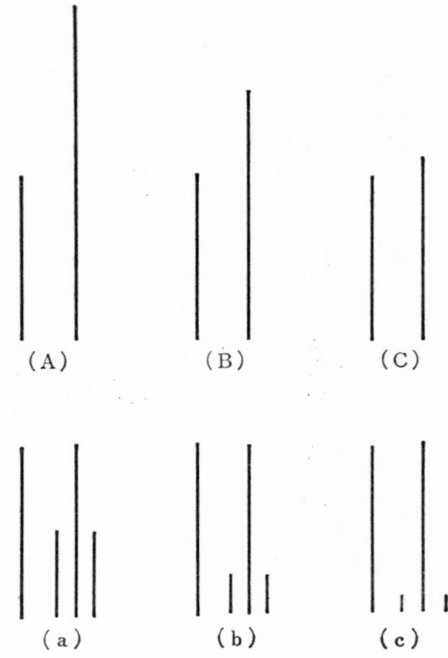

Fig. 9.2 Schematic illustration of relative peak intensities for singly and doubly labeled compounds. A, B, C and a, b, c show the relative peak intensities for singly and doubly labeled compounds at increasing dilutions with the natural abundance compound, respectively.

Fig. 9.2 shows the changes of relative peak intensities on dilution with a natural abundance sample. Fig. 9.2 (A) and (a) illustrate the relative peak intensities of material enriched with singly labeled and doubly labeled precursors. Single labeling gives a peak at least twice as high as the natural abundance one at the same chemical shift. On the other hand, although the intensity of each peak is only one-half of the natural abundance peak, new split peaks appear on both sides of the natural abundance peak due to spin-spin coupling (and the coupled peaks appear as slightly shifted peaks due to isotope shift).

Fig. 9.2 (B) and (b) illustrate relative peak heights when singly and doubly labeled samples are diluted two-fold with a natural abundance sample, respectively. The singly labeled sample gives a peak 50% higher than the natural abundance peak, and the difference of relative peak height in the figure is practically at the detection limit to give reliable results. On the other hand, although the doubly labeled sample gives only one-fourth of the peak height compared with the natural abundance sample, these peaks appear at different positions from the natural abundance peaks, and it is easy to show that the labeled precursors have been incorporated.

Fig. 9.2 (C) and (c) show schematic peak heights when singly and doubly labeled samples are diluted 10-fold with a natural abundance sample, respectively. Although it is impossible to show whether the precursors are incorporated or not for the singly labeled sample, it is easy to confirm the incorporation of labeled precursors for the doubly labeled sample.

Thus, the double labeling method can be applied to biosynthetic studies with both low incorporation and low product yield. Low incorporation of precursors is a particular problem for naturally occurring compounds of plant origin. The metabolic turnover of metabolites in plants is usually slow, and the metabolic pool for final metabolites (usually the target metabolites) is large in plants. These factors result in low total and specific incorporation of the precursors and make it difficult to perform biosynthetic studies with plant metabolites. Double labeling often makes it possible to carry out biosynthetic studies for naturally occurring compounds of plant origin.

The double labeling method is similar to experiments with radioisotopes in the following respects: (i) dilution of the labeled metabolite with natural abundance material is possible, (ii) the total incorporation of precursor is more important than the specific incorporation. In addition, double labeling makes it easy to locate precursor units in the final metabolite. When bond fissions between doubly enriched precursor units occur, spin-spin coupled peaks are no longer observed and only enriched peaks without coupled peaks are observed. Thus, it is easy to detect bond fission or rearrangement during the biosynthetic pathway when doubly enriched precursors are used. Bond fission or rearrangement is not easy to detect by single labeling or by any other method. These advantages are illustrated in biosynthetic studies of ascochlorine,[9)] ovalicin,[10)] and so on.

Application of the double labeling method to a biosynthetic study of cytochalasin B **2** illustrates the above mentioned advantages. [1,2-$^{13}C_2$]

acetate (90% enriched) was diluted 50% with sodium acetate of natural abundance to distinguish the precursor units (^{13}C–^{13}C bonds) from newly formed units (^{13}C–^{12}C bonds) and fed to *Helminthosporium dematioideum*. Fig. 9.3 shows the proton-noise decoupled ^{13}C nmr spectrum of the metabolite, cytochalasin B **2**, in $CDCl_3$-CD_3OD solution (thick lines indicate doubly enriched units). Among 27 carbon signals found in the natural abundance spectrum, 18 showed spin-spin coupled signals, indicating that nine ^{13}C-^{13}C coupled units are incorporated into the metabolite intact. The numbers of protons directly attached to each carbon are easily assigned from the off-resonance spectrum of the natural abundance sample. Functional groups attached to each carbon are assignable from the chemical shift and also from the coupling constants.

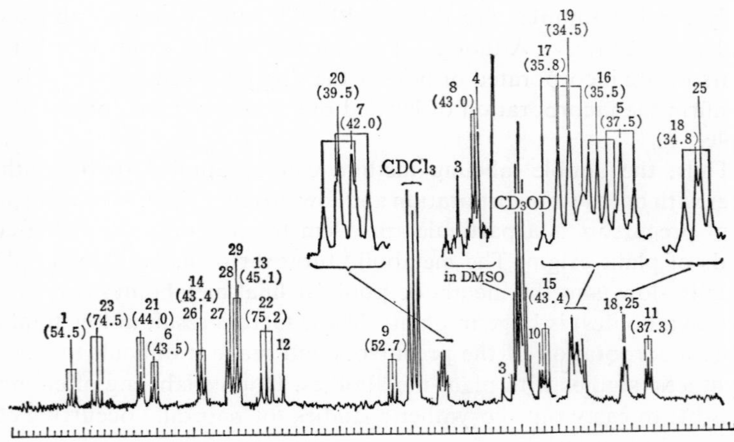

Fig. 9.3 ^{13}C FT-nmr spectrum of cytochalasin B **2** labeled with [1,2-$^{13}C_2$] acetate. [1,2-$^{13}C_2$] acetate was diluted two-fold with natural abundance acetate and fed to *Helminthosporium dematioideum*. The labeled cytochalasin B in $CDCl_3$-CD_3OD (1:1) was analyzed with an XL-100 nmr spectrometer at 25.15 MHz.

There are 30 C–C bond units in cytochalasin B **2** and these can be grouped as follows (numbers are carbon numbers, as shown in Fig. 9.3) according to the combinations of multiplicity due to the numbers of directly attached protons. q-d: 11–5, 25–16. t-t: 17–18, 18–19. t-d: 10–3, 15–14, 15–16, 17–16, 19–20. t-s: 10–26, 12–6. d-d: 3–4, 4–5, 7–8, 8–13, 13–14, 20–21, 21–22, 27–28, 28–29, 29–30, 30–31. d-s: 4–9, 5–6, 7–6, 8–9, 22–23, 27–26, 31–26. s-s: 1–9.

Unequivocal assignments of each carbon signal and biosynthetic deductions are easily possible by grouping C–C bond units according to multiplicities, chemical shift, and coupling constants, as described below. The singlet signal at 176.0 ppm and the singlet peak at 85.8 ppm are accompanied by spin-spin coupled peaks (54.5 Hz and 52.7 Hz, respectively), and are unequivocally assigned to C-1 and C-2. The presence of spin-spin coupling between C-1 and C-9 means that there are no spin-spin couplings between C-9/C-4 or C-9/C-8. Thus, assignment of the carbon signals and biosynthetic studies are easily performed. Selective decoupling is also helpful for the assignment of carbon signals and sometimes coupling patterns (AB or AX type) are also helpful to identify the coupled peaks (for example, AB type for C-16 and C-17).

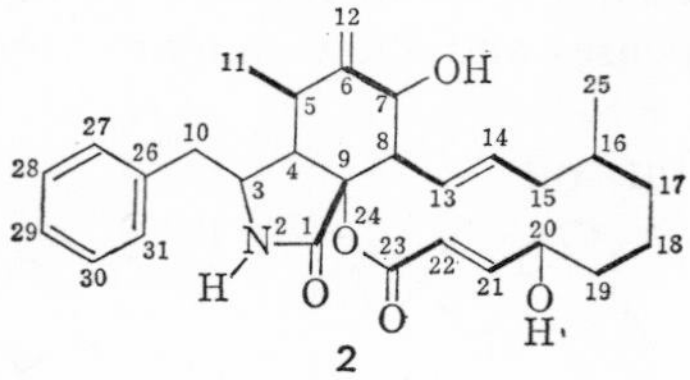

Doubly enriched acetate is thus much more useful than singly enriched acetate for biosynthetic studies of polyketides. Furthermore, doubly enriched acetate is also a good precursor for biosynthetic studies of terpenes, as shown for ascochlorine and ovalicin.[9,10]

Doubly enriched precursors with carbon-13 are extremely useful for biosynthetic studies of carbon skeletons. Other doubly enriched precursors such as ^{13}C-D and ^{13}C-^{15}N are also useful.[1,6] Although singly labeled precursors with deuterium (^{2}H nmr detection) are useful for biosynthetic studies of hydrogen,[11] doubly enriched precursors with ^{13}C and ^{2}H give information about C-H precursor units, rearrangement, and elimination of hydrogen. Doubly enriched precursors with ^{13}C and ^{15}N are useful for biosynthetic studies of nitrogen-containing compounds.

9.3 Isolation and Purification of Labeled Compounds

There are no generalized methods for the isolation and purification of diverse naturally occurring compounds. However, the chemical structures of target compounds for biosynthetic studies are usually known, so it is easier to obtain various derivatives and sometimes to isolate the labeled metabolites with little loss or degradation. The following example illustrates a typical procedure utilizing derivative formation for the isolation of metabolites with similar chemical structures.[12]

Illudol **3** and illudin S **4**, metabolites of *Clitocybe illudens*, are trihydroxylated sesquiterpenses and show similar chromatographic behavior under various conditions. Both metabolites are produced in a culture broth and have been purified by counter-current distribution between chloroform and water. The purification of both metabolites is not easy on a small scale. Among the three hydroxyl groups in both metabolites, three in illudol and two in illudin S are acetylated under mild conditions (acetic anhydride-pyridine). Therefore, it should be easier to separate the metabolites after the acetylation procedure, which should produce a distinct difference in their chromatographic properties. In fact, although it is difficult to get complete separation of illudol **3** and illudin S **4** on thin layer or column chromatography, the acetate derivatives of illudin S and illudol are easily separable under almost any chromatographic conditions. Thus, the preparation of acetate derivatives has the following advantages. (i) Isolation and purification are easier. (ii) Acetylation of hydroxyl groups may stabilize reactive functional groups. (iii) Acetyl groups can be used as an internal standard for natural abundance peaks in the nmr spectra and can be used to check the measurement conditions, i.e., the methyl and carbonyl groups in the acetyl moiety can be used as internal standards for sp^3 carbons with three protons and sp^2 carbons without protons, respectively. When the delay time and pulse width, and the concentrations of samples are adequate, the intensity ratios of methyl and carbonyl groups in the ^{13}C nmr spectra should be the same for both natural abundance (control) and labeled samples. (iv) Acetate derivatives are usually more soluble in chloroform than their parent metabolites. The use of methanol, acetone, or DMSO for polar compounds gives rise to more solvent peaks than chloroform in the ^{13}C nmr spectra, and these peaks often overlap with sample peaks. There is the disadvantage that signals due to acetyl groups are increased, but acetyl signals in the derivatives are quite easy to identify.

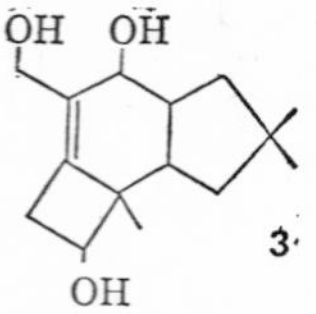

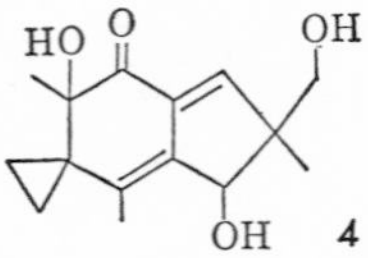

Biosynthetic studies for known naturally occurring substances are usually performed by the reported isolation and purification procedures, but different separation and isolation procedures may be preferred due to recent instrumental and technical developments and/or due to differences in standard methods from laboratory to laboratory.

Helicobasidin **5**, a sesquiterpene pigment of *Helicobasidium mompa*, cannot be purified by the routinely used silica gel column chromatography

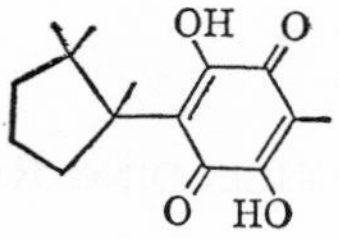

5

due to severe tailing. The pigment which accumulates in mycelia has been purified by drying mycelia followed by direct sublimation from the dried mycelia.[13] As expected from the chemical structure, the pigment is acidic and is quantitatively eluted with benzene from an acidic silica gel column.[14] Although the eluted benzene solution contains a small amount of phenolic compounds as contaminants, helicobasidin is the main component, and can be recrystallized from the eluate to give pure material. It takes only 3–4 hr to obtain pure helicobasidin **5** by an imporoved procedure; namely, direct extraction of helicobasidin **5** with hot acetone from the cultured mycelia, removal of acetone from the extract by evaporation, extraction of helicobasidin with benzene from the residue, acidic silica gel column chromatography, and recrystallization. *H. mompa* is a mushroom and grows very slowly in stationary culture. It takes about three months to obtain cultured mycelia, and the microorganism is not suitable for biosynthetic studies. However, the culture conditions can be improved to obtain faster growth and a better yield of helicobasidin **5** as follows. The microorganism was precultured on wet wood shavings containing 2% glucose and 0.5% yeast extract. The wood shaving culture can be used as a stock culture for several months. The cultured mycelia on the wood shavings were mixed well with the same culture solution, then transferred to a culture solution which contained 2% sucrose, 0.5% corn steep liquor, 0.2% peptone, 0.2% yeast extract, 0.1% potassium phosphate and 0.01% magnesium sulfate. The transferred mycelia covered the bottom of the culture flask within one week. When the mycelia grew on the surface of the culture solution, the production of the pigment, helicobasidin **5**, started and the mycelia turned deep red. Therefore, precursors were added after removing the mycelia from the bottom of the flask and allowing them to float on the surface, and the culture was continued for a further two weeks to obtain deep red mycelia. The yield of helicobasidin increased several times compared with the conditions used previously to give about 150 mg/l of culture solution.

The carbon atoms in helicobasidin **5** can be divided into sp^3 carbons and sp^2 carbons without protons. The acetate derivative of helicobasidin is readily soluble in chloroform, and the methyl and carbonyl groups of the acetyl moieties provide suitable internal standard signals for sp^3 and sp^2 carbons, respectively.[14)]

9.4 Concluding Remarks

It is not easy to generalize on the best kind of precursor to use, how to label and feed the precursor, when the precursor should be administered, and how to extract and purify the metabolites. The present review covers only a small number of recent biosynthetic studies with stable isotopes. It is necessary to select the best method according to the nature of the target metabolites and available instruments in the laboratory.

The data for cytochalasin B, illudol, illudin S, and helicobasidin have not previously been published. The author is grateful to Dr. M. Tanabe (SRI International) for supplying unpublished data and to Mr. M. Imanari (JEOL Co.) for the ^{13}C nmr chart of cholesterol.

References

1) M. Tanabe, *Stable Isotopes in Biosynthetic Studies* in *Biosynthesis* **2** (1973), **3** (1974), and **4** (1975), The Chemical Society.
2) K.T. Suzuki, *Kagaku no Ryoiki Zokan* (in Japanese) **107**, p. 189, Nankodo (1975).
3) M. Yamasaki, Y. Morino, *Kagaku Zokan* (in Japanese) **67**, p. 171, Kagakudojin (1976).
4) A.G. McInnes, J.L.C. Wright, *Acc. Chem. Res.*, **8**, 313 (1975).
5) A.G. McInnes, D.G. Smith, C.T. West, *Chem. Commun.*, **1974**, 281.
6) H. Yamada, M. Hirobe, K. Higashiyama, H. Takahashi, K.T. Suzuki, *J. Am. Chem. Soc.* **100**, 4617 (1978).
7) M. Tanabe, H. Seto, L.F. Johnson, *ibid.*, **92**, 2157 (1970).
8) H. Seto, T. Sato, H. Yonehara, *ibid.*, **95**, 8461 (1973).
9) M. Tanabe, K.T. Suzuki, *Chem. Commun.*, **1974**, 445.
10) M. Tanabe, K.T. Suzuki, *Tetr. Lett.*, **1974**, 4417.
11) Y. Sato, T. Oda, E. Miyata, H. Saito, *FEBS Lett.*, **98**, 271 (1979) and references cited therein.
12) M. Tanabe, K.T. Suzuki, unpublished data.
13) S. Natori, Y. Inoue, H. Nishikawa, *Chem. Pharm. Bull.*, **15**, 380 (1967).
14) M. Tanabe, K.T. Suzuki, W.C. Jankowski, *Tetr. Lett.*, **1974**, 2271.
15) G. Kunesch, C. Poupat, *Biosynthetic Studies Using Carbon-***13** *Enriched Precursors* in *Carbon-13 in Organic Chemistry* (ed. E. Buncel, C.C. Lee) Elsevier (1977) and references cited therein.

CHAPTER 10

Flycidal Constituents from *Tricholoma muscarium* and *Amanita strobiliformis*

Tricholoma muscarium KAWAMURA (Japanese name "*haetorishimeji*") is a species of mushroom belonging to the family Tricholomataceae. The cap of this mushroom is 4 to 6 cm in diameter, colored pale yellow-green or pale yellow, and covered densely with dark green fibers resembling silk. Although the mountain-shaped cap becomes flat as the mushroom grows, the center of the cap is convex and colored green-brown. The interior is white; the folds are white and emerge from the upper part of the stem. The stem is 4 to 7 cm in length, 0.7 to 1.5 cm in diameter, slightly thinner at the upper part and thicker at the lower part, and pointed at the bottom. The interior is largely filled with white material. The spores are ovoid or elliptical, and 4.5 to 5.5 μ in diameter. The mushroom grows on the ground in the chestnut woods of Japan in autumn, and is lethal to flies, though non-toxic to human beings.

Amanita strobiliformis (PAUL.) QUEL. (Japanese name "*ibotengutake*") is a species of mushroom belonging to the family Amanitaceae, discovered and given its Japanese name by Professor H. Matsumoto (Tohoku University) near Aoba Castle, Sendai, during 1921 to 1925. The fruit body is similar to that of *Amanita pantherina*, but much bigger. Characteristically, the cap and the thick lower part are dotted with many pyramidal warts. It is said that this mushroom is non-toxic to human beings and is very tasty, while *Amanita pantherina* is fatally poisonous. Large numbers of *Amanita strobiliformis* grow on the ground in the windbreak zone of *Pinus thumbergii* on the seashore near the outskirts of Sendai in summer and autumn, especially from the end of September to the beginning of October.

10.1 Studies on the Flycidal Constituents of Mushrooms

Tricholoma muscarium has been used for catching flies and similar insects by farmers since olden times, and accordingly it is called "*haetoritake*" or "*haetorikinoko*" (a fly-catching mushroom). However, the name "*haetoritake*" seems to be used generally for the *Amanita* group. For instance, *Amanita pantherina* is called "*haetoritake*," while *Amanita muscaria* is called "*akahaetoritake*." They are used for catching flies in the same way as *T. muscarium*.[3] For the purpose of catching flies, the untreated or slightly heated mushroom is placed in a dish containing a small amount of water. Cane suger may also be added to induce the flies to feed.

In spite of its rapid insecticidal effect, few studies on the active principle of *T. muscarium* have been reported. Ohya[4] studied the principle and reported that the insecticidal constituents were readily soluble in water, were insoluble in organic solvents, were adsorbed by strongly acidic ion-exchange resin in the same way as amino acids, and could be eluted with ammonia water. Further, Ohya observed free amino acids on two-dimensional paper chromatography; there were eighteen ninhydrin-positive spots including spots of aspartic acid, glutamic acid, alanine, valine, serine, glycine, leucine, threonine, either glucosamine or glutamine, and two unidentified yellow-colored amino acids.

Ohya[5] also attempted to isolate insecticidal constituents as follows. The material was extracted with hot water (80° C), decolorized with active carbon, passed through Amberlite IR-120 (H^+), and eluted with 1% ammonia water. The eluate was passed through Amberlite IRC-50, and the effluent was in turn passed through buffered Amberlite IR-45. Fraction B showed activity. This fraction was spotted across paper (40 × 40 cm), developed with *n*-butanol-acetic acid-water (4:1:1), and divided into nine fractions from the top. Each part was extracted and tested for insecticidal activity. The insecticidal activity was found at *Rf* 0.07 to 0.13 (colored light blue with ninhydrin). The fraction was paper-chromatographed with phenol-water (4:1). Two spots of *Rf* 0.25 (red-violet with ninhydrin) and *Rf* 0.6 (blue with ninhydrin) were seen. The former was identified as cystine, but the latter was not identified.

On the other hand, *Amanita strobiliformis* has been known as a flycidal mushroom since olden times (it is generally called "*haetorimodashi*"), but little work has been done on its insecticidal constituents. Many of the mushrooms of Amanitaceae are fatally poisonous. *A. pantherina* and *A. muscaria* are regarded as typical poisonous mushrooms in Japan; they are known to contain muscarine. In addition, poisonous constituents with cyclic peptide structures, such as phalloidin, phalloin, α-amanitin and β-

amanitin have been isolated from *Amanita phalloides*.[6] (These mushrooms are known as "fly agaric" in English and "Fliegenpilz" in German.)

Onda *et al.*[7] reported pantherine as a flycidal constituent of *A. pantherina* and assumed it to be a peptide. Bowden *et al.*[8] isolated a compound called agarin from *A. muscaria* and proposed its chemical structure. However, both substances are in fact the same as a secondary product obtained by the authors from ibotenic acid, a flycidal constituent which the authors isolated from *A. strobiliformis*. Eugster *et al.* noted that *A. muscaria* had a strong CNS (central nervous system) inhibitory action in higher animals (e.g., anesthesia-enhancing action or sedative action), and isolated ibotenic acid[9] and muscimol[10] as effective constituents. They reported that muscimol is not the natural constituent of the mushroom, but a substance produced secondarily from ibotenic acid during the extraction process.[9]

The identification of the flycidal constituents was achieved first by the present authors, partly because of previous experience in isolating the parasiticidal constituents, i.e., kainic acid **1** and domoic acid **2**, from such sea weeds as *Digenea simplex* and *Chondria armata*. Namely, in 1963 the

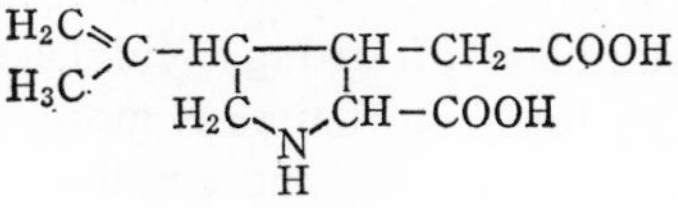

```
H2C≈                                                CH3
     C-HC——CH-CH2-COOH                              C
H3C    |    |                               HC≈  HC——————CH-CH2-COOH
      H2C  CH-COOH                          |     |       |
         N                                  HC≈  H2C     CH-COOH
         H                                     CH     N
                                               |      H
                                              CH
      kainic acid 1                     H3C      COOH      domoic acid 2
```

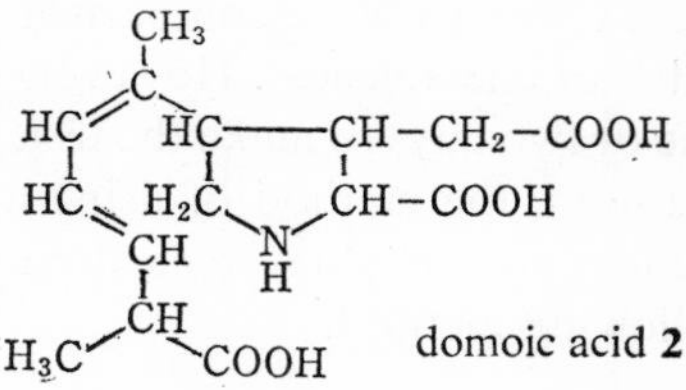

authors succeeded in isolating domoic acid[12] from *Chondria armata* and in establishing its structure after research on the parasiticidal constituent, kainic acid, of *Digenea simplex*. At that time, investigators considered that amino acids were comparatively stable compounds, so when multiple amino acids were produced by acid treatment of a substance, it was considered likely to be a peptide. However, kainic acid and domoic acid were readily isomerized by acid treatment, producing multiple ninhydrin-positive substances.

Furthermore, the authors accidentally found[13] that kainic acid and domoic acid have flycidal properties. When the authors were collecting and drying *Chondria armata* near Yakushima Island, they observed that flies landed on the sea weed, licked it, and immediately died. Thus, when the study of parasiticidal constituents was finished, a study of the flycidal constituents was begun. Tests with captured flies quickly showed that domoic acid itself is the insecticidal constituent of *Chondria armata*. Further, on the basis of screening tests for flycidal constituents, the flycidal action was very strong. Unless the test was a comparative study of the fly-cidal action, observation of a single fly for about an hour was better than the general method,[4] and it was more suitable for research on the effective constituents. In fact, in the case of *T. muscarium*, the effective constituents were obtained in crystalline form in about three months. In the case of *A. strobiliformis*, the isolation of the effective constituents was achieved in about one month. Incidentally, the authors isolated and identified ibotenic acid from *A. pantheria* and *A. muscaria* in addition to *A. strobiliformis.*[15]

10.2 Extraction and Isolation of Flycidal Constituents

Initially, it was important to develop suitable tests for activity. In general insecticidal tests, a definite number of flies is placed in a container. The sample, alone or dissolved in water, is adsorbed on a piece of cotton, which is placed in the container. The insecticidal effect is rated by counting the number of flies which fall and die in a definite time. This method is most suitable for the comparison of samples or for obtaining the threshold value of effectiveness. However, several samples are required and it takes one or two days to make the test. When samples with strong flycidal action are used, the method of using a single fly for a short time is more convenient and reliable. Accordingly, the authors devised and employed the following method.

[Experimental procedure 1] Simple test for insecticidal activity

House flies which had been bred for many generations were employed as test insects. As shown in Fig. 10.1, a glass funnel of 6 cm diameter (the neck of the funnel was closed with a cotton stopper to prevent the flies escaping) was placed upside down on a 20 cm^2 paper. The anesthetized test flies were put in the glass funnel one by one. They were deprived of food

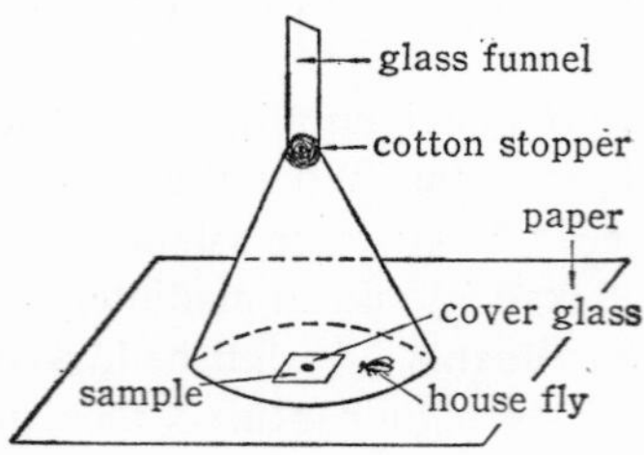

Fig. 10.1 Apparatus for testing insecticidal acitivity.

for about one hour, then the samples were either put in the funnel, or dripped onto a microscope cover glass in the funnel after being mixed with sugar water, honey, and fruit juice. Ethylether or carbon dioxide gas can be used as the anesthetic. However, the authors simply caught the house flies in the flask, stoppered it, and placed it in a refrigerator (low temperature anesthetization). This method does not cause any deaths and is very convenient. The state of the flies was observed in detail after they began to lick the sample. In 4 to 6 min, movement of the fly's legs slowed down, followed by paralysis, a syncopal state, and finally death.

[Experimental procedure 2] Solubilities of the insecticidal constituents[17)]

T. muscarium was collected in Iwate Prefecture in late September, 1960. The mushrooms were carefully dried in an oven below 60° C and then powdered (mushroom powder).

The powdered material was first extracted with petroleum ether (bp 35–50° C) for 4 hr in a Soxhlet extractor, yielding 100 g of light yellow oily petroleum ether extract. Next, the residue was extracted with 15 ml of ether for 3 hr, and 30 mg of light yellow oily ether extract was obtained. The residue of the ether extract was extracted with 15 ml of chloroform for 5 hr, and 30 mg of yellow-brown oily chloroform extract was obtained. Finally, the residue was divided into five equal parts. One part (equivalent to 1 g of mushroom powder) was warmed and soaked four times in 5 ml of 95% ethanol for 1 hr. The combined extracts yielded 100 mg of light brown ethanol extract resembling gluten. Another part of the chloroform

TABLE 10.1 Behavior of insecticidal constituents on treatment with solvents.

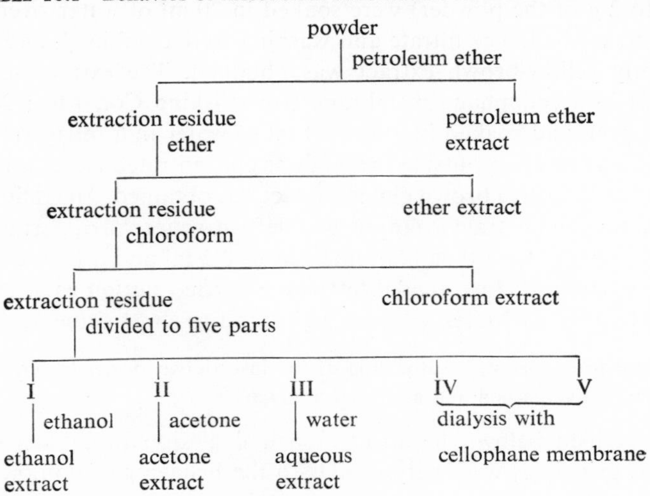

TABLE 10.2 Properties of solvent extracts.

Extract (Ex.)	Properties (yield, mg)	Liebermann reaction	Ninhydrin reaction	Insecticidal activity
Petroleum ether Ex.	light yellow oily substances (100)	+	−	−
Ethylether Ex.	light yellow oily substances (30)	+	−	−
Chloroform Ex.	yellowish-brown oily substances (30)	−	−	−
Ethanol Ex.	light yellow sticky substances (100)	−	−	−
Acetone Ex.	light yellow sticky substances (50)	−	−	−
Aqueous Ex.	brown sticky substances (300)	−	+	+

extract residue was treated with acetone as described for the ethanol extract, and 50 mg of a light yellow acetone extract resembling gluten was obtained. Next, 10 ml of water was added to a third part of the chloroform extract residue. The mixture was cooled and left overnight. Concentration of the filtrate gave 300 mg of a brown aqueous extract resembling gluten. Liebermann's reaction,[18] the ninhydrin reaction,[19] and insecticidal tests were carried out on all the above extracts. Tables 10.1 and 10.2 show the procedure and results, respectively.

[Experimental procedure 3] Permeability through a cellophane membrane

The two remaining parts of the chloroform extract residue (corresponding to 2 g of the powder) were soaked in 20 ml of water overnight, filtered, and washed. The filtrate and washing were combined, and 20 ml of a slightly yellow-brown extract was obtained. The extract solution was placed in a cellophane membrane bag (Visking Co., Ltd.; 2.4 nm pore diameter), and soaked in about 30 ml of water in a refrigerator for three days. The solution outside the bag was concentrated to 0.5 ml under reduced pressure, and a brown gluten extract was obtained. The cellophane bag containing the extract solution was left in water for a further two days. The contents were then concentrated to 0.5 ml and a brown gluten extract was obtained. Insecticidal tests were carried out on each extract, and the extract of the outer solution was found to show activity.

[Experimental procedure 4] Adsorption of the insecticidal principles on ion exchange regin, active carbon and activated alumina

As described below, the adsorption and elution properties of the insecticidal principles were very similar to the behavior of domoic acid, suggesting that the insecticidal principles might be acidic or neutral amino acids.

Mushroom powder (5 g) was soaked in 50 ml of water for about 12 hr. This procedure was repeated twice with 30 ml of water. After filtration, 100 ml of the combined extract solution was passed through a column (1.8 × 20 cm) of Amberlite IR-120 (H^+) 50 ml). Next, the column was washed with water, then eluted with 0.5 N NaOH solution. When the eluate began to show alkaline (about 135 ml), the column was further eluted with 180 ml of alkaline solution. This eluate was passed through a prepared column of Amberlite IRC-50 (H^+) (1.8 × 15 cm) to remove the alkali. The resulting solution and washing were concentrated to 5 ml under reduced pressure (Fraction 1). Next, the Amberlite IRC-50 column was eluted with 2% ammonia water. When the eluate showed alkaline, the column was further eluted with 160 ml of alkaline solution. This eluate was concentrated to 0.5 ml under reduced pressure (Fraction 2).

[Experimental procedure 5] Behavior of insecticidal principles with active carbon

Active carbon (100 mg) was added to half of Fraction 1 obtained above, and the mixture was stirred for about 30 min. The active carbon was filtered off, and washed, then 4 ml of 90% methanol was added. The mixture was stirred for 10 min and filtered. This extraction procedure was repeated 6 times. The resulting methanol extract solution was concentrated to 1 ml under reduced pressure (Fraction 3).

[Experimental procedure 6] Behavior of insecticidal principles with activated alumina

Methanol (22.5 ml) was added to the remaining half of Fraction 1 obtained above and the insoluble matter was filtered off. The filtrate was passed through a column (0.6 × 10 cm) of 4 g of activated alumina, then washed with 40 ml of 90% methanol. The eluate and washing were combined (Fraction 4). Next, the adsorbed material was eluted with 40 ml of water (Fraction 5), followed by 40 ml of 0.1 N NaOH, providing a light yellow eluate (Fraction 6). Fractions 4, 5, and 6 were passed through prepared columns of Amberlite IRC-50 (H^+). After removing inorganic matter, each eluate was concentrated to 1 ml under reduced pressure. Table 10.3 shows the results of these and the previous procedures.

[Experimental procedure 7] Paper electrophoresis of insecticidal constituents

Fraction 3 obtained above (0.05 ml) was spread across paper (Toyo Roshi No. 50; 4 × 40 cm), air-dried, and subjected to electrophoresis in 0.02 M phosphate buffer (pH 7.0) at 400–600 V, 0.5 mA/cm, for 3 hr. The paper was air-dried then 1 cm strips were cut off both edges, and one was

TABLE 10.3 Fraction of insecticidal constituents. + and − indicate the presence and absence of insecticidal activity, respectively.

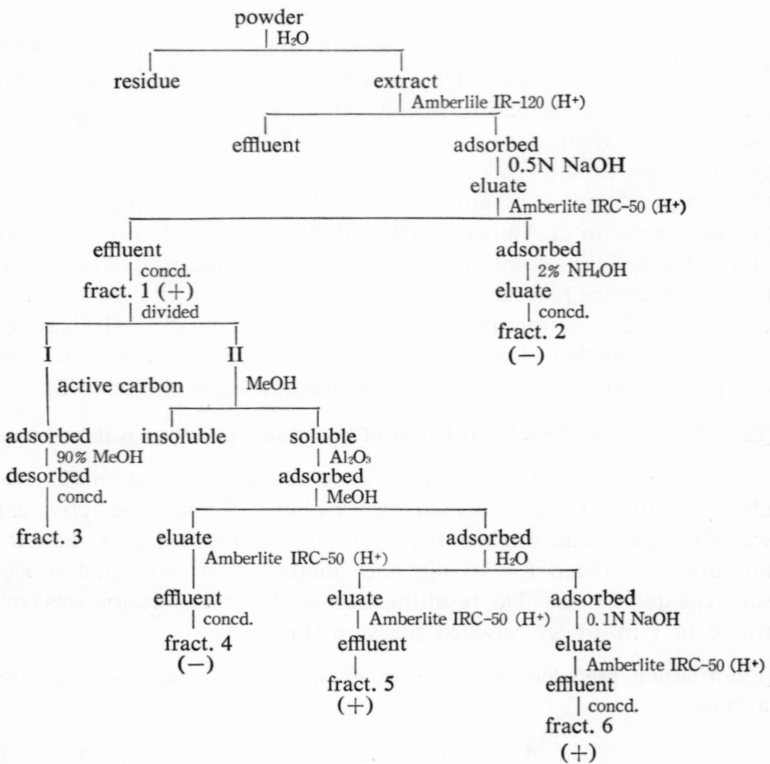

treated with ninhydrin. Color development appeared at electrophoretic migration distances (mm) of −7, +52, +74, +85, and +99. The corresponding parts of the other strip were wetted slightly with cane sugar solution (2%) and the insecticidal properties were tested. Only the zone of +74 migration showed insecticidal activity.

Behavior of Insecticidal Principles with Weakly Basic Ion Exchange Resin

Fraction 1 (7 ml) was passed, through an Amberlite IR-4B column (0.4 × 8 cm), eluting with about 100 ml of water, followed by 0.1 N acetic acid solution, Fractions were concentrated and subjected to paper chromatography paper electrophoresis and insecticidal tests. It was found that the insecticidal principles were adsorbed by Amberlite IR-4B (acetic acid type) and could be eluted gradually with water.

10.3 Isolation of the Flycidal Principle from *T. muscarium*

As described above, and shown in Table 10.4, the insecticidal principles were adsorbed on Amberlite IR-120, and eluted with alkaline solution. The eluate was passed through a column of Amberlite IRC-50, desalted, and concentrated under reduced pressure. Fraction A, with strong insecticidal activity, was obtained. Fraction A was passed through a column of Amberlite IR-4B, and eluted with water. The eluate was concentrated under reduced pressure (Fraction B). A little methanol was

Table 10.4 Isolation of tricholomic acid **3**.

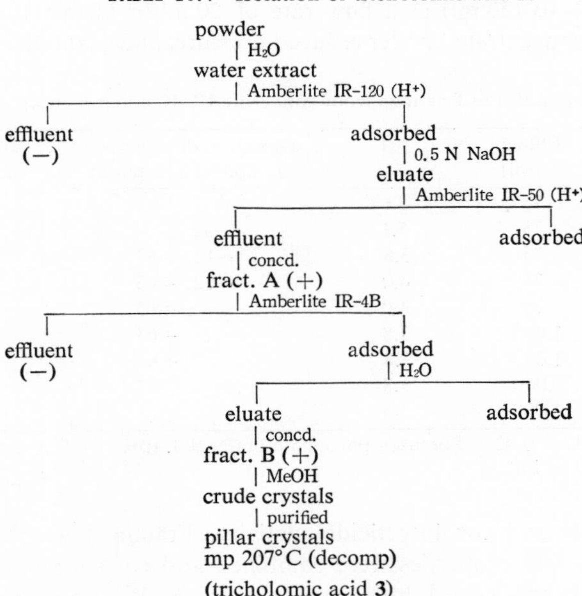

added, and the mixture was allowed to stand, yielding colorless pillar crystals. The crude crystals were recrystallized from water, yielding pillar crystals, mp 207° (decomposition), $C_5H_8O_4N_2$, which were colorless, odorless, and of delicious taste. An aqueous solution (0.01 %) showed strong insecticidal action against house flies. On electrophoresis, it migrated towards the anode in a neutral buffer, and gave a red-violet color with ninhydrin, becoming violet. pKa': near 2, 6.0, and 8.0. The principle was concluded to be a new acidic amino acid, and was named tricholomic acid **3**.

[Experimental procedure 8] Isolation of tricholomic acid

Mushroom powder (80 g) was soaked in 500 ml of water overnight. The residue was again soaked in water (400 ml) for 2 hr (repeated four times), and 1,920 ml of extract was obtained. This was passed through a column of Amberlite IR-120 (H^+ type) (150 ml), and washed with water until Cl^- was no longer detectable in the effluent, then 0.5 N NaOH solution was passed through the column. When the effluent became alkaline, elution was continued with 900 ml of water. The eluate was passed through a column of Amberlite IRC-50 (H^+) (200 ml) and concentrated to 50 ml under reduced pressure. This solution was passed through a column of Amberlite IR-4B (50 ml) at a flow rate of 20 ml/hr (Table 10.5). Each fraction was concentrated under reduced pressure, and examined by paper

TABLE 10.5 Properties of fractions from Amberlite IR-4B column chromatography.

Fraction number	Effluent (ml)	pH	Migration distance (mm) on paper electrophoresis	Insecticidal activity
1	75	5.6	−12	−
2	75	5.1	−12	−
3	75	3.8	−12, +63	±
4	75	4.0	−12, +63	±
5	75	4.0	+63	+
6	1,000	3.8	+63	+
7	1,000	3.8	+63	+
8	1,000	3.8	+63	+
9	1,000	3.8	−	−

Toyo Roshi #50, 0.02 M Sörensen phosphate buffer Sol. (pH 7) 0.5 mA/cm, 2 hrs, detection-ninhydrin

electrophoresis and for insecticidal activity. Fractions which showed strong insecticidal properties were combined and concentrated to 5 ml under reduced pressure. A little methanol was added, and on standing 10 mg of crude crystals was obtained. Recrystallization from water gave colorless pillar crystals, mp 207° (decomposition), $C_5H_8O_4N_2$ (elemental analysis and molecular weight determination), pKa' near 2, 6.0, and 8.6. The product was stained red-violet with ninhydrin, becoming violet. *Rf* 0.10 (Toyo Roshi No. 50, *n*-butanol-acetic acid-water 4:1:1); *Rf* 0.28 (Toyo Roshi No 50, phenol-water 4:1); migration distance (mm) + 63 (Toyo Roshi No. 50, 0.02 M phosphate buffer (pH 7.0), 0.5 mA/cm, 2 hr); migration distance (mm) −5 (Toyo Roshi No. 50, acetic acid-sodium acetate solution (pH 3.7), 0.06 mA/cm, 2 hr).

10.4 Extraction and Isolation of the Flycidal Principle from *A. strobiliformis*

A preliminary examination showed that the flycidal principle of *Amanita strobiliformis* had the following properties.

(1) The active principle was slightly soluble or insoluble in organic solvents, but soluble in water.

(2) It could permeate through a cellophane membrane.

(3) It was adsorbed from an aqueous extract by a strongly acidic ion exchange resin and by active carbon. It could be eluted from active carbon with an alkaline solution or with methanol.

(4) It was adsorbed by alumina and could be eluted with alkaline solution.

(5) It migrated toward the anode in neutral buffer solution during paper electrophoresis.

(b) It was adsorbed by Amberlite IR-4B; it could not be eluted by water but was eluted by 0.1 N acetic acid solution.

The flycidal constituent was stained yellow with ninhydrin, becoming violet on standing. The flycidal principle is clearly more acidic than tricholomic acid, and resembles an amino acid. It was isolated as described below.

[Experimental procedure 9] Isolation of ibotenic acid

Frozen fresh mushroom (800 g) was well ground. Ice-water (1 l) was added to the powder, and left overnight. The residue was extracted with 1 l of water for 1 hr (repeated three times). Viscous and insoluble material was removed from the combined extract with diatomaceous earth (Hyflo Supercel). The filtrate was passed through a column of Amberlite IR-120 (H^+) (300 ml), washed with water and eluted with 0.5 N NaOH. When the effluent became alkaline (1 l of 0.5 N NaOH solution), a further 1.8 l was passed through the column. The eluate was passed through a column of Amberlite IRC-50 (H^+) (500 ml). After desalting, the eluate was concentrated to 50 ml under reduced pressure (neutral and acidic amino acid fraction).

The neutral and acidic amino acid fraction (20 ml) was passed through a column of Amberlite IR-4B (20 ml), eluting with 240 ml of water (neutral amino acid fraction), 560 ml of 0.1 N acetic acid (acidic amino acid fraction a), and finally 300 ml of 2.5 N acetic acid (acidic amino acid fraction b). Each fraction was concentrated under reduced pressure. The residue was extracted with 2 ml of water. The insecticidal properties are shown in Table 10.6.

TABLE 10.6 Fraction of flycidal principles from *Amanita strobiliformis.*

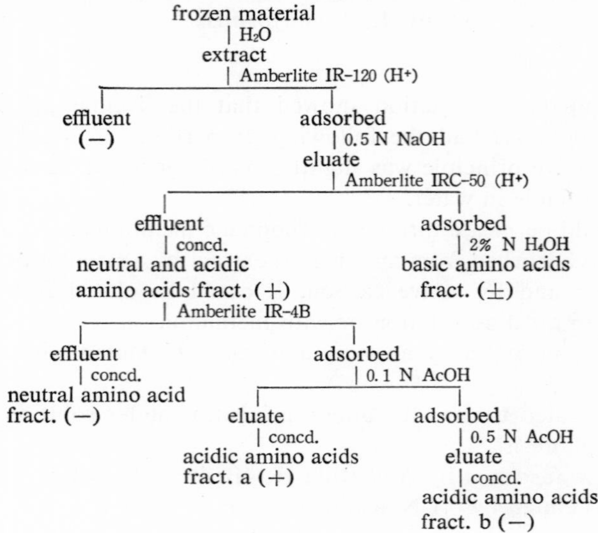

Methanol (90 ml) was added to 10 ml of the neutral and acidic amino acid fraction, and the insoluble material was filtered off. The filtrate was passed through a column (10 g) of activated alumina, and eluted with 150 ml of 90% methanol. The eluates were concentrated to 1 ml under reduced pressure (Fraction A). No insecticidal activity was seen. Next, 200 ml of water was passed through the column (Fraction B). Lastly, 200 ml of 0.2 N NaOH was passed through the column (Fraction C). Fractions B and C were each passed through a column of Amberlite IRC-50 (H^+), and inorganic material was filtered off. The fractions were concentrated to 1 ml under reduced pressure, and the insecticidal properties were tested (Table 10.7). The acidic amino acid fraction (2 ml) was treated with 3 ml of 7% copper acetate and warmed to 60° C. The gray-green precipitate was filtered off, washed with water, and floated on 12 ml of warm water. Copper was removed by saturation with hydrogen sulfide. The filtrate was concentrated under reduced pressure, yielding 5 mg of crude crystals, which were recrystallized from water as colorless pillars, mp 151–152° C (decomposition), $C_5H_6O_4N_2 \cdot H_2O$ (elemental analysis and molecular weight determination), $[\alpha]_D \pm 0$ ($c = 0.900$, H_2O; $c = 4.00$, 2 N NaOH). uv $\lambda_{\text{max.}}^{H_2O}$ 210 nm (log ε 3.84), pK_a' near 2, 5.1, and 8.2. The compound was stained yellow with ninhydrin, becoming violet on standing: *Rf* 0.20 (Toyo Roshi No. 50, *n*-butanol-acetic acid-water (4:1:1); 0.11 (Toyo Roshi No. 50, phenol-water 4:1), migration distance (mm) + 55

TABLE 10.7 Fraction of the flycidal principle by alumina chromatography.

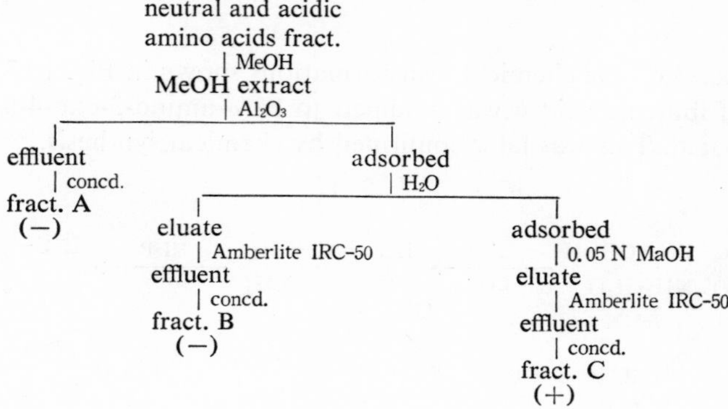

(Toyo Roshi No. 50, 0.02 M phosphate buffer (pH 7), 0.5 mA/cm, 2 hr), −7 (Toyo Roshi No. 50, acetic acid-sodium acetate solution (pH 3.7), 0.06 mA-cm, 2 hr).

10.5 STRUCTURE OF TRICHOLOMIC ACID 3[21]

The chemical properties shown in Fig. 10.2 and other findings indicated **3** to be α-amino-3-oxo-isoxazolidine-5-acetic acid. The structure was confirmed by synthesis[22] at Takeda Chemical Industries.

3 (isoxazolidinone: H_2C—CH(O)—CH(NH_2)—COOH, ring C(=O)–NH–O) —[6 NHCl (105°C, 16 hr)]→ $HOOC-CH_2-CH(OH)-CH(NH_2 \cdot HCl)-COOH$ (**5**); $HOOC-CH_2-CH_2-COOH$; $NH_2OH \cdot HCl$; NH_3

3 —[H_2/PtO_2]→ $H_2NOC-CH_2-CH(OH)-CH(NH_2)-COOH$ (**4**)

4 —[HI-P (160°C, 6 hr)]→ $HOOC-CH_2-CH_2-CH(NH_2)-COOH$ (glutamic acid)

4 —[0.5N HCl, 100°C, 30 min]→ $HOOC-CH_2-CH(OH)-CH(NH_2 \cdot HCl)-COOH$ (**5**)

Fig. 10.2 Reactions of tricholomic acid.

10.6 Structure of Ibotenic Acid 6[23)]

Based on the chemical transformations shown in Fig. 10.3, the structure of ibotenic acid **6** was assumed to be α-amino-3-oxo-4-isoxazolineacetic acid. This was later confirmed by chemical synthesis.[24,25,26)]

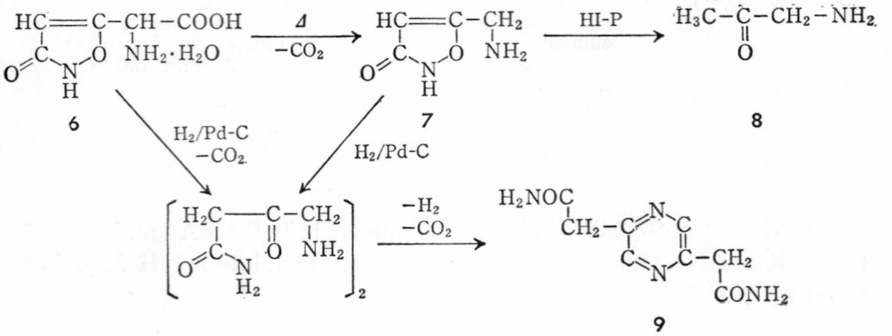

Fig. 10.3 Reactions of ibotenic acid.

[Experimental procedure 10] Pyrolysis of 6 to yield 7

Compound **6** (250 mg) was dissolved in water (10 ml), and heated on a water bath for 4 hr in a tightly closed tube. After opening the tube (evolution of carbon dioxide), the reaction solution was passed through a column of Amberlite IRC-50 (H^+) (50 ml). The column was washed with 1 l of water and **6** was thoroughly removed. Next, 3 l of 0.2% ammonia water was passed through. The eluate was dried and concentrated under reduced pressure. The residue was dissolved in a little water, and the same volume of methanol was added. The resulting light yellow solution was treated with active carbon. Acetone was added dropwise to the filtrate, and precipitated crude crystals (30 mg) were filtered off then recrystallized from water-methanol-acetone as colorless plates, mp 177–178° C (decomposition); $C_4H_6O_2N_2$ (elemental analysis and molecular weight determination). Ninhydrin reaction on the paper gave a yellow color, becoming red-violet. It decolorized potassium permanganate solution.

[Experimental procedure 11] Isolation of 2,5-pyrazinediacetamide 9

Compound **6** (166 mg) was dissolved in 40 ml of water, and 0.1 g of 10% palladium carbon was added. Hydrogenation was carried out for 5 hr (18.4 ml absorbed). The filtrate was dried and concentrated under reduced pressure, yielding crude crystals (24 mg), which were recrystallized from water. $C_8H_{10}O_2N_4$ (elemental analysis and mass analysis); mp 280° C (decomposition); soluble in dilute hydrochloric acid, uv $\lambda_{max}^{H_2O}$: 275.5 nm (log e = 3.88). The substance was negative to ninhydrin. It was still negative after treatment with 6 N hydrochloric acid at 105° C for 16 hr.

Alternatively, compound **7** (156 mg) was dissolved in 30 ml of water, and 30 mg of 5% palladium-carbon was added. Using the procedure described above, 48 mg of crude crystals was obtained. The crude crystals were recrystallized with water, yielding colorless plates, mp 280° C (decomposition). The ir spectrum of the substance was identical with that of **9** obtained above.

References

1) S. Kawamura, *Genshoku Nippon Kinrui Zukan* (in Japanese), **4**, p. 449, Kazamashobo (1954); R. Imazeki, T. Hongo, *Coloured Illustrations of Fungi of Japan* (in Japanese), p. 25, Hoikusha (1957).
2) H. Matsumoto, *Tohoku Kinrui Zufu* (in Japanese), p. 76, Enjushobo (1953).
3) R. Imazeki, T. Hongo, *Coloured Illustrations of Fungi of Japan* (in Japanese), p. 44, Hoikusha (1957).
4) T. Ohya, *Nippon Oyodobutsu Konchu Gakukaishi* (in Japanese), **3**, 41 (1959).
5) T. Ohya, *Kagaku* (in Japanese), **31**, 327 (1961).
6) T. Wieland, *Science*, **159**, 946 (1968).
7) M. Onda, M. Fukushima, M. Akagawa, *Chem. Pharm. Bull.*, 12, 751 (1964).
8) K. Bowden, A.C. Drysdale, *Tetr. Lett.* **1965**, 727.
9) C.H. Eugster, C.F.R. Müller, R. Good, *ibid.*, **1965**, 1813; R. Good, G.F.R. Müller, C.H. Eugster, *Helv. Chim. Acta*, **48**, 927 (1965).
10) G.F.R. Müller, C.H. Eugster, *ibid.*, **48**, 910 (1965).
11) S. Murakami, T. Takemoto, N. Shimizu, *Yakugaku Zasshi*, **73**, 1026 (1953); S. Tatsuoka, *Yukikagaku no Shinpo* (in Japanese), **12**, p. 61, Kyoritsu Shuppan (1958).
12) T. Takemoto, K. Daigo, *Chem. Pharm. Bull.*, **6**, 578 (1958); K. Daigo, *Yakugaku Zasshi*, **79**, 353, 356 (1959).
13) T. Takemoto, T. Nakajima, K. Daigo, *Jap. J. Pharm. Chem.*, **35**, 404 (1963).
14) K. Yasutomi, Y. Inoue, *Eiseigaichu Kujono Rironto Jissai* (in Japanese), p. 123, Hokuryukan (1957).
15) T. Takemoto, T. Nakajima, R. Sakuma, *Yakugaku Zasshi*, **84**, 1233 (1964).
16) T. Takemoto, T. Nakajima, *ibid.*, **84**, 1183 (1964).
17) G. Klein, *Handbuch der Pflanzenanalyse*, I-IV, Julius Springer (1931–1933).
18) K. Yamaguchi, *Shokubutsuseibun Bunsekiho* (in Japanese) I, p. 7,Nankodo (1959).
19) T. Momose, *Yukiteiseibunseki* (in Japanese), p. 136, Hirokawa Shoten (1960).
20) T. Takemoto, T. Yokibe, T. Nakajima, *Yakugaku Zasshi*, **84**, 1186 (1964).
21) T. Takemoto, T. Nakajima, *ibid.*, **84**, 1230 (1964).
22) H. Iwasaki, T. Kamiya, O. Oka, J. Ueyanagi, *Chem. Pharm. Bull.*, **13**, 753 (1965).
23) T. Takemoto, T. Nakajima, T. Yokobe, *Yakugaku Zasshi*, **84**, 1232 (1964); T. Yokobe, T. Takemoto, *ibid.*, **89**, 1236 (1969).

24) A.R. Gagneux, F. Häfliger, R. Meier, C.H. Eugster, *Tetr. Lett.*, **1965**, 2081.
25) K. Shirakawa, O. Aki, S. Tsushima, K. Konishi, *Chem. Pharm. Bull.*, **14**, 89 (1966).
26) Y. Kishida, T. Hiraoka, J. Ide, A. Terada, N. Nakamura, *ibid.*, **14**, 92 (1966); *ibid.*, **15**, 1025 (1967).

CHAPTER 11

Red Tide Toxins: Assay and Isolation of the Toxic Components

Although red tides have recently been attracting considerable attention on environmental grounds, they are not a new phenomenon and have been known since ancient times. The often quoted verse in Exodus," . . . and all the water was changed into blood. The fish died and the river stank, and the Egyptians could not drink water from the Nile,"[1] is said to be the description of a red tide. There is also speculation that American Indians knew of the relationship between toxic shellfish and red tides.

The occurrence of red tides is not necessarily associated with pollution or eutrophication, as suggested by some people. There are a number of cases in which red tides have been observed at places remote from civilization.

The term, red tide, is not a word of choice in the scientific community. The word "bloom"is adopted generally, as exemplified in "The International Conference on Toxic Dinoflagellate Blooms" which has been held twice since 1974. The word "bloom" expresses the explosive multiplication of a single species of plankton. As a broad definition, red tide means the bloom of any microplankton, including bacteria, diatoms, blue-green algae, and dinoflagellates. However, in a restricted sense, it usually refers to the bloom of flagellata, especially dinoflagellates. The reddish color of red tides is due to peridin and related carotinoids.[2]

11.1 The Toxicity of Red Tides

Most plankton including dinoflagellates are not deleterious, and play an important role as the basis of ocean productivity. However, in the case of red tide a large population of algae may not be utilized by other organisms, but may rather have undesirable effects on other organisms. For example, large-scale fish kills often take place due to oxygen deficiency caused by the explosive growth. In reality, few red tide-causing organisms are toxic by themselves, and among the toxic species, only a few are known to be harmful to humans. The species reported to be toxic and their distribution are shown in Table 11.1 and Fig. 11.1.

Table 11.1 List of toxic dinoflagellates.

Organism	Distribution	Toxicity	Toxin
1. *Gonyaulax catenella*	North Pacific (U.S.A., Canada, Alaska, Japan)†2	PSP toxin[3]	saxitoxin[4] gonyautoxins neosaxitoxin[5]
2. *G. tamarensis*†1	North Atlantic (Canada, Great Britain), North Sea, North Pacific (Japan)†2	PSP toxin[6–8]	gonyautoxins neosaxitoxin saxitoxin[9]
3. *G. acatenella*	British Columbia (Canada)	PSP toxin[10]	not known
4. *Pyrodinium bahamense*	South Pacific (Palau, New Guinea)	PSP toxin[11]	not known
5. *Pyridinium phoneus*	North Sea	PSP toxin[12]	not known
6. *G. polyedra*	Southern California	PSP toxin?[13]	not known
7. *G. monilata*	Gulf of Mexico	ichthyo-toxic[14]	not known
8. *Gymnodinium breve*	Gulf of Mexico	lipid-soluble neurotoxin, ichthyo-toxic[14,15]	complex high molecular compounds[16]
9. *Gymnodinium veneficum*	English Channel	neurotoxic[17]	not known
10. *Noctiluca scintillan*	South China Sea (Hong Kong)	ichthyotoxic? toxic shellfish[18]	not known
11. *Exuviaella mariaelebouriae*	Lake Hamana	hepatotoxic clams	venerupin[19]

†1 Some workers claim that the organism responsible is *Gonyaulax excavata*, suggesting that *G. tamarensis* is nontoxic[20], but this is not correct.[21]

†2 The red tide which occurred at Oase Bay, Mie, Japan, in 1975, was tentatively identified as *G. catenella*. On the other hand the species found on the Northern Pacific coast seems to be *G. tamarensis*.

Fig. 11.1 Distribution of red tides in the world (see Tabel 11.1).

Among the toxic dinoflagellates, the damaging are those which cause so-called paralytic shellfish poisoning (PSP). The toxic organisms are taken up by shellfish, and the toxins accumulate in the shellfish bodies. Because of this mechanism, as a rule, filter-feeding bivalves become toxic, while gastropods remain safe.

The toxins are usually concentrated in the hepatopancreas, and after the disappearance of the bloom they dissipate gradually over two or three weeks. However, in the case of Alaskan butter clam (*Saxidomus giganteus*) the toxin remains in the siphons all the year round.[23] The amount of toxins sometimes reaches 30,000 mouse units (see below) per 100 g of meat. The toxicity of *Gonyaulax polyedra*, which occurs along the southern California coast, is unclear. *Gymnodinium breve*, which causes large-scale blooms and fish kills in the Gulf of Mexico, also makes shellfish toxic. However, the nature of *G. breve* toxicity is quite different from PSP. It exhibits mild neurotoxicity to humans, but the hemolytic toxin is fatal to fish. Fig. 11.2 shows diagrams of *Gonyaulax* spp. and *Gymnodinium breve*.

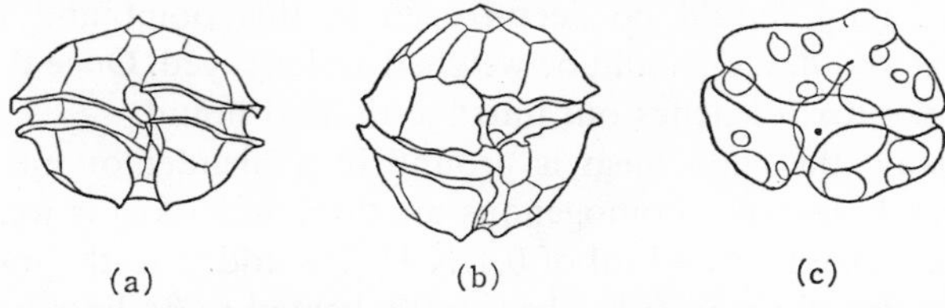

Fig. 11.2 Some toxic dinoflagellates.[22] (a) *Gonyaulax catenella*, (b) *G. tamarensis*, (c) *Gymnodinium breve*.

Toxicity Assay of Paralytic Shellfish Poisons[24)]

The mouse assay which was developed by Sommer and Meyer[4)] and modified by Schantz *et al.* has been adopted as the standard assay for PSP toxins in the United States and in Canada. The unit of toxicity, one mouse unit, was initially defined as the amount required to kill a 20 g white Swiss strain mouse in 15 min, but actually it is the asymptotic point in the dose-to death curve (Fig. 11.3), corresponding to 0.18 μg of saxitoxin dihydrochloride. PSP toxins show only acute toxicity, and if death does not occur within a certain time, a mouse will survive with no apparent effects.

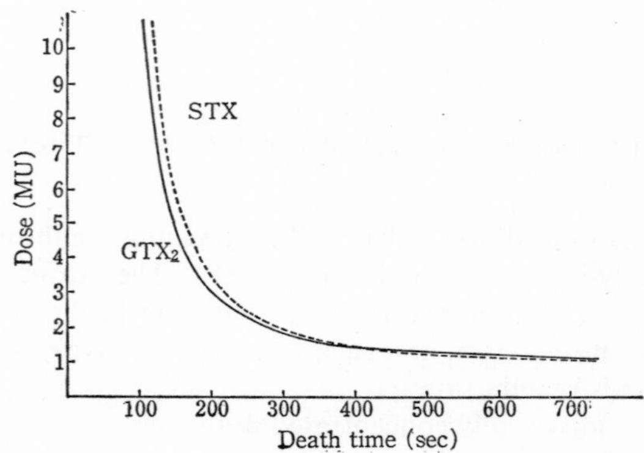

Fig. 11.3 Dose-death time response curves for saxitoxin and gonyautoxin-II.

[Experimental procedure 1] PSP toxin assay procedure using mice

Live clams which have been washed well with fresh water, are opened by cutting the adductor muscles and removed from the shells. No heat or analgesic should be used, and damage to the meat should be avoided. The shucked meat (100–200 g) is drained on a sieve (10 mesh) for 5 min, and the drained water is discarded. When test samples have to be preserved or transported, they should be deep-frozen at this point and the whole thawed meat and liquid should be weighed and assayed. Once the shellfish are frozen the toxin leaches out into the thawed liquid.

For assay, the clam meat is ground in a blender or meat grinder (1/8–1/4 inch holes) to a homogenous mixture, and 100 g is weighed in a tared beaker. One hundred ml of 0.1 N HCl is added to the mixture and the whole is stirred vigorously, then gently heated to boiling, boiled for 5 min and allowed to cool to room temperature. The pH is adjusted to 4.0–4.5 with HCl or NaOH solution, and the total volume is brought to 200 ml by adding distilled water. The supernatant obtained by allowing the solution to stand or, if necessary, by centrifugation or by filtration, is used for mouse assay.

Male mice of 18–22 g (never over 25 g) should be used. The mice have

TABLE 11.2 Relation between death time and number of mouse units (Sommer's table).

Death time†	Mouse units	Death time†	Mouse units†	Death time†	Mouse units
1:00	100	4:00	2.50	9:00	1.16
10	66.2	05	2.44	30	1.13
15	38.3	10	2.38		
20	26.4	15	2.32	10:00	1.11
25	20.7	20	2.26	30	1.09
30	16.5	25	2.21		
35	13.9	30	2.16	11:00	1.075
40	11.9	35	2.12	30	1.06
45	10.4	40	2.08		
50	9.33	45	2.04	12:00	1.05
55	8.42	50	2.00	13:00	1.03
		55	1.96	14:00	1.015
2:00	7.67			15:00	1.000
05	7.04	5:00	1.92	16:00	0.99
10	6.52	05	1.89	17:00	0.98
15	6.06	10	1.86	18:00	0.972
20	5.66	15	1.83	19:00	0.965
25	5.32	20	1.80	20:00	0.96
30	5.00	30	1.74	21:00	0.954
35	4.73	40	1.69	22:00	0.948
40	4.48	45	1.67	23:00	0.942
45	4.26	50	1.64	24:00	0.937
50	4.06			25:00	0.934
55	3.88	6:00	1.60		
		15	1.54	30:00	0.917
3:00	3.70	30	1.48	40:00	0.898
05	3.57	45	1.43	60:00	0.875
10	3.43				
15	3.31	7:00	1.39		
20	3.19	15	1.35		
25	3.08	30	1.31		
30	2.98	45	1.28		
35	2.88				
40	2.79	8:00	1.25		
45	2.71	15	1.22		
50	2.63	30	1.20		
55	2.56	45	1.18		

† min:sec.

to be from an identical colony, and the results should be standardized by using saxitoxin standard solution. One ml of the supernatant is injected peritoneally, and the time to death is recorded. If the death time is shorter than 5 min, appropriate dilutions are made to bring the death time to 5–7 min. At least three mice should be used at each dilution.

The toxicity is computed from the average death time using Sommer's table with appropriate corrections if the mice weigh less than 18 g or more than 20 g (Tables 11.2 and 11.3). The total toxicity can be obtained by multiplying the dilution factor and the total volume (200 ml), and expressed as μg of saxitoxin dihydrochloride equivalent by multiplying by the conversion factor (CF)* obtained with the mouse colony used.

TABLE 11.3 Mouse body weight correction factors.

Mouse weight (g)	Mouse units	Mouse weight (g)	Mouse units
10	0.50	17	0.88
10.5	0.53	17.5	0.905
11	0.56	18	0.93
11.5	0.59	18.5	0.95
12	0.62	19	0.97
12.5	0.65	19.5	0.985
13	0.675	20	1.000
13.5	0.70	20.5	1.015
14	0.73	21	1.03
14.5	0.76	21.5	1.04
15	0.785	22	1.05
15.5	0.81	22.5	1.06
16	0.84	23	1.07
16.5	0.86		

Various attempts have been reported to assay PSP toxins by physicochemical methods, as described below. However, none of them has been proved to be superior to the mouse assay.

[Experimental procedure 2] Bates-Rapoport assay method for PSP toxins

This method involves separation of the toxin on ion-exchange resin and measurement of the fluorescence produced by oxidation with hydrogen peroxide.

* The conversion factor, CF, can be calculated according to the following equation. Standard saxitoxin dihydrochloride solution is obtainable from the U.S. Food and Drug Administration.

$$CF = \frac{\text{satitoxin dihydrochloride } (\mu g) \text{ in standard solution (/ml)}}{\text{MU in standard solution(/ml)}}$$

Frozen clam meat was homogenized with a blender. To 2 g of the homogenate was added 2 ml of 0.5 M trichloroacetic acid. The mixture was stirred with a glass rod, and heated at 85–90° C for 10 min, then cooled with ice to 20° C. The pH was adjusted to 5.0–5.5 with 10% NaOH solution and the mixture was centrifuged at 12,000 xg for 10 min. The supernatant was introduced into a column (6 × 75 mm) packed with 2 ml of Bio-Rex 70 (50–100 mesh) buffered with 0.2 M sodium acetate buffers, pH 5.

The column was washed consecutively with 30 ml of 0.2 M sodium acetate buffer (pH 5.0), 25 ml of distilled water and 1 ml of 0.5 N HCl, and all the eluates were discarded. The column was then washed with 4.0 ml of 0.5 N HCl, and the eluate was divided into two equal portions, and transferred into two separate centrifuge tubes. To one tube was added 2.0 ml of 1.2 M NaOH and 0.05 ml of 10% H_2O_2, and to the other tube (blank) only 2.0 ml of 1.2 M NaOH was added. Both tubes were centrifuged at 1,000 xg. After 40 min, the supernatants were transferred into cuvets and neutralized to pH 5 by adding ca. 0.15 ml of acetic acid. The emission at 330 nm (excitation, 380 mm) was measured using the mixture without H_2O_2 as a blank. For calibration, the Raman peak of pure water (excitation, 330 mm; emission, 371 nm), which is equivalent to 0.017 μg saxitoxin/g meat or 7×10^{-9} M, was used.

The authors of this method claimed that the amounts of toxins in clams were proportional to the fluorescence yields. They also reported that the Bio-Rex 70 column step could be omitted. However, repeated experiments by several groups could not confirm the original claims. The method may be suitable for a sample containing only saxitoxin.

[Experimental procedure 3] Method of Buckley *et al.*[26)]

This method involves separation of a toxin mixture on a thin-layer plate and scanning of the fluorescent spots after spraying hydrogen peroxide. It is applicable only to fairly pure samples, and a considerable amount of toxins may be lost in the purification process. Buckley *et al.* could detect only three toxins out of more than seven toxins present in the clams.

A toxin mixture was spotted on a precoated Silica gel 60 plate, and the plate was developed with a pyridine-ethyl acetate-H_2O-acetic acid mixture. After air-drying, the plate was sprayed with 1% H_2O_2, heated at 100° C for 30 min, and scanned with a tlc fluorescence scanner. The amounts of toxins were calculated based on the peak heights and calibration curves prepared from known concentrations of toxins.

Buckley *et al.*[27)] also proposed an assay method involving measurement of the fluorescence formed by oxidation of PSP toxins with alkaline hydrogen peroxide. It was found that this procedure could not detect gonyautoxin-I, gonyautoxin-IV or neosaxitoxin, but was applicable to saxitoxin, gonyautoxin-II and gonyautoxin-III. Gonyautoxin-V gave an excessively high value.

11.2 Isolation and Purification of Saxitoxin[4)]

Until recently, saxitoxin was the only PSP toxin to have been purified. Although the presence of other toxins in *Gonyaulax tamarensis* had been suspected for some time, the isolation procedure used for saxitoxin was not applicable to these new toxins.[28–30)] The author's group developed a different method applicable to all PSP toxins and have so far purified eight toxins from shellfish and algal samples. Saxitoxin has been found in the Alaskan butter clam, *Gonyaulax catenella,* and crabs from the Amami Islands.[31)] The best material is Alaskan butter clams from certain areas where they accumulate the toxin in the siphons all the year round. The samples used for previous structural and pharmacological studies were derived from this source. A large amount of saxitoxin is said to have been isolated from butter clams for military research.

The purification method is based on the principle that strongly basic saxitoxin can be freed from bulk contaminants simply by adsorbing it on a weakly acidic ion-exchange resin followed by washing with buffer solutions. This was rather fortunate.

[Experimental procedure 4] Isolation and purification of saxitoxin

Sufficient 95% ethanol was added to ground Alaskan butter clam siphons or hepatopancreas of toxic bivalves (toxicity 5,000–10,000 mouse units/100 g) to cover the surface. The mixture was adjusted to pH 2–3 (HCl), mixed with about half its volume of Celite 545, and packed into a large diameter column. The column was then eluted with hydrochloric acid containing (pH 2–3) 15% ethanol. The extract was evaporated down to a small volume, adjusted to pH 5.5 (NaOH), and centrifuged to remove insoluble materials. The supernatant was applied to an Amberlite IRC-50 column (20–50 mesh, Na^+ form, washed until the washing solution reached pH 8.0). The extract was applied until excess toxin started to leach out of the column, and then the column was washed with distilled water and pH 4 acetate buffer (1 N acetic acid + saturated sodium acetate solution). When pH of the eluate reached 4.5, the column was washed again with water. Most of the contaminants and sodium ions were removed by this washing process. Elution with 0.5 N acetic acid gave toxic fractions, which were assayed using mice. It should be noted that acetic acid at high concentrations is toxic to mice and sodium ions can reduce the apparent toxicity.

By this single step, the toxic fraction can be concentrated to a level of ca. 1,000 mu/mg. For further purification, rechromatography on the H^+ form of Amberlite CG-50 or Bio-Rex 70 was very efficient. A sample solution (adjusted to pH 5.5) was charged on the column, and eluted with a gradient of acetic acid starting at 0.05 N. Each fraction was checked by mouse assay and highly toxic fractions were combined, lyophilized, and rechromatographed. Usually saxitoxin can be purified in two chromatographic separations. If neosaxitoxin is also present, it can be separated from saxitoxin by high performance chromatography on Bio-Rex 70 (<400 mesh).[7)]

In Schantz's original method, the final purification of saxitoxin was done by chromatography on acidic alumina using ethanol as an eluting solvent. For the removal of small amounts of impurities, however, chromatography on Bio-Gel P-2 is very efficient.

A Bio-Gel P-2 (Bio-Rad Lab.) column (8 mm × 60 cm) previously washed with dilute HCl solution and glass-distilled water was charged with about 10,000 mouse unit equivalent of toxin in a small amount of glass-distilled water. The column was washed first with water (10 ml) and then with 0.025 N acetic acid. The major fractions were combined and lyophilized. The resulting residue is saxitoxin diacetate; to obtain the dihydrochloride, a small amount of hydrochloric acid may be added before lyophilization.

11.3 Isolation and Purification of the Gonyautoxin Group[6)]

As mentioned earlier, *G. tamarensis* toxins could not be purified by the original saxitoxin isolation method. The majority of toxins passed through an Amberlite IRC-50 (Na^+) column untrapped. The toxin components were also more complex. The newly devised procedure described below is a combination of Sephadex G-15 or Bio-Gel P-2 gel adsorption and high performance ion-exchange column chromatography.

[Experimental procedure 5] Isolation and purification of gonyautoxins

From shellfish contaminated with G. tamarensis: Softshell clams, (*Mya arenaria*) collected at the peak of *G. tamarensis* bloom were shucked, and the hepatopancreas was removed with scissors. Hepatopancreas thus obtained (about 10% of the shucked meat) was homogenized with two volumes of 80% ethanol adjusted to pH 2 with HCl. After centrifugation, the residue was similarly extracted twice more. The combined extracts were reduced to a small volume *in vacuo*, and washed thoroughly with chloroform. The water layer was further reduced to a small volume, adjusted to pH 5.0 with NaOH solution, and applied to a Sephadex G-15 or Bio-Gel P-2 column prepared with distilled water. After washing with distilled water, the adsorbed toxins were eluted with 0.02 N acetic acid. In the presence of a large quantity of contaminants, a part of the toxins may not be adsorbed. In such a case, this procedure can be repeated, or the contaminants (mostly salts) can be removed by precipitation with ethanol. The toxin concentrate thus obtained may be purified further on Sephadex G-15 or Bio-Gel P-2 or subjected to Bio-Rex 70 chromatography. Sephadex G-15 or Bio-Gel P-2 chromatography can only partly fractionate the toxins (Fig. 11.4), and for the separation of individual toxins, chromatography on Bio-Rex 70 is necessary.

Bio-Rex 70 is a methylacrylic polymer, and its carboxyl moieties do not dissociate at low pH, so no adsorption by ion-exchange can occur. On the other hand, at high pH the toxins tend to decompose or to isomerize due to the carboxylate basicity. The chromatography was carried out using a hand-built medium pressure chromatography system, as shown in Fig. 11.5 (a). This system offers excellent performance.

The sample was dissolved in a small amount of water, adjusted to pH 5.5 and introduced into the inlet A. After letting a small volume of water pass, the column was eluted with an acetic acid gradient. Each toxic fraction was assayed using mice. The toxin fractions could be conveniently detected by spot tests on filter paper or tlc (fluorescence detection by 1% H_2O_2 spraying and heating). Monitoring of the toxin uv end absorption at 206 nm was also effective. A typical fractionation pattern is shown in Fig. 11.4 (c). The overlapping fractions were rechromatographed.

The toxin elution sequence is different from that on tlc. The toxins which were not separable on tlc could be separated on the column. Each component was lyophilized to a highly hygroscopic powder. The counter ion was acetate.

Toxin isolation from G. tamarensis cells: Although it is not impossible to collect algal cells directly from sea water, the operation would involve the centrifugation or precipitation of organisms from a vast amount of sea water, which would be inconvenient. It is also difficult to collect a unialgal specimen. Thus, it is more practical to culture the organisms. The culture media and temperature depend on the organism to be cultured. The case of *G. tamarensis* is described here.

For the cultivation of *G. tamarensis,*[32] sea water (salinity: 29–32) which had been kept in a cold room for 2–3 weeks was passed through coconut charcoal and a Millipore filter (0.22 μm) successively. Using this

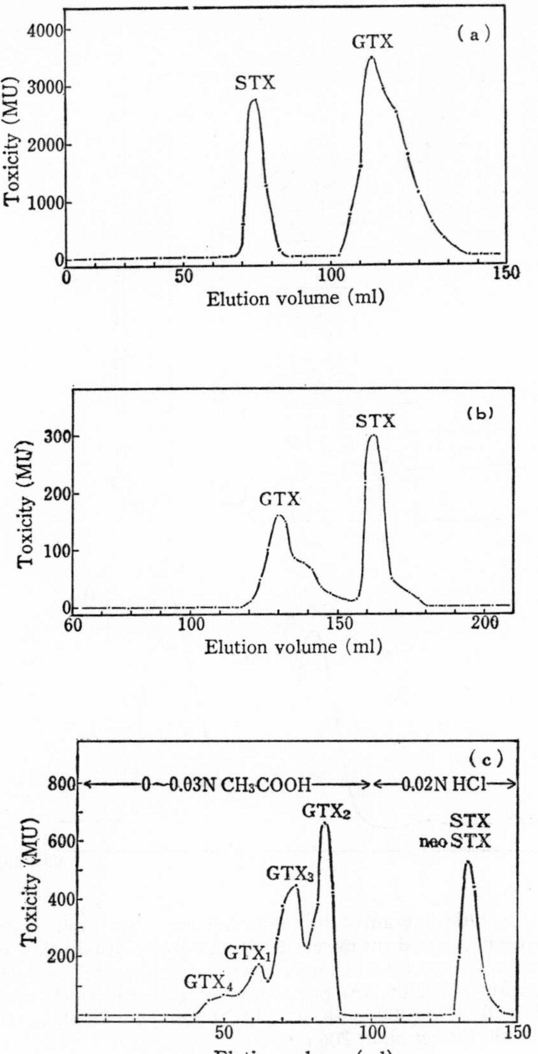

Fig. 11.4 (a) Separation of gonyautoxin mixture (GTX) and saxitoxin (STX) on Sephadex G-15 (60 × 2 cm; void volume 30 ml, elution with 20 ml of H_2O followed by 0.025 N acetic acid); (b) separation of GTX and STX on Bio-Gel P-2 (60 × 1 cm; void volume 20 ml, 0.01 N acetic acid elution); (c) separation of *G. tamarensis* toxins on Bio-Rex 70 H^+.

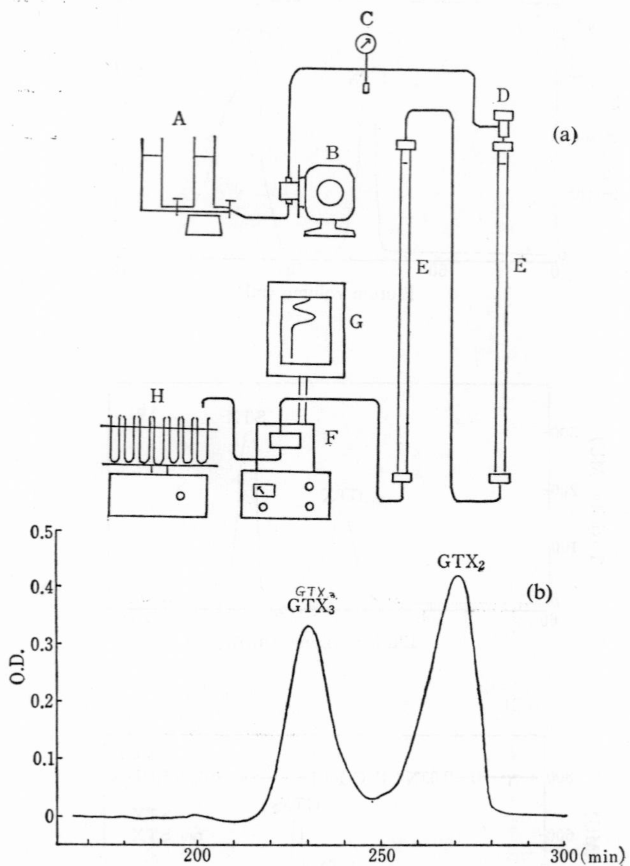

Fig. 11.5 (a) Schematic diagram of the hand-built chromatography system for PSP toxin separation. A, gradient mixer; B, Milton-Roy constant-flow pump; C, pressure gauge and safety valve; D, injection port; E, pressure-resistant glass column; H, fraction collector. (b) Separation of gonyautoxin-II and -III. Column: Bio-Rex 70 H^+, <400 mesh, 6 × 1,000 mm. Mobile phase: 0.02 N acetic acid 1 ml/min. Detection at 206 nm.

sea water as the base, the following medium (Guillard F medium)[3] was prepared.*

A 10 l Pyrex carboy containing 6 l of the medium was inoculated at a

* A mixture of sea water (1000 ml), $NaNO_3$ (150 mg), ferric sequestrene (10 mg), $CuSO_4 \cdot 5H_2O$ (0.0196 mg), $ZnSO_4 \cdot 7H_2O$ (0.044 mg), $COCl_2 \cdot 4H_2O$ (0.360 mg), $Na_2MoO_4 \cdot 2H_2O$ (0.0126 mg) and Tris buffer (2.0 ml; 25%, pH 7.1) was autoclaved, then a mixture of thiamine (0.2 mg), biotin (1.0 μg), vitamin B_{12} (1.0 μg) and $NaH_2PO_4 \cdot H_2O$ (10 mg) was added through a Millipore filter.

ratio of 10,000 cells/l and cultured under fluorescent illumination. In about 4 weeks, the cell population reached 200,000/l. The cells were collected by continuous centrifugation. During the centrifugation, some of the cells were destroyed, and their toxin content leached out into the medium; such leached out toxins are very difficult to recover. The *Gonyaulax* cells are easily destroyed by freezing, alcohol treatment, etc., and the toxin can then be extracted. The easiest extraction method is to heat the cells briefly in dilute acetic acid. After concentration and chlorofom washing, the extract can be processed as in the case of the clam extract.

The toxin composition in the cells is almost identical to that in clams except for the presence of a significant amount of neosaxitoxin which was found only in a trace amount in toxic softshell clams. The toxins so far isolated from various sources are listed in Table 11.4 and their chromatographic and electrophoretic properties[35)] are summarized in Table 11.5.

TABLE 11.5 Chromatographic and electrophoretic behavior of isolated toxins.

Toxins	tlc		Electrophoresis	Elution order from Bio-Rex 70 column
	Rf†1	*Rf*†2	*Rm*†3	
STX	0.62	0.51	1.00	9
GTX_1	0.90	0.70	0.16	4
GTX_4	0.81	0.65	0	3
GTX_2	0.81	0.65	0.56	6
GTX_3	0.69	0.61	0.28	5
GTX_5	0.61	0.52	0.28	2
neoSTX	0.70	0.54	0.47	7
GTX_6	0.57		0.08	1
GTX_7	0.44		0.97	8
$APTX_1$	0.81	0.51	—	2
$APTX_2$	0.70	0.51	—	3
$APTX_3$	0.48	0.50	1.51	1

†1 Silica gel 60; pyridine, ethyl acetate, water, acetic acid (75:25:30:15).
†2 Silica gel GF; *t*-butanol, acetic acid, water (2:1:1).
†3 relative mobility on cellulose acetate strip in tris-HCl buffer, pH 8.7, at 200 V (constant) and 0.2 mA/cm for 1 hr. The value was calculated by taking that of saxitoxin as 1.0.

Neosaxitoxin has properties very similar to those of saxitoxin and its presence was not detected for some time. For its separation, the "saxitoxin fraction" which is strongly adsorbed on Bio-Rex 70 was pooled and rechromatographed on a Bio-Rex 70 column using an acetic acid gradient.[7)] Neosaxitoxin and saxitoxin appeared as overlapping peaks. Further rechromatography was necessary to obtain a pure specimen. Lyophilized neosaxitoxin is a slightly yellow hygroscopic powder. With

Table 11.4 PSP and related toxins isolated from various sources

Source	Purified toxin†[1]											
	STX	GTX_1	GTX_4	GTX_2	GTX_3	GTX_5	neoSTX	GTX_6	GTX_7	$APTX_1$	$APTX_2$	$APTX_3$
Gonyaulax tamarensis†[2]	++	++	+	+++	+	+	+++					
Mya arenaria†[3]	++	++	+	+++	+		+					
Mya arenaria†[4]	++	±		+++	+		+					
Mytilus edulis†[5]	+	+++	+	++	+	++						
Tapes (Amygdala) japonica†[5]	+	+++	+	++	+	++						
Gonyaulax sp.†[6]	+	+++	+	++	+	++						
Saxidomus giganteus†[7]	+++						+					
Mytilus edulis†[8]	+++	+	+	++	+	++		+				
Mytilus edulis†[9]		+++	++	++	++	+	+	+				
Mytilus edulis†[10]	+	+++	+	++	++	+		+				
Placopecten magellanicus†[11]	+	++	+	++	++	+	+++		+			
Gonyaulax catenella†[12]	++	++	+	++	++	+	++		±			
Aphanizomenon flos-aquae†[13]										++	++	++

†[1] Symbols (+, ++, tlc.) indicate approximate amounts isolated.
†[2] Cultured cells.
†[3] Softshell clams collected at Essex, Massachusets, in 1972 and 1974.
†[4] Softshell clams collected at Hampton, New Hampshire, in 1972.
†[5] Mussels and short-necked clams collected at Oase Bay, Mie, Japan, in 1975, exposed to an unidentified *Gonyaulax* sp. bloom.
†[6] A bloom collected at Oase Bay, Mie Japan, in 1975.
†[7] Alaskan butter clams collected in Alaska.
†[8] Mussels originated from Vigo, Spain, caused a massive PSP epidemic in several countries in Western Europe in 1976.
†[9] Mussels collected near Haines, Alaska, in 1976.
†[10] Mussels collected at Elfin Cove, Alaska, in 1977.
†[11] Sea scallops collected in the Bay of Fundy, Nova Scotia, Canada, in 1977.
†[12] Cells from nonaxenic open culture.
†[13] Toxic blue-green alga collected at Kezar Lake, New Hampshire.

hydrogen peroxide it shows a yellowish fluorescence different from that of saxitoxin, gonyautoxin-II, or gonyautoxin-III.

11.4 Pharmacological and Chemical Properties of Paralytic Shellfish Poisons

Pharmacological Properties Like tetrodotoxin from puffer fish, saxitoxin blocks the influx of Na^+ in the excitable membranes. As a result of this blockage, sudden death takes place due to respiratory paralysis. The other PSP toxins are expected to possess similar activity, and this was confirmed for gonyautoxin-I, -II, -III and neosaxitoxin.[37)]

Chemical Properties Saxitoxin is a strongly basic ompound with two guanidinium moieties (Fig. 11.6). It has two p*Ka*'s of 11.5 and 8.24. The later is assigned to the imidazole guanidinium group. The hydrochloride is not crystalline, but the *p*-bromobenzenesulfonate is crystalline and was used for an X-ray diffraction study. Saxitoxin is very stable in acidic solutions, but is easily oxidized at high pH, losing its toxicity. Oxidation of saxitoxin with alkaline H_2O_2 gives aromatized, highly fluorescent aminopurine derivatives (Fig. 11.7).[37)] In view of these degradation products, the nmr spectrum and other physical data, a bridged purine structure was first proposed,[38)] but the correct structure was finally determined as **1** by X-ray diffraction.[39,40)] An unusual structural feature is a hydrated ketone moiety, which is stabilized by the two electron withdrawing guanidinium groups.

saxitoxin I

gonyautoxin-II
gonyautoxin-III
($X = SO_3^-$)

Fig. 11.6 Structures of saxitoxin and gonyautoxin-II and -III and their cmr chemical shifts (*may be reversed).

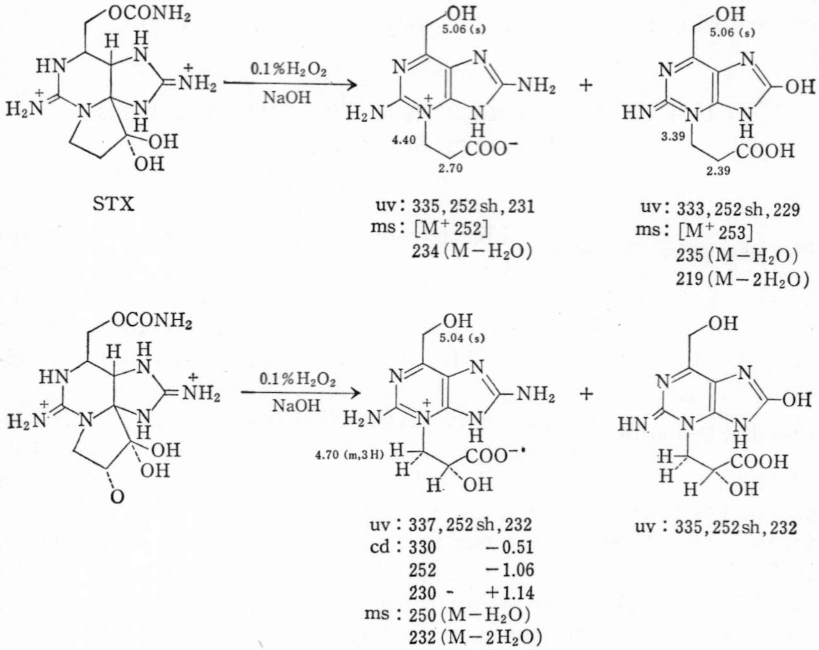

STX

uv: 335, 252 sh, 231
ms: [M^+ 252]
234 ($M-H_2O$)

uv: 333, 252 sh, 229
ms: [M^+ 253]
235 ($M-H_2O$)
219 ($M-2H_2O$)

uv: 337, 252 sh, 232
cd: 330 −0.51
252 −1.06
230 +1.14
ms: 250 ($M-H_2O$)
232 ($M-2H_2O$)

uv: 335, 252 sh, 232

Fig. 11.7 Alkaline hydrogen peroxide degradation of saxitoxin and gonyautoxins.

Gonyautoxin-II and -III were assigned as 11-hydroxysaxitoxin derivatives on the basis of microdegradation, and proton and carbon nmr studies. Recently, however, it was confirmed that they are in the sulfate forms, **2** and **3**.[41] Gonyautoxin-II and -III form an equilibrium mixture on standing. The achievement of equilibrium is drastically accelerated in the presence of a trace of base, e.g. sodium acetate.

The difference of adsorbability of gonyautoxins is due to the presence of a sulfate group.

11.5 *Gymnodinium breve* Toxins

G. breve causes tremendous damage on the Gulf of Mexico coast, especially the Florida coast.

The influx of terrestrial organic and inorganic effluents with rainfall is a suspected but unconfirmed cause of blooms. Massive fish kills represent the heaviest economic loss, but bivalves also become toxic. In some areas, shellfishing is banned throughout the year. The symptoms of poisoning are quite different from those of PSP, and it has been suggested to resemble ciguatera poisoning. *Gymnodinium* blooms have also been observed in the Inland Sea, Japan, but the toxicity has not been fully confirmed.

The nature of *G. breve* toxin is still unclear. Conflicting reports have appeared on the chemical and pharmacological properties of the toxin. The differences are so great that one might question whether the authors are all working on the same organism. Some of the past data are listed in Table 11.6.

TABLE 11.6 Properties of *Gymnodinium breve* toxins reported by various researchers.

Workers	Properties	Molecular weight	Molecular formula	Toxicity (LD_{50} to mice)
Trieff *et al.*[42]	yellow stable amorphous	2,538	$C_{138}H_{226}O_{47}P$	
		1,164	$C_{60}H_{113}O_{14}P$	0.5 mg/kg
		1,630	$C_{92}H_{157}O_{11}P$	
Martin and Chatterjee[43]	faintly yellow solid	1,545	$C_{91}H_{166}O_{57}P$	
Cummins and Stevens[44]	gray solid		$C_{102}H_{157}N$	26 mg/kg
Alam[45]	faintly solid yellow	288 (osmosis) 279 (ms)		0.85 mg/kg
Alam *et al.*[46]	colorless amorphous	725 (ms)		0.25 mg/kg

There is no official assay method for *G. breve* toxin. Some reported assays using fish, brine shrimp, and mice are described below.

[Experimental procedure 6] Assay of *Gymnodinium breve* toxin

Mouse assay:[46] An assay sample was dissolved in a known volume of ethyl ether. An aliquot containing about 200 μg of the sample was mixed with one drop of Polysorbate 80, and ether was removed by evaporation. The residue was homogenized well with 1 ml of saline solution and injected peritoneally into a 20 g white Swiss strain mouse. The mouse was observed for 72 hr. If death did not occur at a dose of 10 mg/kg, the sample was judged nontoxic.

Killifish assay:[15] A killifish (*Fundulus similis*, 3.0–3.5 g, 6–7.5 cm) was placed in a 400 ml breaker containing 50 ml of sea water under standardized conditions (e.g. salinity 32–33; pH 8.0–8.2; 23 $\pm$ 2° C). After the addition of a small amount of methanol or chloroform solution of the test sample, the fish was kept under observation. An amount sufficient to kill the fish in 7–8 min was tentatively called 1 fish-kill unit.

Brine shrimp assay:[47] Four 100 ml beakers were filled with filtered sea water and three individuals of *Artemia salina* (body length 1.2–1.6 cm) were placed in each. A sample solution in methanol or Polysorbate was then added. The sample concentration was adjusted by dilution with sea water. The assay was carried out at 25 $\pm$ 2° C, and the time to the point at which movement of the brine shrimps ceased was designated as the death time. LD_{50}((LC_{50}) was calculated from the death rate (%) at 72 hr.

[Experimental procedure 7] Isolation and purification of *Gymnodinium breve* toxin GB-2[46,16]

The synthetic medium, NH-15[48] of Gates and Wilson was used. The medium was inoculated at 1,000,000 cells/l and kept at 24° C under fluorescent illumination for 18–21 days. The whole culture was then transferred into a separatory funnel and shaken with ether. *G. breve* cells were easily destroyed by this method, and the lipid-soluble toxin was extracted with ether. The ethereal layer was dried over Na_2SO_4 and, after removal of the ether, was immediately chromatographed on silica gel. The fractions eluted with benzene-ethyl acetate were collected and rechromatographed. Further purification was done by preparative tlc (silica gel, benzene-ethyl acetate (1:1)). The toxin is extremely labile to air and light.

Properties of GB–2[16] GB-2 has an LD_{100} of ca. 0.5 mg/kg in mice. It has a rather large molecular weight (over 850). Elemental analysis showed a high content of oxygen and no or very little nitrogen. It shows a very weak absorption due to hydroxyl groups in the ir, suggesting that most oxygens are in cyclic ether form. There are also carbonyl functions, including conjugated ones (ir: 3580, 1743, 1700, 1640, 1100 cm^{-1}; uv: 213 nm). The nmr spectrum showed 5–6 methyl signals and an aldehyde function. The structure is under investigation.

References

1) Old Testament, Exodus 7:20–21.
2) H.H. Strain, W.M. Manning, G.H. Hardin, *Biol. Bull.*, **86**, 169 (1944); H. Pinckard, J.S. Kittredge, D.L. Fox, F.T. Haxo, L. Zechmeister, *Arch. Biochem. Biophys.* **44**, 189 (1953).
3) H. Sommer, K.F. Meyer, *Arch. Pathol.*, **24**, 560 (1937); Association of Official Analytic Chemists, **28**, 319–321 (1975).
4) E.J. Schantz, J.D. Mold, D.W. Stanger, J. Shavel, F.J. Riel, J.P. Bowden, J.M. Lynch, R.S. Wyler, B. Riegel, H. Sommer, *J. Am. Chem. Soc.*, **79**, 5230 (1957).
5) Y. Shimizu, *Toxic Dinoflagellate Blooms* (Taylor and Seliger, eds.), Elsevier-North Holland, Inc., 321 (1979).

6) A.B. Needler, *J. Fisheries Res. Board Can.*, **7**, 490 (1949); A. Prakash, *ibid.*, **20**, 983 (1963); P.C. Wood, *Nature*, **220**, 21 (1968).
7) T. Kawabata, T. Yoshida, Y. Kubota, *Bull. Japan Soc. Sci. Fisheries*, **28**, 344 (1962); I. Kawashiro, H. Tanabe, A. Ishii, T. Kondo, *J. Food Hyg. Soc. Japan*, **3**, 273 (1962).
8) Y. Oshima, W.E. Fallon, Y. Shimizu, T. Noguchi, Y. Hashimoto, *Bull. Japan Soc. Sci. Fisheries*, **42**, 851 (1976).
9) Y. Shimizu, M. Alam, Y. Oshima, W.E. Fallon, *Biochem. Biophys. Res. Commun.*, **66**, 731 (1975).
10) A. Prakash, F.J.R. Taylor, *J. Fisheries Res. Board Can.*, **23**, 1265 (1966).
11) J.L. MacLean, *Pacific Science*, **29**, 7 (1975).
12) H.J. Koch, *Assoc. Franc. Avan. Sci. Paris*, **63rd** *Session*, p. 654 (1939).
13) J. Schradie, C.A. Bliss, *Lloydia*, **25**, 214 (1962).
14) A.M. Sievers, *J. Protozool.*, **16**, 401 (1969).
15) D.F. Martin, A.B. Chatterjee, *Fishery Bulletin*, **68**, 433 (1970).
16) Y. Shimizu, M. Alam, W.E. Fallon, *Proc. 4th Food-Drugs from the Sea Conference*, p. 238 (1974).
17) B.C. Abbott, D. Ballantine, *J. Marine Biol. Assoc. U.K.*, **36**, 169 (1957).
18) J.R. Grindley, A.E.F. Heydorn, *S. African J. Sci.*, **66**, 216 (1970); B. Morton, R.R. Twentyman, *Environ. Res.*, **4**, 544 (1971).
19) M. Nakazima, *Bull. Japan Soc. Sci. Fisheries*, **34**, 130 (1968).
20) L.A. Loeblich, A.R. Loeblich, III, *Proc. 1st Int. Conference on Toxic Dinoflagellate Blooms*, p. 207, The Massachusetts Science and Technology Foundation, Mass. (1975).
21) M.I. Alam, C.P. Hsu, and Y. Shimizu, *J. Fish. Res. Board of Canada*, **36**, 32 (1979).
22) Original picture by Dr. Y. Oshima and Mrs. S. Oshima. Reproduced with their permission.
23) E.J. Schantz, H.W. Magnusson, *J. Protozool.*, **11**, 239 (1964).
24) E.J. Schantz, E.F. McFarren, M.L. Schafer, K.H. Lewis, *J. Assoc. Offic. Agr. Chem.*, **41**, 160 (1958).
25) H.A. Bates, H. Rapoport, *J. Agr. Food Chem.*, **23**, 237 (1975).
26) L.J. Buckley, M. Ikawa, J.J. Sasner, Jr., *ibid.*, **24**, 107 (1976).
27) L.J. Buckley, Y. Oshima, Y. Shimizu, *Anal. Biochem.*, **85**, 157 (1978).
28) E.J. Schantz, *Ann. N. Y. Acad. Sci.*, **90**, 843 (1960).
29) M.H. Evans, *Brit. J. Pharmacol.*, **40**, 847 (1970); *Marine Pollution Bull.*, **1**, 184 (1970).
30) N.E. Ghazarossian, E.J. Schantz, H.K. Schnoes, E.M. Strong, *Biochem. Biophys. Res. Commun.*, **59**, 1219 (1974).
31) T. Noguchi, S. Konosu, Y. Hashimoto, *Toxicon*, **7**, 325 (1969).
32) Y. Shimizu, M. Alam, W.E. Fallon, *Proc. 1st Int. Conference on Toxic Dinoflagellate Blooms*, p. 275, The Massachusetts Science and Technology Foundation, Mass. (1975).
33) R.R.L. Guillard, *Symp. on Marine Microbiol.* (ed. C. Oppenheimer) p. 93, C.C. Thomas Publ. Co. (1963).
34) M. Alam *et al.*, unpublished data.
35) W.E. Fallon, Y. Shimizu, *J. Environ. Sci. Health*, **A12**, 455 (1977).
36) T. Narahashi *et al.*, unpublished data.
37) J.L. Wong, M.S. Brown, K. Matsumoto, R. Oesterlin, H. Rapoport, *J. Am. Chem. Soc.*, **93**, 4633 (1971).
38) J.L. Wong, R. Oesterlin, H. Rapoport, *ibid.*, 7344 (1971).
39) E.J. Schantz, V.E. Ghazarossian, H.K. Schnoes, F.M. Strong, J.P. Springer, J.O. Pezzanite, J. Clardy, *ibid.*, **97**, 1238 (1975).
40) J. Bordner, W.E. Thiessen, H.A. Bates, H. Rapoport, *ibid.*, **97**, 6008 (1975).
41) Y. Shimizu, L.J. Buckley, M. Alam, Y. Oshima, W.E. Fallon, H. Kasai, I. Miura, V.P. Gullo, K. Nakanishi, *ibid.*, **98**, 5414 (1976); G.L. Boyer, E.J. Schantz, and H.K. Schnoes, *Chem. Commun.*, 474 (1978).

42) N.M. Trieff, V. Venkatasubramanian, S.M. Ray, *Texas Rept. Biol. Med.*, **30**, 97 (1972).
43) D.F. Martin, A.B. Chatterjee, *Nature*, **221**, 59 (1969).
44) J.M. Cummins, A.A. Stevens, *Investigations of Gymnodinium breve Toxins in Shellfish*, U.S. Department of Health, Education and Welfare, Public Health Service Publication (1970).
45) M. Alam, Ph.D. Thesis, University of New Hampshire (1972); J.J. Sasner, Jr., M. Ikawa, F. Thunberg, M. Alam, *Toxicon*, **10**, 163 (1972).
46) M. Alam, N.M. Trieff, S.M. Ray, J.E. Hudson, *J. Pharm. Sci.*, **64**, 865 (1975).
47) N.M. Trieff, M. McShan, D. Grajcer, M. Alam, *Texas Rept. Biol. Med.*, **31**, 409 (1973).
48) E.J. Gates, W.B. Wilson, *Limnol. Oceanog.*, **5**, 171 (1960).

CHAPTER 12

Studies on Bromine-Containing Organic Compounds in Marine Algae

It has long been known that bromine is contained in marine algae.[1] Kylin,[2] and Ochi and Takahashi[3] analyzed various marine algae and found that a high content of bromine is generally seen in *Rhodophyta*, especially in *Polysiphonia, Rhodomela* and *Odonthalia.* The content of bromine is relatively low in *Phaeophyta* and *Chlorophyceae.*

It is generally thought that bromine exists in sea water in the form of the ion, Br^-. The average content of bromine is higher than that of iodide, I^-, but is still only 63.8 ppm. It is therefore extremely interesting to investigate the pathway of bromine incorporation and the role of bromine in marine algae.

However, there have been only a few studies on bromine compounds in marine algae. Mastaglin and Augier[4] isolated a compound from a species of *Polysiphonia* in 1949, and gave its chemical composition as $C_6Br_2(SO_3K)_2(OH)(COOH)$. The structure of this compound was not identified. In 1955, Saito and Ando[5] isolated 5-bromo-3,4-dihydroxybenzaldehyde **1** from *Polysiphonia morowii* of the *Rhodomela* family. This was the first bromo compound ever isolated from marine algae and identified.

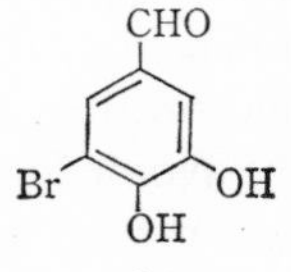

CHO

Br OH

OH

1

Our laboratory initiated a series of studies on the components of marine algae more than ten years ago, focusing on terpenes and bromo compounds. We have investigated various species, and we were successful in isolating laurencin **2** in an early phase of our study, from *Laurencia glandulifer* KÜTZING collected at Oshoro, Hokkaido. It was identified as a

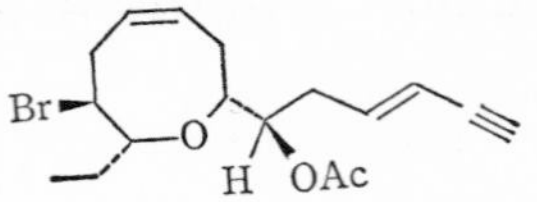

2

bromine-containing eight-membered cyclic ether with a side chain having a conjugated enyne moiety. Therefore, we concentrated our efforts on *Laurencia* and related genuses from an early stage of our investigation. Bromine-containing compounds that we have isolated can be categorized into the following three groups. (1) Phenolic compounds. (2) Terpene compounds. (3) Cyclic ether compounds. Since our early work, many reports have been published especially during the last few years. There is a good review article by Fenical and Faulkner.[7)]

12.1 Collection and Drying of Marine Algae and Isolation of Components

The first problem is the collection of suitable species of algae. Since weight of algae is reduced to one-tenth of the wet weight on drying, and since the contents of terpenes and bromine-containing compounds are very small, it can be very difficult to collect sufficient quantities of certain species of marine algae for this type of study. It is usually difficult to identify a specific species during collection expeditions, especially in the growth stage, because of the marked morphological changes. It may be essential to secure the collaboration of experts on algae during collection.

We have found the drying method has a major effect on the research results with algae. It is particularly important to avoid direct sun light. Better results were usually obtained from algae that were dried in a shaded and breezy place with occasional turning, and that retained some flexibility. Collection of algae should be avoided during periods of high humidity.

During concentration procedures after solvent extraction, it should be considered that many of the algal components may be unstable to heat. The use of a film evaporator is recommended, especially for treating large quantities of extract, to avoid prolonged heating. In other cases, concentration should be carried out under nitrogen at low pressure and temperature.

Algae dried as described above were fractionated as shown in Table 12.1.

TABLE 12.1 Fractionation of components of marine algae.

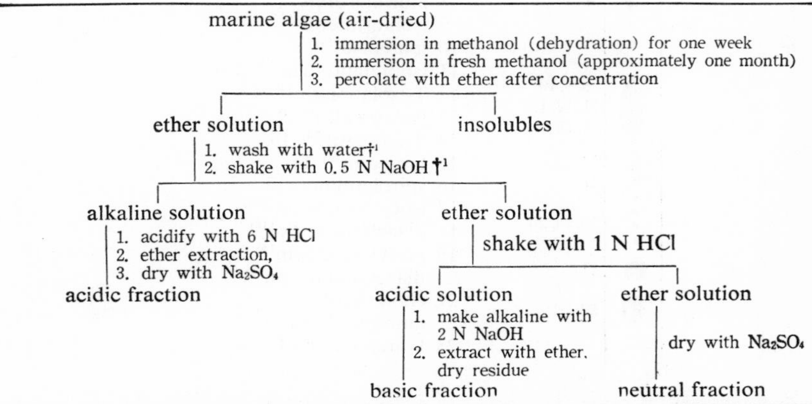

†[1] Components of marine algae tend to be unstable in acid or alkali, so that those procedures should be completed within a short time. Washing with water at step 1 should be exhaustive to avoid emulsification in later steps.

12.2 FRACTIONATION OF *Laurencia nipponica* YAMADA

Air-dried samples of *L. nipponica* YAMADA collected near Hakodate, Hokkaido, in May were fractionated following the procedures described in Table 12.1. The yield from 34 kg of starting material was 340 g of neutral fraction, which gave the silica gel tlc result shown in Fig. 12.1.

Next, rough fractionation of the neutral fraction was attempted by column chromatography. The neutral fraction (36 g) was dissolved in 100 ml of hexane-benzene mixture (1:1), and then passed through a column (5 × 75 cm) filled with 750 g of neutral alumina (Merck). Development was done with hexane, hexane-benzene mixture (1:1), benzene, and then benzene-ethyl acetate mixture (10:1). The portion eluted with 1000 ml of hexane was designated as Fraction 1 (6.5 g, approximately 18 %), that with 2000 ml of hexane-benzene mixture as Fraction 2 (13.3 g, approx. 37 %), that with 1000 ml of benzene as Fraction 3 (1.8 g, approx. 5 %), and that with 2000 ml of benzene-ethyl acetate mixture as Fraction 4 (5.0 g, approx.

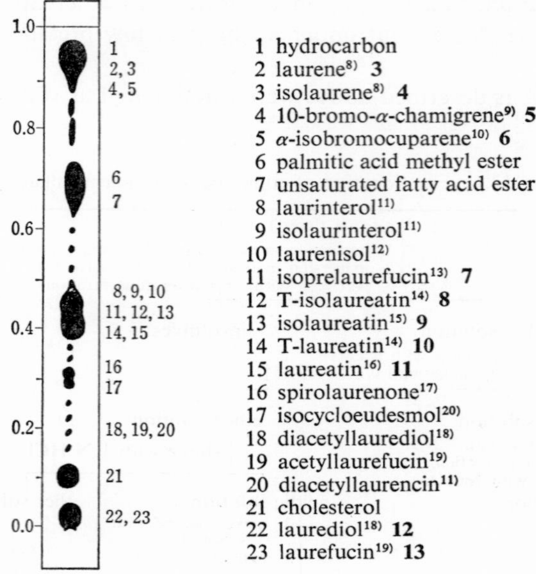

Fig. 12.1 Thin layer chromatography (silica gel) of components of *L. nipponica* YAMADA. Developing reagent: benzene. Coloring reagent: $CeSO_4$ (dissolve 25 g of $CeSO_4$ in 500 ml of 2 N H_2SO_4).

14%). Each fraction was carefully fractionated further. Some of the procedures are described in detail below.

[Experimental procedure 1] Isolation of laurene 3, isolaurene 4, 10-bromo-α-chamigrene 5 and α-isobromocuparene 6

Fraction 1 described above was chromatographed repeatedly on an alumina column. The portion easily eluted with hexane contained compounds **3**–**6**. This portion (20 g) was dissolved in 60 ml of hexane, and passed through a column (5 × 60 cm) filled with 600 g of silica gel (Merck). Development was done carefully with hexane at a rate of one drop of eluent per 5 sec. Laurene **3** (7.2 g, approx. 2%), isolaurene **4** (3.3 g, approx. 0.8%), 10-bromo-α-chamigrene **5** (36 mg, approx. 0.01%) and α-

Br

Br

laurene **3**　　isolaurene **4**　　10-bromo-α-chamigrene **5**　　α-isobromocuparene **6** isobromocuparene **6** (0.2 g, approx. 0.06%) were isolated, all as oily materials. (Yields were calculated based on the amount of neutral fraction.)

[Experimental procedure 2] Isolation of isoprelaurefucin 7, T-isolaureatin 8, isolaureation 9, T-laureatin 10 and laureatin 11

Fraction 2 was rechromatographed on a neutral alumina column. The portion eluted later with hexane-benzene mixture (1:1) was crystallized and contained compounds **7–11**. This portion was chromatographed on 100 times its quantity of silica gel (Merck). Development was carried out

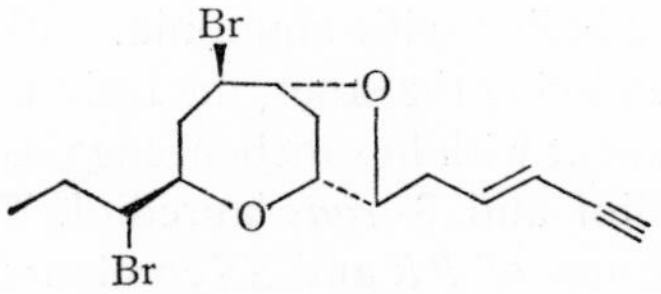

isoprelaurefucin **7**

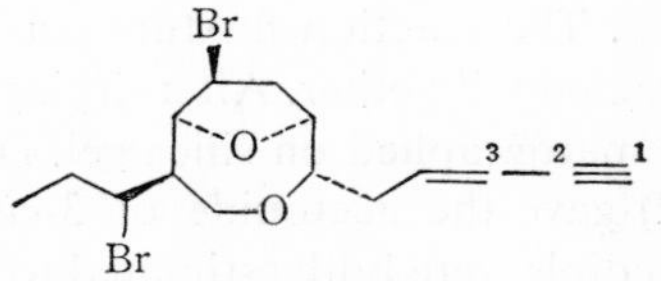

T-isolaureatin **8** (3-*trans*)
isolaureatin **9** (3-cis)

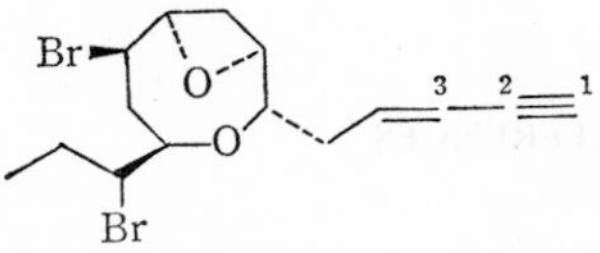

T-laureatin **10** (3-*trans*)
laureatin **11** (3-*cis*)

at a rate of one drop per 5 sec, and the above compounds were obtained in the sequence given, in yields of 0.06%, 0.005%, 6.8%, 0.04%, and 12.8%, respectively. On silica gel tlc, fraction 2 started to separate after several repetitions of development with the 1:1 hexane-benzene mixture; the *Rf* values of these compounds are very similar.

[Experimental procedure 3] Isolation of Laurediol 12 and laurefucin 13

Fraction 4 was separated by silica gel chromatography. Development with benzene-ethyl acetate mixture (10:1) gave an oily substance and a crystalline material, each exhibiting a single spot on tlc. The crystalline substance was recrystallized several times from either ether or isopropylether, yielding laurefucin **13** (1.4%). The crystals obtained from isopropylether were suitable for X-ray diffractometry, as was the case with isolaureatin **9**. The mother liquor of recrystallization and the oily substance described above were subjected to repeated silica gel chromatography. An oily material giving a single spot on tlc was obtained (0.02%). This material was found to be a mixture of stereoisomers having a double bond at the 3-position by mass spectrometry and nmr. The mixture was stirred with 5% phosphomolybdate in acetone solution at room temperature for

3-*cis*-laurediol **12a**
(R, R: S, S = 7: 3)

3-*trans*-laurediol **12b**
(R, R: S, S = 4: 1)

laurefucin **13**

25 hr. The reaction mixture was made alkaline with ammonia, and was extracted with ether. After drying and removing the ether, the residue was chromatographed on silica gel. Development with hexanebenzene mixture (1:2) gave the acetonide of 3-*cis*-laurediol and 3-*trans*-laurediol. These laurediols were both estimated to be mixtures of *RR* and *SS* configuration at the 6- and 7-positions by calculation and comparison with the optical rotations of degradation products of laurencin **2** and laureatin **11**.

References

1) R.A. Lewin, *Physiology and Biochemistry of Algae*, Academic Press (1962); Y. Tsuchiya, *Chemistry of Marine Products*, Volume 2, Kohseisha-Koseikaku (1962); Y. Obata, *Kagaku*, **15**, 403 (1960).
2) H. Kylin, *Z. Physiol. Chem.*, **83**, 171 (1913).
3) S. Ochi, T. Takahashi, *Report from the National Chemical Laboratory for Industry*, **28**, 1 (1933).
4) P. Mastaglin, J. Augier, *Compt. Rend.*, **229**, 775 (1949).
5) T. Saito, Y. Ando, *J. Chem. Soc. Japan*, **76**, 478 (1955).
6) T. Irie, M. Suzuki, T. Masamune, *Tetr. Lett.* **1965**, 1091: *Tetr. hedron*, **24**, 4193 (1968).
7) W. Fenical, *J. Phycol.*, **11**, 245 (1975); D.J. Faulkner, *Tetrahedron*, **33**, 1424 (1977)
8) T. Irie, Y. Yasunari, T. Suzuki, N. Imai, E. Kurosawa, T. Masamune, *Tetr. Lett.*, **1965**, 3619; T. Irie, T. Suzuki, E. Kurosawa, T. Masamune, *ibid.*, **1967**, 3187; T. Irie, T. Suzuki, Y. Yasunari E. Kurosawa, T. Masamune, *Tetrahedron*, **25**, 459 (1969).
9) B.M. Howard, M. Fenical, *Tetr. Lett.*, **1976**, 41; L.E. Wolinsky. D.J. Faulkner, *J. Org. Chem.*, **41**, 597 (1976): M. Suzuki, A. Furusaki, E. Kurosawa, Tetrahedron, **35**, 823 (1979).
10) T. Suzuki, M. Suzuki, E. Kurosawa, *Tetr. Lett.* **1975**, 3057.
11) T. Irie, M. Suzuki, E. Kurosawa, T. Masamune, *ibid.*, **1966** 1837; *Tetrahedron*, **26**, 3271 (1970).
12) T. Irie, A. Fukuzawa, M. Izawa, E. Kurosawa, *Tetr. Lett.* **1966**, 1343.
13) E. Kurosawa, A. Fukuzawa, T. Irie, *ibid.*, **1973**, 4135.
14) T. Irie, A. Fukuzawa, E. Kurosawa, unpublished work (see Refs. 15 and 16).
15) T. Irie, M. Izawa, E. Kurosawa, *Tetr. Lett.* **1968**, 2785; *Tetrahedron*, **26**, 851 (1970); E. Kurosawa, A. Furusaki, M. Izawa, A. Fukuzawa. T. Irie, *Tetr. Lett.*, **1973**, 3857.

16) T. Irie, M. Izawa, E. Kurosawa, *ibid.*, **1968**, 2091; *Tetrehedron*, **26**, 851 (1970); E. Kurosawa, A. Furusaki, M. Izawa, A. Fukuzawa. T. Irie, *Tetr. Lett.*, **1973**, 3857.
17) M. Suzuki, E. Kurosawa, T. Irie, *ibid.*, **1970**, 4995.
18) E. Kurosawa, A. Fukuzawa, T. Irie, *ibid.*, **1972**, 2121.
19) A. Fukuzawa, E. Kurosawa, T. Irie, *ibid.*, **1972**, 3; A. Ferosaki, E. Kurosawa, A. Fukuzawa, T. Irie, *ibid.*, **1973**, 4579.
20) T. Suzuki, H. Kikuchi, E. Kurosawa, *Chem. Lett.*, **1980**, 1267.

CHAPTER 13

Abnormal Secondary Metabolites in Plants

Under natural conditions, plants often suffer damage, e.g., from heavy rain and wind, and attack by fungi, bacteria and viruses as well as insects. Therefore, plants have acquired the ability to recognize injuries and penetrating parasites; they form a wound layer in response to mechanical injury (wounding) and show a defense reaction to parasitic penetration. Principally, wound layer formation and defense reaction involve rapid and large-scale production of secondary metabolites that are not present or hardly present in the healthy plant. Therefore, final and intermediary metabolites which are produced in response to mechanical injury or parasitic penetration are called "abnormal secondary metabolites."

When plants are infected by parasites, antibiotic substances are produced in the infected tissue or non-infected tissue very closely adjacent to the infected region. Müller and Börger[1)] called such substances "phytoalexins," and assumed that phytoalexins participated in the defense reaction.[2)] Such compounds are not produced simply by mechanical injury, but are produced in response to continuing injurious stimuli such as parasitic infection and toxic chemical treatment.[3)] On the other hand, there are also many kinds of abnormal secondary metabolites that are produced by mechanical injury. The metabolites are generally produced in rather larger amounts in response to parasitic infection.

The first phytoalexin to be isolated was ipomoeamaron, which was isolated from *Ceratocystis fimbriata*-infected sweet potato (*Ipomoea batatas*) roots by Hiura.[4)] Later, Kubota and Matsuura[5)] determined its chemical structure, and the compound is now called ipomeamarone (that is, (+)-ngaione).

13.1 Aims of Research on Abnormal Secondary Metabolites

Studies on abnormal secondary metabolites are generally approached from the following points of view, involving basic and applied chemical biology.[6]

(1) Plants have the ability to heal mechanical injury and to provide a defense against parasitic infection. If the chemical and biochemical features of the abnormal secondary metabolites involved can be elucidated, we could develop some practical approaches to improving the yields of various crops.[7]

(2) Until recently, research has focused on the isolation and chemical structure determination of "normal secondary metabolites," and the chemical knowledge thus obtained has been useful in connection with various reactions and syntheses in organic chemistry, and also in chemical systematics and biogenesis studies in phytochemistry. Further, the biological activities of some of these compounds have been of considerable interest. The abnormal secondary metabolites will greatly extend the scope of research in these fields.

(3) Normal secondary metabolites are usually synthesized in healthy plants rather slowly, often synchronized with the growth, although the content is normally high. Therefore, it is difficult to investigate biosynthetic mechanisms in *in vitro* systems from healthy plants, even when radioisotope techniques are used. On the other hand, abnormal secondary metabolites are produced rapidly and in large amounts in response to certain stimuli. Thus, it is relatively easy to follow the enzymatic reactions using *in vitro* systems from injured or diseased plants.

(4) Abnormal secondary metabolites are hardly present in healthy plants. Upon parasitic infection, macromolecular or low-molecular elicitors contained in the parasites may combine with receptors in the plants to induce or activate the enzymes involved in producing the metabolites. Thus, studies of the mechanism of biosynthesis of abnormal secondary metabolites in injured or diseased plants could cast light on the biological control of enzyme induction in plants in general. Further, investigations along this line may give some indication of the biological role of plant hormones in enzyme induction and other important biological events, since the above-mentioned elicitors may participate in enzyme in-

duction or activation in such a way as to stimulate the operation of some specific hormone-like substances.[6)]

13.2 Experimental Considerations

In order to isolate abnormal secondary metabolites from infected or injured plant tissues, healthy plant tissues must first be subjected to parasitic infection or toxic chemical treatment. A high degree of humidity is desirable for successful infection or chemical injury of plant tissues, but the procedures are not difficult. The use of storage organs, such as waterswollen cotyledons or endosperms, tuberous roots or stems, bulbs, underground stems and fruits, is often particularly easy. In a given species of plant, abnormal secondary metabolites are essentially the same among the various organs.

Microbial contamination during the incubation of plant tissues after parasitic inoculation is not usually a major problem. It is usual to select healthy plants without injury or infection, wash the materials thoroughly with tap water, then immerse them in NaClO solution (e.g., 0.05–0.1% for 1 hr and 1% for 10 min for sweet potato roots and dried pea seeds, respectively.) After this sterilization, it is better to incubate the materials under germ-free conditions, but it may be more important to incubate them under conditions of high humidity to ensure effective parasitic infection of the plant tissues. Parasites penetrate easily into healthy plant tissues when there are mechanical injuries, so healthy plant tissues are often cut up with a knife before incubation with the parasites.

Abnormal secondary metabolites are produced in response to the infection of many kinds of parasite, and the metabolites produced are essentially the same. However, the amount produced is related to the type of parasite. Generally, fungi induce larger amounts than bacteria, and the fungi that cause brownish dry-rot are more appropriate than the fungi causing wet rot. Further, dry rot-causing fungi which penetrate easily into the inner part of the plant tissues, but then grow rather moderately should be selected (Fig. 13.1). Spore suspension is usually inoculated on the cut surfaces of plant tissues at a density of 10 to 100 spores per host cell (10^6 to 10^7/ml). Spores that normally germinate in a few hours and penetrate into the host cells soon after germination are preferable. Some kinds of fungus produce large amounts of pectinase or toxins which may be sufficient to kill the host cells, and these often result in loss of the ability to produce abnormal secondary metabolites.

Sometimes, the use of non-pathogenic fungi may enable the plant tissue to produce large amounts of abnormal secondary metabolites and to

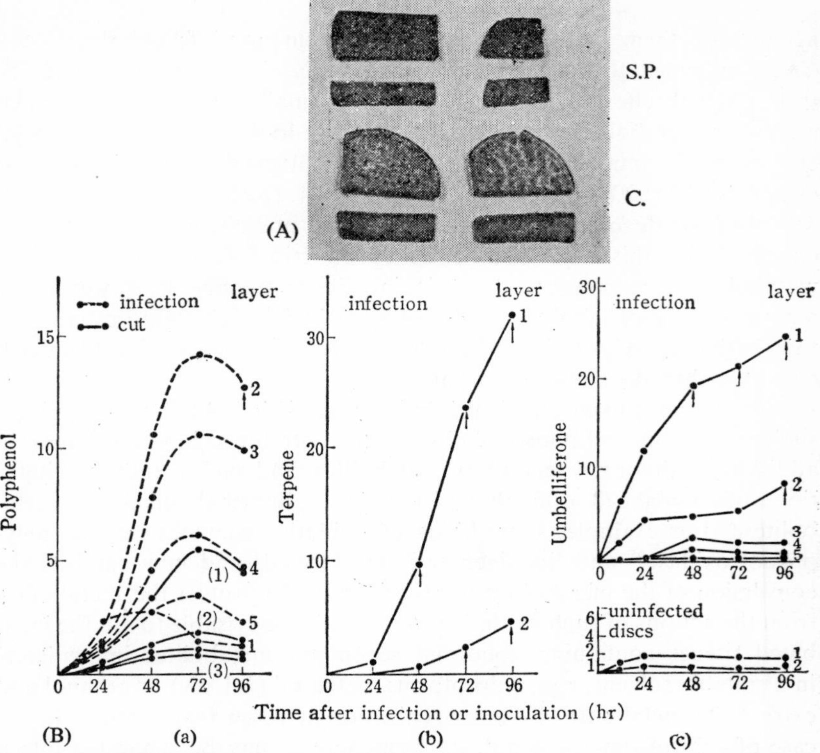

Fig. 13.1. Change in morphology (A) and production of abnormal secondary metabolites (B) in sweet potato root tissue in response to *Ceratocystis fimbriata* infection. (A) There are various strains of *C. fimbriata* which show some difference in parasitism. For example, sweet potato (S.P.) strain preferentially attacks sweet potato, while coffee (C.) strain attacks coffee, but is less effective against sweet potato. Spore suspensions of both strains were inoculated on the cut surface of tissue discs (about 1 cm thick) of sweet potato root. The figure shows the observations on diseased tissue discs incubated at 25°C for 3 days. In the case of S.P. strain, the aerial hyphae grew well and covered all of the cut surface, and the mycelia penetrated into the discs to a depth of about 0.5 to 1.0 mm. The infected region turned brown, and phytoalexins such as ipomeamarone were accumulated heavily there. Polyphenols were accumulated in the adjacent non-infected tissue. In the case of C. strain, the aerial hyphae grew on the cut surface, but not as heavily as in the case of S.P. strain, and the depth of the infected region was about 0.1 to 0.3 mm. The accumulation of phytoalexins and production of polyphenols were less marked than in the case of S.P. strain. (B) Time course of production of polyphenols and phytoalexins in different layers (each 0.5 or 1.0 mm thick) from the cut surface towards the inner part. Since polyphenols are also produced in response to mechanical injury, data for uninfected discs are also shown here. (a) The amounts of polyphenols are shown as chlorogenic acid (mg) per 15 discs (1.0 × 10 mm, 1 mm of thickness and 10 mm of diameter). (b) The amounts of terpenes are shown as ipomeamarone (mg) per 15 discs (0.5 × 10 mm). (c) The amounts of umbelliferone (μg) are shown per 15 discs (0.5 × 10 mm). ↑ : indicates layers infected by the fungus over more than half in their area. 1, . . . 5: five layers (each 0.5 or 1.0 mm thick) from the cut surface towards the inner part.

accumulate them in the infected region or in non-infected tissue very closely adjacent to the infected region. Toxic chemicals such as $HgCl_2$ (0.1 %) can also be used as elicitors, although smaller amounts of abnormal secondary metabolites are generally produced. In the case of such artificial elicitors, it is sometimes necessary to preincubate the plant tissues after cutting for 10 to 20 hr before applying the agent. During the preincubation, carbohydrate decomposition, respiration and polyphenol production are enhanced, and this may assist the rapid production of abnormal secondary metabolites in response to the artificial elicitor treatment. However, the amounts of the metabolites produced by artificial elicitors are usually less than in the case of parasitic infection, and the metabolites, once produced, are often then decomposed in part.[3,6)]

In response to mechanical injury or parasitic infection, compounds such as proteins and phospholipids are also produced in plant tissues, in addition to abnormal secondary metabolites, and such metabolic alterations often make it difficult to isolate the abnormal secondary metabolites.[8)] For example, some kinds of oxidative enzymes such as polyphenol oxidase[9)] and lipoxigenase[10)] are formed, and may catalyze the conversion of the metabolites to other forms. To protect the metabolites from the actions of such oxidative enzymes, it is advisable to dip the incubated tissues containing abnormal secondary metabolites immediately in a suitable solvent, e.g., chloroform: methanol (1:1 v/v) or ethanol and extract the metabolites by homogenization at room temperature. In the case of chloroform–methanol, the homogenate may be separated into a chloroform layer and a methanol:water layer by adding an appropriate volume of water. We can classify abnormal secondary metabolites into two categories: hydrophilic and lipophilic. Hydroxy-phenylpropanoids such as chlorogenic acid belong to the former, but phytoalexins such as ipomeamarone belong to the latter. Ethanol extracts both kinds of metabolites, but chloroform–methanol extracts the latter. However, chloroform: methanol also extracts phospholipids, sterols and sometimes plant resins, such as jalapinic acid which is contained in sweet potato, and plant rubbers which are contained in some kinds of rubber tree.

13.3 Assay Procedures for Abnormal Secondary Metabolites

One of the problems with abnormal secondary metabolites is how to assay them either qualitatively or quantitatively. Generally, polyphenols are produced in large amounts in mechanically injured tissue (Fig. 13.1).

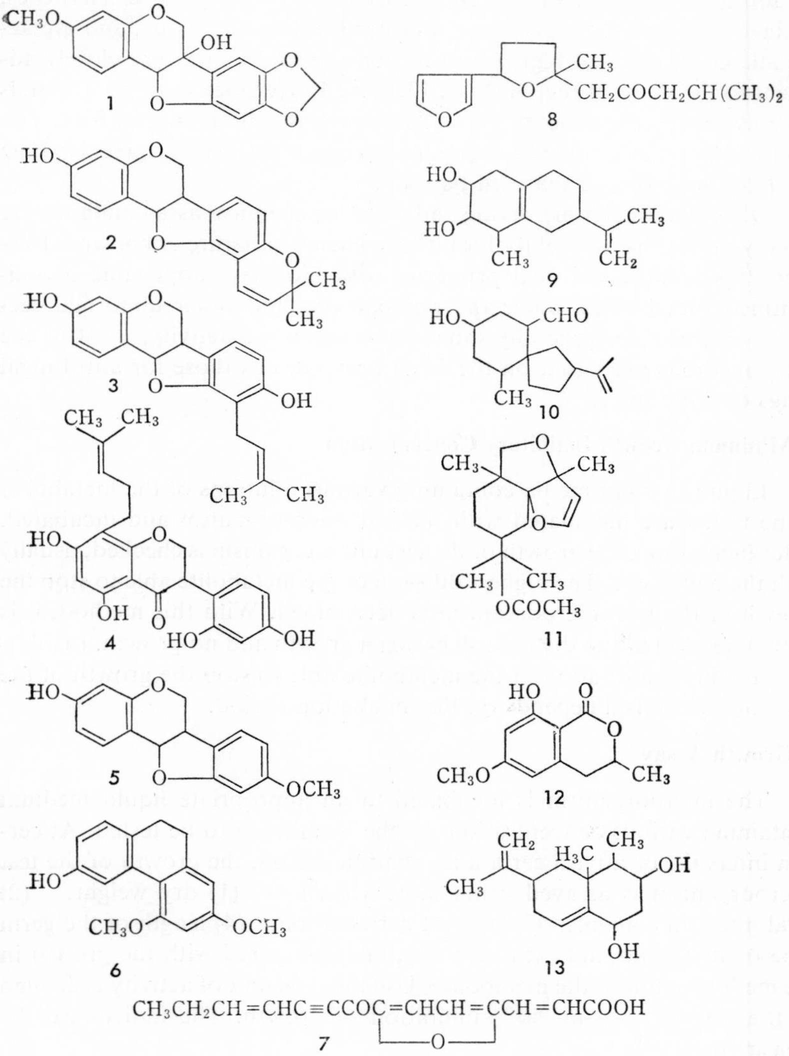

$CH_3CH_2CH{=}CHC{\equiv}CCOC{=}CHCH{=}CCH{=}CHCOOH$ (7)

Fig. 13.2. Main phytoalexins. 1, pisatin (pea); 2, phaseollin (bean); 3, phaseollidin (bean); 4, kievitone (bean); 5, medicarpin (alfalfa, clover); 6, orchinol (orchid); 7, wyeronic acid (broad bean); 8, ipomeamarone (sweet potato); 9, rishitin (white potato, tomato); 10, lubimin (white potato); 11, phytuberin (white potato); 12, 6-methoxymellein (carrot); 13, capsidiol (pepper).[2,11,12]

In addition, phytoalexins which consist mainly of coumarins, terpenes, isoflavonoids and fatty acids are produced in infected tissue, and are accumulated in infected region and in non-infected tissue very closely adjacent to the infected region (Figs. 13.1 and 13.2). Since most polyphenols have uv maxima at around 250 to 330 nm, this can be utilized for assay. On the other hand, some of the coumarins generate a strong greenish-blue to violet fluorescence that can be used.

Of course, such time-saving and accurate chemical assay methods are usually applicable to metabolites whose chemical nature is known. However, this section will deal primarily with bioassays applicable to unidentified metabolites. Since the phytoalexins and other abnormal secondary metabolites generally show some antibiotic activity, most of the assay methods are based on the same principles as those for anti-fungal drugs or antibiotics.[13)]

Minimum Growth-Inhibitory Concentration

Liquid or solid media containing various dilutions of the metabolite to be tested are inoculated with the test microorganism and incubated. After incubation, the growth of the test microorganism is checked, usually with the naked eye. The highest dilution of the metabolite able to stop the growth of the test microorganism is determined. With this method, it is sometimes difficult to distinguish between growth and no-growth. In addition, the highest dilution of the metabolite able to stop the growth of the test microorganism depends on the incubation period.

Growth Assay

The microorganism is incubated in an appropriate liquid medium containing various concentrations of the metabolite to be tested. At certain intervals, or after a certain incubation period, the growth of the test microorganism is assayed using indices such as (1) dry weight, (2) total nitrogen content, (3) percent germination, (4) length of the germ tube or mycelium and (5) turbidity, and compared with the growth in the medium without the metabolite. Usually, one unit of activity is defined as the activity causing 50% inhibition of growth. The indices can be measured as follows.

Dry weight: After incubation, the test microorganism in the medium is collected, washed well and heat- or vacuum-dried. The dried sample is weighed on a micro-balance.

Total nitrogen content: The well-washed microorganism is digested by Kjeldahl's method and assayed for nitrogen content, commonly by Nessler's method.

Germination (percent): Germinated and ungerminated spores are counted under a microscope and the percentage of germinated spores is calculated.

Length of the germ tube or mycelium: The length of the germ tube or mycelium is measured directly using a microscope with a micrometer or by photomicrography. Since there are usually large variations among the lengths of germ tubes or mycelia of the test microorganism, statistical treatment of the data is necessary to get a reliable result.

Turbidity: Turbidity is measured at 430 nm with an uncolored sample or at 660 nm with a colored sample. The proportionality between the turbidity and the growth of the test microorganism should be checked first with each system.

Colony Size Measurement

Small discs of the test microorganism are placed in the center of solid media preparations containing various concentrations of the metabolite, and the media are incubated. After a certain period, the diameters of colonies on the solid media are measured, and compared with those of a control. This method requires neither special apparatus nor reagents, although it has the disadvantage of being slow.

Direct Detection of the Metabolite

Thin layer and paper chromatographies are often used for the purification or detection of phytoalexins and other abnormal secondary metabolites. Some methods have been developed for the direct detection of such metabolites in terms of anti-microbial activity on the chromatography plate or paper. In one method, a uniform layer containing both the medium and spores of the test microorganism is formed with a sprayer on the plate or paper after chromatography, and the plate or paper is incubated for a certain period. The metabolite is located as a growth-inhibited zone or spot on the chromatogram. In another method, the chromatographed plate or paper is plated on a solid medium (preparing a replica) on which the test microorganism is cultured. The solvent used for chromatography is generally toxic to microorganisms. It is therefore necessary in both methods to use developing solvents which can be easily removed after chromatography. It is convenient to use a test microorganism which becomes colored or changes its color after germination, for ease of detection.[14)]

General Remarks

All the bioassay methods mentioned above are based on the inhibitory effect of the metabolites on the germination or growth of the test microorganism. Since germination and growth are not simple physiological events, but complex ones which involve many kinds of metabolism, they are affected by various factors. Moreover, the inhibitory activity of some of these metabolites is not strong. Thus, it is important to discriminate carefully the inhibitory effect of the metabolite to be tested from those of other factors. The germination and growth of the microorganism are profoundly affected by physical, chemical and biological factors such as osmotic pressure, pH and chemical composition of the medium, as well as the age of the test microorganism.

13.4 Isolation of Abnormal Secondary Metabolites from Mechanically Injured Plant Tissues

In response to mechanical injury (wounding), various kinds of abnormal secondary metabolites are produced in plant tissues. The main metabolites are polyphenolic compounds.[6,8,9,15] When sweet potato root tissues are cut and incubated, large amounts of chlorogenic acid and isochlorogenic acid, the most ubiquitous polyphenolics in the plant kingdom, are produced. In the following sections, we will describe our experiments on a possible intermediate of chlorogenic acid biosynthesis that was isolated from injured sweet potato root tissue.

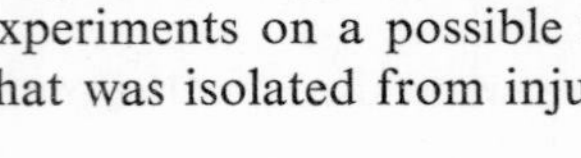

HO, HO–C₆H₃–CH=CHCOOH

caffeic acid

HO, COOH; RO, OH, OH (quinic acid ring, positions 1–6)

chlorogenic acid

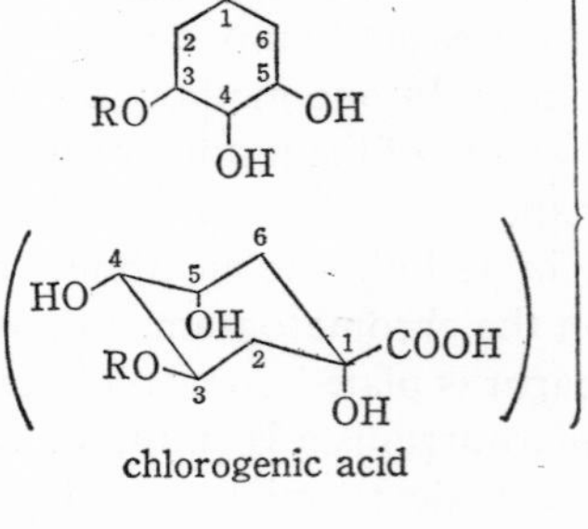

quinic acid

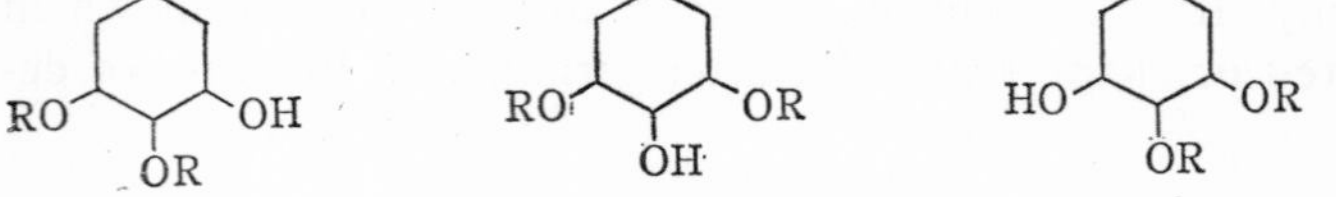

isochlorogenic acid

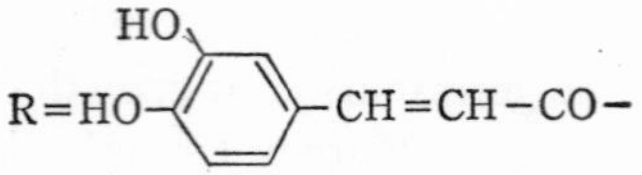

R = HO, HO–C₆H₃–CH=CH–CO–

Detection of a Possible Intermediate of Chlorogenic Acid Biosynthesis[13)]

Tissue discs of sweet potato roots were fed [2-^{14}C]-*trans*-cinnamic acid, which was shown to be the source of the aromatic moiety of chlorogenic acid by *in vivo* experiments. At intervals during the incubation, discs were homogenized with ethanol, and the concentrated extracts were subjected to paper or silica gel column chromatography, to investigate the time course of changes in the distribution of radioactivity during incubation. The results indicated the formation of one compound, tentatively called compound I, which was assumed to be an intermediate in chlorogenic acid biosynthesis, based on the following evidence.

(1) The label from [2-^{14}C] *trans*-cinnamic acid was incorporated into compound I initially, and was then transferred to chlorogenic acid and isochlorogenic acid.

(2) The specific radioactivity of compound I increased prior to that of the fraction containing chlorogenic acid and isochlorogenic acid and decreased subsequently.

(3) The label from ^{14}C-labeled compound I was efficiently incorporated into chlorogenic acid and isochlorogenic acid.

Isolation and Structural Determination of Compound I

The tissue slices (2 mm thick) of sweet potato roots were soaked for about 2 min in a saturated solution of *trans*-cinnamic acid in 5 mM acetate buffer, pH 5.5, then taken out and incubated at 30° C for 12 hr in a moist chamber. The incubated slices were washed with water to remove the unabsorbed *trans*-cinnamic acid, boiled in ethanol for 4 min to inactivate enzymes, then homogenized in a blender. The suspension was mixed with additional ethanol, boiled for 30 min, and filtered with suction. After filtration, the residue was washed three times with ethanol. The filtrate and washings were combined. The extract was dried with Avicel SF powder. The powder was applied to a column of silica gel and chromatographed in a linear gradient formed from 400 ml of cyclohexane: chloroform (3:7 v/v) preequlibrated with 0.25 N sulfuric acid in the mixing chamber and the same volume of *tert*-butanol: chloroform (3:7 v/v) preequilibrated with 0.25 N sulfuric acid in the reservoir. The eluted fractions were monitored in terms of the absorbance at 278 nm, and a fraction containing compound I was obtained. Compound I was crystallized from the concentrated fraction and repeatedly recrystallized from chloroform-ethanol, then from water. Compound I formed white needles when recrystallized from water, mp 152.5–153.5° C, $[M]_D^{12}$–44.3 Anal. Found: C, 57.85; H, 6.15; N, 0. Calcd. for $C_{15}H_{18}O_7$:C, 58.06; H, 5.85; N, 0. MW, 310; uv_{max}^{ETOH} nm (ε): 278 (20,420). Compound I was unstable in alkali and was readily hydrolyzed by 0.5 N sodium hydroxide to give two products.

One proved to be *trans*-cinnamic acid (paper and thin layer chromatographies and uv, ir and ms data). The other product showed exactly the same behavior as D-glucose on paper chromatography using three different solvent systems and on gas chromatography after trimethylsilylation. The nmr spectral data for compound I indicated its structure to be β-1-*O*-*trans*-cinnamoyl-D-glucopyranose **14**.[16] The β-linkage between the D-glucose and *trans*-cinnamic acid moieties was confirmed by the molecular rotation value of the compound and its preferential hydrolysis by β-glucosidase.

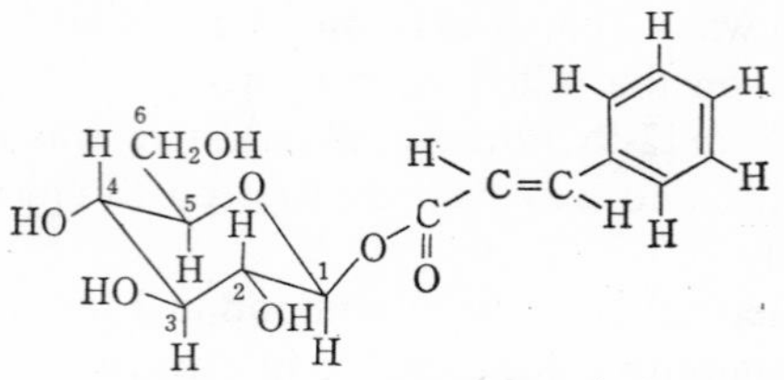

β-1-*O*-*trans*-cinnamoyl-D-glucopyranose **14**

It has been reported that some plants, such as white potato (*Solanum tuberosum*)[17] and *Lathyrus doratus*,[18] contain *p*-coumaroyl-D-glucose and caffeoyl-D-glucose, which are related to *trans*-cinnamoyl-D-glucose in terms of both chemical structure and metabolic pathway. These glucose esters of hydroxycinnamic acids are chemically active and may be intermediates in the biosynthesis of phenylpropanoids. In view of these experimental results,[16–18] we proposed the following biosynthetic pathway of chlorogenic acid in sweet potato roots: *trans*-cinnamic acid is esterified through its carboxyl group with D-glucose at the first step, then two hydroxyl groups are consecutively introduced into the benzene ring to give caffeoyl-D-glucose. Finally, trans-esterification takes place between caffeoyl-D-glucose and quinic acid or activated quinic acid, producing chlorogenic acid.* However, there still remains the possibility that *trans*-cinnamoyl-D-glucose is a by-product of chlorogenic acid biosynthesis. In that case, it may function as a regulatory compound of chlorogenic acid biosynthesis in sweet potato roots.

13.5 Isolation of Phytoalexins from Infected Plant Tissues

As described above, phytoalexins are produced in plant tissues in response to parasitic penetration, and take part in the defense mechan-

* The importance of the glucose esters of hydroxy-cinnamic acids in chlorogenic acid biosynthesis was supported by a recent study by Molderez *et al.* using *Cestrum poeppigil* (M. Molderez, L. Nagels, F. Parmentier, *Phytochemistry*, **17**, 1747 (1978)).

ism. Since the detection of the first phytoalexin from sweet potato (*Ipomoea batatas* LAM.) belonging to Convolvulaceae,[4,5] various kinds of phytoalexins have been isolated from Leguminosae (pea, *Pisum sativam* L.; bean, *Phaseolus vulgaris* L.; soybean, *Glycine max* MERRILL), Solanaceae (white potato, *Solanum tuberosum* L.; tomato, *Lycopersicon esculenum* MILL; pepper, *Capsicum frutesrens* L.), Gramineae (barley, *Hordenum vulgare* L. var. *hexastichon* ASCHERS; oat, *Avena sativa* L.), etc.[2,11,12,19,20] Most of them are terpenes such as ipomeamarone in sweet potato, isoflavonoids such as pisatin in pea, or fatty acids and their derivatives such as wyeronic acid in broad bean (Fig. 13.2). Others include coumarins such as umbelliferone and stilbenes. No alkaloid phytoalexins have yet been detected.

Here, we will describe the isolation of furanoterpenes as phytoalexins in diseased sweet potato.

Furanoterpenes from Infected Sweet Potato Roots

Although coumarins such as umbelliferone **15**, scopoletin **16** and es-

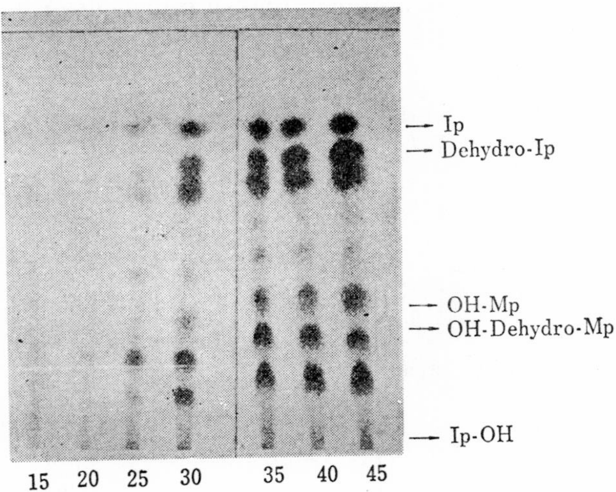

Fig. 13.3 Thin-layer chromatography pattern of terpenes in *Ceratocystis fimbriata*-infected sweet potato roots.

Spore suspension of *C. fimbriata* was inoculated on the cut surfaces of sweet potato root tissue discs (2 to 3 cm thick), and the inoculated discs were incubated at 30° C under high humidity. The samples were taken at appropriate times as indicated, and the infected region was extracted with chloroform: methanol (1:1 v/v). The extracts were applied to a silica gel thin layer, which was then developed with benzene: ethyl acetate (8:2 v/v). After development, the spots on the thin layer were visualized using Ehrlich's reagent.[23]

culetin **17** have been identified as phytoalexins in infected sweet potato roots, the amounts are very low (Fig. 13.1),[6,18] and furanoterpenes are the

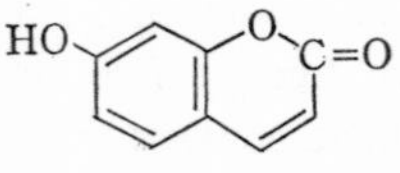

umbelliferone **15** 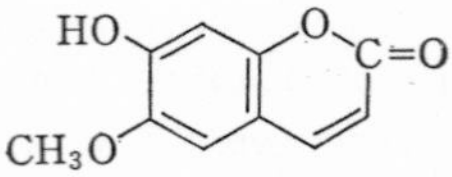 scopoletin **16** 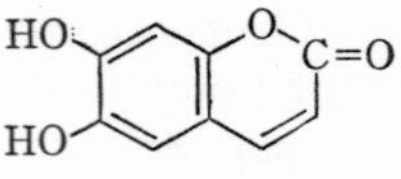 esculetin **17**

main phytoalexins, being present in amounts high enough to protect the tissue from penetration by pathogenic fungi such as *C. fimbriata*. The major component of the furanoterpenes is ipomeamarone, though more than 10 kinds of minor furano-terpenes have been separated on silica gel thin layer plates after chromatography (Fig. 13.3).

The chemical structures of some components have already been determined. They include ipomeamarone (Ip) **8**,[4,5] ipomeamaronol (Ip-OH) **19**,[21,22] dehydroipomeamarone (Dehydro-Ip) **18**,[23] 4-hydroxy-myoporone (OH-Mp) **20**,[24] and 4-hydroxydehydromyoporone (OH-Dehydro-Mp) **21**.[25] Ipomeanine **22**, batatic acid **23** and furan-β-carboxylic acid **24** were also detected in the crude extract.[5] Since they were isolated and purified by distillation with heating, it is quite likely that they were

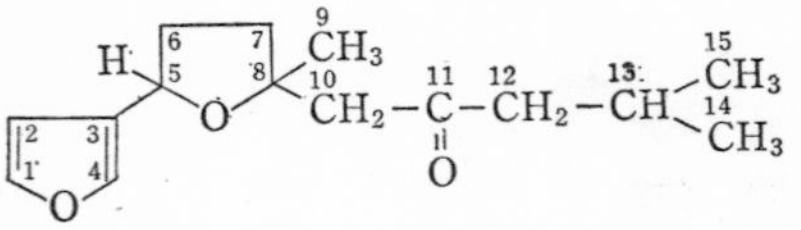

ipomeamarone **8**

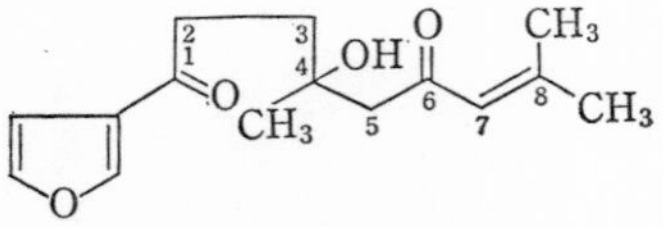

4-hydroxydehydromyoporone **21**

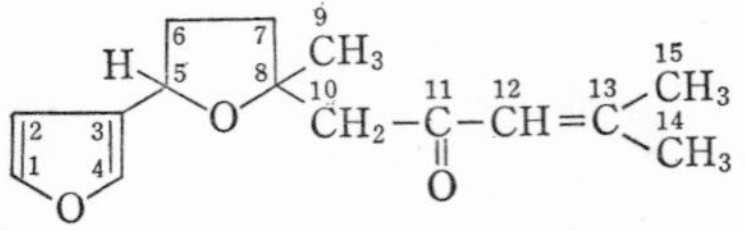

$CO-CH_2-CH_2-CO-CH_3$

ipomeanine **22**

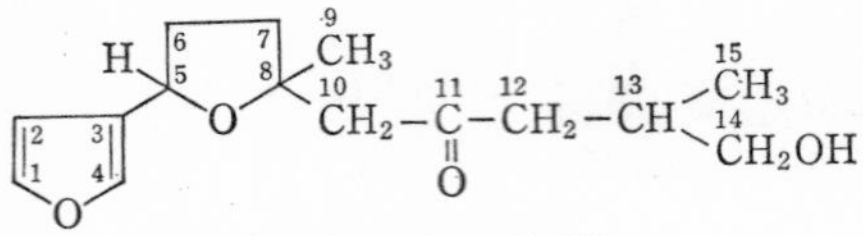

dehydroipomeamarone **18**

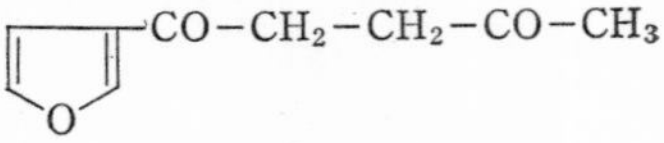

$CO-CH_2-CH_2-CH(CH_3)COOH$

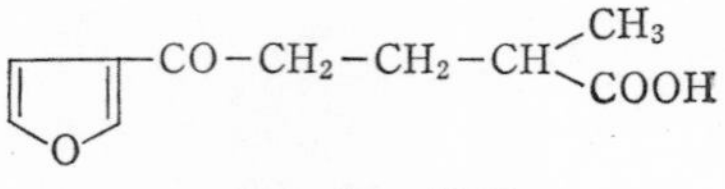

batatic acid **23**

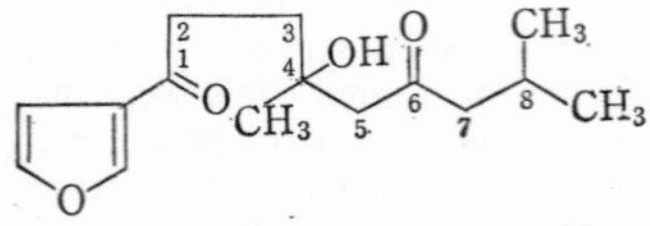

ipomeamaronol **19**

COOH

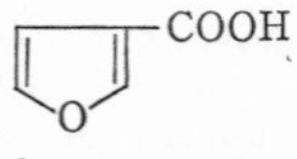

furan-β-carboxylic acid **24**

4-hydroxymyoporone **20**

produced by the decomposition of native phytoalexins during the purification procedure.

Isolation of Ipomeamarone

When sweet potato roots are infected by pathogenic fungi of sweet potato such as *C. fimbriata*, the infected region becomes very bitter (Fig. 13.1). Early research was aimed at the elucidation of the chemical nature of the bitter substances.[4,5] The crude extract was subjected to steam distillation, yielding oily droplets that did not almost dissolve in the distilled water. The aqueous solution containing the droplets was extracted with ethyl ether, and the oily extract was subjected to vacuum distillation (135° C, 3.5 mmHg). A large part of the extract could be distilled by this procedure, and the main component was identified* as Ip.[4,5]

Akazawa *et al.*[30] assumed that Ip was produced in sweet potato roots infected by the larvae of *Cylas formicarius*, sweet potato weevil, since the weevil-infested roots were also bitter. They obtained an oily substance by vacuum distrillation of the crude extract from the above roots, and tried to crystallize the Ip semicarbazone from the substance, but failed to obtain a derivative with a sharp melting point. This suggested that the Ip semicarbazone was contaminated with semicarbazones of other native and artifactual carbonyl group-containing furanoterpenes. The crude ethyl ether extract was subjected to tlc,[31] yielding many spots that were positive to Erlich's reagent (HCl-acidic *p*-dimethylaminobenzaldehyde in ethanol), as shown in Fig. 13.3. The positive reaction indicated the presence of a furan ring in the compounds.

[Experimental procedure 1] Isolation of ipomeamarone 8

The crude ethyl ether extract (5 g) from powder of the air-dried infected region of sweet potato root tissue (corresponding to about 200 to 500 g of the wet infected region) was applied to a silica-gel column (4 × 30 cm), and eluted stepwise with two lipophilic solvents, then one rather hydrophilic solvent: that is, 200 ml of *n*-hexane, 200 ml of *n*-hexane: ethyl acetate (8:2 v/v), and then 95% ethanol. Since the fraction eluted with *n*-hexane: ethyl acetate contained a large amount of Ip, it was subjected to silica gel tlc with *n*-hexane: ethyl acetate. An almost pure Ip sample was obtained. It was distilled *in vacuo* (122° C, 2.5 mm Hg) to obtain the pure material.[31]

The isolation of Ip-OH was performed in a similar way, except for the use of a more hydrophilic solvent such as *n*-hexane: ethyl acetate (1:1 v/v) in place of *n*-hexane: ethyl acetate (8:2 v/v).

* There are two asymmetric carbons in Ip. It was demonstrated that ngaione, which was isolated from *Myoporus* by McDowall[26] is the enantiomorph of Ip.[27] However, the configurations of both Ip and ngaione require confirmation.[28,29]

[Experimental procedure 2] Isolation of dehydroipomeamarone 18

When the crude extract from the infected region of infected sweet potato roots was chromatographed on a silica gel thin layer, we detected a spot as an *Rf* value just smaller than that of Ip, which gave the same color as Ip with Ehrlich's reagent (initially salmon-orange, turning dark violet). Further, [2-^{14}C] acetate was incorporated at about the same rate as into Ip. The results suggested that this compound (Ip′) was very similar in chemical structure to Ip.

The Ip′ fraction was obtained by silica gel column chromatography using the same solvent system as for Ip, and was then applied to a Kieselgel HF_{254} (fluorescent compound-containing silica gel) thin layer plate and chromatographed. The position of Ip′ on the layer was easily detected by means of its fluorescence under a uv lump. The band was scraped off and eluted with ethyl ether. The eluate was gas-chromatographed on a glass columm (3 × 200 mm) of Shimalite (60–80) containing 10% 1,4-butanediol-succinate at 185° C, with He gas flow at a rate of 60 ml per min. Ip′ was obtained in purified form, and was identified as Dehydro-Ip **18**.[23)]*

If N_2 gas was used instead of He gas, or a metal column was used in place of a glass column, dehydro-Ip was gradually decomposed during the gas chromatography.[23)]

Isolation of 4-Hydroxymyoporone and 4-Hydroxy-Dehydromyoporone

Since OH-Mp **20** and OH-Mp′ showed the same color, namely pink, when sprayed with Ehrlich's reagent and the *Rf* values of both compounds on a silica gel thin layer plate were very similar, the chemical structure of OH-Mp′ was assumed to be very similar to that of OH-Mp. After separation, OH-Mp was completely purified by tlc on Kieselgel H or HF.

For the purification of OH-Mp′, tlc was performed repeatedly in an open system for 6 to 8 hr to remove contaminants such as OH-Mp.[25)] It could also be purified by preparative hplc after the isolation by normal tlc. The nmr, ms, uv and ir data, in comparison with those of an authentic sample of OH-Mp **20** kindly provided by Wilson,[32)] indicated that the chemical structure of OH-Mp′ was that of OH-Dehydro-Mp **21**.

It should be noted that OH-Mp appeared unexpectedly at the same position as ipomeanine **22** on silica gel thin layer chromatography and showed the same pink color to Ehrlich's reagent. Ipomeanine **22** was not detected in the process of isolation of OH-Mp, and analyses by nmr and

* It should be noted that the chemical structure of dehydro-Ip was elucidated in part by comparison with that of dehydrongaione, the enantiomorph of dehydro-Ip, which was isolated from *Althanasia crithmifolium* L. by Bohlmann and Rao (F. Bohlmann, N. Rao, *Tetr. Lett.*, **1972**, 1039), and kindly provided by Dr. Bohlmann.

ms indicated that the purified OH-Mp sample was different from ipomeanine. Therefore, it is reasonable to consider that ipomeanine may be produced artificially by the reverse reaction of aldol condensation during heating, as mentioned above. Wilson's group[32] isolated 4-ipomeanol **26** and ipomeanine **22** as well as OH-Mp **20** from sweet potato roots infected by *Fusarium solani*, and suggested that the production of both 4-ipomeanol and ipomeanine might be due to the action of the enzymes of *F. solani*.[24,32]

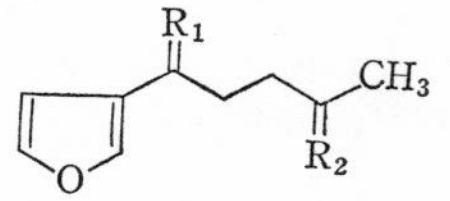

1-ipomeanol	**25**	(R_1 = H, OH	R_2 = O)
4-ipomeanol	**26**	(R_1 = O	R_2 = H,OH)
ipomeadiol	**27**	(R_1 = H,OH	R_2 = H,OH)

According to Wilson's group,[32] *C. fimbriata* has no such enzymes, so sesquiterpenes such as OH-Mp and Ip are not converted to C_9-compounds such as ipomeanine in *C. fimbriata*-infected sweet potato roots.

13.6 Concluding Remarks

Care must be taken when investigating the purity of a specific organic compound by a single technique (tlc, gc or hplc), especially when compounds with very similar chemical structures are present. For example, OH-Mp **20** has the same *Rf* value on silica gel thin layer chromatography as ipomeanine **22**, as mentioned above. Further, the gc retention time of Ip **8** is the same as that of OH-Mp **20**, under the conditions described above for Dehydro-Ip **18**.

It is well known that procedures involving heat treatment should be avoided when isolating terpenes. However, terpenes are usually contaminated with phospholipids and neutral lipids when they are extracted with chloroform: methanol (1:1 v/v) at room temperature. Therefore, it is sometimes necessary to carry out distillation at as low a temperature as possible.

The study of abnormal secondary metabolites continues to offer very wide scope for further research.

References

1) K.O. Müller, H. Börger, *Arb. Biol. Reichsanstalt Landw. Forstw. (Berlin)*, **23**, 189 (1940).
2) I.A.M. Cruickshank, *Kagaku to Seibutsu* (in Japanese), **13**, 168 (1975).

3) I. Uritani, M. Uritani, H. Yamada, *Phytopathology*, **50**, 30 (1960).
4) M. Hiura *Rep Gifu Agric. Coil.* (in Japanese), **50**, 1 (1941).
5) T. Kubota, T. Matsuura, *J. Chem. Soc. Japan* (in Japanese), **74**, 44, 101, 105, 107, 208, 248, 668 (1952); *ibid.*, **75**, 447 (1953).
6) I. Uritani, *Kagaku to Seibutsu* (in Japanese), **12**, 546 (1974).
7) P. Albersheim, *Chem. Eng.*, Apr. 26, 21 (1976).
8) Y. Tanaka, M. Kojima, I. Uritani, *Plant Cell Physiol.*, **15**, 843 (1974).
9) I. Uritani, K. Ôba, *Experimental Methods on Enzymes and Proteins in Plants*, Special Issue of *Protein, Nucleic Acid* and *Enzyme* (in Japanese), **2**, 170 (1976).
10) T. Galliard, *Recent Advances in the Chemistry and Biochemistry of Plant Lipids* (ed. T. Galliard *et al.*), p. 319, Academic Press (1975).
11) K. Tomiyama, *Kagaku to Seibutsu* (in Japanese), **12**, 778 (1974).
12) N. Katsui, A. Matsunaga, T. Masamune, *Tetr. Lett.* **1974**, 4483.
13) M. Kojima, I. Uritani, *Plant physiol.*, **51**, 768 (1973).
14) N.T. Keen, J.J. Sims, D.C. Erwin, E.C. Erwin, J.E. Partridge, *Phytopathology*, **61**, 1084 (1971).
15) Y. Tanaka, *Kagaku to Seibutsu* (in Japanese), **14**, 237 (1976).
16) M. Kojima, I. Uritani, *Plant Cell Physiol.*, **13**, 1075 (1972).
17) K.R. Hanson, *Phytochemistry*, **5**, 491 (1966).
18) J.B. Harborne, J.J. Coner, *Biochem. J.*, **81**, 242 (1961).
19) J. Kuć, *Ann. Rev. Phytopathol.*, **10**, 207 (1972).
20) J.L. Ingham, *Phytopathol. Z.*, **78**, 314 (1973).
21) N. Kato, H. Imaseki, N. Nakashima, I. Uritani, *Tetr. Lett.*, 843 (1971).
22) D.T.C. Yang, B.J. Wilson, T.M. Harris, *Phytochemistry*, **10**, 1653 (1971).
23) I. Oguni, I. Uritani, *Agric. Biol. Chem.*, **37**, 2443 (1973); *Plant Physiol.*, **53**, 649 (1974).
24) L.T. Burka, L. Kuhnert, B.J. Wilson, *Tetr. Lett.*, 4017 (1974).
25) H. Inoue, N. Kato, I. Uritani, *Phytochemistry*, **16**, 1063 (1977).
26) F.H. McDowall, *J. Chem. Soc.*, 731 (1927).
27) A.J. Birch, R. Massey-Westropp, S.E. Wright, T. Kubota, T. Matsuura, M.D. Sutherland, *Chem. Ind.*, 902 (1954).
28) T. Matsuura, M. Nakashima, T. Kubota, *8th Symposium on the Chemistry of Natural Products* (in Japanese), Abstracts, p. 59 (1964).
29) W.D. Hamilton, R.J. Park, F.J. Perry, M.D. Sutherland, *Australian J. Chem.*, **26**, 375 (1973).
30) T. Akazawa, I. Uritani, H. Kubota, *Arch. Biochem. Biophys.*, **88**, 150 (1960).
31) T. Akazawa, *Arch. Biochem. Biophys.*, **90**, 90 (1960).
32) M.R. Boyd, B.J. Wilson, *J. Agr. Food Chem.*, **20**, 428 (1972); *idem*, private communication.

CHAPTER 14

Dormancy Substances: Studies on Germination and Growth Inhibitors in Rice Husks

Dormancy is biologically defined as the situation where germination is not stimulated under suitable physiological conditions. In the case of rice seeds, it is well known that the seeds are not able to germinate immediately after harvest, even when they are kept under favorable conditions for germination.

Many studies have been carried out in attempts to identify the factor(s) causing dormancy.[1] It is well established that seed dormancy is caused by the inhibitory influence of structures covering the embryo. Roberts in 1961 demonstrated the importance of the rice husk[2] by carrying out germination tests on seeds not dehusked (control), seeds dehusked 3 days after harvest, and seeds dehusked immediately before the test. The results are shown in Fig. 14.1.

It is clear that in most seeds the removal of the husk allows the germination of seeds which would otherwise be dormant. Two different hypotheses have been proposed by several research groups. Roberts himself considers that the husk restricts the inward diffusion of gaseous substances such as oxygen necessary for some metabolic process leading to loss of dormancy. The other hypothesis is that the husk contains chemical substances which act as either inhibitors of the growth of the embryo or deactivators of a growth promotor present in the seeds.

Several workers have reported that the inhibitory influence of the husk of higher plants is due to chemical substances present in the husk.[3] In 1944, Kasahara suggested the existence of both growth inhibitors and promotors in rice plant. Since then many groups have attempted to isolate

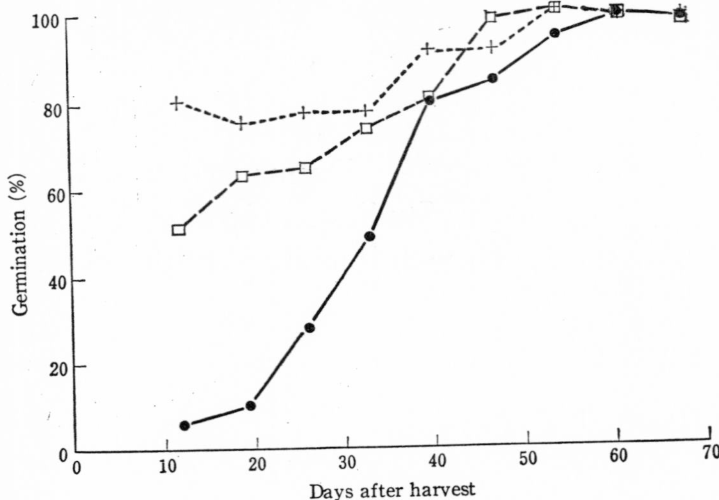

Fig. 14.1 The effect on dormancy of removing the husk. ●——●, intact seed (control); □——□, seed dehusked 3 days after harvest; + ... +, seed dehusked immediately before each test. [from E.H. Roberts, *J. exp. Bot.*, **12**, 432, 1961].

the germination inhibitors from husks of dormant rice seeds. Takeshima, Mikkelsen, Ohta and Takahashi independently reported the discovery of inhibitory substances in the husks of rice seeds. Takahashi also examined the content of the germination inhibitor in husks, rice bran and endosperm and found that the germination inhibitors were mainly present in the husk.[4)]

In 1971, Kato and Takahashi started joint studies of the germination inhibitors in the husk and have characterized several kinds of chemical compounds possessing germination inhibitory activity. The present chapter describes the isolation, characterization and some physiological properties of germination inhibitors in rice husks.

Fig. 14.2 shows how the relative germination ratio differs among several varieties of rice, all of which except for *Surjamukhi*, are widely cultivated in Japan. As is evident from the figure, *Ginmarasi* and *Towada* germinate considerably faster than *Surjamukhi* and *Koshihikari* under the same conditions. Varieties showing lower germination ratios presumably contain higher levels of inhibitors in the husk.

Initial studies on *Surjamukhi*, selected as a rather primitive strain with a high content of inhibitors, yielded insufficient material for structure analysis. We therefore carried out further studies with the more readily available *Koshihikari*.

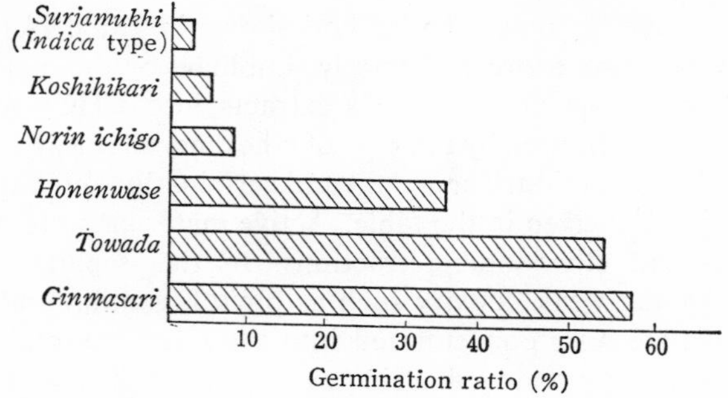

Fig. 14.2 Germination ratios for several varieties of rice (determined at Higashiyama Agricultural Institute).

This change proved to be fortunate, since we later found that *Koshihikari* contains momilactones A and B (the former is predominant), whereas *Surjamukhi* contains momilactone B with only a trace of momilactone A, and momilactone B did not give single crystals suitable for X-ray crystallographic analysis.

14.1 Characterization of the Germination Inhibitors[5)]

Fractionation

Up to 300 kg of rice husks was immersed in MeOH in polyethylene buckets at room temperature for several months. MeOH was then evaporated off with a flash vacuum evaporator at 35° C. Crude extracts of *Surjamukhi* were directly chromatographed on large amounts of silica gel, eluting successively with solvents of different polarity. The active fractions were identified by assaying all the fractions using lettuce seeds. This procedure resulted in the isolation of phytol (**1**), vanillin (**2**) and methyl *p*-hydroxycinnamate (**3**). Although these compounds showed weak inhibition activity towards the germination of lettuce seeds, we could not isolate any compound with strong inhibitory activity.

OH OCH_3 CHO   OH CO_2CH_3

phytol **1**   vanillin **2**   methyl *p*-hydroxycinnamate **3**

After repeated attempts, the method shown in Table 14.1 was developed. The whole procedure was greatly simplified by the first two steps, i.e., separation of the crude MeOH extracts into $CHCl_3$-soluble and insoluble fractions followed by removal of *n*-hexane-soluble materials from the $CHCl_3$-soluble fraction. The weight and germination inhibition activity of each fraction are given in the table. Active materials were obtained in the benzene- and ether-soluble fractions. In this separation scheme, compounds having weak activity such as phytol, vanillin and methyl *p*-hydroxycinnamate were concentrated into the *n*-hexane-soluble fraction and hence were easily removed. Being extremely simple and economical, this separation method seems quite general, and should be valuable for the isolation of physiologically active natural products from whole plants if an established bioassay is available.

TABLE 14.1 Separation of active fractions (values in parentheses show the germination ratio of lettuce seeds at 1000 ppm).

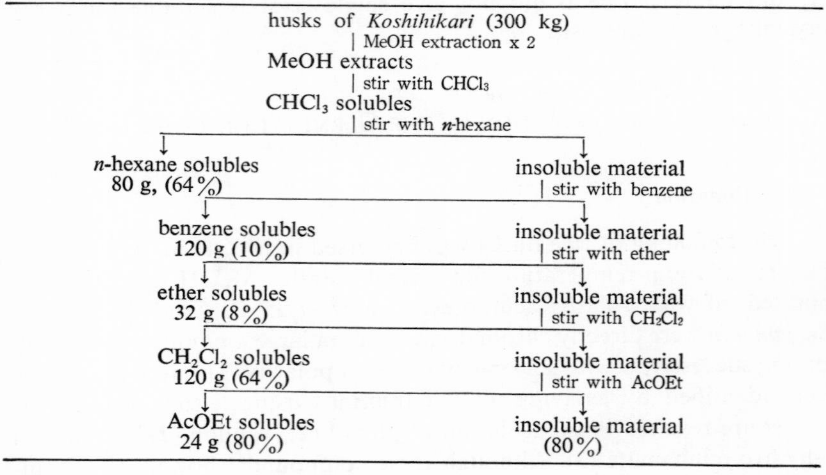

Each fraction was separated into acidic and neutral components, all of which were assayed for germination inhibition activity. The experimental results are summarized in Table 14.2. The neutral materials showed strongest activity, indicating that the germination inhibitors were present in the neutral fractions. The weak activity in the $NaHCO_3$ fraction of the ether-soluble material was attributable to the presence of a weak inhibitor, *p*-coumaric acid, which was isolated by column chromatography on active charcoal, eluting with aqueous acetone. The increased activity of the neutral fraction of *n*-hexane-soluble material compared with the original *n*-hexane-soluble fraction was due to the concentration of weakly active compounds such as phytol, vanillin and methyl *p*-hydroxycinnamate into this fraction.

TABLE 14.2 Germination ratios (%) of acidic and neutral fractions of solvent extracts at 1000 ppm.

solvent / extract	*n*-hexane	benzene	ether	CH_2Cl_2
original extract	64	10	8	64
$NaHCO_3$ solubles	100	32	32	100
NaOH solubles	100	94	100	100
neutral fraction	23	3	6	41

Neutral Components of the Benzene-soluble Fraction

The neutral fraction of the benzene-soluble material was chromatographed on silica gel and each fraction was assayed for inhibitory activity. Fig. 14.3 shows the germination inhibition activities using lettuce seeds while Fig. 14.4 shows the inhibition of the growth of rice seedlings. The same fractions are clearly responsible for both inhibitions. It appears that the dormancy substances might be present in fractions 8 to 12. Repeated column chromatography and recrystallization resulted in the isolation of about 1.0 g of momilactone A from fr. 8 and about 400 mg of momilactone

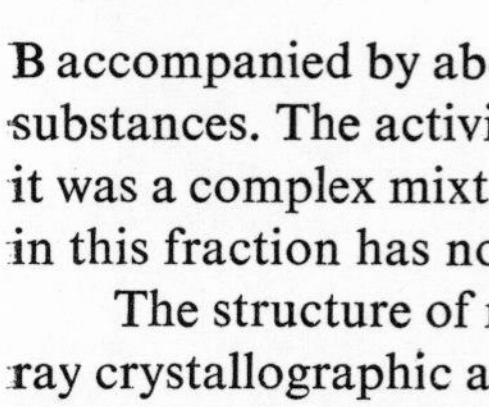

momilactone A **4** 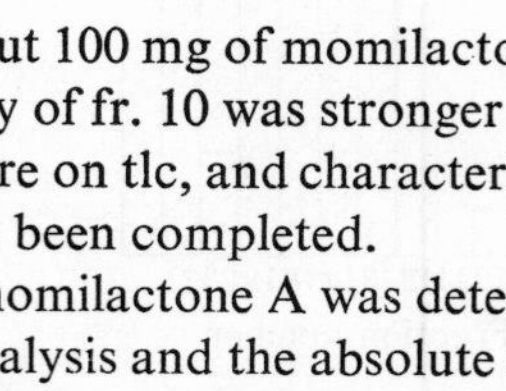 momilactone B **5** 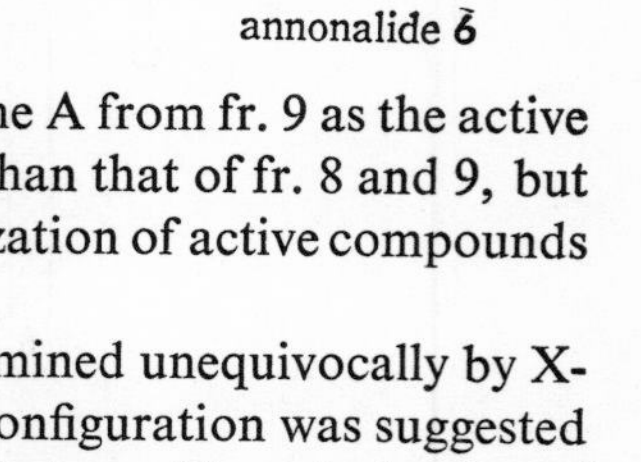 annonalide **6**

B accompanied by about 100 mg of momilactone A from fr. 9 as the active substances. The activity of fr. 10 was stronger than that of fr. 8 and 9, but it was a complex mixture on tlc, and characterization of active compounds in this fraction has not been completed.

The structure of momilactone A was determined unequivocally by X-ray crystallographic analysis and the absolute configuration was suggested by ord and cd measurements of the hydroxy ketone (**7**), a derivative of momilactone A. The structure of momilactone B was deduced on the basis of the nmr data (Table 14.3) and the proposed structures were supported by the results of the chemical degradation reactions summarized in Fig. 14.8.

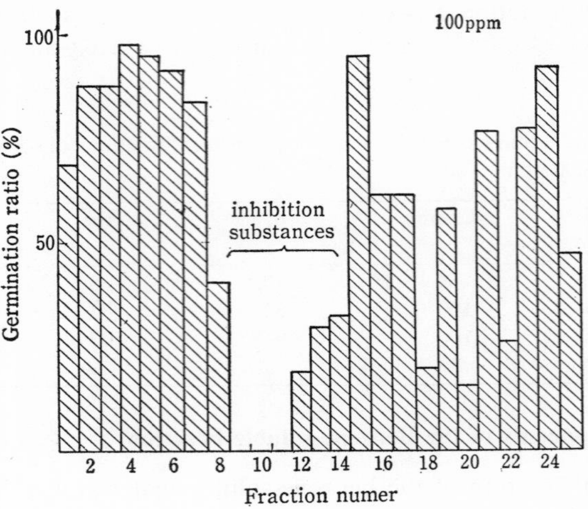

Fig. 14.3 Germination tests with lettuce seeds of fractions of the benzene-soluble neutral material at 100 ppm.

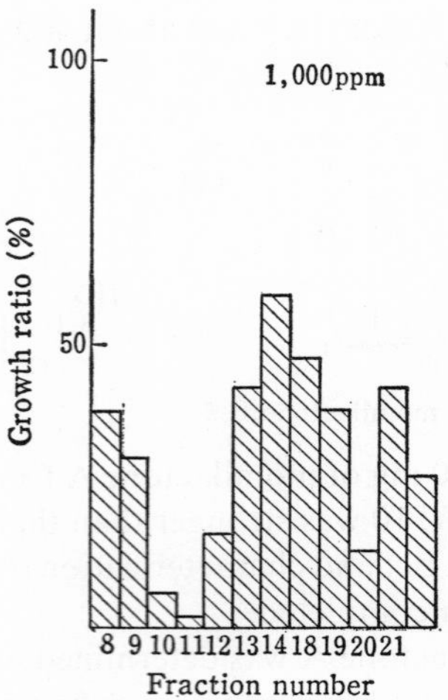

Fig. 14.4 Effect of each fraction of the benzene-soluble neutral material on the growth of roots of rice (*Kinnampu*) at 1000 ppm.

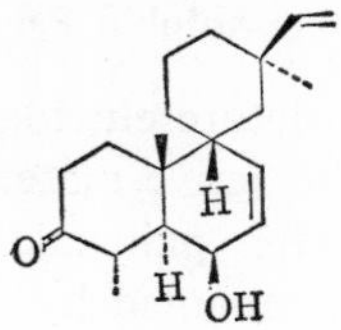

7

TABLE 14.3 Spectroscopic data for momilactones A and B.

pmr

Proton	Momilactone-A Chemical shift, multiplicity, and coupling constants (Hz)	Momilactone-B Chemical shift, multiplicity, and coupling constants (Hz)
5-H	2.32, d, (5.0)	2.20, dd, (7.1 and 2.1)†[1]
6-H†[1]	4.83, t, (5.0)†[1]	4.94, dd, (7.1 and 4.5)†[1]
7-H†[1]	5.70, d, (5.0)†[1]	5.68, d, (4.5)†[1]
14-H_2	2.03 and 2.22, d, (12.5)	overlapped with others
15-H	5.85, dd, (17 and 10.8)	5.83, dd, (17.6 and 10.5)
16-H	4.93, dd, (10.8 and 1.2)	4.92, dd, (10.5 and 1.2)
16-H	4.95, dd, (17 and 1.2)	4.95, dd, (17.6 and 1.2)
17-H_3	0.90, s	0.87, s
18-H_3	1.52, s	1.40, s
20-H_3	1.00, s	3.55, dd, (9.0 and 2.1) 4.07, bd, (9.0)†[1]

†[1] Assignments were confirmed by decoupling.

cmr†[2]

carbon number	Momilactone-A	Momilactone-B	carbon number	Momilactone-A	Momilactone-B
1	34.89†[3]	28.82†[3]	11	24.03	24.81†[3]
2	31.25†[3]	26.45†[3]	12	37.25	37.25
3	205.00	96.59	13	40.16	39.98
4	53.57	50.36	14	47.57	47.44
5	46.67†[4]	43.02†[4]	15	148.95	148.83
6	73.17	73.78	16	110.12	110.18
7	114.06	114.00	17	21.97†[5]	21.90
8	147.98	146.70	18	21.42†[5]	18.99
9	50.24†[4]	44.65†[4]	19	174.25	180.52
10	32.52	30.76	20	21.78†[2]	72.74

†[2] Spectra in $CDCl_3$ taken at 15.359 MHz on a JEOL JNM-TFT-100 spectrometer; chemical shifts in parts per million downfield from TMS.

†[3] Values in each vertical column may be reversed.

†[4] As †[3].

†[5] As †[3].

Neutral Components of the Ether-soluble Fraction

After completion of the structure elucidation of momilactones, our efforts were directed towards the active materials in the neutral fraction of the ether-soluble material. After painstaking repeated procedures of separation and purification accompanied by bioassay, we isolated a few milligrams of a white crystalline substance with inhibitory activity towards germination. We named it ineketone, and attempted to obtain a single crystal suitable for X-ray analysis. However, our attempts failed.

Infrared spectroscopy of ineketone suggested the presence of hydroxyl groups and a conjugated carbonyl group, while nmr showed a multiplet at 3.5 ppm due to proton(s) geminal to hydroxyl group(s). In order to simplify the multiplet, D_2O was added. However, the signal pattern was found to have changed completely. Remeasurement of the ir spectrum and tlc indicated that the compound had decomposed completely, being converted into a complex mixture. Presumably trace amounts of hydrochloric acid generated from D_2O and $CDCl_3$ had destroyed the ineketone. We had lost our whole sample within a few minutes.

Our co-workers, Tsunakawa, Sasaki and Aizawa took up the challenge and started extracting a fresh batch of rice husks. The neutral fraction of the ether-soluble material was divided into three portions by SiO_2 column chromatography, eluting with 10:1 and then 3:1 mixtures of benzene-ethyl acetate followed by ethyl acetate. The first portion obtained from 10:1 benzene-ethyl acetate showed relatively strong inhibition activity. After repeated column and thin layer chromatographies, the activity was shown to be due to the presence of a mixture of momilactones (100 mg). The intermediate portion showed no strong activity. Bioassay of the last portion indicated the existence of substances having stronger activity than momilactones. This was a complex mixture, from which about 15 mg of ineketone (**8**), 3 mg of momilactone C (**9**) and 13 mg of *S*-(+)-dehydrovomifoliol (**10**) were isolated by preparative tlc and high pressure liquid chromatography.

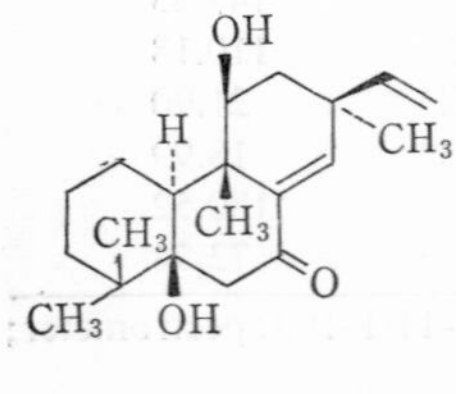

ineketone **8** 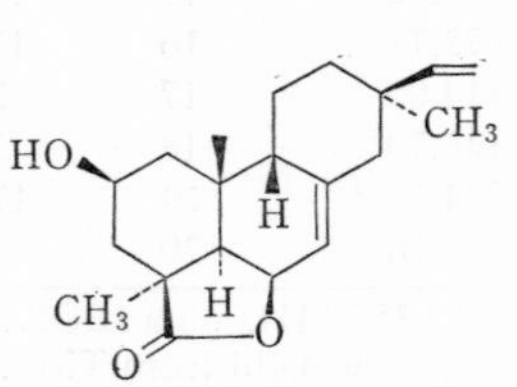 momilactone C **9** 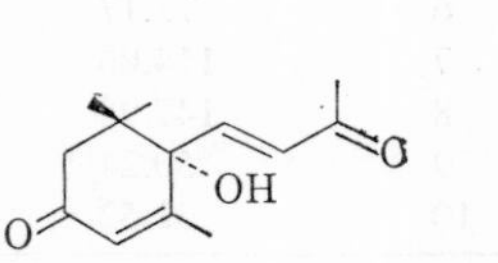 *S*-(+)-dehydrovomifoliol **10**

Thus, we finally characterized three compounds from the neutral fraction of the ether-soluble material. These compounds showed rather weak inhibition activity even at relatively high concentration and hence did not appear to be major factors in dormancy.

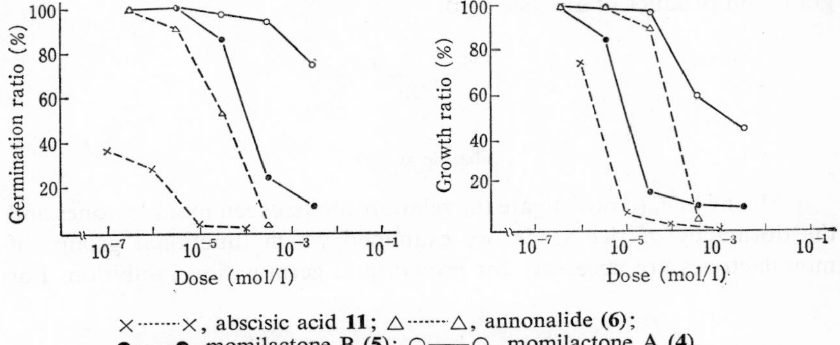

×·······×, abscisic acid **11**; △·······△, annonalide (**6**);
●——●, momilactone B (**5**); ○——○, momilactone A (**4**).

Fig. 14.5 (*left*) Dose Effects of various compounds on the germination of lettuce seeds.
Fig. 14.6 (*right*) Effects of various compounds on the growth of roots of rice.

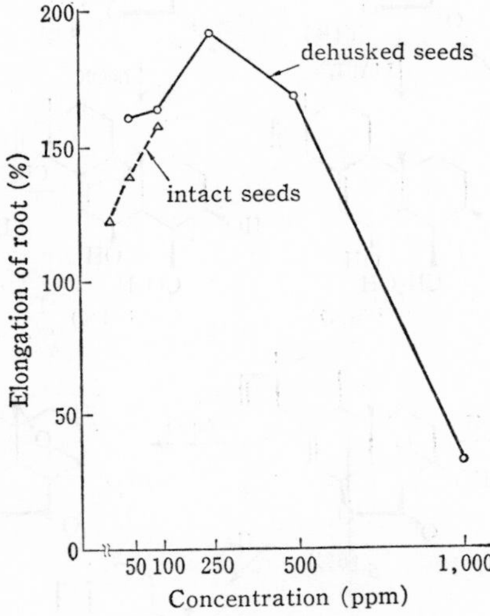

Fig. 14.7 Effect of ineketone on the root elongation of *Koshihikari.*

14.2 Structure-Activity Relationship of Momilactones

All the compounds from rice husks described so far are less active by 10^{-1} to 10^{-3} times as compared with the well known dormancy substance, abscisic acid (**11**). It therefore seemed possible that an abscisic acid analog might be involved in the dormancy of rice seeds, especially since we had identified *S*-(+)-dehydrovomifoliol (**10**), which could be a biodegradation product of abscisic acid.

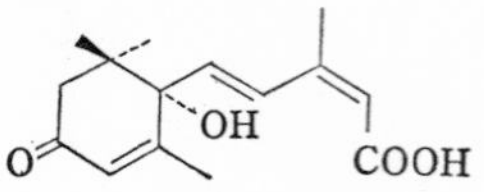

OH
O
COOH

abscisic acid 11

Meanwhile, to investigate the relationship between momilactones and the dormancy of rice seeds, we examined which functional groups of momilactones are necessary for growth and germination inhibition. For

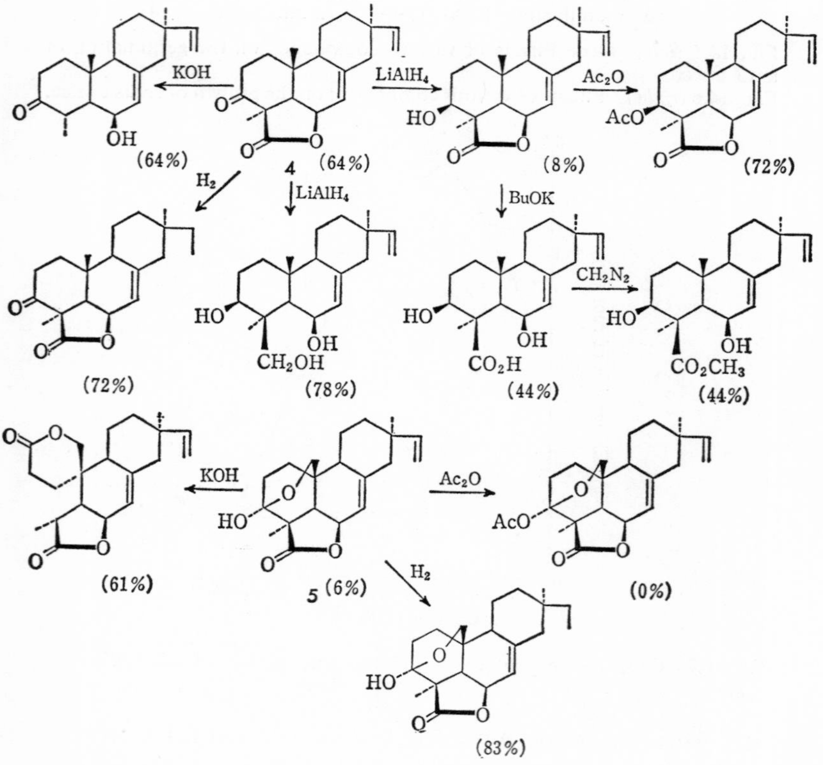

KOH LiAlH$_4$ Ac_2O
O OH (64%) O O 4 (64%) HO O O (8%) AcO O O (72%)
H_2 LiAlH$_4$ BuOK CH_2N_2
O O O (72%) HO OH CH_2OH (78%) HO OH CO_2H (44%) HO OH CO_2CH_3 (44%)
KOH Ac_2O
O O O O (61%) HO O O O 5 (6%) AcO O O O (0%)
H_2
HO O O O (83%)

Fig. 14.8 Chemical degradation of momilactones and germination inhibition activities of the derivatives (values in parentheses show the relative inhibitions at 100 ppm).

this purpose, chemical transformation of momilactones was carried out (Fig. 14.8). Each derivative was tested to determine its effects on the germination of lettuce seeds and the root growth of rice. The relative germination ratio is indicated in parentheses under each derivative.

The results of the bioassay showed that these two physiological activities are linked: that is, all the derivatives which strongly inhibit the germination of lettuce seeds possess strong inhibitory activity towards the root growth of rice. When the carbonyl group of momilactone A was reduced to a hydroxyl group, the activity of the derivative increased to the same level as that of momilactone B. A marked decrease of the activity was observed when the hydroxyl group was protected with an acetyl group. In contrast, acetylation of momilactone B enhanced the inhibition activity.

These results suggest that if the dormancy of rice seeds involves momilactones, they may exist in rice husks as the hydroxy derivative in the case of momilactone A and as esters in the case of momilactone B. Annonalide (**6**),[6] a natural product possessing a partial structure similar to that of momilactone B shows almost the same physiological activity (Figs. 14.5 and 6).

14.3 Phytoalexin Activity of Momilactones

Independently of our study on germination inhibitors, a British group investigated phytoalexins produced by rice blast and found that momilactones identical with ours were present as phytoalexins in rice plants suffering from rice blast.[7] We have also examined the effects of momilactones and two derivatives, momilactone A alcohol (**12**) and momilactone B acetate (**13**) on spore germination, germ tube growth and appressorium formation using several kinds of rice blast strains (*Pyricularia oryzae*).[8]

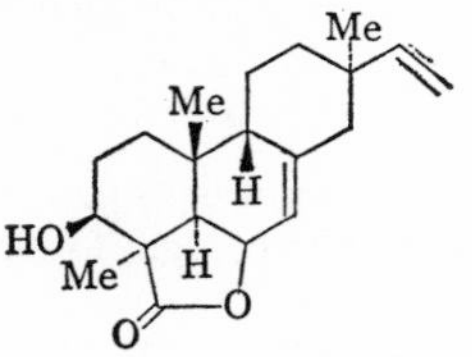

Me, Me, H, HO, Me, H, O, O

momilactone A alcohol **12**

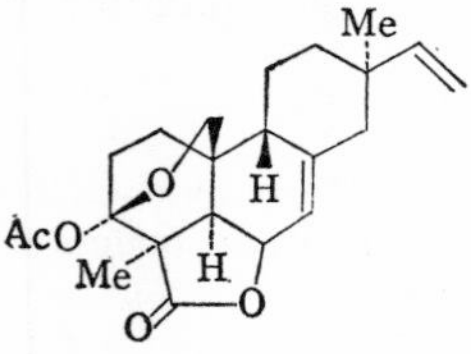

Me, O, H, AcO, Me, H, O, O

momilactone B acetate 13

As shown in Figs. 14.9 and 14.10, all the compounds we examined inhibited spore germination and germ tube growth of the tested strains,

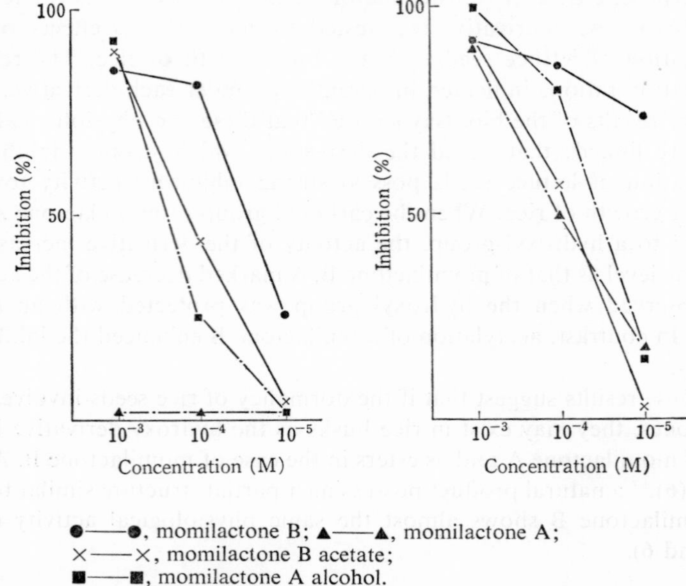

Fig. 14.9 (*left*) Effects of various compounds on spore germination of *Pyricularia oryzae* (F 67–54).
Fig. 14.10 (*right*) Effects of various compounds on germ tube growth of *Pyricularia oryzae* (F 67–54).

as typified by F67–54. The effects of the compounds decreased with time and disappeared almost completely after 9 hr in the case of momilactone A. Momilactone B showed the strongest activity among the four compounds. The activity of momilactone B acetate was next highest, then momilactone A alcohol, and finally momilactone A was least active.

It is worth mentioning that the order of the activity on rice blast is similar to that of germination inhibition of lettuce seeds. As summarized in Table 14.4, momilactone A promoted appressorium formation of the strains.

TABLE 14.4 Effects of momilactones A and B on appressorium formation of *Pyricularia oryzae* (values are percentages of appressorium formation).

Momilactone A

Isolate / Concentration (M)	F67–54	ken60–19	ken62–89
10^{-3}	55.7	66.7	43.2
10^{-4}	33.7	70.8	75.8
10^{-5}	50.3	14.2	8.9
control	14.0	14.0	17.2

TABLE 14.4—*continued*

Momilactone B

Isolate Concentration (M)	F67–54	ken60–19	ken62–89
10^{-3}	22.8	9.4	7.4
10^{-4}	34.9	17.7	15.8
10^{-5}	2.2	9.1	6.4
control	12.9	15.8	13.5

[Experimental procedure 1] Bioassay with lettuce seeds

Each sample is dissolved in MeOH or acetone to make a solution of known concentration. For example, 1 mg of a sample in 1 ml of solvent corresponds to 1000 ppm. A round filter paper 2.8 cm in diameter is placed on the bottom of a Petri dish 3 cm in diameter, and 0.3 ml of the solution is applied. After removing the solvent in the Petri dish by evaporation in a desiccator under reduced pressure, 0.3 ml of aq. Tween 80 prepared by dissolving 0.1 g of commercial Tween 80 in 1 l of distilled water is dropped onto the dried filter paper. Each Petri dish is transferred into a large Petri dish which contains small amounts of aq Tween 80 sufficient to prevent concentration changes of the aq. Tween 80 solution in the small dishes due to evaporation. The whole is kept overnight at room temperature. The next day, fifty seeds of lettuce (*Lactuca sativa* L. cv. Wayahead: Ferry Morse Seed Co., Mountain View, Cal.) are arranged on the filter paper so as to be immersed in the aq. Tween 80 solution. The dishes are kept in the dark at 20° C, and the number of germinated seeds is recorded at 24 hr and 48 hr. The relative germination ratio is calculated based on the number of germinated seeds in a Petri dish which contains 0.3 ml of aq. Tween 80 without any sample (control). To examine the growth-promotive effect of a sample, the germination experiment is carried out at 30° C.

[Experimental procedure 2] Bioassay with rice seeds

First day: A suitable amount of rice seeds (intact or dehusked depending on the nature of the experiments) is immersed in 75% EtOH for 30 sec and then transferred into 0.2% aq. mercuric chloride solution. After 3 min, the seeds are washed with fresh water for 1 hr. The seeds are then sown on a filter paper in a Petri dish and sufficient water is added to immerse the seeds. The seeds are kept in the dark at 30° C for 48 hr.

Second day: A definite amount of sample is dissolved in MeOH or acetone to prepare the solution of known concentration. An aliquot (1 ml) of this solution is put on a filter paper in a tube 2 cm in diameter and 4 cm in height, and the solvent is completely removed *in vacuo*. After adding 1 ml of 100 ppm aq. Tween 80, the stoppered tube is kept at room temperature overnight.

Third day: Uniformly germinated seeds are collected and seven of them are placed on the filter paper in the tube. The tube is kept in the dark for 2 days at 30° C and then for 3 days in the light at 26° C. The second leaf and root lengths of the grown plant are measured and the average values calculated.

[Experimental procedure 3] Separation of active compounds from ether-soluble material

The neutral components (8 g) of the ether-soluble material were column chromatographed on SiO_2 (600 g) and eluted successively with benzene–AcOEt (10:1), benzene–AcOEt (3:1) and finally with AcOEt. The first solvent eluted 200 mg of a mixture of steroids (sitosterol, stigmasterol, campesterol), methyl *p*-coumarate (150 mg) and a mixture of momilactones A and B (100 mg). The 3:1 mixture fraction afforded tricin (20 mg), $C_{17}H_{14}O_7$, mp 288–289° C from MeOH. The AcOEt fraction (1.9 g), after column chromatography on SiO_2 (190 g) and preparative tlc (SiO_2) with benzene-AcOEt (4:1), CH_2Cl_2–MeOH (60:1), and then $CHCl_3$–EtOH (50:1), afforded an active mixture which was again chromatographed on SiO_2 with benzene–AcOEt (4:1) to elute ineketone (**8**) (ca. 10 mg). The active mixture was again chromatographed on SiO_2 and then subjected to preparative tlc, yielding two active fractions, I (50 mg) and II (50 mg). Fraction (I) was separated by hplc (Hitachi gel column, $CHCl_3$ + 2% EtOH) to obtain momilactone C (**9**) [3 mg, mp 227–228° C from *n*-hexane: $CHCl_3$ (1:1)].

Fraction (II) was repeatedly recycled on hplc (μ-Porasil column, 1.6% EtOH in $CHCl_3$) to yield 13.4 mg of *S*-(+)-dehydrovomifoliol (**10**), which corresponds to HLC-3 in Fig. 14.11.

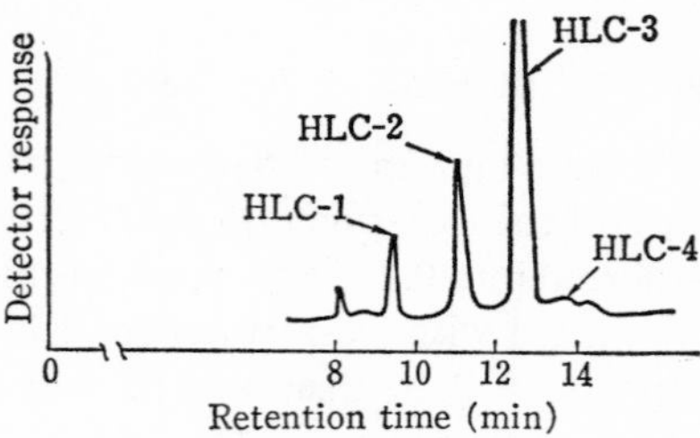

Fig. 14.11 High pressure liquid chromatography of active fraction II.

Two other compounds were also isolated, $C_{15}H_{20}O_3$ (HLC-1) (4.8 mg) and $C_{19}H_{26}O_3$ (HLC-2) (5.9 mg). The structures of the latter two materials were not investigated. *S*-(+)-Dehydrovomifoliol (**10**), $C_{13}H_{18}O_3$, oil, $[\alpha]_D^{25}$ + 142.7°. Ineketone (**8**), mp 206–209°C from *n*-hexane-EtOH. Bioassay data for these two compounds are shown in Fig. 14.12.

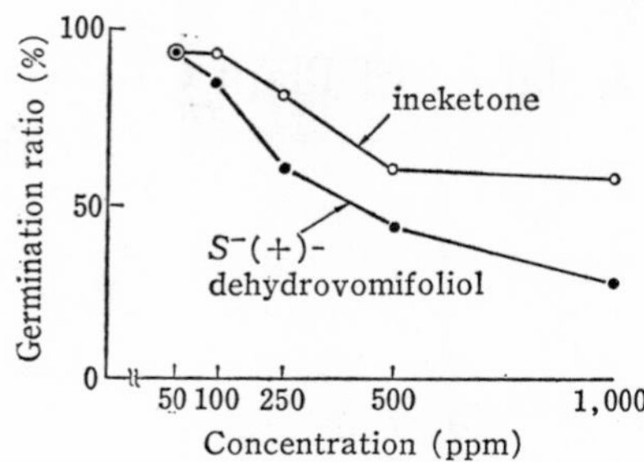

Fig. 14.12 Effects of ineketone and dehydrovomifoliol on the germination of lettuce seeds.

References

1) For a leading reference, see *Chemical Regulation of Higher Plants* [*Shokubutsu no Kagaku Chosetsu* (in Japanese)], **2**, 84 (1967).
2) E.H. Roberts, *J. Exptl. Botany*, **12**, 430 (1961).
3) For example, see T. Ito, *Dormancy and Germination of Higher Plants* [*Shokubutsu no Kyumin to Hatsuga* (in Japanese)], University of Tokyo Press (1975).
4) N. Takahashi, *Rept. Inst. Agr. Res., Tohoku University*, **19**, 1 (1968).
5) T. Kato, C. Kabuto, N. Sasaki, M. Tsunakawa, H. Aizawa, K. Fujita, Y. Kato, Y. Kitahara, N. Takahashi, *Tetr. Lett.*, **1973**, 3861; T. Kato, M. Tsunakawa, N. Sasaki, H. Aizawa, K. Fujita, Y. Kitahara, N. Takahashi, *Phytochemistry*, **16**, 45 (1977); T. Kato, H. Aizawa, M. Tsunakawa, N. Sasaki, Y. Kitahara, N. Takahashi, *J. Chem. Soc. Perkin I*, **1977**, 250; M. Tsunakawa, M. Ohba, C. Kabuto, N. Sasaki, T. Kato, Y. Kitahara, *Chem. Lett.*, **1976** 1157.
6) The annonalide was kindly provided by Professor F. Pelizzoni.
7) C. Cartwright, P. Langcake; R.J. Pryce, D.P. Leworthy, J.P. Ride, *Nature*, **267**, 511 (1977).
8) T. Yamanaka, T. Namai, T. Kato, N. Sasaki, N. Takahashi, *Ann. Phytopath. Soc. Japan*, **46**, 494 (1980).

CHAPTER **15**

Isolation of Plant Growth Inhibitors from Fruits

Naturally occurring plant growth inhibitors, which retard such physiological processes as root and stem elongation, seed germination, bud opening and so on, endogenously regulate plant development and differentiation in cooperation with auxins, gibberellins and/or cytokinins. The inhibitors are also excreted from some plants into the environment to retard the germination and growth of other plants directly or indirectly. In other words, they promote plant succession in wild fields, or cause so-called soil sickness in orchards and farms, playing a part in the regulation of the ecological balance in the plant kingdom. These phenomena are called "allelopathy."

The chemical structures of plant growth promotors isolated from higher plants so far have been restricted to derivatives of indole, *ent*-gibberellane and 6-substituted adenine, though many kinds of compounds, such as aromatic compounds (simple phenols, aromatic acids, phenylpropanes and coumarins), flavonoids, terpenes and fatty acids, are known to have growth inhibitory activity.[2,3] Some of them are thought to be specific secondary metabolites of some plant species, and may be responsible for allelopathic phenomena.

One of the most active inhibitors isolated from plants is abscisic acid,[4] which strongly inhibits the growth of young shoots, leaves, stems and roots, and promotes the abscission of plant organs, dormancy and senescence. One of its most striking effects is its ability to cause the closure of stomata. Abscisic acid is widely distributed in dicotyledons, monocotyledons and pteridophytes, and can thus be regarded as one of the plant hormones. Most plant tissues contain between 20 and 100 μg/kg fresh weight of abscisic acid, but changes in the concentration in an organ occur during the course of development. Fruits are a rich source[2,5] (10 mg/kg fresh weight in avocado fruit pulp)[6].

The substances which cause allelopathic phenomena are known to be present in fruits as well as in roots and leaves. For example, fruits of *Sorbus aucuparia* contain parasorbic acid, which shows growth inhibitory activity towards other plants in the vicinity. This is one reason why plants which do not grow rapidly can proliferate. Many kinds of plant growth inhibitors are expected to occur in fruits.[7]

In this chapter, we describe in detail the isolation procedures for abscisic acid **1**, *trans*-abscisic acid **2**, a new inhibitor, cucurbic acid **3**, which has a cyclopentane moiety, and two cucurbic acid derivatives **4** and **5** from seeds of *Cucurbita pepo* L.[8]

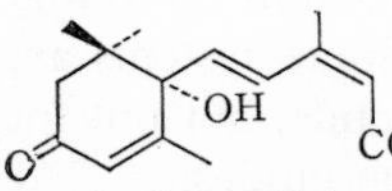

abscisic acid **1**

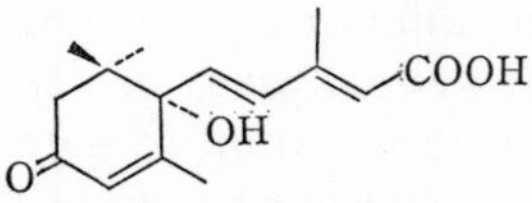

trans-abscisic acid **2**

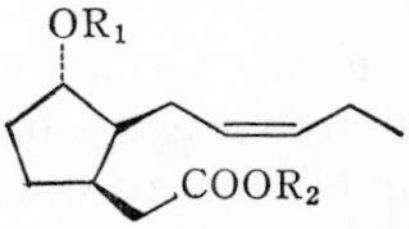

cucurbic acid **3** ($R_1=R_2=H$)
glucosylcucurbic acid **4**
(R_1=glucose, R_2=H)
methyl glucosylcucurbate **5**
(R_1=glucose, $R_2=CH_3$)

15.1 Bioassay

Trace amounts of biologically active compounds cannot be isolated from crude extracts without guidance by appropriate bioassays. Each isolation step must be considered critically, to avoid degradation of the active compounds, because we cannot estimate their chemical properties in advance. Bioassays can generally detect very small amounts of active compounds, which would be insufficient to produce spots or peaks on chromatograms. The presence of functional groups of the active compound can be investigated in terms of the biological activity of the derivatives obtained by acylation, methylation, hydrogenation, bromination and the like.[9]

Many bioassays require precise techniques and a long time to obtain accurate results, and may be the critical steps in the isolation procedures. Bioassays for guiding the isolation of the active compounds should therefore be simple as well as able to detect the activities in many fractions at the same time. Bioassays suitable for guiding the isolation of plant growth inhibitors are summarized in Table 15.1. This table includes assays

TABLE 15.1 Bioassays used for the detection of plant growth inhibitory activities.

Bioassay	Reference	Bioassay	Reference
Avena curvature test	10	*Phaseolus* embryonic axis expansion test	17
Avena straight growth test	11	Cotton explant test	18
Pea hook segment growth test	12	Rice seedling test	19,20
Slit-pea curvature test	13	Azuki bean seedling test	21
Wheat coleoptile growth test	14	*Raphanus* germination test	21
Lettuce hypocotyl growth test	14	Leaf mustard†3 germination test	21
Lepidium†1 germination test	14	Pea stem straight growth test	22
Hedera†2 root formation test	15	Lettuce germination test	23
Raphanus root formation test	16	Cucumber hypocotyl growth test	24

†1 *Lepidium sativum* L.
†2 *Hedera helix* MIQ.
†3 *Brassica juncea* CZERN. et COSS.

for the formation of adventitious roots, since some growth inhibitors promote their formation. No inhibitor can be expected to exhibit inhibitory activity on every bioassay system listed in this table. Some systems are suitable for quantitative determination of pure compounds, but not for crude compounds. These should not be used as isolation guides.

We will describe in detail the isolation procedure for plant growth inhibitors from seeds of *C. pepo*, guided by the rice seedling test, for which two methods are available. One is a dip method,[20] in which the sample is made available to the seedlings through the roots. The other is a microdrop method,[19] in which the sample solution is placed on the surface of leaf buds as a microdrop. The latter is not sensitive to growth inhibitors (abscisic acid, fusaric acid and so on), and therefore is not suitable for their detection. The former is effective for the isolation of growth inhibitors, however. Any variety of rice can be used for detecting inhibitors. Dwarf rice ("*tan-ginbozu*" or "*kotake-tamanishiki*") is recomended for the simultaneous detection of both inhibitors and promotors. At a certain concentration of a mixture of gibberellins and abscisic acid, the second leaf sheath is elongated and the root growth is inhibited, thus indicating the fraction to contain both promotors and inhibitors.

[Experimental procedure 1] Rice seedling test (dip method)

Seeds of rice, after removing the seeds which float on a 7% sodium solution (sp. gr. 1.05), are soaked in ethanol for 5 min, washed with water three times and then dipped in a 1% alkaline sodium hypochlorite solution for 30 to 60 min, followed by washing with running water for 3 hr. The seeds, in a Petri dish filled with water, are kept in a chamber (30 $\pm$ 1°C, 2500 lux) for 3 days. Seven germinated seeds are taken into a glass tube containing a sample solution, and the tube is tightly covered with polyethylene film. The tube is left in the chamber described above for 7 days, then the lengths of the second leaf sheaths are compared with those of seedlings grown in water.

15.2 Extraction and Preliminary Separation of Crude Extracts

Preparation of Sample

The fruits should be harvested at the time when they contain the largest amount of biologically active compounds. If the active compounds are found to be located in a certain organ of the fruit, mechanical separation of the organ from others prior to solvent extraction makes the subsequent isolation procedures much simpler.

Extraction with Solvents

In order to extract biologically active compounds from plant organs, enzymes should first be deactivated by heating or by soaking in appropriate solvents to avoid enzymatic decomposition. Nitsch[25)] investigated the effects of various solvents on the prevention of artichoke tuber browning during auxin extraction. As shown in Table 15.2, methanol yielded white tissues without the slightest trace of browning. Acetone and absolute

Table 15.2 Prevention of tissue browning during auxin extraction.[25)]

Extraction solvent	Browning after 18 hr†[1]
Absolute methanol	0
Absolute ethanol†[2]	+
Acetone	+
Petroleum ether (bp 30–65° C)	+++
Chloroform	+++++
Chloroform + 8-hydroxyquinoline (2 mg/ml)	++++
Ethyl acetate	+++++++
Ether	+++++++
Ether (95%) + absolute methanol (5%)	++++
Ether (70%) + absolute methanol (30%)	+++
Ether (50%) + absolute methanol (50%)	++
Ether (70%) + absolute ethanol (30%)	+++
Ether + $Na_2S_2O_4$ (1 mg/ml undissolved)†[3]	0
Ethyl acetate (50%) + absolute methanol (50%)	++

†[1] Jerusalem artichoke tuber tissues left overnight in an ice-box at about 5°C.
†[2] The tissues are shrivelled after extraction.
†[3] Dissolves when the fresh tissue is put in the mixture.
(Source: ref. 25. Reproduced by kind permission of J.P. Nitsch and Butterworths Scientific Publications.)

ethanol were effective, but tissues in ether and ethyl acetate turned dark brown. Some solvents are known to react with components. For example, α-D-glucosyl abscisate reacts readily with methanol to yield the methyl ester under acidic conditions. Gallic acid is also methylated on leaving it in methanol at room temperature. Acetone readily forms adducts under basic conditions.

In the case of extraction of plant growth substances from a large amount of seeds, methanol is added to the fresh seeds to give a final concentration of 70–75% methanol, which is known to deactivate enzymes. The volume of the methanol required can be calculated from the water content of the seeds (ca. 90%). The fresh seeds of *Lupinus luteus* L. were soaked in methanol without using a blender for several weeks at a low temperature. This extraction procedure was repeated twice. Trituration of the seeds for extraction gave active compounds quantitatively together with a huge amount of inactive components, making the isolation procedure very tedious.[26] The extracts were concentrated in a glass evaporator under reduced pressure.

Fractionation of Extracts

The extracts are generally partitioned into acidic, weakly acidic, neutral and basic fractions. In the case of isolation of abscisic acid from seeds of *L. luteus*,[27] the aqueous concentrate obtained from methanol extracts was extracted with ethyl acetate six times, after adjustment to pH 3 with 6 N HCl. The ethyl acetate layer was partitioned into 1 M phosphate buffer (pH 8) rather than into sodium bicarbonate or sodium carbonate solution since the buffer is easily separated from the ethyl acetate layer, and does not produce carbon dioxide on acidification. This process should be completed as quickly as possible at a low temperature, because some compounds are unstable under basic or acidic conditions. The ethyl acetate layer must be washed thoroughly with water to remove the acid or alkali added. In our isolation procedure described below, the extract was not treated under basic conditions, which might modify phenolic inhibitors.

[Experimental procedure 2] Extraction and fractionation

The fresh seeds of *C. pepo* (50 kg) were soaked in methanol (100 l), and kept in a cold room (5° C) for several weeks. After filtration, the residue was again soaked in methanol (60 l). The extracts were combined and concentrated at 40° C under reduced pressure. The resulting aqueous concentrate (1.4 l), after adjustment to pH 3 with 6 N HCl, was partitioned into the same volume of ethyl acetate four times. The ethyl acetate layer was washed with water until no silver chloride precipitation appeared on addition of silver nitrate, then concentrated under reduced pressure to yield an oil (63 g), which contained acidic and neutral compounds.

15.3 Fractionation of Crude Extract and Detection of Biologically Active Components

It is very important to exclude biologically inactive components as far as possible at the first step of the isolation procedures. The countercurrent distribution method is useful for the exclusion of large amounts of inactive components,[28] but we used column chromatography, which is rather more troublesome, because we were interested in the plant growth inhibitors as well as the promotors. In this case the volume of silica gel added to the column was one-fifth or one-sixth of the usual column volume. Elution was carried out by increasing the polarity of the eluent (chloroform, ethyl acetate and methanol). Fig. 15.1 showed the effect of each fraction on the growth of rice seedlings, indicating the presence of at least 3 growth inhibitors at 20% and 50% ethyl acetate in chloroform and 10% methanol in ethyl acetate, as well as gibberellin-like substances at 80–100% ethyl acetate in chloroform, in the ethyl acetate extract obtained from seeds of *C. pepo*.

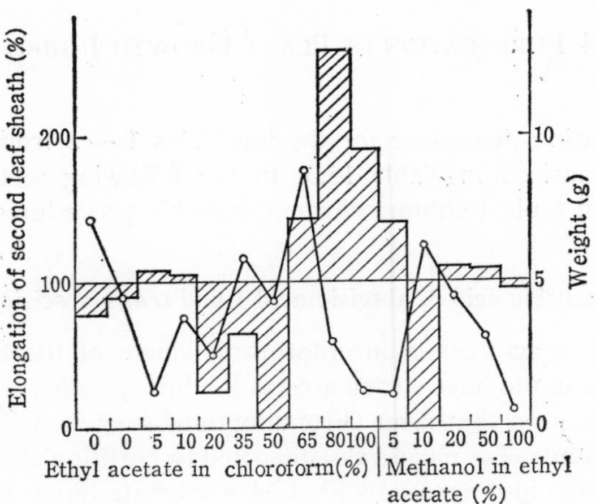

Fig. 15.1. The biological activity on rice seedlings and the weight (O——O) of each fraction obtained by silica gel column chromatography of the ethyl acetate extract.

[Experimental procedure 3] Preliminary fractionation of the crude extract by silica gel chromatography

A mixture of silica gel (600 g, ten times the weight of the sample) and Celite (600 g) suspended in chloroform, the first eluent for column chromatography, was poured into the column (10 × 100 cm) taking care to avoid air bubbles. The commercial Celite should be washed with water and acetone successively, followed by heating for activation. The column was washed with 6 l of chloroform (column volume, 3 l) and then left to settle overnight. Celite was added to speed up the elution. Silica gel (180 g) was added to the sample (63 g, the ethyl acetate extract) dissolved in methanol (40 ml), then the mixture was shaken to homogeneity, and dried under reduced pressure. The sample adsorbed on silica gel was added carefully to a column by slurrying the silica gel with chloroform in a beaker. The sample layer was settled by placing a filter paper or gauze on it.

The column was eluted with chloroform, then successively with chloroform containing 5, 10, 20, 35, 50, 65, 80 and 100% ethyl acetate, and with ethyl acetate containing 5, 10, 20, 50 and 100% methanol in 3 l fractions. The volume of each fraction corresponded to that of the silica gel added to the column, but 6 l of chloroform was used as the first eluent, since the seed extract contained a large amount of lipids. Two ml of each fraction was used for the rice seedling test described above. The results and the weight of each eluate are summarized in Fig. 15.1.

15.4 Purification of Plant Growth Inhibitors

The isolation procedure for the inhibitors **1** ~ **5** from seeds of *C. pepo* is summarized in Table 15.3. In the following sections, we describe in detail typical column chromatographic procedures used in our experiments.

Isolation of cucurbic acid 3, abscisic acid 1 and *trans*-abscisic acid 2

The 50% ethyl acetate in chloroform eluate of the first silica gel column was found to have strong growth inhibitory activity. The behavior of the inhibitors on thin layer chromatograms suggested that they were acidic compounds. The eluate was subjected to partition chromatography on Celite (200 g) impregnated with 1 M phosphate buffer (pH 5.6). The column was eluted in 500 ml fractions with stepwise gradients of *n*-butanol in benzene. The 2.5–5% *n*-butanol eluate (600 mg) showed growth inhibitory activity in the rice seedling test. The procedure for partition chromatography is described in detail in "Experimental procedure 5."

TABLE 15.3 Isolation procedure for plant growth inhibitors from seeds of *Cucurbita pepo*.

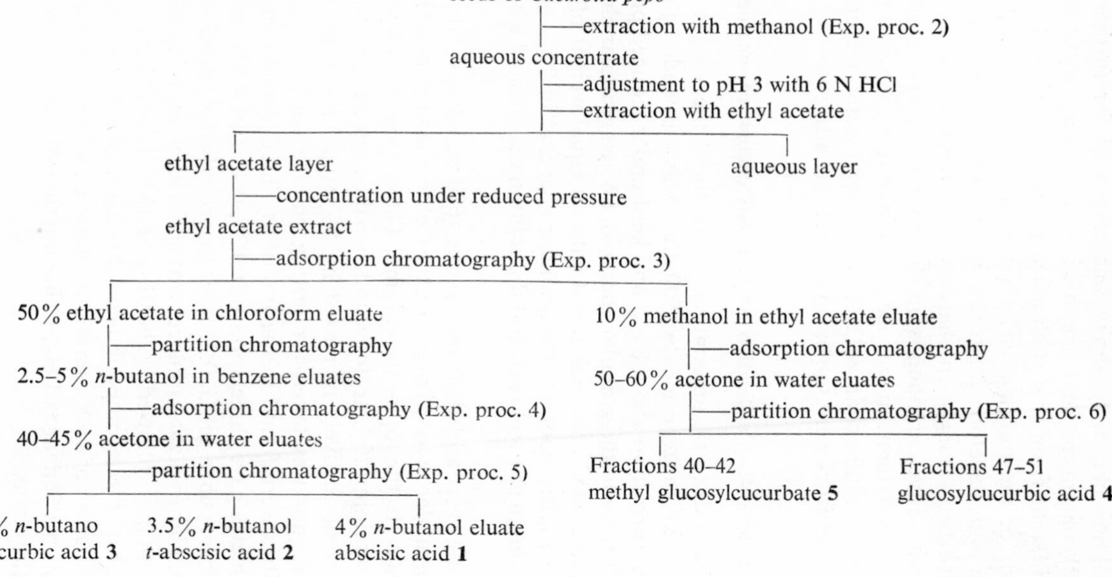

Charcoal column chromatography was employed for further purification. Charcoal column chromatography, in which the adsorption mechanism is quite different from that with silica gel and alumina, has been successfully used for the purification of gibberellins. Elution is usually carried out with water-acetone mixtures, but aromatic compounds are known to be adsorbed strongly on charcoal. It is therefore necessary to test the recovery of active compounds on elution with water-acetone mixtures or other solvent systems. In the case of our inhibitors, the activity was easily eluted from a granular charcoal test column with water-acetone mixtures. Thus, our inhibitors were subjected to charcoal adsorption chromatography.

[Experimental procedure 4] Purification by charcoal column chromatography

Granular charcoal (6 g), amounting to ten times the weight of the 2.5–5% *n*-butanol in benzene eluate (600 mg), was suspended in 20% acetone in water, warmed to about 80° C and kept under reduced pressure to remove air adhering to the granular charcoal. (When granular charcoal is not available, activated charcoal mixed with an equal weight of Celite can be employed.) After cooling, the granular charcoal in 20% acetone was added to a column (20 × 1 cm), and the column was washed with 20% acetone.

Celite (500 mg) was added to the sample (600 mg) dissolved in acetone. After drying under reduced pressure, the Celite was suspended in 20% acetone and applied to the top of the column. The column was eluted with 30 ml of water containing 20% acetone, and then the acetone concentration of the eluent was successively increased by 5% steps. From each fraction, a 0.5 ml aliquot was taken for testing on rice seedlings, and the inhibitory activities were found in the fractions eluted with 40 and 45% acetone. Both eluates gave three spots (*Rf* 0.36, 0.43 and 0.53) on a thin layer chromatogram (silica gel G, 0.25 mm; isopropyl ether: acetic acid = 19:1), detected under uv light as fluorescent spots after spraying with 5% sulfuric acid in ethanol followed by heating at 130° C for 3 min.

The behavior of these compounds on thin layer chromatograms suggested that the isolation of each compound might be achieved by partition chromatography on a Sephadex LH_{20}-Celite mixture; the latter was added to improve the flow rate.

[Experimental procedure 5] Purification by partition chromatography

The solvents (benzene and *n*-butanol) used for this chromatography must be saturated with 1 M phosphate buffer (pH 5.6) before use. Sephadex LH_{20} (50 g) was swollen in the buffer for 2 hr at room temperature. Celite (50 g) was added and mixed thoroughly using a spatula. The Sephadex LH_{20}-Celite mixture was suspended in benzene, the first eluent, and added to the column (70 × 2 cm) in thirty portions. Each portion of the mixture in the column was packed tightly using an appropriate plunger;[29] air did not enter from the bottom of the column, since the column bed was elastic. If necessary, the homogeneity of the column can be tested by eluting a small amount of azobenzene with benzene (described below).

The inhibitory 40–45% acetone eluates (100 mg) obtained in "Experimental procedure 4" were adsorbed on Celite (550 mg) and applied to the top of the column. The column was eluted with benzene containing increasing concentrations of *n*-butanol in 0.5% steps. The last few drops of each eluate were checked on thin layer chromatograms, and if they contained compounds, elution with the same solvent mixture was continued.

Each of the 3, 3.5 and 4% *n*-butanol eluates gave a spot on thin layer chromatography, yielding a new inhibitor **3** (60 mg, an oil) which was named cucurbic acid after the plant name, together with *trans*-abscisic acid **2** (2.5 mg, mp 159–161°C) and abscisic acid **1** (2.5 mg, 162–164°C), respectively.

Isolation of Glucosylcucurbic Acid 4 and its Methyl Ester 5

The 10% methanol in ethyl acetate eluate obtained in "Experimental procedure 3" was further purified by chromatography on granular charcoal. The method is described in "Experimental procedure 4." The 50 and 60% acetone in water eluates showed strong inhibitory activities on the growth of rice seedlings.

Many kinds of partition chromatographies have been described for the separation of gibberellins and related compounds. The partition chromatography which was devised by MacMillan *et al.*[30] was expected to be effective for purifying inhibitors having a carboxyl group. Biphasic solvent systems were devised starting from a mixture of light petroleum (bp 60–80°C), methanol and acetic acid. Ethyl acetate and sufficient water to maintain two phases were added until a suitable partition coefficient ($K_{a/0}$) between the aqueous and organic phases was obtained for a selected standard. The $K_{a/0}$ values were determined by gas-liquid chromatography of methylated aliquots from each phase. This method was applied for the separation of the inhibitors in the 50 and 60% acetone eluates described above.

[Experimental procedure 6] Purification by partition chromatography

A mixture of petroleum ether (bp 60–80°C, 1000 ml), ethyl acetate (400 ml), acetic acid (25 ml), methanol (200 ml) and water (35 ml) was thoroughly shaken in a separatory funnel. Sephadex LH_{20} (50 g) was swollen in the aqueous phase of the solvent mixture for 3 hr at room temperature.

Excess solvent was poured off and the thick slurry was added to a glass column (1.5 × 100 cm) containing the aqueous phase (20 ml). The column was washed with the aqueous phase until the Sephadex formed a firm bed. At this stage, the uniformity of the packed Sephadex could be checked by applying azobenzene in the aqueous phase. The azobenzene was eluted with the aqueous phase. If unsatisfactory, the column could readily be repacked at this stage.

The organic phase was then passed through the column until no more aqueous phase was eluted. The column uniformity was rechecked by carefully adding azobenzene in the aqueous phase (100 μl) onto the column through 3 mm of the organic phase. Elution was then continued with the organic phase. If the band is no wider than 2 cm at the bottom of the column, the column is suitable for the purification of several mg to several hundred mg of a sample.

The 50–60% acetone eluates (230 mg) obtained in "Experimental procedure 3" were dissolved in the aqueous phase, adsorbed on Sephadex LH_{20} (1 g), and placed on top of the column. For smaller amounts of material, the sample in the aqueous phase was applied to the top of the column with a microsyringe. Elution was continued with the organic phase. Fractions (16 ml) were collected and evaporated to dryness after the addition of toluene to remove the acetic acid by azeotropic distillation.

Fractions 40–42 and 47–51 gave one spot each (*Rf* 0.80 and 0.55, respectively) on a thin layer chromatogram (silica gel G, 0.25 mm; $CHCl_3$:MeOH:AcOH = 15:4:1). The former yielded methyl glucosylcucurbate **5** (30 mg, an oil), which might be an artifact derived from the inhibitor **4** by methylation during the isolation procedure. The latter gave glucosylcucurbic acid **4** (150 mg, an oil).

References

1) V.I. Kefeli, C.S. Kadyrov, *Ann. Rev. Plant Physiol.*, **22**, 185 (1971).
2) N. Takahashi, S. Marumo, N. Ōtake, *Seirikassei-Tennenbutsukagaku* (in Japanese), p. 63, Tokyo University Press (1973).
3) K. Yamashita, *Shokubutsu no Seirikassei-Busshitsu* (in Japanese), p. 47, Nankodo (1975).
4) B.V. Milborrow, *Ann. Rev. Plant Physiol.*, **25**, 259 (1974).
5) J.P. Nitsch, *The Biochemistry of Fruits and Their Products* (ed. A.C. Hulme), vol. **1**, p. 445, Academic Press (1970).
6) B.V. Milborrow, *Phytohormones and Related Compounds: A Comprehensive Treatise* (ed. D.S. Letham, P.B. Goodwin, T.J.V. Higgins), vol. 1, p. 307, Elsevier/North Holland Biomedical Press (1978).
7) T. Hemberg, *Encycl. Plant Physiol.*, **14**, 1162 (1961).
8) K. Koshimizu, H Fukui, S. Usuda, T. Mitsui, *Plant Growth Substances* **1973,**

p. 86, Hirokawa Publishing Company, Inc. (1974); H. Fukui, K. Koshimizu, S. Usuda, Y. Yamazaki, *Agric. Biol. Chem.*, **41**, 175 (1977); H. Fukui, K. Koshimizu, Y. Yamazaki, S. Usuda, *ibid.*, **41**, 189 (1977).

9) E. Stahl, *Thin-Layer Chromatography* (ed. E. Stahl), p. 205, Springer-Verlag (1969).
10) K.V. Thimann, *The Action of Hormones in Plants and Invertebrates* (ed. K.V. Thimann), p. 4, Academic Press (1952).
11) J.P. Nitsch, C. Nitsch, *Plant Physiol.*, **31**, 94 (1956).
12) D. Adamson, V.H.K. Low, H. Adamson, *Biochemistry and Physiology of Plant Growth Substances* (ed. F. Wightmann, G. Setterfield), p. 505, The Runge Press (1968).
13) M.S. Smith, R.L. Wain, F. Wightman, *Ann. Appl. Biol.*, **39**, 295 (1952).
14) H.F. Tayler, R.S. Burden, *Proc. Roy. Soc. London*, B **180**, 317 (1972).
15) C.E. Hess, *Régulateurs Naturels de la Croissance Végétale* (ed. J.P. Nitsch), p. 517, C.N.R.S. (1964).
16) M. Mitsuhashi, H. Shibaoka, *Plant & Cell Physiol.*, **6**, 87 (1965).
17) E. Sondheimer, C. Walton, *Plant Physiol.*, **45**, 244 (1970).
18) O.E. Smith, J.L. Lyon, F.T. Addicott, R.E. Johnson, *Biochemistry and Physiology of Plant Growth Substances* (ed. F. Wightman, G. Setterfield), p. 1547, The Runge Press (1968).
19) Y. Murakami, *Botan. Mag.* (Tokyo), **81**, 33 (1968).
20) Y. Ogawa, *Plant & Cell Physiol.*, **4**, 227 (1963).
21) T. Oritani, Y. Yamashita, *Agric. Biol. Chem.*, **38**, 801 (1974).
22) D. Köhler, A. Lang, *Plant Physiol.*, **38**, 555 (1963).
23) D. Aspinell, L.G. Paleg and F.T. Addicott, *Australian J. Biol. Sci.*, **20**, 869 (1967).
24) M. Katsumi, B.O. Phinney, W.K. Purves, *Physiol. Plant.*, **18**, 462 (1965).
25) J.P. Nitsch, *The Chemistry and Mode of Action of Plant Growth Substances* (ed. R.L. Wain, F. Wightman), p. 3, Butterworths Scientific Publications (1956).
26) K. Koshimizu, H. Fukui, T. Kusaki, Y. Ogawa, T. Mitsui, *Agric. Biol. Chem.*, **32**, 1135 (1968).
27) K. Koshimizu, H. Fukui, T. Mitsui, Y. Ogawa, *Agric. Biol. Chem.*, **30**, 941 (1966).
28) N. Ōtake, A. Suzuki, N. Takahashi, N. Murofushi and H. Yonehara, *Busshitsu no Tanri to Seisei* (in Japanese), p. 35 and 218, Tokyo University Press (1976).
29) E.W.H. Edwards, *Chromatographic Techniques* (ed. I. Smith), p. 249, William Heinemann Medical Books (1958).
30) J. MacMillan and C.M. Wels, *J. Chromatogr.*, **87**, 271 (1973).

General References

i) *Annual Review of Plant Physiology*, Annual Reviews Inc.
ii) D.S. Letham, P.B. Goodwin, T.J.V. Higgins (ed.), *Phytohormones and Related Compounds: A Comprehensive Treatise*, vols. 1 and 2, Elsevier/North Holland Biomedical Press (1978).
iii) A.W. Galston, P.J. Davis, *Control Mechanisms in Plant Development*, Prentice-Hall (1970).
iv) Y. Masuda, M. Katsumi, H. Imazeki, *Shokubutsu-Horumon* (in Japanese), Asakura-Shoten (1971).
v) N. Takahashi, S. Marumo, N. Ōtake, *Shokubutsukassei-Tennenbusshitsu* (in Japanese), Tokyo University Press (1973).
vi) K. Yamashita, *Biologically Active Substances produced by Plants* (in Japanese), Nankodo (1975).

CHAPTER 16

Isolation of Gibberellins from Higher Plants

In Japan the rice disease called "*bakanae-byo*" has been known for a long time. Rice cultivation in the warm and humid part of Japan has suffered particularly severe damage from this disease. The disease is caused by a pathogenic fungus called *Gibberella fujikuroi*. When young plants in seedbeds get this disease, they become unusually tall, their green color fades, and in serious cases, they become withered.

Kurosawa showed that this pathogen produces a toxin which causes these symptoms, and later in 1938, Yabuta and Sumiki (at the University of Tokyo) isolated gibberellins A and B as active components able to cause elongation of rice plants. After the war, research on gibberellins was started again among research groups at Imperial Chemical Industries Ltd. (ICI) in England, the Northern Regional Research Laboratory in the U.S.A. and the University of Tokyo in Japan. In 1955[1)], three kinds of gibberellins, A_1, A_2 and A_3, were characterized (abbreviated as $GA_{1,2,3}$ hereafter). In further investigations, new gibberellins ($GA_{4,7,9\sim16,24,25,36,37,40-42,47,54-57}$) were isolated from this fungus and their structures were elucidated. These twenty-five gibberellins are called fungal gibberellins.

The occurrence of gibberellin-like substances in tissues of higher plants was suggested, and in1958, MacMillan *et al.* of ICI isolated $GA_{1,5,6,8}$ from immature seeds of scarlet runner bean (*Phaseolus multiflorus*), while Phinney and West in the U.S.A. isolated $GA_{1,5}$ from those of *Phaseolus vulgaris*. Further, Kawarada and Sumiki of the University of Tokyo identified GA_1 in water sprouts of *Citrus unshiu*. Further investigations showed that gibberellins are distributed widely in higher plants and play important

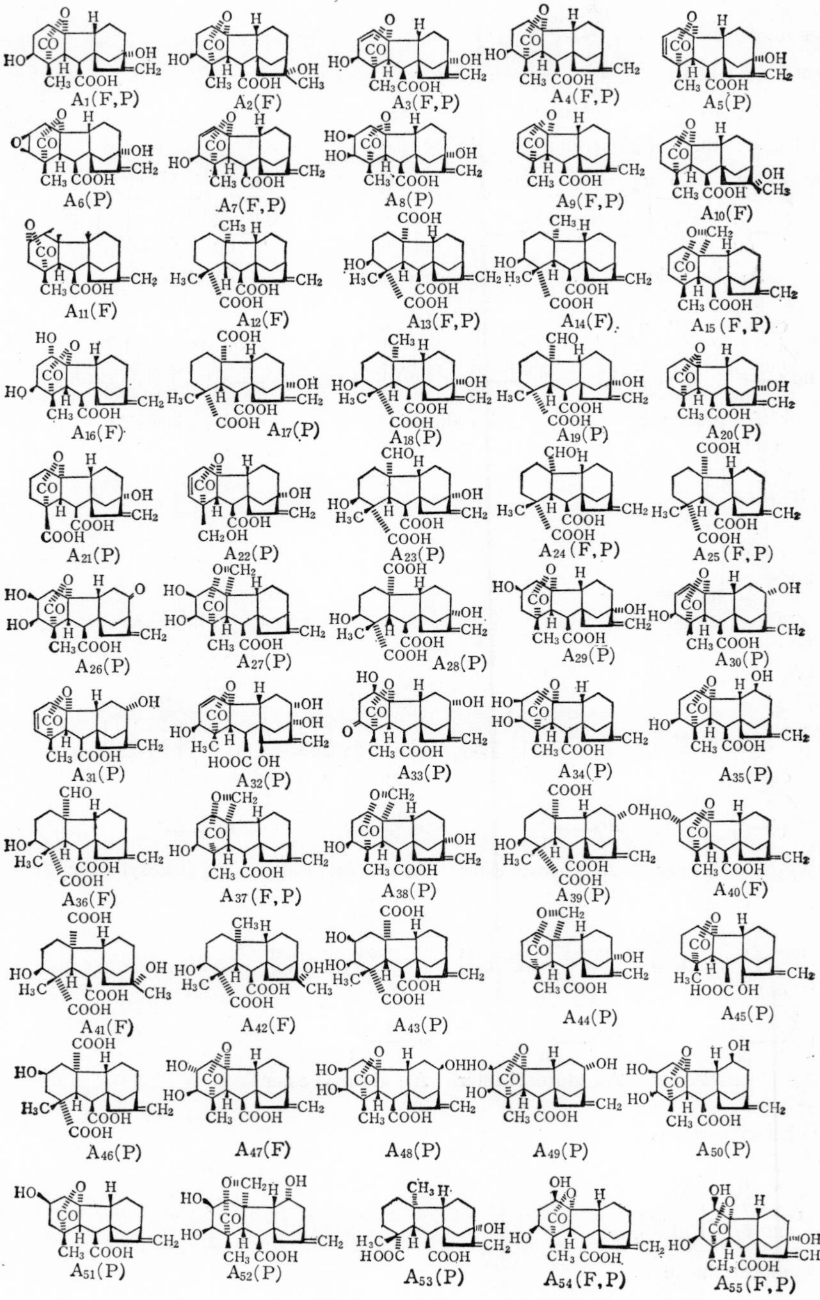

A_1(F,P) A_2(F) A_3(F,P) A_4(F,P) A_5(P)

A_6(P) A_7(F,P) A_8(P) A_9(F,P) A_{10}(F)

A_{11}(F) A_{12}(F) A_{13}(F,P) A_{14}(F) A_{15}(F,P)

A_{16}(F) A_{17}(P) A_{18}(P) A_{19}(P) A_{20}(P)

A_{21}(P) A_{22}(P) A_{23}(P) A_{24}(F,P) A_{25}(F,P)

A_{26}(P) A_{27}(P) A_{28}(P) A_{29}(P) A_{30}(P)

A_{31}(P) A_{32}(P) A_{33}(P) A_{34}(P) A_{35}(P)

A_{36}(F) A_{37}(F,P) A_{38}(P) A_{39}(P) A_{40}(F)

A_{41}(F) A_{42}(F) A_{43}(P) A_{44}(P) A_{45}(P)

A_{46}(P) A_{47}(F) A_{48}(P) A_{49}(P) A_{50}(P)

A_{51}(P) A_{52}(P) A_{53}(P) A_{54}(F,P) A_{55}(F,P)

Fig. 16.1

A_{56}(F) A_{57}(F,P)

Glucosyl ethers of gibberellin

A_1 glucoside A_3 glucoside A_8 glucoside

A_{26} glucoside A_{27} glucoside A_{29} glucoside

GA_{35} glucoside

Glucosyl esters of gibberellin

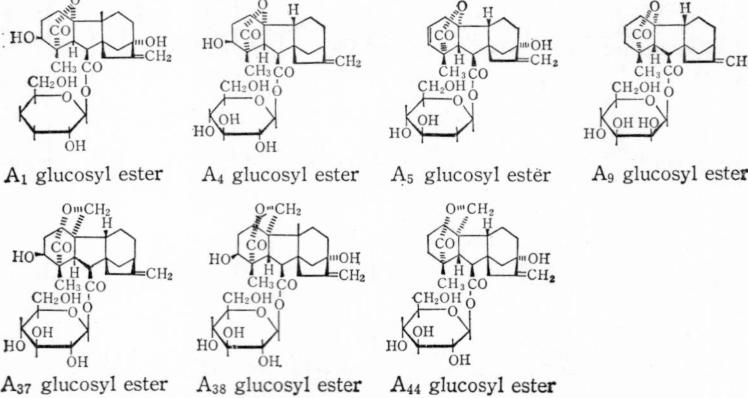

A_1 glucosyl ester A_4 glucosyl ester A_5 glucosyl ester A_9 glucosyl ester

A_{37} glucosyl ester A_{38} glucosyl ester A_{44} glucosyl ester

Other gibberellins

gibberethione

Fig. 16.2

roles in the regulation of their growth. Since 1965, studies to isolate new gibberellins have been conducted by groups at the University of Tokyo and Kyoto University in Japan and the University of Bristol in England. So far, forty-four gibberellins, $GA_{1,3-9,13,15,17-35,37-39,43-46,48-55}$, have been definitely characterized as plant gibberellins. At present, fifty-seven gibberellins in total are known to occur in nature. Their structures are shown in Fig. 16.1, in which the letters (F) and (P) indicate fungal and plant origin, respectively.

These gibberellins are called free gibberellins, since they can be extracted* from an acidic aqueous solution with ethyl acetate. However, the presence of gibberellin-like substances which are more polar than these free gibberellins has been suggested in higher plants.

The group at the University of Tokyo began studies on the chemical nature of the polar gibberellins and succeeded in isolating five polar gibberellins from immature seeds of morning glory (*Pharbitis nil*) and one from immature seeds of yellow broom (*Cytisus scoparius*), and showed that they are glucosyl ethers of gibberellin.[2)] Further, they isolated neutral polar gibberellins from mature seeds of *P. vulgaris* and showed them to be glucosyl esters at the C-6 carboxyl of $GA_{1,4,37,38}$.[3)] Recently, they isolated a new gibberellin derivative called gibberethione[4)] (formerly known as pharbitic acid), which was found to contain 3-oxo-GA_3 and mercaptopyruvic acid. These gibberellin derivatives are called bound gibberellins; their structures are shown in Fig. 16.2.

All free gibberellins contain *ent*-gibberellane (**1**), as a basic skeleton, which may carry many kinds of functionality (fifty-seven gibberellins are known so far).

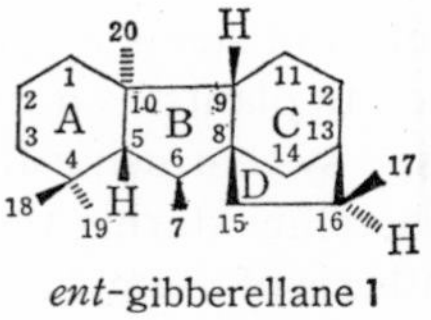

ent-gibberellane **1**

It has been suggested that quantitative and qualitative changes of endogenous gibberellins can be correlated with various physiological phenomena of higher plants. However, the content of endogenous gibberellins in higher plants is very low. Even in the case of immature seeds of Leguminoceae, which are rich sources of gibberellins, the content is usually less than several mg/kg fresh weight. In particular, the content of gibberellins

* GA_{32}, which is classified as a free gibberellin, cannot be extracted with ethyl acetate from an acidic aqueous solution because of its high polarity, resulting from the presence of four hydroxyl groups in the molecule.

in the tissue during vegetative growth is very low; the content of GA_{19} in bamboo shoots has been estimated to be 0.002–0.005 mg/kg.

For the isolation of such a minute amount of gibberellin, it is important to combine several isolation procedures suited to the chemical nature of the gibberellins and to establish bioassay methods for monitoring the behavior of the gibberellins during the isolation process. When the material cannot be collected in an amount sufficient to permit its isolation, identification must be done without isolation by utilizing thin-layer chromatography (tlc), gas-liquid chromatography or, to ensure more definite identification, combined gas-liquid chromatography and mass spectrometry.

In this chapter, the procedures for the bioassay of gibberellins, as well as for their isolation and micro-identification are presented.

16.1 Bioassay and Other Detection Methods for Gibberellins

Gibberellins show many kinds of biological activities in higher plants. They can be summarized as follows.

1) Elongation of shoots of intact plants.
2) Flower induction in plants which require a long day and low temperature for this process.
3) Ending domancy.
4) Stimulation of germination.
5) Acceleration of flowering.
6) Acceleration of *de novo* synthesis of α-amylase in grain.

The elongation of shoots and stems by gibberellins, especially in the case of dwarf plants and young plants, is a unique physiological activity of gibberellins, so a bioassay system using this feature is suitable monitoring the behavior of gibberellins during the isolation process. The stimulation of the *de novo* synthesis of α-amylase in grains can sometimes be used as a bioassay method.

Gibberellin can be visualized as fluorescent spots under uv light (365 nm) on a tlc plate after spraying conc. H_2SO_4 and heating at 120° C for a few minutes.

Bioassay Using Dwarf Plants

Rice and maize are often used as dwarf plants, for bioassay. In Japan, since there is considerable interest in physiological research on rice, many dwarf varieties are known. Those that have been used for bioassay of gibberellins include "*tan-ginbozu,*" "*kotake tamanishiki*" and "*waito-C*", because of their high sensitivity to gibberellins. It should be pointed out that the activity patterns of these varieties for various kinds of gibberellins are not necessarily similar. Among them "*tan-ginbozu*" is most widely used as a test plant because it shows a wide action spectrum with various gibberellins. However, even in the case of bioassay using "*tan-ginbozu*", different action spectra are obtained depending upon the structural variation of the gibberellins. In general, gibberellins with the partial structure (**2**) in ring A, such as $A_{3,7,30,32}$, show very strong activity and those such as $GA_{13,17,24,28,39}$, which contain three carboxyl groups at C–4, –6,–10, and such as $GA_{8,26,27,29,34,40}$, which contain a hydroxyl group at C–2, show very low activity. Glucosyl derivatives of gibberellins show different activity, depending upon the bioassay system used.

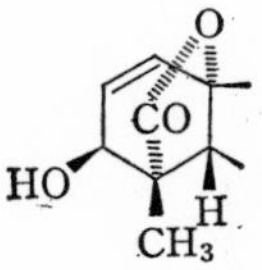

2

[Experimental procedure 1] The dip method

This method is illustrated in outline in Fig. 16.3(a). The seeds of "*tan-ginbozu*" are washed with tap water and placed on filter paper in a Petri dish, and sufficient water is added to just cover the seeds. The Petri dish is kept at 30° C for germination. A test solution (2 ml) is added to a vial (140 x 25 mm) and several seeds that have just germinated to similar extents are placed in it. The open end of vial is covered with polyethylene film. The seedlings are incubated for 5 to 7 days at 30° C under fluorescent light (3000 lux). The length of the second leaf sheath is measured. As little as 0.01 ppm GA_3 can be detected by this method.

[Experimental procedure 2] The micro-drop method

Several seeds that have just germinated are inoculated on plain agar in a vial (55 x 32 mm). Vials are placed in a covered transparent box saturated with water vapor. When the seedlings reach a height of 2 cm (see Fig. 16.3 (b)) after incubation for about 45 hr at 30° C under fluorescent light (3000 lux), a 1 μml aliquot of a test solution is applied between the leaf sheath and the first leaf with a micro syringe. Seedlings are incubated for a further 3 days,then the length of the second leaf sheath is measured.

Another Detection Method

Gibberellins show bluish fluorescence on a tlc plate on spraying 80% sulfuric acid and heating at 120° C for a few minutes. This is very suitable for the detection of gibberellins. Since the fluorescent color tone and inten-

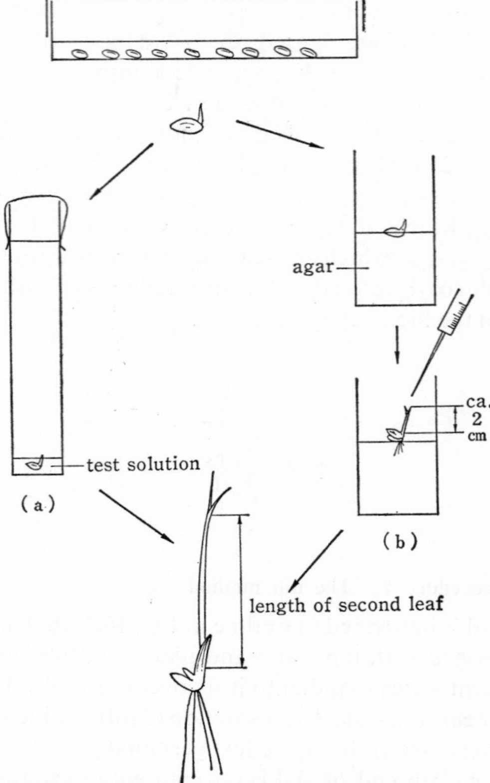

Fig. 16.3. Bioassay of gibberellins with dwarf rice seedlings.
(a) Dip method, (b) microdrop method.

sity are variable, depending on the type of gibberellin, it can be used for estimating the type of gibberellin. For example, GA_3 and GA_7, with the ring A structure shown in **2**, show fluorescence simply on spraying 80% sulfuric acid without heating, while GA_{17} shows weak whitish fluorescence. Since fluorescence can be emitted by other organic compounds on sulfuric acid treatment, it is advisable to combine this method with a bioassay procedure.

16.2 Isolation of Gibberellins from Higher Plant Tissues

Isolation of gibberellins can be carried out by solvent extraction, frac-

TABLE 16.1. Solvent fractionation of gibberellins.

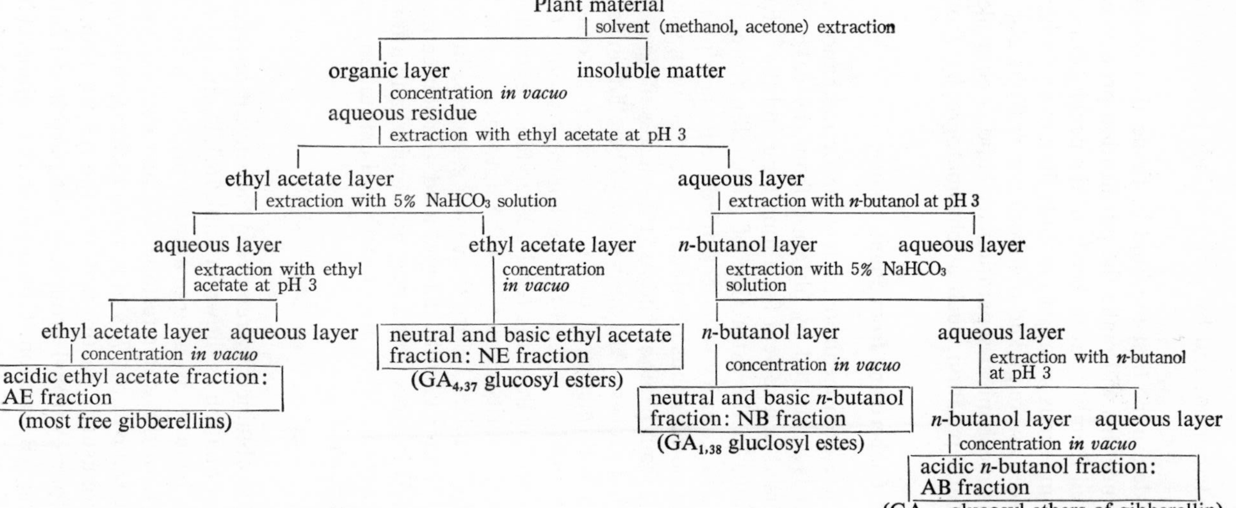

tionation by solvent distribution, and by various kinds of chromatographies.

It is important to determine first the kind of tissue that contains a large amount of GA, since this will make the purification process easier. GA_{32}, which was isolated from immature seeds of the peach (*Prunus persica*), was present in young seeds in larger amounts than in any other part of the fruit. Therefore seeds were picked out from young fruits 2–4 cm in diameter by hand. This mechanical separation by hand is equivalent to several steps of chemical separation, such as chromatography, from the whole fruit.

Solvent Extraction of Gibberellins from Plant Tissues

Generally, tissues are homogenized in acetone or methanol with a blender and the extract is separated by filtration from the cake. Sometimes, tissues are immersed in a solvent for a long time and the solvent is separated by decantation. In the extraction of gibberellin from mature bean seeds (*P. vulgaris*, cv Kentucky Wonder), dried seeds are ground in a mechanical mill and the powder thus obtained is extracted with solvent.

In the case of GA_{19} from bamboo shoots, the success of its isolation rests on the first extraction process. Although most gibberellins in bamboo shoots can be extracted with methanol using a blender, it is technically difficult to apply this method to several tons of bamboo shoots. On hot water extraction, only one-fourth of the total gibberellin is extracted but only a small amount of impurity comes into the hot water layer. Since bamboo shoots are usually boiled before the canning process in order to bleach them, the waste water was collected and used as a source material for further purification. This process was very economical and afforded a much more pure extract than in the case of methanol extraction, making further purification much easier.

Solvent Fractionation

The aqueous residue obtained by removing the solvent *in vacuo* was subjected to solvent fractionation as shown in Table16.1. As in the case of the extract from mature bean seeds, a large amount of fatty material was contained in the extract, so the water residue was first extracted several times with hexane or benzene at pH 7–8 to remove the lipophilic impurities. Free gibberellins are usually partitioned into the AE fraction. However GA_{32}, which has four hydroxyl groups, is not partitioned into the AE fraction but into the AB fraction. Glucosyl esters of gibberellin are partitioned into the NE and NB fractions, and glucosyl ethers into the AB fraction. In the partition process using *n*-butanol, clear-cut separation cannot be achieved.

Counter-Current Distribution Method

The counter-current distribution method is usually used in an early stage of the purification process. Since the p*Ka* values of many free gibberellins are known to be around 4.5, a rough separation can be achieved by the use of ten-transfer counter-current distribution between ethyl acetate and 1 M phosphate buffer of pH 5.8 to 6.0, using separate funnels. After the distribution, the pH of the aqueous layer is adjusted to pH 3 and the acidic water layer is extracted several times with ethyl acetate. The upper layer from each funnel and the ethyl acetate extract obtained from the corresponding lower layer are combined and dried over anhydrous sodium sulfate, then the solvent is removed *in vacuo*.

Charcoal Chromatography

Charcoal chromatography is a very effective purification process. Charcoal for chromatography should have a moderate particle size and be porous (e.q. that from Wako Co.). Charcoal amounting to ten to twenty times the sample weight is mixed with water, and the mixture is heated for a few minutes under reduced pressure to remove air absorbed in the charcoal. The charcoal is packed into column with water, then water is replaced with the elution solvent (acetone-water). A sample dissolved in the solvent is applied to the top of the column and elution is carried out with an increasing concentration of acetone in water. Polar gibberellins are usually eluted earlier and less polar ones later. For example, GA_9 which is the most non-polar gibberellin is eluted in 90% acetone-water solution, while gibberellin glucosides appear in 40–60% acetone solution.

Partition Chromatography

In partition chromatography, silica gel or a Sephadex such as LH-20, G-25 or G-50 is used as a support, equilibrated with 1 M phosphate buffer (pH 5.2), then ethyl acetate-benzene or *n*-butanol-benzene are used for the elution of free acids and *n*-butanol-ethyl acetate for glucosides. Recently the system of silica gel equilibrated with 0.1 M formic acid and eluted with *n*-hexane-ethyl acetate was shown to be effective. At high concentrations of *n*-butanol, inorganic salt from the buffer frequently appears in the eluates. To remove it, the eluate is concentrated and passed through a short charcoal column with a methanol-water solvent system. Salt is eluted with the solvent front, and gibberellins appear later.

Adsorption Chromatography

In the separation of gibberellins by adsorption chromatography, silica gel is generally used as an adsorbent with a benzene-ethyl acetate system (with increasing ethyl acetate content in the range of 30–70%) for free gibberellins, a benzene-methanol (with increasing methanol content in the range of 5–15%), or an acetone-benzene (with increasing acetone content in the range of 30–70%) system for glucosyl esters of gibberellin, and a chloroform-methanol-acetic acid system (90:10:95) for glucosides of gibberellin.

Thin Layer Chromatography

Thin layer chromatography (tlc) is often used for monitoring the behavior of gibberellins during purification and at the final purification step. Table 16.2 shows solvent systems suitable for gibberellins. Paper chromatography was formerly used in gibberellin separation studies, but has now been replaced by tlc.

TABLE 16.2. Solvent systems for the identification and isolation of gibberellins by tlc.

Free gibberellins	ethyl acetate-chloroform-acetic acid (20:8:1) or *n*-butanol-3 N ammonia (5:1)
Methyl esters	isopropyl ether-acetic acid (98:2)
Glucosyl esters	chloroform-methanol (3–4:1) or acetone-benzene (4:1)
Glucosyl ethers	chloroform-methanol (2–3:1)

[Experimental procedure 3] Isolation of GA_{19} from bamboo shoots[5)]

Waste water from 44 tons of boiled bamboo shoots (*Phyllostachys edulis*) during the bleaching process in a canning factory was fractionated to give the AE fraction. This was purified by the process shown in Table 16.3. Finally, 14 mg of GA_{19} was isolated as pure crystals.

[Experimental procedure 4] Isolation of GA_{32} from immature seeds of the peach (*Prunus persica*)[6)]

Immature seeds (35 kg) were removed by cracking young peach fruits (1000 kg). The AB fraction obtained by solvent fractionation was purified by the process shown in Table 16.4 to give the pure GA_{32} (38 mg). It should be pointed out that in the charcoal chromatography, a small amount of an acetonide of GA_{32} was formed from GA_{32} and acetone used as the solvent. The presence of GA_5 in the AE fraction was indicated by gas chromatography and mass spectroscopy after partial purification.

[Experimental procedure 5] Isolation of gibberellins from mature bean seeds (*Phaseolus vulgaris* cv. Kentucky Wonder)[2)]

Powder obtained by grinding the dry bean seeds (100 kg) in a mechanical mill was extracted with a large volume of methanol, which was de-

TABLE 16.3. Isolation of GA_{19} from bamboo shoots.

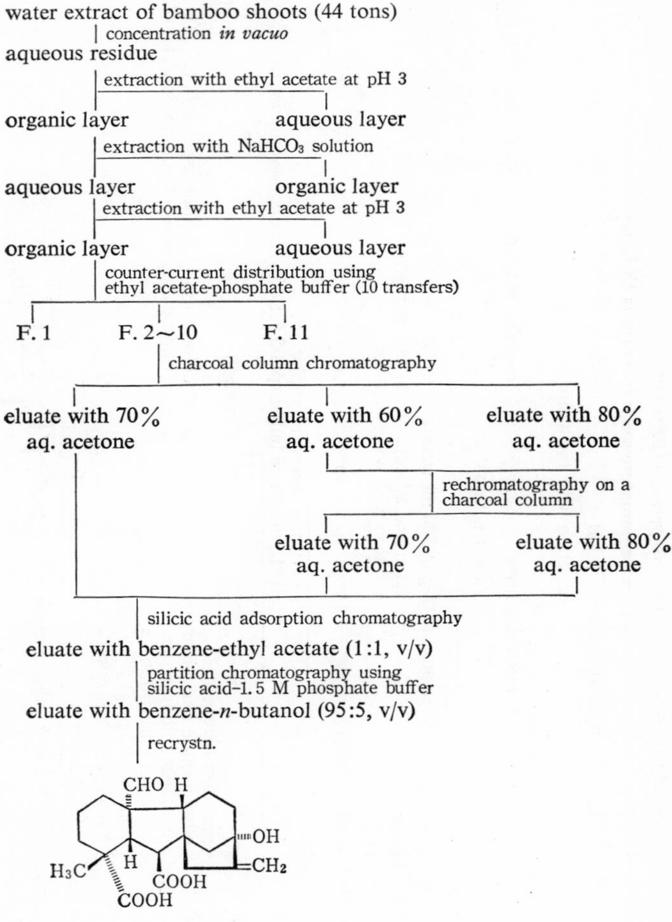

canted off. This process was repeated three times. The aqueous residue obtained by concentrating the combined methanol extracts was extracted with benzene to remove fatty materials, then solvent partition was conducted in the usual way. $GA_{1,8}$ were isolated from the AE fraction and identified. The NE and NB fractions were purified by the processes shown in Table 16.5. A mixture of GA_4 and GA_{37} glucosyl esters was isolated from the NE fraction, while GA_1 and GA_{35} glucosyl esters were isolated from the NB fraction. The NE fraction also contained a very small amount

Table 16.4. Isolation of GA_{32} from immature seeds of the peach.

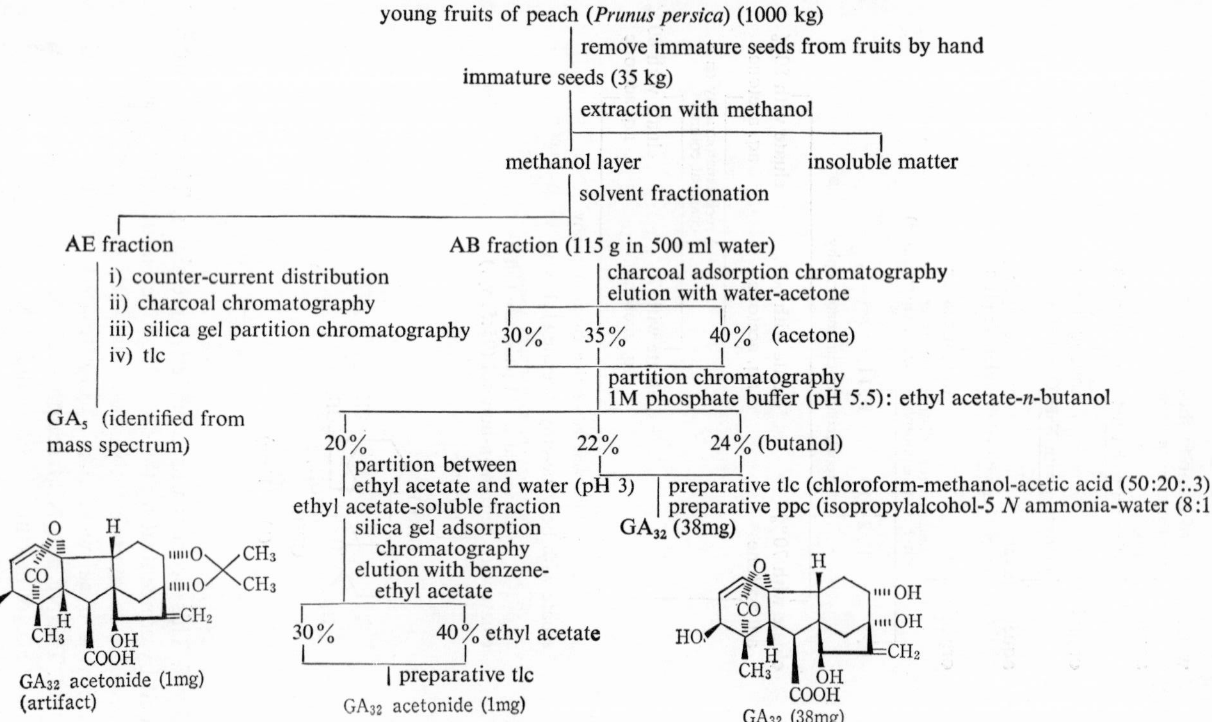

young fruits of peach (*Prunus persica*) (1000 kg)
remove immature seeds from fruits by hand
immature seeds (35 kg)
extraction with methanol
methanol layer
insoluble matter
solvent fractionation
AE fraction
AB fraction (115 g in 500 ml water)
i) counter-current distribution
ii) charcoal chromatography
iii) silica gel partition chromatography
iv) tlc
GA5 (identified from mass spectrum)
charcoal adsorption chromatography
elution with water-acetone
30%
35%
40% (acetone)
partition chromatography
1M phosphate buffer (pH 5.5): ethyl acetate-n-butanol
20%
22%
24% (butanol)
partition between
ethyl acetate and water (pH 3)
ethyl acetate-soluble fraction
silica gel adsorption
chromatography
elution with benzene-
ethyl acetate
30%
40% ethyl acetate
preparative tlc
GA32 acetonide (1mg)
preparative tlc (chloroform-methanol-acetic acid (50:20:.3)
preparative ppc (isopropylalcohol-5 N ammonia-water (8:1:1)
GA32 (38mg)
O
H
CO
HO
CH3
H
OH
COOH
CH3
CH3
CH2
OH
GA32 acetonide (1mg)
(artifact)
GA32 (38mg)

TABLE 16.5. Isolation of glucosyl esters of gibberellins from the NE and NB fractions of the extract of mature bean seeds.

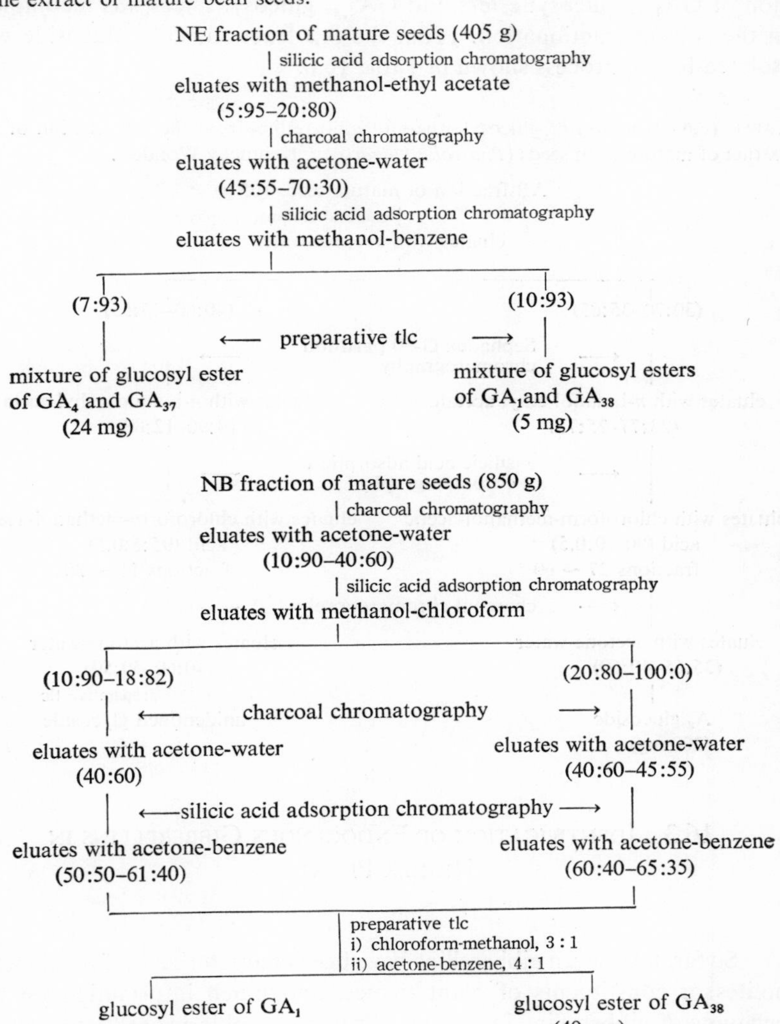

of a mixture of GA_1 and GA_{38} glucosyl esters. This indicates that separation of $GA_{1,38}$ glucosyl esters and $GA_{4,37}$ glucosyl esters was incomplete in the solvent fractionation. From the AB fraction, GA_8 glucoside was isolated by the process shown in Table 16.6.

TABLE 16.6 Isolation of glucosyl ethers of gibberellins from the AB fraction of the extract of mature bean seeds (*Phaseolus vulgaris* cv. Kentucky Wonder).

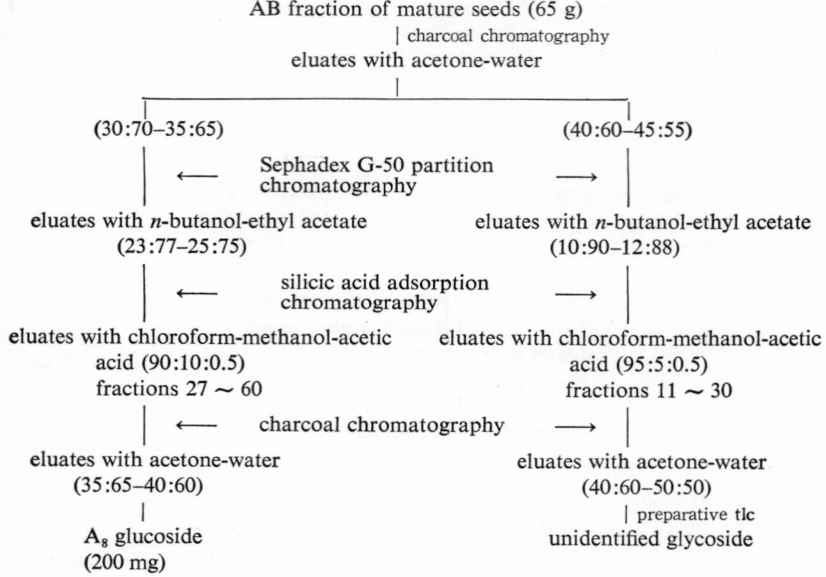

16.3 IDENTIFICATION OF ENDOGENOUS GIBBERELLINS IN HIGHER PLANTS

So far, fifty-seven gibberellins have been characterized as fungal metabolites or constituents of plant tissues, and much information on the behavior of gibberellins in various chromatographic procedures, and on their spectroscopic properties has been accumulated. Identification of endogenous gibberellins in higher plants has provided valuable insight into the roles of gibberellins in the biological processes of higher plants. However it often happens that characterization is impossible due to the difficulty in collecting sufficient material to make isolation possible or due to the complexity of the isolation process. Therefore identification has often been carried out by tlc, paper chromatography, or gas chromatography. Recently new identification methods in which separation is combined with identification have been developed, and the reliability of identification has greatly improved. In particular, gas-liquid chromatography combined with mass spectrometry has been shown to be quite effective for the identification of gibberellins.

Gas-Liquid Chromatography of Gibberellins

Since free gibberellins are not volatile, they are not suitable for gas chromatography without derivatization, usually to their methyl esters, which are formed by treatment with ethereal diazomethane, or to the TMS derivatives of the methyl esters, which are prepared by treatment with silylation reagents (*N*, *O*-bis (trimethylsilyl)-acetamide-trimethylchlorosilane-dry pyridine (2:1:1)). TMS derivatives are particularly useful because no tailing is observed. Ikekawa *et al.* and Covell *et al.* carried out extensive gas chromatographic studies, and recommended the use of SE-30, SE-33, OV-16, QF-1 and XE-60 as stationary phases.

Gas Chromatography-Mass Spectrometry of Gibberellins

MacMillian's group[7)] carried out extensive gas chromatography-mass spectrometric studies of gibberellin methyl esters and their TMS ethers. On the other hand, Takahashi's group[8)] measured the high-resolution mass spectra of many gibberellins and analyzed their spectroscopic features. Based on the data obtained by MacMillan's and Takahashi's groups, the following empirical rules on the structures and mass spectra of gibberellins were derived.

1) In the mass spectra of gibberellin derivatives, molecular ion peaks of moderate intensity are observed. They can thus be used for identification of the species of gibberellins.
2) In addition to M^+, peaks due to $M^+ - 32$, $M^+ - 60$, $M^+ - 62$ are observed with high intensity, and can be used to determine whether an unknown sample belongs to the gibberellin group or not.
3) Mass spectra of TMS derivatives of gibberellins with a C-13 hydroxyl group usually show molecular ion peaks of much higher intensity than gibberellin derivative lacking the C-13 hydroxyl group.
4) Since the mass spectra of gibberellin derivatives (methyl esters and TMS ethers of the methyl ethers) were collected by MacMillan and Takahashi, the identification of known gibberellins can be carried out very quickly.

Since the extracts from higher plants contain many impurities, clean-up is required prior to gas chromatography-mass spectrometry, which is

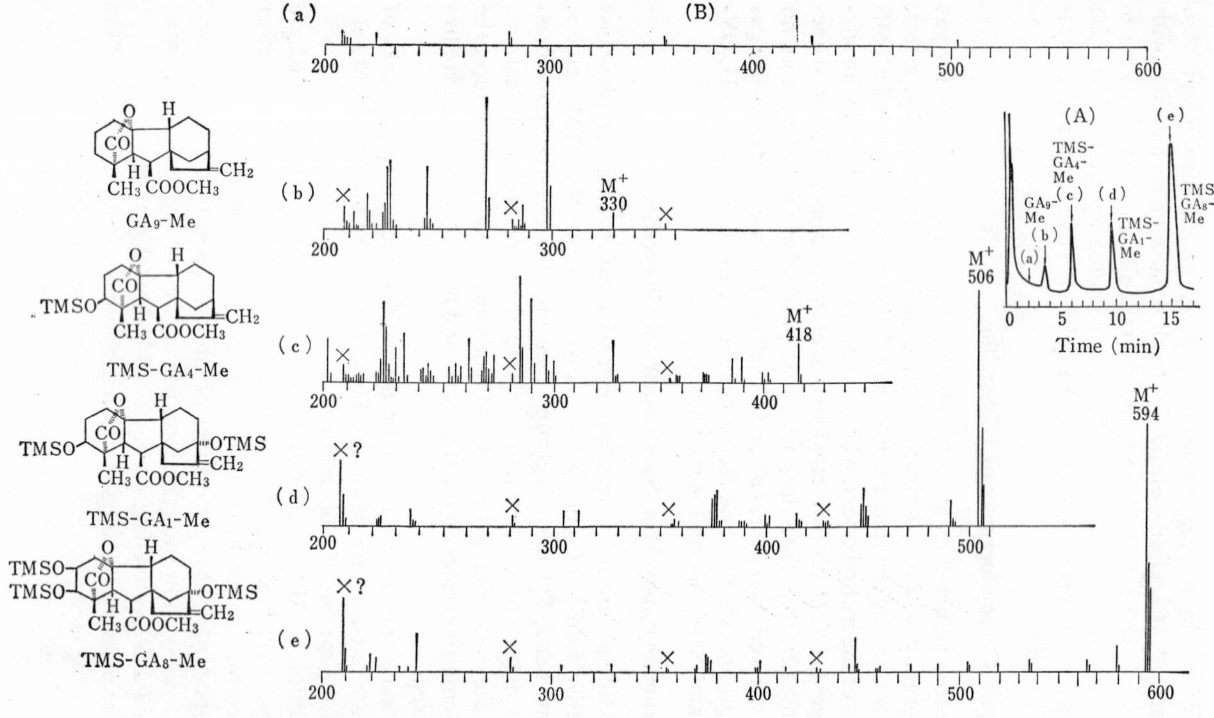

Fig. 16.4. Gas chromatography-mass spectrometry of gibberellin ($GA_{1,4,8,9}$) methyl ester TMS ethers. (a)–(e) indicate the positions of mass spectral determination on the gas chromatogram traces in (A) and their mass spectra in (B). An SE-30 glass column (1.5 m × 3 mm) was used: column temperature, 230°; helium pressure, 18 kg/cm². The symbols × indicate peaks due to the liquid phase.

not suitable for quantitative analyses. For this purpose mass fragmentography (selected ion current monitoring) is recommended.

[Experimental procedure 6] Gas chromatography-mass spectrometry of $GA_{1,4,8,9}$ methyl ester TMS ethers

Fig. 16.4 shows the results for a mixture of $GA_{1,4,8,9}$ methyl ester TMS ethers. Good separation of these gibberellin derivatives (GA_9 methyl ester is not TMS-derivatized because it lacks a hydroxyl group) can be seen, and a good mass spectrum of each peak was obtained.

References

1) S. Tamura (ed.), *Gibberellins* (in Japanese), University of Tokyo Press (1969).
2) T. Yokota, N. Murofushi, N. Takahashi, S. Tamura, *Agr. Biol. Chem.*, **35**, 583 (1971).
3) K. Hiraga, T. Yokota, N. Murofushi, N. Takahashi, *ibid.*, **38**, 2511 (1974).
4) T. Yokota, S. Yamazaki, N. Takahashi, Y. Iitaka, *Tetr. Lett.*, **1974**, 2957.
5) N. Murofushi, S. Iriuchijima, N. Takahashi, S. Tamura, J. Kato, Y. Wada, E. Watanabe, T. Aoyama, *Agr. Biol. Chem.*, **30**, 917 (1966).
6) I. Yamaguchi, T. Yokota, N. Murofushi, N. Takahashi, Y. Ogawa, *ibid.*, **39**, 2399 (1975).
7) R. Binks, J. MacMillan, R. J. Pryce, *Phytochemistry*, **8**, 271 (1969).
8) N. Takahashi, N. Murofushi, S. Tamura, N. Wasada, H. Hoshino, T. Tsuchiya, S. Sasaki, T. Aoyama, E. Watanabe, *Org. Mass Spectrom.*, **2**, 711 (1969).

CHAPTER 17

Insect Antifeedants in Plants

It is well known that special plant constituents function to establish interrelationships between insects and their food plants. An insect reaching or alighting on a plant is stimulated by attractants contained in the plant and inhibited by repellents.

On the other hand, feeding stimulatns elicit a feeding response of insects, whereas insect feeding is inhibited by the presence of feeding deterrents or antifeedants in plants.[1)]

Isolation and structure elucidation of insect antifeedants from plants not only provide basic knowledge about the interrelationships between insects and plants, but may also give indications for the development of agricultural chemicals to protect cultivated plants from their herbivores. Such information may also be useful for the breeding of insect-resistant varieties of crops.

17.1 Bioassay of Insect Antifeedants

Our studies on the extraction and isolation of plant constituents having insect feeding inhibitory activity were monitored by a bioassay procedure established by us. The insect used in the Laboratory of Pesticide Chemistry, Faculty of Agriculture, Nagoya University, was mainly the 3rd instar larvae of "*hasumonyoto*" (*Spodoptera litura* F.). The fresh leaves of sweet potato (*Ipomea batatas* Lam.) were used as the food plant for the larvae. A sample to be tested was dissolved in acetone to give a solution of definite concentration. Leaf disks were punched out with a cork borer (diameter: 20 mm) from the leaves of sweet potato.

Two leaf disks were immersed in an acetone solution of a sample for 1 to 2 min, and were then air-dried, while two other disks, used as a control, were immersed in pure acetone. These four disks were placed crosswise and symmetrically on the bottom of a polyethylene case (12 cm in diameter and 5 cm deep), then ten 3rd instar larvae of *S. litura* were introduced in the center of the case, and the case was closed with a cap containing small holes and placed under room light in a temperature-controlled room (25–28° C). After feeding for 2 hr, the four disks were placed between glass plates and a contact print of the disks was taken. The amounts of the sample disks consumed were expressed as a percentage of those of the control disks. The insect feeding inhibitory activity of a sample was expressed in terms of the concentration (ppm or %) at which the sample showed 80 to 100% inhibition. When the feeding experiments continued for longer than 2 hr, the larvae of *S. litura* died of starvation in some cases.

17.2 Insect Antifeedants of Some Plants

Insect Antifeedant of "*Kamiebi*" (*Cocculus trilobus* DC)

"*Kamiebi*" (Menispermaceae) (*C. trilobus* DC) is the host plant of Japanese fruit-piercing moths, "*akaeguriba*" (*Oraesia excavata* Butler) and "*himeeguriba*" (*O. emarginata* F.), which attack peaches, Japanese medlar, and grapes. These moths attack orchards at night, live near "*kamiebi*" in the day time and lay eggs on the leaves of the host plant. Thus, "*kamiebi*" was considered likely to contain attractants for the fruit-piercing moths. Extraction and isolation of the constituents of "*kamiebi*" and the bioassay of attractants were investigated from 1962 to 1964, but little information was obtained.

However, one day in August 1964, we had been separating the leaves of "*kamiebi*" and other plants gathered from the same place, and noticed that "kamiebi" had scarcely been attacked by the insects, whereas the other plants had been partially eaten. This suggested that "*kamiebi*" might contain insect resistant factors. We therefore attempted to isolate repellents (or antifeedants) from "*kamiebi*" instead of studying attractants for fruit-piercing moths. The bioassay for insect antifeedants was established with the kind cooperation of Dr. T. Saito, Laboratory of Applied Entomology, Faculty of Agriculture, Nagoya University, and an antifeedant (isoboldine **1**) and an insecticide (cocculolidine **2**, a new alkaloid) were

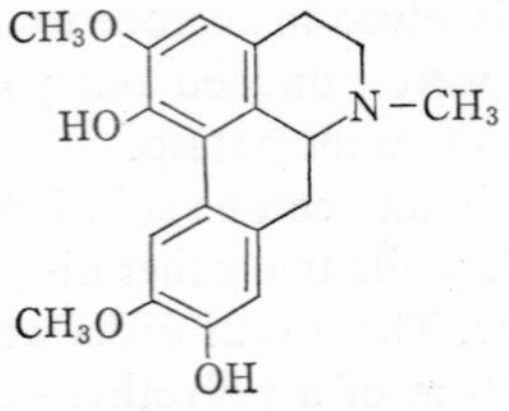

isoboldine **1** 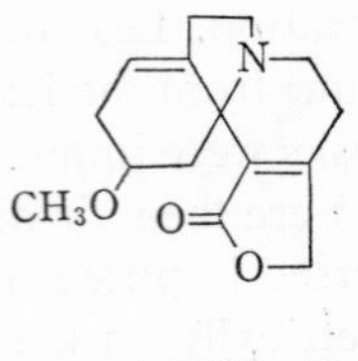 cocculolidine **2**

obtained through extraction and isolation experiments on the insect—resistant factors of "*kamiebi*".[2)]

[Experimental procedure 1] Isolation of insect antifeedant from "*kamiebi*"

The fresh leaves of "*kamiebi*" (15 kg) were cut into thin pieces with a meat slicer and placed in an enameled iron tank fitted with a three-necked cap. Next, 60 l of methanol was added. After refluxing for 1 hr, the hot methanol was decanted, and the same procedure was repeated twice. The combined extracts were then concentrated under reduced pressure at about 50° C. The residual solution (15 l) was acidified with dilute hydrochloric acid and extracted three times with 3 l of ether. The ether extracts were dried over anhydrous sodium sulfate and concentrated to give 19 g of the neutral and acidic fraction. The aqueous layer was made alkaline with ammonium hydroxide and extracted four times with 3 l of chloroform to give 15 g of the basic fraction. The basic fraction showed insect antifeeding activity in the leaf disk test. The chloroform solution of the basic fraction (15 g) was applied to an alumina chromatographic column (3×30 cm) and eluted with a mixed solvent system of chloroform and methanol to give twelve fractions, as shown in Table 17.1. Fractions 5 and 6 showed antifeeding activity, and their main component gave a positive ferric chloride test, suggesting the presence of a phenolic group. Fraction 6 was dissolved in 30 ml of 5% hydrochloric acid. The acidic solution was made alkaline with 20% sodium hydroxide and extracted with chloroform to remove trace amounts of nonphenolic alkaloids. The aqueous layer was made alkaline by the addition of solid ammonium chloride and extracted with chloroform to give 700 mg of oily material.

The material was dissolved in benzene and stored in a refrigerator overnight to give 100 mg of prismatic crystals, mp 99° C. Recrystallization of this antifeeding alkaloid from chloroform yielded prismatic crystals, mp 127° C; this compound was identical with isoboldine **1** by comparison of its physical properties with those of an authentic sample.[2)]

Fractions 1 to 4, having no antifeeding activity, showed only one spot of tlc. Recrystallization of this material from carbon tetrachloride gave

TABLE 17.1 Fractionation of the basic fraction of "*kamiebi*".

Fraction	Solvent (ml)	Eluate (g)	Antifeeding activity	Insecticidal activity
1	$CHCl_3$	1.184	–	++
2	$CHCl_3$ 200	2.503	–	+++
3	$CHCl_3$ 150			
4	$CHCl_3$ 100	0.711	++	++
5	10% CH_3OH–$CHCl_3$ 100	0.398	+++	–
6	10% CH_3OH–$CHCl_3$ 150	0.893	++	–
7	10% CH_3OH–$CHCl_3$ 150	0.393	+	–
8	10% CH_3OH–$CHCl_3$ 100	0.172	+	–
9	50% CH_3OH–$CHCl_3$ 150	0.438	–	–
10	50% CH_3OH–$CHCl_3$ 150	trace	–	–
11	50% CH_3OH–$CHCl_3$ 150	trace	–	–
12	CH_3OH 250	trace	–	–

prismatic crystals, which showed weak toxicity against green rice hopper and "*azuki*" bean weevil. This new insecticidal alkaloid was named cocculolidine and assigned the structure **2**.

Insect Antifeedants from "*Kusagi*" (*Clerodendron tricotomum* THUMB)

While studying the insect antifeedants of "*kamiebi*", we heard that "*kusagi*" (*C. tricotomum*), belonging to *Verbenaceae*, is also not attacked by insects. Antifeeding tests of crude solvent extracts of "*kusagi*" against the larvae of *S. litura* showed that the benzene extract had antifeeding activity.

Two active compounds, named clerodendrin A and B, were found in this plant through the procedure developed for the isolation of the antifeedant from "*kamiebi*". The structures of the compounds were assigned as **3** and **4** from their chemical reactions and spectral data.[3] The absolute

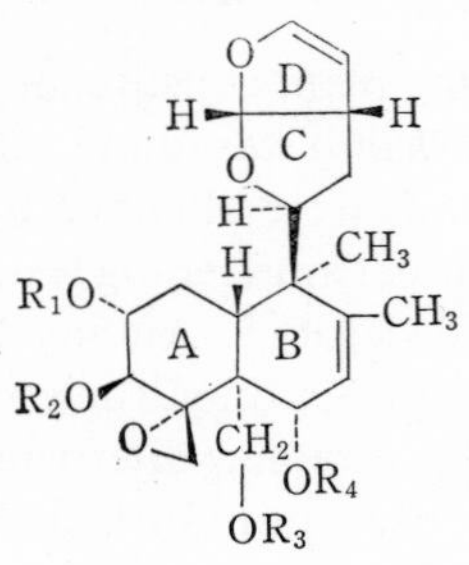

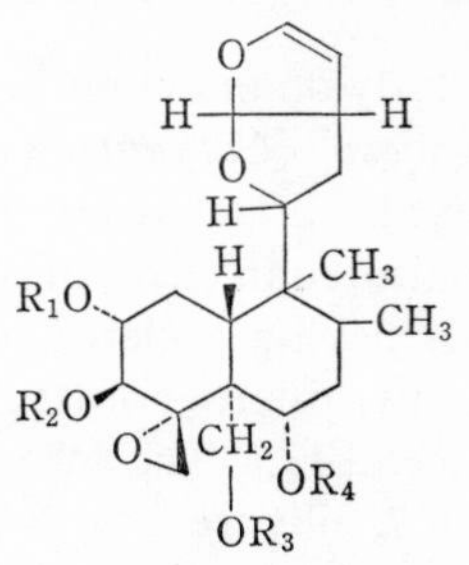

4 (R_1–R_4=same group as in **3**)

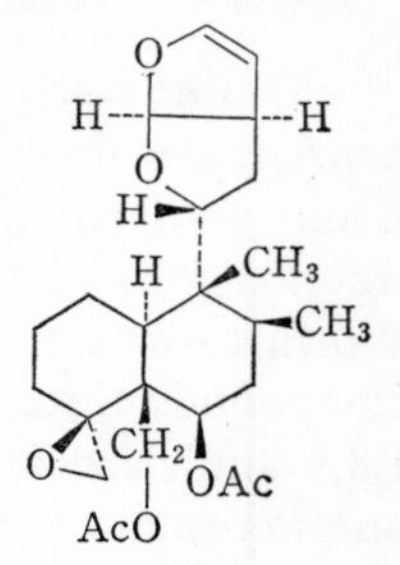

5

3:(R_1=H, R_2=$C_2H_5C(CH_3)(OAc)CO$–, R_3=R_3=Ac)

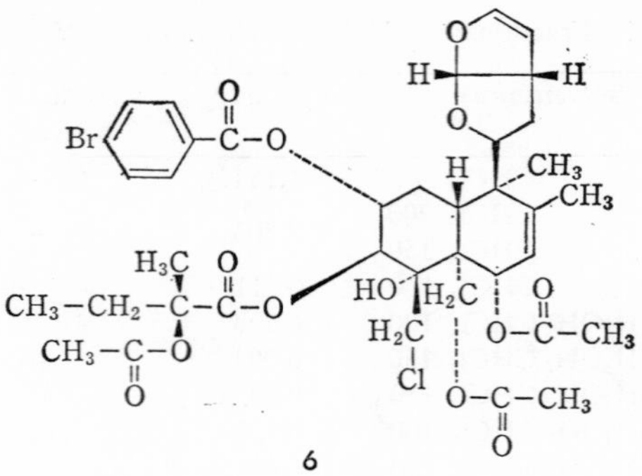

6

configuration of clerodendrin A was determined to be **6** by X-ray crystallography of the *p*-bromobenzoate chlorohydrin derivative.[4] The antifeeding activity was so strong that clerodendrin A and B stopped the larvae of *S. litura* eating at a concentration of 50 ppm; the larvae died, leaving the sample disks unattacked.

[Experimental procedure 1] Isolation of clerodendrin 3 from the leaves of "*kusagi*"

Air-dried leaves of "*kusagi*" (4.7 kg) were extracted 3 times with benzene. The extract was concentrated to 300 ml, and chromatographed on neutral alumina (2 kg). The diterpenoid fraction was eluted by 1% methanol in benzene after washing with benzene. The diterpenoid fraction was further subjected to silica gel column chromatography, and two diterpenoids, clerodendrin A **3** (2 g) and **B 4** (2.5 g) were successively eluted with *n*-hexane-ethyl acetate (5:2). The isolation of these diterpenoids was monitored by tlc and insect antifeeding tests.

Insect Antifeedants from "*Kariganeso*" (*Caryopteris divaricata* Maxim)

During a screening study on insect antifeeding plant constituents, the benzene extract of "*kariganeso*" (*Caryopteris divaricata*) was found to have strong antifeeding activity. Eight active compounds were obtained by extraction and isolation monitored in terms of the insect antifeeding test, and shown to be diterpenoids having a clerodane skeleton **5**. Among these compounds, three were known diterpenoids: clerodin **5**,[5] dihydroclerodin-I **8**,[5] and clerodin hemiacetal **12**.[5] The other five were new diterpenoids, and were named caryoptin **7**, dihydrocaryoptin **9**, caryoptinol **10**, dihydrocaryoptinol **11**, and caryoptin hemiacetal **13**. These diterpenoids and clerodin have been shown physicochemically to share a common absolute stereochemistry[6–8] and were assigned as the antipodes of the structures **5** and **7** to **13**, respectively. It should be noted that the absolute stereochemistries of caryoptin and clerodendrin were recently revised to the structures **7** and **5** by chiroptical methods[9] and X-ray assignment.[10] Thus, the above antifeeding diterpenoids have the absolute configurations **5** and **7** to **13**, respectively.

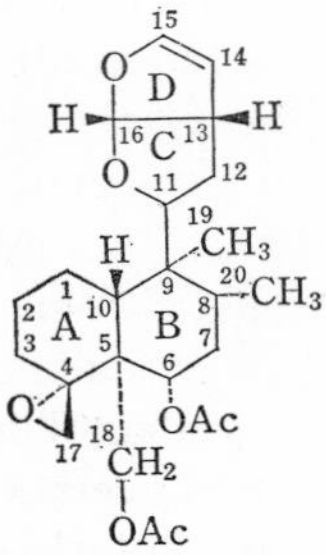

5 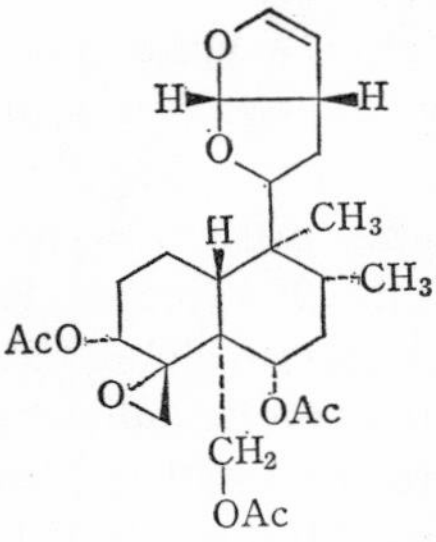 7 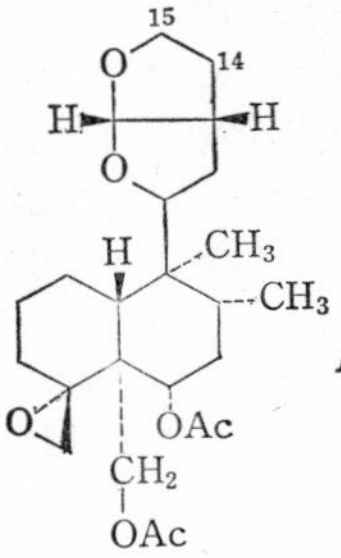 8 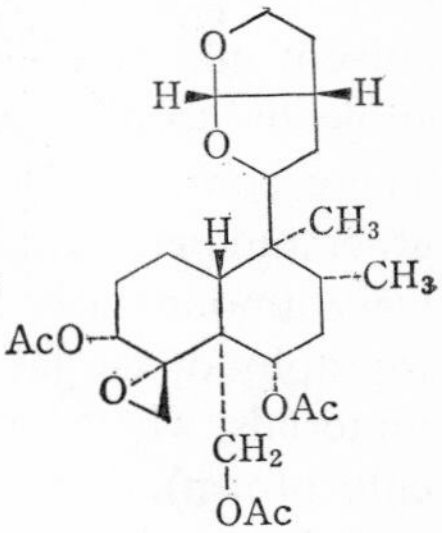 9

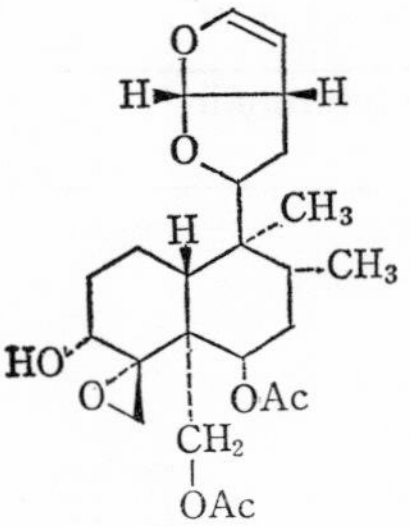

10 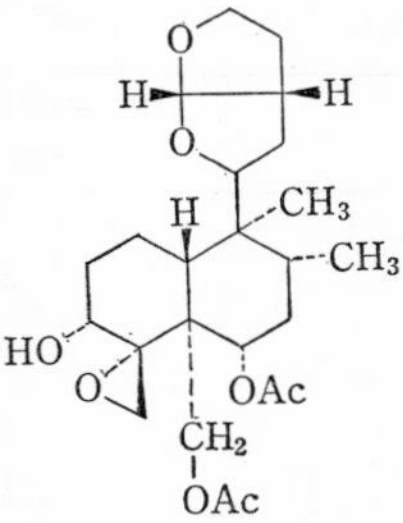 11 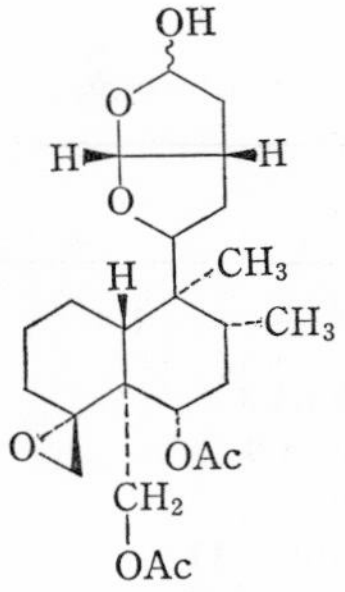 12 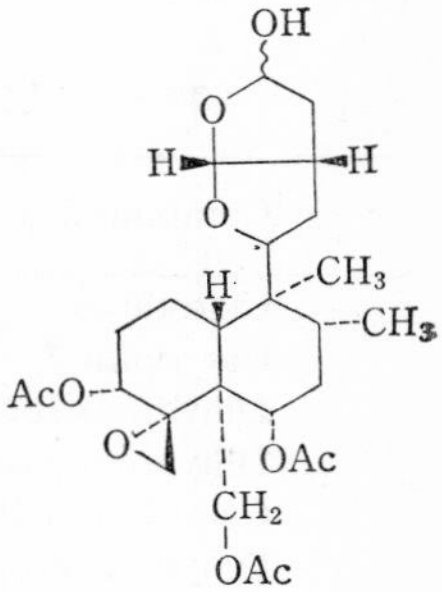 13

[Experimental procedure 3] Isolation of insect antifeedants from "*kariganeso*"

Dried ground leaves and stems (43 kg) were extracted with ether for 3 days at room temperature. This operation was repeated twice.

The filtered ethereal extract was concentrated under reduced pressure to give 850 g of crude gum. This gum (100 g) was chromatographed on neutralized alumina (3 kg; Brockmann grade IV–V; water content, 15–20%) and eluted with benzene, ether-benzene, ethyl acetate-benzene, and ethyl acetate. Each fraction was monitored by means of the insect antifeeding test and tlc. The ether-benzene (1:5) eluate was chromatographed on silica gel, eluting with ethyl acetate-*n*-hexane (1:5). The eluate was crystallized from ether-light petroleum to yield clerodin **5** as needles, mp 164–165° C. The ether-benzene (2:5) fraction was subjected to silica gel chromatography and eluted with ethyl acetate-*n*-hexane (2:5) to give crystalline material showing only one spot on tlc; this was recrystallized from ether to yield colorless prisms, **7**, mp 176–177° C. The ethyl acetate-benzene (1:1) eluate was chromatographed on a silica gel column and eluted with ethyl acetate-*n*-hexane (1:1) to give two crystalline fractions. The first fraction was recrystallized from ethyl acetate—ether-petroleum ether to yield colorless prisms, **8**, mp 169–170° C, and the second gave colorless prisms, **9**, mp 198.5–199.5° C, on recrystallization from the same solvent mixture. The first half of the ethyl acetate eluate was repeatedly subjected to silica gel chromatography, and two compounds were isolated. These compounds were recrystallized from a mixture of ethyl acetate, ether and petroleum ether to yield colorless prisms, **10**, mp 219–220° C, and trace amounts of colorless prisms, **11**, mp 204–205° C. Repeated chromatography of the latter half of the ethyl acetate eluate gave two crystalline materials, which showed *Rf* 0.45 and 0.40 on tlc (silica gel: 5% ethanol in chloroform). These two materials were recrystallized from ethyl acetate-petroleum ether to give colorless prisms, **12**, mp 179–181° C, and colorless prisms, **13**, mp 188–189° C.

TABLE 17.2 Antifeeding activity of the constituents of "*kariganeso*".

Compound	Threshold concentration (ppm)
Clerodin **5**	50
Caryoptin **7**	200
Dihydroclerodin-I **8**	50
Dihydrocaryoptin **9**	80
Caryoptinol **10**	200
Dihydrocaryoptinol **11**	100
Clerodin hemiacetal **12**	50
Caryoptin hemiacetal **13**	200

Antifeedants in Meliaceae Plants

In India, aqueous suspensions of the crushed fruits of *Azadirachta indica* (*Melia azadirachta*), belonging to Meliaceae, used to be sprayed for insect control many years ago. The components in Meliaceae plants have been shown to have antifeeding activity against locusts, *Locusta migratoria* and *Scistoceraca gregaria*. Lavie *et al.* developed a bioassay for locust antifeedants in which fifth mid-stadium hoppers were given filter paper soaked with 0.25 M sucrose and various concentrations of samples.[11] The antifeeding activity was expressed in terms of the weight of sample contained in 1 cm^2 of filter paper. An insect antifeeding triterpenoid, meliantriol, $C_{30}H_{30}O_5$, isolated and identified from *M. azadirachta*, showed 100% antifeeding activity at a level of 3 μg per cm^2.[11] Another group independently isolated a different compound named azadiractin **14**, $C_{35}H_{44}O_{16}$, from the seeds of the same plant using a similar feeding test with the same hopper.[12)]

Azadiractin showed 100% antifeeding activity at a concentration of 1 ng per cm^2. This compound was assigned the structure **14** based on chemical[13)] and spectral data.[14)]

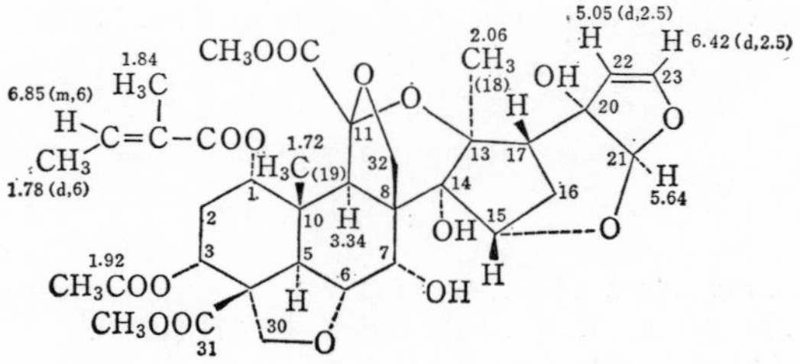

14

17.3 Concluding Remarks

Ten years have passed since we began to study insect antifeeding plant constituents. In the course of studies on antifeedants in various plant species, we found about twenty new compounds, and studied their chemical structures and biological activities against insects. These bioactive compounds belong to many groups, namely, terpenoids, alkaloids, coumarins and lignans. Thus, we had to develop methods appropriate for each of these groups.

Based on our experience, the following methodology is recommended for studying insect antifeedants.

1) Establishment of a method for rearing test insects throughout the year.
2) Survey of insect-resistant plants.
3) Establishment of a leaf disk test.
4) Screening of solvent extracts of plant leaves using the leaf disk test.
5) Collection of large amounts of leaves having antifeeding activity; isolation, monitored by means of the leaf disk test; identification of active compounds.
6) Synthesis of the natural antifeedants and related compounds; field tests of synthetic compounds.

If strongly active antifeedants can be developed, they should be valuable for crop protection from insects. These antifeedants have modes of action different from those of pesticides, and may overcome the environmental problems caused by pesticides.

Insect antifeedants are considered to be one of the important resistance factors in insect-resistant varieties of crops. Identification of the antifeeding substances in a resistant variety will aid the production of new resistant varieties by hybridization and selection, because the contents of the antifeedants in hybrids can be used as an indicator of insect resistance.

References

1) K. Munakata, *Pure Appl. Chem.*, **42**, 57 (1975).
2) K. Wada, K. Munakata, *J. Agr. Food Chem.*, **16**, 471 (1968).
3) N. Kato, M. Shibayama, K. Munakata, *J. Chem. Soc. Perkin Trans. I,* **1973**, 712.
4) N. Kato, K. Munakata, C. Katayama, *ibid.*, **1973**, 69.
5) D.H.R. Barton, N. T. Cheung, A. D. Cross, *J. Chem. Soc.*, **1961**, 5061.
6) S. Hosozawa, N. Kato, K. Munakata, *Phytochemistry*, **12**, 1833 (1973).
7) S. Hosozawa, N. Kato, K. Munakata, *ibid.*, **13**, 1019 (1974).
8) S. Hosozawa, N. Kato, K. Munakata, *Tetr. Lett.*, **1974**, 3753.
9) N. Harada, H. Uda, *J. Am. Chem. Soc.*, **100**, 8022 (1978).
10) D. Rogers, G. G. Unal, D. J. Williams, S. V. Ley, *Chem. Commun.*, **1979**, 97.
11) D. Lavie, M. K. Jain, S. R. Shpan-Gabrielith, *ibid.*, **1967**, 910.
12) J. H. Butterworth, E. D. Morgan, *ibid.*, **1968**, 23.
13) J. H. Butterworth, E. D. Morgan, G. R. Percy, *J. Chem. Soc., Perkin Trans. I,* **1972**, 2445.
14) P. R. Zanno, I. Miura, K. Nakanishi, D. L. Elder, *J. Am. Chem. Soc.*, **97**, 1975 (1975).

General References

i) C. Hirano, *Konchu to Kishushokubutsu* (in Japanese), Kyoritsu Shuppan (1971).
ii) D. L. Wood, R. M. Silverstein, M. Nakajima (eds.) *Control of Insect Behavior by Natural Products*, Academic Press (1970).
iii) K. Yamashita, *Biologically Active Substances Produced by Plants* (in Japanese), Nankodo (1975).

CHAPTER 18

Isolation of Biologically Active Substances from Ichthyotoxic Plants

The roots of *Derris elliptica* (Leguminosae), which contain rotenone, a naturally occurring insecticide, have been used for hundreds of years as a stupefying agent for catching fish. Many kinds of plants besides *Derris* roots are known to be used as fish stupefying agents, so-called fish poisons. Today this method of catching fish is generally prohibited, but the custom remains as a traditional ceremony or recreation in some tropical areas.[1] The plants used traditionally for catching fish (ichthyotoxic plants) were surveyed and summarized by Burkill[2] and Altschul.[3]

These plants range over various families, and the toxic parts may include the roots, seeds, fruits, bark, latex and/or leaves.

Since the discovery of rotenone as an insecticide, extensive chemical studies on the constituents of ichthyotoxic plants have been carried out. However, it is difficult to isolate minor and non-crystallizing bioactive constituents without monitoring the bioactivity.

Recently it has been found that some ichthyotoxic plants contain not only piscicidal or insecticidal substances but also various other bioactive substances such as plant growth inhibitors (*Juglans mandshurica*), insect antifeedants (*Melia azadirachta, M. azedarach*), antitumor agents (*Stephania hernandiifolia, Daphne mezereum*), and co-carcinogenic or irritant substances (*Croton tiglium*).

The author has been investigating the constituents of some species of ichthyotoxic plants by monitoring various bioactivities.[4] Among the plants investigated, the leaves of *Viburnum awabuki* (Japanese name, *sangoju*), which were used as a fish poison in Okinawa Prefecture, are taken here as an example, and the experimental procedures used to isolate a piscicidal substance, vibsanine A and a plant growth inhibitor, vibsanine B, are described. Vibsanine A, $C_{25}H_{38}O_4$, and vibsanine B, $C_{25}H_{36}O_5$, have the structures **1** and **2**, i.e., β,β-dimethylacrylic acid esters of novel

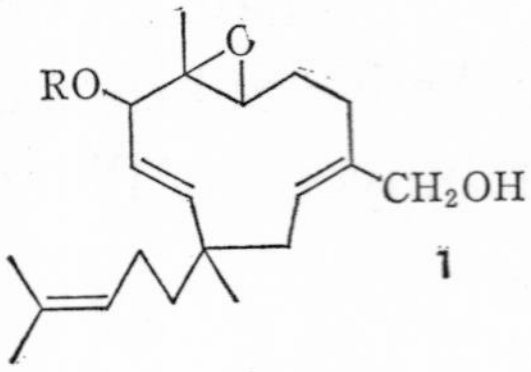

RO OH O CH₂OH **1**   RO OH O CH₂OH **2**

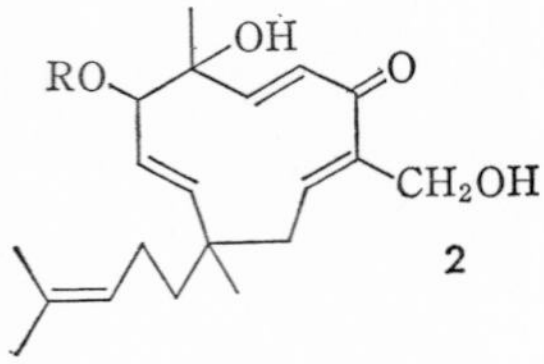

R= $(CH_3)_2C{=}CH{-}CO{-}$

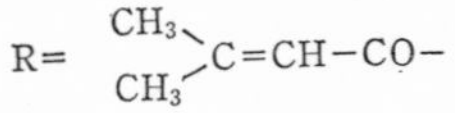

diterpenes with the humulene skeleton. The piscicidal activity of vibsanine A, in terms of the TLm (24 hr) value against the killie-fish (*Oryzias latipes*) is 0.1 μg/ml, and the plant growth inhibitory activity, IC_{50}, for rice seedling growth, of vibsanine B is 60 μg/ml.[5)]

18.1 Bioassay

In order to isolate bioactive substances from ichthyotoxic plants it is desirable to have available as many kinds of bioassay method as possible. These bioassay methods should be simple and rapid. The piscicidal test, the rice seedling growth test, and the *Cochliobolus miyabeanus* conidium germination test[6)] were used as fractionation guides for isolating various pesticidal substances. More recently a new test method using *Drosophila melanogaster* (a kind of feeding test) has been devised[7)] for insect development inhibitory activity.

Piscicidal Test

Piscicidal activity was tested by a contact method, *i.e.*, by examining whether fish are killed by a compound poured into the water. In this method a compound tested is considered to be absorbed through the gills directly into the blood.

In the standard method for evaluating the toxicity of commercial insectides in Japan, the young of *Cyprinus carpio* are used, but this species can be replaced by *Pseudorsbora parvae* or *Oryzias latipes*.

Since the killi-fish (*Oryzias latipes*) is commercially available as an ornamental fish, and is cheap, the author used this as a test fish. In the standard method, ten fish were used in 10 l of water, but the activity of constituents fractionated at each step of the isolation procedures was examined by observing five fish in 150 ml of water after 24 hr, because of the small amounts of constituents available.

[Experimental procedure 1] Piscicidal test of the concentrate of the methanol extract of *Viburnm awabuki.*

Killi-fish, *Oryzias latipes*, averaging 350 mg in weight and 3–3.5 cm in length were purchased and reared in a water tank and were not fed for two days before use in a test. A methanol solution (1 ml) of the methanol concentrate at various concentrations (dilution ratio, 2) was added with vigorous stirring to water (150 ml) in a beaker (200 ml) aerated for two days prior to use by means of an air-pump. The solution prepared by adding only the solvent served as a control. Five fish were introduced into the test solution, which was maintained between 18° and 19° C during the test. Fish that died during the test were immediately removed from the solution, and the number of survivors in each container was recorded at 24 hr after their introduction. If more than one fish died in the control, the results were considered unreliable, and the test was repeated. The result is represented by minimum lethal concentration, and that of the methanol concentrate was 60 μg/ml.

If a test compound is poorly soluble in methanol, acetone or dimethyl sulfoxide can be used.

Rice Seedling Growth Test

The test can be done with any cultivar of rice insofar as the seeds germinate. It is convenient to use a dwarf rice, e.g., "*kotaketamanishiki*" or "*tanginbōzu*", for detecting gibberellins too. Most exogenous inhibitors inhibit only root growth, while abscisic acid inhibits the growth of the root and leaf sheath. Growth inhibitors are surveyed by measuring the length of the second leaf sheath and the weight of roots in comparison with those of a control.[8)]

[Experimental procedure 2] Rice seedling growth inhibitory test of fractions from charcoal column chromatography of the ethyl acetate-soluble material

Rice seeds were soaked in 99% ethanol, washed three times with water, then soaked in 1% sodium hypochlorite solution for 1 hr and thoroughly washed with water. These seeds, soaked in water in a Petri dish, were kept at 30° C under illumination (2,000–6,000 lux) to germinate. After two days, seeds with ca. 1 mm of plumule were selected for test. Concentrate (1 mg) from each chromatography fraction dissolved in ethyl acetate (0.5 ml) was added to a glass tube (3 cm × 13 cm). The solvent was removed by blowing air followed by evacuation in a desiccator. Water

(2 ml) was added to the tube and mixed with the compound and 7 germinated seeds were placed on the bottom. The tube was covered with a polyethylene film, and kept for a week under the conditions mentioned above. After removing water on the surface of the seedlings with a filter paper, the length of the second leaf sheath was measured. Root weight was obtained as the difference between the total weight of the 7 seedlings and that of the seedlings from which the roots had been cut with a scissors. The ratios of the length of the second leaf sheath and the root weight to those of the control are shown in Table 18.1. Activity was observed in the water-acetone (20:80) fraction, and the root growth was more markedly inhibited than the leaf sheath growth. Since the ethyl acetate-soluble material before this chromatography inhibited root growth by 5% compared to the control at a dose of 6 mg/2 ml, the activity was enhanced 6 times by this purification procedure.

TABLE 18.1 Rice seedling growth test of fractions from charcoal chromatography (dose 1 mg).

Fraction	Water-acetone				
	40:60	30:70	20:80	10:90	0:100
Second leaf sheath elongation (% of control)	87	88	52	95	94
Root weight (% of control)	52	60	22	74	70

18.2 EXTRACTION AND ISOLATION

An aqueous solution prepared by soaking finely shredded leaves (2.78 g) of *Viburnum awabuki* in water (1 l) overnight showed piscicidal activity. This activity was also observed when an extract was prepared with methanol. Therefore, the leaves were extracted with methanol. The piscicidal activity in the methanol extract could be transferred to benzene or ethyl acetate by extraction with these solvents, but the plant growth inhibitory activity was not completely transferred to the benzene phase.

In an initial experiment the concentrate of the methanol extract was partitioned with ethyl acetate. At every stage of the purification procedures, bioassays should be carried out to identify the fraction(s) containing active principle. The bioassay was performed at three successive dilutions (dilution ratio, 2), beginning from half the minimum effective concentration at the previous step of purification.

If the activity was not enhanced proportionally with the decreasing

TABLE 18.2 Isolation procedures for bioactive constituents from the leaves of *Viburnum awabuki.*

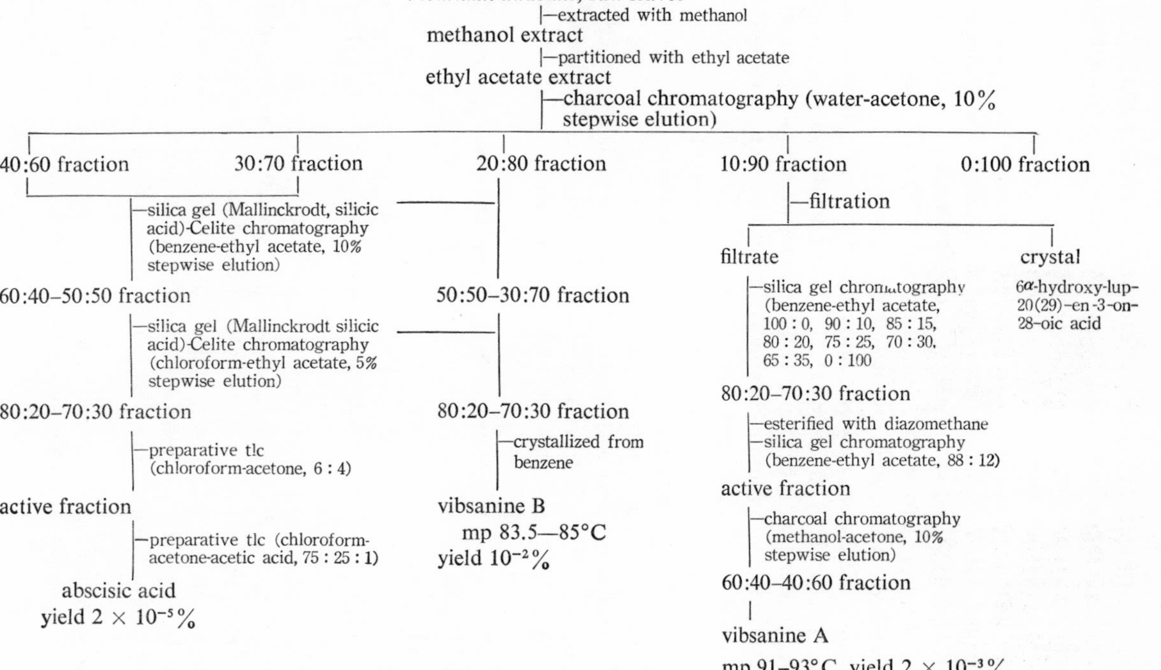

amounts of constituents in the active fraction, the purification procedure might be inadequate, or more than one active principle might participate synergistically in the activity.

Ethyl acetate-soluble material was chromatographed on charcoal, eluting with water-acetone to give a plant growth inhibitor in the water-acetone (40:60, 30:70, 20:80) fractions and a piscicidal substance in the water-acetone (10:90, 0:100) fractions. Each fraction was further purified as shown in Table 18.2 to afford abscisic acid, vibsanine B (a plant growth inhibitor) and vibsanine A (a piscicidal substance).

As both vibsanine A and vibsanine B were soluble in benzene, for large-scale preparations the concentrate of the methanol extract of the leaves was extracted with benzene, and the benzene-soluble material was subjected to charcoal column chromatography. This requires both less time and less charcoal. The procedure was as follows. Raw leaves (5 kg) were soaked in methanol in a polyethylene bottle (20 l) for a month. The methanol extract was concentrated in a 3 l three-mouthed round-bottomed flask below 40° C under a nitrogen atmosphere. On one mouth of the flask, a Dimroth condenser was mounted, through which ice-cooled water was circulated, and the extract was supplied automatically through the center mouth. This apparatus enabled us to concentrate the methanol extract at a rate of 1 l/hr.

A flash evaporator with steam heating can concentrate the material three times faster than this apparatus, but results in the formation of artifacts due to the high temperature.

In order to remove a large amount of pigments contained in the ethyl acetate-soluble fraction, this fraction was subjected to charcoal chromatography, eluting stepwise with water-acetone.

The ethyl acetate-soluble material caused 100% mortality in 24 hr at a dose of 3 mg, and inhibited rice seedling growth at a dose of 2 mg. Therefore, each fraction from the charcoal chromatography was assayed for piscicidal and rice seedling growth inhibitory activities at these doses. As shown in Table 18.3, these two activities were fortunately found in separate fractions. From the water-acetone (40:60–30:70 and 20:80)

TABLE 18.3 Piscicidal and rice seedling growth inhibitory activities of fractions from charcoal chromatography of ethyl acetate-soluble material.

Fraction	Water-acetone				
	40:60	30:70	20:80	10:90	0:100
Rice seedling growth test†[1] (dose 2 mg)	+	+	+	—	—
Piscicidal test†[2] (dose 3 mg)	—	—	+	+	+

†[1] +, root weight is less than 30% of the control.

†[2] +, no survivors after 24 hr.

fractions, abscisic acid and vibsanine B were isolated by repeated chromatography and preparative thin layer chromatography, respectively.

The isolation procedures for vibsanine A from the water-acetone (10:90) fraction are described in detail below.

[Experimental procedure 3] Charcoal chromatography

The benzene-soluble material of the methanol concentrate was fractionated by charcoal columns (6.2 cm × 45 cm). The benzene-soluble material (360 g) was divided into 4 parts, each of which was applied to a column. Charcoal (240 g) for chromatography was suspended in water-acetone (1:1) in a 2 l round-bottomed flask and the suspension was degassed by evacuation. Celite 545 (10 g) was applied with water-acetone (1:1) to a glass column tube (6.2 cm × 60 cm) fitted with a glass filter (No. 2) and a stopcock on the bottom. The charcoal suspension was poured in, and water-acetone (1:1) was passed several times through the adsorbent to settle the column. Charcoal (40 g) was added to the benzene-soluble material (90 g) dissolved in acetone (100 ml) and then water (66 ml) was added to adjust the ratio of water to acetone to 40:60. The sample adsorbed on the charcoal was placed gently on the top of the column with water-acetone (40:60) and eluted with five solvents containing stepwise-increasing ratios of acetone. Each fraction was assayed for piscicidal activity at doses of 1 mg and 0.5 mg. The dry weights eluted in each fraction and their piscicidal activities are listed in Table 18.4.

TABLE 18.4 Amount and piscidal activity of fractions from charcoal chromatography.

Fraction		Water-acetone				
		40:60	30:70	20:80	10:90	0:100
Amount (g)		1	5	16	15	8
Piscicidal activity†	dose, 1 mg	−	−	+	+	+
	0.5 mg	−	−	−	+	−

† +, no survivors after 24 hr.

Among the fractions from the charcoal column chromatography the water-acetone (10:90) fraction, which showed the strongest piscicidal activity, was concentrated to give crystals. These crystals, mp 267–268° C, showed no piscicidal activity on recrystallization from ethyl acetate and were identical with 6α-hydroxy-lup-20(29)-en-3-on-28-oic acid **3**, isolated by Ageta[9)] from the leaves of this plant.

[Experimental procedure 4] Silica gel chromatography of the water-acetone (10:90) fraction from the charcoal column.

The remainder (55 g) after removal of the crystalline triterpene carboxylic acid from the water-acetone (10:90) fraction (80 g) was adsorbed on silica gel (Merck, 0.2–0.5 mm) (60 g) with benzene. A slurry of silica gel (750 g) in benzene (1.2 l) was put in a glass tube (6.2 cm × 60 cm) with a glass filter and a stopcock, and one volume of benzene was passed through the column. The sample, adsorbed on charcoal, was applied to the top of the column, which was eluted with 8 portions of 1.5 l of benzene and ethyl acetate in ratios of 100:0, 90:10, 85:15, 80:20, 75:25, 70:30, 65:35, and 0:100. Amounts of material eluted in each fraction and their piscicidal activities are listed in Table 18.5. Comparison of the thin layer chromatograms of each fraction (Fig. 18.1) suggested that the piscicidal constituent corresponded to spot I.

TABLE 18.5. Amount and piscicidal activity of fractions from silica gel chromatography.

Fraction	Benzene-ethyl acetate							
	100:0	90:10	85:15	80:20	75:25	70:30	65:35	0:100
Amount (g)	0.4	0.5	2.2	5.1	11.2	9.5	9.6	12
Piscicidal activity† (dose 0.3 mg)	−	−	−	+	+	+	−	−

† +, no survivors after 24 hr.

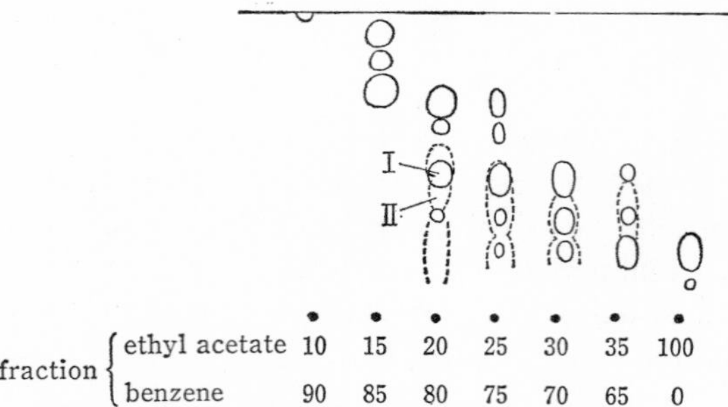

Fig. 18.1. Thin layer chromatogram. Adsorbent, silica gel G + GF_{254} (2:1); developing solvent, benzene-ethyl acetate (2:1); detection under ultraviolet light or with vanillin-sulfuric acid. Dotted line, uv non-absorbing spot.

The tailing spot II, which does not absorb ultraviolet light, and overlaps with spot I, is evidently that of the triterpene carboxylic acid. It was expected that the methyl ester of the latter would have a considerably higher *Rf* value than the acid itself, so the acid could probably be separated as its methyl ester from the piscicidal constituent. Since methylation of the active fraction with diazomethane was confirmed not to decrease the piscicidal activity of the fraction, the active fractions (benzene-ethyl acetate, 80:20–70:30) were esterified with diazomethane. The methylated product showed, as expected, an ultraviolet light non-absorbing spot at a higher *Rf* value on the thin layer chromatogram (Fig. 18.2). This methylated mixture was developed with benzene-ethyl acetate (2:1) on a silica gel

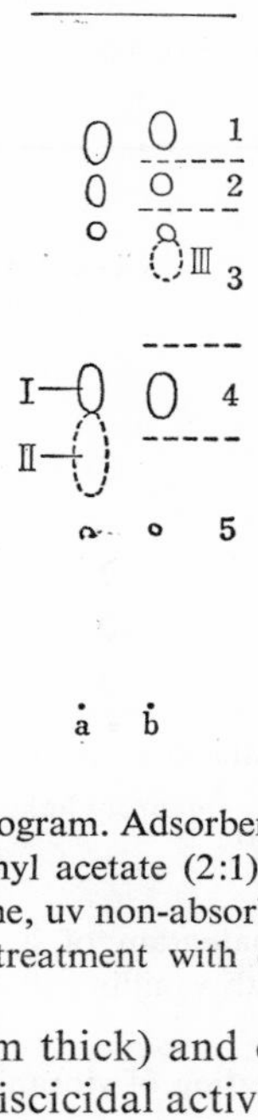

Fig. 18.2. Thin layer chromatogram. Adsorbent, silica gel G + GF_{254} (2:1); developing solvent, benzene-ethyl acetate (2:1); detection under uv or with vanillin-sulfuric acid. Dotted line, uv non-absorbing spot. a, Before treatment with diazomethane; b, After treatment with diazomethane.

G + GF_{254} (2:1) plate (0.5 mm thick) and divided into five fractions as shown in Fig. 18.2 to test for piscicidal activity. Table 18.6 shows that the fourth fraction was active. Thus, the uv-absorbing spot, which gave a

TABLE 18.6. Piscicidal activity of fraction from thin layer chromatography.

Fraction	1	2	3	4	5
Amount eluted (mg)	1.0	6.4	0.5	0.5	0.9
Dose (mg)	1.0	1.6	0.5	0.5	0.9
Activity†	—	—	—	+	+

† +, no survivors after 24 hr.

purple color after spraying with 5% vanillin in sulfuric acid was considered to correspond to the active constituent. This fraction was chromatographed on silica gel using benzene-ethyl acetate (88:12) as an eluent, and collecting small fractions. The active fractions were combined on the basis of thin layer chromatography results.

Though the piscicidal fraction appeared as one spot on tlc (silica gel; benzene-ethyl acetate (1:2)), its color tone against the spraying reagent, vanillin-sulfuric acid, suggested that the piscicidal constituent was overlapping with an analogous constituent. Further tlc conditions were explored (Fig. 18.3), and it proved possible to separate one spot from the other. Preparative thin layer chromatography under these conditions showed that the upper spot was due to the piscicidal constituent and the lower one to the analogous compound. However, preparative tlc is not suitable for large-scale preparation of unstable compounds.

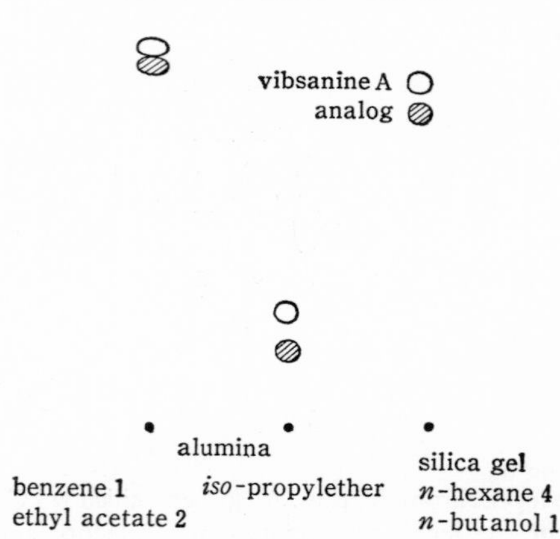

Fig. 18.3. Thin layer chromatogram of a mixture of vibsanine and its analog. Detection by uv or with vanillin-sulfuric acid.

[Experimental procedure 5] Isolation of vibsanine A an and analogous compound by charcoal chromatography

A difference in the absorptivities of these compounds on charcoal was anticipated on the basis of their different uv absorption intensities (253 nm). Thus, charcoal column chromatography using methanol-acetone was carried out.

The combined active fraction from silica gel column chromatography was used. The crude active constituent (3 g) was dissolved in methanol and adsorbed on Celite 545 (6 g). A degassed suspension of charcoal (for chromatography) (180 g) in water-acetone (50:50) was put in a glass tube (ϕ 4.8 cm) and methanol was passed through the column to settle it. The sample, adsorbed on Celite, was placed on the top of the column. Stepwise elution in which the ratios of acetone were increased by 10% steps afforded vibsanine A in the methanol-acetone (60:40–40:60) fraction, and the analogous compound in the (30:70–0:100) fraction. The tlc of each eluate is shown in Fig. 18.4. The yields of fractions containing vibsanine A and its analog were 2 g and 1 g, respectively.

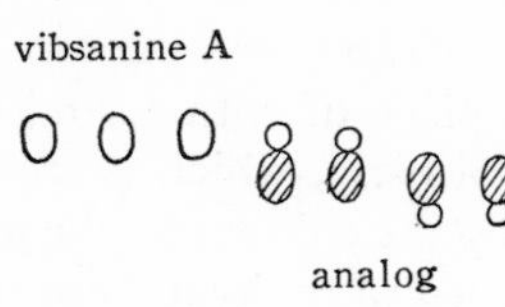

fraction								
	acetone	40	50	60	70	80	90	100
	methanol	60	50	40	30	20	10	0

Fig. 18.4. Thin layer chromatogram of fractions from charcoal chromatography. Silica gel G + GF_{254} (2:1); solvent, *n*-hexane-*n*-butanol (4:1); detection by uv or with vanillin-sulfuric acid.

[Experimental procedure 6] Separation of vibsanine A and its analog by dry column chromatography

Vibsanine A and its analog obtained in the fractions from "Experimental procedure 5" were still mutually contaminated. For the purpose of separating the mixture, silica gel dry column chromatography was employed. An earlier method[10] proposed the use of a thin film nylon tube packed with a dry fluorescent adsorbent. The column was loaded with a mixture to be fractionated, developed, marked under ultraviolet light, and then sliced. However, it is difficult to pack such a thin film tube evenly and tightly with adsorbents. As all the bioactive constituents of the leaves of *Viburnum awabuki* absorb ultraviolet light, a quartz tubing (30 cm long) was used and marked under short-wavelength ultraviolet light (253 nm). By the use of tapering tubing, it is possible to push out the adsorbent towards the wider end of the tubing. The method is as follows.

Silica gel PF_{254} (Merck, no binder) (50 g) was deactivated by mixing thoroughly with water (7.5 g). A mixture (180 mg) of vibsanine A and its analog dissolved in a small amount of benzene was adsorbed on the deactivated silica gel (760 mg).

In a quartz tubing (2.1 cm × 30 cm) with one end plugged with a rubber stopper was put the compound-silica gel mixture. The deactivated silica gel equilibrated with 10% *n*-hexane-*n*-butanol (4:1) was poured in until the tube was two-thirds full, at which point the adsorbent was compacted in the tube by tapping the rubber stopper end on a desk. The tube was then filled to the top and compacted. The top of the column was pressed with a filter paper and cotton. The column tube was inverted, the stopper was removed, and then the top of the column was covered with a filter paper and a piece of cotton, and connected to a solvent reservoir. *n*-Hexane-*n*-butanol (4:1), which is transparent to short-wavelength uv, was used to develop the column by the descending technique. The column was examined under ultraviolet light and the desired zones were successively removed with a spatula and extracted with ethyl acetate.

Vibsanine A (40 mg) and its analog (130 mg) were each isolated in a pure state.

REFERENCES

1) Masatake Matsubara, *Kikan Jinruigaku* (in Japanese), **1**, 129 (1970).
2) I. H. Burkill, *A Dictionary of the Economic Products of the Malay Peninsula,* Oxford University Press (1935).
3) S. von Reis Altschul, *Drugs and Foods from Little-known Plants,* Harvard University Press (1973).
4) Kazuyoshi Kawazu, *J. Synth. Org. Chem Japan* (in Japanese), **30**, 615 (1972).
5) Kazuyoshi Kawazu, Tetsuo Mitsui, *Symposium Papers, the 18th Symposium on the Chemistry of Natural Products*, p. 77, (1974).
6) Shigeyasu Akai, Kiichi Katsura (eds.), *Syokubutsubyōgaku Jikken Nōtsu* (in Japanese), p. 233, Yōkendō (1974).
7) Kazuyoshi Kawazu, Makoto Ariwa, Yoshiaki Kii, *Agric. Biol. Chem.*, **41**, 223 (1977).
8) Masayuki Takeuchi *et al.* (eds.), *Syokubutsusoshikibaiyō* (in Japanese), p. 445, Asakurashoten (1972).
9) Hiroyuki Ageta, Reiko Kobayashi, *Abstracts of Papers, the* **90***th Annual Meeting of the Pharmaceutical Society of Japan* (in Japanese), p. II-182 (1970).
10) B. Loev, M. M. Goodman, *Chem. Ind.*, **1967**, 2026.

CHAPTER 19

Biosynthetic Studies on Secoiridoid Glucosides

The term "secoiridoid glucoside" refers to substances bearing the carbon skeleton formed by the fission of the C-7,8 bond of iridoid glucosides, as shown in Fig. 19.1, and the number of known glucosides of this class amounts to more than fifty.[1,2]

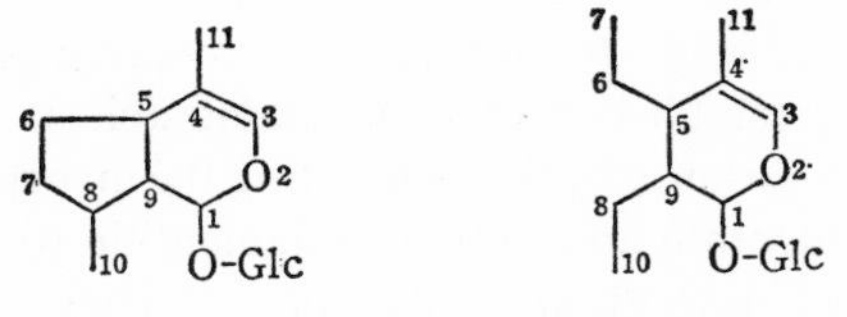

iridoid glucosides secoiridoid glucosides

Fig. 19.1

Secoiridoid glucosides are classified according to their structures into: (i) sweroside **1** type having a vinyl group at C-9, (ii) morroniside **2** type which might be formed by addition of an oxygen function to the vinyl side chain of a sweroside **1**, (iii) oleuropein **3** type, having an ethylidene or hydroxyethylidene group at C-9, and (iv) alkaloidal glucosides, such as vincoside **4**, formed by condensation of tryptamine or tryptophan with a secoiridoid glucoside.

The biosynthesis of secoiridoid glucosides has attracted much attention since these glucosides were recognized to have the same carbon skeleton as the non-tryptophan portion of a number of indole alkaloids. Initially, several hypotheses were presented concerning the origin of the non-tryptophan portion of indole alkaloids. However, in 1965–1967, Scott, Arigoni, Battersby, Leete and their associates confirmed the incorporation of mevalonic acid or geraniol into *Vinca rosea* alkaloids such as vindoline.[3,4] Thus, it was established that the non-tryptophan portion of the indole alkaloids is of terpenoid origin.

Regarding the biosynthesis of secoiridoid glucosides, Coscia *et al.*[5] and Inouye *et al.*[5] demonstrated in 1967 the incorporation of mevalonic acid into sweroside **1**, swertiamarin **5** and gentiopicroside **6**, revealing the monoterpenoid origin of these glucosides.

In this chapter, an outline of recent studies on the biosynthesis of secoiridoid glucosides will be presented, followed by some representative examples of experimental procedures used in these studies. Lack of space, however, does not allow us to deal with the alkaloidal glucosides of group (iv) which should more appropriately be dealt with as indole alkaloids.

19.1 Biogenesis of Secoiridoid Glucosides

The Biosynthetic Route from Mevalonic Acid to Secologanin

Though these glucosides, as mentioned above, are terpenoids biosynthesized from mevalonic acid, it is noteworthy that randomization of the label between C-3 and C-11 was observed in these glucosides when [2-^{14}C]mevalonic acid was administered. The research groups of both Arigoni and Battersby established that mevalonic acid is incorporated into these glucosides via geraniol by demonstrating the incorporation of labeled geraniol into foliamenthin and dihydrofoliamenthin (both are derivatives of secologanic acid) in *Menyanthes trifoliata*.[7,8] They have also examined the incorporation of 10-hydroxygeraniol, 10-hydroxycitronellol, 10-hydroxylinalool, etc., into loganin **8** and the alkaloids of *Vinca rosea*, and it was found that only 10-hydroxygeraniol and 10-hydroxynerol were metabolized as precursors.[9,10] Arigoni *et al.* further proposed that, of these two substances, 10-hydroxynerol, a supposed direct precursor, could cyclize as shown later in Fig. 19.3, resulting in randomization of the label of [2-^{14}C]-mevalonic acid.

Regarding the biosynthetic processes after the formation of the iridoid skeleton, the present authors demonstrated in 1969 the intermediacy of deoxyloganic acid **9** in the biosynthesis of many iridoid and secoiridoid glucosides from a survey of the structures of many iridoid glucosides and from results obtained by feeding labeled 7-deoxyloganic acid **9** to various plants.[11] The observation that **9** is incorporated into loganin **8**[11a,11c] and jasminin, a glucoside of the oleuropein type,[12] provided the first evidence that 7-deoxyloganic acid **9** is hydroxylated at C-7 to give loganic acid **10** or loganin **8** and that **9** is also a precursor of a secoiridoid jasminin. Later, Battersby *et al.* also succeeded in the incorporation of 7-deoxyloganin **11** into **8** and indole alkaloids.[13]

Incorporations of **8** and **10** labeled with ^{14}C or ^{3}H into secologanin **12**[14,15] (or secologanic acid **7**[15]), gentiopicroside **6**[11a,12b,16], jasminin[12] and morroniside **2**[12b,17] provided further direct evidence for the intermediacy of loganin **8** or loganic acid **10** in the biosynthesis of secoiridoid glucosides. Using morroniside **2** and secologanin **12**, it was confirmed by the authors and also by Battersby's group that the proton at C-7 in loganin **8** is retained after cleavage of the cyclopentane ring.

Thus, it has been established that 7-deoxyloganic acid **9** (or 7-deoxyloganin **11** and loganic acid **10** (or loganin **8**) intervene between alicyclic terpenes and secoiridoid glucosides as the key biosynthetic intermediates. The next problem is the mechanism of cleavage of the cyclopentane ring during the conversion of loganin **8** to secologanin **12**. The ring cleavage could be considered to take place at one of the following four stages, (i) 7-dehydrologanin **13**, (ii) the 7,8-dihydroxy compound (for example, gentioside **14**), (iii) 10-hydroxyloganin **15** and (iv) loganin **8** itself. The first

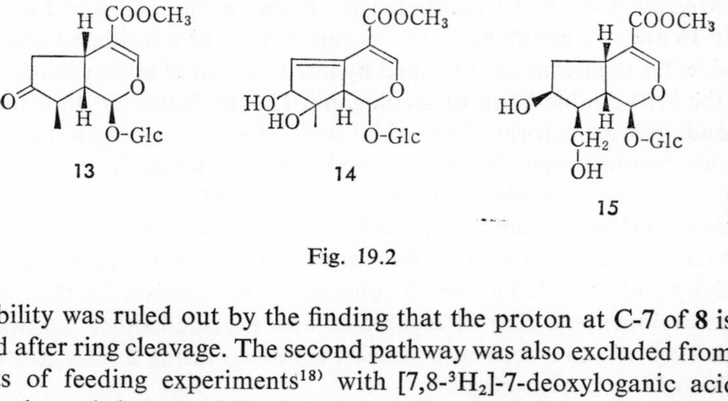

Fig. 19.2

possibility was ruled out by the finding that the proton at C-7 of **8** is retained after ring cleavage. The second pathway was also excluded from the results of feeding experiments[18] with [7,8-$^{3}H_2$]-7-deoxyloganic acid **9**, which showed the retention of both tritium atoms in secologanin **12** and morroniside **2**. Although cleavage of the cyclopentane ring could thus occur at stages (iii) or (iv), the details have not yet been elucidated. This remains one of the most important unsolved problems in the biosynthesis of secoiridoid glucosides.

The Biosynthesis of Each Group of Glucosides from Secologanin

Sweroside-type Glucosides: Besides the glucosides **1, 5, 6, 7** and **12,** substances such as secologanoside (a C-7 carboxylic acid congener of secologanic acid **7**) and its 11-methyl ester, foliamenthin, dihydrofoliamenthin, bakankosin (a lactam corresponding to **1**), amarogentin, amaroswerin, amaropanin, trifloroside and centapicrin (these are ester derivatives of **1** or **5**) belong to this type of secoiridoid glucosides, and are widely distributed in Gentianaceae. The incorporation of [2-^{14}C]mevalonic acid into **1**, **5** and **6** in *Swertia japonica* and *Gentiana triflora* was observed by the authors, and the biosynthetic sequence **1** → **5** → **6** was suggested based on a comparison of the ratios of specific incorporation into **1** and **5**, which occur together in *Swertia japonica*, and from a consideration of their structural features. This view was confirmed by the high incorporation of [10-^{14}C]sweroside **1**, prepared by chemical modification of sweroside **1**,[19] into gentiopicroside **6** on administration to *Gentiana scabra*. Accordingly, it was clear that reduction of the aldehyde group at C-7, followed by lactonization, hydroxylation, dehydration, esterification, etc., could lead to other glucosides such as **1**, **5**, **6**, amarogentin, amaroswerin and amaropanin. Coscia *et al.* carried out a stereochemical investigation of the biosynthesis of gentiopicroside **6** from mevalonic acid by examining the incorporation of 4R, [4-^{3}H]-, 4S, [4-^{3}H]-, 2R,[2-^{3}H]- and 2S, [2-^{3}H]-mevalonic acid administered simultaneously with [2-^{14}C]mevalonic acid to *Swertia caroliniensis* (see Fig. 19.3).[20]

Morroniside-type Glucosides: The glucosides morroniside **2** and kingiside **16** are two members of this group. It may be considered that morroniside **2** is the hemiacetal formed by introduction of an oxygen function into the 8,10-double bond of secologanin **12**, and kingiside **16** is the corresponding lactone formed by oxidation of the hemiacetal group of **2**. In fact, we demonstrated the incorporation of [7-^{3}H]-loganin **8** and [carbo-^{14}C-methoxy]secologanin **12** into morroniside **2** by feeding these substances to *Gentiana thunbergii* and *Cornus officinalis*, respectively.[12b,17] In particular, as mentioned above, the tritium atom of [7-^{3}H]-**8** is retained intact at the C-7 position of **2**. This result suggested a mechanism for the opening of the cyclopentane ring of loganin **8**, and excluded other pathways in which kingiside **16** is formed first and then morroniside **2** is formed by reduction of the lactone ring.

Oleuropein-type Glucosides: Substances belonging to this group are derivatives of oleoside or 10-hydroxyoleoside and are generally esterified at the C-7 carboxyl group with *p*-hydroxy-or 3,4-dihydroxyphenethylalcohol. In addition to oleuropein **3**, glucosides such as ligustroside **17**, 10-hydroxyligustroside **18**, 10-acetoxyligustroside **19**, 10-acetoxyoleuropein **20**, jasminin and nuezhenide belong to this type. No glucoside of this type has so far been isolated from plants belonging to any family other than Oleaceae. As briefly described above, 7-deoxyloganic acid **9** and lo-

mevalonic acid

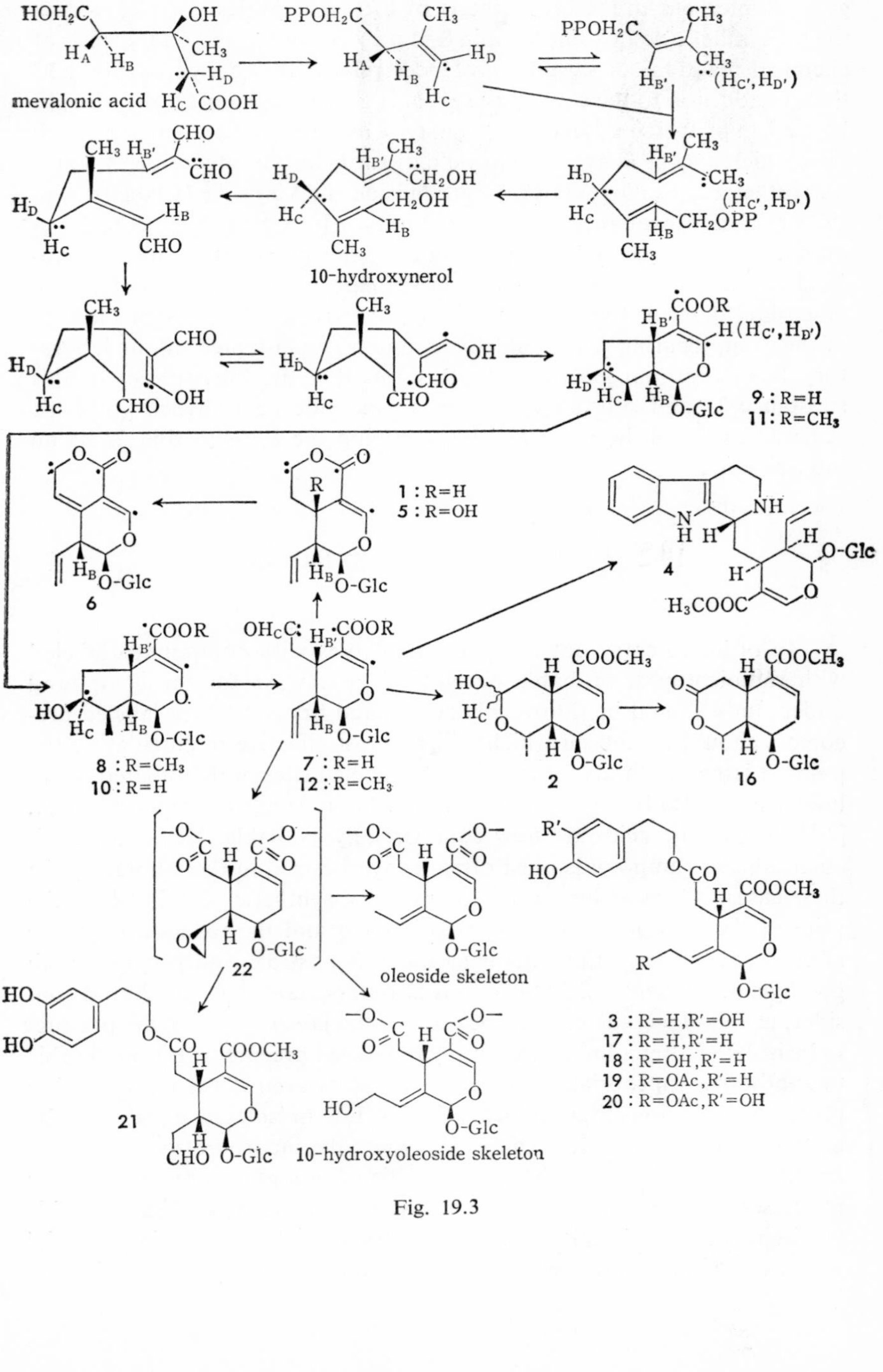

10-hydroxynerol

9 : R=H
11 : R=CH_3

1 : R=H
5 : R=OH

6

4

8 : R=CH_3
10 : R=H

7 : R=H
12 : R=CH_3

2

16

22

oleoside skeleton

3 : R=H, R′=OH
17 : R=H, R′=H
18 : R=OH, R′=H
19 : R=OAc, R′=H
20 : R=OAc, R′=OH

21

10-hydroxyoleoside skeleton

Fig. 19.3

ganin **8** intervene in the biosynthesis of these glucosides also. Regarding the biosynthetic route after loganin **8**, it may be considered on the basis of chemical reactions of several substances related to 7-dehydrologanin **13** that, in addition to the route passing through secologanin **12**, there might be a route by which a Baeyer-Villiger-type reaction of **13** gives 8-epikingiside, which may be dehydrated to yield oleuropein-type glucosides **3**. Thus, we synthesized secologanin **12**, 8-epikingiside and kingiside **16** labeled with ^{14}C or ^{3}H and examined their incorporation into oleuropein **3** and jasminin. It was concluded that glucosides of this group might also be biosynthesized via secologanin **12**, as in the case of the other two types of glucosides described above.[12)] Recently, ligustroside **21**, another member of the group bearing a C-10 aldehyde moiety, was isolated in our laboratory from *Ligstrum japonicum*.[21)] It seems that the biosynthesis of this substance **21** and substances having an oleoside or 10-hydroxyoleoside skeleton may well be explained by assuming the 8,10-epoxide **22** as an intermediate.

19.2 Synthesis of Labeled Precursors

In order to carry out biosynthetic studies with compounds labeled with radioisotopes, probable biosynthetic pathways for the compound under study should be thoroughly considered first, and then labeled precursors should be chosen which will be most effective in elucidating the point at issue. Methods for locating the labeled sites in the isolated metabolites must also be considered carefully. Many labeled compounds, e.g., [2-^{14}C]mevalonic acid, are now commercially available. However, particular labeled compounds must often be synthesized in the laboratory. In these cases, it is most important to develop a synthetic plan in which the yield of the isotope-introducing step is good and the reaction sequence after introduction of the label is as short as possible and proceeds with good yield. As mentioned above, it is also important that one should consider, at this stage, the ease of location of the labeled sites in the product to be isolated after administration of the labeled precursor. It is preferable to synthesize appropriately labeled precursors even if their preparation proves to be tedious than to locate the labeling in isolated metabolites by multistage degradations, because, generally, the incorporation of precursors into metabolites in higher plants is fairly low and lengthy degradation processes may lead to considerable loss of radioactivity. Thus, suitable synthetic routes to the labeled target substance from some starting material are not necessarily the same as the most convenient route in a cold run.

Bearing these considerations in mind, the synthesis of some labeled substances employed in our studies on the biosynthesis of secoiridoid glucosides are described below.

[Experimental procedure 1] Synthesis of [10-^{14}C] sweroside 1[19]

As mentioned on page 274, it had been inferred from the results obtained from the feeding experiment with [2-^{14}C]mevalonic acid that gentiopicroside **6** could be biosynthesized via sweroside **1** and swertiamarin **5**. Labeled sweroside **1** was required not only to verify this, but also to examine the incorporation of **1** into indole alkaloids. On the basis of the requirements discussed above, it was considered advantageous to introduce ^{14}C into the C-10 position of sweroside **1** by a chemical modification (Fig. 19.5). Osmium tetroixde oxidation of sweroside tetraacetate **23** gave the diol **24**, which was cleaved to the aldehyde **25** with sodium periodate. Compound **25** was allowed to react with [^{14}C]methylene-triphenylphosphorane to give [10-^{14}C]-**23**, which underwent the Zemplén reaction to furnish the desired [10-^{14}C]sweroside **1**.

Synthesis of triphenyl-[^{14}C]methyl-phosphonium bromide from [^{14}C] methyl bromide and triphenyl phosphine: A breakable ampoule A containing [^{14}C]methyl bromide (0.5 mCi) was connected to a manifold as shown in Fig. 19.4. Carrier methyl bromide (150 mg) was placed in the vessel B, and a benzene solution of triphenyl phosphine (300 mg, dissolved in 2 ml of anhydrous benzene) in vessel C. Stopcocks (b), (c), (d) and (e) were

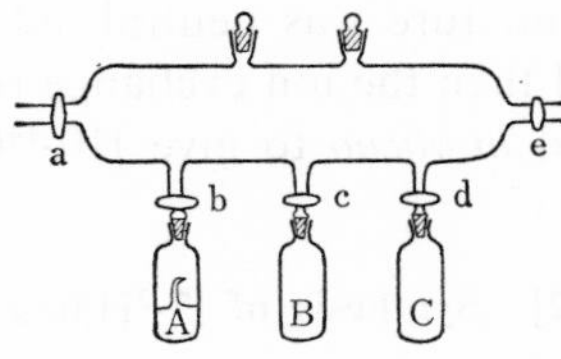

Fig. 19.4

closed and the apparatus was evacuated to a pressure of 1×10^{-2} mm Hg through stopcock (a) using a mercury diffusion pump. Stopcock (a) was then closed, and vessel B was cooled with liquid nitrogen. The ampoule A was then broken and the stopcocks (b) and (c) were opened for about 30 min. By this procedure, [^{14}C]methyl bromide was transferred to vessel B. After closing stopcocks (b) and (c) and then warming vessel B to room temperature, stopcocks (c) and (d) were opened while vessel C was kept cooled with liquid nitrogen. This procedure causes diluted [^{14}C]methyl bromide in vessel B to be transferred into vessel C. Finally, stopcocks (c) and (d) were closed and the reaction mixture was allowed to stand at room temperature for 2 days. Precipitates formed in vessel C were collected by centrifugation to give triphenyl-[^{14}C]methyl-phosphonium bromide (160 mg, 1.37 × 10^8 dpm/mmol).

Throughout all the experiments described below, radioactivity was measured in a liquid scintillation counter with samples dissolved in a scintillation mixture consisting of 2,5-diphenyloxazole (PPO) (50 mg), 2,2′-pheneylenebis-5-phenyloxazole (POPOP) (3 mg) and toluene (10 ml) or of PPO (70 mg), POPOP (5 mg),naphthalene (1 g) and dioxane (10 ml).

Synthesis of [10-^{14}C]sweroside **1**: A suspension of triphenyl-[^{14}C] methyl-phosphonium bromide (59 mg, specific activity 1.37 × 10^8 dpm/mmol) and carrier (60 mg) in anhydrous ether (7 ml) was treated with an ethereal 0.8 N *n*-BuLi solution (0.35 ml) under nitrogen, and the mixture was stirred at room temperature for 2.5 hr. The aldehyde **25** (110.9 mg) dissolved in anhydrous tetrahydrofuran (5 ml) was then added, and the reaction mixture was stirred for a further period of 2 hr. The solvent was removed *in vacuo*, carrier **23** (11.0 mg) was added to the residue and the mixture was purified by chromatography on silica gel. After recrystallization from ethanol, [10-^{14}C]sweroside tetraacetate **23** (21.5 mg) was obtained as colorless needles, mp 165–166° (specific activity 1.87 × 10^7 dpm/mmol). This material was diluted with the carrier, and a solution of this substance (73.9 mg, specific activity 2.34 × 10^6 dpm/mmol) in anhydrous methanol (4 ml) was allowed to react with methanolic 0.1 N sodium methoxide (0.15 ml) solution under reflux for 2 min. After cooling with ice-water, the reaction mixture was neutralized with Amberlite IRC-50 resin (COOH form) and then the ion exchange resin was filtered off. The filtrate was concentrated *in vacuo* to give [10-^{14}C]sweroside **1** (52.8 mg) as a colorless syrup.

[Experimental procedure 2] Synthesis of [7-^{3}H]loganin 8[11c)]

It was important to know whether or not the C-7 proton of loganin **8** is retained in morroniside **2** not only for elucidation of the biosynthetic sequence, loganin **8** → secologanin **12** → morroniside **2**, but also to understand the mechanism of ring cleavage of loganin **8** to yield secologanin **12**. Therefore, the authors synthesized [7-^{3}H]loganin **8** according to the scheme shown in Fig. 19.5. 7-Dehydrologanin tetraacetate **27** obtained from asperuloside tetraacetate via several steps was reduced with NaB^3H_4 to [7-^{3}H]-7-epiloganin tetraacetate **28**. The tosylate **29** of this compound was subjected to Walden inversion and the resulting [7-^{3}H]loganin pentaacetate **30** was deacetylated to give [7-^{3}H]loganin **8**. The experimental procedures for the synthesis of [7-^{3}H]loganin **8** from **27** were as follows: a solution of 7-dehydrologanin tetraacetate **27** (147 mg) in dioxane (18 ml)

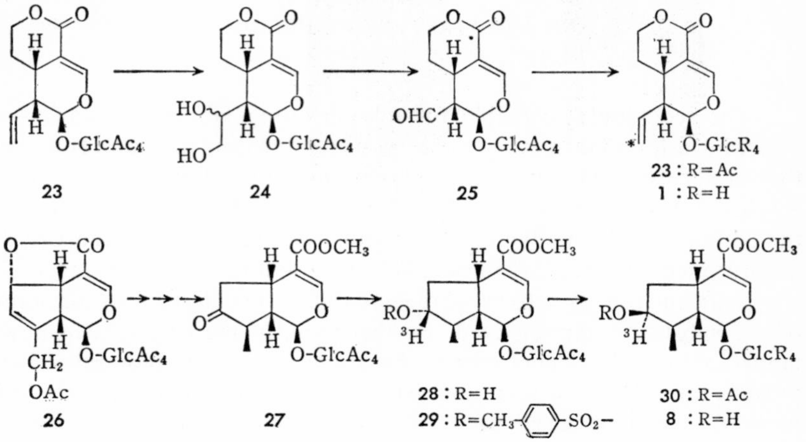

Fig. 19.5

was mixed with NaB^3H_4 (5 mCi; 102 mCi/mmol), $NaBH_4$ (38.3 mg) and water (0.5 ml), and stirred at room temperature for 1 hr. After addition of acetic acid (3 drops), the solvent was removed *in vacuo*. (The evolved tritium gas was converted to ^3H_2O by passing it through a heated CuO tube and the resulting ^3H_2O was trapped by passing it through a calcium chloride tube cooled with liquid nitrogen.) Methanol (5 ml) was added to the residue and the solvent was again removed *in vacuo*. This manipulation was repeated three times. The residue was extracted with chloroform. The chloroform extract was washed with water, dried over anhydrous magnesium sulfate and concentrated *in vacuo* to give [7-^{3}H]-7-epiloganin tetraacetate **28** (138 mg) as colorless needles. A portion of this substance (65.6 mg) was tosylated with *p*-toluenesulfonyl chloride in pyridine. The reaction product was recrystallized from ethanol to give the tosylate **29** (56.0 mg) of [7-^{3}H]-7-epiloganin tetraacetate as colorless needles, mp 116–117°. Tetraethylammonium acetate (120 mg) was then added to a solution of the tosylate (50.9 mg) in anhydrous acetone (8 ml) and the reaction mixture was refluxed for 22 hr. The solvent was removed *in vacuo* and the residue was purified by chromatography on silica gel and recrystallized from ethanol to give [7-^{3}H]-loganin pentaacetate **30** (16.4 mg) as colorless needles, mp 139–139.5°. This substance (14.0 mg) was diluted with carrier (19.0 mg) and treated with methanolic 0.1 N sodium methoxide to give [7-^{3}H]loganin **8** (13.2 mg) as a colorless syrup, specific activity 3.41×10^9 dpm/mmol.

19.3 Administration of Labeled Compounds to Plants and Isolation of the Products

The techniques available for administering labeled substances to plants include the cotton wick, hydroponic and coating methods. However, the former two methods or a combination of them are most frequently employed. When the cotton wick method is adopted, it is advisable to wash the cotton threads in advance with diethyl ether in order to improve the absorption of an aqueous solution. When the solution is absorbed from the root, care must be taken to avoid the possibility of bacterial transformation of the substance administered. Furthermore, in the case of chemically labile compounds, decomposition before arrival at the biosynthetic site must be taken into consideration. Although labeled compounds are administered as an aqueous solution, substances insoluble in water are usually administered as emulsions by adding a small amount of detergent, for example, Tween 80. Water should be added whenever necessary to the container of the labeled substance to permit maximum absorption of the remaining radioactive material by the plant. At the end of administration, the radioactivity administered is calculated by subtracting the radioactivity remaining in the container from the initial activity. The administration period can range from several hours to about one week, depending upon the circumstances. It is advisable to purify radioactive constituents at the final stage by recrystallization. If a compound cannot be crystallized, it can be purified after conversion to a crystalline derivative. Although substances are usually isolated only in small amounts, if there is a sufficient amount of radioactivity, dilution with carrier makes the purification procedure easier. Various kinds of chromatographies can be used. However, the use of complicated apparatus such as counter-current distribution apparatus should be avoided as it may result in contamination by the radioactive substances. Total radioactivity of the isolated substance can be calculated from the specific activity after attaining a constant value and the amount of substance which can be regarded as pure on the basis of several criteria, including tlc.

[Experimental procedure 3] Administration of [10-^{14}C]sweroside 1 to *Gentiana scabra* and the isolation of gentiopicroside 6[19]

[10-^{14}C]-Sweroside **1** (8.7 mg, specific activity 6.90×10^5 dpm/mmol) was dissolved in water (1.5 ml) and the aqueous solution was administered by the cotton wick method to nine *G. scabra* plants (8–10 cm in height). Four days after beginning the administration, the whole plants (wet weight, 34.5 g) were cut into pieces and extracted with four 150 ml portions of methanol. The methanolic extracts were combined and concentrated *in vacuo*. The residue was extracted with water. After washing with ethyl acetate, the aqueous extract was concentrated to about 20 ml and then extracted with *n*-butanol. After washing with a small quantity of water, the *n*-butanol layer was concentrated *in vacuo* to give crude gentiopicroside **6** (800 mg) as a pale yellow powder. A portion (337 mg) of the crude glucoside was acetylated with acetic anhydride-pyridine and the reaction product was recrystallized from ethanol to give gentiopicroside tetraacetate **31** (173.6 mg) as colorless needles, mp 142–143°. This substance was recrystallized from ethanol to a constant activity of 2.96×10^3 dpm/mmol. The total incorporation ratio amounted to 40%.

[Experimental procedure 4] Administration of [7-^{3}H]loganin 8 to *Gentiana thunbergii* and the isolation of morroniside 2[12b)]

Roots of five clumps of *G. thunbergii* (washed carefully so as not to damage the root hairs) were soaked in a solution of [7-^{3}H]loganin **8** (9.9 mg, specific acitivity 3.41×10^9 dpm/mmol) in water (3 ml) and allowed to stand for 5 days, the water being replenished as necessary. The whole plants (6.5 g) were then extracted four times with 50 ml portions of hot methanol. The extracts were combined and concentrated *in vacuo*. The residue was dissolved in methanol (5 ml) and the insoluble material was filtered off. The filtrate was concentrated *in vacuo* and the residue was purified by chromatography on silica gel (chloroform-methanol) to give crude morroniside **2** (12.7 mg). This material was acetylated with acetic anhydride-pyridine and the reaction product, after purification by chromatography on silica gel (diethyl ether), was recrystallized from a mixture of diethyl ether and petroleum ether to give morroniside pentaacetate **32** (5.6 mg) as colorless needles, mp 149–155°. This substance was diluted with carrier **32** (101 mg) and recrystallized from the same solvent to a constant specific activity of 1.26×10^7 dpm/mmol (value before dilution). The total incorporation ratio was 0.13%.

19.4 Determination of Labeled Sites by Degradation

Even if the radioactivity of the substance administered is found in the isolated metabolite, this finding in itself is insufficient to indicate whether the administered substance was incorporated via the supposed route or after decomposition. Accordingly, it is essential for a definitive experiment on biosynthesis to determine the site of the label in the isolated constituent. As mentioned above, successful location of the label on the radioactive metabolite depends upon the appropriateness of the labeling method. As it is essential to carry out degradation reactions on small amounts of radioactive materials, practice with small quantities of non-radioactive sample is advisable.

[Experimental procedure 5] Degradation of [10-^{14}C]gentiopicroside tetraacetate 31[19]

^{14}C-Labeled gentiopicroside tetraacetate **31** (78.9 mg, specific activity 2.96 × 10^3 dpm/mmol) obtained in "Experimental procedure 3" and carrier (70.2 mg) were dissolved in anhydrous chloroform (20 ml). The solution was cooled in a dry ice-methanol bath and a stream of ozone was passed through the solution for 6 hr. The residue, after addition of water (5 ml), was distilled. During the distillation, water (50 ml) was added dropwise to maintain the volume of water in the flask constant. The distillate was collected in a flask containing dimedone (72.9 mg) in acetate buffer (pH 4.5, 20 ml). The resulting white precipitate was collected by centrifugation and dried under reduced pressure over phosphorus pentoxide at room temperature to give formaldehyde-dimedone (42.2 mg), mp 191–193°. This material was further purified by distillation, bp_1 150°. The specific activity was 2.88 × 10^3 dpm/mmol (value before dilution) corresponding to 98% of the activity of the original ^{14}C-labeled **31**. Thus, it was confirmed that ^{14}C in the administered substance was located at C-10 of gentiopicroside **6**.

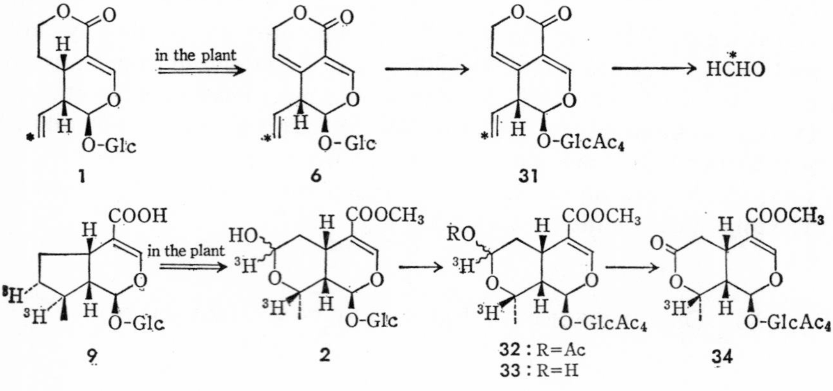

in the plant
H
O-Glc
1
6
O-GlcAc₄
31
H$\overset{*}{C}$HO
COOH
in the plant
HO
RO
^{3}H
COOCH$_3$
O-GlcAc$_4$
9
2
32 : R=Ac
33 : R=H
34

Fig. 19.6

[Experimental procedure 6] Degradation of [7,8-$^3H_2$]-morroniside pentaacetate 32[18]

Radioactive morroniside **2** obtained from *G. thunbergii* to which [7,8-

$^3H_2$]-7-deoxyloganic acid **9*** had been administered was converted to the pentaacetate **32** in the usual way. A portion of this material (12.8 mg, specific activity 2.77 × 10^7 dpm/mmol) diluted with the carrier (54.0 mg) was dissolved in ethanol (10 ml). An aqueous solution of potassium bicarbonate (15 mg/2.1 ml) was added to the solution and the mixture was allowed to stand at 35° for four days. The reaction mixture, after addition of water, was extracted with chloroform. The chloroform extract was concentrated to give crude morroniside tetraacetate **33** (8.9 mg), which was dissolved in acetone (2 ml). Jones reagent (0.02 ml) was added. After stirring for 1 hr under ice cooling, the mixture was diluted with water and extracted with chloroform. The chloroform extract was washed with water, dried over anhydrous magnesium sulfate and concentrated *in vacuo*. The residue was purified by chromatography on silica gel (chloroform) to give kingiside tetraacetate **34** (6.5 mg). This material was recrystallized from ethanol to a constant specific acitivity of 1.47 × 10^7 dpm/mmol (value before dilution), which amounts to 53% of that of the original morroniside pentaacetate **32**. The results described above may be explained as shown in Fig. 19.6 by assuming that both tritiums in the precursor are retained in the morroniside **2** isolated and that kingiside tetraacetate bearing a tritium only at C-8 is obtained by partial deacetylation of **32** at C-7, followed by Jones oxidation. If [10-^{14}C]-**9** could be readily synthesized, this experiment could be more easily performed by the so-called double-labeling tracer method, where this material would be administered combined with [7,8-$^3H_2$]-**9**. However, because of the difficulties in synthesizing the ^{14}C-labeled compound we were compelled to restrict our experiments to tritium labeling only.

19.5 Concluding Remarks

Many problems concerning the biosynthesis of secoiridoid glucosides remain unsolved. For example, the mechanism of the ring opening of loganin **8** to secologanin **12** has not yet been clarified. In addition, the precise mechanism of cyclization of alicyclic terpenes to iridoids has not been established. It was observed in our laboratory that C-2 of mevalonic acid incorporated into C-3 but not into C-11 of asperuloside, lamioside, etc.,

* This substance **9** was obtained by catalytic reduction of deoxygeniposidic acid tetraacetate over Pd-C using tritium gas, followed by deacetylation. However, catalytic reduction with tritium gas can only be performed at certain specialized commercial facilities. Administration of this material to the plant was carried out as in the case of "Experimental procedure 4". Therefore, the synthetic method and administration procedure are not dealt with here.

shwoing that the mechanism of the cyclopentane ring formation in the biosynthesis of these glucosides differs from that of secoiridoid glucosides.[22] Moreover, both groups are frequently biosynthesized via common intermediates, deoxyloganin **11** and loganin **8**. It is of great interest to examine the relationship between these phenomena and the taxonomical positions of the plants containing iridoid and/or secoiridoid glucosides. Studies so far have been mostly concerned with the biosynthetic routes, but next each step will be studied at the enzyme level. In this connection, biosynthetic studies using cell cultures should prove useful. If improvement in the incorporation of precursors into the glucosides can be achieved by using cell cultures, the use of stable isotopes such as ^{13}C and ^{2}H will be possible, resulting in faster progress in the field.

References

1) V. Plouvier, J. Favre-Bonvin, *Phytochemistry*, **10**, 1697 (1971).
2) H. Inouye, S. Ueda, Y. Takeda, *Heterocycles*, **4**, 527 (1976).
3) A. I. Scott, *Accounts Chem. Res.*, **3**, 151 (1970).
4) G. A. Cordell, *Lloydia*, **37**, 219 (1974).
5) a) C. J. Coscia, R. Guarnaccia, *J. Am. Chem. Soc.*, **89**, 1280 (1967); b) C. J. Coscia, L. Botta, R. Guarnaccia, *Biochemistry*, **12**, 5036 (1969).
6) H. Inouye, S. Ueda, Y. Nakamura, *Tetr. Lett.* **1967**, 3221; b) *Chem. Pharm. Bull.*, **18**, 2043 (1970).
7) P. Loew, Ch. von Szczepanski, C. J. Coscia, D. Arigoni, *Chem. Commun.*, **1968**, 1276.
8) A. R. Battersby, A. R. Burnett, G. D. Knowles, P. G. Parsons, *ibid.*, **1968**, 1277.
9) S.E.P. Loew, D. Arigoni, *ibid.*, **1970**, 823.
10) A. R. Battersby, S. H. Brown, T. G. Payne, *ibid.*, **1970**, 827.
11) a) H. Inouye, S. Ueda, Y. Aoki, Y. Takeda, *Tetr. Lett.*, **1969**, 2351; b) H. Inouye, S. Ueda, Y. Takeda, *ibid.*, **1970**, 3351; c) H. Inouye, S. Ueda, Y. Aoki, Y. Takeda, *Chem. Pharm. Bull.*, **20**, 1287 (1972).
12) a) H. Inouye, S. Ueda, K. Inoue, Y. Takeda, *Tetr. Lett.*, **1971**, 4073; b) *Chem. Pharm. Bull.*, **22**, 676 (1974).
13) A. R. Battersby, A. R. Burnett, P. G. Parsons, *Chem. Commun.*, **1970**, 826.
14) A. R. Battersby, A. R. Burnett, P. G. Parsons, *J. Chem. Soc. (C)*, **1969**, 1187.
15) a) R. Guarnaccia, C. J. Coscia, *J. Am. Chem. Soc.*, **93**, 6319 (1971); b) R. Gauarnaccia, L. Botta, C. J. Coscia, *ibid.*, **96**, 7079 (1974).
16) D. Gröger, P. Simchen, *Z. Naturforsch.*, **24b**, 356 (1969).
17) H. Inouye, S. Ueda, Y. Takeda, *Tetr. Lett.*, **1971**, 4069.
18) Y. Takeda, H. Inouye, *Chem. Pharm. Bull.*, **24**, 79 (1976).
19) a) H. Inouye, S. Ueda, Y. Takeda, *Tetr. Lett.*, **1968**, 3453; b) *Chem. Pharm. Bull.*, **19**, 587 (1971).
20) a) R. Guarnaccia, L. Botta, C. J. Coscia, *J. Am. Chem. Soc.*, **91**, 204 (1969); b) C. J. Coscia, L. Botta, R. Guarnaccia, *Arch. Biochem. Biophys.*, **136**, 498 (1970).
21) K. Inoue, T. Nishioka, T. Tanahashi, H. Inouye, *Abstract Paper of the 96th Annual Meeting of the Pharmaceutical Society of Japan*, Part II, p. 207 (1976).
22) a) H. Inouye, S. Ueda, S. Uesato, *Phytochemistry*, **16**, 1669 (1977); b) H. Inouye, S. Ueda, S. Uesato, K. Kobayashi, *Chem. Pharm. Bull.*, **26**, 3384 (1978).

CHAPTER 20

Isolation of Triterpenoidal Saponins

Glycosides which produce a foaming aqueous solution are generally called saponins. They also have a hemolytic action and characteristically form precipitates with cholesterol in alcohol. Based on the chemical structure of the sapogenin (aglycone), saponins are classified into two large groups named steroidal and triterpenoidal saponins.

Among natural organic compounds, saponins have comparatively large molecular weights and high polarity, so that they have been considered to be difficult to isolate in a pure state. Thus, most studies have been carried out on steroidal or triterpenoidal sapogenins obtained by hydrolysis of saponin mixtures. Since many sapogenins obtained by acid hydrolysis are now known to be artifacts, new isolation and hydrolysis procedures for saponins are required. In addition to the Smith degradation and De Mayo's method, new hydrolysis methods using enzymes or soil bacteria have been developed.

Many saponins are present in higher plants, and saponin-containing plants have been used for medical purposes. Among the Oriental drugs there are many herbal drugs which contain saponins as their principal constituents. Besides expectorant, antitussive, tonic and diuretic activities, the saponins contained in crude drugs are also expected to show some specific biological activities. Recently, comparative investigations of purified saponins isolated from Oriental drugs have revealed that saponins consisting of the same sapogenin and different sugar sequences have different biological activities. The methods of administration of saponins into test animals depend on their solubility, but recent studies on the biological activities of pure saponins have demonstrated antiinflammatory activities, central nervous system action, prevention of stress ulcers, antifatigue activity, antibacterial action, stimulation of lipid metabolism, promotion of nucleic acid and protein syntheses, antitumor effect, and other actions. The relationships between biological activities and chemical structures have also been investigated, but much more work is required.

20.1 Separation and Purification of Saponins

The separation and purification of saponins was described by Tschesche.[1] The main problem in the isolation of saponins is the exclusion of phenolic coloring matter or coexisting saponins having similar chemical structures. Accordingly, it is necessary to establish methods to fractionate the crude saponins and then to purify each saponin.

Extraction Procedures

Care is required in the treatment of saponins, because during the course of extraction, esterification of acidic saponins, saponification of ester linkages and migration of acyl groups often occur in the alcohol used. Before extraction, particular care must be taken in the case of fresh material to prevent enzymatic action on the aglycone, sugar moiety and acyl groups of labile saponins. In general, the materials are extracted with methanol or ethyl alcohol. In the case of seeds or fruits which contain large amounts of fatty oil, it is desirable to defat the material with a suitable organic solvent or by means of a press. Leaves are extracted with cyclohexane or carbon tetrachloride to remove chlorophyll. Direct extraction with hot water is sometimes employed for stable saponins.

Fractionation of Crude Saponins from the Extract

After removal of the solvent under reduced pressure, the extract is suspended in water and extracted with an organic solvent, that is ethyl ether, hexane, benzene or chloroform. Less polar saponins dissolve in chloroform, and the water layer is extracted several times with *n*-butanol saturated with water. Each extract should be checked by thin layer chromatography (tlc), because the constituents of saponins in each extract are sometimes variable. The most polar saponins left in the water layer can be separated from sugars and inorganic compounds by sephadex gel filtration. To remove contaminants from saponins, the following methods are available. (i) the alcoholic solution of cholesterol is added to a solution of crude saponins to form a precipitate of complex compounds (separation of saponin fraction). (ii) A solution of saponins in a small amount of methanol is poured into a large amount of ethyl ether or acetone to form a precipitate (removal of nonpolar substances). (iii) Filtration on a small quantity of alumina (removal of phenolic substances and coloring matter). The filtration and separation by column chromatography can be carried out at the same time by packing a small amount of alumina on top of the adsorbent. If a colorless saponin mixture is obtained by these preliminary treatments, the efficiency of subsequent column chromatography to isolate each saponin will probably be increased. Only a few saponins can be isolated by recrystallization or reprecipitation, and saponin mixtures sometimes crystallize more easily than the pure saponin. Thus, it is desirable to examine the purity of crystalline saponin rigorously by tlc.

Isolation of Saponin from a Saponin Mixture

The usual methods used to isolate each saponin from a saponin mixture are column chromatography and preparative tlc. Recently, the application of droplet counter-current chromatography (dcc) has also been reported, as described elsewhere in this book.

Silica gel, silicic acid (Mallinckrodt Co. Ltd.), alumina, cellulose, Sephadex, etc., are suitable adsorbents. The solvent systems used for development and elution include $CHCl_3$–MeOH–H_2O, AcOEt–MeOH–H_2O, BuOH–EtOH–H_2O, BuOH–AcOEt–H_2O, $CHCl_3$–AcOEt–MeOH–H_2O, BuOH–EtOH–NH_3, toluene-BuOH and other solvents. Although successful separation and purification of saponin by chromatography are usually possible, some loss of saponin can usually not be avoided. By careful pre-examination and by the use of combinations of several kinds of chromatographic systems, the isolation of a saponin can usually be carried out effectively, however.

As an example of the problems that may occur, an acidic saponin isolated by column chromatography on silica gel forms a crystalline salt, and an additional procedure is necessary to liberate free acidic saponin. Therefore, in the case of acidic saponins, careful selection of chromatographic conditions and chemical derivation methods are very important. Of course the selection of a wet or dry packing method is important, but the key point for successful separation of saponins is the packing technique. A solvent system used for detection of saponins by tlc is not always suitable for separation by column chromatography.

It is clear that recent advances in saponin chemistry are due largely to the application of modern separation techniques. Some examples are described below.

According to Tschesche *et al.*, triterpenoidal saponins can be classified as follows.

I. Monodesmosides
   1. Neutral glycosides
   2. Esterified saponins
   3. Acidic glycosides based on uronic acid
   4. Acidic glycosides based on aglycone
   5. Acidic glycosides based on uronic acid and aglycone
   6. Acylglycoses

II. Bisdesmosides
   I. Neutral glycosides
   2. Acidic glycosides

The following examples are described in accordance with this classification.

[Experimental procedure 1] Neutral glycoside (I-1): isolation of saikosaponins a, b_1, b_2, b_3, b_4, c, d[2)]

The dried and sliced roots of *Bupleurum falcatum* L. ("*mishima-saiko*" in Japanese) (4 kg) were extracted with ether (2 × 10 l) and then with methanol (3 × 8 l) at room temperature. Material from the methanolic extract (310 g) was dissolved in *n*-BuOH (3 l) and the solution was washed with water. The aqueous layer was extracted with *n*-BuOH. The combined *n*-BuOH layer was washed with sodium chloride solution, dried (Na_2SO_4), and concentrated *in vacuo*. The residue (112 g) was extracted with acetone (2 × 600 ml) to leave the crude saponin (56 g). The saponin was chromatographed on silica gel with $CHCl_3$–CH_3OH (5:1) to give saikosaponins a and d (fraction A) (20 g), saikosaponin b (fraction B) (6.4 g), and saikosaponin c (10 g). Fraction A was rechromatographed on silica gel (solvent, EtOAc–EtOH–H_2O, 8:2:1) to give saikosaponins a (12.4 g) and d(6.6 g). Fraction B (1 g) was separated by preparative tlc (silica gel; solvent, $CHCl_3$–MeOH–H_2O, 30:10:1) into saikosaponin b_2, *Rf* 0.58 (830 mg), and saikosaponin b_4, *Rf* 0.61 (120 mg).

Saikosaponin a described above is a mixture of saikosaponin a and saikosaponin b_3 (ratio 9:1). Although saikosaponin a showed a single spot on tlc, its peracetate showed two spots. The isolation of both saikosaponins was carried out via the peracetates. Saikosaponin a (500 mg), obtained from fraction A described above, was dissolved in acetic anhydride (10 ml) and pyridine (20 ml), and heated at 90° C for 5 hr. The mixture was poured into ice-water and extracted with ethyl acetate to give an acetate mixture (560 mg). The mixture (70 mg) was separated by preparative tlc (silica gel; benzene-acetone, 9:1; triple development) into saikosaponin a octa-acetate, a white powder (60 mg), *Rf* 0.18, and saikosaponin b_3 nona-acetate, a white powder (5 mg), *Rf* 0.14. Saikosaponin b_3 nona-acetate (510 mg) was dissolved in 2% sodium hydroxide in methanol (5 ml) and

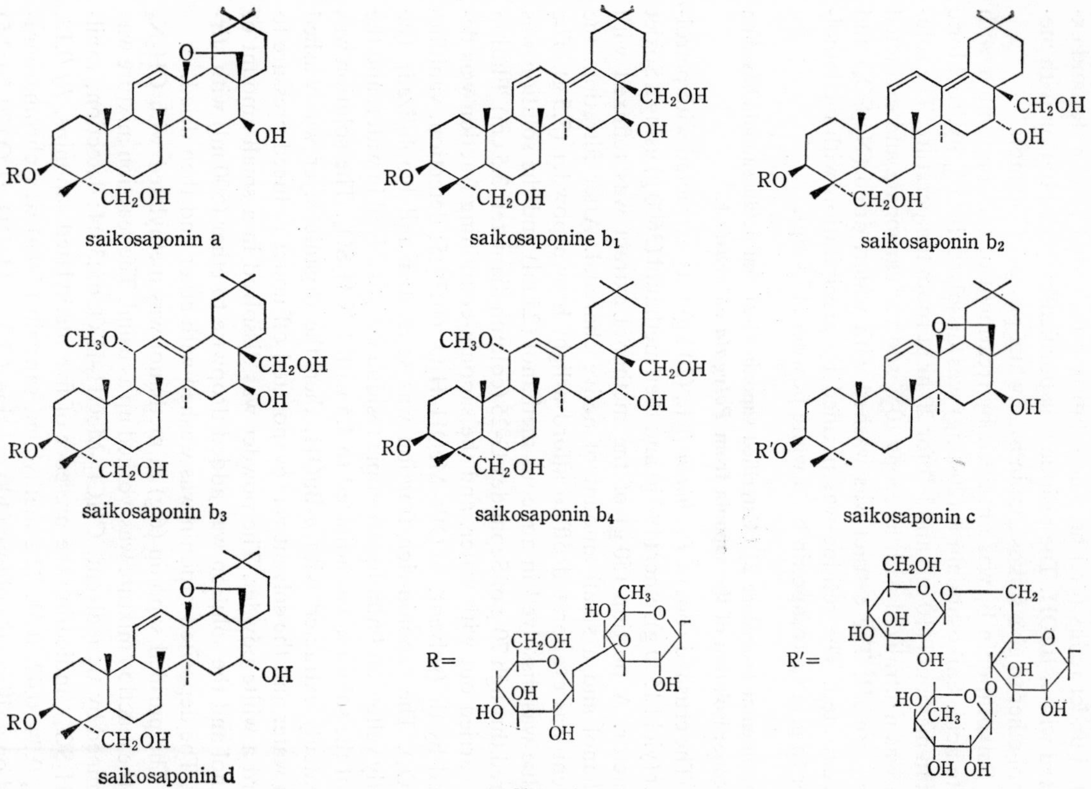

RO CH$_2$OH OH O

saikosaponin a

RO CH$_2$OH CH$_2$OH OH

saikosaponine b$_1$

RO CH$_2$OH CH$_2$OH OH

saikosaponin b$_2$

CH$_3$O RO CH$_2$OH CH$_2$OH OH

saikosaponin b$_3$

CH$_3$O RO CH$_2$OH CH$_2$OH OH

saikosaponin b$_4$

R'O O OH

saikosaponin c

RO CH$_2$OH O OH

saikosaponin d

R= CH$_2$OH HO OH OH CH$_3$ HO O O OH

R'= CH$_2$OH O O CH$_2$ HO OH OH O OH OH HO O O CH$_3$ OH OH

Chart 1

heated under reflux for 1 hr. The solution was diluted with water and extracted with *n*-BuOH. The residue was purified by precipitation with methanol-ether to give saikosaponin b_3, a white powder (390 mg).

Saikosaponin b_1 was obtained by treatment of saikosaponin a with dilute acid. Saikosaponin a (200 mg) was dissolved in 5% hydrochloric acid in methanol (10 ml) and left for 20 hr at room temperature. The solution was neutralized with aqueous 10% sodium carbonate and extracted with *n*-BuOH. The extract was washed with water, dried (Na_2SO_4) and concentrated. The residue was purified by precipitation with methanol-ether to give saikosaponin b_1, a white powder (170 mg).

[Experimental procedure 2] Esterified saponin (I-2) and acidic saponin based on aglycone: isolation of the saponin from *Polygala chinensis* L.[3)]

The crushed root of *P. chinensis* L. (20 kg) was extracted with petrol-ether (yield 1520 g), ether (190 g) and then methanol (3600 g) using a Soxlet extractor. A portion (50 g) of the methanol extract was refluxed with methanol and a small amount of active charcoal. After filtration, the solvent was evaporated off to afford a light brown powder (3.5 g). The residue was dissolved in aqueous methanol (25 ml) and the solution was filtered through 70 g of Sephadex G-25 (column diameter, 3.5 cm). Elution was carried out with water, and the saponin-containing fraction was detected by tlc (solvent, AcOEt–MeOH–H_2O, 60:30:5; detection, vanillin–H_2SO_4). The saponin-rich fraction was separated and lyophilized. The faintly yellow and non-hygroscopic residue was dissolved in water, and the pH of the solution was adjusted to 4.5 with 1 N H_2SO_4. The solution was repeatedly extracted with *n*-BuOH, then the organic layer was washed with water and the solvent was evaporated off under reduced pressure to afford a white powder. The powder was dissolved in a small amount of ethanol and the solution was added dropwise to ether (500 ml) with stirring. The deposited saponin was washed with ether and then dried.

The purified saponin (6 g) in methanol was methylated with CH_2N_2 and the reaction mixture was worked up as usual. The reaction mixture was examined by tlc (solvent, $CHCl_3$–MeOH–H_2O, 65:30:4; detection, vanillin–H_2SO_4), indicating the presence of five methylated saponins, *Rf* 0.18, 0.25, 0.34, 0.42, 0.51. The methylated saponin (7.0 g) was chromatographed on a silica gel column (400 g) using $CHCl_3$–MeOH–H_2O (70:20:2.5) to separate each saponin. Impure fractions were purified by rechromatography. The fraction containing saponin corresponding to *Rf* 0.25–0.42 was rechromatographed on a silica gel column (80 g) with $CHCl_3$–MeOH–H_2O (65:25:1) and the eluted fraction was further chromatographed on a silica gel column (40 g) with AcOEt–MeOH–H_2O (65:25:5) to afford the pure dimethyl ester of the major saponin (680 mg), mp 245–248° C, $[\alpha]_D^{20}$–

2° (c = 1.0, MeOH). The purity of the main saponin was confirmed by chromatography, and the yields of the dimethyl esters of the saponins were 80 mg (*Rf* 0.18), 320 mg (*Rf* 0.34), 200 mg (*Rf* 0.42) and trace (*Rf* 0.51).

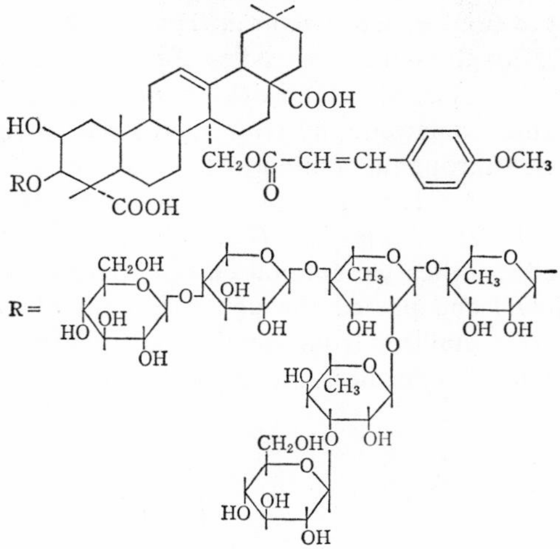

Chart 2

[Experimental procedure 3] Acidic glycoside based on uronic acid (I-3): isolation of soyasaponins I, II and III from soybeans[4,5)]

Crushed seeds of *Glycine max* (10 kg) were defatted by extraction with hot hexane, and the residue was refluxed with MeOH. The MeOH extract (1.4 kg) was partitioned in *n*-BuOH-water (1:1) mixture. The isoflavone mixture (*ca.* 30 g) was separated out as a light yellow precipitate during this procedure. Repeated fractional recrystallization of the isoflavone mixture from MeOH gave genistin, mp 253–255°, and daidzin, mp 234–236°. The *n*-BuOH layer was concentrated under reduced pressure to give a residue which was dissolved in a small amount of MeOH and poured into a large quantity of ether. The precipitated crude saponin was collected by filtration and passed through a column of charcoal (200 g, Tokusei-Shirasagi, Takeda Chem. Ind.)-Celite 535 (200 g, Wako Pure Chem. Ind.) with the aid of MeOH to give a saponin mixture (30 g), which was washed with *n*-BuOH and weak aq. alkali. The white insoluble portion (10 g) collected by centrifugation was then chromatographed on silica gel (500 g), eluting with $CHCl_3$–MeOH–water (7:3:1, lower layer), to give soyasaponins I (5.5 g), II (0.5 g) and III (0.1 g), and a mixture of I and II (2 g).

Soyasaponin I (1.2 g) obtained above, ir ν_{max}^{Nujol} cm^{-1}: 3350 (br, OH),

1720 (br, COOH), 1610 (br, COO^-), was hardly soluble in MeOH and EtOH, but was dissolved in warm, weakly acidic aq. MeOH (50 ml, pH 3–4) and filtered to remove the insoluble portion. Examination by tlc before and after acidic MeOH treatment gave an identical saponin chromatogram, except for the change of the carboxylate form to the acidic form. The crystals (0.95 g) separated out from the filtrate were recrystallized from MeOH to give colorless needles of pure soyasaponin I, mp 238–240°, $[\alpha]_D^{14}$ −8.5° (c = 1.0, MeOH). Soyasaponin II (100 mg) obtained by silica gel column chromatography (*vide supra*) was dissolved in MeOH (50 ml) and passed through a column of Dowex 50w × 8 (H^+, 20 g) and recrystallized from MeOH to give pure soyasaponin II as colorless fine crystals of mp 212–215°, $[\alpha]_D^{29}$−9.6° (c = 0.5, MeOH). Soyasaponin III (50 mg) obtained by silica gel column chromatography (*vide supra*) was dissolved in MeOH and passed through a column of Dowex 50w × 8 (H^+, 10 g), then recrystallized from MeOH to give a pure sample of soyasaponin III as colorless needles, mp 215–216°, $[\alpha]_D^{29}$ + 15.0° (c = 0.5, MeOH).

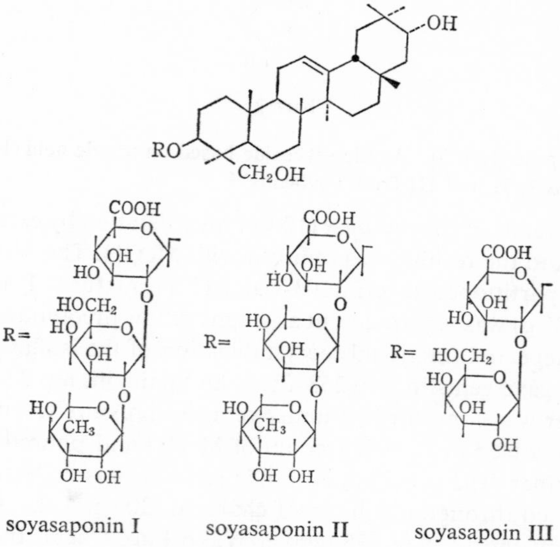

soyasaponin I soyasaponin II soyasapoin III

Chart 3

[Experimental procedure 4] Acidic glycoside based on aglycone and uronic acid (I-5): isolation of spinasaponins A and B from *Spinacia oleracea* L.[6]

Kofler[7] first reported the hemolytic activity of the leaves of *Spinacia oleracea* L., but Dafert[8] could not confirm this, though he found very strong hemolytic activity due to crude saponin from the root. Tschesche *et al.* isolated both saponins and elucidated their chemical structures.

The fresh roots (15 g) washed with water were frozen at -30° and crushed. The material was extracted three times with MeOH-water (5:1) and the extract was concentrated. The aqueous solution was acidified with 2 N HCl and extracted three times with butanol. The butanol layer was washed with water and then concentrated to afford a residue (92.5 g), which was chromatographed on silica gel (8.5 g), eluting with $CHCl_3$–MeOH–water (65:25:10, lower layer) to give non-polar substances (33 g) and a crude saponin mixture (54 g). The crude saponin mixture (2 g) was suspended in MeOH and stirred with 2 N H_2SO_4 (20 ml) at 40° (bath temperature) for 10 min. The reaction mixture was concentrated under reduced pressure to remove methanol and the residue was diluted with water (200 ml). The aq. solution was extracted three times with butanol, and the butanol layer was washed with water then concentrated. The residue was dissolved in methanol containing a small amount of $CHCl_3$, and excess CH_2N_2 solution in ether was added. A thin-layer chromatogram (solvent, $CHCl_3$–MeOH–water, 65:19:10) of the reaction mixture indicated the presence of a small quantity of non-polar impurity, two main saponins and a minor saponin.

The crude saponin (23 g) contained 14 g of the above esterified compounds, and chromatography on a silica gel column (2.5 kg) with $CHCl_3$–MeOH–water (65:15:10, lower phase) afforded chromatographically pure spinasaponin A dimethyl ester (1.16 g), softening point 176–179° C, $[\alpha]_D^{20}$ + 54° (c = 0.68, MeOH), spinasaponin B dimethyl ester (0.96 g), softening point 195–198° C, $[\alpha]_D^{20}$ + 100° (c = 0.42, MeOH), and a minor saponin (0.2 g).

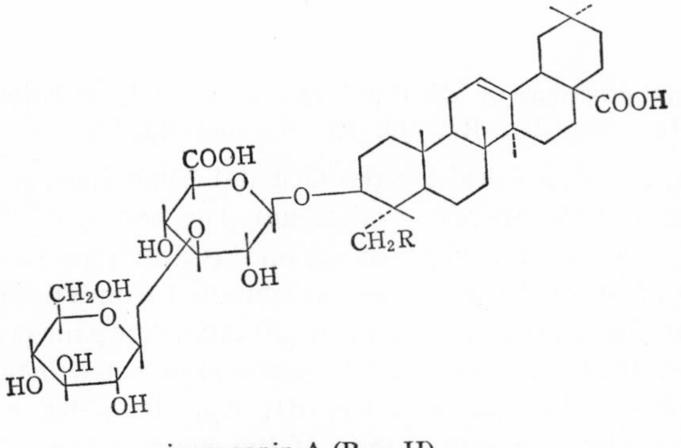

spinasaponin A (R = H)
spinasaponin B (R = OH)

Chart 4

[Experimental procedure 5] Acylglycose(I-6): isolation of mollugo glycoside A from the roots of *Mollugo spergula*[9]

Mollugo spergula (Ficoidaceae), a herbaceous plant commonly found at the base of the Eastern Himalayas, is reputed in Indian medicine to have antiseptic and antidermatitic properties.

The crushed roots were extracted with light petroleum, chloroform and alcohol successively. The alcohol extract was worked up for the isolation of saponins by the butanol method. The butanol extract containing saponins was washed (H_2O) and concentrated under reduced pressure. The brown syrup was taken up in a minimum amount of MeOH and precipitated with Et_2O, and this process was repeated until an almost colorless solid was obtained (yield 0.7%). On tlc (solvent, $CHCl_3$–MeOH–water, 10:4:1) it showed a streak with 8 spots. It did not react with CH_2N_2. It was chromatographed on a column of silica gel, but most of the fractions consisted of mixtures (tlc). The fractions eluted with $CHCl_3$–MeOH (84:16) contained only two compounds, a major and a minor one (tlc). The first one, called mollugo glycoside A, was obtained by preparative tlc (solvent, $CHCl_3$–MeOH–water = 10:4:1; yield, 0.0025 %). Mollugo glycoside A, mp 220–225°, $[\alpha]_D + 30°$ ($c = 0.61$, pyridine).

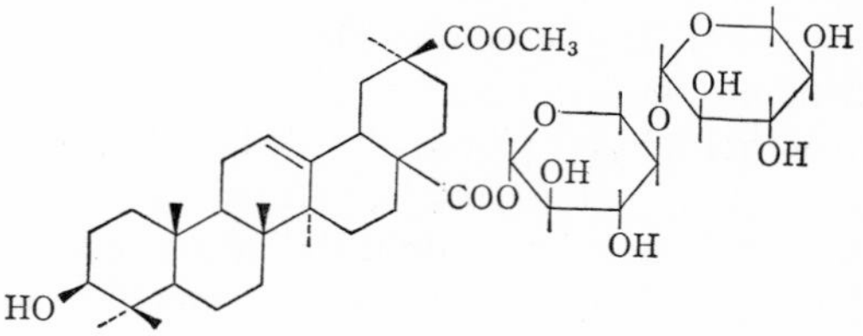

COOCH₃ OH OH O O OH COO OH HO OH

mollugo glycoside A

Chart 5

[Experimental procedure 6] Neutral bisdesmoside (II-1): isolation of ginsenoside-Ro, -Ra, -Rb$_1$, -Rb$_2$, -Rc, -Rd, -Re, -Rf, -Rg$_1$ and -Rg$_2$[10,11]

Ginseng root, a world-famous Oriental crude drug, is prepared from *Panax ginseng* C.A. MEYER (Araliaceae). The history of chemical, pharmacological and biochemical studies on this drug goes back many years. Shibata *et al.* studied the sapogenins obtained from crude saponins and established the chemical structures of 20S-protopanaxadiol and 20S-protopanaxatriol. He also named the saponins detected on tlc: ginsenoside-Ro, -Ra, -Rb$_1$, -Rb$_2$, -Rc, -Rd, -Re, -Rf, -Rg$_1$, -Rg$_2$, -Rg$_3$ and -Rh.[12] The structure of ginsenoside-Rg$_1$, one of the main saponins, has been established by Shibata *et al.*[13] using material isolated via the acetate.

Shoji *et al.* have established a method for the isolation of each ginsenoside, and they obtained amounts of the pure saponins sufficient for chemical and biological studies. The dried roots of *Panax ginseng* C.A. MEYER (3 kg), cultivated in Nagano prefecture, Japan, were crushed and extracted five times with 2 l of hot methanol. After removal of the solvent under reduced pressure, 530 g of brown residue was obtained (yield from the dried roots, 17.7 %). The extract was suspended in water and extracted with BuOH saturated with water. The BuOH layer was concentrated *in vacuo* to afford 103 g of crude saponin (yield from the dried roots, 3.4 %). The BuOH–soluble fraction was analyzed by tlc (silica gel plates; solv. 1, $CHCl_3$–MeOH–H_2O = 65:35:10, lower phase; solv. 2, BuOH–AcOEt–H_2O = 4:1:5, upper phase) and shown to contain ginsenoside-Ro, -Ra, -Rb_1, -Rb_2, -Rc, -Rd, -Re, -Rf, -Rg_1, and -Rg_2 (Fig. 20.1). The *Rf* value of each ginsenoside reflects the oligosaccharide structure of each saponin. Recently, 20-glucoginsenoside-Rf and ginsenoside-Rb_3 (not shown in the figure) have been obtained in very small quantities and their chemical structures have been established.[14)]

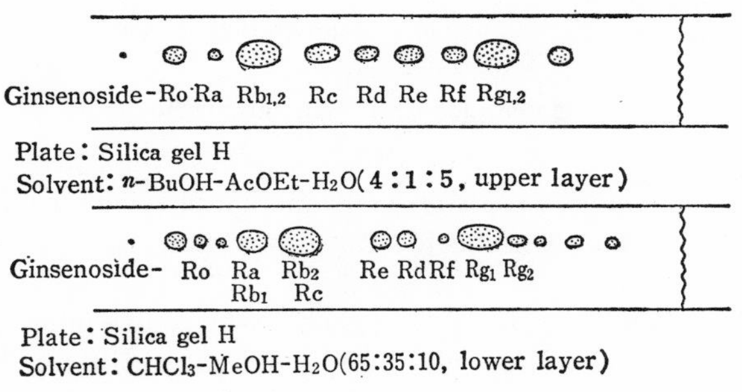

Fig. 20.1

The crude saponin (30 g) was dissolved in a small amount of methanol and the solution was mixed with silica gel (60 g). The mixture was dried at room temperature to remove the solvent, and the dried powder was applied to the top of a silica gel column (1 kg) packed by the dry method. The column was developed with $CHCl_3$–MeOH–H_2O (65:35:10, lower phase) and elution was continued with the same solvent. The eluate was examined by tlc at intervals and separated into five fractions (Fr. 1–Fr. 5) (Table 20.1).

Fraction 1 appeared to contain a very polar, acidic saponin, which might form a salt during silica gel chromatography. Fr. 1 (0.5 g) was treated by column chromatography on silicic acid (100 g) with $CHCl_3$–MeOH–H_2O (65:35:10, lower phase) to afford a free acidic saponin, gin-

TABLE 20.1 Extraction and isolation of ginseng saponins

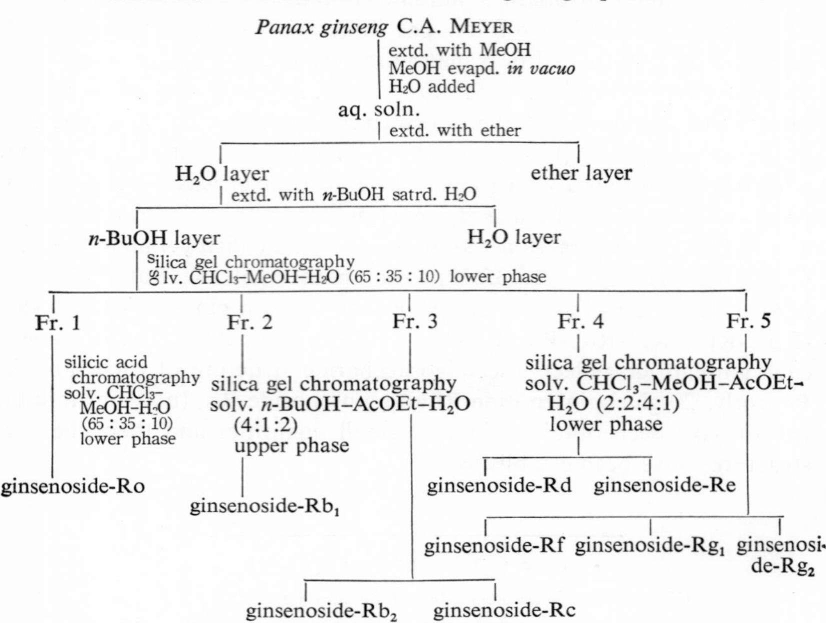

senoside-Ro, as colorless needles from methanol (yield from BuOH extract, 0.8 %). This pure ginsenoside-Ro showed no carboxylate absorption in the IR spectrum.

Fractions 2 and 3 (each 4 g) were each dissolved in a minimum amount of MeOH and each solution was mixed with silica gel (8 g). The mixture was dried at room temperature and the dried powder was applied to the top of a silica gel column (500 g). The columns were developed and eluted with BuOH–AcOEt–H_2O (4:1:2, upper phase) to afford ginsenoside-Rb_1 from Fr. 2 and ginsenoside-Rb_2 and -Rc from Fr. 3.

Fractions 4 and 5 were each subjected to column chromatography on silica gel using $CHCl_3$–MeOH–AcOEt–H_2O (2:2:4:1, lower phase) to afford ginsenoside-Rd and -Rc from Fr. 4 and ginsenoside-Rf, -Rg_1 and -Rg_2 from Fr. 5. The isolation procedure is shown in Table 20.1 and the general properties of each saponin are listed in Table 20.2.

[Experimental procedure 7] Acidic bisdesmoside (II-2): isolation of senegins I, II, III and IV from Senegae Radix

The dried roots of *Polygala senega* LINNE var. *latifolia* TORREY et GRAY (500 g) cultivated in Hyogo prefecture, Japan, were extracted three times with hot methanol (3 l). The extract was concentrated under reduced

TABLE 20.2 Properties of various ginsenosides.

Ginsenoside	Properties	mp (°C)	$[\alpha]_D^{22}$ (*c* in MeOH)	Formula	IR (KBr) cm^{-1}
Ro	colorless needles (MeOH)	239–241	+15.33° (0.91)	$C_{48}H_{76}O_{19}$	3400 (OH), 1740 (COOR), 1728 (COOH)
Rb_1	white powder (EtOH–BuOH = 1:1)	(197–198)	+12.42° (0.91)	$C_{54}H_{92}O_{23}$	3400 (OH), 1620 (C=C)
Rb_2	white powder EtOH–BuOH = 1:5)	(200–203)	+3.05° (0.98)	$C_{53}H_{90}O_{22}$	3400 (O=H), 1620 (C=C)
Rc	white powder (EtOH–BuOH = 1:5)	(199–201)	+1.93° (1.03)	$C_{53}H_{90}O_{22}$	3400 (OH), 1620 (C=C)
Rd	white powder (EtOH–AcOEt = 1:1)	(200–209)	+19.38° (1.03)	$C_{48}H_{82}O_{18}$	3400 (OH), 1620 (C=C)
Re	colorless needles (50% EtOH)	201–203	0––1.00° (1.00)	$C_{48}H_{82}O_{18}$	3380 (OH), 1620 (C=C)
Rf	white powder (acetone)	(197–198)	+6.99° (1.00)	$C_{42}H_{72}O_{14}$	3380 (OH), 1620 (C=C)
Rg_2	colorless needles (EtOH)	187–189	+5.00– −6.00° (1.00)	$C_{42}H_{72}O_{13}$	3400 (OH) 1620 (C=C)

pressure to afford a syrupy residue (220 g, yield 44%). The residue was suspended in water and extracted with benzene (benzene-soluble fraction, 16 g). The aqueous layer was extracted three times with BuOH saturated with water and the extract was concentrated *in vacuo* to afford a pale yellow powder (95 g, yield 19%). The crude saponin was subjected to column chromatography on silica gel using mixtures of ethyl acetate saturated with water and MeOH to yield four fractions [5% MeOH (Fr. 1), 10% MeOH (Fr. 2), 20% MeOH (Fr. 3), and 50–100% (Fr. 4)].

Fraction 1, eluted with 5% MeOH, was a resinous brown substance (yield 8.8%). Fr. 2, eluted with 10% MeOH, was a pale yellow powder (yield 17.6%, phenolic glycosides). Fr. 3 (20% MeOH) was a brown powder (yield 5.0%, 1,5-anhydro-D-sorbitol) and Fr. 4 (50–100% MeOH) was a yellow powder (yield 39.8%, crude saponin). The tlc pattern (Kieselgel H plates treated with oxalic acid; solvent, $CHCl_3$–MeOH–H_2O = 65:35:10, lower phase) of Fr. 4 showed the presence of four saponins which were named senegins I (*Rf* 0.075), II (*Rf* 0.063), III (*Rf* 0.042) and IV (*Rf* 0.021) in order of increasing polarity (Fig. 20.2).

Fr. 4 (200 mg) was subjected to column chromatography on silicic acid (200 g, Mallinckrodt Co. Ltd., 100 mesh) packed by the dry method using $CHCl_3$–MeOH–H_2O (7:3:1, lower phase). The eluent was examined by tlc, and the elution was also followed by measuring the absorbance at 315 nm. The yields of senegins I–IV based on the dried roots were 0.39%,

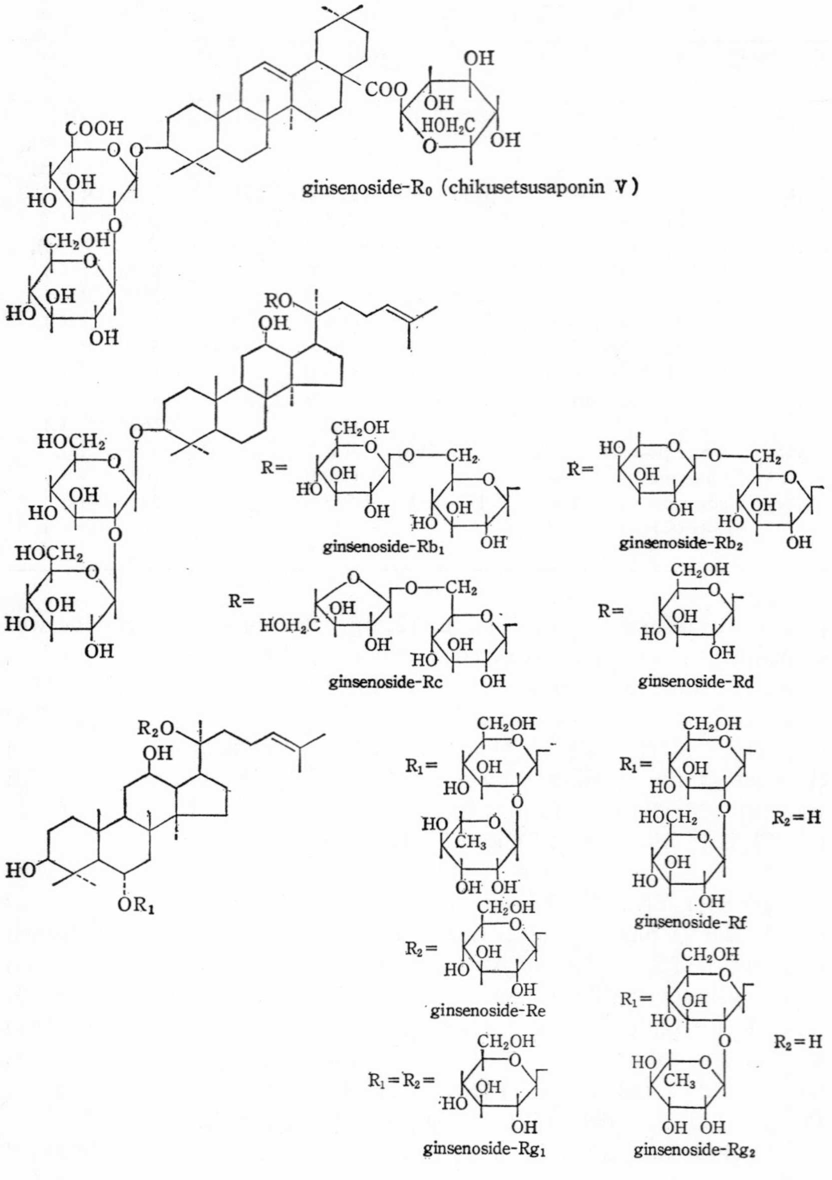

ginsenoside-R_0 (chikusetsusaponin V)

ginsenoside-Rb_1

ginsenoside-Rb_2

ginsenoside-Rc

ginsenoside-Rd

R_2=H

ginsenoside-Rf

ginsenoside-Re

ginsenoside-Rg_1

R_2=H

ginsenoside-Rg_2

Chart 6

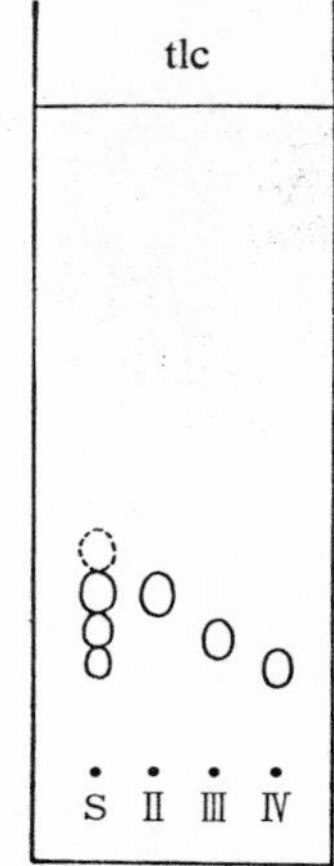

Fig. 20.2 tlc of senegins.

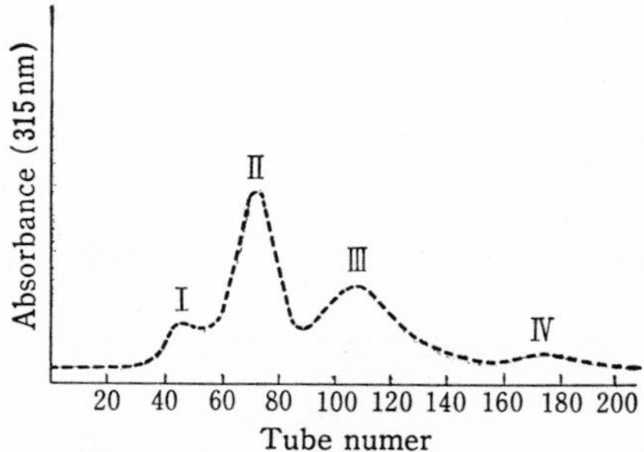

Fig. 20.3 Elution of senegins. Column, silicic acid; solvent, $CHCl_3$–MeOH–H_2O (7:3:1) lower phase.

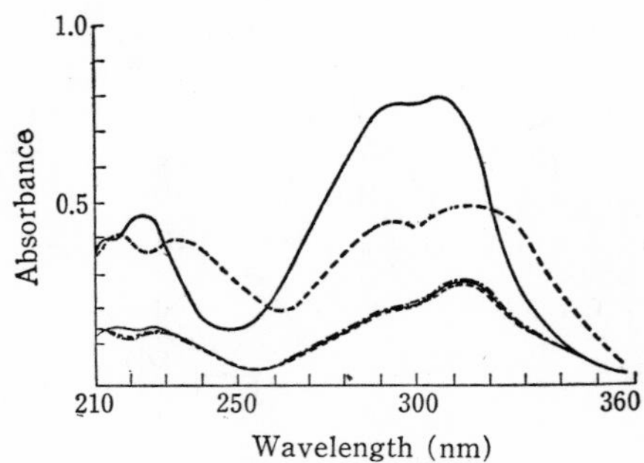

Fig. 20.4 Absorption spectra of senegins.
——, 4-methoxy cinnamic acid; - - - -, 3,4-dimethoxy cinnamid acid; ·······, senegin II; –·–·–, senegin III; ——, senegin IV.

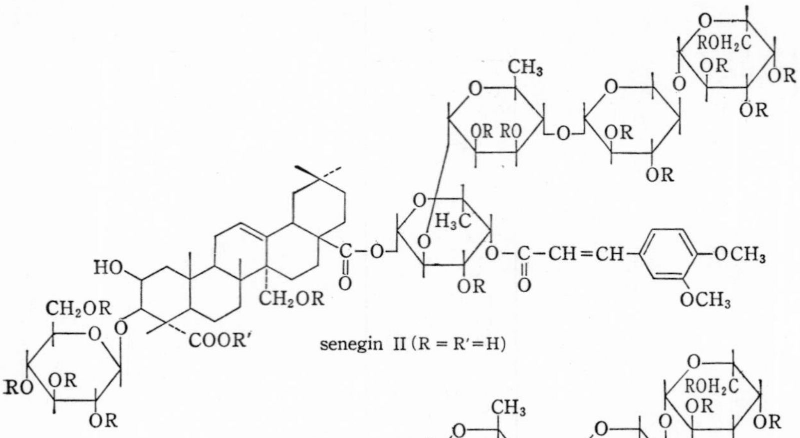

senegin II (R = R′ = H)

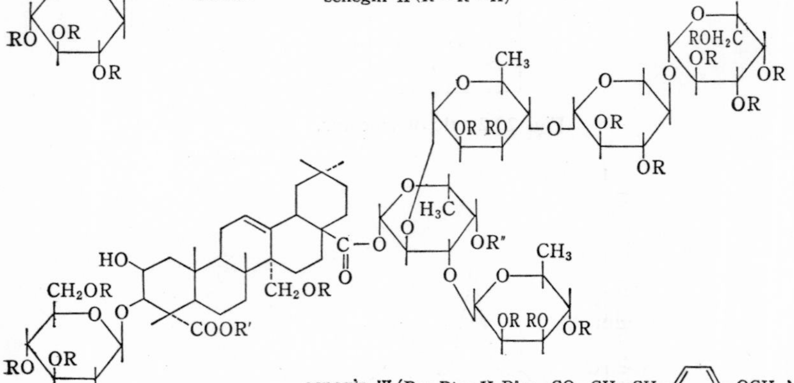

senegin III (R = R′ = H, R″ = –CO–CH=CH–C_6H_4–OCH_3)

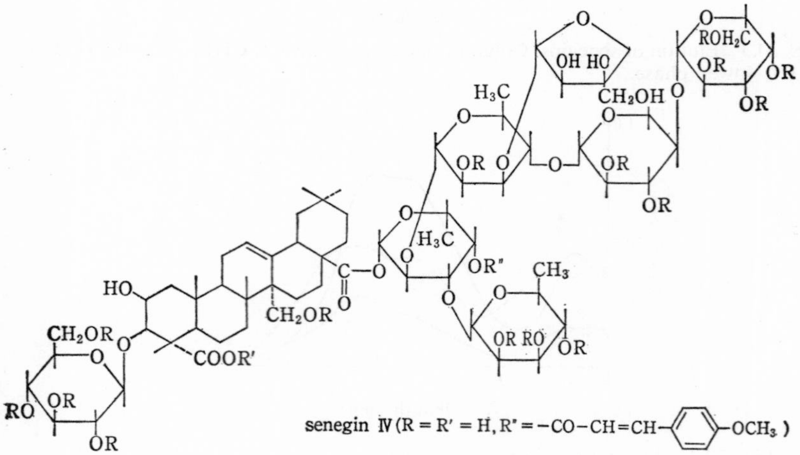

senegin IV (R = R′ = H, R″ = –CO–CH=CH–C_6H_4–OCH_3)

Chart 7

2.4%, 2.4% and 1.1%, respectively. The structures of senegins II, III and IV have been established.

REFERENCES

1) R. Tschesche, G. Wulff, *Fortschritte der Chemie Organischer Naturstoffe*, vol. 30, p. 461, Springer-Verlag (1973); R. Tschesche, *Kagaku no Ryoiki*, **25**, 571 (1971); T. Kawasaki, *Sogorinsho*, **16**, 1053 (1967); K. Hiller, M. Keipert, B. Linzer, *Pharmazie*, **21**, 713 (1966); M. Steiner, H. Holtzem, *Triterpene und Triterpene-Saponine*, *Paech-Tracey, Moderne Methoden der Pflanzenanalyse*, vol. 3, p. 58, Springer (1955); H. D. Woitke, J. P. Kayser, *Pharmazie*, **25**, 133, 213 (1970); S. K. Agarwal and R. P. Rastogi, *Phytochemistry*, **13**, 2623 (1974); R. S. Chandel and R. P. Rastogi, *Phytochemistry*, **19**, 1889 (1980).
2) A. Shimaoka, S. Seo, H. Minato, *J.C.S. Perkin I*, **1975**, 2043.
3) C. H. Brieskorn, W. Kilbinger, *Arch. Pharmaz.*, **308**, 824 (1975).
4) I. Kitagawa, M. Yoshikawa, I. Yoshioka, *Chem. Pharm. Bull.*, **24**, 121 (1976).
5) I. Kitagawa, M. Yoshikawa, Y. Imakura, I. Yoshioka, *ibid.*, **22**, 1339 (1974).
6) R. Tschesche, H. Rehkämper, G. Wulff, *Liebigs Ann. Chem.*, **726**, 125 (1969).
7) L. Kofler, *Wien. Klin. Wochschr.*, **44**, 852 (1931).
8) O. Dafert, *Z. Lebensm.-Untersuch. Forsch.*, **60**, 408 (1930).
9) V. Hariharan, S. Rangaswami, *Phytochemistry*, **10**, 621 (1971).
10) S. Sanada, N. Kondo, J. Shoji, O. Tanaka, S. Shibata, *Chem. Pharm. Bull.*, **22**, 421 (1974).
11) S. Sanada, N. Kondo, J. Shoji, O. Tanaka, S. Shibata, *ibid.*, **22**, 2407 (1974).
12) S. Shibata, O. Tanaka, T. Ando, M. Sado, S. Tsushima, T. Ohsawa, *ibid.*, **14**, 595 (1966).
13) Y. Nagai, O. Tanaka, S. Shibata, *Tetrahedron*, **27**, 8811 (1971).
14) S. Sanada, J. Shoji, *Chem. Pharm. Bull.*, **26**, 1694 (1978).
15) N. Kondo, Y. Marumoto, J. Shoji, *ibid.*, **19**, 1103 (1971).
16) J. Shoji, S. Kawanishi, Y. Tsukitani, *Yakugaku Zasshi*, **91**, 198 (1971).
17) Y. Tsukitani, S. Kawanishi, J. Shoji, *Chem. Pharm. Bull.*, **21**, 791 (1973).
18) Y. Tsukitani, J. Shoji, *ibid.*, **21**, 1564 (1973).

CHAPTER 21

Detection and Isolation of Steroid Saponins

Saponins are a group of plant constituents,* which are the glycosides of complex alicyclic compounds, and which show characteristic properties in aqueous solution, for example, foaming, toxicity towards fish, hemolysis and complex formation with cholesterol.[1]

The saponins were formerly divided into neutral and acidic saponins, but since the aglycones (sapogenins) of the former were found to be either steroid or triterpenoid and those of the latter to be triterpenoid compounds, they are now classified into steroid and triterpenoid saponins. Since 1962, a number of typical steroid saponins, namely steroid glycosides having the afore-mentioned characteristic (saponic) properties, e.g., dioscin **1**, gracillin **2**, parillin **19**, digitonin **3**, F-gitonin **17**, have been successively isolated in a pure state and their structures have been elucidated in detail. It was shown that they have structural features different from those of other non-saponic steroid glycosides, such as sterol, cardiac, and digitanol glycosides. That is, they are all spirostan-3β-ol derivatives having a branched-chain oligosaccharide linked with the hydroxyl group at C-3. However, another type of spirostanol glycosides (atypical ones) has also been discovered. They contain a monosaccharide or a straight-chain oligosaccharide, and, in some cases, have the sugar moiety linked to a hydroxyl group other than that at C-3. In addition, molecules having two sugar moieties and glycosides of spirostan-3α-ol or modified spirostane derivatives, e.g., yononin **4**, avenacoside-B **5** and trillenoside-A **38**, have been

* Saponins from animals such as sea-cucumbers and starfish are also known.

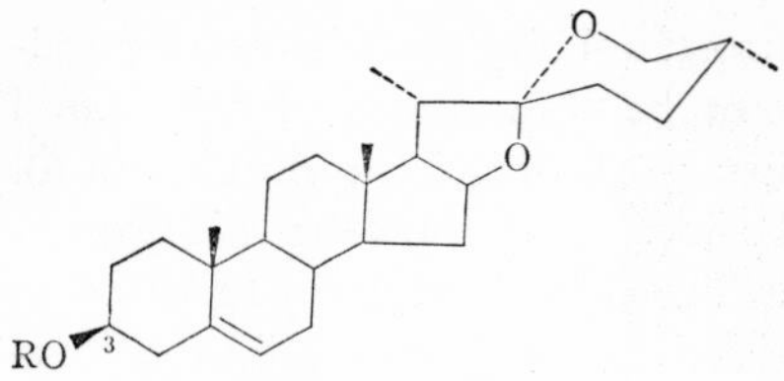

1 : R= Rha-(2)-[Rha-(4)]-Glc-

2 : R= Rha-(2)-[Glc-(3)]-Glc-

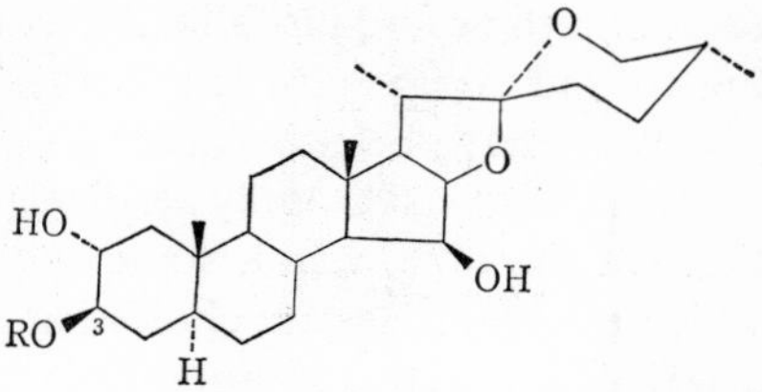

3 : R= Glc-(3)-Gal-(2)-[Xyl-(3)]-Glc-(4)-Gal-

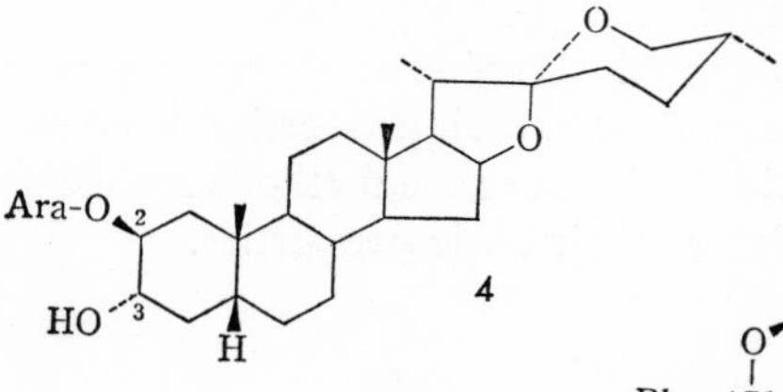

4: Ara-O (C-2), HO (C-3), H

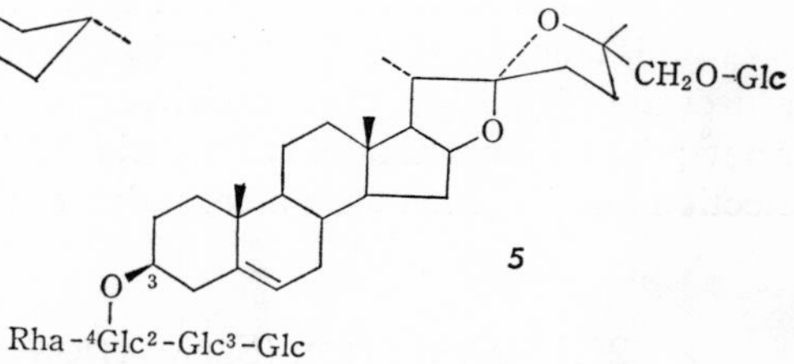

5: Rha-[4]Glc[2]-Glc[3]-Glc (O at C-3); CH_2O-Glc

6: Rha-(2)-[Rha-(4)]-Glc (O at C-3); HO at C-22; O-Glc at C-26

7: Rha-[2]Glc (O at C-3); OCH_3

8: Rha-(2)-[Rha-(4)]-Glc (O at C-3); H, N

9: Rha-(2)-[Glc-(3)]-Gal (O at C-3); N

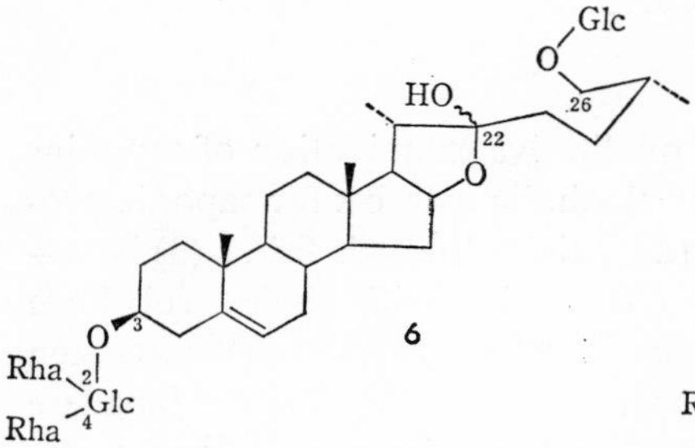

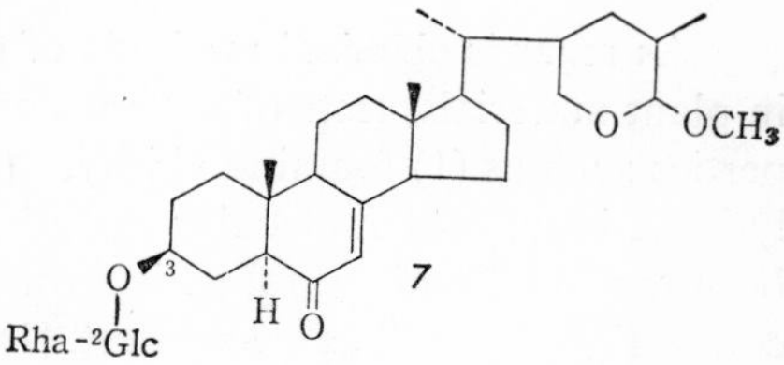

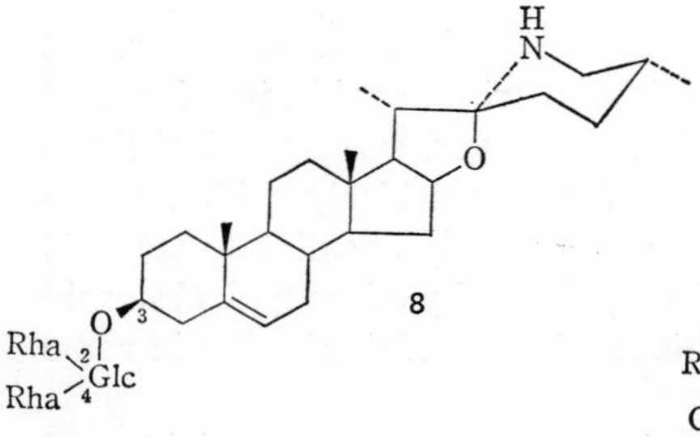

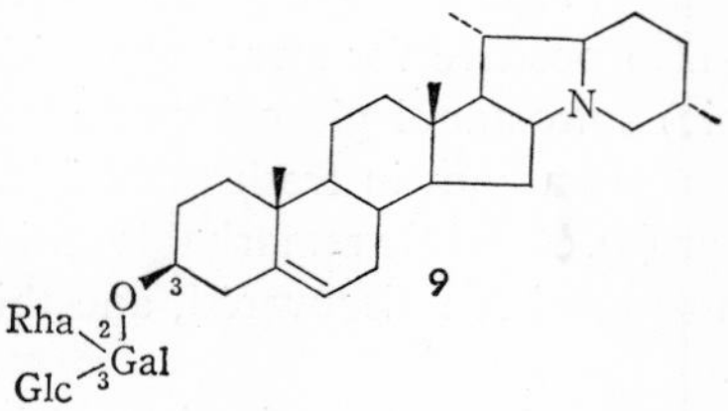

isolated and characterized. Following the discovery in 1966 by Schreiber *et al.* of jurbine and jurbidine, several furostanol bisglycosides corresponding to and being regarded as the prototype compounds of spirostanol glycosides have been isolated* (e.g., sarsaparilloside **22**, protodioscin **6**). Almost all the structurally atypical glycosides fail to show saponic properties. On the other hand, some steroid glycosides have been reported in which the aglycones are cholestane derivatives, but they exhibit the characteristic properties of saponins (e.g., polypodosaponin methylacetal **7**). A series of basic steroid glycosides called solanum alkaloids (e.g., α-solamargine **8**, α-solanine **9**) are closely related to the typical steroid saponins in structure as well as in properties. Their aglycones (alkamines) are the *N*-analogs of spirostane derivatives, and the sugar moieties are similar to or, occasionally, identical with those of typical steroid saponins.

In this chapter, the term "steroid saponins" refers to the spirostanol or furostanol glycosides in general, irrespective of their properties (saponic or not); the solanum alkaloids are excluded because the methods for their detection and isolation are different due to their basic properties.

21.1 Detection (Separation and Qualitative Examination)

As regards classical methods of qualitative examination of saponins in plant materials, tests of aqueous or alcoholic extracts for saponic properties such as (1) foaming ability, (2) toxicity towards fish, (3) hemolytic activity, (4) molecular complex formation with cholesterol, have been reported, and (5) the Liebermann-Burchard reaction has also been conventionally used as a color test. (1) and (5) are adopted in the Japanese Pharmacopea as identification methods for crude drugs containing saponins (e.g., Platycodi Radix, Anemarrhenae Rhizoma). However, (1) through (3) are common to both steroid and triterpenoid saponins, and (5) is also positive for sterol glycosides. Moreover, some atypical (non-saponic) spirostanol glycosides and the furostanol bisglycosides, sometimes predominant in fresh materials, are naturally negative to tests (1) through (4). Recently remarkable anti-fungal activity of the typical steroid saponins has been discovered, and this can be a useful criterion in some cases.

In general, the saponins have relatively large molecular weights and

* A kryptogenin glycoside **25**, in which the aglycone has both the E and F rings of the spirostane skeleton opened, has been isolated.

are mixtures of various kinds of homologs and analogs which differ slightly in the structures of the aglycone and/or sugar moieties. Furthermore, they are often accompanied by very polar substances such as saccharides and coloring matter, and are hence hygroscopic and not easily crystallized. Accordingly, suitable methods for the detection of individual saponins were not available until the development and application of paper partition chromatography and subsequently of thin-layer chromatography and other modern techniques.

Paper partition chromatography (ppc) played a very important role in early studies on steroid saponins of *Dioscorea* and *Convallaria* plants. The solvent systems found to be useful were, for example, BuOH–AcOH–water (4:1:5)[2)] or MeCOEt saturated with water[3)] for free saponins, and toluene –$CHCl_3$–water (10:2:5)[2)] for saponin acetates. For visualizing the spots,[2)] the Sannie reagent, which was known to be useful for the detection of steroid sapogenins on ppc, was mainly used; namely, (a) $SbCl_3$ in $CHCl_3$ (15%), (b) cinnamic aldehyde or anisaldehyde in EtOH (1%), (c) Ac_2O–H_2SO_4 (2:1) were sprayed in the order b, c and a, followed by heating at 60–70° C for a few min. The saponins usually appeared on the chromatograms as yellow spots, and those having unsaturated aglycones (sapogenins) gave a reddish-orange yellow or pink color with reagent a alone. However, since the colorations of saponins are due to the sugar moieties as well as the aglycones, care is required before assuming the presence of a steroid saponin simply on the basis of the color of a spot. Spraying 25% trichloroacetic acid solution in $CHCl_3$ gave a fluorescent spot of saponin under a uv lamp, and blood provided a faded spot on a colored background due to hemolysis, but these reagents were not very sensitive, and the latter is only applicable to typical steroid saponins. The *Rf* values on ppc depended mainly on the number of component monosaccharides of a saponin, suggesting that the approximate number of sugar units might be deduced from the *Rf* value relative to those of reference compounds. However, those which are slightly different in structure, for instance, having the same sugar moiety but homologous or analogous aglycones, could not be distinguished. This was one of the most serious weaknesses of ppc, but Tschesche *et al.*[4)] overcame this problem by an improved procedure using formamide-impregnated paper, which made it possible for the first time successfully to separate and detect several component saponins in the seeds of *Digitalis purpurea.* This new method, however, in our experience, showed relatively poor reproducibility due to the rather intricate procedure and its susceptibility to moisture in the chromatography chamber.

In the meantime, thin-layer chromatography (tlc) on silica gel, initiated by Stahl[5)] in 1958, became available as a simple and fast method with high resolution and sensitivity for very small amounts of organic compounds. Therefore, as an alternative to ppc, we[6] investigated the application of this method to steroid saponins, and found that a solvent system consisting of chloroform, methanol and water (for example, in ratios of 65:35:10, 70:30:10, 70:30:5, etc.) was very effective in separating saponins according not only to differences of the sugar moieties but also to slight structural divergencies of the aglycones. As color reagents, $SbCl_3$ (15%), anisaldehyde (0.5 ml) + EtOH (9 ml) + H_2SO_4 (0.5 ml) (+ 0.1 ml AcOH) and 10% H_2SO_4(in the latter cases, spraying followed by heating) gave a satisfactory result. The last one was a simple but useful reagent, because the color change of a spot, on heating and then cooling, was essentially characteristic of the kind of aglycone, while the second reagent was convenient to distinguish steroid saponins (yellow, in general) from sitosterol glucosides (purple) and cardiac glycosides such as digitoxin (dark blue). A mixture of saponin peracetates or permethylates was also separated by using $CHCl_3$–MeOH (100:2) or benzene–MeCOMe (80:20), respectively, as a developing solvent, and the size of the sugar moiety could be deduced from the *Rf* value. BuOH–AcOH–water (50:10:40) and BuOH–EtOH–96% NH_3 (30:60:50)[7] as solvents and chlorosulfonic acid–AcOH (1:2), 30% H_2SO_4, 1% ceric sulfate in 10% H_2SO_4,[8] and cobalt chloride[9] as detectors for free steroid saponins were later reported. As for the furostanol bisglycosides* (prototype saponins, e.g. protodioscin **6**), we[10] found that the Ehrlich reagent (1 g of *p*-dimethylaminobenzaldehyde + 50 ml of 36% HCl + 50 ml of EtOH; spraying and heating) gave a red color and was useful for their detection on tlc (those (e.g. **34**) corresponding to the glycosides of pennogenin (bearing a hydroxyl group at C-17) and a kryptogeninglycoside **25**, exceptionally, were negative to the reagent).

Thus, by means of tlc, the detection of individual saponins in an extractive, the monitoring of fractionation and separation of crude saponins into components as described later, examination of the purity of a sample, and its identification with an authentic specimen became relatively easy and fast. It should be mentioned, however, that some other steroid glycosides and triterpenoid saponins sometimes show misleading behavior on tlc similar to that of steroid saponins, and that, for the purpose of unambiguous identification, direct comparison with an authentic sample and confirmation of the characteristic absorptions of the spiroketal skeleton in the ir spectrum[11] of the isolated glycoside (if it is a furostanol bisglycoside,

* These are converted on refluxing with methanol to the less polar 22-OCH_3 derivatives, which regenerate the original 22-OH compounds on treatment with boiling water or dilute dioxane. Treatment with enzyme (glucosidase) cleaves the 26-*O*-glucosidic linkage to yield the corresponding spirostanol glycosides, which are negative to the Ehrlich reagent.

after treatment with enzyme to remove the 26–*O*-glucosyl moiety, yielding a spirostanol glycoside) are required.

A mixture of a few steroid saponins of relatively small size can be separated by gas-liquid chromatography[12] of their permethylates and pertrimethyl silyl ethers, but this method has limitations; for instance, it is not suitable for crude extractives or complex saponins.

21.2 Extraction from Plant Materials and Separation from Other Constituents of Crude Saponins

Plant materials (powdered, cut or sliced) are usually extracted with hot water or aqueous alcohol (MeOH or EtOH) repeatedly, but having regard for the subsequent procedures of fractionation and separation, percolation or soaking at room temperature is preferable. The extracts generally foam while distilling off a solvent containing water, so about 90% alcohol for air-dried materials and at least 95% alcohol for fresh materials should be used for extraction; in order to prevent violent foaming, a higher aliphatic alcohol or a silicone polymer can be added if necessary. The extractives are subsequently defatted with petroleum ether, hexane, benzene or ether, and if excessive amounts of chlorophylls are present, they are removed with cyclohexane, chloroform, dichloromethane or carbon tetrachloride. It should be taken into account that in these procedures considerable amounts of the less polar saponins may also be transferred to the ether, chloroform and dichloromethane layers. Formerly, the crude saponin fraction was obtained from the defatted extractives or water-soluble and -insoluble portions thereof by one or several of the following procedures: (a) passage through a charcoal or charcoal–celite column so as to remove the coloring matter, (b) crystallization from appropriate solvents, (c) dissolution in alcohol followed by addition of water, ether or acetone to precipitate the saponins, (d) dissolution in ethanol and admixture with an ethanol solution of cholesterol to yield saponin cholesterides, followed by treatment with pyridine–ether or dimethylsulfoxide–hexane[13] to decompose the complexes and liberate the free saponins.

In 1952, Rothman *et al.*[14] reported a simple but useful method for the separation of crude saponins from extractives; that is, the extractives were dissolved or suspended in water and shaken with butanol saturated with water. This is effective in the separation of the less polar saponins from highly polar ones together with water-soluble contaminants. We[15] employed a similar method, partition between BuOH–AcOEt and water layers to separate highly polar saponins such as furostanol bisglycosides from others (see "Experimental procedure 4"). The carbohydrates and salts accompanying the saponins in the aqueous layer and the phenolic compounds in the organic layer can be removed, respectively, by passage through Sephadex and inactivated (with water) alumina columns. Non-saponic contaminants can also be removed by acetylation of a crude fraction and subsequent chromatography on alumina using benzene–$CHCl_3$ as an eluent. The eluate is saponified to regenerate a saponin mixture. Suitable combinations and repetitions of the above procedures afford several fractions which each consist of saponins similar in size and structure, although a homogeneous saponin cannot necessarily be obtained.

21.3 Separation of a Saponin Mixture into the Individual Components (Isolation of Individual Saponins)

It was formerly a painstaking task to isolate component saponins in as pure a state as possible from a complex mixture, but new chromatographic techniques have made it rather easier.

We[16] succeeded in obtaining pure dioscin **1** and gracillin **2** by extraction of the air-dried rhizomes of several kinds of *Dioscorea* plants with 90% MeOH (EtOH) followed by treatment of the extractives with water and subsequent column chromatography of the water-insoluble portion on alumina, eluting with $CHCl_3$–MeOH 5:1, 1:1, and MeOH. Kimura *et al.*[3] successfully isolated seven homogeneous saponins from 50% MeOH extractives of the flowers of a lily-of-the-valley (*Convallaria keisukei* Miq.) by means of treatment with solvents and successive column chromatographies on active charcoal, alumina, Sephadex G-15, and celite-Florisil (*cf.* "Experimental procedure 1"). Tschesche *et al.*[4] obtained pure digitonin **3** by column chromatography over formamide-impregnated cellulose powder, as an extension of the improved ppc method devised by them.

Our column chromatography on alumina is not able to separate a mixture of saponins which are only slightly different in structure (for instance, different aglycones with the same sugar moiety) into the pure component saponins. On the other hand, column chromatography on silica gel with $CHCl_3$–MeOH–water (in various ratios) has been developed for preparative separation. Thus, two saponins with the same sugar moiety but different aglycones, F-gitonin **17** and desgalactotigonin **18**, were isolated in a pure state ("Experimental procedure 2").[17] Later, CH_2Cl_2–MeOH–water (in various ratios),[18] 5%, 10% and 15% MeOH–AcOEt saturated with water[19] and other solvent systems were devised for silica gel column chromatography, which now has wide applicability.

Among other preparative separation methods, preparative tlc, chromatographies on silica gel (eluting with hexane–AcOEt) for a mixture of peracetates,[18] and on a silicic acid–celite column (eluting with $CHCl_3$–MeOH)[20] are noteworthy. The classical counter-current distribution method is also available, but the recently exploited droplet counter-current chromatography (dcc) is far easier to operate and is theoretically superior. Its further development and wider application seem certain both as an alternative to and a supplement for silica gel chromatography. The report by Tanimura *et al.*[21] on the preparative separation of commercial "digitonin" into its component saponins by this method is interesting.

21.4 Examination of Purity, and Identification

Examination of purity and identification with an authentic sample of an isolated saponin are, at present, the areas still requiring development.

The saponins have relatively large molecular weights and cannot always be obtained as crystals. Even if they can be crystallized, some water of crystallization is usually included which cannot easily be removed completely. They melt in many cases accompanied by decomposition. Therefore it is difficult to decide the purity and identity of samples simply from the melting point, the optical rotation value, and other physical constants. The ir spectra are not very informative concerning the detailed structures, while the nmr (of free saponins and their derivatives)[22] and mass spectra (of the derivatives)[23] are rather complicated, and may be further complicated by the presence of impurities. Generally, the purity of a sample is judged by examining whether it shows a single spot on several types of tlc, developed and stained under a variety of conditions, and identity with another sample is assessed by comparison of the *Rf* values and colorations, and by co-chromatography.

It has recently become possible to determine the molecular weight of a saponin fairly reliably by vapor pressure osmometry of a solution of free saponin or by mass spectrometry[23] of the peracetate or permethylate. The molecular formula of a saponin can now be determined on the basis of the molecular weight thus obtained and the results of elemental analysis together with qualitative and quantitative analysis data for the aglycone and component mono-saccharides obtained by complete hydrolysis of the sample, and also quantitative determination (chemical or NMR spectrometric) of acetyl groups of its peracetate. Unequivocal determinations of purity and identity are based on the molecular formula, the findings during the procedures to determine the formula, physical constants, and chromatographic and spectral data. In identification, comparisons should be made directly with an authentic sample obtained under the same conditions of recrystallization, drying, tlc, hydrolysis and measurements of physical constants and spectra.

Recently Tanaka *et al.*[24] reported that ^{13}C-nmr spectrometry was a very useful technique for structure elucidation and identification of the glycosides of kaurene-type diterpenoids and dammarane-type triterpenoids.

[Experimental procedure 1] Isolation of steroid saponins from the flowers of *Convallaria keiskei* Miq. (Liliaceae)[3]

Dried and powdered materials (9.9 kg) were extracted three times with 50% MeOH (30 l each) at room temperature for 48 hr. The extracts were combined, a few drops of Toshiba Silicone TS-984-E were added, and the mixture was concentrated *in vacuo* at 45–50° C. The residue was treated with ethanol and the soluble part (480.9 g) was dissolved in water and applied to an active charcoal (980 g) column. The eluate (163.1 g) with $CHCl_3$–MeOH (4:1) was dissolved in water (500 ml) and successively extracted with $CHCl_3$ (300 ml each, 4 times) and $CHCl_3$–EtOH (2:1) (300 ml each, 4 times). The combined $CHCl_3$ and $CHCl_3$–MeOH layers and the water layers were each fractionated as shown in Tables 21.1 and 21.2, monitoring by paper chromatography (solvent, MeCOEt saturated with water; Kedde reagent for detection of cardiac glycosides) to give convallasaponin-A **10,**

TABLE 21.1. Fractionation procedure

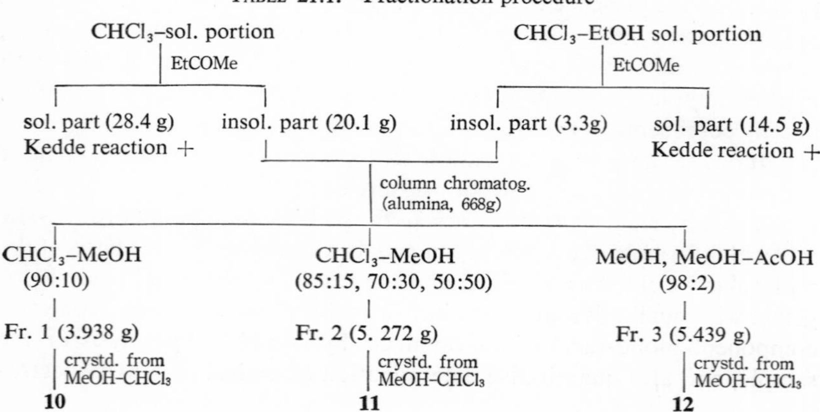

TABLE 21.2.

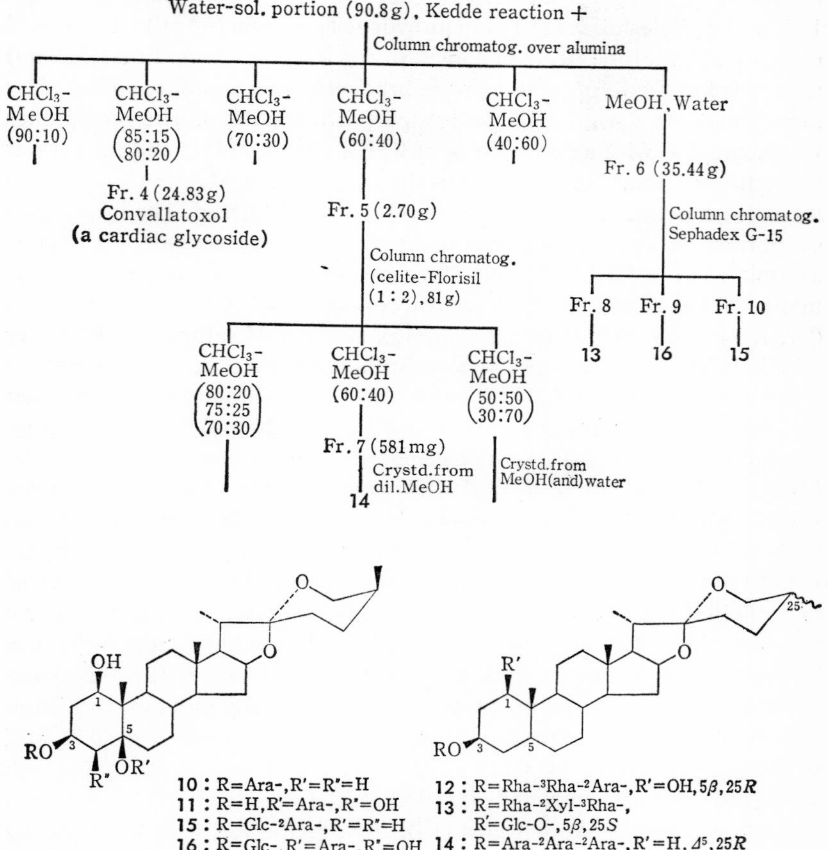

-B **11**, -C **12**, -D **13**, -E **14**, and glucoconvallasaponin-A **15** and -B **16**, all in a pure state.

[Experimental procedure 2] Isolation of F-gitonin 17 and desgalactotigonin 18[17]

The butanol extractives (50 g) of the residue in the manufacture of cardiac glycosides from *Digitalis purpurea* leaves were dissolved in MeOH (700 ml), boiled with active charcoal for 30 min and filtered. The filtrate was concentrated *in vacuo* and the residue was treated with $CHCl_3$ (200 ml). The insoluble portion was dissolved in 90% EtOH (500 ml) and a solution of cholesterol (10 g) in 99% EtOH (150 ml) was added. The mixture was heated on a water-bath for 10–20 min and then left to stand in a refrigerator overnight. The precipitates (saponin cholesterides) were collected by filtration, washed with EtOH, ether and dried in a desiccator (yield 10–15%). The cholesterides (10 g) in anhydrous pyridine (60 ml) were heated on a water-bath for 1 hr, then cooled to room temperature. Ether (600 ml) was added to the solution to provide precipitates, which were collected by filtration, washed with ether and dried (yield 65–70%). The precipitates were extracted with a hot mixture of $CHCl_3$–MeOH (1:1) and the extract was concentrated. The residue was dissolved in hot MeOH (250 ml), refluxed with active charcoal (3 g) for 30 min, and filtered. The filtrate was evaporated to dryness *in vacuo* to provide a crude saponin. The crude saponin (3 g) in $CHCl_3$–MeOH (1:1) was applied to an alumina (Brockmann, 300 g) column and eluted successively with $CHCl_3$–MeOH (1:1) (Fr. 1, 340 mg), MeOH (Fr. 2, 290 mg), and BuOH saturated with water (Fr. 3, 1330 mg). Fraction 3 was rechromatographed in the same way to give Fr. 1′ (196 mg), 2′ (230 mg) and 3′ (890 mg). Fraction 3′ was further separated in a similar manner to afford Fr. 1″ (220 mg) and 2″ (570 mg). Fraction 1 showed a single spot (saponin A) on paper chromatography (Toyo-Roshi No. 50, BuOH–AcOH–water (4:1:5)), but on tlc (silica gel G Merck, $CHCl_3$–MeOH–water (65:35:10, bottom layer)) two spots of saponin A-1 (= desgalactotigonin **18**) and A-2 (= F-gitonin **17**). Fraction 2″ gave one spot (saponin B) on PC and two spots of saponin B-1(= tigogenin glycoside)*1 and B-2(= gitogenin glycoside)*1 on tlc. Saponin A (150 mg) was separated by Tschesche's method[4] (on a formamide-impregnated cellulose powder column) to give **18** (20 mg) and **17** (60 mg). The crude saponin (10 g), obtained by precipitation with cholesterol followed by regeneration with pyridine-ether, was dissolved in 20% EtOH (3 l) containing AcOH (3 ml) and toluene (2 ml) and incubated with Takadiastase A or hemicellulase (20 g) at 30° C for 8 days.*2 The reaction mixture was diluted with water (1 l), the precipitates were collected by filtration and extracted with $CHCl_3$–MeOH (1:1). The extracts were concentrated and left to stand, to yield saponin A (3.9 g). This compound was also obtained when the original BuOH extractives in MeOH were refluxed with charcoal, filtered, concentrated, dissolved in water, incubated with the enzyme and treated as described above. Saponin A (500 mg) was subjected to column chromatography on silica gel (Kanto-Kagaku, 100–200 mesh, 250 g), eluting with $CHCl_3$–MeOH–water (70:30:10, bottom layer). Four hundred and sixteen fractions (5 ml each) were taken, among which fractions 266–300 (116 mg) were crystallized from dilute MeOH to give

*1 These are presumed to be pentaglycosides.
*2 A terminal sugar unit of saponin B is split off to give saponin A.

18, while fractions 338–390 (178 mg) were crystallized from dilute MeOH or BuOH saturated with water to provide **17**.

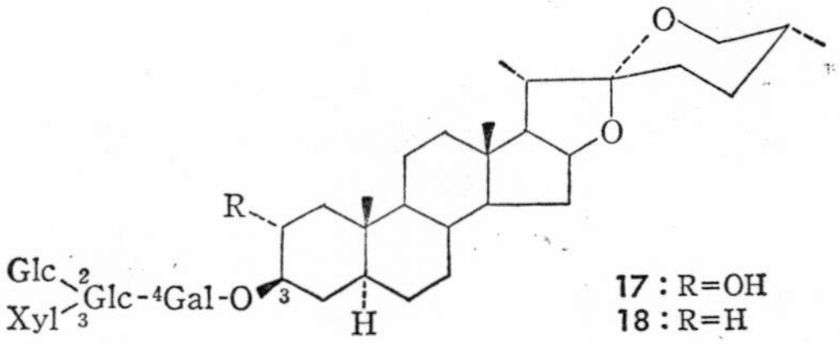

Glc-2, Xyl-3 > Glc-4Gal-O 3

17 : R=OH
18 : R=H

[Experimental procedure 3] Isolation of parillin 19, desglucoparillin 20, desgluco-desrhamnoparillin 21 and sarsaparillosides 22 and 22′[25]

Vera-Cruz sarsaparilla (Sarsaparillae Radix) (commercial material) (3.74 kg) was cut up and percolated at 42° C with MeOH and 95–90% MeOH to give extractives I (with 1 l MeOH, 120 g), II (4 l MeOH, 820 g), III (4 l MeOH, 193 g), IV (4 l 95% MeOH, 581 g) and V (4 l 90% MeOH, 915 g). Extractives III–V were combined, dissolved in water (4.5 l) and shaken with BuOH (2 l each, 3 times). The BuOH layer was concentrated to provide a crude saponin (23.4 g), which was combined with extractives I and II. This crude saponin fraction (925 g), which showed eleven spots (those of parillin **19** and less polar saponins being major) on tlc (silica gel G Merck, $CHCl_3$–MeOH–water (65:35:10, bottom layer)), was dissolved in MeOH (700 ml), diluted with water (400 ml), and the mixture was allowed to stand for two days. The precipitates (82.6 g) were collected by filtration, and washed with $CHCl_3$. The residue (72.2 g) in hot 99% EtOH (800 ml) was mixed with a saturated hot solution of cholesterol (64 g) in ether (350 ml). The mixture was left to stand for 24 hr and the precipitates were collected by filtration, then washed with ether to give cholesterides (21 g) of **19** and other saponins. The filtrate and washings were combined, concentrated, and the residue (115 g, a mixture of **19**, **20**, **21**, other saponins, and cholesterol) was treated repeatedly with ether in order to remove cholesterol and other soluble substances, providing a saponin mixture (31.9 g). This was dissolved in hot 99% EtOH (400 ml) and diluted with water (200 ml) to yield crystalline precipitates consisting principally of desglucoparillin **20** and desgluco-desrhamnoparillin **21**. The supernatant contained mainly **19**. The precipitates and the residue after concentration of the supernatant were each subjected to the above procedure (addition of water to the EtOH solution) repeatedly to yield a fraction (9.18 g) of almost homogeneous **19** and a fraction (16.5 g) consisting of **20** and **21**, respectively. The latter was recrystallized twice from EtOH–water (1:1),

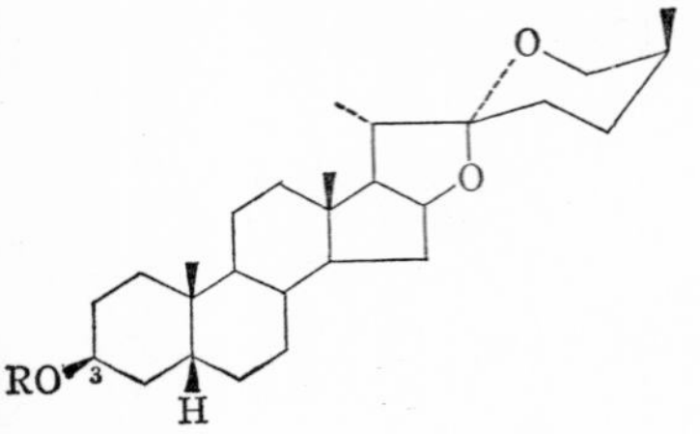

RO 3 H

19 : R= Glc-2(Rha-4)(Glc-6)Glc-

20 : R= Rha-4(Glc-6)Glc-

21 : R= Glc-6Glc-

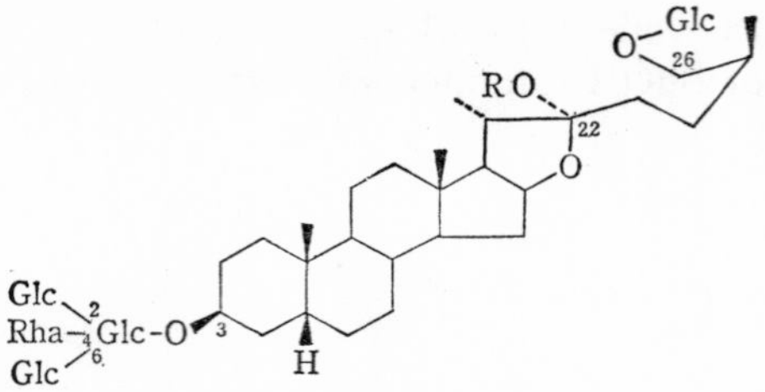

RO 22 O Glc 26

Glc-2(Rha-4)(Glc-6)Glc-O 3 H

22 : R = H

22′ : R = CH_3

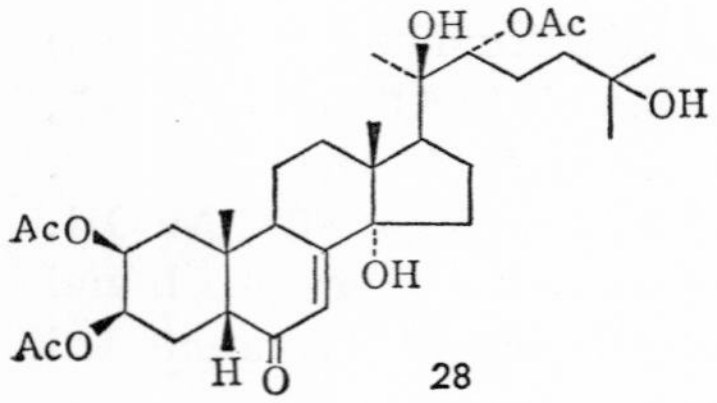

OH OAc OH

AcO AcO H O OH

28

HO 20 O 17 O

RO 3

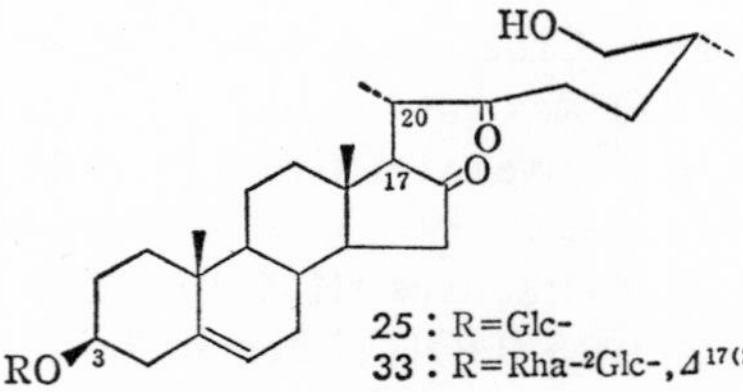

25 : R=Glc-

33 : R=Rha-2Glc-, $\Delta^{17(20)}$

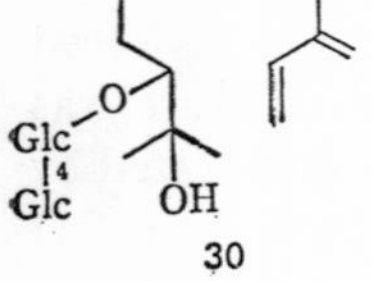

Glc 4 Glc O OH

30

O R′ 17 O RO 3

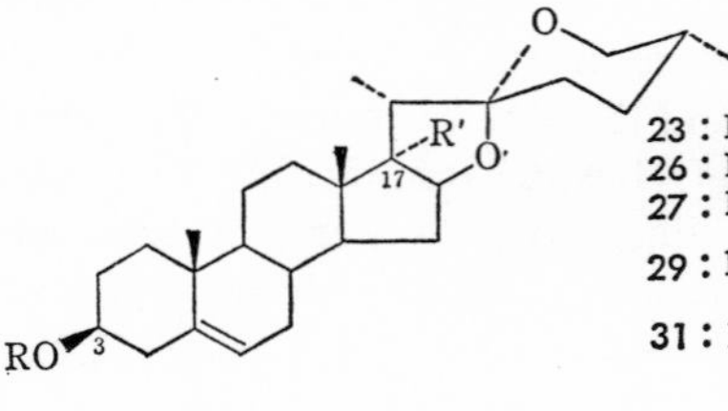

23 : R=H, R′=OH

26 : R=Rha-2Glc-, R′=H

27 : R=Rha-2Glc-, R′=OH

29 : R=Rha-2(Rha-4)Glc-, R′=OH

31 : R=Rha-2(Rha-4Rha-4)Glc-, R′=OH

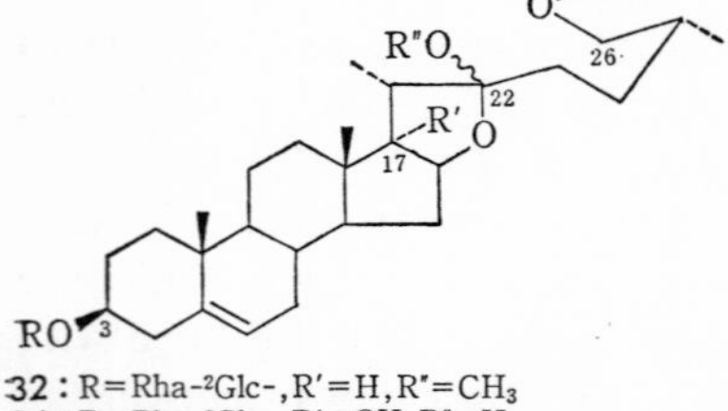

O-Glc R″O 22 26 R′ 17 O

RO 3

32 : R=Rha-2Glc-, R′=H, R″=CH_3

34 : R=Rha-2Glc-, R′=OH, R″=H

36 : R=Rha-2(Rha-4)Glc-, R′=OH, R″=H

39 : R=Rha-2(Rha-4Rha-4)Glc-, R′=OH, R″=H

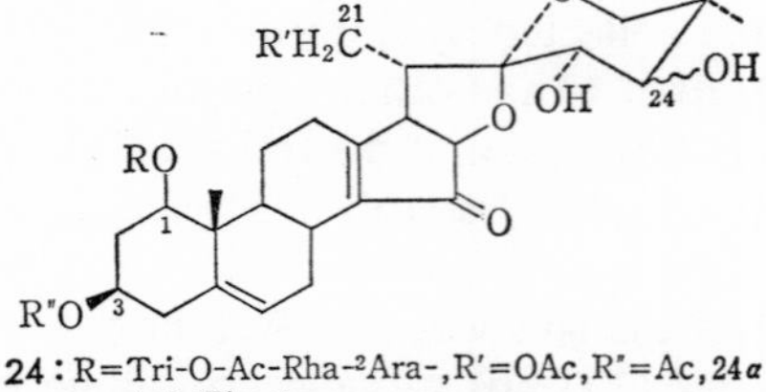

21 R′H_2C O O OH 24 OH RO 1 O R″O 3

24 : R=Tri-O-Ac-Rha-2Ara-, R′=OAc, R″=Ac, 24α

35 : R=Api-3Rha-2(Xyl-3)Ara-, R′=R″=H, 24β

37 : R=Rha-2(Xyl-3)Ara-, R′=OH, R″=H, 24β

38 : R=Api-3Rha-2(Xyl-3)Ara-, R′=OH, R″=H, 24β

TABLE 21.3.

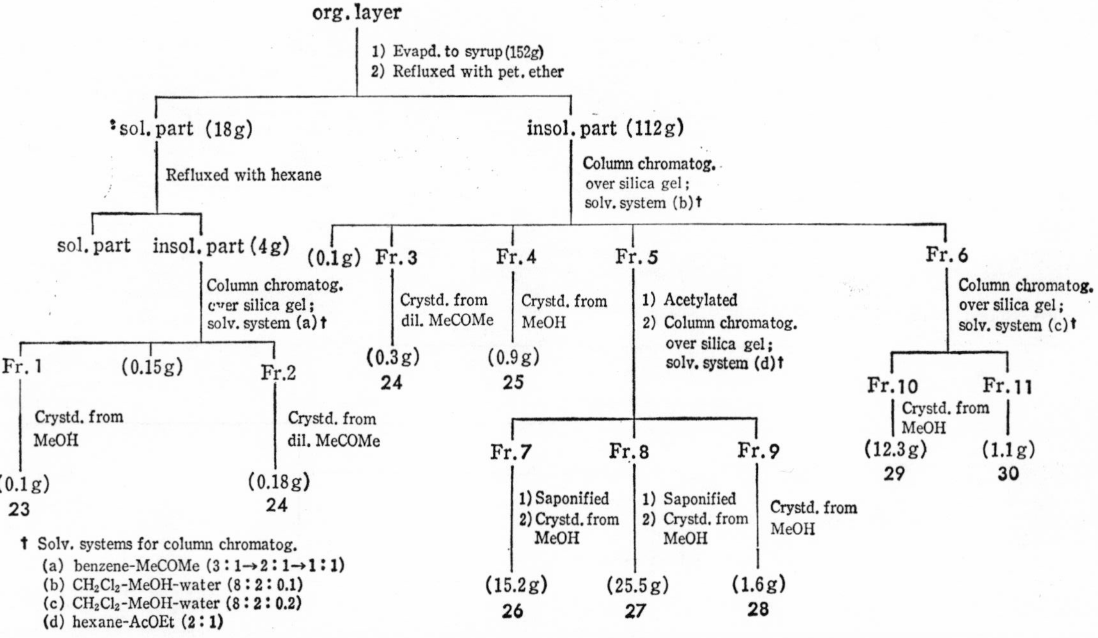

† Solv. systems for column chromatog.
(a) benzene-MeCOMe (3 : 1→2 : 1→1 : 1)
(b) CH_2Cl_2-MeOH-water (8 : 2 : 0.1)
(c) CH_2Cl_2-MeOH-water (8 : 2 : 0.2)
(d) hexane-AcOEt (2 : 1)

TABLE 21.4.

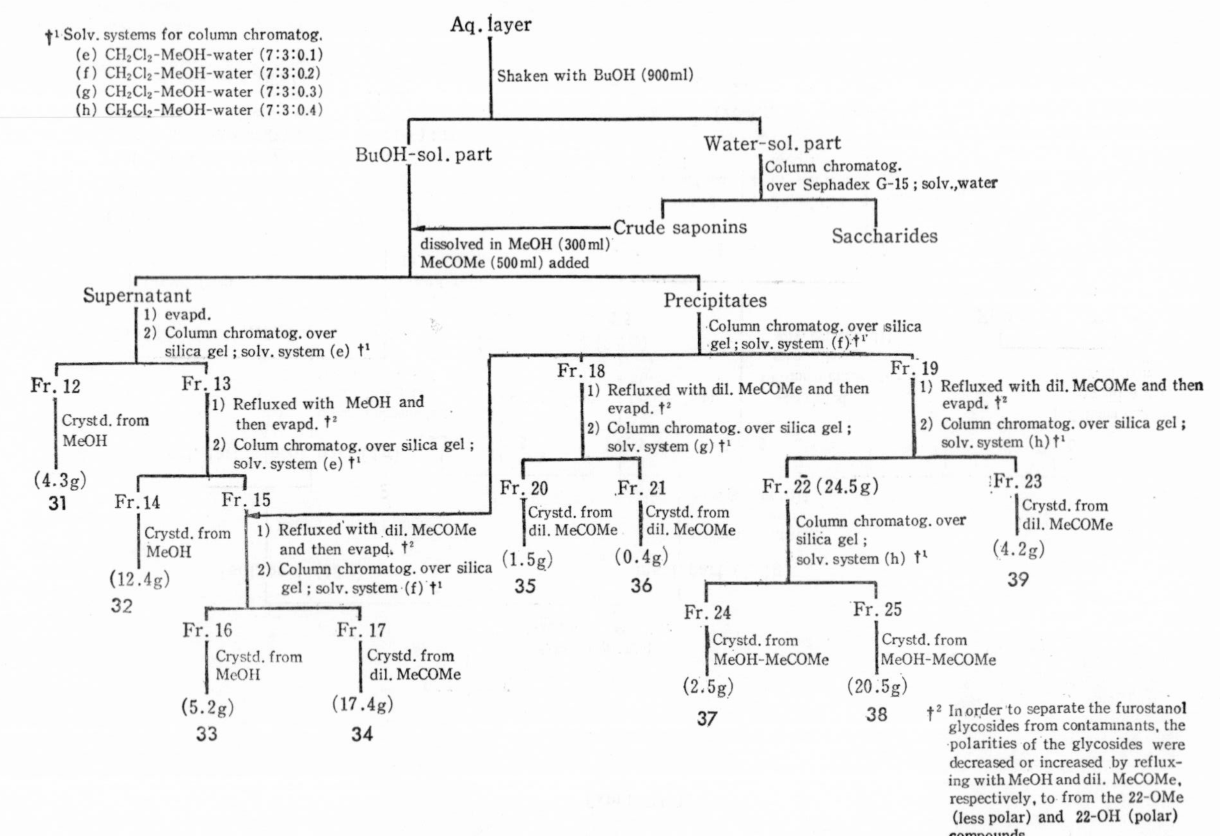

†1 Solv. systems for column chromatog.
(e) CH_2Cl_2-MeOH-water (7:3:0.1)
(f) CH_2Cl_2-MeOH-water (7:3:0.2)
(g) CH_2Cl_2-MeOH-water (7:3:0.3)
(h) CH_2Cl_2-MeOH-water (7:3:0.4)
Aq. layer
Shaken with BuOH (900ml)
BuOH-sol. part
Water-sol. part
Column chromatog. over Sephadex G-15 ; solv.,water
Crude saponins
Saccharides
dissolved in MeOH (300ml) MeCOMe (500ml) added
Supernatant
1) evapd.
2) Column chromatog. over silica gel ; solv. system (e) †1
Precipitates
Column chromatog. over silica gel ; solv. system (f) †1
Fr. 12
Crystd. from MeOH
(4.3g)
31
Fr. 13
1) Refluxed with MeOH and then evapd. †2
2) Colum chromatog. over silica gel ; solv. system (e) †1
Fr. 14
Crystd. from MeOH
(12.4g)
32
Fr. 15
1) Refluxed with dil. MeCOMe and then evapd. †2
2) Column chromatog. over silica gel ; solv. system (f) †1
Fr. 16
Crystd. from MeOH
(5.2g)
33
Fr. 17
Crystd. from dil. MeCOMe
(17.4g)
34
Fr. 18
1) Refluxed with dil. MeCOMe and then evapd. †2
2) Column chromatog. over silica gel ; solv. system (g) †1
Fr. 20
Crystd. from dil. MeCOMe
(1.5g)
35
Fr. 21
Crystd. from dil. MeCOMe
(0.4g)
36
Fr. 19
1) Refluxed with dil. MeCOMe and then evapd. †2
2) Column chromatog. over silica gel ; solv. system (h) †1
Fr. 22 (24.5g)
Column chromatog. over silica gel ; solv. system (h) †1
Fr. 24
Crystd. from MeOH-MeCOMe
(2.5g)
37
Fr. 25
Crystd. from MeOH-MeCOMe
(20.5g)
38
Fr. 23
Crystd. from dil. MeCOMe
(4.2g)
39
†2 In order to separate the furostanol glycosides from contaminants, the polarities of the glycosides were decreased or increased by refluxing with MeOH and dil. MeCOMe, respectively, to from the 22-OMe (less polar) and 22-OH (polar) compounds.

and a solid (13 g) that separated out was chromatographed on a silica gel (Herrmann, 2 kg) column, eluting with $CHCl_3$–MeOH–water (65:25:10), to give six fractions; Fr. 1 (3.32 g), 2 (4.40 g, **21**), 3 (1.25 g, **21** + **20**), 4 (2.30 g, **20**), 5 (0.73 g) and 6 (1.12 g, **19**). Fractions 2 and 4 were crystallized from 80% EtOH to give pure **21** (3.2 g) and **20** (1.7 g), respectively.

The afore-mentioned fraction (9.18 g), consisting largely of **19**, and Fr. 6 (1.12 g) of the above chromatography were combined and subjected to column chromatography on silica gel (2 kg), eluting with $CHCl_3$–MeOH–water (65:35:10, bottom layer) to yield a homogeneous **19** (1.27 g), which was crystallized from 80% EtOH. The aqueous layers on partition of extractives III–V between water and butanol were concentrated and each residue was again shaken with water (4.5 l) and BuOH (2 l). This procedure was repeated three times. The BuOH phase contained **19** together with other weakly polar saponins, while the aqueous phase was concentrated *in vacuo* to yield a residue consisting of **22**, its 22–methyl ether **22′**, some coloring matter and oligosaccharides. The residue (20 g) was applied to a Sephadex G-25 (350 g) column (40 × 8 cm) and eluted with water to provide a mixture (4–5 g) of **22** and **22′** contaminated with some pigments and other highly polar substances. The mixture (10 g) of crude **22** and **22′** and silica gel (50 g) were laid on a silica gel (50 g) column and eluted with $CHCl_3$–MeOH–water (65:35:10, bottom layer) to remove the bulk of the contaminants. The saponin fraction (ca. 6.5 g) consisting of **22** and **22′** was again mixed with silica gel (ca. 25 g) and placed on a silica gel (700 g) column. Elution with the same solvent provided three fractions; Fr. 1 (1.2 g, **22′** + **22** (ca. 10%)), Fr. 2 (2.2 g, **22′** + **22** (ca. 50%)), Fr. 3 (1.1 g, **22′** + **22** (ca. 85%)).

[Experimental procedure 4] Isolation of steroid saponins from the fresh rhizomes to *Trillium kamtschaticum* Pall. (Liliaceae)[18)]

The finely sliced rhizomes (10.75 kg) were soaked in MeOH (12.5 l) at room temperature for 8 days. The extracts were concentrated *in vacuo* and the residue (435 g) was shaken with (partitioned between) BuOH–AcOEt (1:5) (1.2 l) and water (1 l). The organic and aqueous layers were each fractionated as shown in Tables 21.3 and 21.4, monitoring by tlc (silica gel G Merck, $CHCl_3$–MeOH–water (65:35:10, bottom layer) and (70:30:5), CH_2Cl_2–MeOH–water (8:2:0.2)). Silica gel used for columns was from Merck (0.05–0.2 mm) or Kanto-Kagaku (100–200 mesh) in 30- to 50-fold excess relative to the applied materials. In addition to pennogenin **23**, ecdysterone acetate **28** and a sesquiterpene glycoside **30**, the following fourteen steroid saponins were isolated, all in a pure state: Ta **26**, Tb **27**, Tc **29**, Td **32**, Te **34**, Tf **33**, Tg **31**, Th **39**, Tj **36**, Tk **25**, y_1

(= trillenoside-A **38**), y_2 (= trillenoside-B **37**), y_3 (= isotrillenoside-C–PA **24**) and y_4 (= deoxytrillenoside-A **35**).

Addendum

New methods for the separation of crude saponins from the water-soluble portion of extractives using Amberlite XAD-2 column and for the isolation of individual saponins from a mixture by reversed phase high performance liquid chromatography have recently been reported.[26]

The molecular weight of a free saponin was directly determined and the useful informations concerning the structure were obtained by field desorption mass spectrometry.[27]

References

1) R. Tschesche, G. Wulff, *Fortschritte der Chemie organischer Naturstoffe* (ed. W. Herz, H. Grisebach, G. W. Kirby), vol. 30, p. 461, Springer-Verlag (1973); J. Elks, *Rodd's Chemistry of Carbon Compounds* (ed. S. Coffey), vol. II_E, p. 1, Elsevier Publishing (1971); Supplements to vol. II_{C-E} (ed. M. F. Ansell), p. 205, Elsevier Scientific Publishing (1974); T. Kawasaki, *Methodicum Chimicum* (ed. F. Korte, M. Goto), vol. 11, part 3, p. 88, Academic Press, Georg Thieme Publishers, Maruzen Co. (1978).
2) T. Tsukamoto, T. Kawasaki, A. Naraki, T. Yamauchi, *Yakugaku Zasshi* (in Japanese), **74**, 1097 (1954).
3) M. Kimura, M. Tohma, I. Yoshizawa, *Chem. Pharm. Bull.*, **14**, 50 (1966); *ibid.*, **16**, **25**, 2191 (1968).
4) R. Tschesche, G. Wulff, *Tetrahedron*, **19**, 621 (1963).
5) E. Stahl, *Chemiker—Ztg.*, **82**, 323 (1958); *Pharmaz. Rundsch.*, **2**, 1 (1959).
6) T. Kawasaki, K. Miyahara, *Chem. Pharm. Bull.*, **11**, 1546 (1963): T. Kawasaki, *Thin-Layer Chromatography* (in Japanese) (ed. S. Hara, O. Tanaka, S. Takiya), vol. 1, p. 31, Nankodo (1964).
7) E. Stahl, *Arch. Pharm.*, **306**, 693 (1973).
8) I. Yoshioka, K. Imai, Y. Morii, I. Kitagawa, *Tetrahedron*, **30**, 2283 (1974).
9) B. Pasich, *Planta Medica*, **11**, 16 (1963).
10) S. Kiyosawa, M. Hutoh, T. Komori, T. Nohara, I. Hosokawa, T. Kawasaki, *Chem. Pharm. Bull.*, **16**, 1162 (1968).
11) M. E. Wall, C. R. Eddy, M. L. McClennan, M. E. Klumpp, *Anal. Chem.*, **24**, 1337 (1952); C. R. Eddy, M. E. Wall, M. K. Scott, *ibid.*, **25**, 266 (1953).
12) T. Kawasaki, T. Yamauchi, *Chem. Pharm. Bull.*, **16**, 1070 (1968).
13) C. H. Issidorides, I. Kitagawa, E. Mosettig, *J. Org. Chem.*, **27**, 4693 (1962).
14) E. S. Rothman, M. E. Wall, H. A. Walens, *J. Am. Chem. Soc.*, **74**, 5791 (1952).
15) R. Higuchi, K. Miyahara, T. Kawasaki, *Chem. Pharm. Bull.*, **20**, 1935 (1972); T. Kawasaki, T. Komori, K. Miyahara, T. Nohara, I. Hosokawa, K. Mihashi, *ibid.*, **22**, 2164 (1974).
16) T. Tsukamoto, T. Kawasaki, A. Naraki, T. Yamauchi, *Yakugaku Zasshi* (in Japanese), **74**, 984 (1954).
17) T. Kawasaki, I. Nishioka, *Chem. Pharm. Bull.*, **12**, 1311 (1964); T. Kawasaki, I. Nishioka, T. Yamauchi, K. Miyahara, M. Enbutsu, *ibid.*, **13**, 435 (1965).
18) T. Nohara, K. Miyahara, T. Kawasaki, *ibid.*, **23**, 872 (1975); T. Nohara, A. Na-

kano, K. Miyahara, T. Komori, T. Kawasaki, *Tetr. Lett.*, **1975**, 4381; S. Imamura, T. Nohara, T. Kawasaki, *Ann. Mtg. Pharm. Soc. Japan* (in Japanese), *Abstracts* II, p. 257 (1975); *ibid.*, *Abstracts* II, p. 175 (1976); N. Fukuda, T. Nohara, T. Kawasaki, *ibid.*, p. 176 (1976); T. Nohara, T. Komori, T. Kawasaki, *Chem. Pharm. Bull.*, **28**, 1437 (1980); N. Fukuda, N. Imamura, E. Saito, T. Nohara, T. Kawasaki, *ibid.*, **29**, in press (1981).

19) H. Kato, S. Sakuma, A. Tada, S. Kawanishi, J. Shoji, *Yakugaku Zasshi* (in Japanese), **88**, 710 (1968).

20) H. Sato, S. Sakamura, *Agr. Biol. Chem.*, **37**, 225 (1973).

21) T. Tanimura, J. J. Pisano, Y. Ito, R. L. Bowman, *Science*, **169**, 54 (1970); T. Tanimura, H. Otsuka, Y. Ogihara, *Kagaku-no-Ryoiki* (in Japanese), **29**, 895 (1975).

22) K. Miyahara, T. Kawasaki, *Chem. Pharm. Bull.*, **22**, 1407 (1974).

23) T. Komori, Y. Ida, Y. Muto, K. Miyahara, T. Nohara, T. Kawasaki, *Biomed. Mass Spectr.*, **2**, 65 (1975).

24) R. Kasai, J. Asakawa, O. Tanaka, *Ann. Mtg. Jap. Soc. of Pharmacognosy* (in Japanese), *Abstracts*, p. 35 (1975); H. Kanda, T. Kobayashi, K. Yamasaki, R. Kasai, O. Tanaka, K. Nishi, *Ann. Mtg. Pharm. Soc. Japan* (in Japanese), *Abstracts* II, p. 273 (1976); M. Yahara, R. Kasai, O. Tanaka, *ibid.*, p. 274 (1976); K. Yamasaki, H. Kohda, T. Kobayashi, R. Kasai O. Tanaka, *Tetr. Lett.*, **1976**, 1005; H. Kohda, R. Kasai, K. Yamasaki, K. Murakami, O. Tanaka, *Phytochemistry*, **15**, 981 (1976); K. Yamasaki, H. Kohda, T. Kobayashi, N. Kaneda, R. Kasai, O. Tanaka, K. Nishi, *Chem. Pharm. Bull.*, **25**, 2895 (1977); I. Sakamoto, K. Yamasaki, O. Tanaka, *ibid.*, **25**, 3437 (1977); J. Asakawa, R. Kasai, K. Yamasaki, O. Tanaka, *Tetrahedron*, **33**, 1935 (1977); S. Yahara, O. Tanaka, I. Nishioka, *Chem. Pharm. Bull.*, **26**, 3010 (1978); O. Tanaka, S. Yahara, *Phytochemistry*, **17**, 1353 (1978); J. Kim, K. Han, K. Yamasaki, O. Tanaka, *ibid.*, **18**, 894 (1979).

25) R. Tschesche, R. Kottler, G. Wulff, *Liebigs Ann. Chem.*, **699**, 212 (1966); R. Tschesche, G. Lüdke, G. Wulff, *Tetr. Lett.*, **1967**, 2785; *Chem. Ber.*, **102**, 1253 (1969).

26) N. Imamura, R. Imanari, N. Yokota, K. Kudo, K. Miyahara, T. Kawasaki, *Ann. Mtg. Pharm. Soc. Japan* (in Japanese), *Abstracts*, p. 184 (1980); T. Komori, K. Sakamoto, Y. Itakura, H. Nanri, I. Maetani, T. Kawasaki, *The 23rd Symposium on the Chemistry of Natural Products* (in Japanese with English summary), *Symposium Papers*, p. 482 (1980).

27) H.-R. Schulten, T. Komori, T. Kawasaki, *Tetrahedron*, **33**, 2595 (1977); H.-R. Schulten, T. Komori, T. Nohara, R. Higuchi, T. Kawasaki, *ibid.*, **34**, 1003 (1978); T. Komori, M. Kawamura, K. Miyahara, T. Kawasaki, O. Tanaka, S. Yahara, H.-R. Schulten, *Z. Naturforsch.*, **34c**, 1094 (1979); T. Komori, I. Maetani, N. Okamura, T. Kawasaki, T. Nohara, H.-R. Schulten, *Liebigs Ann. Chem.*, in press (1981).

CHAPTER 22

Cleavage of Glycoside Linkages

Glycosides are widely distributed in nature, especially in the plant kingdom. Chemically, they are the acetal derivatives of carbohydrates. The hemiacetal hydroxyls in their cyclic pyranose or furanose structures can be substituted with a variety of alkyl or aromatic residues. The ether linkages formed between these hemiacetal hydroxyls and the various kinds of residues are known as glycosidic linkages. The carbohydrate portions of glycosides often consist of oligosaccharides containing one or more kinds of monosaccharide constituents. The ether linkages binding the monosaccharide constituents in the oligosaccharide are chemically equivalent to those in the glycoside linkages.

On complete hydrolysis of a glycoside with acid or by any other procedure, the glycoside linkage is cleaved to liberate the component monosaccharides (one or more kinds) in addition to the non-carbohydrate moiety, which is usually called the aglycone or genin. In the case of saponin, the non-carbohydrate portion is termed a sapogenol or sapogenin.

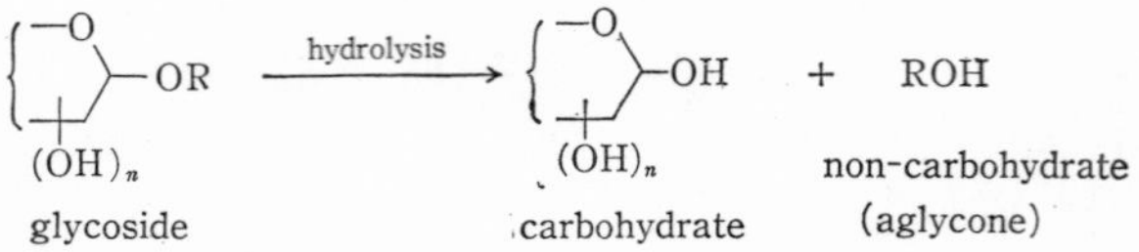

$$\text{glycoside } [(OH)_n \text{ ring}-OR] \xrightarrow{\text{hydrolysis}} \text{carbohydrate } [(OH)_n \text{ ring}-OH] + \underset{\text{non-carbohydrate (aglycone)}}{ROH}$$

Most natural glycosides are *O*-glycosides, as shown above, in which the oxygen atom of the hemiacetal hydroxyl at C-1 of the carbohydrate is linked with the aglycone. There are also some other types of glycosides occurring in nature, such as *S*-glycosides, *C*-glycosides, and *N*-glycosides. In *S*-glycosides, e.g., the mustard oil glycosides, the glycosidic linkage is a thioether bond which links the aglycone and the carbohydrate constituent through the sulfur atom of a 1-thio sugar. *C*-Glycosides are common among flavonoid derivatives, and the aglycone is directly attached to C-1 of the carbohydrate constituent through a C-C bond. The glycoside linkage in *N*-glycosides is a C-N bond which is formed between the nitrogen atom of the aglycone and C-1 of the carbohydrate, as seen in nucleosides.

As regards the carbohydrate constituents in glycosides, hexoses such as D-glucose, D-galactose, D-mannose (rare in glycosides, but often found in polysaccharides), L-rhamnose, D- and L-fucose, D-quinovose, D-fructose (rare), D-glucuronic acid, and D-galacturonic acid (rare), and pentoses such as D-xylose, L-arabinose, and D-ribose, are generally known. As special examples, desoxy-sugars and *O*-methyl sugars have been identified as the carbohydrate constituents of cardiac glycosides, and D-apiose, a branched carbohydrate, is an uncommon carbohydrate constituent of oligoglycosides.

Thus, although the kinds of carbohydrate constituents in glycosides are relatively few, the structures of aglycones are chemically diverse.[1,2] This chapter deals with cleavage methods for glycoside linkages, which are essential for chemical investigations on the structures of *O*-glycosides.

22.1 Cleavage of Glycoside Linkages

In order to elucidate the chemical structures of glycosides, the following points must be clarified: (1) the structure of the genuine aglycone, (2) the composition and sequence of the component monosaccharides in the carbohydrate moiety of the glycoside, and (3) the location and configuration of the linkage between the aglycone and the carbohydrate moiety. In the initial stage of structural studies of glycosides, cleavage of the linkage between the aglycone and carbohydrate moiety is a common approach.

Various methods have been devised for the cleavage of glycoside linkages. They can be divided roughly into chemical methods (using acid or some other reagent), biochemical methods (using an enzyme), and other procedures. Among the chemical methods, acid hydrolysis is the most common.

In this section, chemical cleavage methods for the glycoside linkage using acids and other reagents are described. In addition, some other methods for the cleavage of glycoside linkages are considered, including procedures for liberating the genuine aglycones or sapogenols from glycosides or saponins.

Acid Hydrolysis[3)]

In glycosides, the glycosidic linkage between an aglycone and a furanose-type monosaccharide is readily hydrolyzed by acid as compared with that of the corresponding pyranoside. For example, the rate of hydrolysis of aldofuranoside is known to be 50–200 times higher than that for aldopyranoside. The chemical structure of the aglycone also affects the hydrolysis rate. Phenolic glycoside linkages are generally more readily hydrolyzed than terpenoidal and steroidal glycosides.

For acid hydrolysis, glycosides are usually heated for a certain period with sulfuric or hydrochloric acid at a suitable concentration in a polar solvent such as water, ethanol, or methanol (or sometimes dioxane). Afterwards, the reaction mixtures are concentrated under reduced pressure to remove the organic solvent and diluted with water. The aqueous mixtures are then extracted with organic solvents to take up the liberated aglycones or are filtered to collect the aglycones as precipitates.

The aqueous portions are neutralized either with alkali or with basic ionic resin and concentrated under reduced pressure to yield syrupy residues which are subjected to paper partition and gas-liquid chromatographic analyses to determine the sugar compositions in the glycosides. When sulfuric acid is used for hydrolysis, barium hydroxide or barium carbonate may be employed for neutralization of the total hydrolysate.

In some cases, such chromatographic analysis of component sugars can give very complicated chromatograms due to concomitant degradation of the saccharides during the hydrolysis procedure. In these cases, it is desirable to treat standard monosaccharides in advance under the same acidic hydrolysis conditions to provide guidelines in the chromatographic analysis.

In acid hydrolysis methods, heating under acid conditions can cause secondary chemical transformations of the aglycones. If this occurs, the liberated aglycones are artifacts, and various devices for obtaining the genuine aglycones and sapogenols must be attempted, as will be described later.

Acetolysis[4)]

When the carbohydrate portion of a glycoside is an oligosaccharide comprising more than two kinds of monosaccharides, acetolysis is often undertaken in order to elucidate the sugar chain structure.

In acetolysis, the glycosidic linkage between the sugar chain and aglycone is cleaved in addition to one or more of the linkages connecting the component monosaccharides, and acetylated derivatives of the aglycone, monosaccharides, and/or oligosaccharides are obtained.

In these cases, the acetylium cation CH_3CO^+ attacks the acetal bond in the glycoside to liberate the acetylated products. Usually, a combination of acetic anhydride with sulfuric acid, perchloric acid, boron trifluoride, or zinc chloride, or a mixture of trifluoroacetic anhydride and acetic acid is employed for acetolysis.

[Experimental procedure 1] Acetolysis of *Bupleurm* root saponin derivative[5]

BF_3-etherate (2 drops) was added to a solution of **1** (100 mg) in glacial AcOH (1 ml) and Ac_2O (0.5 ml), and the mixture was heated at 100° for 1 hr in an oil bath. After cooling, the reaction mixture was poured into ice-water and extracted with CH_2Cl_2. The CH_2Cl_2 layer was taken and washed successively with water, aq. 5% $NaHCO_3$, and water, then dried over Na_2SO_4. Removal of the solvent gave a residue (102 mg) which was purified by preparative tlc (silica gel G, benzene-AcOEt = 4:1) to afford crude heptaacetyl disaccharide **2** (36 mg, 64%) (from *Rf* 0.05) and crude tetraacetyl sapogenol **3** (42 mg, 76%) (from *Rf* 0.59). Recrystallization from MeOH gave **2**, mp 162–166°, as colorless needles, while recrystallization

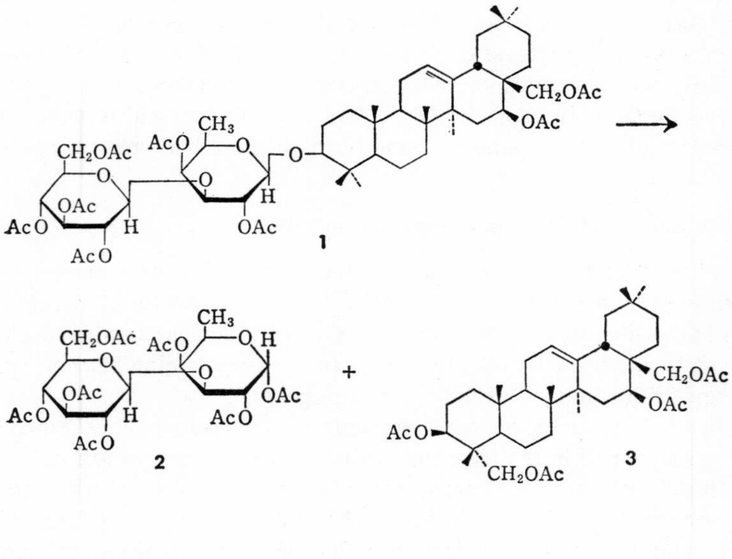

CH3 CH2OAc AcO O O OAc AcO H OAc AcO CH2OAc OAc **1**

CH3 CH2OAc AcO O H O OAc AcO H OAc AcO AcO **2** + AcO CH2OAc CH2OAc OAc **3**

from ether and petr. ether gave **3**, mp 277–279.5°, as colorless needles. Acetylated monosaccharide was not formed in this acetolysis.

Other Chemical Methods

Since glycosides are generally stable in dilute alkali at room temperature, alkali is usually not suitable for cleavage of the glycoside linkage. However, depending upon the chemical structure of the aglycone, some glycosidic linkages can be hydrolyzed with alkali.[1,6] In addition, ester-type glycoside linkages, in which the carboxyl residue of the aglycone is glycosylated, are hydrolyzed by alkali, as expected.

Among special cleavage methods for glycosidic linkages, the following examples can be cited: hydrogenolysis for cleavage of the glycoside linkage in benzyl alcohol derivatives, chlorinolysis by treatment with chlorine under acidic conditions, photolysis for the cleavage of glycoside linkages connected to aromatic aglycones in which the photochemical energy is initially absorbed in the aglycone moiety and then transferred to split the glycoside linkage, and γ-ray irradiation.[1]

22.2 Methods for Obtaining Genuine Aglycones

Acid hydrolysis often results in the concomitant modification of aglycones (sapogenols or sapogenins), and this constitutes one of the serious impedimenta in structural studies of saponins which contain oligosaccharides. In order to avoid such reactions and to obtain the genuine aglycones or sapogenols, various chemical devices and biochemical methods using enzymes and microorganisms have been proposed. Some examples are given below.

Mild Acid Hydrolysis (in a two-phase medium)

When cyclamin, a saponin from the root of *Cyclamen europaeum*, is hydrolyzed by heating with aqueous 10% sulfuric acid for 12 hr, cyclamiretin D **4** is liberated as the sapogenol. On the other hand, when cyclamin is hydrolyzed with ethanolic hydrochloric acid containing benzene, cyclamiretin A **5** is obtained as the major sapogenol.[7]

In the latter case, the reaction medium is inhomogeneous. Since benzene is suspended in the aqueous medium, the liberated sapogenol is readily transferred into the benzene phase, so that secondary modification of the sapogenol by acid may be reduced. Acid treatment of cyclamiretin A **5** with heating furnishes cyclamiretin D **4**, so that A is the genuine sapogenol and D is a secondary product. Cyclamiretin A **5** contains an acid-labile 13β, 28-oxide moiety in its oleanane skeleton, while cyclamiretin D **4** has an ordinary olean-12-ene skeleton.

cyclamin $\xrightarrow{H^+/EtOH\text{-benzene}}$ cyclamiretin A **5** $\xrightarrow{H^+}$ cyclamiretin D **4**

Although some genuine sapogenols may be obtained in such inhomogeneous (two-phase) media, aqueous alcoholic acid hydrolysis with an immiscible organic solvent (benzene in most cases) in this way does not always prevent the formation of secondary products.

[Experimental procedure 2] Isolation of cyclamiretin A (5) from cyclamin[7)]

Benzene (200 ml) and conc. HCl (3.3 ml) were added to a solution of cyclamin (3 g) in a mixture of water (200 ml) and 96% EtOH (200 ml) (thus making *ca.* 0.1 N HCl solution). The total mixture was heated at 100° for 6 hr. After cooling, the benzene layer was separated and the solvent was evaporated off to give a residue which was crystallized from ether and then from aq. 96% EtOH to afford cyclamiretin A **5** (40 mg).

Chemical Modification of Aglycones prior to Hydrolysis

During the hydrolysis of glycosides, secondary conversion of the genuine aglycone can sometimes be avoided by chemical modification of the aglycone moiety prior to acid hydrolysis.

On acid hydrolysis, ginsenoside Rb,c, a saponin fraction from the root of *Panax ginseng*, furnishes panaxadiol (20S + 20R) **6**, an epimeric mix-

ginsenoside $R_{b,c}$ → protopanaxadiol (20S,20R) **7**; ginsenoside $R_{b,c}$ $\xrightarrow{H_2/PtO_2}$ dihydro deriv. $\xrightarrow{H^+}$ **8**; protopanaxadiol **7** $\xrightarrow{H_2/PtO_2}$ **8**; protopanaxadiol **7** $\xrightarrow{H^+}$ panaxadiol **6**; ginsenoside $R_{b,c}$ $\xrightarrow{H^+}$ panaxadiol **6**

ture of artifact sapogenols. The tetrahydropyran ring in **6** is secondarily formed through acid-catalyzed ring closure between the C-20 hydroxyl and C-24,25 double bond in protopanaxadiol **7**, the genuine sapogenol of ginsenoside Rb,c.

However, when the acid hydrolysis is carried out after catalytic hydrogenation of the C-24,25 double bond in ginsenoside Rb,c, the undesired ring closure cannot occur, but instead a dihydro derivative of the genuine sapogenol, named dihydroprotopanaxadiol **8**, is obtained.[8)]

[Experimental procedure 3] Dihydroprotopanaxadiol (8) from ginsenoside Rb,c[8)]

A solution of ginsenoside Rb,c (2 g) in a mixture of EtOH–MeOH–glacial AcOH was hydrogenated over Adams catalyst (41 ml of H_2 was consumed in 50 min). After removing the catalyst by filtration, the filtrate was concentrated under reduced pressure. The residue was dissolved in a mixture of conc. HCl (10 ml), water (20 ml), and EtOH (20 ml), and the total mixture was heated under reflux for 4 hr. White precipitates thus formed were taken up with ether and the product obtained by usual work-up of the ether extract was crystallized from MeOH to furnish dihydroprotopanaxadiol **8** (350 mg) as colorless needles.

On acid hydrolysis of furostanol-type steroidal saponins, spirostanol-type sapogenols are liberated. In order to clarify the relationship, a furostanol-type saponin **9** was catalytically hydrogenated prior to acid hydrolysis; in this case, the conversion to a spirostanol-type sapogenol **10** cannot occur, and instead a C-22 desoxy derivative of the furostanol-type sapogenol **11** is formed.[9)]

Glc
HO O
RO
H
6 (R= sugar chain)
i) H_2/PtO_2-MeOH
ii) 3 NHCl
H^+
HO
H
10
H_2
H_2/PtO_2-AcOH
H HO (S)
HO
H
11

Periodate Degradation

In order to liberate the genuine aglycone from the parent glycoside, several degradation methods by which the carbohydrate portion is specifically decomposed have been investigated. In the Smith degradation method for polysaccharides, periodate is employed to cleave initially the α-glycol moiety in the carbohydrate. Successive reduction with sodium borohydride followed by mild acid treatment finally decomposes the carbohydrate chain.[10)]

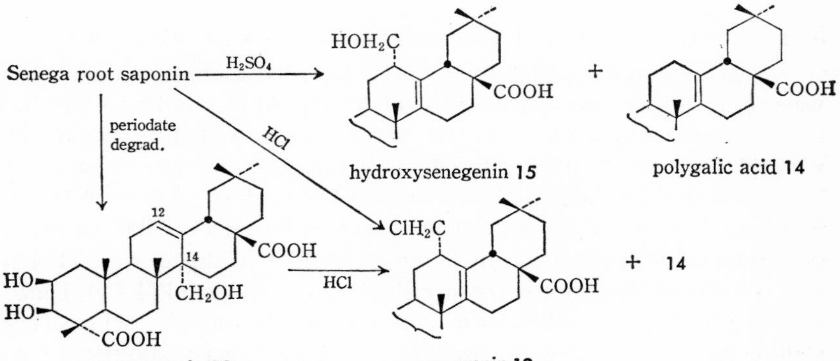

Senega root saponin $\xrightarrow{H_2SO_4}$ hydroxysenegenin **15** + polygalic acid **14**

Senega root saponin $\xrightarrow{HCl}$ senegenin **13** + **14**

Senega root saponin $\xrightarrow{\text{periodate degrad.}}$ presenegenin **12** $\xrightarrow{HCl}$ senegenin **13** + **14**

By utilizing a modified Smith degradation method, Dugan and de Mayo were successful in the liberation of presenegenin **12**, which is the genuine sapogenol of saponins from the root of *Polygala senega*.[11)] On acid treatment, presenegenin **12** is readily convertible to senegenin **13** and polygalic acid **14**, or **13** and hydroxysenegenin **15**, depending upon the mineral acid used. Presenegenin **12** is never obtained by direct acid hydrolysis of the parent saponins.

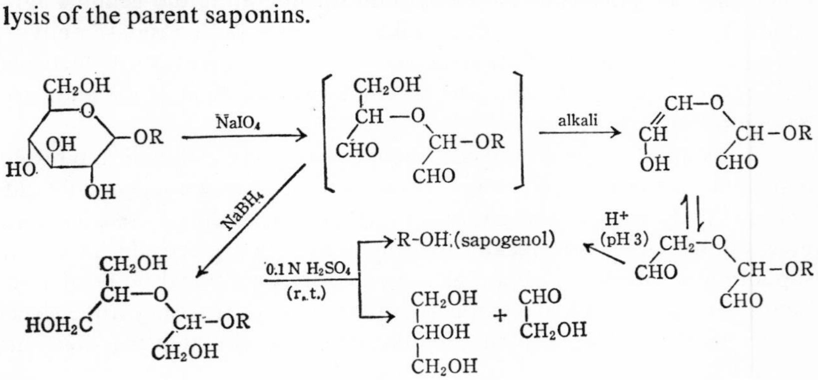

Glycoside (–OR) $\xrightarrow{NaIO_4}$ [dialdehyde: CH_2OH–CH–O–CH(OR)–CHO, CHO] $\xrightarrow{\text{alkali}}$ CH(OH)=CH–O–CH(OR)–CHO ⇌ CH_2(CHO)–O–CH(OR)–CHO $\xrightarrow{H^+\ (pH\ 3)}$ R-OH (sapogenol)

Dialdehyde $\xrightarrow{NaBH_4}$ HOH_2C–CH(CH_2OH)–O–CH(OR)–CH_2OH $\xrightarrow{0.1\ N\ H_2SO_4\ (r.t.)}$ R-OH (sapogenol) + CH_2OH–CHOH–CH_2OH + CHO–CH_2OH

The reaction pathway in the periodate degradation has been explained in terms of the following mechanism.

The following genuine sapogenols have been isolated from their parent saponins by the periodate degradation method: sapogenols of the root saponins from *Bupleurum falcatum*,[12] *Panax ginseng*,[13] and *Primula sieboldi*.[14] However, it should be noted that the scope of the method seems to be rather limited because of the reactivity of periodate. It may react with a sapogenol possessing an α-glycol or chemically equivalent moiety in its molecule. The yield of the degradation product (or genuine sapogenol) is usually unsatisfactory.

[Experimental procedure 4] Presenegenin (12) from *Senega* root saponin[11]

$NaIO_4$ (41 g) was added to an ice-cooled solution of saponin (30 g) in water (1 l) with stirring over a period of 30 min. After standing in the dark at room temperature for 24 hr, the stirred reaction mixture was treated with KI (20 g) and then with $NaAsO_2$ until the color of iodine disappeared. The mixture was neutralized with solid KOH and an additional 50 g of KOH was added. The stirred solution was then heated at 100° under an N_2 atmosphere for 1 hr, cooled, and carefully acidified with HCl to pH $\sim$ 3. Extraction with ether furnished crude sapogenol (4.5 g). Diazomethane methylation followed by chromatography on silica gel gave crystalline presenegenin dimethyl ester (1.5 g). Several recrystallizations from EtOAc gave the pure substance.

Enzymatic Hydrolysis

Glycosidase, which catalyzes the biosynthesis of glycoside, is also capable of hydrolyzing the same glycoside. Some hydrolysis methods have been reported in which this reversibility of the intracellular enzyme is utilized.[15] On the other hand, hydrolysis methods using glycosidases from other kinds of organisms are also useful for liberating the genuine aglycones from glycosides. Emulsin, which possesses β-glucosidase activity, is a well-known example. Since emulsin specifically hydrolyzes β-glucosidic linkages, it is commonly used for preliminary assessment of the anomeric configuration in a glucoside.

In the case of oligoglycosides such as saponins, although the results depend upon the kinds of component sugars, emulsin is not generally suitable for hydrolysis of the glycoside linkages. In addition, because pure glycosidases which specifically hydrolyze a certain glycoside linkage in saponin are difficult to obtain at present from a practical point of view, increasing numbers of enzymatic hydrolysis procedures using crude glycosidase mixtures have been successfully reported, although the yields are not yet satisfactory.

Some examples include the hydrolysis of *Panax ginseng* saponin using crude hesperidinase,[16] the hydrolysis of holotoxins, the antifungal oligoglycosides from the sea cucumber *Stichopus japonicus*, with a cellulase, crude naringinase, or Takadiastase A preparation,[17] and the hydrolysis of Mi-saponin A, the seeds saponin of *Madhuca longifolia* L., with a crude hesperidinase or Takadiastase A preparation.[18]

Another interesting example is the hydrolysis of saponin from the starfish *Acanthaster planci* (crown of thorns) using a glycosidase mixture extracted from the marine gastropod *Charonia lampas*, which is known to be the natural enemy of *A. planci*.[19]

[Experimental procedure 5] Hydrolysis of holotoxin A (18) with a Takadiastase A preparation[17]

Powdered crude Takadiastase A (60 g, Sankyo Co.) was extracted with dist. water (420 ml) for 3 hr under ice-cooling, then a small amount of Celite 535 was added to the aqueous extract with stirring and the mixture was filtered. The filtrate was then treated with aq. 1 M $Ca(OAc)_2$ (102 ml) and adjusted to pH 7.0 with aq. 5 N NaOH to give a suspension, which was filtered with the aid of Celite 535. Cold acetone was added dropwise to the ice-cooled filtrate up to 40% concentration by volume. The precipitates were collected by centrifugation and dissolved in AcOH–AcONa buffer (pH 5.1, 200 ml); this solution was used as the Takadiastase A preparation.[17a]

For the hydrolysis of holotoxin A **18**,[17b] a mixture of holotoxin A **18** (950 mg) in a solution of the Takadiastase A preparation (AcOH–

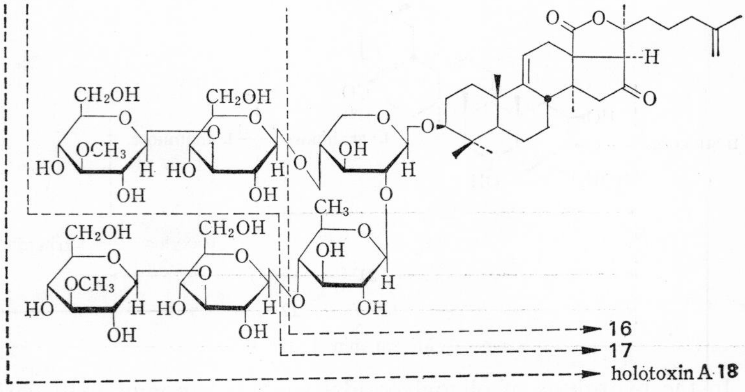

AcONa buffer solution, pH 5.1, 200 ml) was stirred at 31° for 6 days. After the addition of *n*-BuOH (200 ml) and warming for a while, the whole was centrifuged to separate the *n*-BuOH layer. The resulting precipitate was collected and washed with *n*-BuOH and a small amount of MeOH. The washings were combined with the *n*-BuOH layer and the combined solution was concentrated under reduced pressure to give the *n*-BuOH extractive (2.85 g). The extractive (2.8 g) was successively subjected to column chromatography (silica gel 70–230 mesh, 60 g), developing with $CHCl_3$–MeOH–H_2O = 7:3:1 (lower layer), and to medium pressure column chromatography (silica gel 80 g; column preparation at 5 kg/cm^2; elution at 3 kg/cm^2; column size 2.5 × 60 cm; flow rate 50 ml/hr; developing solvent, $CHCl_3$–MeOH–H_2O = 13:3:1 (lower layer) → 10:3:1 (lower layer)) to furnish crude **16** (60 mg), crude **17** (223 mg), and crude **18** (510 mg recovered). Recrystallization from MeOH gave pure samples of **16** (28 mg), mp 281–282°, and **17** (56 mg), mp 274–276°.

[Experimental procedure 6] Hydrolysis of Mi-saponin A (19) with crude Hesperidinase[18)]

A solution of Mi-saponin A **19** (2 g) in Na_2HPO_4-citric acid buffer solution (pH 4.0) (250 ml) was treated with crude hesperidinase (Tanabe Pharmaceutical Co., Lot No. N-30) (250 mg). The total mixture was stirred gently at 31–33° for 70 hr, then extracted with *n*-BuOH. The *n*-BuOH solution was washed with water and concentrated under reduced pressure to give a product (1.1 g). Column chromatography of the product on silica gel (70 g), developing with *n*-BuOH saturated with water, furnished protobassic acid (the genuine sapogenol) (87 mg), **20** (24 mg), **21** (621 mg), **22** (22 mg), and mixtures of these compounds (total, ca. 200 mg).

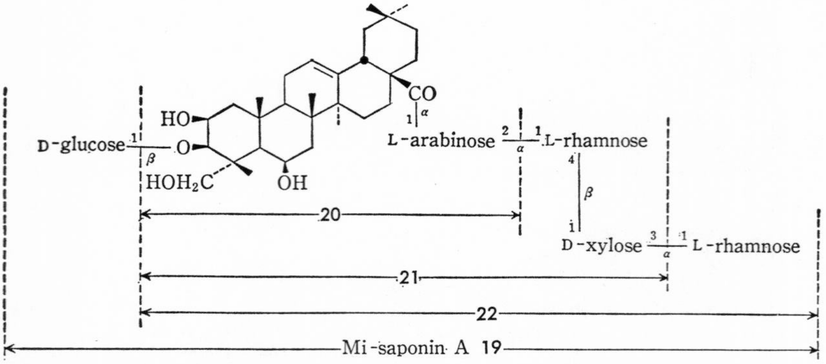

HO, D-glucose —1 (β)— O, HOH_2C, OH, CO —1 (α)— L-arabinose —2 (α) 1— L-rhamnose —4 (β) 1— D-xylose —3 (α) 1— L-rhamnose

←20→ ←21→ ←22→ ←Mi-saponin A 19→

In the hydrolysis of oligoglycosides such as saponins with various crude glycosidases, various partial hydrolysates of the oligoglycoside may be produced in addition to the genuine sapogenol. These partial hydrolysates are valuable for the structure elucidation of the parent saponin. If pure and specific glycosidases become readily available in the future, the utility of the enzymatic hydrolysis method would be substantially increased.

Microbial Hydrolysis

Microbial hydrolysis is considered to involve combinations of enzymatic hydrolyses, and the microbial hydrolysis method is essentially an extension of the enzymatic method employing crude glycosidase mixtures.

When a fungus of *Corynespora* sp. or *Alternaria* sp. is cultured for 7 days on the Czapek Dox medium, which contains saponin from *Agave sisalana* as the sole carbon source, the sapogenol hecogenin is liberated almost quantitatively in the medium.[20)] Sapogenols of *Medicago sativa* can be obtained from the saponin mixture in a similar fashion by cultivation of *Aspergillus* sp., *Rhizopus* sp., or *Mucor* sp.[21)] Diosgenin has been successfully obtained on a semi-industrial scale by microbial hydrolysis of the saponin from *Dioscorea tokoro* using *Aspergillus terreus*.[22)]

The above-mentioned findings are examples in which the microorganisms are expected to utilize glycosides such as saponins in the medium as a carbon source. In other words, these are examples of microbial transformations. Since pure cultures of microorganisms are employed for hydrolysis in these cases, extensive preliminary screening tests are always necessary to find a suitable microbial strain.

[Experimental procedure 7] Isolation of diosgenin from *Dioscorea* rhizomes[22)]

A mixture of air-dried and powdered rhizomes of *Dioscorea tokoro* (5 g), rice-hulls (0.6 g), $CaCO_3$ (0.01 g), and dist. water (5 ml) was mixed well and sterilized at 100° for 1 hr in a sterilized 100 ml Erlenmeyer flask. After cooling, a spore suspension (1 ml) which was prepared by the addition of sterilized water (5 ml) to one-week pre-cultured mycellia of *Aspergillus terreus* D_4 was added, and the mixture was allowed to stand at 30° for 48 hr. The total culture was then treated with sterilized water (40 ml) (adjusted to pH 4.5 with HCl beforehand) and toluene (1 ml), and left to stand at 37° for 96 hr with occasional shaking to effect the hydrolysis. After the reaction, the solid material was collected by filtration, dried, and extracted with hexane in a Soxhlet extractor to obtain diosgenin. Since the insoluble residue in the extractor was rich in saponin, it was again subjected to microbial hydrolysis. In all, 113 mg of diosgenin was obtained.

From 10 kg of the rhizomes, 374 g of the crude crystalline sapogenol was obtained, and recrystallization from MeOH furnished 145 g of diosgenin.

Hydrolysis by Soil Microorganisms[23)]

In this method, a suitable microbial strain is selected from various soil samples, which contain many kinds of microorganisms, including ones which may hydrolyze glycosides. In this method, the selection of microorganisms is made from soil samples, rather than by screening tests with pure cultures.

In principle, when soil samples are cultured repeatedly on a synthetic medium containing a glycoside as the only carbon source, any microorganism which can grow on the medium is selected. The microorganisms thus selected are expected to consume the carbohydrate which is liberated through enzymatic hydrolysis of the glycoside or saponin and to liberate the genuine aglycone or sapogenol in the medium.

Soil samples generally contain various microorganisms: bacteria of *Pseudomonas* sp., *Clostridium* sp., *Bacillus* sp., *etc.*, as well as actinomycetes, fungi, yeasts, etc. Therefore, when a selection culture is started with various soil samples, many kinds of microorganisms are potentially available. Athough the microorganisms may also transform the sapogenol or aglycone moiety during the cultivation, this can usually be avoided by adjusting the cultivation period.

In general, soil samples (collected at 20–30 different places) are incubated at 31° in test tubes containing 3 ml of a synthetic medium. Microorganisms which grow well (as judged from the turbidity of the medium) are selected and cultivated repeatedly on the same synthetic medium. This selection culture is carried out at 2- to 3-day intervals, although this depends on the growth of the microorganism. By these procedures, a microorganism, able to hydrolyze the glycoside and adapted to the synthetic medium, is selected.

After repeated selection culture (generally four selections are enough), the microorganism is grown in stationary culture on a larger scale. The microorganism possessing the hydrolysis activity is also cultured on an ordinary medium (e.g., meat-bouillon agar for bacteria) for identification. The appropriate period for cultivation varies depending upon the glycoside in the medium and is determined by tlc monitoring of the total culture broth. The culture broth obtained in large-scale cultivation is extracted (after concentration *in vacuo* if necessary) with a suitable organic solvent (ether, chloroform, ethyl acetate, or *n*-butanol, etc.) and the extractive is purified to furnish the genuine sapogenol or aglycone.

[Experimental procedure 8] Presenegenin (12) from *Senega* root saponin[24)]

Thirty soil samples collected at different places were grown in stationary culture at 31° in test tubes each containing 3 ml of a synthetic medium prepared with $(NH_4)_2HPO_4$ (4 g), KH_2PO_4 (1 g), NaCl (1 g), $MgSO_4 \cdot 7H_2O$ (0.7 g), $FeSO_4 \cdot 7H_2O$ (0.03 g), Senega root saponin (3 g) and distilled water (1 l), adjusted to pH 6.0 with dil. HCl, and sterilized at 120° (in an autoclave) for 30–40 min. The highly turbid tubes were selected and cultured repeatedly in the same medium at 2- to 3-day intervals. By tlc monitoring, three bacterial strains suitable for hydrolysis were obtained.

For the isolation of presenegenin **12**, one of the selected bacterial strains was grown in stationary culture at 31° for 16 days in nine 2 l Erlenmeyer flasks (each containing 0.5 l of medium), with 13.5 g of Senega root saponin in total. The combined culture broth was extracted with ether and the crystalline residue obtained by removal of the ether was recrystallized from MeOH to furnish presenegenin **12** (1.72 g) as colorless leaflets. The microorganism used for the hydrolysis was cultured in meat-bouillon medium and identified as *Pseudomonas* sp.

For glycosides which are unstable during thermal sterilization, the cultivation medium is prepared by aseptic addition of the glycoside to an inorganic salts medium thermally sterilized in advance.[25)] In the case of less soluble glycosides, occasional shaking of the culture is desirable.[26)]

By this method, genuine triterpenoid sapogenols of the following plant materials were obtained: *Panax japonicum* rhizome,[24)] *Aesculus turbinata* seeds,[27)] *Sanguisorba officinalis* root,[26)] *Styrax japonica* pericarps,[28)] *Panax ginseng* root,[29)] and *Madhuca longifolia* seeds.[30)] The genuine steroidal sapogenols from *Metanarthecium luteo-viride*,[31)] the diterpenoid aglycone (steviol) of stevioside from *Stevia rebaudiana*,[32)] and monoterpene aglycones from *Paeonia albiflora* and *Scrophularia buergeriana*[25)] were also successfully obtained.

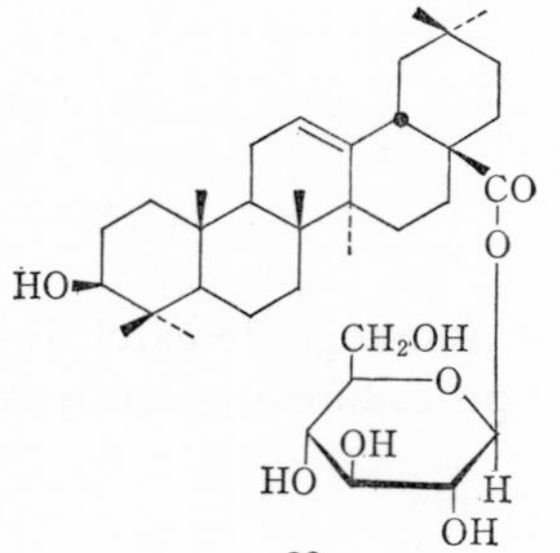

HO, CO, O, CH_2OH, O, OH, HO, H, OH

23

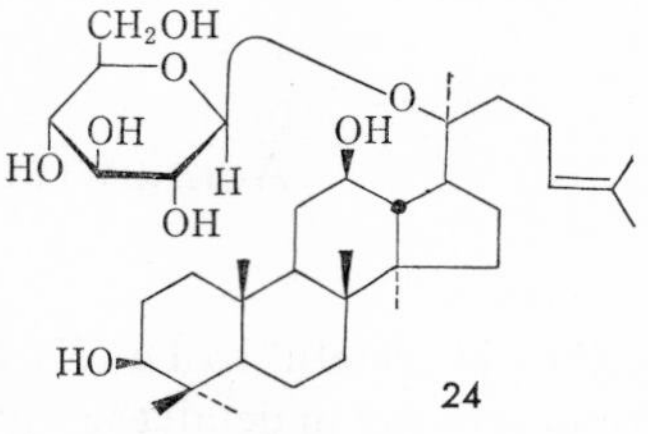

CH_2OH, O, O, OH, OH, HO, H, OH, HO

24

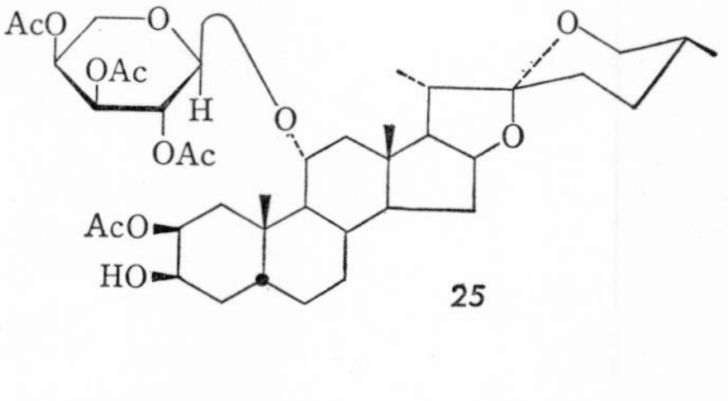

AcO, O, O, OAc, H, OAc, O, O, AcO, HO

25

Among the studies mentioned above, partial hydrolysis products **23**,[24) **24**,[29) and **25**,[31) which retain a sugar moiety attached to a hydroxyl on a carbon other than C-3 of the triterpenoid or steroid skeleton, were obtained in the cases of *P. japonicum*,[24) *P. ginseng*,[29) and *M. luteo-viride*.[31) These findings provided valuable evidence for the structural elucidation of the parent saponins.

Since hydrolysis by selection culture of soil microorganisms can be carried out on a small scale, it may be applicable for the examination of a genuine sapogenol or aglycone which has been isolated from the parent material by some other hydrolysis method. The procedure is very simple, and preservation of a particular culture of microorganisms is not normally necessary, since fresh isolation of a suitable strain from soil can be carried out as required.

22.3 Concluding Remarks

Various procedures for liberating the intact non-carbohydrate moiety from a glycoside have been described above.

In addition, the authors have recently developed four selective degradation methods for the glucuronide linkage.[33,38) They are: (1) photolysis,[34) (2) lead tetraacetate degradation,[35) (3) acetic anhydride-pyridine degradation,[36) and (4) anoidic oxidation.[33,39) These investigations were carried out using oligoglycosides as substrates as part of a search for selective chemical cleavage methods for specific glycoside linkages. They have been shown to be useful as model studies for finding new chemical approaches to the structural elucidation of polysaccharides.[37)

Acknowledgement

The author is grateful to Dr. K. Igarashi of Shionogi Research Laboratory for describing in detail the experimental procedure for acetolysis.

References

1) W. G. Overend, *The Carbohydrates, Chemistry and Biochemistry* (ed. W. Pigman, D. Horton), vol. 1A, p. 279, Academic Press (1972).
2) J. E. Courtois, F. Percheron, *ibid.*, vol. IIA, p. 213, Academic Press (1970).
3) J. N. BeMiller, *Advan. Carbohydrate Chem.*, **22**, 25 (1967).
4) R. D. Guthrie, J. F., McCarthy, *ibid.*, **22**, 11 (1967).
5) K. Igarashi, unpublished data.
6) C. E. Ballon, *Advan. Carbohydrate Chem.*, **9**, 59 (1954).
7) R. Tschesche, F. Inchaurrondo, G. Wulff, *Ann. Chem.*, **680**, 107 (1964).
8) S. Shibata, O. Tanaka, T. Ando, M. Sado, S. Tsushima, T. Ohsawa, *Chem. Pharm. Bull.*, **14**, 595 (1966).
9) R. Tschesche, G. Lüdke, G. Wulff, *Chem. Ber.*, **102**, 1253 (1969).
10) I. J. Goldstein, G. W. Hay, B. A. Lewis, F. Smith, *Methods Carbohydrate Chem.*, **5**, 361 (1965).
11) J. J. Dugan, P. de Mayo, *Can. J. Chem.*, **43**, 2033 (1965).
12) T. Kubota, H. Hinoh, *Tetrahedron*, **24**, 675 (1968); N. Aimi, H. Fujimoto, S. Shibata, *Chem. Pharm. Bull.*, **16**, 641 (1968).
13) M. Nagai, T. Ando, N. Tanaka, O. Tanaka, S. Shibata, *Chem. Pharm. Bull.*, **20**, 1212 (1972).
14) I. Kitagawa, A. Matsuda, I. Yosioka, *ibid.*, **20**, 2226 (1972).
15) M. M. Krider, M. E. Wall, *J. Am. Chem. Soc.*, **76**, 2938 (1954).
16) H. Kohda, O. Tanaka, *Yakugaku Zasshi*, **95**, 246 (1975).
17) a) I. Kitagawa, T. Sugawara, I. Yosioka, *Chem. Pharm. Bull.*, **24**, 275 (1976); b) I. Kitagawa, H. Yamanaka, M. Kobayashi, T. Nishino, I. Yosioka, T. Sugawara, *ibid.*, **26**, 3722 (1978).
18) a) I. Kitagawa, A. Inada, I. Yosioka, *Chem. Pharm. Bull.*, **23**, 2268 (1975); b) I. Kitagawa, K. Shirakawa, M. Yoshikawa, *ibid.*, **26**, 1100 (1978).
19) I. Kitagawa, M. Kobayashi, T. Sugawara, *ibid.*, **26**, 1852 (1978); I. Kitagawa, M. Kobayashi, *ibid.*, **26**, 1864 (1978).
20) C.H. Hassall, B.S.W. Smith, *Chem. Ind.*, **1957**, 1570.
21) W. A. Lourens, M. B. O'Donovan, *S. African J. Agr. Sci.*, **4**, 293 (1961) [*C. A.*, **56**, 10196 (1962)].
22) Y. Nagai, M. Sawai, Y. Kurosawa, *Nippon Nogei Kagaku Kaishi*, **44**, 15 (1970).
23) *Natural Products Chemistry* (ed. by K. Nakanishi *et al.*), vol. 1, p. 380, Kodansha /Academic Press (1974).
24) I. Yosioka, M. Fujio, M. Osamura, I. Kitagawa, *Tetr. Lett.*, **1966**, 6303.
25) I. Yosioka, T. Sugawara, K. Yoshikawa, I. Kitagawa, *Chem. Pharm. Bull.*, **20**, 2450 (1972).
26) I. Yosioka, T. Sugawara, A. Ohsuka, I. Kitagawa, *ibid.*, **19**, 1700 (1971).
27) I. Yosioka, K. Imai, I. Kitagawa, *Tetr. Lett.*, **1967**, 2577.
28) I. Yosioka, S. Saijoh, I. Kitagawa, *Chem. Pharm. Bull.*, **20**, 564 (1972).
29) I. Yosioka, T. Sugawara, K. Imai, I. Kitagawa, *ibid.*, **20**, 2418 (1972).
30) I. Yosioka, A. Inada, I. Kitagawa, *Tetrahedron*, **30**, 707 (1974).
31) I. Yosioka, K. Imai, Y. Morii, I. Kitagawa, *ibid.*, **30**, 2283 (1974).
32) I. Yosioka, S. Saijoh, J. A. Waters, I. Kitagawa, *Chem. Pharm. Bull.*, **20**, 2500 (1972).
33) I. Kitagawa, M. Yoshikawa, *Heterocycles*, **8**, 783 (1977).
34) I. Kitagawa, M. Yoshikawa, Y. Imakura, I. Yosioka, *Chem. Pharm. Bull.*, **22**, 1339 (1974).
35) I. Kitagawa, M. Yoshikawa, K. S. Im, Y. Ikenishi, *ibid.*, **25**, 657 (1977).
36) I. Kitagawa, Y. Ikenishi M. Yoshikawa, K. S. Im, *ibid.*, **25**, 1408 (1977).
37) B. Lindberg, J. Lönngren, S. Svensson, *Advan. Carbohydrate Chem. Biochem.*, **31**, 185 (1975).

38) I. Kitagawa, M. Yoshikawa, K. Kobayashi, Y. Imakura, K. S. Im, Y. Ikenishi, *Chem. Pharm. Bull.*, **28**, 296 (1980).
39) a) I. Kitagawa, T. Kamigauchi, H. Ohmori, M. Yoshikawa, *ibid.*, **28**, 3078 (1980); b) I. Kitagawa, T. Kamigauchi, K. Shirakawa, Y. Ikeda, H. Ohmori, M. Yoshikawa, *Heterocycles*, **15**, 349 (1981).

CHAPTER 23

Toxic Alkaloids and Diterpenes from Euphorbiaceae (including Daphniphyllaceae)

Though there are many toxic plants in the family Euphorbiaceae, only limited knowledge was available on their toxic components in 1965, when we started research on these plants. Though it has now been suggested that Daphniphyllaceae should be separated from Euphorbiaceae to form a distinct family, Daphnipyllaceae, they were then classified as Euphorbiaceae (There are only three species in Japan: *Daphniphyllum macropodum*, *D. teijsmanni*, and *D. humile*.). We began our research by studying alkaloids of *Daphniphyllum macropodum*.

We were also interested in the possibility of discriminating between *D. macropodum* and *D. humile* chemotaxonomically by studying the alkaloids of those species. However, this approach was not fruitful due to the similarity of the main alkaloid components of the two species, the large seasonal variation in the contents of the alkaloids, and the variety of the alkaloids contained.

It is interesting that about thirty kinds of alkaloid have been isolated from only three species of *Daphniphyllum* existing in Japan, and, in particular, that the skeletons of these alkaloids are quite different from those from other plant species. Classification of these alkaloids into six groups has been carried out on the basis of skeletal structure.

23.1 Alkalods of Daphniphyllaceae[2,3)]

About 35 toxic alkaloid components of Daphniphyllaceae have been

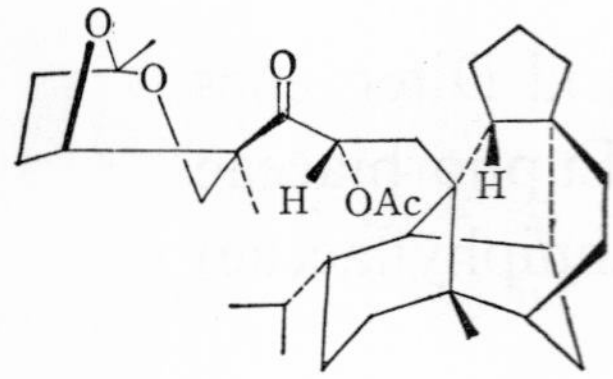

O
O
O
H
OAc
H
daphniphylline 1 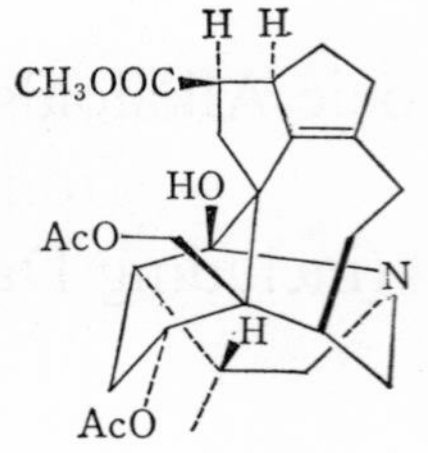
H H
CH3OOC
HO
AcO
N
H
AcO
yuzurimine 2 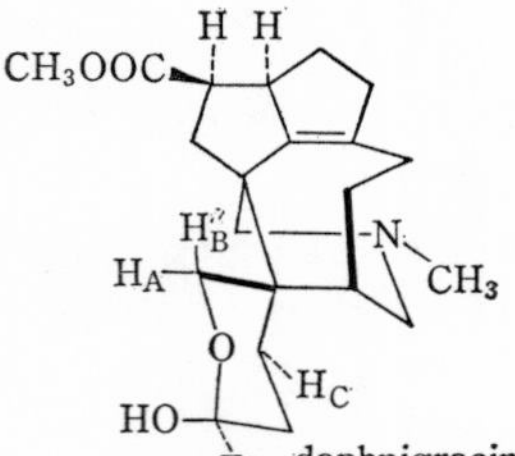
H H
CH3OOC
HB
HA
N
CH3
O
HC
HO
Et
daphnigracine 3 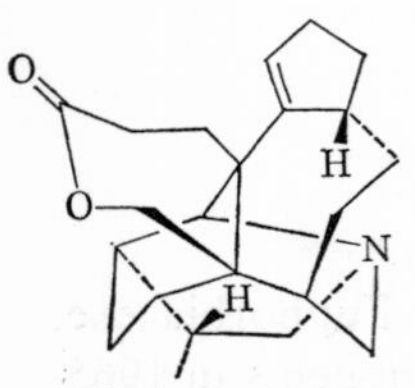
O
O
H
N
H
daphnilactone B 4 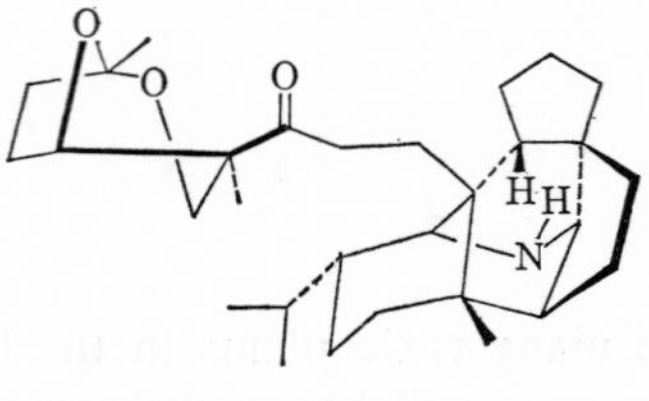
O
O
O
H
H
N
secodaphniphylline 5 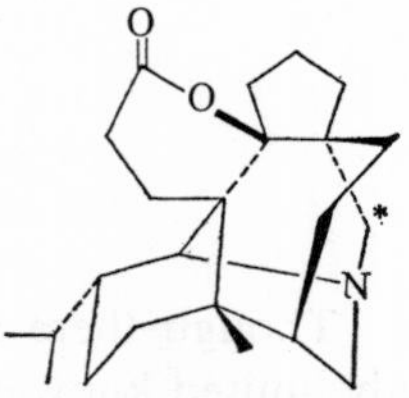
O
O
N
*
daphnilactone A 6
O
O
O
H
N
7
codaphniphylline 7
CH3OOC
H
N
methyl homodaphniphyllate 9
2
CH3OOC
H H
N
5
secodaphniphyllate 8
squalene
CH3OOC
NH
[A]
6
CH3OOC
H
NH
[B]
4

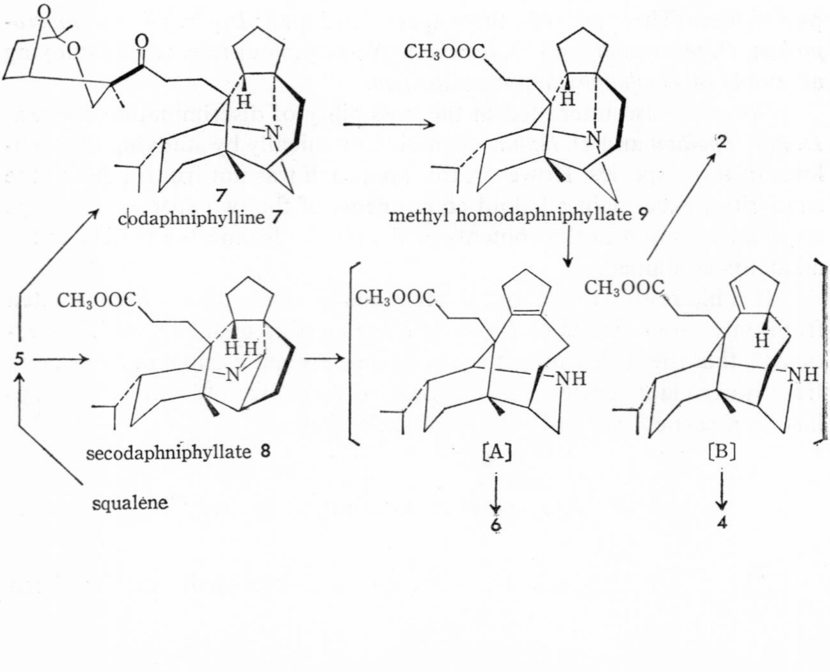

isolated so far, and the structures of most of them have been determined. About 30 of these have been isolated from leaves and cortices, and about half of the compounds have the same structural skeletons as the main components, daphniphylline **1** and yuzurimine **2**. Alkaloids (5 kinds) from a related plant from New Guinea (*Daphniphyllum gracile* GAGE), in contrast, are of yuzurine type **3**, and no alkaloids corresponding to the main components, **1** and **2**, have been detected. Ten alkaloids have so far been isolated from the fruits, and the main component was identified as daphnilactone B **4**, which is not present in the leaves or cortices. This lactone is considered to occupy an intermediate possition between daphniphyllin **1** and yuzurimine **2** in the biosynthetic pathway.

The alkaloids from Daphniphyllaceae can be grouped into six categories, as described above. Representative compounds are the alkaloids **1**, **2**, **3** and **4** described above, and secodaphniphylline **5** and daphnilactone **6**. Secodaphniphylline occupies an important position in relation to the biosynthesis of Daphniphyllaceaealkaloids. Daphnilactone is a very minor component (0.00001 %), and has a skeleton with 23 carbon atoms, i.e., one more carbon than yuzurimine and daphnilactone. (The carbon marked * is considered to have been introduced into the baisc carbon skeleton.)

The following schematic diagram illustrates the interrelations of these alkaloids based on the results of biosynthetic experiments carried out so far. Firstly, squalene gives secodaphniphylline **5**, and then codaphniphylline 7. These give methyl homosecodaphniphyllate **8** and methyl homodaphniphyllate **9** after oxidative elimination of the ketal portion. (Methyl homodaphniphyllate was synthesized during the structural study of daphniphyllin, and was subsequently isolated from the fruits of Daphniphyllaceae.) The intermediates A and B have not been isolated. Yuzurimine **2**, daphnigracine **3**, and daphnilactone A **6**, and B **4** are derived from these hypothetical intermediates. The results of various studies of the compounds are summarized briefly below.

Reduction of the Carbonyl Group to a Methylene Group

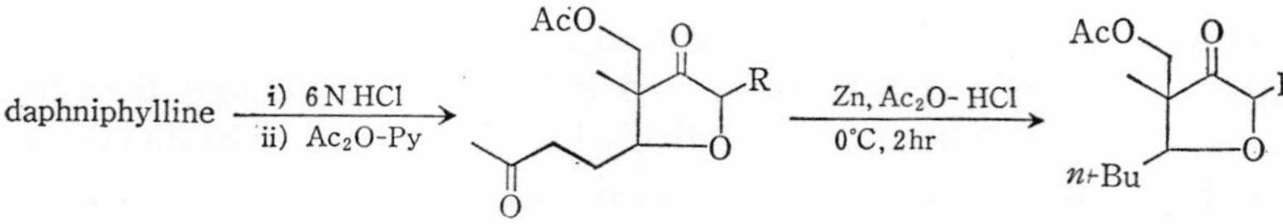

This reaction was explored as a model for a general reaction for the reduction of a carbonyl group to a methylene group under mild conditions, and has been applied to many steroids. The reducing reaction under mild conditions is considered to proceed by the same mechanism as Clemmensen reduction.[4)]

Discovery of Compounds that are Contrary to Bredt Rule

An imino compound $C_{23}H_{35}O_2N$, was obtained from methyl homosecodaphniphyllate by oxidation with $Pb(OAc)_4$ in anhydrous benzene, and was reduced to the original compound by treatment with $NaBH_4$. Two possibilities were considered for the structure of the imino body: a bicyclo [3.3.1] and a bicyclo [2.2.2] compound with a bridge-head double bond. Model building showed that these structures, both of which are contrary to Bredt rule, are feasible. Only one anti-Bredt imine has previously been isolated. As expected, the C = N bonding in the compound was found to have a simpler in pattern (ν_{max} (KBr) 1,589 cm^{-1}) than ordinary imines.

Biosynthesis

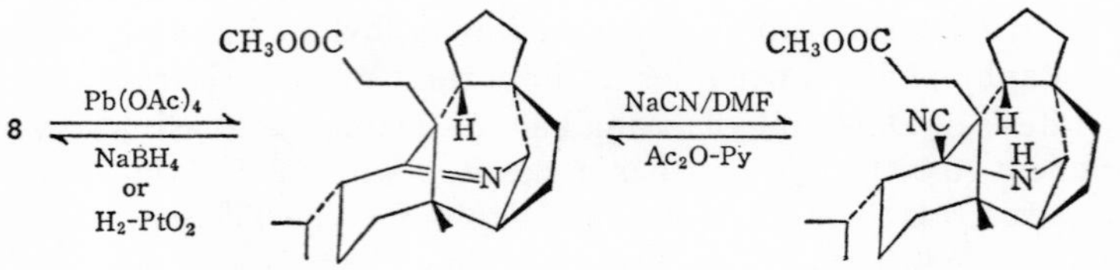

8 ⇄ [$Pb(OAc)_4$ / $NaBH_4$ or H_2-PtO_2] ... ⇄ [NaCN/DMF / Ac_2O-Py] ...

CH_3OOC H N CH_3OOC NC H H N

The routes of biosynthesis of these new compound were investigated by isotopic labeling studies. As mentioned above, an unusual specific cyclization reaction was apparently involved in the biosynthetic pathway from squalene.[2,3)]

23.2 Isolation of *Daphniphyllum* Alkaloids

Most of the *Daphniphyllum* alkaloids do not crytallize readily, so that their isolation and purification was rather difficult. After initial extraction, each alkaloid was isolated by column chromatography, monitored by tlc.

Bark of *Daphniphyllum* was used at first, according to the literature. However, collecting bark from standing trees would damage them, and it was not easy to find cut trees for bark collection. Fortunately the content of alkaloids was found to be relatively high in leaves and branches of the tree. Some alkaloids were unique to the fruits, and the fruits of *D. teijsmanne* Zollinger were collected on the shore line of Yakushima island.

[Experimental procedure 1] Extraction of Bark of *Daphniphyllum*

Finely chopped bark of *D. macropodum* Miquel (15 kg) was immersed in methanol (60 l) for two weeks. The methanol filtrate (Dragendorff-positive) was concentrated under reduced pressure to 5 l, then an equal volume of water was added and the whole was filtered. This filtrate was used as a source of alkaloids. Conc. HCl was added to bring the pH to 2 or or lower, and the solution was shaken with a large amount (5 l) of ether to eliminate chlorophyll, acidic material and neutral material. The water phase (containing alkaloids) was treated with 4 N NaOH to give pH 9–13, and was again extracted with a large amount (20 l) of ether. The ether phase was extracted with 0.1 N HCl (1 l), and the acidic water phase was carefully brought to pH 9–13 with 4 N NaOH, and again extracted with ether. The ether phase was washed with water (1 l), and was dried with anhydrous sodium sulfate. Elimination of ether from this solution under reduced pressure gave an oily crude mixture of alkaloids (30 g). The results of tlc of this oily substance are shown in Fig. 23.1.

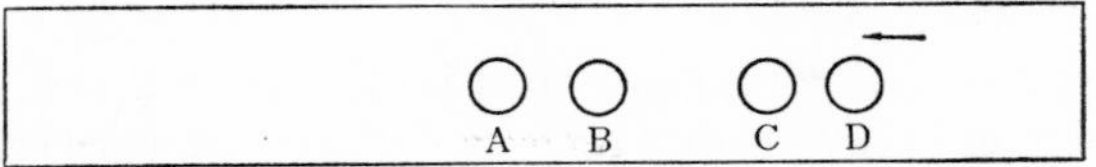

Fig. 23.1. Thin layer chromatography of crude alkaloids.
Silica gel G (Merck), ether-hexane-diethylamine (20:15:3.5).
A, daphniphylline, codaphniphylline; B, neoyuzurimine, yuzurimine A; C, yuzurimine, neodaphniphylline; D, yuzurimine B.

This oily substance was separated into three fractions by chromatography on 200 g of alumina, eluting with chloroform and chloroform-methanol mixtures. Elution with chloroform (1 l) gave about 10 g of oily material, but this was a mixture of hydrocarbons (Fr. 1). Elution with chloroform-methanol mixture (100:1, 1.5 l) gave about 10 g of oily material (Fr. 2); this was dissolved in chloroform-ether and saturated with HCl. Elimination of the solvent under reduced pressure provided a colorless oily material. White crystals (2.6 g) were obtained by crystallization from chloroform-ethyl acetate. Recrystallization from chloroform-ether gave crystals of the HCl salt of yuzurimine **2** (mp 247.5–249° C). Elution of the column with chloroform-methanol (20:1, 1.5 l), gave 10 g of an oily material (Fr. 3) and this was rechromatographed on 300 g of alumina. Elution with benezene-diethylamine (100:0.45) gave about 5 g of oily material, which was soluble in chloroform-ether saturated with HCl gas. Elimination of the solvent under reduced pressure provided an oily material that was crystallized from chloroform-ether to yield about 1 g of crystals. Recrystallization from chloroform-ether gave needle crystals of codaphniphylline **7**, mp 238–240° C and $[\alpha]_D + 43.7°$.

[Experimental procedure 2] Extraction of leaves and branch of *Daphniphyllum*

Crude alkaloid (630 g) was obtained as an oil by extracting 1000 kg of leaves and branches with 400 l of methanol and treating the extract as described above.

This oily material was chromatographed on 3 kg of alumina, and fractionated into five portions by successive elution with hexane, hexane-benzene (1:1), benzene, chloroform, chloroform-methanol (1:1).

Hexane fraction: Oily material (10 g) eluted with hexane was rechromatographed on alumina with hexane, and gave three alkaloids in pure form; alkaloid A_1 (1 g), yuzurine (1.1 g), and methyl secodaphniphyllate (1.7 g).

Hexane-benzene (1:1) fraction: Oily material (about 10 g) was obtained, and was rechromatographed on 100 g of alumina. Secodaphniphylline **5** (1.13 g) and daphnilactone A **6** (0.18 g) were obtained by elution with hexane and hexane-benzene (1:1).

Benzene fraction: Oily material (15.0 g) was obtained, and was chromatographed on 1.5 kg of silica gel with hexane-benzene (1:1) as an eluant, providing 100 g of daphniphylline **1** and 0.3 g of codaphniphyllin **7**.

Chloroform fraction: Oily material (about 250 g) was obtained, and was separated into 1 g of yuzurimine A, 150 g of yuzurimine **2**, and 0.5 g of yuzurimine B by chromatography on 1.5 kg of silica gel with chloroform and chloroform-methanol (1–10%) as eluants.

Chloroform-methanol (1:1) fraction: About 50 g of oily material was obtained, and was chromatographed on 500 g of silica gel with chloroform-methanol (10%) to yield 1.4 g of yuzurimine.

[Experimental procedure 3] Extraction of fruits of *D. teijsmanni* Zollinger

Fruits of the plant (10 kg) were crushed and extracted with a large amount of methanol (10 l). Crude alkaloids (12 g) were obtained by the procedure used in the case of bark. The crude oily material (12 g) was chromatographed on 60 g of basic alumina (manufactured by Nakarai Chemical Co., 300 mesh), eluting with hexane, hexane-benzene (1:1), benzene, benzene-chloroform (1:1), and chloroform, to give six fractions (Fig. 23.2)

The first three fractions were combined (5.3 g) and recrystallized from hexane-benzene to yield pure daphnilactone B **4** (4.8 g). The mother liquor of recrystallization was chromatographed on 1 g of alumina. Elution with hexane and hexane-benzene (1:1) resulted in the separation of methyl homosecodaphniphyllate **8** (0.02 g), methyl homodaphniphyllate **9** (0.01 g), and daphniphylline **1** (0.05 g) in that order. It is interesting to note that the HCl salf of daphniphylline is quite difficult to crystallize unless it is rather pure. We have not been able to crystallize daphniphylline itself at all.

The following two fractions from the previous chromatography were

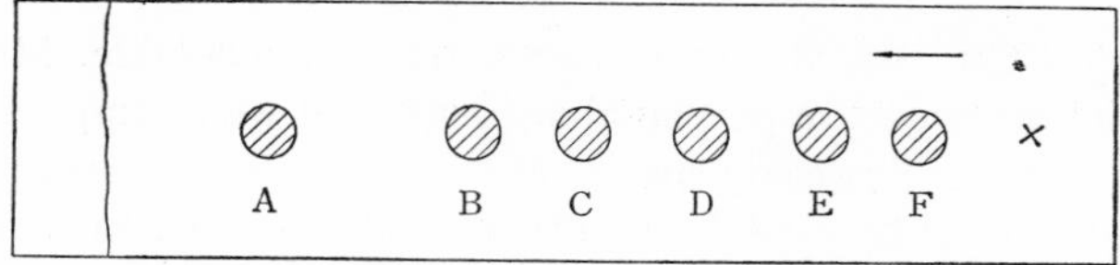

Fig. 23.2. Silica gel thin layer chromatography of crude alkaloids. A, methyl homosecodaphniphyllate; B, methyl homodaphniphyllate; C, daphniphylline; D, daphnilactone B; E, yuzurimine; F, yuzurimine B.

combined (about 2 g), and chromatographed on 4 g of alumina. Elution with benzene and benzene-chloroform (1:1) yielded three alkaloids; daphnilactone B **4** (0.5 g), yuzurimine **2** (0.1 g), and yuzurimine B (0.1 g) in that order.

The final eluate with chloroform gave a crystalline material after rechromatography, eluting with ethyl acetate-methanol (9:1). Recrystallization from water-acetone gave a product of mp 247–248° C. It was a hydrolysis product of daphnilactone B **4**, and was a zwitterion alkaloid.

23.3 Isolation of Toxic Diterpenes from Eurphorbiaceae[5–8]

Little is known of the toxic components of Euphorbiaceae. The milky liquid of Euphorbiaceae plants other than Daphniphyllaceae is negative to the Dragendorff coloring reaction, in spite of its toxicity, and its real nature was totally obscure. The plant body above the ground of *Euphorbia adenochlora* was collected from Kohzu Island, but studies were unsuccessful because its toxicity was unexpectedly weak. Finally, we found Dragendorff-positive material in *E. millii* Ch. des Moulins, Euphobiaceae. We isolated three components in pure form (milliamines A, B, and C) as the toxic components of the plants, and determined their structures. We concluded that the toxic components of Euphorbiaceae plants in general might be diterpene derivatives with high oxygen contents.

We next studied toxic components of the root of *E. jolkini* Boiss., *E. kansui* Liou. and the root of *E. fischeriana* Steudel. (The latter two were obtained as crude medicines from China.) No clear common characteristics were found, however, and isolation of the components was therefore not straight forward. A high-speed chromatography technique was introduced at the final stage of this study, and was extremely effective. Subsequent studies dealt with the structure-toxicity relationships, biosynthetic pathways and reactions of the isolated compounds.

Materials Used

E. millii CH. DES MOULINS is commercially available. Most of the commercial products are perpetual variants, and we attempted to obtain the original spring flowering species as far as possible (an established original strain should be used for studies on plant components, since strain variation often affects the components present). The author is grateful to Messrs. Gotoh and Matsuoka of the Kyoto Farm of Takeda Chemical Industries, Ltd. for assistance in collecting the samples.

E. jolkini BOISS. grows along craggy shores in warm regions of Japan such as Kushimoto in Wakayama Prefecture and Yakushima Island in Kagoshima Prefecture. The roots usually penetrate into rocks, and excavating them is extremely difficult. Some people are sensitive to the milky liquid of the plant. The milky liquid in the portion above ground is turbid and white, while that from the underground portion is yellowish and strongly toxic.

"*Kansui*" (*E. kansui* LIOU.) obtained in Japan in the past showed weak toxicity, and the material used in this study was kindly provided by Dr. Hon-Yuan Shui of Taiwan.

Roots of *E. fischeriana* STEUDEL used in the study were kindly supplied by the Kyoto Farm of Takeda Chemical Industries, Ltd. and were from Hong Kong.

Isolation and Structure Determination of Toxic Components and Related Substances

Ingenol derivatives, milliamines A, B, and C, and C, and also a compound, that appeared to be a diterpene derivative, $C_{20}H_{28}O_6$ **13**, were isolated from the root of *E. millii* CH. DES MOULINS.

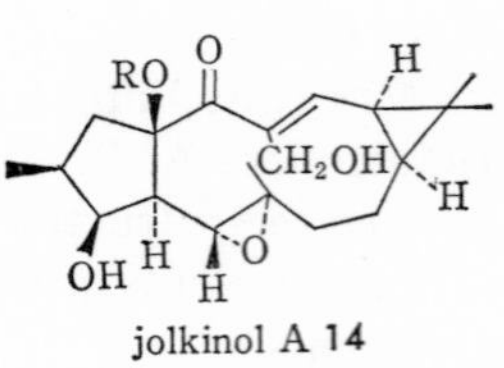

jolkinol A **14**

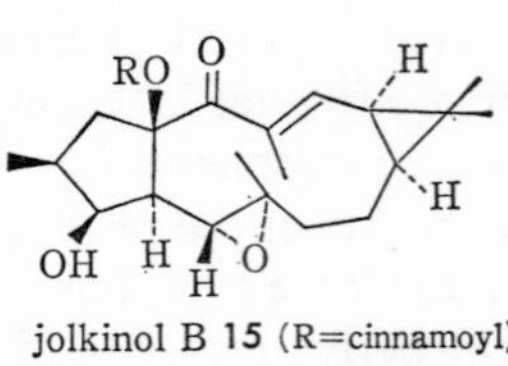

jolkinol B **15** (R=cinnamoyl)

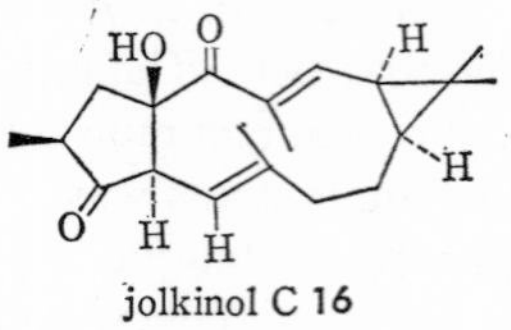

jolkinol C **16**

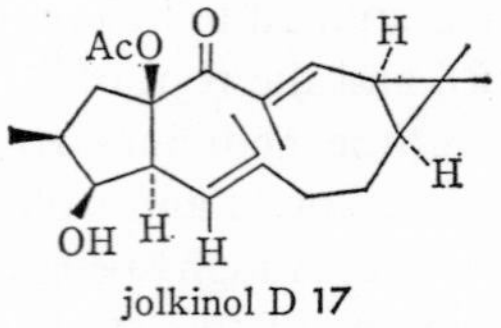

jolkinol D **17**

Ingenol, esterified at the 3-OH with 2,4,6,8,10-tetradecapentaenoic acid, was also identified as the toxic component of the root of *E. jolkini*

BOISS. Jolkinols A **14**, B **15**, C **16** and D **17** were also isolated and their structures determined. Diterpenes with abietane skeletons, though they are not toxic components, were also isolated and identified (**18–22**).

Toxic components of the Chinese crude drug "*Kansui*" (*E. kansui* LIOU) were isolated and identified as ingenol derivatives and a 13-oxyingenol derivative. Kansuinine A **23**, and B **24** were also isolated, and their structures determined.

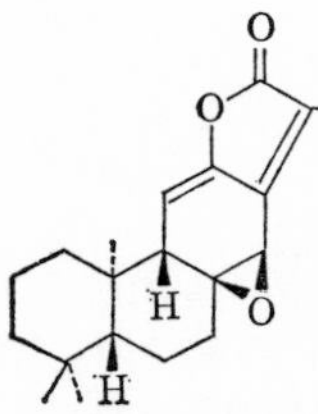

jolkinolide A **18** 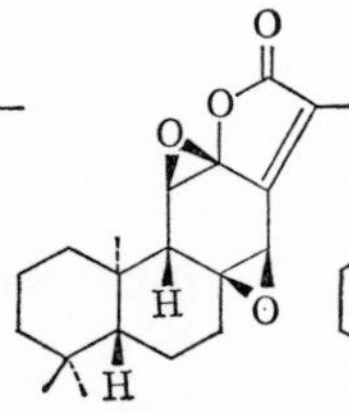 jolkinolide B **19** 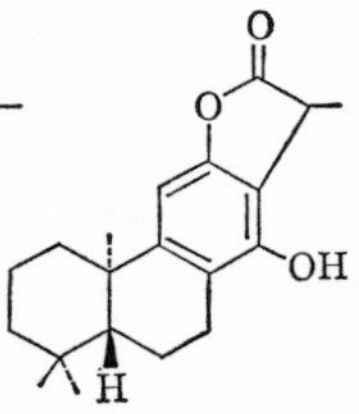 jolkinolide C **20** 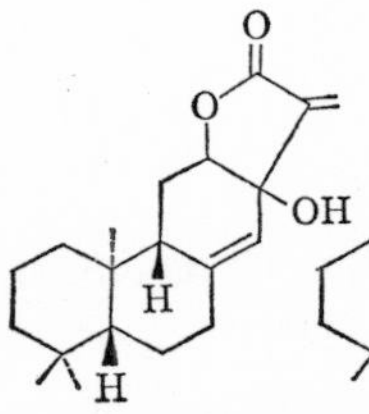 jolkinolide D **21** 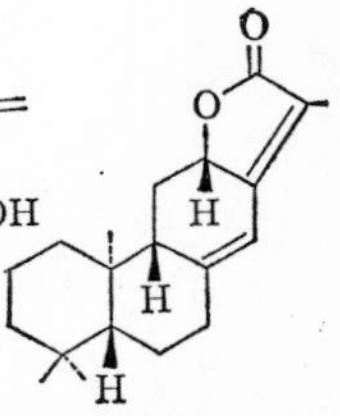 jolkinolide E **22**

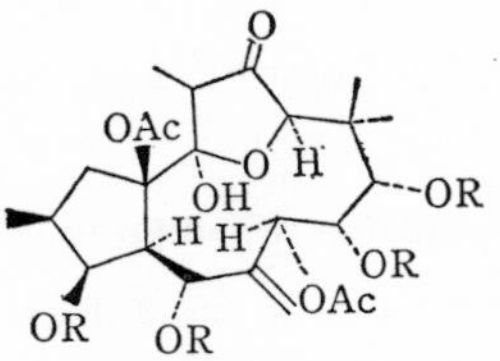

kansuinine A **23** 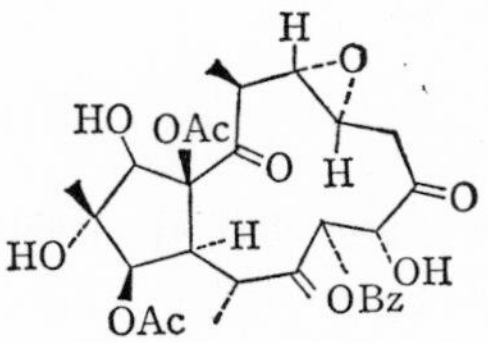 kansuinine B **24**

O-Acetyl–*N*-(*N'*-benzoyl–L-phenylalanyl)–L-phenylalaninol, which is not toxic, was isolated from the root of *E. fischeriana* STEUDEL obtained from China.

Except for kansuinine A and B, the toxic components are derivative. of ingenol and of 13-oxyingenol, while ingenol itself has little toxicity. Toxicity appears to depend on the position of esterification and the type of acid. During the structure determination of the toxic components of *E.*

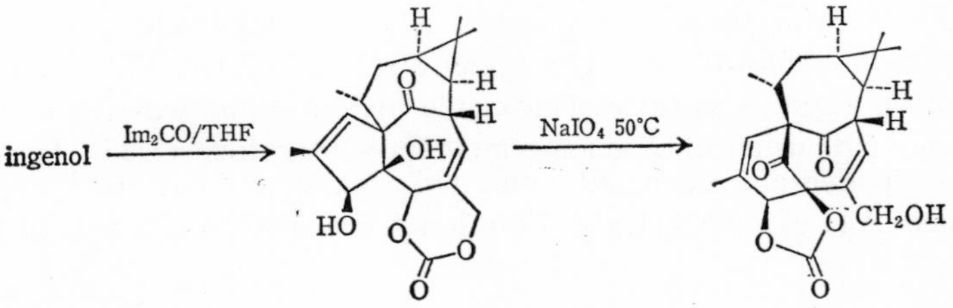

ingenol $\xrightarrow{Im_2CO/THF}$ $\xrightarrow{NaIO_4\ 50°C}$

jolkini BOISS., and in the reactions of kansuinine A and B, some interesting chemical transformations were found.

[Experimental procedure 4] Isolation of kansuinine A and B

The isolation of kansuinine A and B is as a typical example of the separation of toxic components of Euphorbiaceae plants.

"*Kansui*" (8 kg) was immersed in 20 l of ethanol at room temperature for several days. The ethanol filtrate was concentrated under reduced pressure. Extraction in this way with ethanol and concentration were repeated twice, and the concentrates were combined (859 g). This was treated with 500 ml of benzene and 500 ml of water. The water phase was separated and washed with benzene. Both benzene phases were combined, dried with anhydrous sodium sulfate, and concentrated. The residue (20.0 g) was chromatographed on 1.5 kg of silica gel (100 mesh, Mallinckrodt Co.) with benzene. Benzene was passed until compounds of low polarity, such as hydrocarbons and terpenes, become undetectable. Elution was then continued with chloroform, monitored by silica gel tlc (developing solution, 3% methanol-chloroform). The material (3.2 g) eluted with methanol-chloroform (2:98) was toxic. This crude toxic fraction was separated by preparative tlc using silica gel and methanol-chloroform (3:33). The silica gel was divided and extracted with ethyl acetate, and two fractions were obtained, A (1 g), and B (1.3 g). Fraction A was dissolved in 5 ml of ether.

On adding 5 ml of petroleum ether and leaving the solution to stand, 350 mg of kansuinine A was obtained as crystals.

Fraction B was separated by high speed liquid chromatography: 3.3 mg was subjected to recycling chromatography on Micro Pak NH_2 with dichloromethylene-isooctane (1:2), and ten recycles yielded pure kansuinine B and 0.2 mg of unidentified substances. This pure kansuinine B was crystallized by the above method.

Applicability of High-Speed Liquid Chromatography

High speed liquid chromatography with Permaphase ETH could separate 13-oxyingenol derivatives effectively (Fig. 23.3), but did not give good separation in the case of kansuinine. It is also not suitable for preparative use because of the small column capacity.

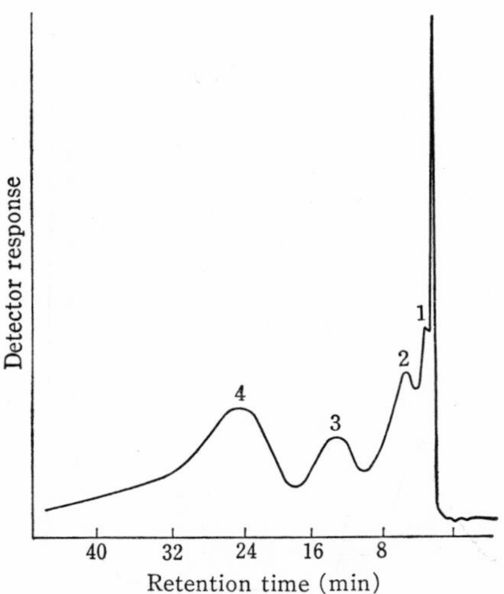

Fig. 23.3. High speed liquid chromatogram of toxic components of "*Kansui*".
1, kansuinine A; 2, kansuinine B; 3,4, 13-oxyingenol derivatives.

Homogeneous and porous column packings of silica gel, Micro Pak Si-10, could successfully separate kansuinine (Fig. 23.4), though kansuinine was first crystallized, as mentioned above, after the elimination of minor components by recycling chromatography (ten times) with Micro Pak NH_2 (Fig. 23.5).

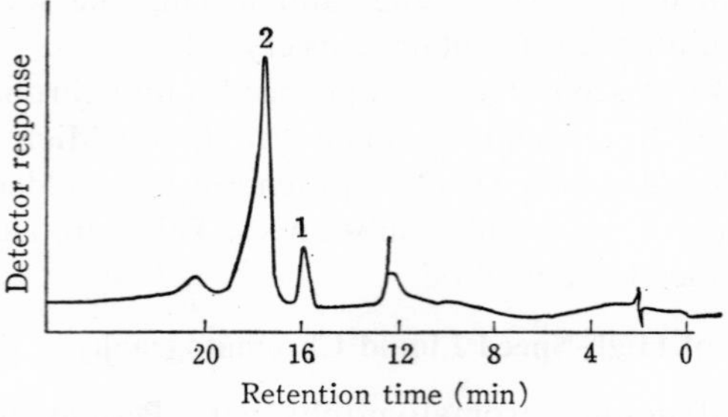

Fig. 23.4. Separation of kansuinine A and kansuinine B on Micro Pak Si-10.
1, kansuinine A; 2, kansuinine B. Column: Micro Pak Si-10 (50 cm × 2 mm), moving phase: A, 33.3% dichloromethylene in isooctane; B, 5–10% gradient of methanol in dichloromethylene, 0.5%/min. Flow rate: 40 ml/hr.

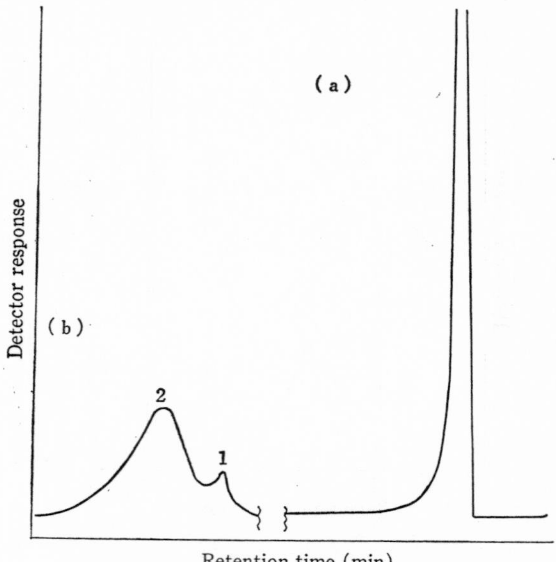

Fig. 23.5. Purification of kansuinine B by recycling chromatography. (a) first cycle, (b) tenth cycle.
1, unknown substance (about 0.2 mg); 2, kansuinine B (2.1 mg).
Column: Micro Pak NH_2 (25 cm × 2 mm inside diameter) × 4.
Solvent: dichloromethylene-isooctane (1:2) at 40 ml/hr v (2,000 psi). Room temperature, with uv (254 nm) detection.

Fig. 23.6 shows the separation of ingenol unsaturated esters, **2** and **3** 3-ester and 20-ester, respectively, which are unstable to light and acid.

Fig. 23.7 shows the separation of jolkinolides A and B, which are unstable in acid and alcohol. There has been no previous report on the

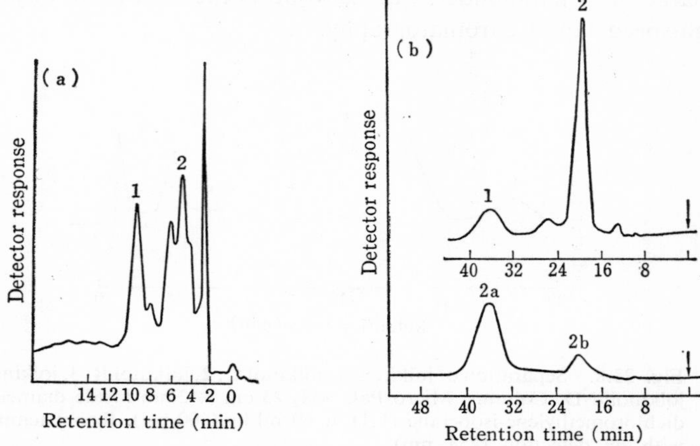

Fig. 23.6. (a) Separation of **1** and **2** on Zipax-Carbowax 600 Column: 100 cm × 1.8 mm inside diameter, 1.2% Carbowax 600 (Zipax), with 2.5% THF in isooctane at 40 ml/hr. Carried out at 28° C with uv detection (254 nm). (b) Separation of **1** and **2** on Perisorb A Column: 200 cm × 1.8 mm inside diameter, Perisorb A, with 20% THF in isooctane at 20 ml/hr (100 psi), 25° C.

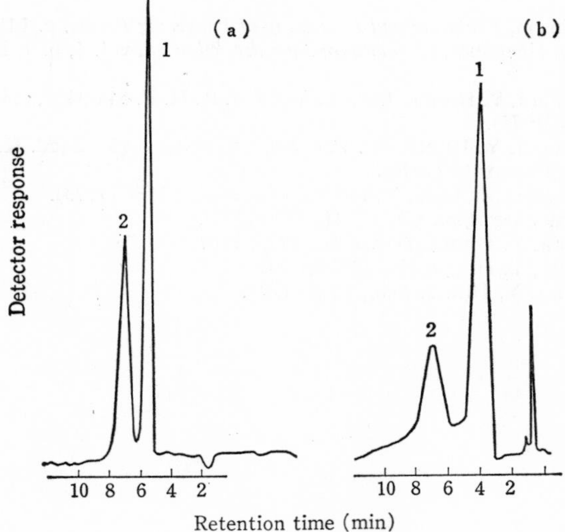

Fig. 23.7. Separation of jolkinolides A and B.
1, jolkinolide A; 2, jolkinolide B.
(a) Column: 100 cm/1.8 mm inside diameter, Bio Sil A (25–35 μm) with 20% THF in isooctane at 60 ml/hr (2,200 psi). Carried out at 35° C with uv detection (254 nm). (b) Column: 50 cm × 1.8 mm inside diameter, Permaphase ETH, with 25% ethanol in water at 160 ml/hr (2,600 psi). Carried out at 29° C with uv detection 254 nm).

separation of jolkinolides. Fig. 23.8 shows the separation of jolkinols by high-speed liquid chromatography.

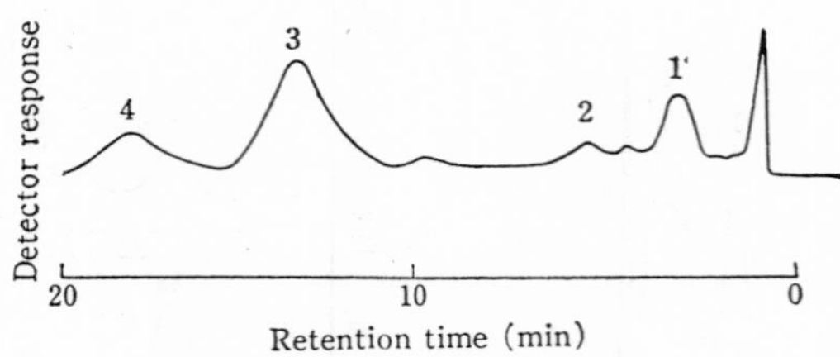

Fig. 23.8. Separation of jolkinos. 1, jolkinol C; 2, jolkinol B; 3, jolkinol A; 4, jolkinolid D. Column: Micro Pak NH_2 25 cm × 2mm inside diameter with dichloromethylene-isooctane (1:1) at 60 ml/hr (800 psi). Room temperature with uv detection (254 nm).

REFERENCES

1) J. Hutchinson, *Evolution and Phylogeny of Flowering Plants,* p. 141, Academic Press (1969): R. Hegnauer, *Chemotaxonomie der Pflanzen*, vol. 4, p. 9, Birkhaeuser, Basel (1966).
2) S. Yamamura, Y. Hirata, *The Alkaloids* (ed. R. H. F. Manske), vol. 15, p. 41, Academic Press (1975).
3) S. Yamamura, Y. Hirata, *Int. Rev. Sci., Alkaloids Series* 2 (ed. K. Wiesner), vol. 9, p. 161, Butterworths (1976).
4) S. Yamamura, M. Toda, Y. Hirata, *Org. Syn.*, **53**, 86 (1973).
5) Y. Hirata, *Pure Appl. Chem.*, **41**, 175 (1975).
6) D. Uemura, Y. Hirata, *Tetr. Lett.*, **1975**, 1967.
7) D. Uemura, Y. Hirata, *ibid.*, **1975**, 1701.
8) D. Uemura, Y. Hirata, *ibid.*, **1975**, 1705.

CHAPTER 24

Indole Alkaloids of Japanese Plants

In this chapter, the term indole alkaloids is taken to include indole and dihydroindole alkaloids, such as yohimbine **1** and strychnine **2**, as well as oxindole alkaloids such as rhynchophylline **3**.[1] The former two types account for 80% of all the indole alkaloids.[2–4]

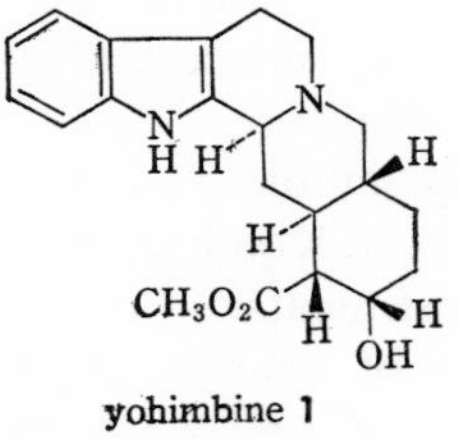

yohimbine **1** 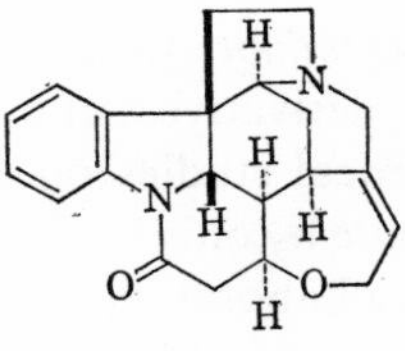 strychnine **2** 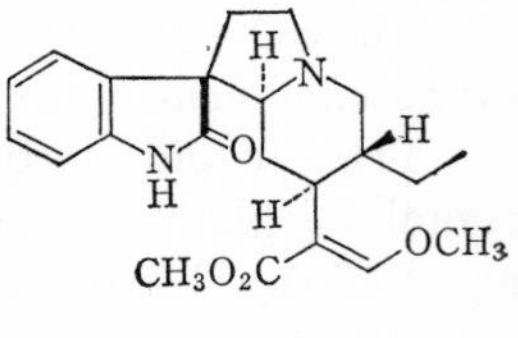 rhynchophylline **3**

Up to the present time, about 800 kinds of indole alkaloids have been isolated and most of their structures have been elucidated. This number amounts to nearly one-fifth of all the known nitrogen-containing plant bases (alkaloids). Since around 1950, spurred by early work on the alkaloids of *Rauwolfia serpentina*, elaborate structural studies on reserpine (an antihypertensive) or ajmaline (an anti cardiac arrythmic) were carried out and rapid development in the study of indole alkaloids began. The finding of an antileukemic bisindole alkaloid, vincristine, which is composed of indole and dihydroindole moieties, in *Catharanthus roseus* (*Vinca rosea*) confirmed the importance of indole alkaloids as objects of pharmaceutical study. From a purely chemical standpoint, many characteristic reactions have been found in this field, making the study of indole alkaloids of special interest. However, most of the plants which contain indole alkaloids grow in the tropical regions of the world, and the present chapter deals with only a limited number of species that appear in Japan.

24.1 Indole Alkaloids and Plants

The first discoveries of indole alkaloids were the results of chemical studies on folk medicines or crude drugs used in various parts of the world, such as India (*Rauwolfia serpentina*), Europe (*Catharanthus roseus*) and Japan (*Uncaria rhynchophylla*). This approach proved to be a very effective one for natural products chemists. The question then arises, what kinds of plants contain indole alkaloids? This can be answered in part with the help of established plant taxonomy.[5,6] The accumulated results of early researchers indicate that plants which contain monoterpenoid indole alkaloids mostly belong to three families of Gentianales, i.e., Apocynaceae (e.g. *Catharanthus roseus*; vincristine), Loganiaceae (e.g. *Strychnos nux-vomica*; strychnine **2**) and Rubiaceae (e.g. *Uncaria rhynchophylla*; rhynchophylline **3**). A few other kinds of indole alkaloids are also found in Rutaceae (*Evodia rutaecarpa*; evodiamine), Leguminosae (*Physostigma venenosum*; eserine), Calycanthaceae (*Meratia praecox*; chimonanthine), Alangiaceae (*Alangium platanifolium*; tuburosine), Simaroubaceae (*Picrasma ailanthoides*; canthione derivatives) and other families.

Prior to our study, chemical studies on Japanese indole alkaloids-containing plants had been made by only two groups; Prof. Kondo and his group (Tokyo University; Itsuu Laboratory) on the constituents of *Uncaria* spp. (Rubiaceae) and Prof. Kimoto (Kyoto College of Pharmacy) on *Amsonia elliptica* (Apocynaceae). The present author started work on indole alkaloids of *Gardneria* sp., which grows naturally in Chiba Prefecture, from about 1964 in collaboration with Prof. Haginiwa of Chiba University.

One reason for selecting this genus for study was its taxonomical position in Loganiaceae. Earlier taxonomists had divided the family of Loganiaceae to two subfamilies, Loganioideae and Buddleioideae. The former subfamily contains the genera *Gelsemium* and *Strychnos*, both of which are known to contain indole alkaloids. However plants belonging to the latter subfamily, such as *Buddleja venerifera*, did not show any indication of the presence of alkaloids in our screening tests. It is interesting to note that a recent edition of "Syllabus der Pflanzen" by Engler (1964)[6] treats Buddlejaceae and Loganiaceae as independent families, in accord with this result. As is well known, the trees of *Strychnos* are tropical plants and do not grow naturally in Japan. On the other hand, the plants of *Gardneria* spp. of Loganiaceae are reasonably common in Japan, and so we were able to collect the plant material relatively easily.

Loganiaceae (18 genera, 500 species)

Gelsemium sempervirens ATLANT, which inhabits the southern parts of North America, is known to contain gelsemine and other related oxindole alkaloids in the roots, while *G. elegans* found in southern parts of China and Malaysia also contains similar indole alkaloids. *Strychnos nux-vomica* contains strychnine **2** and burcine in the seeds. These plants, however, do not grow in Japan. The plants of this family growing in Japan are limited to two genera, *Mitrasacme* (e.g. *M. pygmaea*) and *Gardneria* (e.g. *G. nutans*), taking Buddlejaceae as a separate family. The plants of *Mitrasacme* have a great disadvantage as sources of indole alkaloids, i.e., the smallness of the plants. *Gardneria* species are evergreen climbers found in places south of Chiba.

Five species of this genus are known in Japan and Formosa. We collected all five and made a comparative study. From a morphological point of view, these plants could be separated into two groups. A representative species of group A is *Gardneria nutans* and one of group B is *G. multiflora*, which grows mainly in the Kinki and Chugoku areas.

TABLE 24.1. Alkaloid-containing *Gardneria* spp. growing in Japan.

Group A	*Gardneria nutans* SIEB. et ZUCC.	"*Horaikazura*"†
	G. insularis NAKAI	"*Eishukazura*"
Group B	*G. multiflora* MAKINO	"*Chitosekazura*"
	G. shimadai HAYATA	"*Taiwanchitosekazura*"
	G. liukyuensis HATSUSHIMA	"*Ryukyuchitosekazura*"

† Japanese name

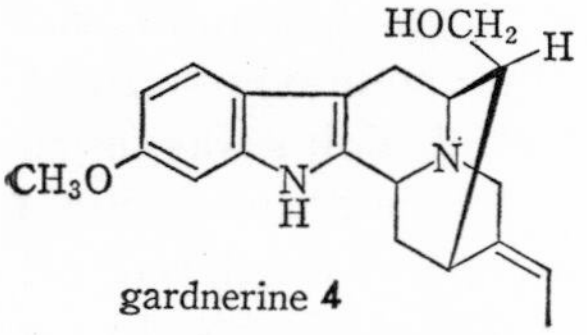

HOCH$_2$ H CH$_3$O N H N

gardnerine **4**

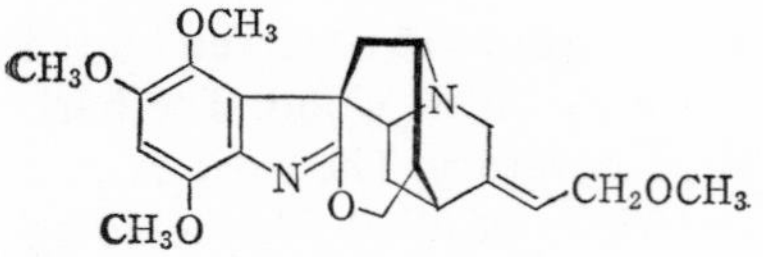

OCH$_3$ CH$_3$O N N O CH$_2$OCH$_3$ CH$_3$O

gardneramine **5**

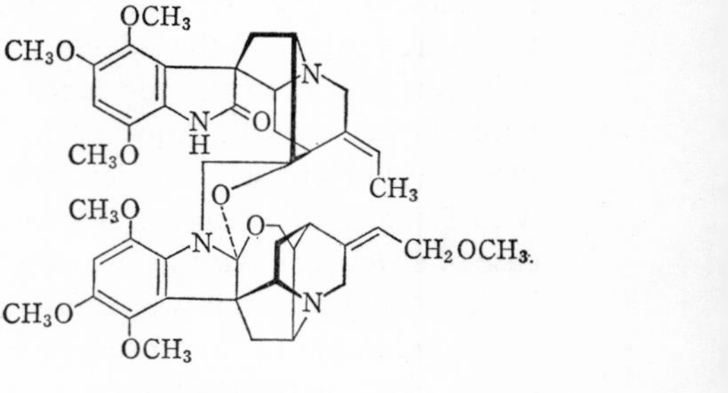

OCH$_3$ CH$_3$O N N H O CH$_3$O CH$_3$ CH$_3$O O N O CH$_2$OCH$_3$ CH$_3$O N OCH$_3$

gardmultine **6**

The plants belonging to group A contained gardnerine **4** and a masked oxindole alkaloid, gardneramine **5**, as the main bases. On the other hand the plants of group B were found not to contain gardnerine **4**, and their main bases were gardneramine **5**, the sole common base of the genus, and a bisindole alkaloid, gardmultine **6**. These results provide an example of the coincidence of morphological taxonomy with chemotaxonomy.

Apocynaceae (200 genera, 2,000 species)

This family is divided to three subfamilies, Plumerioideae, Cerebroideae and Apocynoideae. Indole alkaloids are only found in the plants of Plumerioideae, of which a representative member is *Rauwolfia*. As Japanese plants of this subfamily, only *Amsonia elliptica*, *Ochrosia nakaiana* and another *Ochrosia* species are known. The subfamily Apocynoideae includes the genus *Strophanthus*, which is well known to contain cardiac glycosides such as strophanthin. The Japanese plants *Trachelospermum asiaticum* and *Anodendron affine* are also members of this subfamily. These genera, like *Nerium*, contain no alkaloid.

We carried out several surveys and collections on the Bonin Islands (Ogasawara Islands) after 1968. The trunk bark of *Ochrosia Nakaiana* Koidz. (Apocynaceae), which sometimes has a trunk 30–40 cm in diameter, were collected. On these occasions we planned to collect alkaloid-containing plants of Apocynaceae, Rubiaceae and other families. For this purpose we carried the following materials; chloroform, sodium carbonate, tartaric acid, Meyer reagent (a reagent for the detection of alkaloid; prepared in a small polyethylene eye washer), a filtering funnel, a separatory funnel (50 ml) and watch glasses. Small pieces of collected plants were cut with a knife as finely as possible and macerated in 5–10% tartaric acid overnight in a tea cup. The filtered solution was carefully made basic (pH test paper) with solid sodium carbonate. Extraction was carried out with chloroform. The chloroform layer was put in a tea cup and concentrated under an electric fan. The concentrated solution was then transferred to a watch glass and the solvent was removed. A dilute solution of tartaric acid was added to the residue and the presence of alkaloids was tested with Meyer reagent. Another convenient method for detecting alkaloid uses the juice pressed from leaves or bark onto a filter paper which is tested with Dragendorff reagent. This method, however, is not reliable. In the case of *Ochrosia nakaiana*, a strongly positive reaction was observed for the bark but the reaction was almost negative for the heart wood. The bark was dried in the sun and brought back to the laboratory. The basic extract of the plant was a mixture of a great many kinds of alkaloids including vobasine **7**, a typical 2–acylindole alkaloid, and several yohimbinoid

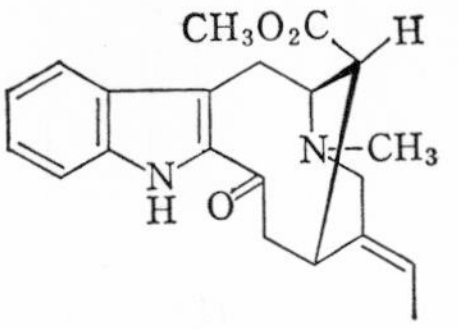

vobasine **7**

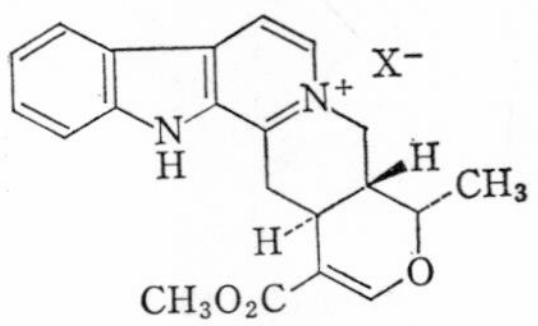

serpentine **8**

alkaloids. As for water-soluble alkaloids, serpentine **8** was obtained through the Reinecke salt.

β-Yohimbine was reported as a component of an Apocynaceae plant (*Amsonia elliptica* ROEM. et SCHULT) by Prof. Kimoto. We continued the study of this plant and identified the secamines **9** as another main component in the roots. We also found characteristic minor alkaloids, pleiocarpamine **11** and antirhine **12**, together with some yohimbinoid alkaloids. The secamines can be considered to be produced from the corresponding presecamines **10** by the action of acid in the course of extraction. The secamines were first found by Smith in a *Rhazya* sp. (not present in Japan), a genus close to *Amsonia* taxonomically. *Amsonia elliptica* is found even in Ominato, Aomori Prefecture, in the north of

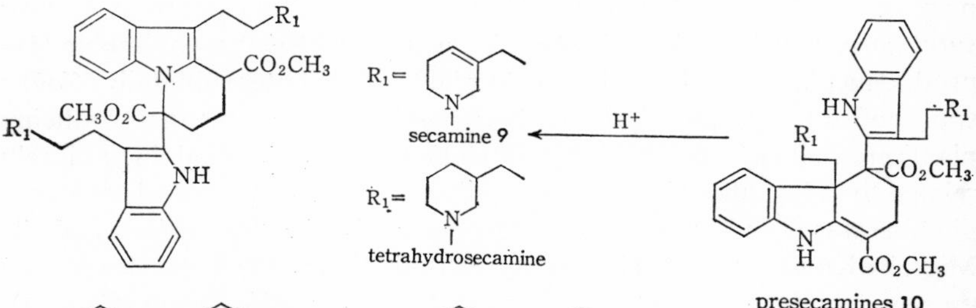

secamine 9

tetrahydrosecamine

presecamines **10**

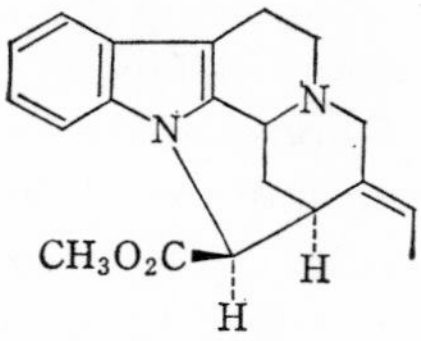

pleiocarpamine **11**

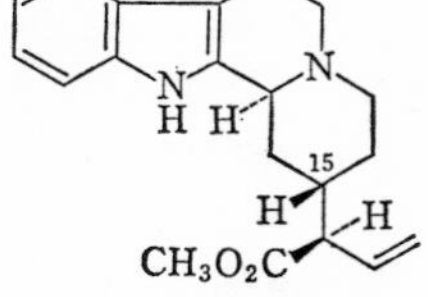

antirhyne **12**

Honshu. *Amsonia elliptica* may thus be regarded as an exceptional plant, since other indole-containing plants are generally distributed in tropical or warm regions of the world. Recently we examined the constituents of the seeds and found that the components were quite different from those of the roots. The seeds contained tabersonine **13**, an aspidosperma alkaloid, as

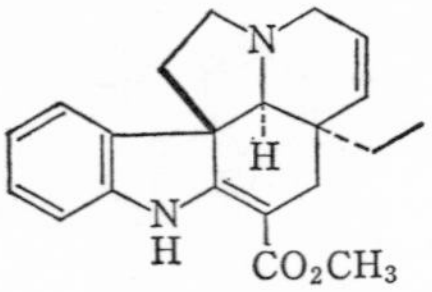

tabersonine **13**

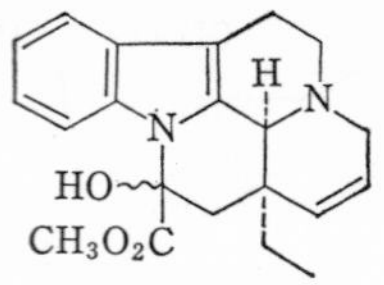

Δ^{14}-vincamine **14** (β-OH)
16-epi-Δ^{14}-vincamine **15** (α-OH)

the main base together with minor bases, such as vincamine analogs **14**, **15**.

Thus we could isolate additional constituents from the same material studied previously. Undoubtedly this was due to the development of new analytical methods such as gas chromatography (gc), thin layer chromatography (tlc), counter current distribution and high performance liquid chromatography (hplc) since the early work. The isolation of the alkaloids was mainly carried out in our laboratory by standard procedures, e.g. column chromatography on alumina or silica gel and preparative tlc. No special technique was employed.

Rubiaceae (450 genera, 6,000 species)

This family is divided to two subfamilies, Cinchonoideae and Rubioideae (Coffeoideae). All the indole or oxindole alkaloids—containing plants of this family belong to the former subfamily, for example, *Cinchona succirubra, Pausinystalia yohimbe* (*Corynanthe yohimbe*; a yohimbine (**1**)–producing plant) and *Uncaria* spp. As plants of the latter subfamily, *Coffea* spp. (coffee plant) and *Cephaelis ipecacuanha* (ipecac; emetine-producing plant) are well known. Emetine, though an isoquinoline alkaloid, is closely related to indole alkaloids biogenetically. As already mentioned, rhynchophylline was found in a Japanese Rubiaceae plant, *Uncaria rhynchophylla* Miq., by Kondo in 1928. The structure of this alkaloid was clarified by two groups, Kondo and his colleagues, and Marion and his group (Canada). *Uncaria rhynchophylla* is the only species of this genus in Japan, and a characteristic feature of this plant is the presence of hooks at the bottom of the leaves on the young branches. Botanically, these hooks are deformed twigs and they are used as a Chinese medicine called "*Cho-to-ko*" for sedative purposes. A large quantity of tannins is contained in this plant and so usual extraction with methanol gives a very poor yield of alkaloids. A good result was obtained when benzene extraction was carried out with plant material pretreated with calcium hydroxide. This procedure is often used for the extraction of caffeine from tea leaves.

[Experimental procedure 1] Extraction of alkaloids from *Uncaria rhynchophylla*

The plants were collected at Kiyosumi, Chiba Prefecture, in November, 1974. The bark of the underground parts of the plants was dried and powdered. The material was extracted in two portions as follows: 7.5 kg was well macerated with $Ca(OH)_2$ (1.5 kg) in water (2 l) and the whole was heated in a stainless steel extractor (Minami extractor) for 5 hr. Benzene (40 l) was added to the material and extraction was continued for 2 days under reflux. Extraction was repeated three times using fresh benzene. The second portion (remaining material) (7.0 kg) was treated in the same way.

All the benzene extracts were concentrated to ca. 1.5 l *in vacuo* and the basic fraction was extracted with 2% H_2SO_4 (3 l). The acid layer was made basic with NH_4OH and the freed alkaloids were taken into $CHCl_3$ containing 10% MeOH. The precipitates that separated during the above process were sulfates of alkaloids, and the free bases were extracted with $CHCl_3$ after adding NH_4OH. The $CHCl_3$ layers were combined, washed with water, dried and the solvent was evaporated off to give 87.5 g (0.60%) of crude base fraction. The alkaloids detected therein are shown here.

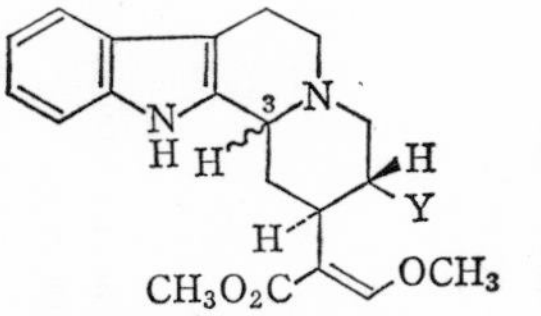

hirsutine **16** (3*R*, Y=CH_2CH_3)
dihydrocorynantheine **17** (3*S*, Y=CH_2CH_3)
hirsuteine **18** (3*R*, Y=$CH=CH_2$)
corynantheine **19** (3*S*, Y=$CH=CH_2$)

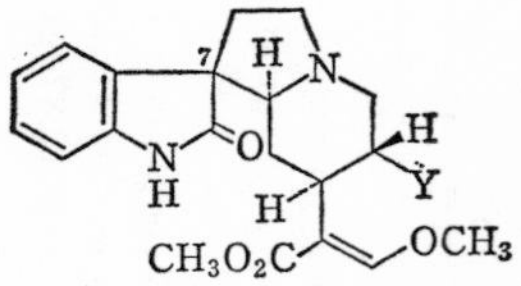

rhynchophylline **3** (7*R*, Y=CH_2CH_3)
isorhynchophylline **20** (7*S*, Y=CH_2CH_3)
corynoxeine **21** (7*R*, Y=$CH=CH_2$)
isocorynoxeine **22** (7*S*, Y=$CH=CH_2$)

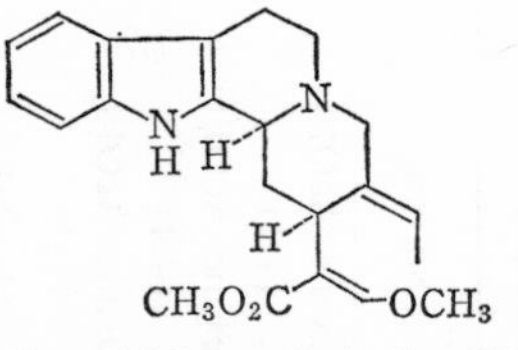

geissoschizine methyl ether **23**

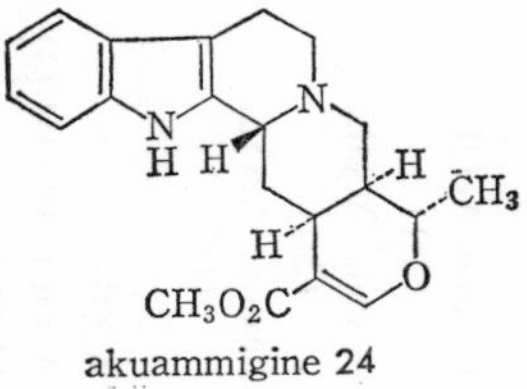

akuammigine **24**

[Experimental procedure 2] Separation of alkaloids of *Uncaria rhynchophylla*

The material plants were collected in November, 1971. The crude base (20.0 g; 0.15%) was obtained from the aerial and subterranean parts (2–3 cm diameter; bark and heart wood) (13 kg) in the manner described above. Brockmann-type alumina (activity II-III) (Merck) and Silica gel 60 (finer than 230 mesh) (Merck) were used for column chromatography. Silicagel GF_{254} (Merck) was used for tlc (Tables 24.1 through 24.7).

In column chromatography, all fractions were checked for alkaloids by tlc (silica gel, $CHCl_3$–Me_2CO (5:4); detected under a uv lamp or by

Table 24.2. Chromatography A (crude base 20g, alumina 450 g).

Fraction No.	Eluant (ml/fr.)	Solvent	Separated crystal. base	Noncrystalline part (g)	Next step
1– 4	300	hexane–AcOEt (2:1)		3.54	chromatography B
5– 9	300	hexane–AcOEt (2:1)	hirsutine, hirsuteine mixed cryst. (9.01 g)	0.92	chromatography C
10–18	300	hexane–AcOEt (1:2)	isorhynchophylline–$HClO_4$ isocorynoxeine–$HClO_4$ mixed cryst. (2.87 g)	0.97	chromatography C
19–21	300	AcOEt		0.11	
22–24	300	AcOEt–MeOH (3%)	rhynchophylline, corynoxeine	0.66	
25	1000	MeOH	mixed cryst. (0.63 g)	0.48	

Table 24.3. Chromatography B (from chromatography A, fr. 1–4 noncrystalline part) (alumina 100 g).

Fraction No.	Eluant (ml/fr.)	Solvent	Separated crystal. base	Noncrystalline part (g)	Next step
1	300	hexane–AcOEt (4:1)		1.21	chromatography D
2	300	hexane–AcOEt (4:1)		1.46	chromatography E
3,4	300	hexane–AcOEt (4:1)		0.16	chromatography E
5–10	300	hexane–AcOEt (2:1) MeOH	hirsutine, hirsuteine mixed cryst. (0.46 g)	0.28	

TABLE 24.4. Chromatography C (from chromatography A, fr. 5–18 noncrystalline part) (alumina 100g).

Fraction No.	Eluant (ml/fr.)	Solvent	Separated crystal. base	Noncrystalline part (g)
1	200	hexane–AcOEt (3:1)		0.02
2–10	200	hexane–AcOEt (3:1)	hirsutine, hirsuteine mixed cryst. (0.74 g)	0.46
11–12	200	hexane–AcOEt (2:1) MeOH	isorhynchophylline–$HClO_4$ isocorynoxeine–$HClO_4$ mixed cryst. (0.57 g)	0.11

TABLE 24.5. Chromatography D (from chromatograph B, fr. 1 noncrystalline part) (alumina 80 g).

Fraction No.	Eluant (ml/fr.)	Solvent	Noncrystalline part (g)	Next step
1	200	hexane–AcOEt (6:1)	0.05	
2,3	200	hexane–AcOEt (6:1)	0.33 (dihydrocorynantheine, corynantheine mixture)	
4–6	200	hexane–AcOEt (3:1)	0.74	chromatography F

TABLE 24.6. Chromatography E (from chromatography B, fr. 2–4 noncrystalline part) (alumina 110 g).

Fraction No.	Eluant (ml/fr.)	Solvent	Noncrystalline part (g)	Next step
1– 4	200	hexane–AcOEt (6:1)	0.64	chromatography F
5– 7	200	hexane–AcOEt (3:1)	0.82	

TABLE 24.7. Chromatography F (from chromatography D, fr. 4–6 and E fr. 1–4) (silica gel 70 g).

Fraction No.	Eluant (ml/fr.)	Solvent	Noncrystalline part (g)	Remarks
1	150	$CHCl_3$	0.13	
2–4	150	$CHCl_3$–acetone (7:1)	0.127 (akuammigine **24**)	isolated by preparative tlc ($CHCl_3$–acetone 5:4)
5–9	150	$CHCl_3$–acetone (7:1) MeOH	1.09	

TABLE 24.8. Chromatography G (from chromatography F, fr. 5–9) (silica gel 60 g).

Fraction No.	Eluant (ml/fr.)	Solvent	Separated cryst. base (g)	Noncrystalline part (g)
1–42	20	$CHCl_3$–acetone (8:1)		0.08
43–173	20	$CHCl_3$–acetone (8:1–6:1)	0.40 (geissoschizine methyl ether **23**)	0.23
174	200	MeOH		0.21

heating after spraying 1% $Ce(SO_4)_2$ in 10% H_2SO_4). Gas chromatography (Shimadzu GC–3BM; 1.5% OV–1 on Shimalite W, 1.5 m; column temp. 260° ; detector temp. 300° ; N_2 40 ml/min.) was also used when necessary. The retention times were between 4.0 and 8.5 min. Separations of rhynchophylline (**3**) from corynoxeine (**21**) and corynantheine (**19**) from dihydrocorynantheine (**17**) were achieved by preparative silica gel tlc. On the other hand, hirsutine (**16**) and hirsuteine (**18**) were separated by column chromatography on silica gel with $CHCl_3$–Me_2CO (7:1–1:1). These pairs of compounds have the same structures except for a difference of the side chains (vinyl and ethyl), and therefore when only the ethyl derivatives were required as starting materials for chemical conversion the crystalline mixtures were catalytically reduced, usually with 10% Pd-C in ethanol. When the materials were colored due to impurities, pre-treatment was carried out with active charcoal.

TABLE 24.9. Alkaloids of *Uncaria rhynchophylla*.

Part	Climbing stems (aereal and subterrean) 13 kg	Bark of the subterrean parts 14.5 kg	Commercial "*Cho-to-ko*" 4.8 kg
Crude base	20.0 g	87.5 g	12.0 g
Separated base			
Rhynchophylline **3**, Corynoxeine **21**	0.63 g (3.2%)	0.83 g (1.0%)	1.55 g (12.9%)
Isorhynchophylline **20**, Isocorynoxeine **22**	2.64 g (13.2%)	3.12 g (3.6%)	3.93 g (32.8%)
Hirsutine **16**, Hirsuteine **18**	10.80 g (54.0%)	64.00 g (73.1%)	2.63 g (21.9%)
Dihydrocorynantheine **17**, Corynantheine **19**	0.33 g (1.7%)	0.82 g (0.9%)	trace
Geissoschizine methyl ether **23**	0.40 g (2.0%)	1.82 g (2.1%)	0.015 g (0.1%)
Akuammigine **24**	0.13 g (0.7%)	0.20 g (0.2%)	not detected

Some results for *Uncaria rhynchophylla* are shown in Table 24.8. It is interesting to note that while oxindole alkaloids **3**, **20**, **21**, **22**, i.e., alkaloids at a high oxidation level, are abundant in the aerial parts, indole alkaloids **16**, **17**, **18**, **19**, **23**, **24**, i.e., alkaloids at a lower oxidation level, are mainly present in the subterranean parts. A particularly great difference was observed between the constituents of the hooks (a commercial crude drug named "*Cho-to-ko*"; usually imported from Hong Kong) and those of the bark of the underground parts.

As described above, many indole alkaloids have been isolated from plants in which only oxindole alkaloids **3**, **20** were previously found. This again, is due to the development of superior separative and analytical techniques in recent years.

In connection with the study on *Uncaria rhynchophylla*, we collected a certain amount of *Uncaria kawakamii* in Formosa in 1968 in the center of the island and to the north. At the same time we purchased Formosan "*Cho-to-ko*." Previous studies on the constituents of this plant had been made by Kondo. We confirmed the presence of heteroyohimbine oxindole alkaloids, formosanine (uncarine B) and isoformosanine (uncarine A). At Heng-chun, to the south of Formosa, another *Uncaria* species (*U. florida*) was found. From this plant pteropodine (uncarine C) **25**, an isomer of formosanine at C-19 and C-20, and its C-7 epimer, isopteropodine (uncarine E), were isolated in high yield (ca. 0.3% in crystalline form). For the separation of these two alkaloids counter current distribution was effective. In our laboratory an automatic apparatus made by Quickfit (England) was used. This apparatus consists of 120 tubes which contain 20 ml each of the upper and lower layers. Both the layers can be transferred in opposite directions. About 9 g of a mixture of the above two bases was separated cleanly with 98% recovery using benzene-0.02 N AcOH as the solvent system. This separation was carried out in a single operation, but it took about one week to work up the separated fractions. This ease of separation of the epimers may arise because protonated pteropodine can form a hydrogen bond while the other epimer cannot. In fact, pteropodine **25** has a larger p*Ka* value than isopteropodine.

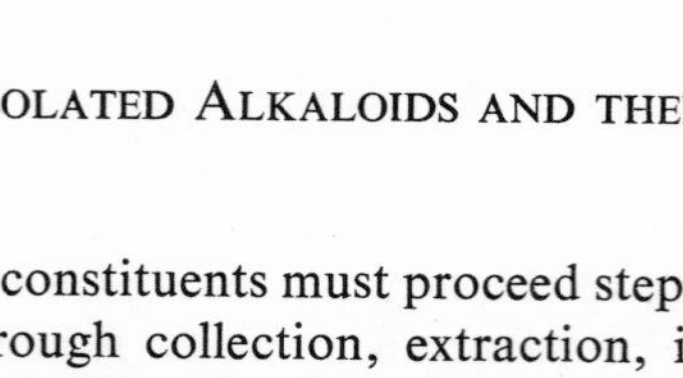

pteropodine **25**

24.2 The Isolated Alkaloids and their Chemistry

A study of plant constituents must proceed stepwise from selection of a material plant, through collection, extraction, isolation of the components, and structure determination. Enormous amounts of data have accumulated during the last 20 years in the indole alkaloid field and now in many cases the structures can be tentatively assigned by a combination of mass and nmr spectroscopy. The absolute configuration may be determined by ord or cd spectroscopy. Of course, it is not always possible to deduce the structure in this way. In our laboratory the relatively complicated structure of gardneramine **5** was fully clarified only after X-ray structural analysis.

Biosynthesis and Biosynthetic Transformations of Indole Alkaloids

No biosynthetic study has been done in this laboratory. Many reports have shown that the indole alkaloids of the three families, Loganiaceae, Apocynaceae and Rubiaceae, are biosynthesized by a general route which begins with the condensation of tryptamine **26** and secologanin **27** and proceeds through strictosidine **28**.

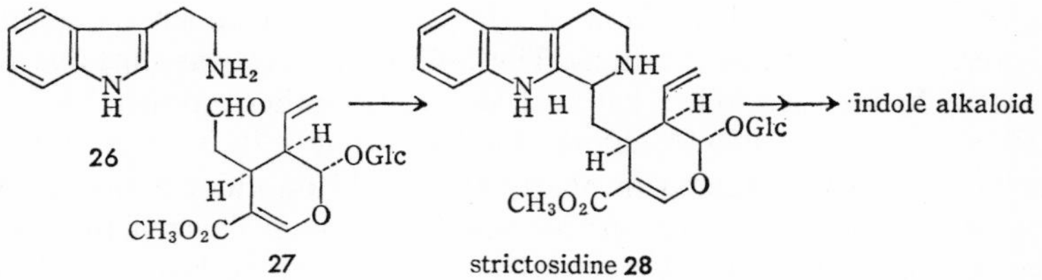

26 + **27** → strictosidine **28** → indole alkaloid

Recently interesting transformations were reported by Brown *et al.*, who imitated *in vitro* the final stages of natural biosynthesis.[7]

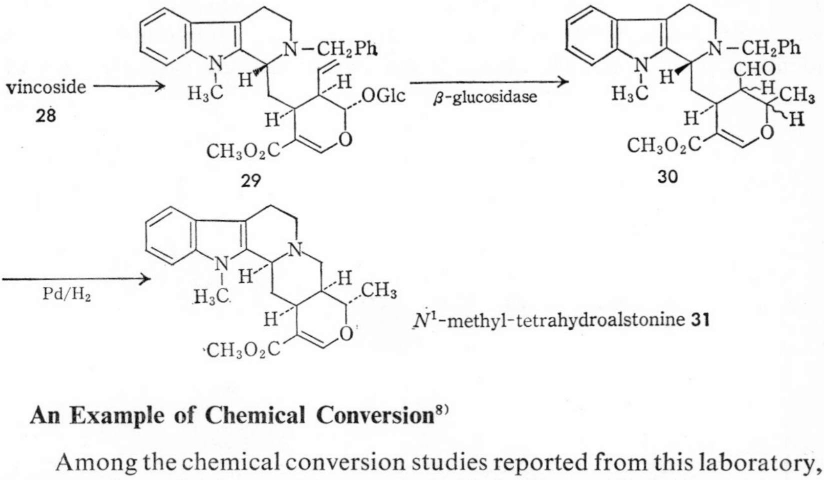

vincoside **28** → **29** —β-glucosidase→ **30** —Pd/H2→ N^1-methyl-tetrahydroalstonine **31**

An Example of Chemical Conversion[8]

Among the chemical conversion studies reported from this laboratory, a conversion of geissoschizine methyl ether **23**, an alkaloid of *Uncaria rhynchophylla*, to alkaloids of the pleiocarpamine group (pleiocarpamine **11** was found in *Amsonia elliptica* in this laboratory) will be described here as an example. Hydrogen chloride gas was introduced into an acetone solution of geissoschizine methyl ether **32a** = **23** to saturation and the solution was left to stand at –13°C for 24 hr. Demethylation took place. After protection of the resulting enol moiety by treatment with ethyl chloroformate, the product was treated with freshly distilled (reduced pressure) cyanogen bromide in chloroform containing ethanol. The resulting C/D ring cleaved product **33a** was treated with sodium hydroxide-methanol to remove the protecting group, giving **33b** (44% from **32b**). Cleavage of the C/D ring system to a ten-membered ring was done to enhance the conformational flexibility, permitting the formation of a bond from C-16 to C-7 or N-1. We next planned to chlorinate **33b** at C-7 using tertiary butyl hypochlorite and to form a bond between C-16 and C-7 by base catalysis for the formation of **37**. Chlorination under ice cooling, however, provided three products. The 7–Cl derivative was unstable and was easily hydrolyzed to a 3–keto derivative. The other products were the 16–Cl and 16–deformyl-16–chloro **33c** derivatives. It is interesting to note that **33c** was obtained in nearly 80% yield on chlorination of **33b** with the same reagent under dry ice/acetone cooling. This reaction probably proceeded by a radical mechanism. The attempted bond formation between C-7 and C-16 by removal of hydrogen chloride to yield a picraline type alkaloid **37** failed under various conditions of base catalysis. However, with sodium hydride in warm DMSO, bond formation between C-16 and N-1 occurred and **34** having 16β-H was obtained in ca. 40% yield. Formation of the epipleiocarpamine **35b** ring system was achieved by treatment with AcOH–AcONH$_4$ under heating. Compound **35b** was identical with a sample obtained by base-catalyzed isomerization of

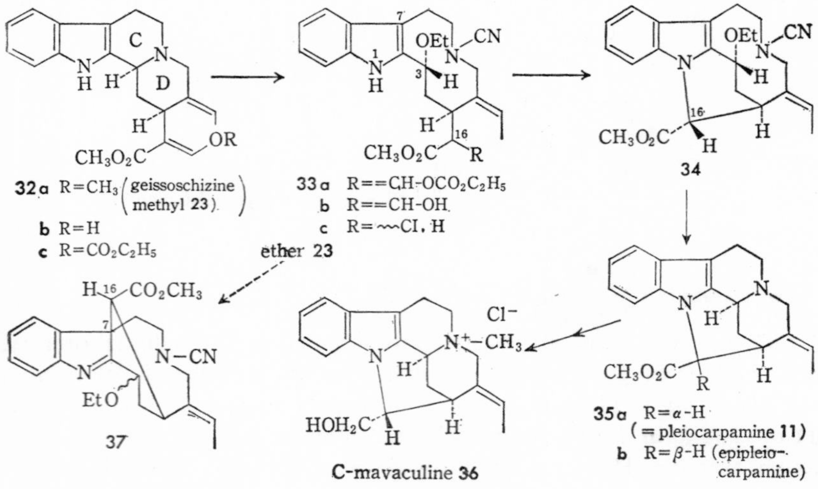

32a R=CH_3 (geissoschizine methyl 23)
b R=H
c R=$CO_2C_2H_5$

33a R==CH-$OCO_2C_2H_5$
b R==CH-OH
c R=~~Cl, H

34

ether 23

37

C-mavaculine **36**

35a R=α-H (= pleiocarpamine **11**)
b R=β-H (epipleiocarpamine)

natural pleiocarpamine **35a** = **11**, which was isolated from *Amsonia elliptica*. C-Mavacurine **36**, a base of *Strychnos* spp., was obtained from **35b** by Schmid and Hesse and thus we also succeeded in the formal partial synthesis of **36**.[9]

In the above and other chemical conversions we looked for biogenetically patterned routes where possible, but this was not always successful. Recently many researchers on indole alkaloids have become interested in this kind of partial synthetic work. A partial synthesis of an antitumor bisindole alkaloid, vincristine, is a particularly important accomplishment.[10]

24.3 Concluding Remarks

As described in the begining of this article, recent studies on indole alkaloids have adopted various kinds of approaches. They include simple synthesis of useful medicinal drugs, total synthesis of alkaloids, and the exploitation of new synthetic reactions. There is also great interest in mimicking biogenetic routes in the laboratory. This is important, since a slight structural difference can often cause a reaction to go in an entirely different direction from that expected. Such problems are generally not completely susceptible to theoretical analysis as yet, so experimental work on chemical transformations still has much to offer.

Collection of the plant materials described here was made under the guidance of Prof. Haginiwa, division of pharmacognosy, this faculty.

References

1) S. Sakai, *Heterocycles*, **4**, 131 (1976).
2) J. S. Glasby, *Encyclopedia of the Alkaloids*, vols. 1, 2, Plenum Press (1975); *see also* R.H.F. Manske, *The Alkaloids*, vols. 1–17, Academic Press (1950–1979); J. A. Joule, *Indole Alkaloids*, *Specialist Periodical Reports*: *The Alkaloids*, vols. 1–8, The Chemical Society (1971–1978); N. Neuss, *Physical Data of Indole and Dihydroindole Alkaloids*, Eli Lilly Co. (1960–1974).
3) R. F. Raffauf, *A Handbook of Alkaloids and Alkaloid-Containing Plants*, John Wiley & Sons (1970).
4) M. Hesse, *Indol alkaloid in Tabellen*, vols. 1, 2, Springer-Verlag (1964, 1968); M. Hesse, *Indol alkaloide, Fortschritte der Massenspektrometrie* (ed. H. Budzikiewicz), vol. 1, Verlag Chemie (1974).
5) A. Engler, *Die Natürlichen Pflanzenfamilien*, IV, Abt. 1, 4, 5, Wilhelm Engelmann (1891, 1897); Y. Satake, *Shokubutsuno Bunrui*, (in Japanese), Daiichi Hoki Shuppan (1965).

6) *A. Engler's Syllabus der Pflanzenfamilien*, II. Band, Gebrüder Borntrageger (1964); K. Watanabe, *Shokubutsu Bunruigaku* (in Japanese), Kazama Shoten (1966); H. Ito, *Shin Kotoshokubutsu Bunrui Hyo*, (in Japanese), Hokuryu Kan (1968).
7) R. T. Brown, C. L. Chapple, *Chem. Commun.*, **1973**, 886; *ibid.*, **1974**, 740; *ibid.*, **1974**, 756; *ibid.*, **1974**, 929; *ibid.*, **1975**, 295.
8) S. Sakai, N. Shinma, *Heterocycles*, **4**, 985 (1976).
9) M. Hesse, W. V. Philipsborn, D. Schumann, G. Spiteller, M. Spiteller-Friedmann, W. I. Taylor, H. Schmid, P. Kerrer, *Helv. Chim. Acta*, **47**, 878 (1964).
10) P. Potier, N. Langlois, Y. Langlois, F. Guéritte, *Chem. Commun.*, **1975**, 670; J. P. Kutney, A. H. Ratcliffe, A. M. Treasurywala, S. Wunderly, *Heterocycles,* **3**, 639 (1975); N. Langlois, P. Potier, *Tetr. Lett.*, **1976**, 1099; J. P. Kutney, J. Balsvich, G. H. Bokelmann, T. Hibino, I. Itoh, A. H. Ratcliffe, *Heterocycles*, **4**, 997 (1976); N. Langlois, F. Guéritt, Y. Langlois, P. Potier, *J. Am. Chem. Soc.*, **98**, 7017 (1976).

CAHPTER 25

Isolation of Higenamine, a Cardiac Component of *Aconitum*

Monkshood, *Aconitum*, is a perennial plant of the family Ranunculaceae. The mother root and daughter root of the plant are called "*Uzu*" and "*Bushi*", respectively, in oriental medicine, and are stated to have cardiaconic, diuretic, excitation and analgesic activities. They are also well known to be strongly toxic, and have been used as an arrow poison used among the Ainu. "*Bushi*" is said to be the natural drug used as a final measure in oriental medication; it occupies an important position comparable to that of camphor injection in western medicine. Thus, "*Bushi*" is frequently used as a drug in spite of its high toxicity.

Research on the composition of the root of monkshood in the past has been mostly on the aconite alkaloids as toxic components in the root, and several tens of alkaloids have so far been isolated and their structures determined. There has, however, been little research on its beneficial components.

A pharmacological study on "*Bushi*" as a crude medicine was first reported in 1958 by Yakazu.[1] He attempted to isolate the cardiac component from "*Bushi*", and fractionated the methanol extract into a petroleum ether-soluble fraction, alkaloids, and a water-soluble fraction. Cardiac activity was tested by the method of Yagi using isolated heart of toad (*Bufo bufo japonicus*), and was found in the water-soluble fraction, from which alkaloids had been eliminated. The active component of monkshood has been further fractionated into the chloroform-insoluble portion of the above water-soluble fraction. "*Bushi*" produced in China was also found to have a cardiac effect on various animals.[2] Takahashi *et al.*[3] attempted the separation of the active component by chromatography on silica gel and ion exchange resin, but isolation and identification of the substance were not achieved. In view of the importance of monkshood as a natural drug, the author attempted the isolation and structure determination of the cardiac component, and obtained the active principle, named Higenamine. This chapter deals with the isolation of Higenamine.

25.1 Techniques for the Isolation of Physiologically Active Components from Natural Materials

There are two approaches in isolating physiologically active components from natural sources: one is to separate as many components as possible based on their chemical nature and to screen each component for specific physiological activity by bioassay, while the other is to apply bioassay to all fractions obtained through the separation procedures and to concentrate on the fraction that shows the strongest activity until the active component is isolated. The former apporach is suitable when the component is stable and is present in the raw material in relatively high amount, but is inappropriate when the component is unstable or is present in only a small amount, since it may often be overlooked or lost during the isolation. This approach has, however, been widely used in the study of natural products since it is not necessary for the researchers who isolate the materials to carry out extensive bioassays.

The latter approach is obviously better than the former as a technique for isolating physiologically active components. However, most natural products studies have been carried out by organic chemists who are unfamiliar with bioassay techniques. It is now possible to utilize simple bioassay procedures in many cases, and advice may be taken from experts in pharmacology regarding the simple bioassay which is most appropriate as an indicator for the target active component.

25.2 The Bioassay Used in the Present Study

Cardiac activity has been tested by Straub's method using the hearts of leopard frogs (*Rana nigromaculata*) according to Yakazu and Lao, by Yagi's method using the hearts of toads (*Bufo bufo japonicus*), and by Langendorff's method using the hearts of guinea pig (*Cavla porcellus*) or rabbits (*Oryctolagus cuniculus*). A variation of Yagi's method using the hearts of *Rana nigromaculata* was selected in the present study. Hearts were taken from leopard frogs by the usual procedure, and the contractual force of the heart muscle was recorded on a chimograph. Test materials were applied after dissolving in Linger's solution. The minimum activation concentration (MAC), the lowest concentration of the sample clearly enhancing the contractual force, was taken as the indicator of cardiac activity.

25.3 Isolation of Higenamine

Selection of Monkshood Plants

There are many species within the genus *Aconitum*, and the number of species existing in Japan is said to be about fifty. The taxonomy of the plants is not yet complete. Types of diterpenes and alkaloids vary by species, collection site, and collection time.

We looked for a species of *Aconitum* with a high content of active component as a starting material, using the isolation method of Yakazu (Table 25.1).

TABLE 25.1. Isolation method for chloroform-insoluble substance according to Yakazu.

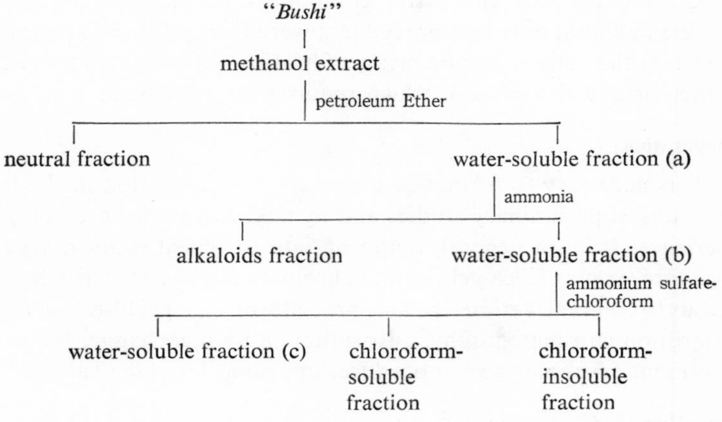

The "chloroform-insoluble substance" obtained in this method is the fraction that is salted out by ammonium sulfate from the water-soluble fraction after eliminating alkaloids from the methanol extract. For convenience, the cardiac activity was tested using the water-soluble fraction (b). The results of cardiac activity tests showed poor reproducibility among different lots of "*Shirakawa Bushi*" and "*Ho Bushi*" (commercial products), and also *Bushi* obtained from Katsuyama City in Fukui Prefecture and from Shizuoka Prefecture at different collection times and sites. However, *Aconitum japonicum* THUMB collected at the end of June in Aikawa Town, Sadogun, Niigata Prefecture always gave stable cardiac activity. Only one species of *Aconitum* is grown in Sado, and forms a very large colony. Thus, the species is convenient for collection, and was judged to be suitable for the study.

Estimation of the Properties of the Cardiac Component

The results of various tests with the water-soluble fraction (b) led us to consider that the active component had the following properties.

(1) Soluble in water, methanol, acetic acid, and pyridine: insoluble in chloroform, and acetone.
(2) Extremely labile in alkali. The activity is lost immediately on treatment with caustic alkali or alkali carbonate.
(3) Relatively stable to acid and heat.
(4) Strongly adsorbed or decomposed on commonly used separating agents such as alumina, silica gel, active charcoal, cellulose powder, and Sephadex G, and difficult to elute efficiently. It became evident that these separating reagents were inappropriate. Sephadex LH-20 had no effect on the activity, and appeared to be suitable.

Considering that the active component was soluble in water and unstable in alkali, only two procedures were thought to be applicable for separating the active component; gel filtration with Sephadex LH-20 and countercurrent distribution, which requires no adsorbent.

Separation

It is necessary to formulate a strategy for separation in the light of the results of preliminary studies and by trial and error based on general experience. It is, in general, inappropriate to repeat a given separation procedure such as silica gel column chromatography. It is usually advantageous to combine various feasible procedures, e.g., gel filtration followed by partition chromatography, adsorption and ion exchange. However, in the present case, only two procedures appeared feasible (Table 25.2).

Isolation of Higenamine

Roots of aconitum (700 kg) collected at the end of June in Sado were dried (150 kg), ground, and extracted three times with methanol under reflux. The extract was concentrated under reduced pressure, yielding a dark brown viscous material (25 kg). The methanol extract was dissolved

TABLE 25.2. Isolation procedures for higenamine.†

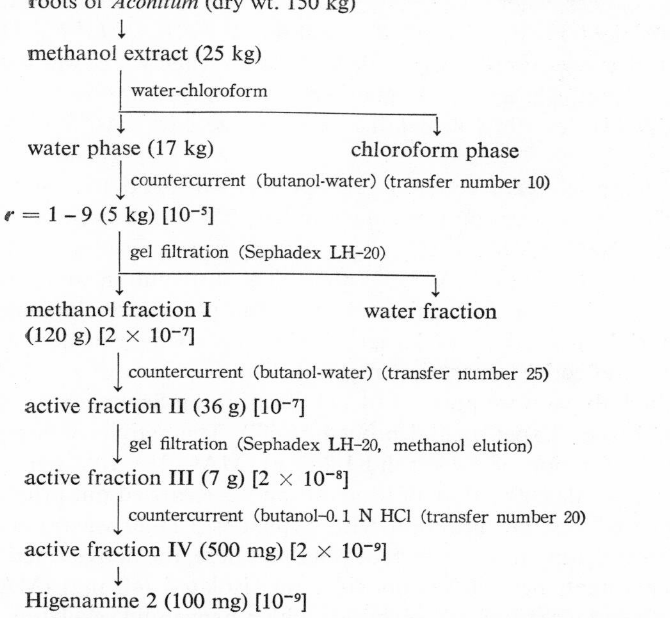

† Values in parentheses are yields; those in square brackets are MAC values (g/ml).

in 5 volumes of water, and then extracted with the same volume of chloroform. The water phase was concentrated under reduced pressure to give the water extract (17 kg). This was subjected to countercurrent distribution fractionation (transfer number, 10) with an equilibrated solvent system of butanol-water, and was separated into 11 fractions of $r = 0$–10. Here $r = 0$ was the fraction that was more soluble in water and remained in the first distribution tube when the upper phase was the moving phase, and $r = 10$ was the fraction that was more soluble in butanol and moved to the eleventh distribution tube. Cardiac activity was recognized in the fractions of $r = 1$–9 (5 kg) (MAC 10^{-5} g/ml), but not in $r = 0$ or $r = 10$. This procedure successfully removed most of the inorganic substances and sugars ($r = 0$), and lipid-soluble substances ($r = 10$).

The fractions of $r = 1$–9 (5 kg) were purified by gel filtration on a column of Sephadex LH-20 prepared with water (see "Experimental procedure 1"). Elution was done first with water and then with methanol. Cardiac activity was concentrated in the methanol-eluted fraction I (120 g) (MAC/2×10^{-7} g/ml). This separation is not an ordinary gel filtration, but utilizes the weak adsorptive ability of Sephadex and the strong adsorbability of the active component. Usual gel filtration is applied for the separation of substances where $1 \geq Kd \geq 0$, where Kd is the value of the partition coefficient in the theoretical equation, $V_e = V_o + Kd.V_i$. The Kd value of the active component was larger than 3 when water was used as an eluting solvent, while it was smaller than 2 when methanol was used. (This procedure is described in detail later). The activity was increased 50–fold.

The methanol-eluted fraction I (120 g) was subjected to counter-current distribution separation (transfer number, 25) with butanol-water (see "Experimental procedure 2"). The activity was collected in the fractions of $r = 5$–18 (36 g) (MAC 10^{-7} g/ml). The distribution coefficient of the active fraction was approximately 1.0. The procedure eliminated inactive material in fractions of larger distribution coefficient, and was found to be an effective separation method. The fractions of $r = 5$–18 (36 g) described above were purified by gel filtration on Sephadex LH-20 with methanol (see "Experimental procedure 3"). The activity was concentrated in the fraction of Kd around 1.2 (7 g) (MAC $2 = 10^{-8}$ g/ml).

When large-scale extraction and separation were carried out prior to the above procedures, gel filtration with Sephadex LH-20 was applied repetitively to the same fraction as that obtained here, the active fraction III. One component, named Yokonoside, was isolated (40 mg) (MAC 2×10^{-8} g/ml) and identified as a glycoside with a benzanilide skeleton, **1**.

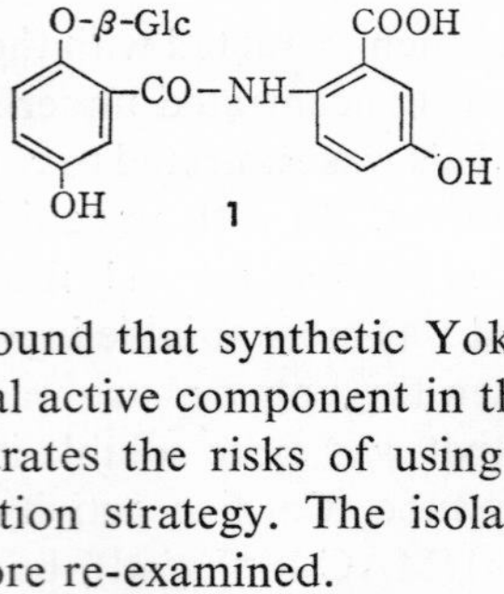

O-β-Glc COOH
CO–NH
OH OH
1

However, it was later found that synthetic Yokonoside had no cardiac activity, and that the real active component in that fraction was a minor contaminant. This illustrates the risks of using mere repetition of one procedure as a purification strategy. The isolation procedures beyond fraction III were therefore re-examined.

Since the active fraction III was assumed to contain yokonoside and other glycosides in relatively high amounts, separation was attempted after derivatizing the fraction into lipid-soluble form by acetylation. The active fraction (300 mg) was treated with acetic anhydride-pyridine, and the chloroform-soluble portion (200 mg) was subjected to silica gel chromatography. This yielded a large amount of acetate fraction, Ac–A (131 mg), on elution with chloroform, and a small amount of another fraction, Ac-B (14 mg), on elution with chloroform-methanol. Cardiaic activity was not found in Ac-A, but was present in Ac-B to some extent (MAC 10^{-3} g/ml). Both fractions were hydrolyzed with hydrochloric acid. A large amount of glucose was identified in the hydrolysate of Ac-A, but none in that of Ac-B. No activity was detected in the hydrolysate of Ac-A, while activity was found in the hydrolysate of Ac-B, almost equal to that of Ac-B itself.

Thus, the following conclusions were reached.

(1) The amount of the actual active component is less than about 10% of the total amount of the active fraction III.

(2) The active component is not a glycoside since no glucose was detected in the hydrolysate of Ac-B.

(3) The active component is fairly stable to acid, or the hydrolysis products of the active component also have strong activity, since no decrease in activity occurred on hydrolysis of Ac-B with HCl.

The isolation of the active component thus required removal of the innactive glysocide, assumed to represent more than 90% of the active fraction III.

Glycosides in active fraction III were hydrolyzed to aglycone and sugar on acid treatment. The distribution coefficient of the aglycone in the solvent system of butanol-water became higher than that of the glycoside, while that of the sugar was smaller than 0.1. The distribution coefficient of the active component was, on the other hand, larger than 1.0, and, consequently, separation of the aglycone, sugar and the active component should be possible, assuming the distribution coefficient of the active component is unaltered.

The active fraction (600 mg) was hydrolyzed with HCl and subjected to the countercurrent distribution (transfer number, 20) with butanol-water. The active component was obtained in the fractions of $r = 4$–6; this was active fraction IV (42 mg) (MAC 2×10^{-9} g/ml). Colorless platelet crystals (8 mg) were fortunately obtained from this active fraction.

The distribution coefficient of the active fraction was near 0.5, and was smaller than that of active fraction III, 1.0. This led us to consider the possibility of chemical modification or the formation of the hydrochloride salt of the active component. Countercurrent distribution (20 transfers) with butanol-0.1 N hydrochloric acid was carried out to check this. The activity was collected in the fraction of $r = 4$–6, and identical crystals were obtained. Recrystallization from methanol gave chlorine-containing colorless platelet crystals of mp 260° C (decomposition). When hydrobromic acid was used in place of hydrochloric acid bromine-containing colorless platelet crystals of mp 267° C were obtained in the same manner. Thus, the former crystals were the hydrochloride salt of the active component, and the latter the hydrobromide salt. The crystals showed activity even at such a high dilution as 10^{-9} g/ml, and no activity change was observed after recrystallization three times. Thus, the substance was judged to be the cardiac component of the root of *Aconitum*, and was named Higenamine.

Higenamine was identified[4,5] as *dl*-demethyl coculaurine **2** on the basis of spectroscopic studies and comparisons with various samples.

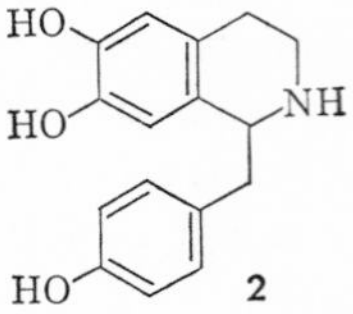

HO HO NH HO **2**

[Experimental procedure 1] Sephadex LH-20 gel filtration

Sephadex LH-20 (230 g) sufficiently swollen in 1.5 l of methanol (requiring about 5 hr) was filtered by suction, then washed three times with equal volumes of methanol. The methanol was replaced by water in the following manner. After adding 1.5 l of distilled water, the suspension was stirred vigorously, filtered by suction, and washed with an equal volume of water. The gel was then transferred into a pear-shaped 3 l flask, and 1.5 l of water was added. Degassing was accomplished by reducing the inside pressure with an aspirator, with occasional shaking (20 min). The degassed gel was carefully packed into a column, through which purified water was passed for 5–7 hr to stabilize the column bed. For fine separation, the eluting solvent was also degassed. The volume of the column bed used here was approximately 950 ml (90 mm × 150 mm).

The fractions of $r = 1$–9 (125 g) were dissolved in 60 ml of purified water and applied to the top of column. Elution was carried out with purified water at a flow rate of 5 ml/min. The eluent was changed to methanol after passing 1700 ml of water, and 1500 ml of methanol was passed. The methanol fraction was concentrated under reduced pressure, yielding 3.0 g of dark brown viscous material. The cardiac activity of this material was equivalent to MAC 2×10^{-7} g/ml.

The *Kd* value of the active portion was calculated from the relation

$$V_t = V_0 + V_i + V_g,$$

where V_t is the total column bed volume, V_0 is the void volume (interparticular clearance volume), V_i is the intraparticular clearance volume, and V_g is the real volume of gel particles. V_0 was calculated to be 237 ml, since $V_t = 950$ ml, $V_g = 230$ ml, and $V_i = 230 \text{ ml} \times 2.1 = 483$ ml.

The solvent volume required to elute the active component, V_e, is given by

$$V_e = V_0 + Kd\ V_i$$

V_e was greater than 1700 ml with water as a solvent, and $Kd > 3.0$. Similarly, V_e for methanol elution was smaller than 1500 ml – 237 ml = 1263 ml, and $Kd < 2.0$.

[Experimental procedure 2] Separation of methanol fraction I by countercurrent distribution

Butanol and water were well shaken in a separating funnel, and the upper and lower phases were separated after standing overnight. Methanol fraction I (10 g) was dissolved in 100 ml of the lower phase. The solution was placed in the first distribution tube (200 ml separating funnel), and the upper phase (100 ml) was added. Distribution was carried out by shaking the funnel gently. Vigorous shaking would result in the formation of an emulsion, which would require a long period for phase separation. A long, cylindrical separation funnel is effective, and distribution can be performed by repeated inversion. On completing the first distribution, the content of the tube was separated into upper and lower phases by standing. The lower phase was then transferred into the next distribution tube, in which the upper phase had already been placed, and fresh lower phase (100 ml) was added into the first distribution tube. Each distribution tube was shaken as before. The procedures were repeated 25 times, and corresponding fractions were obtained. The solvent in each fraction was removed under reduced pressure, and the residue was weighed precisely after drying. Fig. 25.1 shows the weight distribution (solid line) and cardiac activity (broken line).

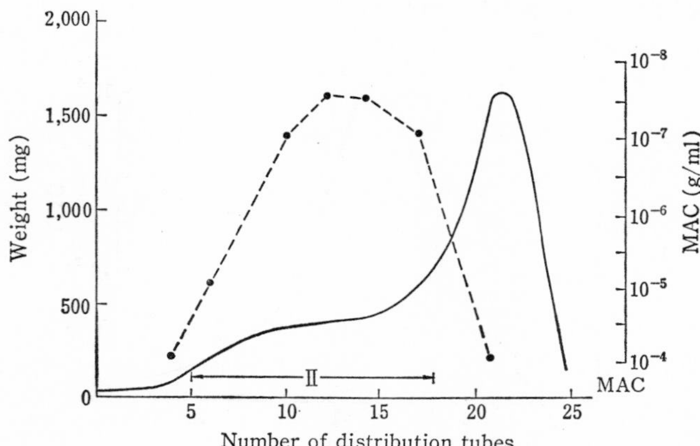

Fig. 25.1. Separation of methanol fraction I by countercurrent distribution. The broken line indicates cardiotonic activity (MAC).

[Experimental procedure 3] Separation of active fraction II

Sephadex LH-20 (400 g) was swollen in methanol (3 l), and packed into a column after washing and degassing. The column bed was stabilized by passing methanol for 7 hr. The bed volume was 1750 ml (45 mm × 1100 mm). Methanol solution of active fraction II (6.6 g in 20 ml of MeOH) was applied to the top of column, and methanol was passed at a flow rate of 0.57 ml/min. After passing 1000 ml, fractions of 20 ml were collected with a fraction collector. The solvent was eliminated from each fraction, and the residue was precisely weighed after drying. The weight distribution (solid line) and cardiaic activity (broken line) are shown in Fig. 25.2. *Kd* of the active fraction III obtained here was calculated to be approximately 1.2 in the manner described above.

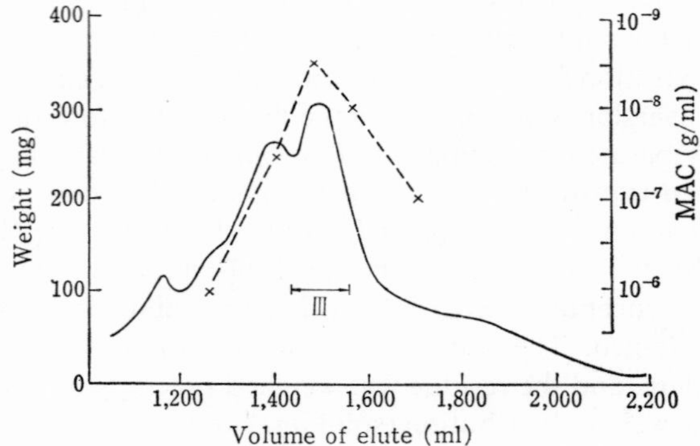

Fig. 25.2. Separation of active fraction II by gel filtration. The broken line indicates cardiotonic activity (MAC).

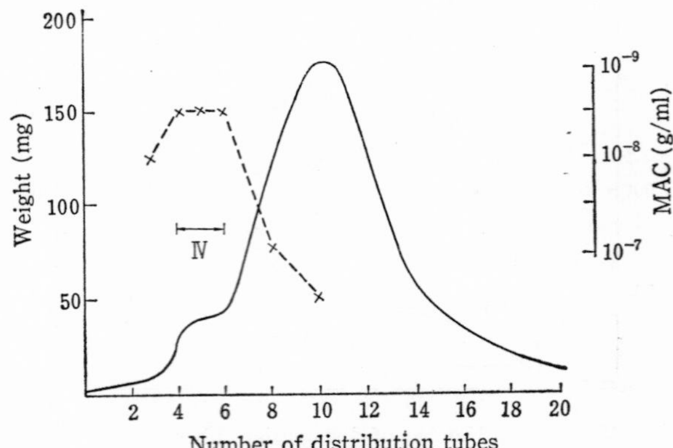

Fig. 25.3. Separation of active fraction III by countercurrent distribution. The broken line indicates cardiotonic activity (MAC).

[Experimental procedure 4] Separation of active fraction III by countercurrent distribution

Fully automated countercurrent distribution equipment (manufactured by Mitamura Riken) with distribution tubes of 50 ml was used with a butanol-0.1 N hydrochloric acid system (10 ml in each phase; transfer number, 20). The solvent was removed from each fraction after completing the distribution, and the residue was precisely weighed. Fig. 25.3 shows the weight distribution (solid line) and cardiaic activity distribution (broken line).

25.4 Concluding Remarks

Fundamental considerations in preparing for the isolation of a physiologically active component from natural material have been discussed, using our study on the cardiaic component in the root of monkshood, *Aconitum*, as an example. The isolation of Yokonoside provides an excellent illustration of the pitfalls that can arise in attempting to isolate active components.

In general, tests of homogeneity must be treated with care, and it is rather important to obtain the active component in crystalline form, if possible. Even though the present study took over 10 years, the author feels confident that the isolation of even an unstable compound present in extremely small quantities can be achieved with the aid of experience and modern analytical and separative techniques.

References

1) S. Yakazu, *J. Pharm. of Japan,* **54**, 895 (1958).
2) M. Lao, *Acta Pharma aceutica Sinica*, **13**, 195 (1966).
3) S. Takahashi, *Annal Rept. Tokyo Coll. Pharm.*, **12**, 180 (1962).
4) T. Kosuge, *Farumashia*, **11**, 103 (1975).
5) T. Kosuge, M. Yokota, *Chem. Pharm. Bull.*, **24**, 176 (1976).

CHAPTER 26

The Constituents of Plant Tissue Cultures and their Biosynthesis

Until recently, organic natural products chemists were not very familiar with tissue cultures of higher plants. However, recent advances in chromatography, especially in high-performance liquid chromatography, have made the isolation of pure chemical substances from very small quantities of natural materials easier. Structure determination methods have also improved, with computer-assisted systems such as FT-nmr, FT-ir, etc. Workers engaged in the chemical synthesis of natural products have been aiming at more complex structures, and also utilizing biomimetic synthesis, modeled on the reactions in biological systems. In addition, interest has developed in the biological significance of low molecular weight organic compounds in living cells form the viewpoint of physiological activity or biosynthetic consequence. As a part of this process, greater interest has appeared in plant tissue culture methods.

Many investigations on plant tissue cultures have been carried out with the aim of utilizing them as a source of food or medicine, and reviews have been published.[1] For medical applications, the cultured tissue of medicinal plants should contain the same components as the intact plants. Sometimes the quantities of various compounds in cultured cells differ significantly from those of the intact plants, or compounds not present in the original plants may even be formed. Variation of the physical or chemical environment of cell cultures often also causes a remarkable variation in chemical constituents. The arising problems here would be the biogenesis of these compounds and each biosynthetical steps mediated by enzymes, and further the mechanisms of control on the expression of these steps, should be the extended subjects to investigate. Biosynthetic studies can be made using intact plants, but callus cultures are attractive in terms of convenience for culture under artificially controlled conditions.

Another aspect of interest to natural product chemists is the application of plant tissue culture techniques to chemotaxonomy, which is closely related to the biosynthesis of secondary products. The callus is a mass of dedifferentiated cells, and appropriate culture methods often cause redifferentiation to whole plants or parts of plants. Dedifferentiated cells of plants closely related by systematic classification should have very similar features in secondary metabolism. Investigations of the changes in metabolism accompanying the redifferentiation to different forms could provide much knowledge about chemical systematics, because chemotaxonomy is based on the principle that peculair morphological forms of plants are reflected in their metabolic pathways. However, these lines of investigation still await improved culture methods able to control the differentiation of cells more effectively.

One more advantage of plant tissue cultures is that chemical investigations on plants which are scarce may be made much easier by this method.

We describe here the results of investigations on callus cultures of *Glycyrrhiza* spp., the plants which yield the crude drug licorice. They provide an example of tissue culture studies, including the isolation of new compounds from the callus, feeding experiments leading to the finding of a previously unknown route of biogenesis, and enzymological investigations with a cell-free extract of the callus.

26.1 Constituents of the Cultured Cells of *Glycyrrhiza echinata*

Crude drug licorice available in Japan is mostly imported from China and is widely used for medicinal purposes, and also as a sweetening agent. The plants of origin are several species of the genus *Glycyrrhiza*. The sweet principle is glycyrrhizin, a well-known triterpenoid saponin. Phenolic components may also be pharmacologically significant, since fractions containing mainly flavonoids, but with a low content of glycyrrhizin, obtained from the extract of licorice root have been shown to have an antiulcer effect. Shibata's group has thoroughly investigated the flavonoids of several kinds of commercial crude licorice, resulting in the isolation of many new flavonoids.[2)] They demonstrated the existence of characteristic compounds for each kind of licorice, affording a good example of chemotaxonomy. Thus, comparison of the thin layer chromatogram pattern of flavonoids is especially useful for identifying the species of origin of a drug, and also for the evaluation of crude drugs. We have been engaged in studies on tissue cultures of various plants for medicinal purposes including licorice. Our initial interest was investigation of the chemical constituents of callus tissue, and then of the differences in constituents between the original palnts and tissue cultures under various conditions.

The alcoholic extract of *G. echinata* callus subcultured in the dark on White's medium containing yeast extract showed many fluorescent spots on tlc, indicating the existence of several phenolic compounds. They were mainly flavonoids, and a compound with marked green fluorescence was the major one; the content of this in the original plant was very low. This compound, named echinatin **1**, was shown to be a new chalcone with an unusual substitution pattern of oxygen functional groups. Echinatin 4-*O*-glucoside was also obtained from the water-soluble fraction. The other compounds in the organic layer were two flavones, 7,4′-dihydroxyflavone **2**, licoflavone A **3**, and an isoflavone, formononetin **4**. Afterwards, in the course of studies on echinatin biosynthesis, further investigation of the minor components of the callus resulted in the isolation of licodione **5**, a dibenzoylmethane derivative of rare natural occurrence.[10]

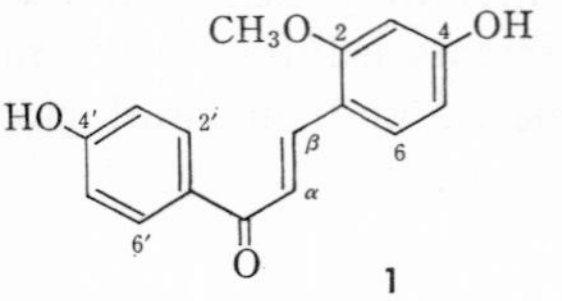

1

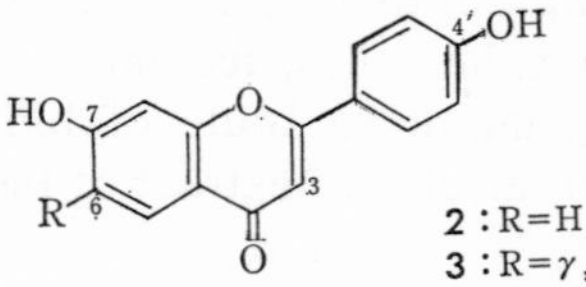

2 : R=H
3 : R=γ,γ- dimethylallyl

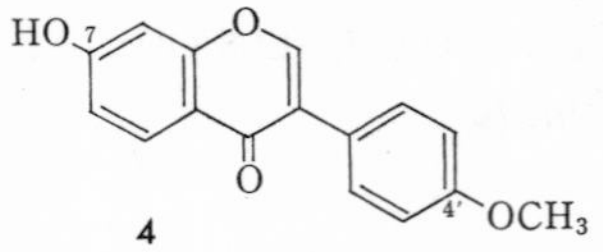

4

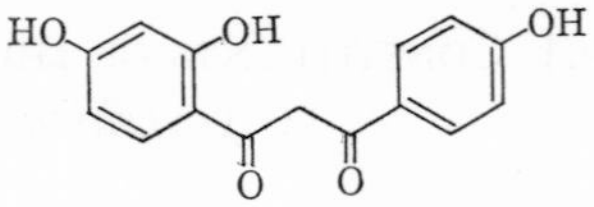

5

Echinatin **1**,[3] yellow needles, mp 209.5–212° (decomp.), was obtained by successive column chromatography on silica gel of the ethyl acetate-soluble fraction of ethanol extract of the callus. The molecular formula was determined to be $C_{16}H_{14}O_4$ by mass spectrometry, and uv and ir spectra clearly indicated the chalcone skeleton with two hydroxy and one methoxy groups. The absence of phenolic hydroxy groups at the 2′ and 6′ positions in ring A, with the hydroxy groups at these positions chelated to carbonyl oxygen, was suggested by the observation that no change occurred in the uv spectrum after the addition of aluminum chloride. Mass spectrometry showed that ring B has a methoxy group in addition to a hydroxyl at C-4, and the characteristic fragment ion $(M - OCH_3)^+$ demonstrated that a methoxyl is present at C-2. The nmr spectrum was also in agreement with this structure, showing an A_2X_2 system of ring A protons and an ABX system arising from ring B protons. On degradation of **1** with alkali, *p*-hydroxyacetophenone and 2-methoxy-4-hydroxybenzaldehyde were afforded. Synthesis of **1** was carried out by the condensation of these two substances under alkaline conditions.

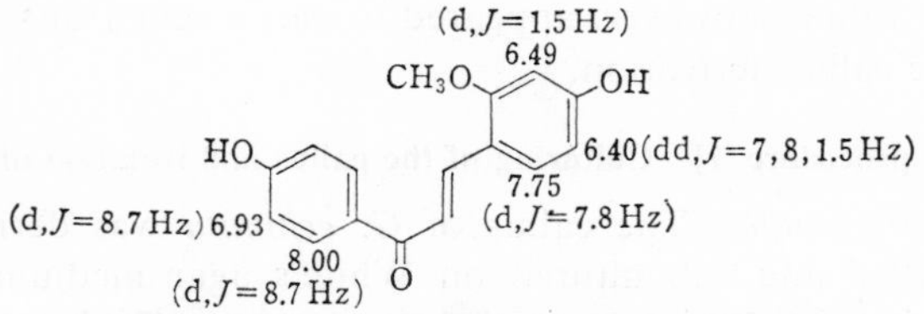

The nmr signals of aromatic protons of echinatin **1**

Licodione **5**,[4] $C_{15}H_{12}O_5$, deep yellow needles, mp 152–153° (decomp.), was also obtained from the ethyl acetate extract, and was shown to be a tetrahydroxychalcone. In fact, the uv spectra in the presence and absence of alkali revealed a chalcone pattern with phenolic hydroxy groups at the *para* position to the C_3 linkage. A tautomeric form of β-diketone C(OH)= CH–C =O was suggested by the IR absorptions at 1620 (sh.) and 1600 (br.) cm^{-1}, and the positive β-diketone color reaction (green color after treatment with phenylhydrazine and reduction with sodium, followed by addition of conc. sulfuric acid and $NaNO_2$) indicated that **5** is a dibenzoylmethane derivative. The predominant peaks in mass spectrometry were due to mono- and dihydroxybenzoyl groups. On alkaline hydrolysis of **5**, *p*-hydroxyacetophenone, resacetophenone, β-resorcylic acid and *p*-hydroxybenzoic acid were obtained, and **5** was readily cyclized by acid with the loss of water to give 7,4′-dihydroxyflavone. The nmr spectrum showed an equilibrium mixture of diketo and keto-enol forms in solution.[11]

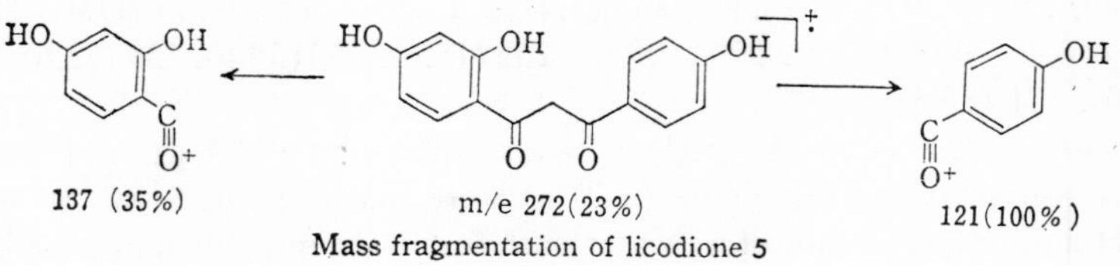

Mass fragmentation of licodione **5**

The existence of echinatin and licodione in the extract of callus or intact plant was easily detected on tlc, since they both showed characteristic colorations. A green fluorescent spot of echinatin became orange-red (visual) on heating after spraying with dil. H_2SO_4, while licodione gave a yellowish fluorescent spot which turned blue under a uv lamp after spraying with H_2SO_4 and slight heating. In addition, tlc confirmed the existence of both compounds in an ethyl acetate extract of the root of the original plant.

The contents of those flavonoids in the cultured cells are 0.01% for echinatin (maximum), of the order of 0.001% for the flavone and isoflavone derivatives, and 0.0003–0.0004% for licodione per fresh weight of callus. These values, however, appeared to vary slightly during subculturing from the callus derivation.

[Experimental procedure 1] Culturing of the callus and isolation of constituents

Culturing of the callus The callus of *G. echinata* was derived from a seedling in 1965 and subcultured on White's agar medium (see Table 28.4) containing 2,4–D (0.1 ppm) and yeast extract (0.1%). It was transferred to fresh medium each 5–6 weeks and maintained in the dark at 26°. When 100 ml Erlenmeyer flasks containing 40–50ml of medium were used, ca. 2 g of callus grew to ca. 10 g per vessel.

Isolation of echinatin **1** The homogenized callus (2.7 kg) was extracted at room temperature and under reflux (twice) with ethanol. To this EtOH extract (76.6 g), water and ethyl acetate were added and the mixture was shaken to yield AcOEt extract (3.3 g). This was subjected to column chromatography on silica gel (410 g) and eluted with chloroform mixed with methanol ($2 \rightarrow 5$%). The fractions (480 mg) containing echinatin (*Rf* 0.31 on silica gel tlc with $CHCl_3$–MeOH, 9:1, 0.24 with benzene–ethyl acetate–methanol–petroleum ether (6:4:1:3) (BEMP)) were further subjected to column chromatography on silica gel (170 g), eluting with chloroform, then a final chromatography on silica gel (80 g; eluent, benzene–AcOEt ($0 \rightarrow 100$%)) gave fractions which showed a single spot on tlc. On recrystallization from ethanol-water, 5.6 mg of yellow needles, mp. 209–212° (decomp.), was obtained. Molecular formula, $C_{16}H_{14}O_4$ (M^+ 270.0896); uv λ_{max}^{EtOH} nm (log ε); 237 (3.79), 312 (3.94), 370 (4.20); $\lambda_{max}^{EtOH-EtONa}$ nm (log ε); 252 (3.79), 271 (3.80), 435 (4.41); ir ν_{max}^{KBr} cm^{-1}: 3400 (OH), 1635 (C = C–C =O); ms *m/e* 270 (100%), 255 (18.6), 240 (58.6), 239 (82.6), 177 (32.0), 121 (85.3).

Synthesis of Echinatin **1** *p*-Hydroxyacetophenone (0.55 g) and 2-methoxy–4-hydroxybenzaldehyde (0.50 g) were dissolved in 8 ml of 70% KOH and heated on a boiling water bath for 10–15 min, then poured into ice-water followed by slow addition of dil. HCl, yielding a yellow precipitate. The mixture was left to stand overnight, then the precipitate was collected by suction filtration and recrystallized from EtOH–H_2O. Yellow needles (0.30 g) were obtained.

Isolation of Licodione **5** The callus (17.0kg), cultured and harvested for 6 months, was homogenized and extracted with cold and then hot MeOH (twice). Ethyl acetate extract (20.6 g) from the MeOH extract was chromato graphed on silica gel (2.2 kg), eluting with $CHCl_3$–MeOH (1 → 50%). Fractions containing licodione (*Rf* 0.39 on silica gel plates with $CHCl_3$–MeOH (9:1), 0.28 with BEMP) were again subjected to silica gel column chromatography, eluting with benzene–AcOEt (0→100%). Fractions containing licodione (75 mg) were collected and recrystallized from EtOH–H_2O to yield deep yellow needles (53.7 mg), mp 152–153° (decomp.). Molecular formula, $C_{15}H_{12}O_5$ (M^+ 272.0692), uv λ_{max}^{MeOH} nm (log ε): 285 (4.3), 376 (4.6), $\lambda_{max}^{MeOH-MeONa}$ nm (log ε): 242 (4.1), 342 (4.7), ir ν_{max}^{KBr} cm^{-1}: 3380 (OH), 1625 (sh) and 1600 (br) (C(OH) = CH–C = O), MS*m/e* 272 (23.3%), 255 (6.4), 254 (5.3), 163 (4.5), 137 (34.6), 121 (100).

26.2 Biosynthesis of Retrochalcone

Echinatin **1** and licochalcones A **6** and B **7**,[5] isolated from Sinkiang licorice by Shibata *et al.*, have unusual oxygen substitution patterns without O-functional groups at C-2′ in ring A and with *meta* substitution of the resorcinol type in ring B.

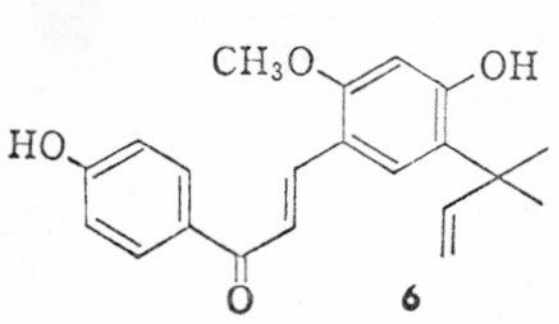

CH₃O OH HO O **6**   OH CH₃O OH HO O **7**

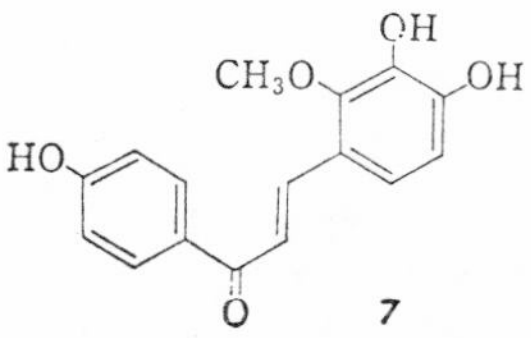

Flavonoids are known to be biosynthesized via a C_{15} intermediate, generated by the condensation of one molecule of *p*-coumaroyl CoA and three molecules of malonyl CoA, and ring A is formed via the acetate-malonate pathway, having hydroxyls at the *ortho* and *para* positions of C_3 linkage, while ring B arises from the shikimate pathway, having oxygens at *para* and occasionally *meta* positions. However, these substitution patterns are reversed in the chalcones **1, 6, 7**, which were designated as retrochalcones, in which the origins of the two rings are opposite to those of normal flavonoids, with the inversion of the ketone and olefin of the C_3 moiety after the C_{15} intermediate stage. This proposal for retrochalcones was confirmed by tracer studies carried out by us and Shibata's group.[6]

In biosynthetic experiments, the acquisition of suitable plant materials is important. The intact original plant of Sinkiang licorice, which grows in China, is not available in Japan. The fresh root of *G. echinata* contains only a small amount of echinatin. In contrast, *G. echinata* callus produces echinatin as a major phenolic component under certain culture conditions. Moreover, the famous studies of Hahlbrock[7] on the enzymes involved in flavone glycoside biosynthesis using cultured cells of parsley suggests the possibility of future studies on retroflavonoid biosynthesis with cell-free systems from callus.

Cinnamic acid (biosynthetically 2 steps before *p*-coumaroyl CoA) labeled with ^{14}C at the carboxyl group (C-1) or at C-3 was fed to suspension cultures of *G. echinata* callus and after shaking for 6 days the callus was extracted, and radioactive echinatin was isolated. A part of this was used for counting radioactivity and the rest was degraded with alkali to determine the location of the labeled atom in the molecule.

Table 26.1 shows clearly that the carboxyl carbon of cinnamate was incorporated into the β-position of echinatin while the carbon at C-3 of cinnamate was incorporated into the carbonyl carbon of echinatin. This result implies that the C_3 unit of echinatin is biosynthetically reversed compared with normal chalcones.

TABLE 26.1 Incorporation of ^{14}C-labeled cinnamic acid into echinatin and the distribution of radioactivity (^{14}C) in echinatin.

	Precursors	
	[3-^{14}C] Cinnamic acid (0.05 mCi)	[1-^{14}C] Cinnamic acid (0.05 mCi)
Echinatin:		
specific activity (dpm/mM)	1.46×10^6	5.22×10^6
total incorporation ratio (%)	0.08	0.52
specific incorporation ratio (%)	0.0013	0.056
Echinatin:		
specific activity (dpm/mM)	1.46×10^6	5.22×10^6
distribution (%)	100	100
p-Hydroxyacetophenone:		
specific activity (dpm/mM)	1.82×10^6	nil
distribution (%)	125	0
2-Methoxy-4-hydroxybenzaldehyde:		
specific activity (dpm/mM)	0.09×10^6	5.53×10^6
distribution (%)	6	106

A tracer experiment with tritium-labeled isoliquiritigenin, a normal chalcone corresponding to an earlier stage of flavonoid biosynthesis, was carried out by Shibata's group in the same manner. The results (Table 26.2) indicates that ring B of isoliquiritigenin was converted into ring A of echi-

TABLE 26.2. Incorporation of [3,5-T] isoliquiritigenin into echinatin and the distribution of radioactivity (T) in echinatin.

Isoliquiritigenin: total activity	8.56 mCi
Echinatin: specific activity (dpm/mM)	2.63×10^8
total incorporation ratio (%)	0.11
specific incorporation ratio (%)	1.38
Echinatin†: specific activity (dpm/mM)	4.55×10^5
distribution (%)	100
p-Hydroxyacetophenone:	
specific activity (dpm/mM)	3.89×10^5
distribution (%)	85.5
2-Methoxy-4-hydroxybenzaldehyde:	
specific activity (dpm/mM)	nil
distribution (%)	0

† Diluted with carrier.

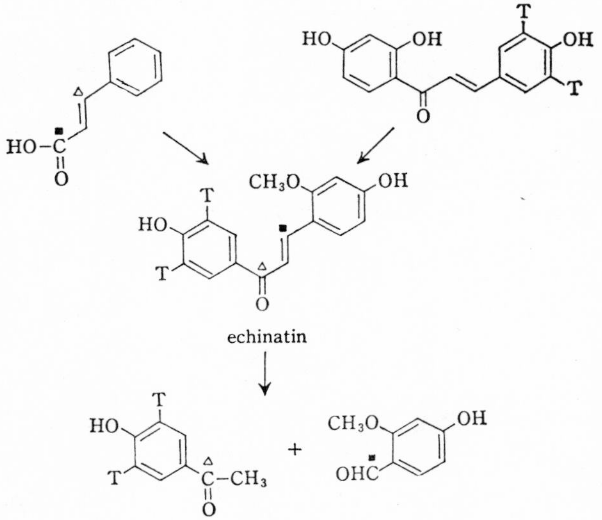

Fig. 26.1 Feeding experiments with labeled precursors. △, 3-^{14}C cinnamic acid; ■, [1-^{14}C] cinnamic acid; T, [3,5-^{3}H] isoliquiritigenin.

natin, suggesting that echinatin is biosynthesized from the same C_{15} intermediate as other normal flavonoids (Fig. 26.1).

All flavonoids of *G, echinata* callus have oxygen substitutions *para* to the C_3 linkage in one ring and at *ortho* and *para* positions in the other. Bearing this in mind. and locating licodione as the precursor of echinatin, the biogenetic relationship of flavonoids in *G. echinata* cultured cells can be illustrated as shown in Fig. 26.2. The role of dibenzoylmethanes in flavonoid biosynthesis is often thought to be as a precursor of flavones, especially those having a prenyl group at C-3,[8] because prenylation of the active methylene of dibenzoylmethane seems a reasonable process.

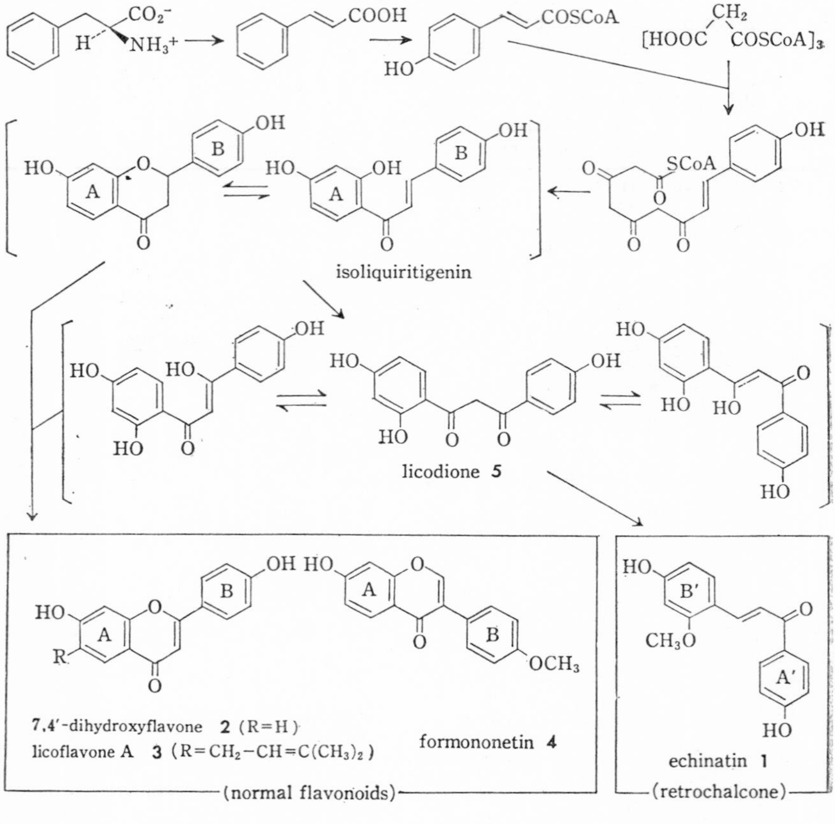

Fig. 26.2 Biosynthetic relationships among flavanoids in *Glycyrrhiza echinata* callus.

G. echinata callus produces echinatin as the major component only when cultured on White's medium containing yeast extract in the dark. We could detect neither echinatin nor licodione in the extract of callus cultured in the light or on media not containing a suitable concentration of yeast extract, whereas normal flavonoids were present regardless of chemical regulation or the presence of light. This suggests a close relationship between dibenzoylmethane and retrochalcone, and tracer studies as well as cell-free experiments to demonstrate this relationship are in progress.

[Experimental procedure 2] Radioisotopic experiments on the biosynthesis of echinatin

Feeding of Radioactive Precursors Each [^{14}C] cinnamic acid (0.05 mCi) was dissolved in a small quantity of EtOH and divided equally into ten 1 l Erlenmeyer flasks. The solvent was removed by passing a nitrogen stream, and then liquid medium (400 ml) was poured into each flask and sterilized in an autoclave (ca. 1 atm, 20 min). Static callus (60–70 g) subcultured for 6 weeks was transferred and shaken in the dark at 28°. After 6 days, the callus was collected by filtration of the medium through nylon cloth and extracted with MeOH and then AcOEt. The AcOEt extract was subjected to column and preparative layer chromatography. Radioactive echinatin (16.8 mg) was isolated from 644 g of callus after feeding [3-^{14}C] cinnamate, and 29.7 mg of radioactive echinatin was obtained in the experiment using [1-^{14}C]cinnamate as the precursor.

Degradation of Radioactive Echinatin [Carbonyl-^{14}C]echinatin (7.2 mg) diluted with cold echinatin to give 9.7 mg, was dissolved in 3 ml of 40% KOH and refluxed on an oil bath for 4 hr. After cooling to room temperature, the mixture was made weakly acidic with 6 N HCl and extracted with AcOEt. The products were separated by preparative layer chromatography of the ethyl acetate extract, using silica gel plates (0.5 mm thick) and developing with petroleum ether (bp 35°)–Et_2O(1:1). *p*-Hydroxyacetophenone (4.5 mg) (*Rf* 0.26) and 2–methoxy–4-hydroxybenzaldehyde (4.2 mg) (*Rf* 0.16) were obtained. Degradation of [β-^{14}C]echinatin was carried out in the same manner, and 8.4 mg of *p*-hydroxyacetophenone and 9.6 mg of 2-methoxy–4-hydroxybenzaldehyde were obtained from 18.5 mg of echinatin (after dilution with carrier).

Determination of Radioactivity A suitable amount of radioactive material was dissoved in toluene scintillant containing one-half volume of Triton X-100, and the radioactivity was counted with a liquid scintillation counter. Counting efficiency was calculated using quenching correction curves preparted by standard counting of a known activity of [^{14}C] sodium acetate with various amounts of quenchers, echinatin or acetone. Radioactive compounds were diluted with carrier and recrystallised until a constant value (less than 5% variation) of radioactivity was obtained. Recrystallization 3–5 times for each compound gave constant radioactivity. The solvents used for recrystallization were as follows: 40% ethanol for echinatin, chloroform-*n*-hexane (1:1) for *p*-hydroxyacetophenone and 5% ethanol for 2-methoxy–4-hydroxybenzaldehyde.

26.3 Metabolic Control and Cell-Free Experiments in *G. echinata* Cultured Cells

Glycyrrhiza echinata callus on White's medium in the dark is yellow, the color of echinatin, but when it is transferred onto Murashige and Skoog's medium it becomes colorless. The main difference in additives in these media is yeast extract (White) and the stock solution 5 (MSV; see Table 26.4) of Murashige and Skoog. On the other hand, the effect of light on flavone glycoside biosynthesis in parsley cell cultures was studied by Hahlbrock's group. They clearly demonstrated the induction of enzymes involved in the biosynthesis by light, and their studies have recently been extended to the biogenesis of mRNAs responsible for these enzymes.[7] It generally appears that the physical impositions, such as tissue damage or irradiation with light increase the amounts of phenylpropanoids in plants, and from the viewpoint of plant physiology, phenylalanine ammonia-lyase (PAL), a key enzyme at a branching point from the shikimate pathway to proteins and phenylpropanoids, plays a significant role in this phenomenon. We have been interested in the regulation of the biosynthesis of retrochalcone, and we examined the effects of additives and of light, in conjunction with cell-free experiments intended to identify the enzymes involved in this biosynthetic pathway.

The changes of echinatin accumulation, PAL activity and the growth of suspension cultures in White's and in Murashige and Skoog's liquid media in the dark and the light are shown in Fig. 26. 3. It was demonstrated that Murashige and Skoog's is a better medium for growth, but the production of echinatin is greatly facilitated in cells in White's medium in the dark. The timing of the increase is several days after the maximum of PAL activity. Under conditions of low echinatin content, PAL activity remains, which is consistent with the observation that normal flavonoids exist in the callus regardless of chemical or physical reguation (Table 26. 3). It is also noteworthy that an increase of PAL activity in cultures in Murashige and Skoog's medium in the light was observed, presumably due to the induction of this enzyme by light. Yeast extract was also shown to be a positive effector. In White's medium not containing yeast extract but with MSV, the cells grew to about two-thirds of the maximum weight of cells in the medium with yeast extract, but the content of echinatin decreased markedly (to 10–20%). The cells in Murashige and Skoog's medium without MSV but with yeast extract showed good growth and contained some echinatin.

Such metabolic control experiments with alterations of the chemical

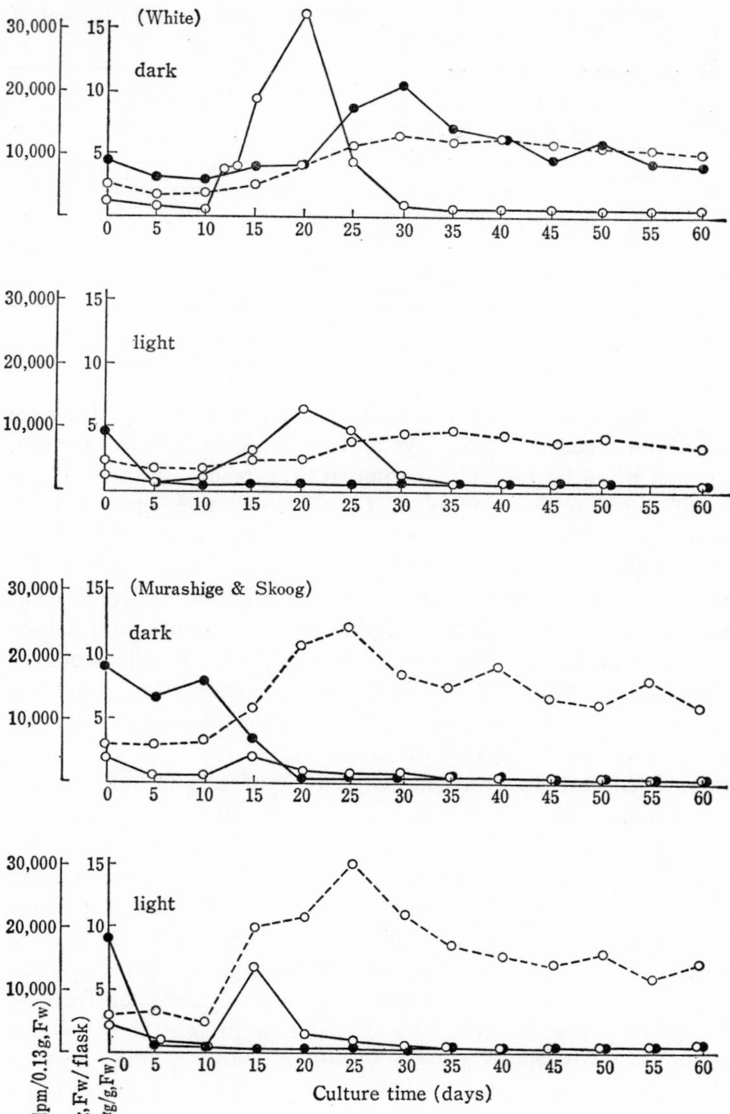

Fig. 26.3 Time courses of growth, echinatin content and PAL activity. ●—●, Echinatin content; ○—○, PAL activity; ○---○, growth.

TABLE 26.3. Presence of various flavonoids in the call us and the original plant of *Glycyrrhiza echinata*.

Culture conditions: Medium		Licodione	Retrochalcone: Echinatin	Normal flavoroids: 7,4′-Dihydroxy flavone	Normal flavoroids: Licoflavone A	Normal flavoroids: Formononetin
White	dark	+	+	+	+	+
	light	±	±	+	+	+
Murashige and Skoog	dark	—	—	+	+	+
	light	—	—	+	±	+
Original plant (roots)		+	+	+	+	+

and physical environment of tissue cultures can provide considerable information about metabolic pathways and their regulations in the cells. However, it can often be difficult to interpret the results in view of the complexity of the processes involved. In addition, the significant effector is not always the obvious one. In fact, the effect of light on echinatin biosynthesis in *G. echinata* cultured cells, though clear in the experiments using suspension cultures, as mentioned above, is not so clear in static cultures on agar media; the callus in media containing a high concentration of kinetin turns green in the light and occasionally produces large amounts of echinatin. Accordingly, further knowledge about the biosynthetic pathway and its regulation is necessary, involving enzymatic studies on the individual steps of the pathway.

We have been studying retrochalcone biosynthesis in cell-free extracts of *G. echinata* cultured cells in recent years, and some interesting results have been obtained. We have detected the activities of PAL, TAL (tyrosine ammonia-lyase) and cinnamate 4-hydroxylase, enzymes related to general phenylpropanoid biosynthesis, and licodione *O*-methyltransferase,[9] presumably the enzyme specifically involved in retrochalcone biosynthesis. The properties of licodione *O*-methyltransferase suggest strongly that licodione is the precursor of echinatin, as expected from its structural similarity to echinatin. We consider at present that the last steps of echinatin biosynthesis are the methylation to licodione and reduction of the diketone in methyllicodione to a keto-alcohol, followed by enzymatic or non-enzymatic dehydration to yield the retrochalcone.

[Experimental procedure 3] Suspension cultures and assay of enzymes

Suspension cultures The static callus (ca. 3 g) subcultured for 6 weeks under the conditions described in "Experimental procedure 1" was transferred into a 100 ml Erlenmeyer flask containing ca. 30 ml of liquid medium and shaken at 26° in the dark or the light (5,000 lux). The composition of the medium was the same as that shown in Table 28.4, but without agar. The cells were harvested by filtration of the medium through nylon cloth. The fresh weight was measured, then the cells were stored in a deep freeze (—20°) until use for enzyme assay and determination of echinatin content.

Assay of phenylalanine ammonia-lyase Callus (0.5 g or 1 g) was sus-

TABLE 26.4. Compositions of the media used.

Murashige and Skoog		White	
Stock solution	Preparation method	Stock solution	Preparation method
Solution 1	Solution 1 20(ml)	Solution 1	Solution 1 10(m*l*)
NH_4NO_3 165 g	2 10	KNO_3 4g	2 10
KNO_3 190 g	3 10	KCl 3.25 g	3 10
KH_2PO_4 17g	4 10	$NaH_2PO_4 \cdot H_2O$ 825 mg	$Fe_2(SO_4)_3$
H_3BO_3 620 mg	5 5	$MnSO_4 \cdot 4H_2O$ 350 mg	2.5 mg
$MnSO_4 \cdot 4H_2O$ 2,230 mg	Agar 9 g	$ZnSO_4 \cdot 7H_2O$ 150 mg	Agar 8 g
$ZnSO_4 \cdot 4H_2O$ 860 mg	Sucrose 30 g	H_3BO_3 75 mg	Sucrose 20 g
KI 83 mg	Add water to make	KI 37.5 mg	Yeast
$Na_2MoO_4 \cdot 2H_2O$ 25 mg	1 l, adjust to pH	Na_2SO_4 10 g	extract 1 g
$CoCl_2 \cdot 6H_2O$ 2.5 mg	5.7–5.8 with 5%	Dissolve in 500 ml	Add water to
Add water to	KOH	of water (× 100)	make 1 l
make 2 l (× 50)		Solution 2	
Solution 2		$Ca(NO_3)_2 \cdot 4H_2O$ 15 g	
$CaCl_2 \cdot 2H_2O$ 44 g		Dissolve in 500 ml	
Dissolve in 1 l		of water (× 100)	
of water (× 100)		Solution 3	
Solution 3		$MgSO_4 \cdot 7H_2O$ 36 g	
$MgSO_4 \cdot 7H_2O$ 37 g		Dissolve in 500 ml	
Dissolve in 1 l		of water (× 100)	
of water (× 100)			
Solution 4			
Na_2-EDTA 3.73 g			
$FeSO_4 \cdot 7H_2O$ 2.78 g			
Dissolve in 1 l			
of water (× 100)			
Solution 5 (MSV)			
Myo-inositol 20 g			
Nicotinic acid 100 mg			
Pyridoxine HCl 100mg			
Thiamine HCl 20 mg			
Dissolve in 1 l			
of water (× 200)			

pended in 0.5 or 1 ml of Tris-HCl buffer (pH 8.8) and homogenized at 0° for 5 min. After centrifugation (10,000 g, 10 min), the supernatant was used as crude enzyme. The reaction mixture was as follows.

0.05 M Tris-HCl buffer, pH 8.8	1.00 (ml)
[^{14}C(U)] L-phenylalanine, 0.1 mCi/10 ml	0.01
crude enzyme	0.50
water	0.49
total volume	2.00

The mixture was incubated at 30° for 2 hr, then 1 N HCl (0.5 ml) was added, followed by extraction of cinnamate with 3 ml of peroxide-free ether. An aliquot of this ether layer (1 ml) was transferred into a vial and the radioactivity was counted with a liquid scintillation counter to determine the enzyme activity.

Determination of echinatin content This was carried out by gas-liquid chromatography. Echinatin isolated by the method described previously (successive extraction of the callus with methanol and ethyl acetate followed by preparative chromatography on silica gel plates) was converted to the TMS ether by the addition of 20–30 μl of Pz-kit and slight warming on a water bath. The peak at *Rt* 5.6 min on a glass column (2 m) packed with 1% SE-30 on Gas-Chrom Q (60–80 mesh) at CT 240°, DT 255°, with nitrogen as a carrier gas, was due to echinatin.

Assay of Licodione methyltransferase The buffer used was 0.1 M phosphate, pH 8.0 containing 14 mM mercaptoethanol. Buffer (2 ml) was added to the callus (2 g), and the mixture was homogenized with a Teflon homogenizer at 0° for 5 min. After centrifugation (1,000 g, 10 min) the supernatant was used as the crude enzyme. The assay solution was as follows.

licodione (5.7 μmol)	0.10 (ml)
crude enzyme	0.50
$MgCl_2$ (0.1 μmol)	0.01
buffer	0.82
S-adenosyl–L-[Me-^{14}C]-methionine (SAM) (*ca.* 100,000 dpm)	0.02
total volume	1.45

After incubation at 30° for 2 hr, ethylene glycol monomethyl ether (0.05 ml) and acetic acid (0.02 ml) were added and the mixture was extracted with 4.0 ml of ethyl acetate. The radioactivity in the ethyl acetate layer was divided by the radioactiviy of SAM added and this conversion ratio was defined as the enzyme activity.

References

1) T. Furuya, *Biotech* (in Japanese), **2**, 113 (1971); T. Furuya, *The Tissue Culture* (in Japanese), **1**, 111 (1975); M. Tabata, N. Hiraoka, *ibid*, **1**, 120 (1975).
2) T. Saitoh, H. Noguchi, S. Shibata, *Chem. Pharm. Bull.*, **26**, 144 (1978) and references cited therein; S. Shibata, T. Saitoh, *Metabolism and Disease* (in Japanese), **10**, 157 (1973).
3) T. Furuya, K. Matsumoto, M. Hikichi, *Tetr. Lett.*, **1971**, 2567.
4) T. Furuya, S. Ayabe, M. Kobayashi, *ibid.*, **1976**, 2539.
5) T. Saitoh, S. Shibata, *ibid.*, **1975**, 4461.
6) T. Saitoh, S. Shibata, U. Sankawa, T. Furuya, S. Ayabe, *ibid.*, **1975**, 4463.
7) J. Schröder, F. Kreuzaler, E. Schäfer, K. Hahlbrock, *J. Biol. Chem.*, **254**, 57 (1979); K. Hahlbrock, H. Grisebach, *The Flavonoids* (ed. J. B. Harborne, T. J. Mabry, H. Mabry), p. 866, Chapman and Hall (1975).
8) K. Venkataraman, *The Flavonoids* (ed. J. B. Harborne *et al.*), p. 277, Chapman and Hall (1975).
9) S. Ayabe, T. Yoshikawa, M. Kobayashi, T. Furuya, *Phytochemistry*, **19**, 2331 (1980).
10) S. Ayabe, M. Kobayashi, M. Hikichi, K. Matsumoto, T. Furuya, *ibid*, **19**, 2179 (1980).
11) S. Ayabe, T. Furuya, *Tetr. Lett.*, **31**, 2965 (1980).

Note added in proof: An intensive study on licodione *O*-methyltransferase (LMT) revealed that in LMT catalyzed reaction only the hydroxy group at *ortho* to C_3 linkage is methylated. This fact strongly supports the hypothesis about the last steps of echinatin biosynthesis described in p. 391 (for details, see ref. 9). Furthermore, a feeding experiment with ^{14}C-labeled licodione has recently provided a direct evidence that licodione is a precursor of echinatin (S. Ayabe and T. Furuya, to be published).

CHAPTER 27

Components of Marijuana: Analysis and Isolation of Cannabinoids

Marijuana (marihuana, ganja) is the dried leaves or immature spikes of hemp (*Cannabis sativa* L.). Resin of hemp is generally called hashish or charas. Hemp is said to have originated in Central Asia or the Baikal region, and the plant is currently cultivated and grows wild all over the world. It must have migrated with human beings since hemp is one of the oldest materials used for making textiles. Abuse of marijuana has become a serious social problem in western countries only in recent years. The use of marijuana for hedonistic and religious purposes has occurred since ancient times, and it also has a long history as a medicine. Consequently, publications on the pharmacology and chemistry of marijuana had already appeared in the nineteenth century. It is known to cause euphoria, melancholy, enhancement of suggestibility, paropsis, visual hallucination, and effects due to abnormality in thinking, perception and emotional activity.

Its effects in rats and mice[1] also indicate psychotropic action. The active component is the hallucinogen tetrahydrocannabinol (THC).

27.1 Structure, Classification and Numbering System of Cannabinoid

Since the early studies,[2,3] many cannabinoids have been isolated and their structures determined. The term cannabinoid was first proposed by

Mechoulam *et al.*[4] for the C_{21} compounds in marijuana, related compounds and conversion products. The word is currently interpreted in a wider sense, as a general term for compounds structurally related to THC. The names and acronyms of all cannabinoids isloated from or confirmed to be present in marijuana are listed in Table 27.1, and their chemical structures are shown in Fig. 27.1. These cannabinoids can be cate-

TABLE 27.1. Cannabinoids isolated from or identified in marijuana and their acronyms

I. Pentylcannabinoid	
A-a. Original Cannabinoid Acid	
1. tetrahydrocannabinolic acid[5]	(THCA)
2. tetrahydrocannabinolic acid-B[6]	(THCA-B)
3. cannabidiolic acid[7,9]	(CBDA)
4. cannabichromenic acid[8]	(CBCA)
5. cannabigerolic acid[9]	(CBGA)
6. cannabigerolic acid monomethyl ether[10]	(CBGAM)
A-b. Secondary Cannabinoid Acid	
1. cannabinolic acid[10]	(CBNA)
2. cannabicyclolic acid[11]	(CBLA)
3. cannabielsoic acid-A[12]	(CBEA-A)
B. Neutral Cannabinoid	
1. tetrahydrocannabinol[13]	(THC)
2. cannabidiol[14]	(CBD)
3. cannabichromene[13]	(CBC)
4. cannabigerol[13]	(CBG)
5. cannabigerol monomethyl ether[15]	(CBGM)
6. cannabidiol monomethyl ether[16]	(CBDM)
7. cannabinol[17]	(CBN)
8. cannabicyclol[13]	(CBL)
9. cannabicitran[18]	(CBT)
10. $\Delta^{1(6)}$-tetrahydrocannabitriol[19]	($\Delta^{1(6)}$-THCT)
II. Propylcannabinoid	
A-a. Original Cannabinoid	
1. tetrahydrocannabivarinic acid[20]	(THCVA)
2. cannabidivarinic acid[20]	(CBDVA)
3. cannabichromevarinic acid[20]	(CBCVA)
B. Neutral Cannabinoid	
1. tetrahydrocannabivarin[21]	(THCV)
2. cannabidivarin[22]	(CBDV)
3. cannabichromevarin[23]	(CBCV)
4. cannabigerovarin[23]	(CBGV)
5. cannabivarin[24]	(CBV)
III. Methylcannabinoid†	
B. Neutral Cannabinoid	
1. tetrahydrocannabiorcin[25]	(THCO)
2. cannabidiorcin[25]	(CBDO)
3. cannabiorcin[25]	(CBO)

† Confirmed by gc-ms.

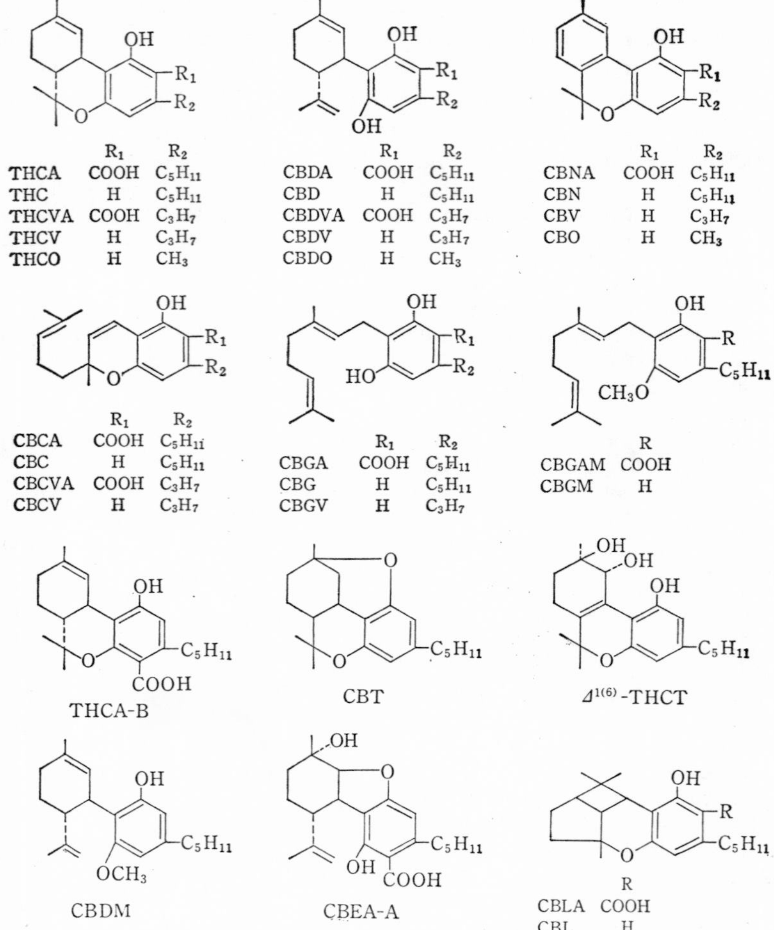

Fig. 27.1. Structural formulae of cannabinoids.

gorized in terms of the alkyl side chain as pentylcannabinoids [I], propylcannabinoids [II], and methylcannabinoids [III]. Each of those categoires includes cannabinoid acids (CA) [A] and neutral cannabinoids (NC) [B]. CA is still grouped into native cannabinoids CA [A-a], and secondary cannabinoids CA [A-b].

Many numbering systems have been used. Natural THC is, for example, described as Δ^1-THC, Δ^2-THC or Δ^9-THC. Numbering systems currently in use are shown in Fig. 27.2. (1) is based on monoterpenes and is often used in chemical structure research, while (2) is based on di-

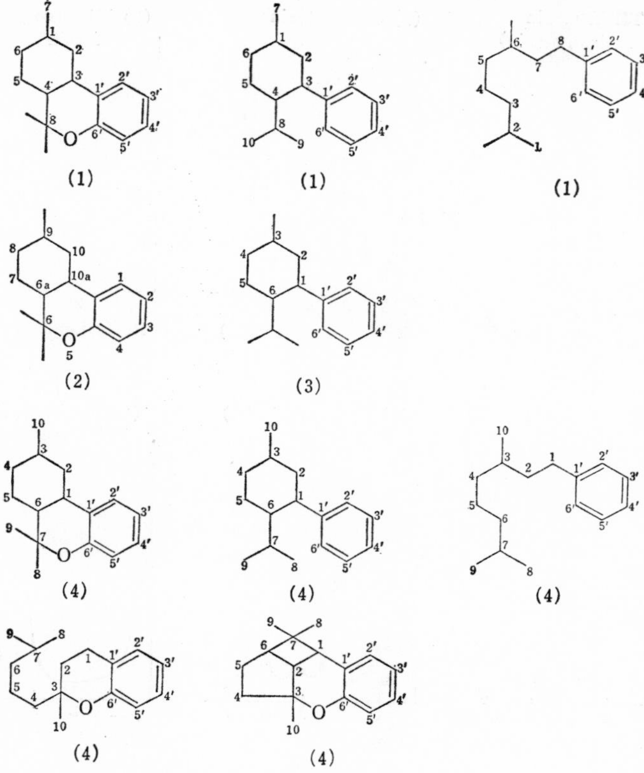

Fig. 27.2 Numbering of cannabinoids.

benzopyran, and is used in the literatures on chemical synthesis and among pharmacological workers in the U.S.A. (3) is based on diphenyl, and is only used for CBDs. (4) is one used by the author based on the biosynthesis.

27.2 Selection of Source Materials

Physiological Species of Hemp[26)]

There are many physiological species of *Cannavis sativa* L. with specific cannabinoid patterns (Fig. 27.3). The strains containing pentyl cannabinoid are the strain with THCA as the main cannabinoid, THCA strain (Mexican strain, Minami Oshihara (Tochigi prefecture, Japan) strain), the strain with CBDA as the main cannabinoid, and CBDA strain, and addtionally intermedial type generated by cross hybrid of those strain.

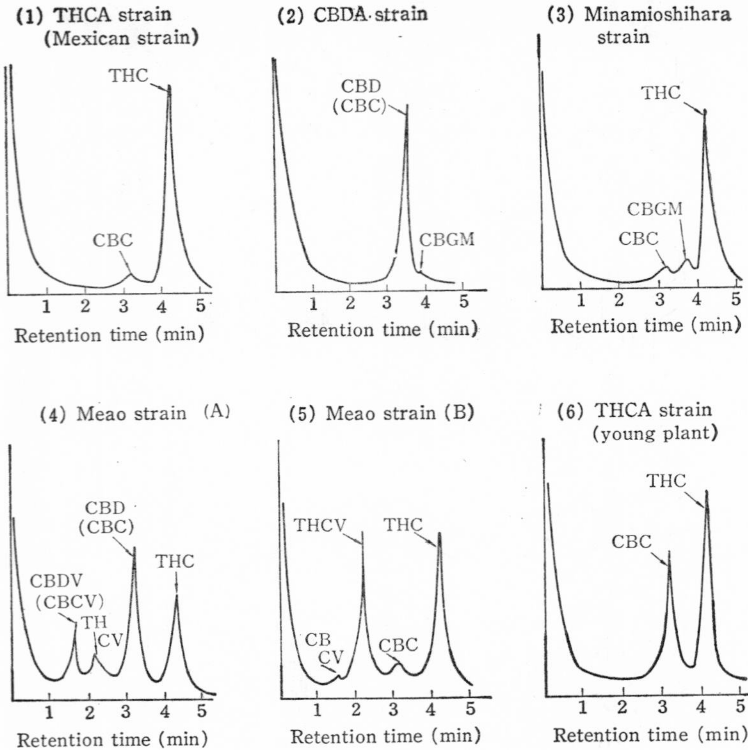

Fig. 27.3. Cannabinoids patterns of various hemps.

Most hemp strains currently grown are the dominant THCA strain. No pure strain containing propylcannabinoid has so far been discovered, but there are THCVA and CBDVA strains corresponding to the pentylcannabinoid-containing THCA and CBDA strains. Currently grown strains include crossed ones giving extremely complex cannabinoid patterns (Fig. 27.3, (4), (5): Meao variant A, B). The existence of methylcannabinoids has been confirmed by gc-ms, but the contents are extremely low in currently available strains.

Variations of Cannabinoids Content in the Plant Body[27)]

Fig. 27.4 shows examples of analysis of various leaves from a fully grown hemp body, and clearly indicates that the younger leaves are, the higher is the content of cannabinoid. A young leaf was found to contain almost three times more cannabinoid than an old leaf. This sug-

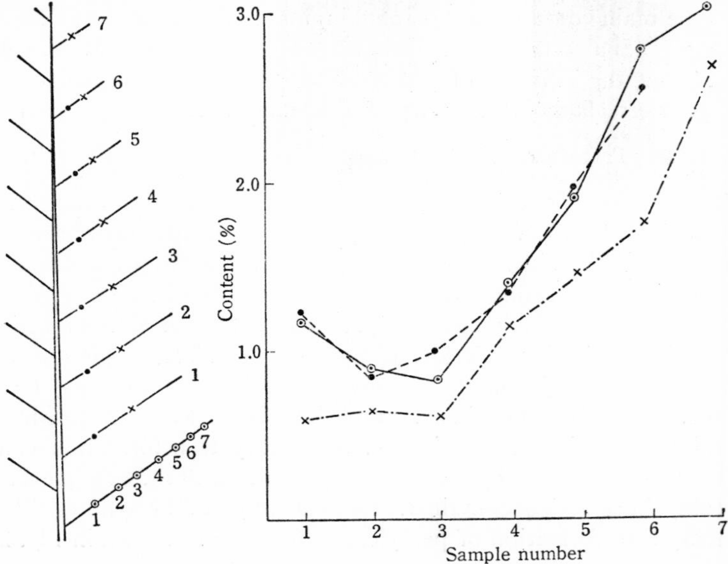

Fig. 27.4. THC(A) content of leaves from various parts of the plant.

gests that the portion of plant used should be clearly described when the contents of cannabinoids from different samples are compared. It is also desirable to use the leaves from the middle protion for determining the content when the contents in various individuals are to be compared.

Variation of Cannabinoids Content during Growth[27)]

The pattern of cannabinoids varies with the growth of hemp. As shown in Fig. 27.5, the main cannabinoid in THCA strain is CBCA until

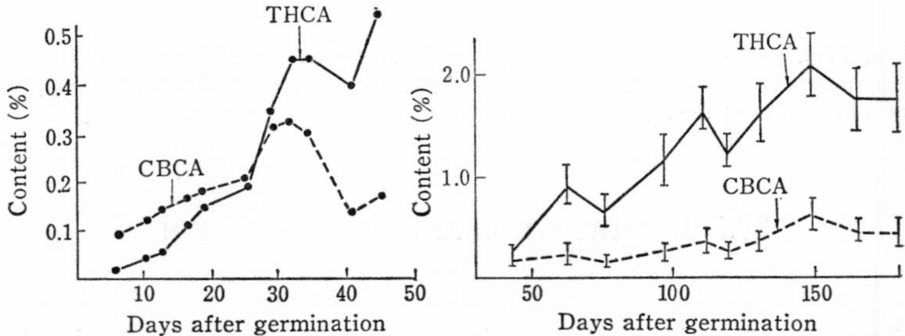

Fig. 27.5. Variation of the cannabinoids contents during the growth of hemp.

30 days after germination. THCA increases rapidly after that and becomes the main component of cannabinoids, until the ratio of THCA to CBCA reaches a certain level. Similar behavior is also seen in the CBDA strain. Thus, the isolation of CBCA and its derivatives is relatively easy when younger plants are used as starting material.

Secondary Transformation of Cannabinoids

The number of native cannabinoid acids is no more than about ten, as indicated in Table 27.1. They are fairly unstable and are transformed into secondary cannabinoids. The main transformation reaction is decarboxylation, and CAs are readily transformed into NCs. This reaction proceeds rapidly and quantitatively on heating. Another transformation reaction is autoxidation, which is accelerated by light and heat. THC(A) is sometimes not detected in marijuana after storage for several months; only CBN is detected. Thus, a fresh marijuana sample is desirable as the starting material for intact original cannabinoids. Secondary transformation can be avoided to some extent by storing marijuana in a dark, cold place, and preferably in a sealed container. The pathways of biosynthesis[28] and secondary transformation of pentylcannabinoids are shown in Fig. 27.6.

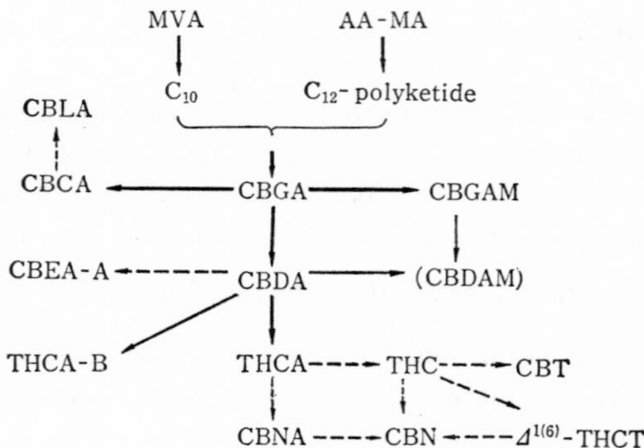

Fig. 27.6. Pathways of biosynthesis and secondary transformation of pentylcannabinoids.——, pathways of biosynthesis;——, unconfirmed pathways of biosynthesis;·······, pathways of secondary transformation.

27.3 Identification and Determination

The Beam reaction and its variants and the Duquenois reaction have been developed and utilized for many years for detecting components of marijuana. Thin-layer chromatography (tlc) and gas chromatography (gc) are mainly used at present for identification and determination.

Coloring Reactions

The Beam reaction[29] Addition of five drops of an alcoholic solution of 5% potassium hydroxide to a petroleum ether extract of marijuana develops a purple color. Dilution of the solution with water changes the color to blue. There are several variations of the method. CBD(A) and CBG(A) are positive in this reaction, but THC(A)s are negative. This reaction is, therefore, not entirely appropriate as a method to identify marijuana, though it is a well-known classical method.

The Ghamrawy reaction[29] Petroleum ether extract of marijuana is treated with a solution of *p*-dimethylaminobenzaldehyde (1 g) in 5 ml of sulfuric acid and 1 ml of water. Heating the mixture on a water bath for one minute develops a reddish-brown color, which turns purple on cooling. Addition of water to the solution changes the color to blue, and the blue color is retained even after further addition of water.

The Duquénois reaction[29] Petroleum ether extract of marijuana is treated with 2 ml of a solution of 0.4 g of vanillin and 0.06 ml of acetaldehyde dissolved in 20 ml of 95% ethanol. On adding several drops of hydrochloric acid to the mixture, a blue color develops.

Diazotized benzidine reagent[30] Benzidine (2.5 g) is dissolved in 7 ml of water, then 500 ml of water is added. This solution is mixed with 10% nitrous acid immediately before use. (The mixture can be used for 10 days if it is stored in refrigerator.) This is a coloring reagent for phenolic compounds, but it develops specific coloring with each cannabinoid (Table 27.2) and is a suitable coloring reagent for tlc.

Echt Blau Salz B (fast blue salt B) reagent[31] Echt Blau Salz B (Merck) di-*o*-anisidine tetrazolium chloride) is dissolved in N/10 sodium hydroxide (15mg/20ml). This is also a coloring reagent for phenols, and gives a specific coloring reaction with each cannabinoid (Table 27.2). The color developed by this reagent is more stable than that with diazotized benzidine, and it is a suitable coloring reagent for tlc.

Thin-layer Chromatography

Identification of cannabinoids by tlc was first attempted by Korte *et al.*,[31] who reported the isolation of CBDA, CBD, THC, and marijuana extract, but their method was complex and is rarely used now. Aramaki and his co-workers[32] have examined several solvents for CBD, THC, and CBN analysis on silica gel thin-layer plates, and they obtained good results with hexane-benzene-diethylamine (25:10:1). The author succeeded in isolating

TABLE 27.2. Values of R_f (tlc) and R_t(gc) and coloring of cannabinoids.

	R_f †1			R_t	coloring†2	
	1	2	3	(TMS derivative) (min)	1	2
THCA	0.59	s	s	4.1(7.4)	red	reddish brown
CBDA	0.67	s	s	3.3(4.8)	orange	orange brown
CBCA	0.24	s	s	3.3(7.7)	reddish brown	purple brown
CBGA	0.67	s	s	5.0	orange yellow	—
CBGAM	0.59	s	s	3.6(9.1)	red	—
CBNA	0.20	s	s	5.0(9.4)	reddish purple	purple brown
CBLA	0.65	s	s	2.7	red	—
THCVA	0.30	s	s	2.5	red	—
CBDVA	0.40	s	s	1.9	orange	—
CBCVA	0.18	s	s	1.9	reddish brown	reddish brown
THC	t	0.57	0.30	4.1	orange red	orange
CBD	t	0.60	0.34	3.3	orange	reddish purple
CBC	t	0.44	0.17	3.3	reddish brown	orange red
CBG	t	0.37	0.20	5.0	orange yellow	—
CBGM	t	0.82	—	3.6	orange	—
CBDM	t	0.95	—	2.4	orange yellow	—
CBN	t	0.53	0.25	5.0	reddish purple	reddish purple
CBL	t	0.60	0.32	2.7	orange red	—
$\Delta^{1(6)}$-THCT	0.40	s	s	—	purple	—
THCV	t	0.48	—	2.5	red	—
CBDV	t	0.53	—	1.9	orange	—
CBCV	t	0.32	—	1.9	reddish brown	—
CBGV	t	0.30	—	2.9	orange	—
CBV	t	0.48	—	2.9	reddish purple	—

†1 Silica gel thin-layer plates:
1, hexane-ethyl acetate (1:1); s, starting point; t, top
2, benzene
3, hexane-benzene-diethylamine (25:10:1)
†2 Spot test:
1, diazotized benzidine reagent
2, Echt Blau Salz B reagent.

CAs and $\Delta^{1(6)}$-THCT with hexane-ethyl acetate (1:1) and NCs with benzene on silica gel thin-layer plates.

Gas Chromatography

Gas chromatography is now widely applied for the analysis of cannabinoids, especially for quantitative determination. CAs are usually decarboxylated in the column and are detected as the corresponding NCs. However, preliminary trimethylsilylation makes it possible to separate and to identify CAs from NCs. CBD(A) and CBC(A) can be identified by combining gc with tlc.

[Experiment procedure 1] Identification of cannabinoids in marijuana[33]

Marijuana (100 mg) was extracted with 10ml of benzene with heating. The benzene extract was dissolved in 0.2–1.0 ml of acetone, and 2 μl of the solution was injected into a Shimadzu 5A gas chromatograph (FID) (column, 1.5% SE 52, Chromosorb W, DMCS (60–80 mesh); 2.25 m × 4 mm i.d.; column temperature 240° C; detector temperature 240° C; sample injection gate temperature 250° C; nitrogen 25 ml/min, hydrogen 50 ml/min, air 500 ml/min). The retention times are shown in Table 27.2.

[Experimental procedure 2] Determination of THC(A) in marijuana[27]

Marijuana (200 mg) was immersed in 20 ml of benzene. The extract was evaporated to dryness and the residue was dissolved in 0.5–2.0 ml of acetone to prepare a sample solution. An aliquot (0.1 ml) of the sample solution was mixed well with 0.1ml of cholestane (*Rt*, 11 min.) solution (1.292 mg/ml), and 2 μl of the mixutre was subjected to gc under the conditions described above. The peak area was calculated by the half band width method, and the content was obtained from the following equation using a standard curve. The standard deviation for THC (3.88 mg) was ± 0.04 mg.

$$\text{Content (\%)} = \frac{1.292 \times W \times V}{M} \times 100$$

where W is the weight ratio of THC to cholestane obtained from the standard curve, V is the volume of acetone (ml) used to dissolve the benzene extract, and M is the weight of marijuana (mg).

The standard curve was prepared by placing 64.14 mg of cholestane in a 50 ml volumetric flask and dissolving it in a small amount of hexane. The solution was made up to 50 ml with acetone (1.292 mg/ml). THC solutions of various concentrations, in hexane were also prepared, then 0.1 ml of the THC solution and 0.1 ml of the cholestane solution were mixed, and 2 μl of the mixture was injected into the gas chromatograph. The standard curve was prepared by plotting the area ratio (ordinate) versus the weight ratio (abscissa). The peak area was calculated by the half band width method.

High-Speed Liquid Chromatography[33]

The author was the first to apply high-speed liquid chromatography to the identification and determination of cannabinoids, obtaining good results with a "Permaphase" ODS column and a moving phase of 63% methanol (Fig 27.7). This method was found to have the advantage that the separation of CBD from CBC and of CAs from NCs was possible, whereas it could not be achieved by gas chromatography. High-speed liquid chromatography should become a powerful tool in the identification and determination of cannabinoids.

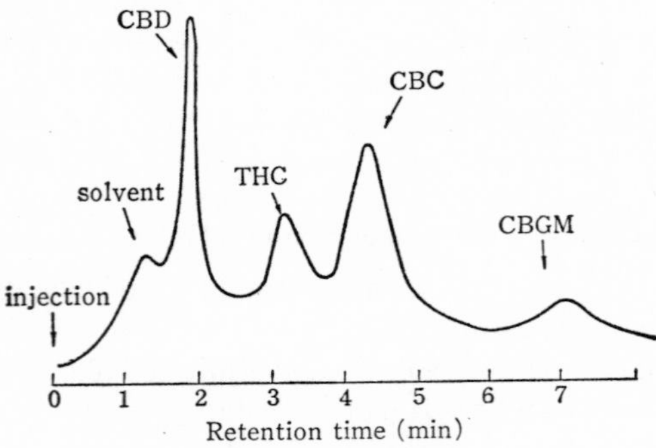

Fig. 27.7. Separation of neutral-cannabinoids by high-speed liquid chromatography. Du Pont 830 machine with a Permaphase ODS column (1 m × 2 mm i.d.) using 63% methanol (v/v). Temperature 40°C, column pressure 1700 psig, UV detector (254 nm).

Gas Chromatography-Mass Spectrometry

Analysis by gc-ms has made the identification of unknown cannabinoids and the determination of micro amounts of cannabinoids possible. It has been used to study methylcannabinoids[25] and to detect THC in the blood of marijuana smokers.[35]

[Experimental procedure 3] Identification of THCV and CBDV[33]

Marijuana (Meao strain from Thailand) (100 mg) was immersed in 10 ml of warm ether, and the ether extract was subjected to gc-ms using a Shimadzu LKB 9000 machine (1.5% OV 17, Chromosorb W (60–80 mesh); 2 m × 3 mm i.d.; helium 32 ml/min; column temperature 240° C; sample injection gate temperature 280° C; ion source temperature 290° C; separator temperature 240° C; ionizing voltage 70 eV; trap current 60 mA; acceleration voltage 3.5 kV).

Peak A; *m/e* 286 (M^+), 271, 243, 218, 203 (base peak). Peak B; *m/e* 286 (M^+), 271, 243, 203 (base peak). The fragment peaks of both at *m/e* −28 and their relative intensities, were consistent with those of CBD and THC, and A was identified as CBDV, and B as THCV.

27.4 Purification and Isolation

Cannabinoids such as THC and THCA were generally purified by distillation until the 1940s, since they are syrupy materials and can not be purifed by recrystallization. CAs were not detected such preparations, since decarboxylation was unavoidable. Separation and purification of cannabinoids are now carried out mostly by combinations of various types of chromatography, and in a few cases by the countercurrent method. Separtation methods for CAs and NCs are outlined in Figs. 27.8 and 27.9. The fractions from chromatography must be monitored constantly by tlc since the composition of cannabinoids varies from sample to sample.

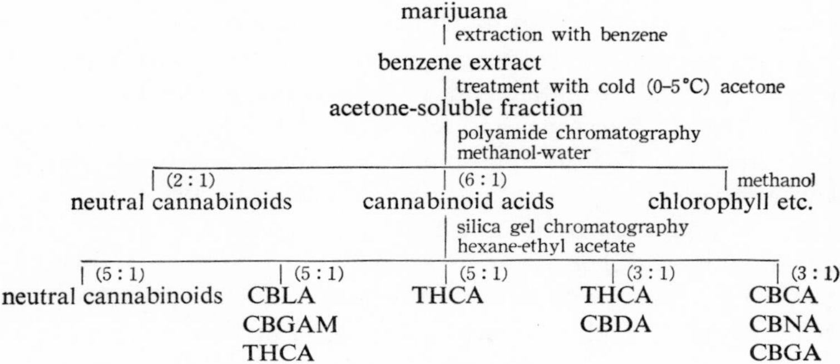

Fig. 27.8. Isolation procedure for cannabinoid acids.

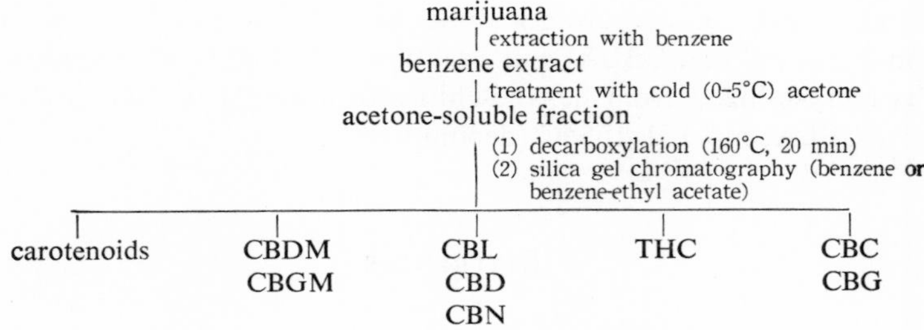

Fig. 27.9. Isolation procedure for neutral cannabinoids.

[Experimental procedure 4] Isolation of THC from marijuana

Marijuana (Mexican strain) (100 g) was extracted with 1000 ml of cold benzene. The solvent was removed from the extract under reduced pressure, leaving a residue (6.8 g). This residue was treated with 3 volumes of cold acetone (0–5° C). The acetone-soluble portion was heated in an oil bath at 160° C for 20 min, then subjected to column chromatography on silica gel (140 g) using hexane-ethyl acetate (50:1) as the eluant. Monitoring by tlc, the elute was fractionated into Fraction 1 (3.8 g, THC and a small amount of CBN) and Fraction 2 (CBC and a small amount of CBG). Fraction 1 was dissolved in 20 ml of methanol and left to stand for 5 hr, then filtered. The filtrate was concentrated under reduced pressure, yielding 3g of residue. The residue was chromatographed on a silica gel (150 g) column, eluting with hexane-ethyl acetate (50:1). Colorless oily THC (1.8 g) was obtained. The purity was confirmed by gc and tlc.

[Experimental procedure 5] Isolation of CBLA from marijuana

Marijuana (obtained in Japan; collected at the early nutritional phase and stored for 4 months) (2.2 kg) was immersed in 22 l of cold benzene. Removal of the solvent under reduced pressure gave 200 g of residue. This residue was treated with 3 volumes of cold acetone (0–5° C) and insoluble material was filtered off. (Alternatively, the residue could be dissolved in 600 ml of acetone, and left to stand for 3 hr at 0–5° C, followed by filtration.) The acetone-soluble portion was chromatographed on a polyamide column (850 g), eluting with ethanol-water (2:1 → 6:1). The elute was monitored by tlc and fractionated into Fraction 1 (10 g, NCs) and Fraction 2 (21 g, CAs). Fraction 2 was chromatographed on silica gel (210 g), eluting with hexane-ethyl acetate (5:1 → 3:1). The elute was monitored by tlc, and was fractionated into Fraction 2–1 (very small amount, NCs), Fraction 2–2 (470 mg, CBLA and THCA), Fraction 2–3 (7 g, THCA), Fraction 2–4 (450 mg, THCA and CBCA) and Fraction 2–5 (2 g, CBCA, CBNA, and CBGA). Fraction 2–2 was rechromatographed on a silica gel (47 g) column, eluting with hexane-ethyl acetate (5:1), to yield Fraction 2–2–1 (60 mg, CBLA) and Fraction 2–2–2 (THCA). Fraction 2–2–1 was recrystallized from hexane-chloroform (50:1) to give colorless prisms of CBLA, mp 152–155° C (decomposed).

References

1) S. Ueki, M. Fujiwara, N. Ogawa, *Physiol. Behav.*, **9**, 585 (1972).
2) K. Takeda, *Kagaku no Ryoiki*, **2**, 170 (1946).
3) Z. Krejči, F. Šantavý, *Acta Univ. Palackianea Olomuc.*, **6**, 59 (1955) (*Chem. Abstr.*, **50**, 12080 (1956)).
4) R. Mechoulam, Y. Gaoni, *Fortschr. Chem. Org. Naturstoffe*, **25**, 174 (1967).
5) T. Yamauchi, Y. Shoyama, H. Aramaki, T. Azuma, I. Nishioka, *Chem. Pharm. Bull.*, **15**, 1075 (1967).
6) R. Mechoulam, Z. Ben-Zvi, B. Yagnitinsky, A. Shani, *Tetr. Lett.*, **1969**, 2339.
7) Z. Krejči, M. Horák, F. Šantavý, *Pharmazie*, **14**, 349 (1959).

8) Y. Shoyama, T. Fujita, T. Yamauchi, I. Nishioka, *Chem. Pharm. Bull.*, **16**, 1157 (1968)
9) R. Mechoulam, Y. Gaoni, *Tetrahedron,* **21**, 1223 (1965).
10) Y. Shoyama, T. Yamauchi, I. Nishioka, *Chem. Pharm. Bull.*, **18**, 1327 (1968).
11) Y. Shoyama, R. Oku, T. Yamauchi, I. Nishioka, *ibid.*, **20**, 1927 (1972).
12) A. Shani, R. Mechoulam, *Tetrahedron*, **30**, 2437 (1974).
13) Y. Gaoni, R. Mechoulam, *J. Am. Chem. Soc.*, **93**, 217 (1971).
14) R. Mechoulam, Y. Shvo, *Tetrahedron*, **19**, 2073 (1963).
15) T. Yamauchi, Y. Shoyama, Y. Matsuo, I. Nishioka, *Chem. Pharm. Bull.*, **16**, 1164 (1968).
16) Y. Shoyama, K. Kuboe, I. Nishioka, T. Yamauchi, *ibid.*, **20**, 2072 (1972).
17) R. Adams, B. R. Baker, R. B. Wearn, *J. Am. Chem. Soc.*, **62**, 2204 (1940).
18) C. A. L. Bercht, R. J. J. Ch. Lousberg, F. J. E. M. Küppers, C. A. Salemink, *Phytochemistry*, **13**, 619 (1974).
19) F. Yamato, G. Nonaka, A, Okabe, Y. Shoyama, 1. Nishioka, *Abstract of the* **94*th*** *Annual Meeting of the Pharmaceutical Society of Japan*, II, p. 239 (1974).
20) Y. Shoyama, H. Hirano, H. Makino, N. Umekita, I. Nishioka, *Chem. Pharm. Bull.*, **25**, 2306 (1977).
21) E. W. Gill, *J. Chem. Soc.*, **1971**, 579.
22) L. Vollner, D. Bieniek, F. Korte, *Tetr. Lett.*, **1969**, 145.
23) Y. Shoyama, H. Hirano, M. Oda, T. Somehara, I. Nishioka, *Chem. Pharm. Bull.*, **23**, 1894 (1975).
24) F. W. H. M. Merkus, *Nature,* **232**, 579 (1971).
25) T. B. Vree, D. D. Breimer, C. A. M. Van Ginneken, J. M. Van Rossum, *J. Pharm. Pharmacol.*, **24**, 7 (1972).
26) I. Nishioka, *Abstract of the 11th Phytochemistry Symposium*, p. 11 (1975).
27) H. Kushima, Y. Shoyama, I. Nishioka, *Chem. Pharm. Bull.*, **28**, 594 (1980).
28) Y. Shoyama, M. Yagi, I. Nishioka, T. Yamauchi, *Phytochemistry*, **14**, 2189 (1975).
29) H. Asahina, *Narcotics* (in Japanese), p. 82, Nankodo (1960).
30) J. E. Koch, W. Krieg, *Chem. Zentr.*, **62**, 140 (1938).
31) F. Korte, H. Sieper, *J. Chromatog.*, **13**, 90 (1964).
32) H. Aramaki, N. Tomiyasu, H. Yoshimura, *Chem. Pharm. Bull.*, **16**, 822 (1968).
33) I. Nishioka, *Abstract of the First Symposium on the Analysis of Crude Drugs*, p. 26 (1972).
34) Unpublished
35) S. Agurell, B. Gustafsson, B. Holmstedt, K. Leander, J. E. Lindgren, I. Nilsson, F. Sandberg, M. Asberg, *J. Pharm. Pharmacol.*, **25**, 554 (1973).

CHAPTER 28

Chinese Drug Constituents: Isolation of the Biologically Active Principles

The earliest medicines were naturally occurring substances. Greek medicine, from which modern Western medicine is believed to have derived, has now practically disappeared, but traditional Chinese medicine and Indian Ayurvedic medicine still exist as established therapeutical systems. Their philosophical basis is quite different from that of Western medicine, and has attracted renewed interest in recent years. "*Kanpo*" or Chinese medicine was practiced more than 2000 years ago, and was apparently first systematized in a book called *"Shang Han Tsu Ping Lun"* said to have been compiled by Chang Chung-Ching in the 2nd century A.D., and in *"Shin-Nung Pen T'sao Ching"*, also compiled in the same century by an unknown author. Today, only revised editions exist: the originals have been lost.

The 10 volumes of "*Shang Han Lun*" and the 3 volumes of *"Chin Kuei Yao Lueh"* which are said to be a recompiled and revised edition by Wan Su-Ho (210–285) of the original 16 volumes of "*Shang Han Tsu Ping Lun*" by Chang Chung-Ching (142–210), explained how to prescribe medicine according to the syndromes observed in the course of development of a fever disease. The book includes 113 prescriptions wherein about 80 kinds of crude drugs are used.

"*Shin-Nung Pen T'sao Ching*", recompiled and revised by Tao Hung-Ching (456–536), is the earliest book of medicinal herbs *(Pen T'sao)* that still exists today, and includes 365 kinds of crude drugs which are classified into three classes, i.e. superior, general and inferior.

According to the book, those belonging to the superior class (120 in number) work on "life" itself, are non-toxic, cause no undesirable effect even if taken in excess, improve health, heighten the spirit, and bring about extension of the life span. Those belonging to the general class (120) work on the human body, may be non-toxic or toxic, heal disease, and improve physical well-being. Those belonging to the inferior class (125) act on the diseases themselves, are often toxic and should not be taken in quantity, and cure accumulated illnesses, fevers and chills.

Later in the Tang dynasty, a number of medical books were published, such as "*Chien Chin Fang*", "*Chien Chin I Fang*", "*Wai Tai Mi Yao*" and others. By that time medicine had become more influenced by Taoism, while the number of prescription formulas and the number of crude drugs used had greatly increased.

In Japan, the traditional medicine according to "*Shan Han Lun*" is called "*Ko-ho*" (classical method), whereas that newly devised in the Chin and Yuan eras is called "*Gosei-ho*" (new method). "*Kanpo*" or Chinese medicine, was first brought to Japan through the Korean peninsula before the Nara era. Over 60 kinds of crude drugs still remain among the offerings to the Great Buddha in Todaiji Temple in Nara, dedicated by Empress Komyo in 735 A.D., and kept in Shosoin Treasure House.

In the early Heian Era, (982–984) Yasuyori Tamba issued 30 volumes of "*Ishin-po*", in which the essence of Chinese medicine practised in the Tang dynasty in China was described. Later, Todo Yoshimasu (1703–1773) advocated "*Shang Han Lun*" and wrote three books, "*Ruijuho*", "*Hokyoku*" and "*Yakucho*". A Chinese medical practitioner, Yodo Odai (1798–1870) published "*Ruijuho-Kogi*" and "*Juko-Yakuchō*", which are revised editions of the above books by Todo Yoshimasu.

Since then, Chinese medicine practitioners in Japan have mostly stuck to "*Koho*", whereas in China the original techniques have continued to be developed; there were great advances in the Min and Qing dynasties.

In European medicine, prescriptions are made on the basis of pathological observations on the cells, tissues or organs, aiming at the restoration of their original functions. In Chinese medicine, prescriptions are chosen from a number of established formulas according to the overall physical state of the patient. In Chinese medicine, a diagnosis is normally given in terms of the prescription formulas which the doctors think are appropriate for the patient and will remove or mitigate the syndromes.

Thus patients with different pathological symptoms, i.e., presumably having different kinds of diseases, may be treated with the same prescription formulas if their physical conditions are the same, or patients with the same disease may be treated with different prescription formulas if their physical states are not the same. Unlike ordinary folk medicine, it is rare in Chinese medicine that the prescriptions consist of only one or two crude drugs: several are usually used. This makes the pharmacological analysis of Chinese medicine even more difficult.

The first modern chemical studies on the constituents of Chinese drugs was the work on Moutan Cortex Radicis and Ephedra by Nagai in the early Meiji Era. This was followed by many similar studies.

However, most of these workers merely achieved the isolation or separation of the major component(s) in the material, or those easiest to separate, and paid little attention to their biological activities.

A major aim of our investigations on Chinese drugs is to find a modern scientific basis for the apparent efficacy of traditional Chinese medicine, in the hope of finding new biologically active substances.

28.1 Pharmacological and Biochemical Studies on Chinese Drugs

The techniques used in chemical studies on the constituents of Chinese drugs are essentially the same as those used in the chemistry of natural products. However, there are two approaches in studies on the pharmacological effects of Chinese drugs. One is to test the crude drugs constituting certain formulas for particular pharmacological activities deduced from the usage of the formulas in medical parcitice. The other is to carry out general pharmacological screening.

The first approach is quite useful in the analysis of Chinese medicine. However, crude drugs used in Chinese medicine normally have low toxicity and mild biological activity, and since they are not used singly but always in combination, the compositions or the usage of formulas may not clearly suggest the possible activity of the constituents of the formulas. Sometimes two compounds isolated from crude drugs contained in one formula have pharmacologically antagonistic or quite different activities: Chinese medicine relies on the net results of such complex effects. Generally, a preliminary pharmacological analysis of the constituents may be helpful before carrying out actual biological assays. Todo Yoshimasu described in his book *"Yakucho"* the expected or possible pharmacological effects of some 50 kinds of crude drugs included in *"Shang Han Lun"*.

For example, in the book *"Yakucho"*, Ephedra is described as "mainly cures coughing and edema", and Pueraia Radix as "mainly overcomes stiffness of the neck". His conclusions may not always be persuasive, but his analysis was partly based on his own clinical experience, and is even today regarded as a useful reference on the pharmacological activities of crude drugs. Harada[2)] attempted a more precise pharmacological evalua-

TABLE 28.1. Content of Paeony root in prescriptions containing it.

Prescriptions	Content of Paeony root (%)
"*Shakuyaku-Kanzo-to*"	50
"*Haino-san*"	<43
"*Shokenchu-to*"	<35
"*Keishi-ka-Shakuyaku-to*"	35
"*Toki-Shakuyaku-san*"	35
"*Shigyaku-san*"	25
"*Shinbu-to*"	25
"*Bushi-to*"	21
"*Keishi-Bukuryo-gan*"	20
"*Oogon-to*"	20

tion of Paeony root or Licorioe, by scoring its effects (on a scale of ten) on varions symptoms. Table 28.2 shows an example where the analysis of Paeony root was attempted, using 10 formulas (Table 28.1) that contain more than 20% Paeony and rating their effectiveness on the symptoms.

The main pharmacological effect of Paeony root was found to be relief of pain and muscle spasm in the abdomen.

The results obtained coincide with the description given in "*Yaku-cho*", so examination for analgesic and sedative effect, or anti-inflammatory effect, may be sufficient in studies aimed at isolating the active component(s).

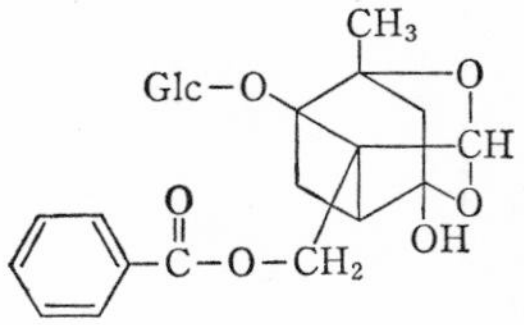

Glc–O, CH_3, O, CH, O, OH, O, C–O–CH_2

Harada eventually showed that a monoterpene glycoside, paeoniflorin, prepared from Paeony root had sedative, analgesic, temperature-lowering, anti-inflammatory and vasodilative effects, and prolonged hexobarbital-induced sleep, though some of these activities were observed only when large quantities of paeoniflorin were administered. The toxicity of paeoniflorin is so low (LD_{50} 3,530 mg/kg i.v., 9,530 mg/kg i.p.), however, that large quantities may be safely administered. Generally speaking, paeoniflorin seems to have sedative, mild central nervous system suppressing and anti-inflammatory activities.

Shibata[3)] and Lee[4)] also investigated formulas containing Bupleuri Radix and Platycodi Radix, respectively, as their main component, and

TABLE 28.2. Symptoms for which prescriptions containing Paeony root are adopted.

Prescriptions \ Symptoms	Excitation, Spasm	Tiredness	Suppurative swelling	Anemia	Bleeding	Edema	Neuralgia, myalgia, stiffness of muscles	Feeling of cold in extremities	Hot feeling in the head	Dizziness	Palpitation	Heavy feeling of the epigastrium	Stiffness on region along the lower rib.	Abdominal pain, stiffness of abdomal muscle.	Heavy feeling of the abdomen	Diarrhea	Pollakuria	Pain on pressing specific region of navel	Menstrual disorder	Total
"*Shakuyaku-Kanzo-to*"	1						4.5							4.5						10
"*Oogon-to*"												3		4		3				10
"*Shigyaku-san*"								1				1	3	3		2				10
"*Haino-san*"			10																	10
"*Keishi-ka-Shakuyaku-to*"														4.5	2.5	3				10
"*Keishi-Bukuryo-gan*"					1.5				1.5					1				3	3	10
"*Shinbu-to*"		1				1.5		1.5		1.5	1.5			1.5		1.5				10
"*Bushi-to*"						2	4	2						2						10
"*Shokenchu-to*"	1	3.5									1			3.5			1			10
"*Toki-Shakuyaku-san*"				1.5		1.5		1.5						2				1.5	2	10
Total (= %)	2	4.5	10	1.5	1.5	5	8.5	6	1.5	1.5	2.5	4	3	26	2.5	9.5	1	4.5	5	100

tried to find a correlation between the crude drugs and the usage of such formulas, as shown in Table 28.3.

TABLE 28.3. Symptoms for which Bupleuri Radix and Platycodi Radix are adopted.

Bupleuri Radix		Platycodi Radix	
Inflammation	42	Inflammation	63
Pyrexia	40	Cough	16
Pain	21	Pain	11
Muscle spasm	21	Pyrexia	9
Neurosis	7	Hypertension	4

Such preliminary analyses are usually helpful in deciding the nature of the biological examinations to be performed on the crude drug or the various fractions derived therefrom in more detailed studies. However, some of the activities may not be detected in such an analysis.

The second method is the so-called general blind screening,[5] wherein samples are subjected to various biological examinations, such as tests on the effects on the nervous system, behavior, breathing, blood pressure, cardiac functions, the small intestine, the stomach, and the uterus, as well as tests for anti-inflammatory, anti-anaphylaxis, and antihistaminic activities, and so on.

When Bupleuri Radix was tested by this method, its saponins were found to show sedative anti-inflammatory and analgesic effects, and to be effective on coughing or pyrexia. Platycodin, a saponin from Platycodi Radix was later shown to possess these and other activities. Thus, the sa-

ginsenosides Rx

Rb_1 R = D-Glc β(1→2)D-Glc
R′ = D-Glc β(1→6)D-Glc

Rb_2 R = D-Glc β(1→2)D-Glc
R′ = L-Ara(pyr) α(1→6)D-Glc

Rc R = D-Glc β(1→2)D-Glc
R′ = L-Ara(fur) α(1→6)D-Glc

Rd R = D-Glc β(1→2)D-Glc
R′ = D-Glc

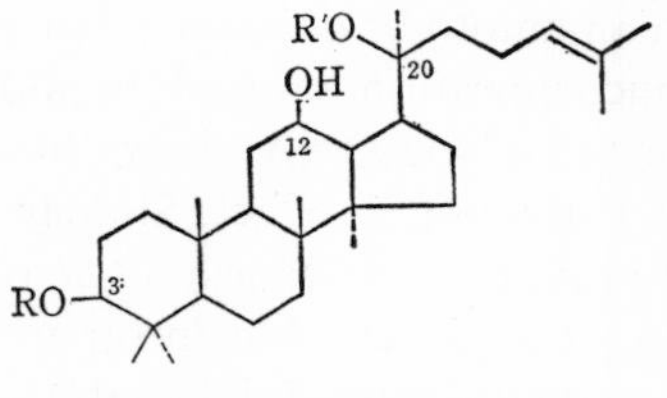

R=R′=H (20S)-protopanaxadiol

Re R = L-Rha α(1→2)D-Glc
R′ = D-Glc

Rf R = D-Glc β(1→2)D-Glc
R′ = H

Rg_1 R = D-Glc
R′ = D-Glc

Rg_2 R = L-Rha α(1→2)D-Glc
R′ = H

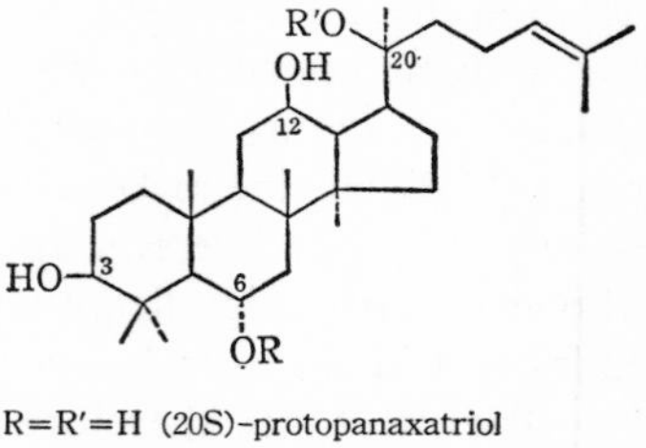

R=R′=H (20S)-protopanaxatriol

ponins, which have generally been regarded simply as anti-tussive agents, were shown to have anti-inflammatory and sedative activities, as could have been predicted from the usage of the crude drugs in Chinese medicinal formulas.

In East Asian countries, Ginseng is widely regarded as a cure-all medicine, a tonic, or even as a drug for longevity. Accordingly, it has attracted the attention of many investigators in the field of natural products chemistry.[6–9] Shibata *et al.*[8] isolated over 10 kinds of dammarane-type saponins (ginsenoside Rx) and determined their structures.

General blind screening of Ginseng saponins showed that ginsenoside Rb_1 has central nervous system (CNS)-suppressing and tranquillizing effects, while ginsenoside Rg_1 has a CNS-activating effect. These physiological activities observed with the isolated components are not inconsistent with the description of Ginseng in "*Shin Nung Pen T'sao Ching*", which says "Panax ginseng is used for repairing the five viscera, quieting the spirit, curbing emotion, overcoming agitation, removing noxious influences, brightening the eyes, enlightening the mind and increasing wisdom".

Ohura and Hiai attempted biochemical screenings of aqueous extracts of various crude drugs by measuring their effects on the incorporation of ^{3}H- and ^{14}C-labelled crotonic acid into nuclear RNA of rat liver cells. Ginseng was found to have remarkable activity in this test, and the so-called prostisol fraction (corresponding to the ginsenoside Rx fraction) was found to be, responsible for this effect.

Potentiation of nucleic acid and protein biosynthesis, or of carbohydrate and lipid metabolism, by ginsenoside Rx was further investigated by Ohura, Hiai *et al.*,[10] Yamamoto *et al.*,[11] and Higashi *et al.*[12] For example Yamamoto *et al.* found that ginsenosides Rb_2, Rc and Rg_1 increased the incorporation rate of thymidine *in vivo* into rat bone marrow cells (an index of the DNA biosynthetic rate), while ginsenosides Rb_1 and Re apparently had no effect. Similar tendencies were observed in protein and lipid biosynthesis. Higashi *et al.* observed that when rats were given 5 mg/100 g of a ginsenoside intraperitoneally ginsenoside Rb_1 increased the incorporation rate of [^{3}H]orotic acid into nuclear RNA of liver cells after 4 hr, while Rg_1 had very little effect and Rc rather reduced the rate. The acceleration effect of Rb_1 seems to be an indirect effect, because no increase in RNA polymerase activity was observed in *in vitro* experiments.

Higashi *et al.*[12] also observed that ginsenosides Rb_1 and Rc increased the rate of [^{14}C] leucine incorporation into serum protein by 2.2 and 2.4 times, respectively, while Rg_1 apparently had no effect. Ginsenoside Rb_1 was shown to have a significant effect on the biosynthesis and metabolism of cholesterol in rat liver *in vivo*. However, this effect was not observed *in vitro*. The effect may therefore again be considered to be indirect. The saponins in Gingseng were thus clearly shown to have very interesting biological activities, though the correlation between those biochemical experimental results and actual pharmacological effects produced in patients remains to be investigated in detail.

28.2 Identificatin of Samples

Chinese crude drugs comprise mainly parts of higher plants, with some higher fungi, materials of animal origin and minerals to lesser extents, and those available in Japan are mostly imported from China or south-east Asia. Thus, the biological identification of the crude drugs is extremely difficult and laborious, and not always successful, except for those that are cultivated in Japan.

Study of the internal and external morphological characteristics may not always lead to correct identification, but determination of the profiles of the constituents may sometimes aid identification.

In the market, materials considered to be identical may sometimes be called under different names, depending on the region of production, quality, etc., which makes definite botanical identification of the material extremely difficult or even impossible. Equally, crude drugs of different botanical origin may sometimes be called by the same name. These often belong to the same family and the same genus, but not always. A part of every specimen should always be retained for reference in case further identification studies should subsequently be desirable.

For example in the Japanese market, "*Tu-Huo*" (*Dokkatsu*), a crude drug believed to relieve or mitigate headache, leg pain, etc., is the root of *Aralia cordata* (Araliaceae), while in the Chinese market, it is the root of *Angelica pubescens* (*A. shishiudo*), *A. hemsleyganum*, *A. laxiflora*, or *A. megaphylla* (Umbelliferae).

From Japanese "*Tu-Huo*" (root of *Aralia cordata*) and the root of cultivated *A. racemosa*, which provides a cough mixture or medicine for stomach troubles according to European folk medicine, the authors[13] isolated diterpenes such as (–)-pimaric acid and its derivatives, and (–)-kaurenoic acid derivatives, whereas from *Angelica pubescens* (Chinese "*Tu-Huo*"), Kariyone and Hata obtained coumarin derivatives such as grablalactones. Thus, according to the chemical constituents isolated so far, the two seem to be quite different.

Another example is "*Wujiapi*", a crude drug used as a tonic. Root cortex of various *Acanthopanax* plants (Araliaceae) such as *A. gracilistylus* ("*Nan-Wujiapi*"), *A. senticosus* (= *Eleutherococcus senticosus*) is sold under the

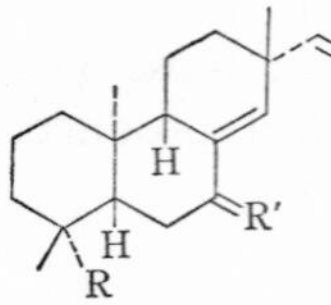

(CH₂OH, OH, H, COOH structural formulas)

R = COOH, R′ = H_2	(−) pimara-8(14)15-dien-19-oic acid	
R = COOH, R′ = <H, OH	7-α-hydroxy-(−)-pimara-8(14)15-dien-19-oic acid	
R = COOH, R′ = O	7-keto-(−)-pimara-8(14)15-dien-19-oic acid	
R = COOH, R′ = <OH, H	7β-hydroxy-(−)-pimara 8(14)15-dien-19-oic acid	
R = CH_2OH, R′ = H	(−)pimara-8(14)15-dien-19-ol	

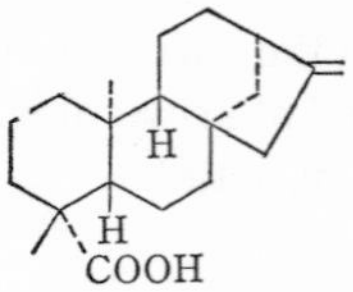

(−)kaur-16-en-19-oic acid

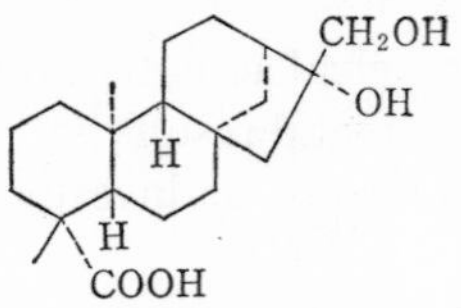

16,17-dihydroxy-16β(−)-kauran-19-oic acid

name *"Wujiapi"*, though in Japan most of the *"Wujiapi"* imported is from *Periploca sepimu* (Ascrepiadaceae) (*"Bei-Wujipai"*). Elyakov *et al.*[14] isolated eleutherosides A, B, B_1, C, D, E and F from the root of *E. senticosus* (Araliaceae). The aglycone moiety of eleutherosides B, D and E is a lignan, silingaresinol. According to Brekhman, these glycosides possess antifatigue or antistress activity (so-called "adaptogen" activity). On the other hand, Shoji *et al.*[15] isolated from the root bark of *Periploca sepium* a small amount of cardiac glycoside, periplocin, along with several steroid glycosides of the pregnane series. Here again, crude drugs having different chemical components are apparently used for the same purpose, under the same name.

28.3 Separation, Identification, and Determination of the Constituents of Chinese Crude Drugs

As mentioned above, crude drugs are conventionally identified according to their external and internal morphological characteristics. However, the chemical profiles of the constituents may often be a more reliable indicator for identification. Some examples will be given here of the use of modern separative methods for the identification of constituents of crude drugs.

Droplet Counter-Current Chromatography[16]

Droplet counter-current chromatography (DCC) was originally devised by Tanimura and Pisano.[17] It is based on the difference in the partition coefficients of substances between two mutually immiscible solvents, and thus, the method is in principle similar to the conventional counter-current distribution method. The droplet method, however, has several advantages: the apparatus is smaller, does not use fragile distribution tubes, does not require mechanical shaking, uses only a small quantity of solvents, etc.

With some modifications, this method was found to provide a very useful means for semi-micro or micro quantitative and qualitative analysis, and for the separation or purification of the components contained in crude drugs. It is particularly suitable for processing compounds such as saponins that foam easily. The method is also advantageous in that it does not involve the use of a solid absorbent or carrier, and heating, which may cause decomposition, it unnecessary. The compounds are not in contact with air during the process, and may easily be kept in the dark, so that oxygen-sensitive compounds may be safely processed.

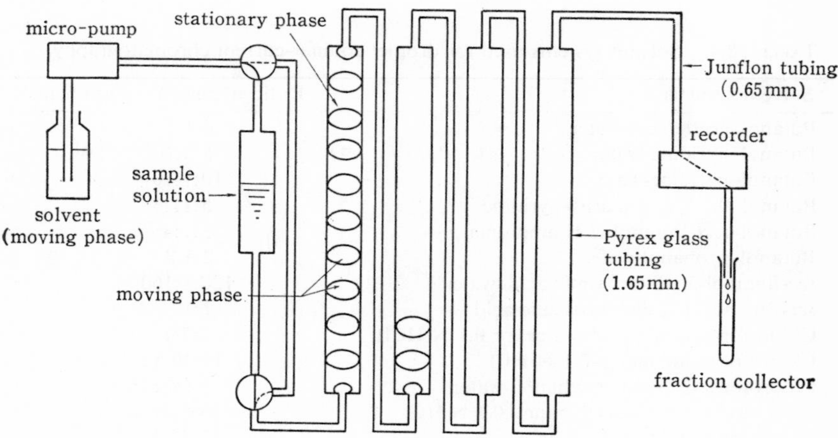

Fig. 28.1. Droplet counter-current chromatography apparatus.

The apparatus (Fig. 28.1) comprises 500–1,000 pyrex glass tubes (40 cm × 1.65 mm) connected by narrow plastic tubes ("Junflon", F_4-ethylene F_6-ethylene copolymer) (55 cm × 0.65 mm), which are stable to acid, alkali or usual organic solvents.

In the ascending method, a lower phase solvent is pumped into all the glass tubes with a micro-pump to produce a stationary phase. The sample to be analyzed is dissolved in a small quantity of a 1:1 mixture of upper phase and lower phase solvents, which is charged into the sample chamber. The upper layer solvent (moving phase) is pumped into the tubes, through which it moves upwards as a series of small droplets. The components contained in the original sample are subjected to separation during the movement of the droplets through the stationary phase solvent, according to their partition coefficients. At the far end of the apparatus a fraction collector is used to collect 3–5 ml fractions of the eluate.

In the descending method, the solvent system is chosen so that a lighter upper phase solvent constitutes the stationary phase while a heavier lower phase solvent moves from the upper end of the tube to the bottom as small droplets.

Collected fractions are analyzed either spectrophotometrically after color development or gravimetrically after evaporation to dryness. Recovery is quantitative.

Solvent systems are chosen from among those used for the development of ordinary tlc that procude droplets of suitable size. Examples are given in Table 28.4.

TABLE 28.4. Solvent systems used for droplet counter-current chromatography.

Solvent systems	Ratio of solvent components
Butanol–acetic acid-water	4:1:5
Butanol–pyridine-water	5:2:10
Butanol–pyridine-water	10:1:10
Butanol–0.1% acetic acid–pyridine	5:11:3
Butanol–*tert*-Butanol–2 N ammonia	3:1:4
Butanol–propanol-water	2:1:3
sec-Butanol–Trifluoroacetic acid–water	120:1:160
sec-Butanol–1% dichloroacetic acid	1:1
Chloroform–acetic acid–water (or 0.1 N HCl)	2:2:1
Chloroform–methanol–0.1 N HCl	19:19:12
Chloroform–benzene–methanol–water	15:15:23:7
Chloroform–benzene–methanol–0.1 N HCl	10:5:11:4
Chloroform–methanol–water	35:65:40
Dichloroethylene–methanol–water	10:5:5
Hexane–ethanol–water–ethyl acetate	5:4:1:2
Ethyl acetate–propanol–water	4:2:7

The size of droplets is affected by the surface tension, the difference of specific gravity between the two liquids, and the diameter and quality of the glass tubes used in the apparatus. For example, in a hexane–water system, the surface tension is large enough to result in the formation of droplets with a diameter of 5 mm, which is too large to pass through the glass tubes as a droplet: movement occurs by plug flow. Thus, such a system is not suitable.

[Experimental procedure 1] Separation of "*Suan-T'sao-Jen*" saponins

Methanol extract of "*Suan-T'sao Jen*" (*Zizyphus jujuba* seeds) was dissolved in water and treated with ether to remove lipids. Butanol extraction provided the saponin fraction. The crude saponin fraction (25 mg) was subjected to DCC separation by the ascending method, using 509 tubes with chloroform–methanol–H_2O (5:6:4) as a solvent. Fractions of 4 ml were collected, and determined spectrophotometrically after phenol treatment with a modified solvent system [chloroform-methanol-propanol-water (45:60:5:40)] Jujuboside A was obtained as crystals (Fig. 28.2).

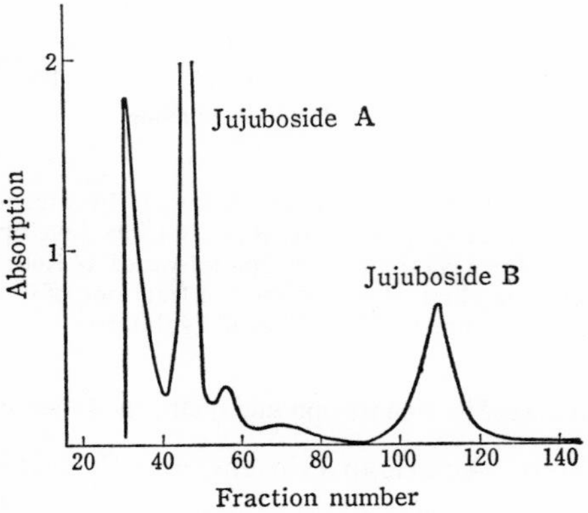

Fig. 28.2. Separation of jujubosides A and B of the saponin of "*Suan-T'sao Jen*" (*Zizyphus jujuba* seeds).
The saponin fraction (25 mg) was developed by ascending DCC using chloroform–methanol–water (50:60:40) over 509 Pyrex glass tubes. Fractions of 4 ml of the moving phase eluates were collected and measured colorimetrically using phenol–H_2SO_4 as a color reagent.

[Experimental procedure 2] Components of *Rheum*[16]

A mixture of 3 mg each of sennosides A, B and C, which are purgative bianthraquinonyls of *Rheum officinalis,* was subjected to DCC with chloroform:methanol:propanol:water (45:60:10:40). Fractions of 4 ml were collected and evaporated to dryness. Five ml of 50% ethanol was added to each residue and the extinction at 260 nm was determined. Sennoside B was successfully separated from its geometrical isomers, though the separation of sennosides A and C was not successful due to their similar retention times (Fig. 28.3).

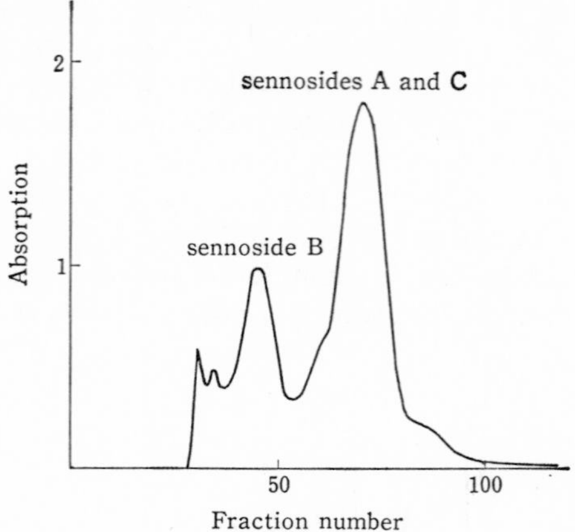

Fig. 28.3. Separation of sennosides A, B and C by ascending DCC. A mixture (3 mg each of sennosides A, B and C) was developed by ascending DCC using chloroform-methanol-*n*-propanol-water (45:60:10:40) over 509 Pyrex glass tubes. The eluate was collected in fractions of 4 ml and measured colorimetrically at 260 nm in 50% ethanolic solution.

[Experimental procedure 3] Separation and quantitative estimation of saponins[16]

Korean Ginseng, American Ginseng and San-Chi Ginseng contain dammarane saponins as their main constituents, and these ginsenosides are known to have interesting physiological activities. However, their separation and purification are not easy.

Powdered white Ginseng (50 g) was treated twice with 200 ml each of benzene for 5 hr to remove lipid-soluble fractions and then extracted three times with 200 ml each of methanol for 5 hr to yield 6.83 g (13.7%) of methanol extract. The extract was suspended in water and treated several times with butanol to yield 1.43 g (2.9%) of butanol-soluble fraction, i.e. saponin fraction, which was then subjected to DCC using 509 separation tubes with chloroform–propanol–methanol–water (45:6:60:40), using the lower phase as the stationary phase. Exactly 25 mg of the sample was dissolved in 3 ml of the stationary phase and injected into the sample chamber. The moving phase was pumped into the apparatus by means of a micro pump, and fractions of 3 ml of the eluate were collected at the far end of the apparatus. Fractions No. 1–36 contained stationary phase solvent only. Fractions No. 37–100 were evaporated to dryness and dissolved in 6 ml each of distilled water; 2 ml aliquots were subjected to analysis. Fractions from No. 101 onward were evaporated to dryness and each was dissolved in 2 ml of distilled water. These samples were analyzed colorimetrically by adding 1 ml of 5% (w/w) phenol and then 5 ml of conc. H_2SO_4 in one portion. After standing at room temperature for 30 min, and then in cold water, the O.D. values were determined at 490 nm (Fig. 28.4). Quantitation was done by reference to calibration curves prepared with authentic ginsenoside samples.

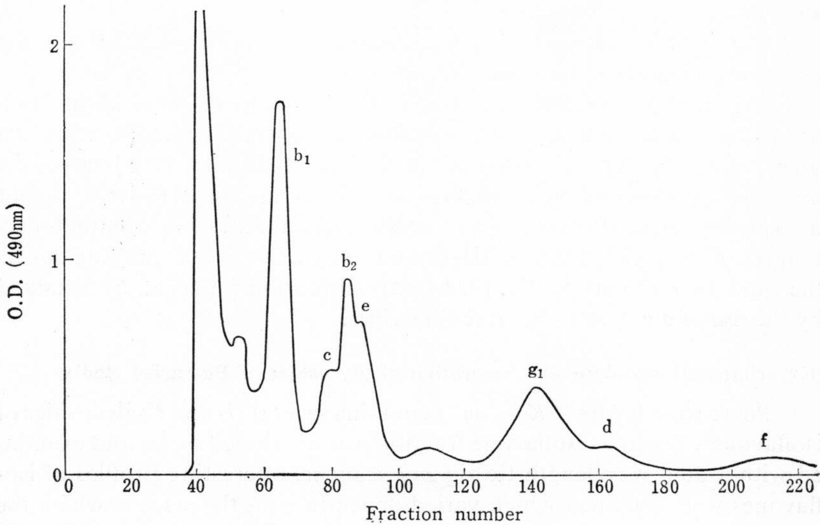

Fig. 28.4. Separation of saponins (ginsenosides R_X) of Ginseng by ascending DCC using chloroform–propanol–methanol–water (45:6:60:40)) over 1,000 Pyrex glass tubes. Measured colorimetrically using phenol–H_2SO_4 as a color reagent. The contents of ginsenosides R_X in the saponin fraction (yield: 2.9%) of White Ginseng produced in Shinshu were as follows:

Ginsenoside R_X	b_1	b_2	c	d	e	f	g_1
%	20	9	5	1	7	4	17

Dual Wavelength Chromatogram Scanning Method

Recent advances in chromatographic techniques have permitted the microchemical investigation of many naturally occurring substances by means of chromatographic analysis. However, simplicity and ease of operation make thin layer chromatography (tlc) outstanding. For quantitative analysis by means of tlc, it is usual to scrape the plate at the region of interest, extract the resulting material with an appropriate solvent and spectrophotometrically examine the extract directly or after color development with appropriate reagents. However, quantitative recovery of the material from the adsorbent, or exact estimation of the recovery, is not easy.

Direct densitometric analysis of the spots on a thin layer chromatogram is more reliable, but suffers from the disadvantage that linear calibration plots are not always obtainable. This may arise from (1) error due to the thin layer plate film itself (background error), (2) errors due to distorted shape of the spots, as the scanning is performed in one direction only, and (3) a non-proportional relationship between the size of the spot and the quantity of the material.

The dual wavelength tlc chromatoscanner largely overcomes these problems, offering reliable and sensitive measurement.

The method may be characterized as follows. (1) One of the two beams, of wavelength λ_S, is used for measurement of the sample while the other, of wavelength λ_R, is used for background scanning. Thus, the results obtained are all based on the difference of the extinction of the two, so that background error is eliminated or at least minimized. (2) Scanning is carried out slowly with a lateral zig-zag movement; thus the scanning covers the spots two-dimensionally. (3) Linear calibration plots can be obtained by the use of a calibration curve linearizer.

[Experimental procedure 4] Separation of glycosides of Puerariae Radix

Puereariae Radix (*"Kakkon"*) contains several *O*- and *C*-glycosides of isoflavones. When its isoflavone fraction was developed on tlc and examined with a dual wavelength tlc zig-zag scanner, interesting profiles of isoflavones were obtained which varied depending on the place at which the material had been obtained. The resulting profiles may be used for the identification of the crude drug (Fig. 28.5).

The above results show clear qualitative differences among various *"Kakkon"* specimens. Thus Korean and Japanese *"Kakkon"* specimens were similar while those of Chinese and Thai origin were apparently different. The white variety of Thai *"Kakkon"* resembles Japanese and Korean *"Kakkon"* in part, but is not exactly the same.

When a scanning wavelength of 250 nm (λ_S) is used for a tlc plate on which an extract of Chinese *"Kakkon"* has been developed, with 320 nm background compensation instead of 400 nm (the usual wavelength for background scanning), a characteristic invested peak is observed. This is caused by a compound which has stronger absorption at 320 nm than at 250 nm.

Thus, plots obtained at various wavelengths can provide as much information on these compounds as the spectrogram measured with the isolated compounds.

In the above case (Figs. 28.5 and 28.6) the main component, puerarin, emits fluorescence when irradiated with uv light, so that the amount estimated is only half of the amont actually present.

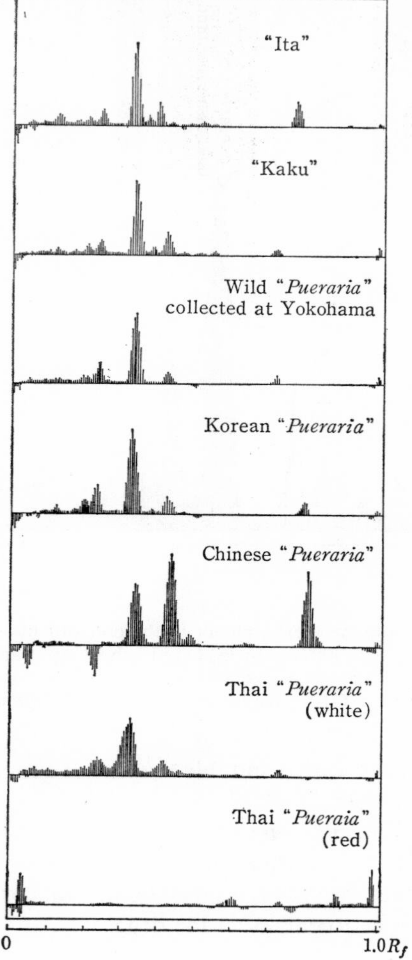

Fig. 28.5. Profile analysis of extracts of "*Pueraria*" root by tlc, using precoated plates of silica gel 60F-254 (Merck) with the lower phase of chloroform–methanol–water (65:35:10). λ_S 250 nm; λ_R 320 nm.

When *"Kakkon"* extract was developed on tlc with the lower phase of chloroform–methanol–ethyl acetate–water (2:4:4:1), the puerarin spot (Fig. 28.6) was resolved into two compounds, i.e., puerarins and another unidentified compound (Table 28.5).

When the compounds of interest have no uv absorption, the chroma-

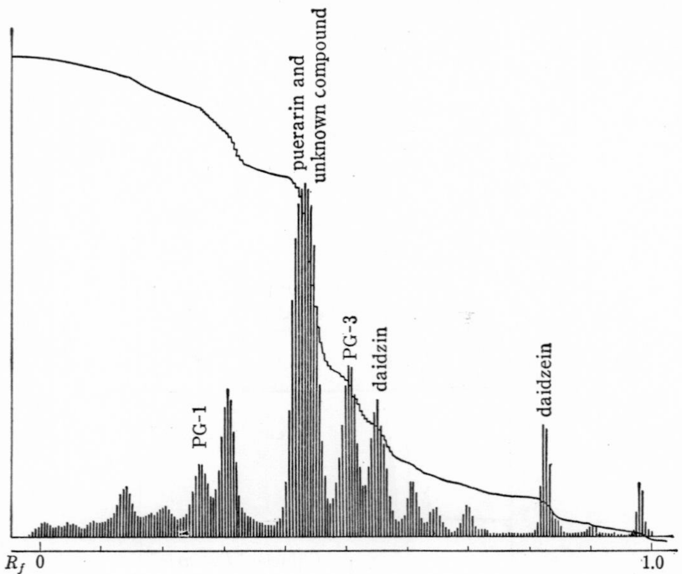

Fig. 28.6. Quantitative measurement of the constituents of "*Pueraria*" root with a Shimadzu CS-900 dual-wavelength chromatogram zig-zag scanner. Precoated silica gel 60F-254 plates (Merck) with the lower phase of chloroform–methanol–water (65:35:10). Wavelengths: 280 and 400 nm.

togram may be treated with a suitable reagent or conc. H_2SO_4 before densitometric measurement.

The profiles of absorption at various wavelengths may provide quantitative and qualitative information on unknown compounds, and may also indicate whether the spot is a single compoents or not, which is useful in surveys of the constituents of crude drugs.

ACKNOWLEDGEMENTS

The pharmacological data cited in this article are taken from papers reported by Prof. K. Takagi, Prof. M. Harada, Dr. M. Shibata and Dr. E. B. Lee. The DCC separation and the dual wavelength chromatogram scanning of Chinese drug components were performed by Dr. H. Otsuka, University of Tokyo, and Prof. T. Saitoh, Teikyo University, respectively, when they were in my laboratory at the University of Tokyo. I wish to

TABLE 28.5 Constituents of "*Pueraria*"†[1]

I R=H
II R=Glc

III R=H
IV R=OH
V R=OCH_3

	Daidzein (I)	Daidzin (II)	Puerarin (III†[2])	Puerarin (III†[3])	Pg-1 (IV)	Pg-3 (V)
Chinese "*Pueraria*"	0.1	0.2	0.3	0.1	0	0
Korean "*Pueraria*"	1.3	0.6	5.5	2.4	2.3	0.7
"*Ita*"	0.3	0.2	1.9	0.9	0.8	0.4
"*Kaku*"	2.0	1.0	6.9	3.1	2.1	1.1
Yokohama	1.4	0.6	6.9	3.9	2.1	0.4
Thai (White)	0.0	0.0	0.3	0.2	0	0
Thai (Red)	0	0	0	0	0	0

†[1] Figures are gravimetric % of the components in the dried "*Pueraria*" roots.
†[2] Measured at 250 nm.
†[3] Measured at 365 nm (fluorescence).

thank them for their cooperation. I am indebted to Dr. (Mrs.) M. Watanabe for the preparation of the English text of this article.

REFERENCES

1) H. Nishikawa, *Ruijuho-kogi, Hoki, Hokyoku, Juko-Yakucho* (in Japanese), Ikeuchi-shoji (1969).
2) M. Harada, *Yakugaku Zasshi*, **89**, 899 (1969).
3) M. Shibata, *ibid.*, **90**, 398 (1970).
4) E. B. Lee, *ibid.*, **93**, 1188 (1973).
5) K. Takagi, H. Saito, H. Nabata, *Japan J. Pharmacol.*, **22**, 245 (1972); K. Takagi, H. Saito, Y. Higuchi, A. Yamaguchi, *Oyoyakuri*, **5**, 5 (1971).
6) I. I. Brekhman, I. V. Dardmov, *Ann. Rev. Pharmacol.*, **9**, 419 (1969).
7) G. B. Elyakov, L. I. Strigna, N. K. Kochetkov, *Khim. Prirodn. Soedin.*, **3**, 149. (1965); G. B. Elyakov, N. I. Uvarova, R. P. Gorshkova, Yu. S. Ovodov, N. K. Kochetkov, *Dokl. Akad. Nauk. USSR*, **165**, 1309 (1965); G. B. Elyakov *et al.*, *Tetr. Lett.*, **1966**, 141; *Tetrahedron*, **24**, 5483 (1968) and references cited therein.
8) Y. Nagai, O. Tanaka, S. Shibata, *ibid.*, **27**, 881 (1971); S. Sanada, N. Kondo, J. Shoji, O. Tanaka, S. Shibata, *Chem. Pharm. Bull.*, **22**, 421, 2407 (1974).

9) H. Saito, Y. Yoshida, K. Takagi, *Japan J. Pharmacol.*, **24**, 119 (1974) and references cited therein.
10) S. Hiai, H. Oura, *Tanpaku-Kakusan-Koso*, **18**, 13 (1973) and references cited therein; H. Oura, S. Hiai, *Wakanyaku*, **10**, 564 (1973) and references cited therein.
11) M. Yamamoto, *Wakanyaku*, **10**, 581 (1973) and references cited therein.
12) K. Sakakibara, Y. Shibata, T. Higashi, S. Sanada, J. Shoji, *Chem. Pharm. Bull.*, **23**, 1009, (1975); T. Higashi *et al.*, *ibid.*, **24**, 2400 (1976).
13) S. Shibata, S. Mihashi, O. Tanaka, *Tetr. Lett.*, **1967**, 5241; S. Mihashi, I. Yanagisawa, O. Tanaka, S. Shibata, *ibid.*, **1969**, 1683.
14) Yu. S. Ovodov, R. G. Ovodova, T. F. Solovera, G. B. Elyakov, N. K. Kochetkov, *Khim. Prirodn. Soedin.*, **1**, 3 (1965); Yu. S. Ovodov, G. M. Frolova, L. A. Elyakova, G. B. Elyakov, *Izu. Akad. Nauk SSSR Ser. Khim.*, **11**, 2065 (1965); Yu. S. Ovodov, G. M. Frolova, M. Yu. Nefedova, G. B. Elyakov, *Khim. Prirodn. Soedin.*, **1**, 63 (1967).
15) J. Shoji, S. Kawanishi, S. Sakuma, S. Shibata, *Chem. Pharm. Bull.*, **15**, 720 (1967); *ibid.*, **16**, 326 (1968) and references cited therein.
16) Y. Ogihara, O. Inoue, H. Otsuka, K. Kawai, T. Tanimura, S. Shibata, *J. Chromatog.*, **128**, 218 (1976).
17) T. Tanimura, J. J. Pisano, Y. Ito, R. L. Bowman, *Science*, **169**, 54 (1971).
18) H. Otsuka, Y. Morita, Y. Ogihara, S. Shibata, *Planta Medica*, **32**, 9 (1977).
19) T. Saitoh, S. Shibata, *Abstr. Ann. Meeting of Japan Soc. Pharmacog.*, p. 23 (1975).

CHAPTER 29

Studies on the Constituents of Crude Drugs by Bioassay: Bioactive Components of Rhei Rhizoma and *Artemisiae capillaris* Flos

Research on the constituents of crude drugs has developed greatly with recent advances in analytical techniques, and many compounds have been isolated from crude drugs. However, the biological activities of many compounds have not been fully studied in spite of the medicinal use of the source crude drugs. Recently, new bioassays have been developed for the investigation of new active substances or estimation of the quality of crude drugs.

The authors have applied bioassays for many years to estimate the effectiveness of crude drugs or to investigate new bioactive substances.[1)] In the course of our research, for example, it was shown that the antiinflammatory effect of Coptidis Rhizoma and Phellodendri Cortex was attributable to berberine, which is the major alkaloid of both drugs. New hypocholesterolemic triterpenes, alisols, were also isolated from Alismatis Rhizoma.

Although derivatives of oxyanthraquinone and their glycosides were thought to be the purgative principles of Rhei Rhizoma until the nineteen-sixties, more active compounds, sennosides, were detected by Zwaving in 1965. The authors also investigated in detail the bioactive constituents of this drug with the aid of bioassay, and showed that strong activity always appeared in the fractions which contained sennosides. Six analogs of sennoside were subsequently isolated.

In the screening of substances which inhibit the antibiotic action of chloramphenicol against *E. coli,* the authors isolated an active compound, named capillarisin, from *Artemisiae capillaris* Flos, used as an important cholepoietic in Chinese medicine, and showed that capillarisin was one of the cholepoietic components of the crude drug.

29.1 Studies on the Purgative Components of Rhei Rhizoma

Rhei Rhizoma (*daio* in Japanese) is the dried rhizome of *Rheum* spp. (Polygonaceae), and has been used as a purgative or a digestive. In Chinese medicine, it is included in many important prescriptions such as "*dai-saiko-to*" or "*daio-kanzo-to*". The quality of the drug varies markedly according to the species of the original plant. The useful drugs, in general, are available from the *Palmata* section such as *R. palmatum* L., *R. tanguticum* Max., *R. officinale* Baillon or *R. coreanum* Nakai and are named "*kinmon daio*", "*to daio*", etc. in Japanese markets. On the other hand, the drugs originating from the *Rhapontica* section, such as *R. rhaponticum* L. or *R. undulatum* L., and named "*wa daio*" or "*do daio*" in the market are less effective as a purgative. Although the best drug is produced principally in China, a kind of "*kinmon daio*" has been produced in Japan recently.

The well-known constituents of Rhei Rhizoma include derivatives of oxyanthraquinone such as chrysophanol, physcion, emodin, aloe-emodin and rhein, and their glycosides. Fairbairn and Matsuoka independently examined the purgative activity of crude drugs and oxyanthraquinones, and concluded that active components were present. In 1965, Zwaving detected sennosides A, B and C in rhizomes of *R. palmatum* L. by paper partition chromatography (ppc). Miyamoto *et al.* also recognized that the acidic fraction of rhizomes of *R. coreanum* Nakai showed strong purgative activity using the mouse assay method, and they succeeded in isolating sennoside A in 1967. Sennosides A and B were first isolated as the purgative principles of *Cassia angustifolia* Vahl by Stoll in 1949, then sennosides C and D were isolated from the same plant later.[2,3] In the course of chemical and biological studies of a hybrid between *R. coreanum* Nakai and *R. palmatum* L., which was named "*Shin-shu daio*", the authors isolated sennosides A, B, C, D and two new analogs, sennosides E and F (Table 29.1).[4,5]

Assay of the Purgative Activity of Rhei Rhizoma

Grot *et al.*, Miller *et al.*, Lou *et al.*, and Matsuoka used rats or mice for the assay of purgatives. In 1969, Fujimura *et al.* also designed a valuable method of assay: the purgative was administered orally, then ED_{50} was calculated statistically from the rate of diarrhea. This method showed good reproducibility, and was used in our study.[6]

TABLE 29.1. Structures of sennosides and oxyanthraquinones.

Compound	R_1	R_2	Stereostructure (10–10′)
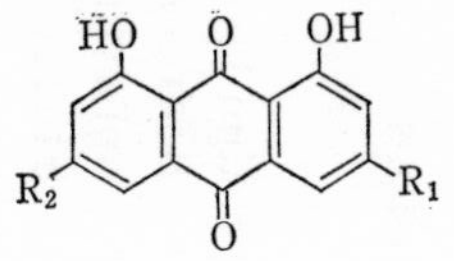 chrysophanol	CH_3	H	
physcion	CH_3	CH_3O	
emodin	CH_3	OH	
aloe-emodin	CH_2OH	H	
rhein	COOH	H	
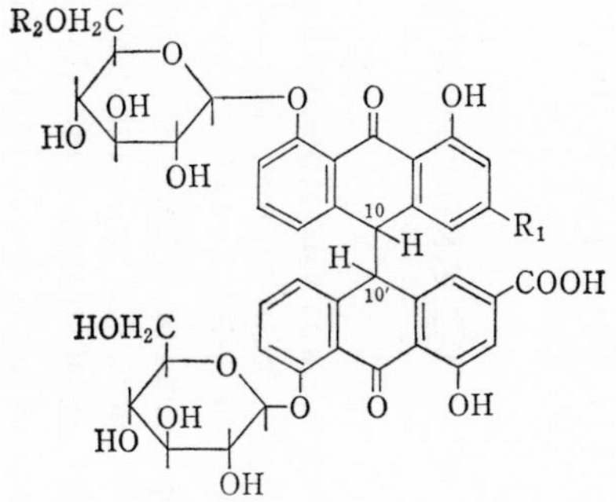 sennoside A	COOH	H	*threo*
sennoside B	COOH	H	*erythro*
sennoside C	CH_2OH	H	*threo*
sennoside D	CH_2OH	H	*erythro*
sennoside E	COOH	COCOOH	*threo*
sennoside F	COOH	COCOOH	*erythro*

[Experimental procedure 1] Bioassay of purgative activity

Three hundred mg of ground crude drug, Rhei Rhizoma (below 50 mesh), was weighed accurately, then suspended in 5% gum arabic solution to 20 ml. A portion of the suspended solution was diluted 1.2–fold with the gum arabic solution, and 6 suspensions were prepared. In this test, 4-week-old $CF_{\#1}$ mice were used. The mice received diet and water *ad lib.* until 1 hr before the test, and mice producing normal feces were selected. The suspension was orally administered to the mice with a 1 ml tuberculin syringe at 0.2 ml/10 g.

Sixty mice were used to assay a purgative. The mice were transferred individually to small cages, and the floors were covered with filter papers. Food and water were not given during the test. The feces were observed at hourly intervals for 8 hr after administration. The feces were classified as normal (0), swollen (0.5) and adhering to the paper, and fluid (1.0), staining the paper. The purgative rate was calculated from total index of 10 mice in a group. The ED_{50} value and 95% confidence limits (C.L.) were also calculated by the method of Litchfield and Wilcoxon.

The results were: 300 mg/kg 10/10 (100%), 250 mg/kg 9/10 (90%), 208 mg/kg 6/10 (60%), 174 mg/kg 3/10 (30%), 145 mg/kg 0.5/10 (5%), 121 mg/kg 0/10 (0%). ED_{50} (95% C.L.) = 196 (175–220) mg/kg.

[Experimenal procedure 2] Purgative activity of sennoside A

Sennoside A (25mg) was suspended in 5% gum arabic solution to 20 ml. A portion of the suspended solution was then diluted 1.3 fold with the gum arabic solution, and the suspensions were administered by the method described above.

The results were: 25.0 mg/kg 10/10 (100%), 19.2 mg/kg 9.5/10 (95%), 14.8 mg/kg 7/10 (70%), 11.4 mg/kg 2/10 (20%), 8.8 mg/kg 1/10 (10%), 6.7 mg/kg 0/10 (0%). ED_{50} (95% C.L.) = 13.5 (12.2–15.0) mg/kg.

29.2 Analyses of Sennosides

For the qualitative analysis of sennosides, ppc, thin layer chromatography (tlc) or paper electrophoresis is convenient. Both sennosides and oxyanthraquinones are visible as yellow-colored spots on a chromatogram. However, these spots can be distinguished easily, since the color of the former changes to enhanced yellow and that of the latter to orange or violet on spraying a solution of sodium hydroxide. In addition, when the paper is placed between a uv lamp and a sheet of Lumicolor plate, spots of sennosides are visible as dark violet, and spots of oxyanthraquinones as orange through the transmitted beam.

Photometric and polarographic methods have been proposed for the quantitative analyses of sennosides. High-performance liquid chromatography gave good results for sennoside A, the main compound, but it was difficult to clean up the starting materials sufficiently to make it possible to determine the other sennosides.

[Experimental procedure 3] Paper partition chromatography of sennosides

The extract of Rhei Rhizoma and authentic compounds were spotted on paper which had been impregnated with 0.2 M citrate buffer (pH 6.2) and dried. The paper was developed overnight using the upper phase of BuOH–EtOH–the buffer (2:1:2) by the ascending method. After drying the paper, the spots were detected by the Lumi-color or alkali method.

Rf values of sennosides (Toyo Roshi No. 51 paper) were: A,0.37; B,0.18; C,0.60; D, 0.54; E, 0.24; F, 0.11.

[Experimental procedure 4] Thin layer chromatography of sennosides

Thin layer chromatography was performed with silica gel plates, eluting with PrOH–AcOEt–H_2O (4:4:3). After developing for more than 10 cm, each plate was dried, then sprayed with NaOH solution. A broad spot which consisted of sennosides B, E and F was observed near the base line, and sennoside A was located close by. Sennosides C and D could not be distinguished from oxyanthraquinones because of the complexity of the spot pattern. This method is useful because of its speed and simplicity, but it is not wholly reliable, because the *Rf* values of sennosides depend on the concentrations or purities of test samples.

[Experimental procedure 5] Paper electrophoresis of sennosides

After spotting the extract of Rhei Rhizoma or authentic compounds on the cathode side of the paper, electrophoresis was performed in 0.04 M citrate buffer (pH 6.2) for 1 hr at 900 V, 1.6 mA/cm. The paper was dried and the distance of each sennoside from the starting point was measured.

The migration distances of sennosides (Toyo Roshi No. 51 paper) were: A, 69 mm; B, 82 mm; C, 58 mm; D, 64 mm; E,95 mm; F, 135 mm.

[Experimental procedure 6] Quantitative analysis of sennosides by high-performance liquid chromatography

Ground Rhei Rhizoma (1 g) (below 100 mesh) was added to 10 ml of phosphate buffer (pH 6.0). After shaking for 30 min, the suspension was centrifuged. One-half ml of supernatant solution was adsorbed on top of a column (1 × 3 cm), which was packed with cellulose powder. After washing with 30 ml of AcOEt, the sennosides were eluted with 70% MeOH solution, then the eluate was concentrated below 40° C *in vacuo*. The residue was dissolved again in 10 ml of phosphate buffer, and 5 μl of this solution was subjected to high-performance liquid chromatography on a Hitachi 634 unit: Hitachi gel 3010 (2.1 × 1,000 mm), 60° C, 0.6 ml/min, 280 nm. The retention times and amounts of sennosides were: (a) Mc-Ilvaine buffer (pH 2.2). THF (4:1): A, 14 min 0.71%; mixture of B and C, 10 min 0.65%; E, 24 min 0.31%. (b) Clark-Lubs phosphate buffer (pH 6.0)-THF (9:1): C, 11 min 0.11%; mixture of A, B and E, 5–5.5 min. Sennosides D and F were not detected in this experiment since they were present in very small amounts.

[Experimental procedure 7] Isolation of sennosides from Rhei Rhizoma (1)

Fractions which contain sennosides should generally not be treated above 60° C, and it is desirable that the temperature during extraction and concentration should not exceed 50° C. When ground crude drug, for example, is suspended in water and heated at 100° C for 1.5 hr, the activity of the drug falls to about half. The target compounds should be checked regularly by bioassay and qualitative analysis during the isolation procedure for sennosides, because sennsodies gradually tend to degrade to inactive brown substances. The isolation procedure adopted was therefore as follows.

A mixture of 490 g of ground crude drug (ED_{50}, 250 mg/kg) and 1 l of saturated NaCl solution was washed three times with 2 l of THF. The last washing showed almost no coloration. Although the substances soluble in THF showed activity as high as ED_{50} 200 mg/kg, these consisted mainly of oxyanthraquinones without a carboxyl moiety in the structure. The muddy residue was acidified with 50 g of oxalic acid and again extracted three times with 2 l of THF. The residue no longer showed activity. The THF solution was dried over Na_2SO_4, then concentrated to dryness *in vacuo* below 40° C to afford 75 g of an almost black mass (ED_{50} 40 mg/kg). The mass was dissolved in 200 ml of MeOH, then 4.9 g of yellow crystals consisting of sennosides A, C and E (ED_{50} 14 mg/kg) precipitated gradually. The crude crystals were recrystallized twice from 70% acetone. The yellow crystals were almost pure sennoside A. After concentrating the mother liquor, the residue was treated with a small volume of 70% MeOH, then neutralized with $NaHCO_3$ and dissolved entirely. The solution was subjected to Sephadex LH-20 column (7 × 100 cm) chromatography, eluting with 70% MeOH. Each fraction was concentrated below 50° C until MeOH was removed, then the residue was acidified to pH 1.5 with HCl and extracted with BuOH. The BuOH solution was concentrated to dryness below 50° C. The fractions which contained sennoside E, A or C were recrystallized from 70% acetone.

The mother liquor from which the crude crystals of sennosides A, C and E had been isolated was still active (ED_{50}, 50 mg/kg). It was subjected to silica gel column (6 × 50 cm) chromatography, and sennosides were eluted with 99% acetone. Sephadex LH-20 column chromatography of this fraction yielded sennosides F, E, B and A successively. Sennoside D was not separated from oxyanthraquinones. Yields of individual sennosides with respect to the raw material were: A, 0.772%; B, 0.014%; C, 0.037%; E, 0.067%; F, 0.0006%.

[Experimental procedure 8] Isolation of sennosides from Rhei Rhizoma (2)

After extracting 1 kg of ground crude drug three times with 4 l of 70% THF, all extracts were combined and concentrated to dryness. The extract (495 g) was active (ED_{50}, 110 mg/kg), but no activity remained in the residue. The extract was dissolved in a heterogenous mixture of 2 l each of THF and saturated NaCl solution. The aqueous phase was acidified to pH 1.5 with HCl, then extracted twice with 2 l of THF. After drying over Na_2SO_4, the THF solution was concentrated *in vacuo* below 40° C affording 40 g of a black mass (ED_{50}, 35 mg/kg). The black mass was dissolved in MeOH and left for several hours; 8.2 g of dark yellow crystals consisting of sennosides A, C and E precipitated. Next, 300 ml of 10% Pb (OCO-$CH_3)_2$ solution was added to the mother liquor, and the brown amorphous precipitate was filtered. After removing oxyanthraquinones by washing with MeOH, the amorphous material was added to 300 ml of 5% Na_2SO_4 solution and stirred vigorously. Sennosides were dissolved in the aqueous solution, and insoluble substances were filtered off. The red solution was acidified to pH 1.5 with HCl, then extracted with BuOH, and BuOH was removed *in vacuo* to afford a brown powder (ED_{50} 20 mg/kg). When Sephadex LH-20 column chromatography was applied as described in "Experimental procedure 7", sennoside D was also obtained in a yield of 0.0002%.

29.3 Isolation and Activity of Oxyanthraquinones from Rhei Rhizoma

The solubilities of oxyanthraquinones in Rhei Rhizoma vary widely. Chrysophanol, for example, is soluble in petroleum ether, but rhein glycoside is soluble only in polar solvents, like sennosides. Although all the glycosides are glucosides, eight kinds of analogs have been isolated from

Rhei Rhizoma. Nishioka *et al.*[7] reported the isolation and purification of oxyanthraquinones. The free oxyanthraquinones can be isolated easily by the following procedure. After moistening the crude drug with 5% hydrochloric acid solution, it is extracted with ethyl acetate. The extract is fractionated by silica gel column chromatography using petroleum ether, ethyl acetate, acetone, methanol, water or various mixtures of these solvents. Chrysophanol, physcion, emodin, aloe-emodin and rhein are isolated successively.

The glycosides of oxyanthraquionones can be isolated by the following procedure. The 60% methanol extract of the crude drug is dissolved in a mixture of butanol and acidic water. The butanol soluble substances are fractionated by silica gel column chromatography, eluting with ethyl acetate–methanol–water (95:5:5). Each fraction, if necessary, is purified by Polyamide column chromatography with methanol. 8-Glucosylrhein is generally isolated together with sennosides.

Oxyanthraquinones can be analyzed effectively on silica gel thin layer plates. The following mixed solvents are useful as developers: (a) petroleum ether-ethyl acetate (9:1), (b) benzene–ethyl formate–formic acid (75:25:2), (c) benzene–ethyl formate-ethyl acetate–formic acid (75:24:0.8:0.2), (d) ethyl acetate–methanol–water (100:16.5:13.5), (e) propanol–ethyl acetate–water (4:4:3). The polyamide–methanol system is often useful.

Sennosides have the strongest purgative activities among the constituents of Rhei Rhizoma. All sennosides have comparable activities in the mouse assay method. The activities of oxyanthraquinones, on the other hand, vary widely. Aloe-emodin and rhein are fairly active, but chrysophanol, physcion and emodin have little activity (Table 29.2).

TABLE 29.2. Purgative activities of constituents of Rhei Rhizoma.

Compound	ED_{50} (mg/kg)	Compound	ED_{50} (mg/kg)
Sennoside A	13.5	chrysophanol	$>$ 500
Sennoside B	13.9	physcion	$>$ 500
Sennoside C	13.3	emodin	$>$ 500
Sennoside D	15.8	aloe-emodin	59.6
Sennoside E	13.5	rhein	97.5
Sennoside F	16.1		

There remain many ambiguous factors regarding the correlation between active components and the overall activity of rhubarb. A mixture of sennoside A and inactivated crude drug, for example, shows stronger activity than the same dosage of sennoside A in mice. However, it seems likely that the main active components are sennosides, because the activity of the crude drug correlates roughly with its sennosides content.

29.4 Studies on Cholepoietic Components of Artemisiae Capillaris Flos

Artemisiae capillaris Flos, (*inchinko* in Japanese), has been used as a cholepoietic in Chinese medicine, and corresponds to *Artemisiae capillaris* Thunb. or *A. scoparia* Waldst. et Kit. (Compositae). In Japan, the dried spikes of the former are commonly used, and it is the main drug in such Chinese prescriptions as *"inchinko-to"* or *"inchin-gorei-san"* which have been used as a remedy for jaundice. Dimethylesculetin is known to be the effective component of the crude drug.[8)]

In the course of investigation of the effects of extracts of crude drugs on microorganisms, the authors noted that the effect of chloramphenicol on *E. coli* was inhibited by the extract of Artemisiae capillaris Flos, and a new active compound, named capillarisin, was isolated. It was assumed on the basis of the medicinal use of the source material that capillarisin would show cholepoietic action, and in fact, it showed stronger activity than dimethylesculetin in bioassays with rats, and dogs. The authors also isolated four analogs of capillarisin, and elucidated their chemical structures from the physico-chemical data, together with synthetic studies. These compounds have a 2-phenoxychromone moiety, which is a novel structure among natural products (Table 29.3).[9,10)]

Table 29.3. Structures of capillarisins.

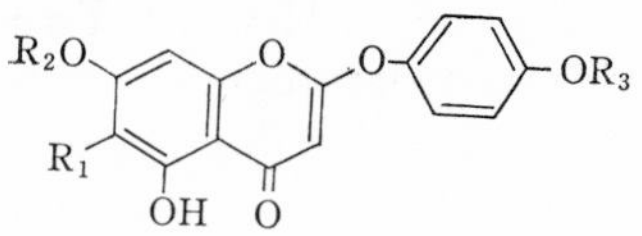

Compound	R_1	R_2	R_3
capillarisin	CH_3O	H	H
7-methylcapillarisin	CH_3O	CH_3	H
4′-methylcapillarisin	CH_3O	H	CH_3
6-desmothoxycapillarisin	H	H	H
6-desmethoxy-4′-methyl-capillarisin	H	H	CH_3

Inhibition of Chloramphenicol Activity by Extracts of Crude Drugs

Some substances are known to be antagonistic to antimicrobial substances, e.g., PABA to sulfa drugs. The authors found that several extracts of crude drugs suppressed or enhanced the antibiotic action of chloramphenicol towards *E. coli,* but it was difficult to decide on a possible mechanism.

The authors examined 1,700 kinds of plant extracts to seek the physiologically active substance(s). The activity was observed in the extracts of crude drugs such as diuretics, cholepoietics and antifebriles, and specifically in Rhamnaceae, Zingiberaceae and Polypodiaceae with high frequency. The 77 kinds of extracts which showed activity were examined again by Dye's agar medium method, and 22 kinds were selected for study.

[Experimenal procedure 9] Screening with a liquid medium

(a) Bouillon medium, (b) bouillon medium with chloramphenicol (2.26 ppm) and (c) bouillon medium with both chloramphenicol (2.26 ppm) and extract (0.05%) were prepared, and 2 ml portions were poured into test tubes. *E. coli* IFO 3044 was cultured in bouillon medium until the turbidity at 630 nm reached 1.0, then it was diluted to 1,000 volumes. Portions (2 ml) of the diluted solution were added to the test tubes containing media (a), (b) and (c). The test tubes were shaken for 15 hr at 30° C, then the turbidities were measured and the rates of increase were calculated using the following equation:

$$\text{rate of increase } (\%) = (C - B)/A \times 100$$

where A, B and C correspond to the turbidities of media (a), (b) and (c) respectively. An extract which gave a rate of more than 30% was regarded as positive.

Rates of increase were: Artemisiae capillaris Flos, 35%; Alismatis Rhizoma, 44%; Rhamni japonicae Cortex, 51%; Zingiberis Rhizoma, 37%.

[Experimental procedure 10] Screening with agar medium

Agar plates were prepared by pouring the bouillon agar medium containing *E. coli* IFO 3044 into a square Petri dish (7.5 × 22.5 cm). Two filter paper strips (Toyo Roshi No. 51, 5 mm wide), one impregnated with 0.1% chloramphenicol EtOH solution and dried, and the other impregnated with 10% solution of extract, were placed on the surface of the medium crosswise. The plate was incubated for 15 hr at 30° C, then the width of the inhibition zone was measured. The width of the chloramphenicol inhibition zone was ca. 15 mm under these conditions. A decrease of more than 2 mm in the width of the inhibition zone by the extract was regarded as positive.

Decrease of the inhibition zone: *Artemisiae capillaris* Flos, 5.5 mm; Alismatis Rhizoma, 3.0 mm; Rhamni japonicae Cortex, 4.5 mm; Zingiberis Rhizome, 1.5 mm.

[Experimental procedure 11] Bioassay of cholepoietics in rats

In the assay of cholepoietic activity, various animals such as rats or dogs can be utilized, but rats are preferable when many samples are to be examined efficiently.

Wistar rats (300–400 g) were anesthetized with Nembutal, then the biliary duct was cannulated. The secretion of bile was observed for more than 2 hr. If the secretion was stationary, the drug suspended in 5% gum arabic-saline was administered from the jugular vein. The volume of bile was recorded every 30 min for 3 hr (Fig. 29.1).

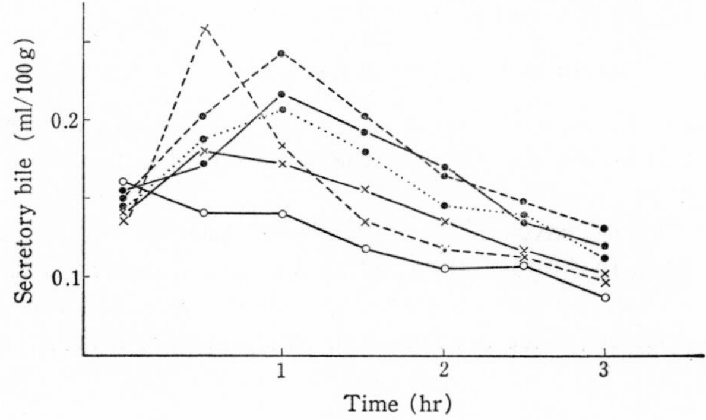

Fig. 29.1. Cholepoietic activities of constituents of *Artemisiae capillaris* Flos. ○——○, Control; ●----●, capillarisin; ●——●, 7-methylcapillarisin; ●······●, 4′-methylcapillarisin; ×······×, sodium dehydrocholate; ×——×, dimethylesculetin. Dose: 70 mg/kg, i.v.

[Experimental procedure 12] Isolation of capillarisin and its analogs from *Artemisiae capillaris* Flos

The extract of the crude drug was fractionated systematically on the basis of cholepoietic activity; the fraction containing capillarisin showed the strongest activity. Four analogs which gave a red-violet color, like capillarisin, with ferric chloride reagent were isolated from this fraction. Various flavonoids such as genkwanin, cirsimaritin, cirsilineol and rhamnocitrin were also isolated[10)] (not described here).

The spikes of *A. capillaris* THUNB. (15 kg) were extracted twice with 100 l of MeOH at room temperature. After concentration to 1.8 l, the extract was poured into 10 l of acetone, and the precipitate was removed by decantation, then the acetone solution was concentrated to dryness. The residue was dissolved again in 15 l of MeOH and washed twice with 4 l of hexane. The fraction soluble in hexane showed no activity. On leaving the MeOH solution to stand overnight, 22 g of crystals of dimethylesculetin precipitated. The mother liquor, which showed strong activity at a dose of 200 mg/kg, was subjected to silica gel column (15 × 130 cm) chromatography, eluting with $CHCl_3$ and a mixture of $CHCl_3$ and MeOH. The ratio of MeOH was increased from 0.05%. The eluate was examined by tlc using Kieselgel 60 F_{254} (Merck) with $CHCl_3$–ether (10:1), and the substances which colored with ferric chloride were isolated.

Yields and *Rf* values on tlc of capillarisin and its analogs were: capillarisin, 13 g 0.17; 7-methylcapillarisin, 0.75 g 0.28; 4′-methylcapillarisin, 1.4 g 0.57; 6-desmethoxycapillarisin, 0.05 g 0.15; 6-desmethoxy-4′-methylcapillarisin, 0.5 g 0.54 (c.f. dimethylesculetin, 33 g 0.72).

Seasonal Variation of Capillarisin Contents

As the constituent contents of a plant, in general, are seasonal it is necessary to harvest the plant at the most suitable time to obtain good crude drug.

In the case of leaves of *A. capillaris* THUNB., the capillarisin contents reached a peak in August, then fell as the ears appeared (Fig. 29.2). The spikes, moreover, contained more capillarisin than any other parts of the plant, as expected, since the spikes of this plant are utilized as the crude drug.

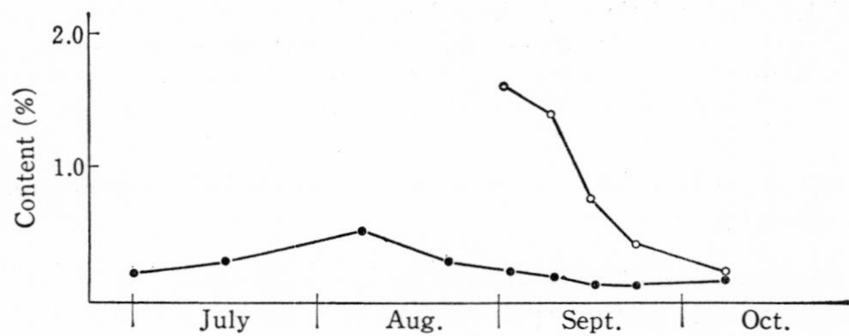

Fig. 29.2. Seasonal variation of capillarisin content. ○—○, Spike; ●—●, leaf.

[Experimental procedure 13] Quantitative analysis of capillarisin

The crude drug (1 g) was extracted three times by refluxing in 20 ml of MeOH for 1 hr. After removing MeOH by evaporation, the residue was dissolved again in 2 ml of MeOH, then 50 μl of the solution was applied to a silica gel plate 5 cm wide and developed with $CHCl_3$–MeOH (10:1). The zone of silica gel including capillarisin was removed and extracted with THF. The optical density of THF solution at 289 nm was estimated, and the capillarisin content was calculated from a calibration curve (Fig. 29.2).

29.5 Concluding Remarks

A crude drug contains many constituents, and even the bioactivity may be due to several components. In addition, different sorts of bioactive components, which act in opposite ways are often contained in a crude drug. Further, it is difficult to prepare such crude drugs of uniform quality, because they often differ as regards the original plants, habitats, harvesting time, parts of the plants, and methods of preparation. In addition, some consituents of a crude drug may be unstable under certain conditions, such as high temperature.

It is therefore necessary to develop suitable bioassays for the target physiologically active substances and for quality estimation of the crude drugs.

References

1) M. Goto, series entitled as *Studies on crude drugs and their prescriptions by bioassays*, *Yakugaku Zasshi* (in Japanese), **75**, 1180 (1955); *J. Takeda Res. Lab.* (in Japanese), **35**, 146 (1976).
2) A. Stoll, B. Becker, *Fortschr. Chem. Org. Naturstoffe*, **7**, 248 (1950).
3) W. Schmid, E. Angeliker, *Helv. Chim. Acta*, **48**, 1911 (1965); J. Lemli, J. Cuveele, *Pharm. Acta Helv.*, **40**, 667 (1965).
4) M. Miyamoto, S. Imai, M. Shinohara, S. Fujioka, M. Goto, T. Matsuoka, H. Fujimura, *Yakugaku Zasshi* (in Japanese), **87**, 1040 (1967); *CA*, **68**, 921e (1968).
5) H. Oshio, S. Imai, S. Fujioka, T. Sugawara, M. Miyamoto, M. Tsukui, *Chem. Pharm. Bull.*, **22**, 823 (1974).
6) M. Tsukui, M. Yamazaki, Y. Toyosato, T. Matsuoka, H. Fujimura, *Yakugaku Zasshi* (in Japanese), **94**, 1095 (1974); *CA*, **82**, 38706r (1975).
7) H. Okabe, K. Matsuo, I. Nishioka, *Chem. Pharm. Bull.*, **21**, 1254 (1973).
8) H. Mashimo, K. Shimizu, G. Chihara, *The Saishin-Igaku* (in Japanese), **18**, 1430 (1963).
9) T. Komiya, M. Tsukui, H. Oshio, *Yakugaku Zasshi* (in Japanese), **96**, 841 (1976); *CA*, **86**, 89536a (1977).
10) T. Komiya, Y. Naruse, H. Oshio, *Yakugaku Zasshi* (in Japanese), **96**, 855 (1976); *CA*, **86**, 55239r (1977).

CHAPTER 30

Isolation of Blepharismone and Blepharmone, Two Conjugation-Inducing Gamones of *Blepharisma japonicum* var. *intermedium*

Conjugation of ciliates is induced by interaction between cells of complementary mating types.[1] In *Blepharisma japonicum* var. *intermedium* (Fig. 30.1) complementary mating types I and II communicate by conjugation signals or "gamones" excreted into the medium, as shown diagrammatically in Fig. 30.2.[2,3] Type I cells excrete gamone 1 (step 1). This gamone gives specific information to type II cells (step 2) and transforms them so that they can unite (step 3), at the same time inducing them to excrete gamone 2 (step 4). This gamone gives specific information to type I cells (step 5) and transforms them so that they can unite (step 6). The transformed cells unite to form pairs (step 7).

The union between transformed cells can occur in any combination of mating types. Therefore, in mixtures of the two cell types, both heterotypic pairs (type 1–type II) and homotypic pairs (type 1–type I, type 2–type II) are formed. If cells of only one type are treated with gamone of the other type, homotypic pairs are formed. Using this cellular response, gamone can be detected and assayed (Fig. 30.1).

The excretion of gamone 1 begins autonomously when type I cells are suspended in a nutrient-free medium. The rate of excretion reaches a maximum after 1 day and then declines, but the excretion continues for a week or even longer, until the cells die from starvation. During this time of reduced excretion, gamone 2 can enhance the excretion. Therefore, the reaction chain consisting of steps 1, 2, 4, and 5 in Fig 30.2 is a positive feedback cycle.

Some cultures of type II cells (*non-augex* form) do not autonomously

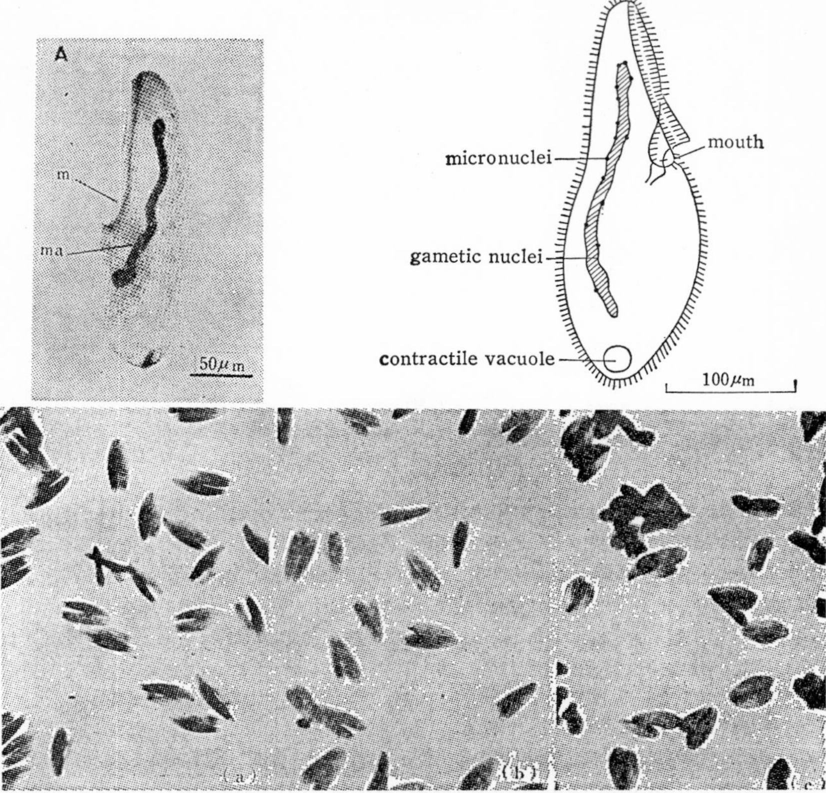

Fig. 30.1. Cells of *Blepharisma japonicum* var. *intermedium* in pairing. × 25. (a) Heterotypic pairs between mating types I and II. The darker looking cell of a pair is mating type II, 20 hr after mixing the two mating types. (b) Homotypic pairs of mating type I induced by cell-free fluid of mating type II with gamone II activity, 24 hr after beginning the treatment. (c) Homotypic pairs of mating type II induced by cell-free fluid of mating type I with gamone I activity, 2 hr after beginning the treatment.

excrete gamone 2. They excrete gamone 2 only when gamone 1 is present. Others (*augex* form) autonomously excrete gamone 2. Like the gamone 1 excretion by type I cells, gamone 2 excretion by *augex* changes according to the nutritive conditions. At times of diminished excretion, gamone 1 enhances the excretion.

Gamone 2 was isolated first, identified as calcium–3-(2′-formylamino–5′-hydroxybenzoyl) lactate (Fig. 30.3) in 1973[5] and chemically synthesized (1973).[4] Pure gamone 2 (blepharismone) at a concentration of 1 ng/ml can induce homotypic pairs in type I cells suspended at a density of 500–1000 cells/ml. At 4 ng/ml, it can induce pairs in nearly all cells in about 2 hr

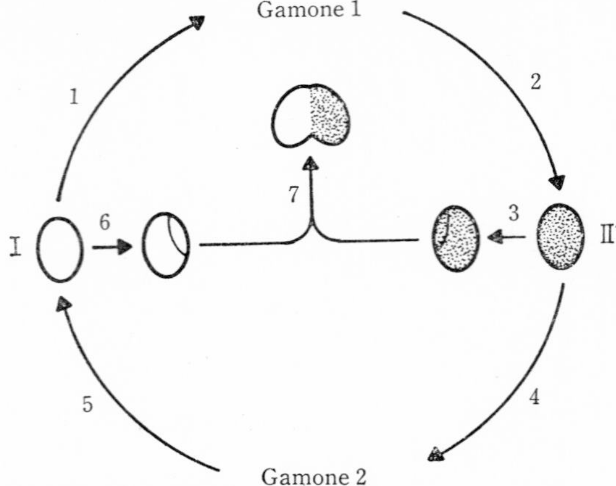

Fig. 30.2. Roles of gamones 1 and 2.

under the same conditions. Synthetic blepharismone is approximately half as active as the natural compound. This can be explained by assuming that one of the enantiometric forms (L) is several times more biologically active than the other form (D). This assumption is supported by the finding that L-isomers of gamone 2 inhibitors are more effective than their D-isomers. The chemical structure of the gamone 2 molecule suggests that a precursor of this gamone is tryptophan.[5] Indeed, type II-cells incorporate [^{14}C] tryptophan into gamone 2.[6]

In the isolation of the more unstable gamone 1 (blepharmone), the finding that serum albumin is a potent protector of this gamone played a critical role. Thus, all the processes of concentration and purification were carried out in the presence of bovine serum albumin until albumin was finally removed in the last step of purification.[7] Gamone 1 was identified as a glycoprotein with a molecular weight of 20,000.[7,8] It has characteristically high contents of tyrosine, aspartic acid, threonine, and serine (13, 26, 17 and 19 residues, respectively, out of a total of 175 amino acids excluding tryptophan), a low content of glutamic acid (7 residues), and 5% of sugars (3 glucosamine, 3 mannose).[9] Purified gamone 1 (blepharmone) at a concentration of 0.06 ng/ml can induce homotypic pairs in type II cells suspended at a density of 500–1000 cells/ml.

When type II cells (*non-augex*) were treated with gamone 1, they started producing gamone 2 after a time lag of 1–2 hr.[8] If the gamone 1 treatment was discontinued after 1, 1.5 or 2 hr by washing the cells, gamone 2 production began at nearly the same time and continued for 3–4 hr as the same rate as in the control, which was continuously exposed to gamone 1 (A. Miyake, unpublished). Induction was also achieved by a

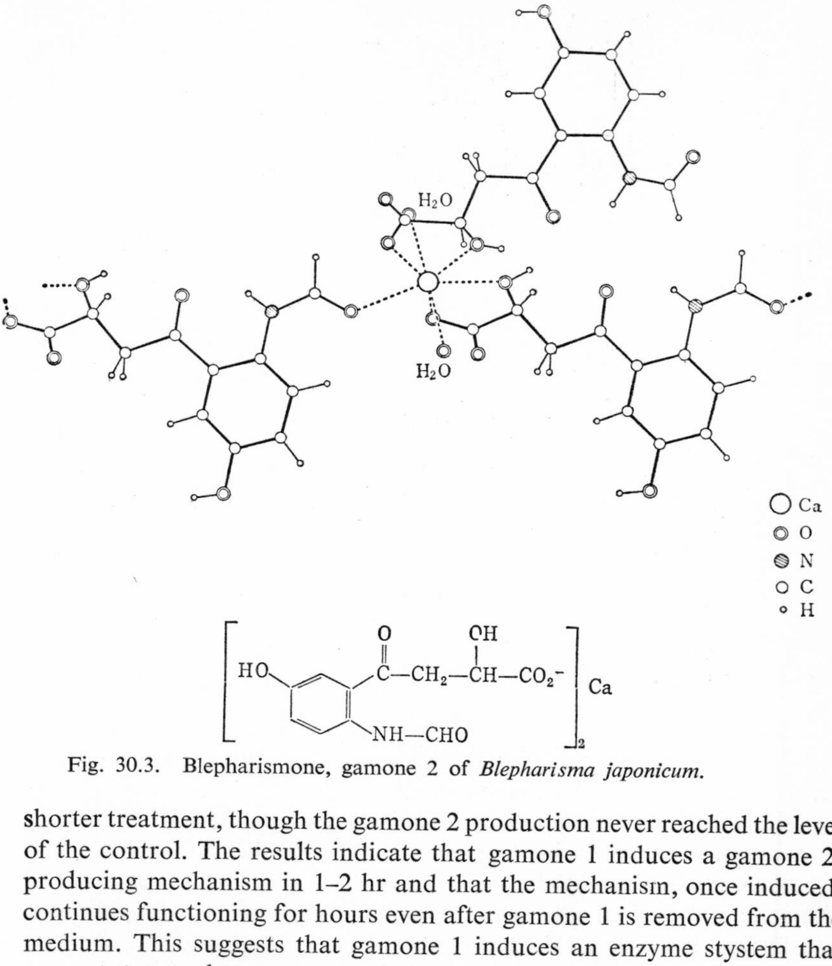

Fig. 30.3. Blepharismone, gamone 2 of *Blepharisma japonicum.*

shorter treatment, though the gamone 2 production never reached the level of the control. The results indicate that gamone 1 induces a gamone 2-producing mechanism in 1–2 hr and that the mechanism, once induced, continues functioning for hours even after gamone 1 is removed from the medium. This suggests that gamone 1 induces an enzyme stystem that converts tryptophan to gamone 2. In the induction of gamone 1 by gamone 2, a time lag of 1–2 hr was also observed,[8] but this phenomenon has not been further analyzed.

Blepharmone and blepharismone are the only conjugation signals so far isolated in ciliates. Comparative studies on four species of the genus *Blepharisma* indicate that they all share blepharismone as gamone 2, while each species has a species-specific gamone 1.[6,18] The significance of a striking chemical difference between a pair of gamones that provoke similar responses in target cells was discussed recently.[6,7,10,18]

A hypothetical molecular mechanism of cell interaction in *Blepharisma* suggested by Miyake[11,18] is shown in Fig. 30.4.

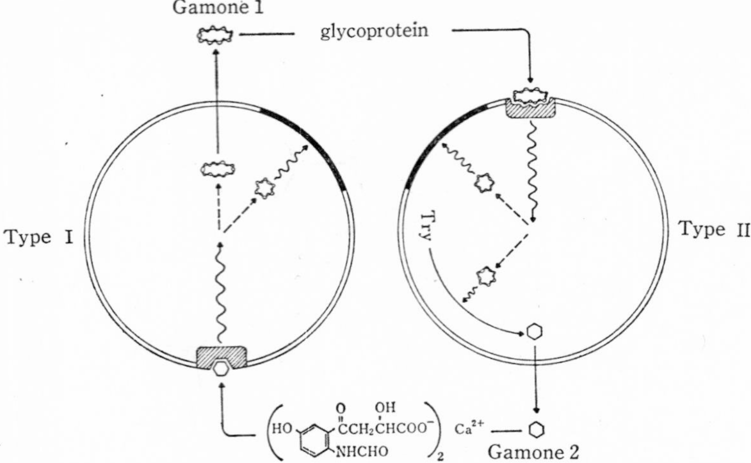

Fig. 30.4. Hypothetical molecular mechanism of cell interaction in *Blepharisma.*

30.1 Culture of *Blepharisma japonicum*

Culture Medium[8,12]

General techniques for culture and handling were the same as those described by Sonneborn[12] for *Paramecium*. The culture medium was a modified Ca-poor fresh lettuce medium.[2] Fresh lettuce leaves (10 kg) were washed twice with tap water and deionized water, boiled for 5 min and homogenized with a Waring blender (CB-5) for 5 min at speed "Lo". The homogenate was filtered through 10 sheets of gauze at 4° C. The filtrate was diluted to make a final volume of 20 1 and frozen. After thawing, the juice was filtered through a thin layer of cotton wadding. One liter of this filtrate was diluted to 15 1, buffered with 2 mM Na-phosphate buffer, pH 6.8, and autoclaved. This medium was incubated with *Aerobacter aerogenes* for one day before use. Culture and handling, as well as experimental procedures, were performed at $25 \pm 1°$C.

Culture

To start a cell culture, a single cell was isolated from the stock culture, "washed"[12] if necessary, and placed in a few drops of culture medium in a slide depression that could hold about 1 ml of medium. Culture medium was added every other day so as to fill the depression in about a week. Thereafter the culture was first grown in a test tube with an aluminum cap and then in a wide mouth Erlenmeyer flask with a cotton plug or in a 20 l bottle until the culture reached an appropriate size. The amount of culture placed in each flask was limited to half the capacity of the flask. In the 20 l bottle up to 16 l of culture was grown with continuous aeration using sterilized air. Fission rates during tube, flask, and 20 l bottle culture were adjusted to about one fission per day by doubling the volume of culture each day by adding freshly prepared culture medium. All these procedures were performed under aseptic conditions. One day after feeding, the culture had usually "cleared", i.e., the cells had consumed most of the food (bacteria) and the culture became transparent. If a culture failed to clear within two days after feeding, usually as a result of contamination by unsuitable microorganisms, a single cell was washed with freshly prepared culture medium and grown as described above.

The cells were concentrated by centrifugation and washed with SMB (1.5 mM NaCl, 0.05 mM KCl, 0.4 mM $CaCl_2$, 0.1 mM $MgSO_4$, 2×10^{-3} mM EDTA, 2 mM sodium phosphate buffer of pH 6.8). If necessary, collected cells were put in a plastic cylinder, the end of which was covered with a Nylon net of 50 μm mesh, immersed in SMB solution of 1–1.5 cm depth, as shown in Fig. 30.5. The cells went through the net into the SMB solution after 2–3 hr and impurities (fine particles of lettuce, excreta of *Blepharisma* and clusters of bacteria) remained on the net.

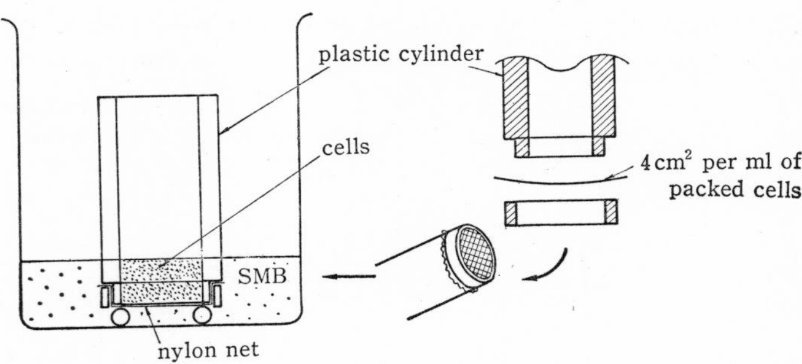

Fig. 30.5. Culture of *Blepharisma* (see the text).

At the final stage, the cells were suspended in SMB containing 32 units of gamone 2 per milliliter and 0.01% albumin (bovine serum albumin, Behriogwerke) at a density of 0.01 ml of packed cells per milliliter (about 10^4 cell/ml). Albumin was added to protect the gamone activity of blepharmone.[7,8] After keeping the suspension for 1 day at 25°C, the cell-free fluid was obtained by removing cells by centrifugation. The cells were resuspended as described above, and the process was repeated for several days. The gamone 1 activity of the cell-free fluid was 0.8×10^4 to 12.8×10^4 unit/ml; the cell-free fluid could be stored frozen for several months without change in the activity.

30.2 Purification of Blepharmone (Gamone 1)[7)]

The cell-free fluid (20 l) having an average activity of 3.2×10^4 units /ml was concentrated under reduced pressure to 1 l at 25° C or below, and lyophilized. The lyophilized material was dissolved in 100 ml of H_2O, and centrifuged at 10,000 g for 30 min. The supernatant was divided into five portions, each of which was separately chromatographed on Bio-Gel P-150 (Fig. 30.6 a). The fractions in the gamone activity peaks of the five protions were pooled, lyophilized, and chromatographed again as described above. The fractions in the activity peak were lyophilized, dissolved in SMB (pH 5.8) from which NaCl had been omitted and phosphate buffer increased to 4 mM, and dialyzed for 1 day at 4° C against this buffer. The dialysate (20 ml) was chromatographed on carboxymethyl cellulose (Fig. 30.6 b). Blepharmone was eluted in two peaks, but about 90% of the activity was found in the second peak. The fractions in this peak were lyophilized, dissolved in SMB (pH 7.6), and dialyzed for 1 day at 4° C against this buffer. The dialysate (10 ml) was applied to a column (2.5 by 34 cm) of

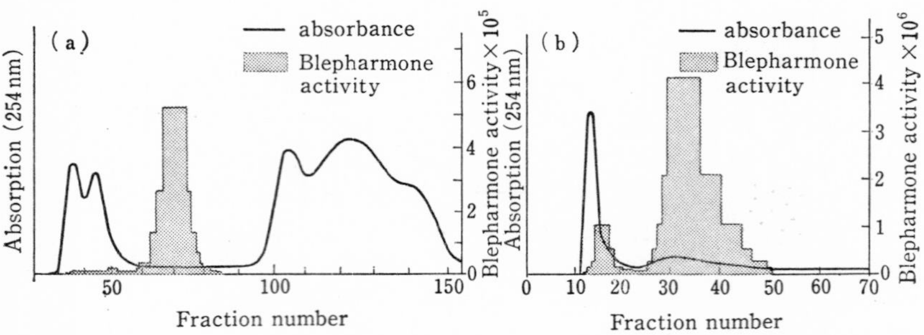

Fig. 30.6. Elution patterns of crude blepharmone samples. Chromatography at 4°C; fraction size 6 ml. (a) Bio-Gel P-150 (100 to 200 mesh) was used. The column size was 5 × 40 cm, and the elution medium contained SMB plus 0.01 % albumin. (b) Carboxymethyl cellulose (Whatman, CM-52) was used. The column size was 2.5 × 35 cm. The equilibration medium was SMB, pH 5.8 (NaCl was omitted, and the phosphate buffer was increased to 4 mM). The elution medium consisted of the equilibration medium plus 0.01% albumin. Before sample application, the column was eluted with 2 l of elution medium.

diethylaminoethylcellulose (Whatman DE 52) equilibrated with SMB (pH 7.6) and eluted with the same buffer. Blepharmone was eluted at the front with a small absorption peak recored at 254 nm. The blepharmone sample obtained by lyophilizing the fractions in this peak contained 2.0 mg of protein and showed an activity of 3.2×10^7 units. On the assumption that the real activity of blepharmone in the starting material was 8×10^7 units, the recovery was 40%. Blepharmone is very unstable in a diluted state, and successful purification was achieved only in the continuous presence of albumin. Thus, the Bio-Gel and carboxymethyl cellulose columns were both eluted with a buffer containing albumin. When albumin was removed in the last step of putification, blepharmone was no longer so unstable, probably because its concentration was sufficiently high.

Assessment of Purity[7)]

This sample was subjected to acrylamide gel electrophoresis by a modification of the method of Gordon and Louis[13)] and by the method of Davis,[14)] which gave different band patterns for a less purified sample of blepharmone. For each run, 1/40 of the purified sample described above was used. In both methods, a single band was strongly stained by amido black (Fig. 30.7 a), and most of the gamone activity was found there. The rest of the activity was at the origin. If stained with the periodic acid-Schiff (PAS) reagent for the detection of carbohydrate,[15)] a single band was seen at the position stained by amido black. Comparing the intensity of the PAS reaction of the band with that of orosomucoid and β_2-glycoprotein 1 (Behringwerke), the carbohydrate in the electrophoresed material was estimated to correspond to 5% of the protein. The sample was also subjected to isoelectric focusing. For each run, 1/40 of the sample was used. Only a single band was strongly stained. In addition, two or three barely visible bands in the more acidic region were observed (Fig. 30.7 b). The peak of the blepharmone activity was at the position of the strongly stained band. By measuring the pH and blepharmone activity of a gel cut into 3-mm thick slices, the isoelectric point of blepharmone was determined to be pH 7.5. This value is consistent with that obtained by the chromatographic filtration method. Blepharmone dissolved in SMB passed freely through the Amicon PM-30 membrane but was retained (about 95%) by the PM-10 membrane. To avoid absorption of blepharmone on the membrane, a preliminary run was made for 5 min with SMB containing 1% albumin, which did not appreciably change the flow rate. Since these membranes retain globular molecules with a molecular weight larger than 3×10^4 and 1×10^4, respectively, the molecular weight of blepharmone should be in the range of 1×10^4 to 3×10^4. This value is consistent with the previous value of 2×10^4 measured by the gel filtration method.[8)]

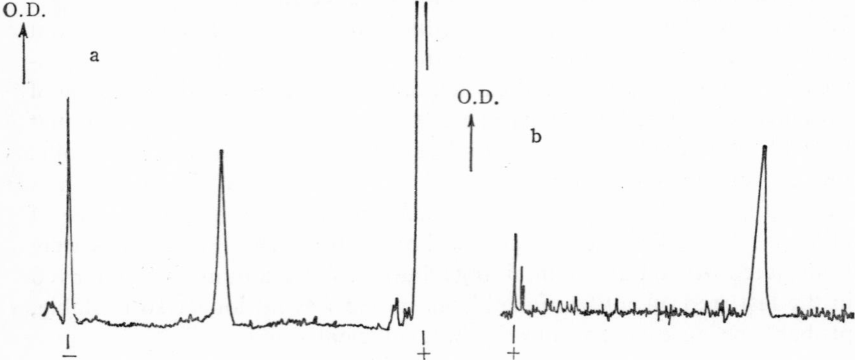

Fig. 30.7. Electrophoresis and isoelectric focusing of purified blepharmone. The optical density (O.D.) was measured with a microdensitometer (Joyce, Loebl MK IIIC) with a red filter (No. 620). The symbols + and – indicate the anodal and cathodal ends of the gel. (a) Electrophoresis was carried out in a polyacrylamide gel column (5 × 62 mm) with a modification of the system of Gordon and Louis.[13] The desalted sample was applied in 0.2 ml of 0.05 M tris · HCl buffer, pH 6.8, containing 25 % Ficoll (Pharmacia). Fixation and staining: 7 % acetic acid, 1 % amido black. (b) Isoelectric focusing in a polyacrylamide gel column (5 × 100 mm) was carried out with a modification of the system of Righetti and Drysdale[16]; 4 % Ampholine (LKB), pH 3 to 10, was used. After fixation and removal of Ampholine, the gel was stained with 0.05% Coomassie blue in 10% trichloroacetic acid.[17]

30.3 Purification of Blepharismone (Gamone 2)[5]

Preliminary Experiment by TLC

Cell-free gamone 2 solution was first examined by tlc (Merck, silica gel G, 25 × 25 cm). When 75% aqueous ethanol was used as the developing solvent, the chromatograms shown in Fig. 30.8 were obtained [(a) under

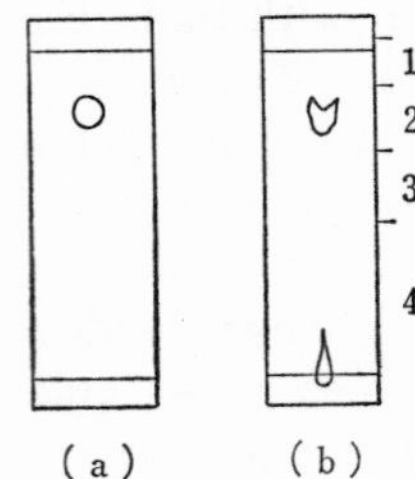

Fig. 30.8. Results of a preliminary tlc study (see the text for details). Visualized under uv at 365 nm (a) and by exposure to iodine vapor (b).

uv (365 nm) ; (b) iodine vapor]. The silica gel layer was divided into four parts (Fig. 30.8). Each part was scratched off, and extracted with aq. ethanol (75% for 1–2 and 50% for 3–4). Bioassay revealed that the gamone 2 activity was concentrated in part 2, suggesting that the active substance might be fluorescent (uv: λ_{max} 234 and 340 nm).

Pretreatment

Gamone 2 is heat-stable, and concentration of the cell-free preparation could be performed on a rotatory evaporator at a bath temperature below 50° as readily as by freeze-drying. Being considerably polar (as seen from the tlc behavior), the active component could not be extracted by organic solvents (e.g. continuous extraction with chloroform was ineffective). In order to free it from inorganic salts and carbohydrate, the concentrated fluid was treated with ethanol [pure gamone is moderately soluble in water (~2 mg/ml) and insoluble in most organic solvents, including ethanol]. The processing scheme for 1 l of the cell-free fluid is illustrated in Table 30.1. The fractions (1)–(4) were tested for gamone activity with the following results; (1) 1.0; (2) 0.03; (3) 0.08; (4) 20. Thus, most of the activity was concentrated in fraction (4), which still contained a large amount of salts.

TABLE 30.1. Initial fractionation of blepharismone.

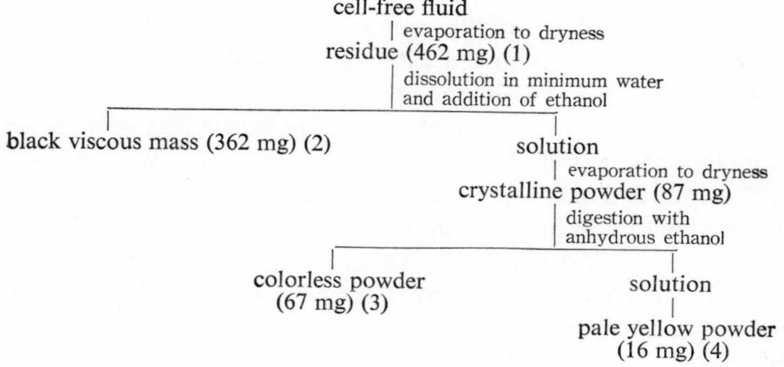

Paper Chromatography

Further separation of fraction (4) above was conveniently done by paper chromatography. Of the developing solvents tested, those containing a base generally gave better separation. Thus, several fluorescent spots were observed in the chromatogram (Fig. 30.9). [In later work, tlc using a cellulose plate (e.g. Merck, DC Fetigplatten, Cellulose, 0.1 mm, or Macherey Nagel & Co. CEL 300, 0.25 mm) proved to be a suitable alternative.]

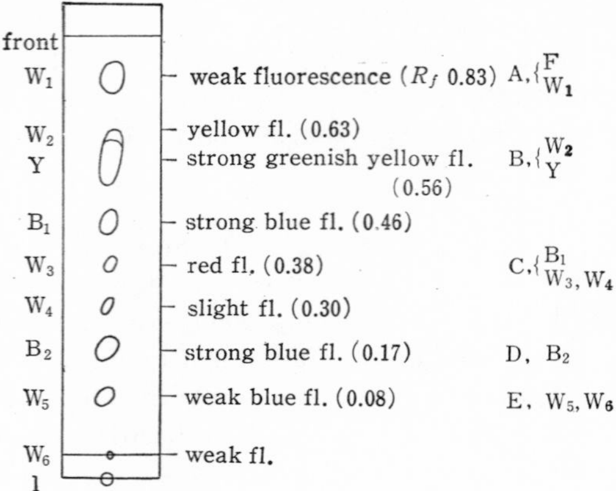

Fig. 30.9. Paper chromatography of fraction (4) from cell-free fluid of *Blepharisma*. See the text for details. Toyo Roshi No. 51 paper; *t*-butyl alcohol-conc. NH_4OH-water (3:1:1).

When the groups A-E of spots were tested for biological activity, only the extract from the spot corresponding to group B was found to be effective. Next, fraction (4) obtained from 3.5 l of the cell-free fluid was separated by descending paper chromatography using seven sheets of filter paper (Toyo Roshi No. 131, 40 x 40 cm) and the same developing solvent as above. The B region was cut off and extracted with water. The extract was then paper chromatographed using γ-collidine-ethanol-water (4:3:2) as a developing solvent, giving the chromatogram shown in Fig. 30.10. The gamone activity was mainly concentrated in the spot B-Y; spot G-Y was only 4% as active. The spot B-Y was again extracted and the extract was paper chromatographed with *n*-butanol-acetic acid-water (12:3:5). Extraction of the yellow spot finally afforded the active substance, which showed a single spot in paper chromatographic analysis with several different solvent systems. When this material was dissoved in a small amount of water and kept in a refrigerator, pale yellow prisms mixed with an oily substance were observed. The separated crystals exhibited gamone activity at a concentration of 0.005 mg/ml. They appear to be gamone 2, and were designated as blepharismone.

Separation and Purification of Blepharismone

For the isolation of blepharismone on a larger scale, a procedure involving column chromatography with cellulose powder was designed.

The cell-free fluid with gamone 2 activity (activity 80 units/ml, 48 l)

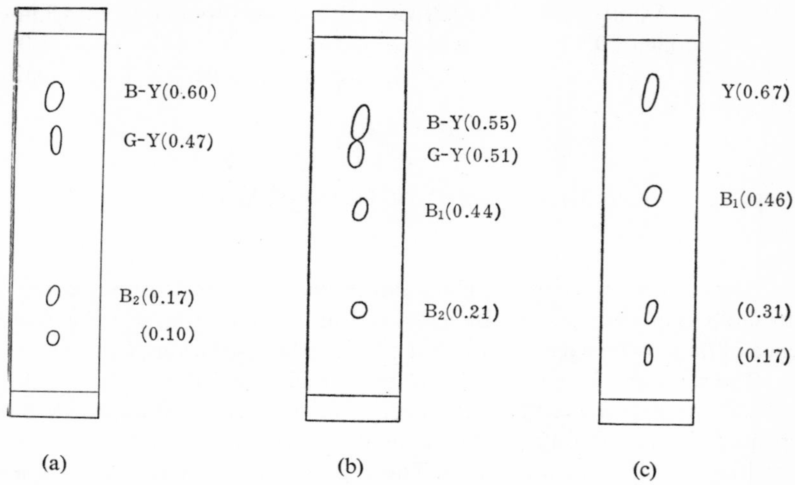

Fig. 30.10. Paper chromatography of the group B material from Fig. 30.9. (a) γ-collidine–ethanol–water 4:3:2, (b) pyridine–butanol–water 1:1:1, (c) butanol–acetic acid–water 12:3:5.

was concentrated *in vacuo* and the deposited red pigment (Type II cells contains a hypericin-like pigment, blepharismin, which is pink in color) was removed by filtration. The filtrate was concentrated further to 200–300 ml. An equal amount of ethanol was added and the deposited viscous material was separated by decantation. It was digested several times with 50% aq. ethanol. The combined supernatant solution and washings was evaporated to dryness and the crystalline residue was extracted several times with warm absolute ethanol. The combined extract solution was concentrated to give a crystalline residue (163 mg), which was dissolved in 75% *n*-propanol (5 ml) and chromatographed on a column of cellulose powder (Toyo Roshi, 100–200 mesh, 100 g). Elution was performed with the same solvent, and the movement of the bands was followed by means of a uv lamp. Fractions of 50 ml were collected. Fractions 4–7 were found to contain the active factor (monitoring by paper chromatography) and were further purified by preparative paper chromatography. The residue left after removal of the solvent was dissolved in water (2.4 ml) and subjected to preparative paper chromatography using 18 sheets of filter paper (40 × 40 cm, Toyo Roshi No. 131). [It was found later that tlc with cellulose plates, e.g. Merck DC Fertigplatten, or Macherey-Nagel & Co., CEL 300–50 (0.5 mm), was convenient for this purpose.] The band with light blue fluorescence was cut off and extracted with water. The crystals obtained were recrystallized twice from water to give blepharismone (3.68

mg). Since 16 units of gamone activity corresponds to a concentration of blepharismone of 0.02 mg/ml, and the original cell-free fluid contains 0.02 $\times 10^{-3}$ mg $\times$ 48 $\times 10^{3} \times 80/16 = 4.8$ mg, the recovery in this isolation procedure is 75%.

30.4 Measurement of Gamone Activity[8)]

The gamone activity of a given sample was obtained by comparing its threshold dilution, i.e., the largest dilution that could induce the occurrence of face-to-face pairs, with that of a standard solution of gamone.

The standard for gamone 1 was prepared in the following way. About 4×10^{6} one-day-starved A5 cells were homogenized in 10 ml of SMB by repeated ejection through a plastic medical syringe without a needle then centrifuged at 25,000 g for 15 min. The supernatant was chromatographed on a Bio-Gel P-150 column under the conditions described in Fig. 30.6. Five fractions with the strongest gamone 1 activity in the second activity peak were pooled, diluted 4 times with SMB, and frozen.

The unit of gamone activity was defined as the smallest amount of gamone able to induce the formation of face-to-face pairs in about 500 one-day-starved cells suspended in 1 ml of SMB. Since a 2^{13}-fold dilution of the standard for gamone 1 induced pairing in highly responsive Type II but a 2^{14}-fold dilution did not, its activity was determined to be 2^{13} or 8.2×10^{3} units/ml.

Similarly, the activity of the gamone 2 standard, which was a 2 μg/ml solution of purified crystallized gamone 2, was determined to be 16 units /ml.

References

1) A. Miyake, *Current Topics Microbiol. Immunol.*, **44**, 69 (1974).
2) A. Miyake, *Proc. Japan Acad.*, **44**, 837 (1968).
3) A. Miyake, *Current Topics in Developmental Biology*, vol. 12, p. 37, Academic Press (1978).
4) T. Tokoroyama, S. Horii, T. Kubota, *Proc. Japan Acad.*, **49**, 461 (1973).
5) T. Kubota, T. Tokoroyama, T. Tsukuda, H. Koyama, A. Miyake, *Science*, **179**, 400 (1973).
6) A. Miyake, L. K. Bleyman, *Genet. Res.*, **100**, 31 (1976).
7) A. Miyake, J. Beyer, *Science*, **185**, 621 (1974).
8) A. Miyake, J. Beyer, *Exp. Cell Res.*, **76**, 15 (1973).
9) V. Braun, A. Miyake, *FEBS Letters*, **51**, 131 (1975).
10) D. L. Nanney, *Microbal Interactions* (ed. T. Reissig), Chapman and Hall (1977).

11) A. Miyake, *25 Mosbacher Colloquima. Gessel. Biolog. Chemie, Biochemistry of Sensory Function*, p. 299 (1974).
12) T. M. Sonneborn, *Methods in Cell Physiology* (ed. D.M. Prescott), vol. 4, p. 241, Academic Press (1970).
13) A. H. Gordon, I. N. Louis, *Anal. Biochem.*, **21**, 190 (1967).
14) B. J. Davis, *Disc Electrophoresis*, Distillation Products Division, Eastman Kodak Co. (1962).
15) K. Flgenhauer, *Clin. Chim. Acta*, **27**, 305 (1970).
16) P. Righetti, J. W. Drysdale, *Biochem. Biophys. Acta*, **236**, 17 (1971).
17) A. Chrambach, M. Wyckoff, J. Zaccardi, *Anal. Biochem.*, **28**, 150 (1967).
18) A. Miyake, *Current Topics in Developmental Biology*, **12**, 37 (1978).

CHAPTER 31

The Wing-raising Pheromone of the German Cockroach

At present, about 3,500 species of cockroaches are known, most of which have outdoor havitats such as tropical forests, and only a few of which live indoors in association with man. In Japan there are several species of domestic cockroaches including the smokybrown cockroach, *Periplaneta fuliginosa,* the Japanese cockroach, *P. japonica,* and the German cockroach, *Blattella germanica.*

Extensive studies on the mating behavior of cockroaches have been made by several workers to discover the mechanisms which bring them together for mating.[1] It is now known that females of several species produce volatile sex pheromones which play important roles in attracting conspecific males from a distance and in eliciting the characteristic courtship behavior. Of these volatile sex pheromones of cockroaches, periplanone B,[2] isolated from virgin females of the American cockroach, *Periplaneta americana,* is the first to have been isolated and identified chemically.

Females of the German cockroach, *Blattella germanica,* may not produce a volatile sex pheromone but they contain in the cuticular wax a nonvolatile one which exerts its pheromonal effect by contact chemoreception through the antnneae. This chapter describes biological and chemical study of this pheromone.

31.1 Biological Study

The Mating Behavior of the German Cockroach

As observed by Roth and Willis,[3] when an adult male of the German cockroach comes in contact with a sexually mature female of the same species, the male shows the characteristic sequence of mating behavior illustrated in Fig. 31.1: (1) "sparring" with the antennae for several seconds, (2) rapid horizontal wagging of his abdomen, followed by turning 180° and simultaneously raising the wings, (3) producing a secretion from the tergal glands for the female to lick, (4) pushing the abdomen under the female and grasping her genitalia, (5) eventually copulating. This sequence of mating behavior is always observed between sexually mature males and females of the German cockroach, and copulation has never been observed without it.

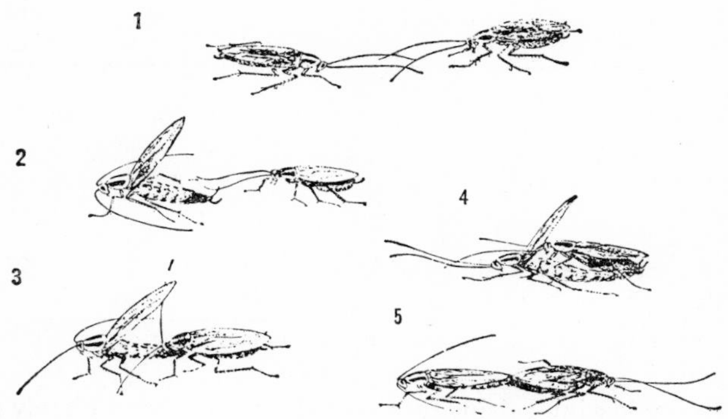

Fig. 31.1. The sequential mating behavior of the German cockroach.

Occurrence of the Wing-Raising Pheromone

It was demonstrated by Roth and Willis[3] and also by Ishii[4] that the male behavior items 1 and 2 (raising the wings and turning) can be elicited even by artificial contact of the antennae with an isolated female's antenna. As shown in Fig. 31.2, a small glass rod carrying an antenna cut from a sexually mature female could produce the first step of the normal mating behavior of a living adult male. This suggests that communication through antennal contact plays a decisive role in species and sex discrimination by the adult male, as well as stimulating the courtship behavior.

As shown in Table 31.1, the ability of the isolated female's antenna to

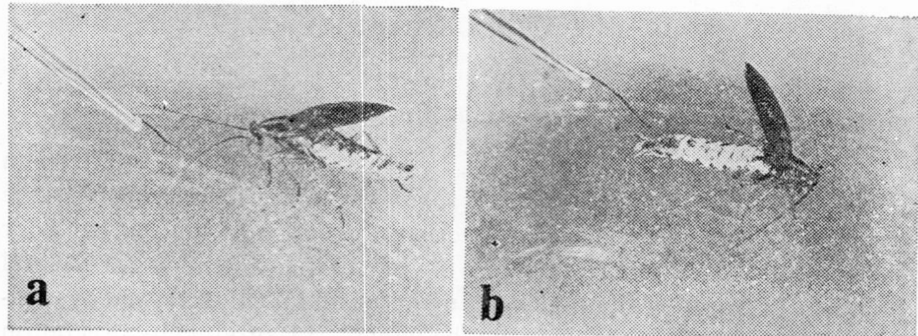

Fig. 31.2. The behavior of the male German cockroach in response to artificial contact with an isolated female's antenna
(a) The male responds to an isolated female's antenna by raising his wings,
(b) and turning 180°.

TABLE 31.1. Changes in the wing-raising pheromone activity on various treatments of isolated antenna.

Treatment	Response (%) of Living Males†
Female's antenna	
untreated	100
washed with water (25° C)	100
washed with hexane	0
Male's antenna	
untreated	0
coated with hexane washings of female's antenna	97.5 ± 6.2
coated with hexane washings of male's antenna	0

† The data are means of four bioassays using 30 males each.

produce the male wing-raising response disappeared completely on washing with hexane before use. Of course, an isolated male's antenna does not elicit the wing-raising response from living males. However, an isolated male's antenna coated with the hexane extract of a female antenna showed the ability to produce wing-raising behavior in a male to the same extent as the female antenna itself. These experimental results suggest that communnication through antennal contact is principally due to contact chemoreception of male antennae with a substance(s) present in cuticular wax on the surface of the antennae of sexually mature females. The substance(s) responsible for eliciting the first step of the male courtship behavior is tentatively named the wing-raising pheromone of the German cockroach.

It is noteworthy that the wing-raising pheromone of the German cockroach shows some unusual characteristics and functions as compared with sex pheromones of other species studied so far. Since the classical study by Butenandt *et al.*[5] on the isolation and identification of Bombykol, the sex pheromone, produced by the female silkworm, *Bombyx mori,* a large number of sex pheromones, especially among Lepidoptera, have been identified. In general, most of them are released by a female into the air, conveyed downwind as a vapor or aerosol to distant conspecific males, and there stimulate the males sexually, causing them to locate the pheromone source: the female. In contrast to these lepidopterous sex pheromones the wing-raising pheromone of the German cockroach is less volatile, displaying activity only on direct antennal contact between the male and female. Another special feature of the wing-raising pheromone of the German cockroach is that artificial contact of an isolated female antenna with the antennae of a living male causes the male to raise his wings and simultaneously to turn, but the subsequent behavioral responses of the male, which are always observed with a living female (Fig. 31.1) are not released; the raised wings are brought down after several seconds and the male begins walking again as usual. Thus, the pheromonal activity of the wing-raising pheromone is limited to eliciting only the initial step of the male behavioral reponses, whereas the sex pheromones of Lepidoptera usually elicit every response of the conspecific male, leading to copulation.

Bioassay Method for the Wing-Raising Pheromone

In order to design an isolation procedure for a biologically active compound, it is normally essential to establish a bioassay method which can monitor the active compound in every step of the isolation procedures. The bioassay method should be a simple system able to analyze the amount of the active compound as sensitively and precisely as possilbe. A bioassay method for the wing-raising pheromone of the German cockroach was established by us on the basis of the following observations.[6,7] (1) An antenna isolated from an adult male is inactive when brought into contact with the antennae of a living male. (2) An isolated male antenna coated with the hexane extract of antennae of sexually mature females, however, elicits the behavioral response of wing-raising from a living male on contact with antennae of the male. It would be more convenient to use some other material, such as nylon fishing line, fine wire, *etc.*, to carry the compound tested. In practice, however, such materials do not appear to be suitable for eliciting a positive response from males. It is likely that some physical property of the holder, for example, surface fine structure, elasticity, flexibility, etc., may have an important influence on the activity of the wing-raising pheromone.

As shown in Fig. 31. 3, the ability of the male to respond to the wing-raising pheromone varies so markedly during one week after the adult ecdysis that males reared for at least 10 days must be used for the assay.

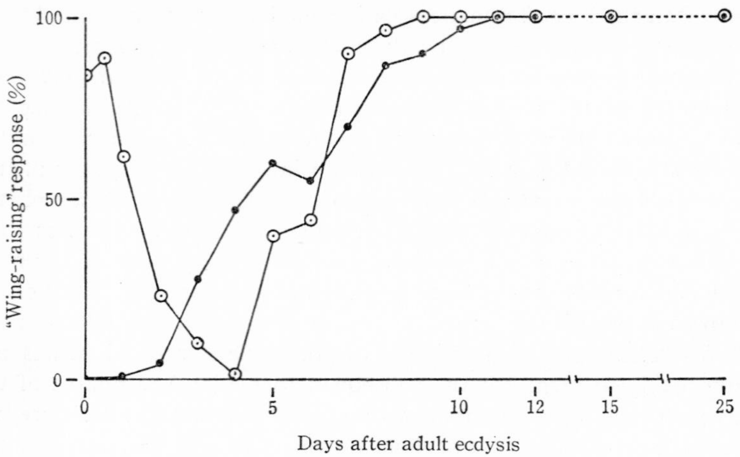

Fig. 31.3. The change of pheromone activity after adult ecdysis. ●——●, Change of the ability of the male to respond to a standard pheromone sample; ⊙——⊙, Change in the pheromonal activity of isolated female's antennae after adult ecdysis.

Figure 31.3 also shows that the pheromone content on the surface of the female antenna varies with age after the adult ecdysis. This suggests that a better yield of the wing-raising pheromone can be obtained by the extraction of German cockroach females reared for more than 10 days, as mentioned later.

[Experimental procedure 1] Bioassay for the wing-raising pheromone

An antenna was cut off close to its base from an adult male German cockroach just before use, attached to a small glass rod with paste, dipped in a carbon tetrachloride solution containing a definite concentration of the test material for 2–3 sec, and kept in the air for about 30 sec to allow the solvent to evaporate off. This treated antenna was brought into contact manually with antennae of 10 successive males which had been segregated from females immediately after their adult ecdysis and confined in glass pots (11 cm in diameter, 7 cm in height, 2 individuals/pot) for about 2 weeks. The bioassay was carried out at about 25° C and was considered positive if wing-raising behavior of the male was observed within 30 sec. The activity was expressed as:

$$\frac{\text{Number of males which responded positively}}{\text{Number of males tested}} \times 100$$

This assay method is suitable for monitoring the active compound(s) during the isolation procedures. For more detailed discussion, however, it was necessary to estimate the BR_{50} value of the candidate compound by the following procedure: carbon tetrachloride solutions of the test compound were prepared at concentrations of 1×10^{-n}, 2×10^{-n}, 3×10^{-n}, and 5×10^{-n} mol/ml near the roughly estimated critical concentration for wing-raising activity which had previously been determined by preliminary trials with ten-fold dilutions of the compound. Each solution was tested with 30 males as described above, and the procedure was repeated four times. The mean values of percentage response were plotted against concentration on a probit scale section paper to determine the BR_{50} value of the compound.

It should be noted that the BR_{50} value is indicated in terms of the concentration of the solution in which the isolated male antenna is dipped, because the quantity of the test compound on the antenna itself cannot be estimated.

31.2 Chemical Study

Source Material and Extraction of the Wing-Raising Pheromone

It is evident, as mentioned above, that the chemical factor responsible for the wing-raising behavior of the male German cockroach is contained in the cuticular wax on the surface of the antennae of sexually mature females and is transmitted to males by direct contact chemoreception *via* the antennae.

Fortunately, however, the pheromone is found not only in the cuticular wax of antennae but also on the body surfaces of sexually mature females of the German cockroach. This was confirmed by the following experimental result: 50 female German cockroaches were segregated from males after their adult ecdysis, reared for two weeks, and then separated into antennae and bodies. Each part was extracted with hexane and gave, upon evaporation, a waxy material in yields of about 0.5 mg and of 12 mg, respectively. Bioassay results showed wing-raising pheromone activity amounting to 100 F.E. (female equivalent)/ml for the antennae extract and 10 F.E./ml for the bodies extract.

This suggests that the cuticular wax on the surfaces of female bodies is a better source material for the isolation of the pheromone, because even though the pheromone content per unit area is much higher on the antenna than on the body surface, the total surface area of the antenna is much smaller than that of the body.

As shown in Fig. 31.3, the quantity of the wing-raising pheromone varies with the age of the female after adult ecdysis, decreasing at the early stage and then increasing rapidly to a plateau after about 10 days. Accordingly the females were segregated from the males just after the adult ecdysis, reared for more than two weeks, and "rinsed" with hexane to isolate the pheromone. Extraction with hexane for a long period could fail to yield pheromone activity, possibly due to the extraction of some chemicals that mask the activity. Quick rinsing of the body surface with hexane overcomes this problem.

[Experimental procedure 2] Preliminary extraction[6,8]

Virgin females (1,600) of the German cockroach which had been segregated from the males immediately after adult ecdysis were reared at about 25° C for 15–30 days. They were anesthetized with carbon dioxide and then rinsed three times with a total of 1,600 ml of hexane in a Buchner funnel within 3 min. The residue was then extracted with 800 ml of ether for 2 min, followed by extraction twice with 400 ml portions of a mixture of dichloromethane: methanol (1:1) by immersion in the mixture for two days. The activity was found only in the hexane rinse (404 mg), which was then fractionated into acidic, basis, and neutralfractions. Each fraction was concentrated under reduced pressure, and 10 and 100 F.E./ml solutions were prepared by dilution with carbon tetrachloride for bioassay. The activity was found only in the neutral fraction. This fraction (381 mg) was chromatographed on a silicic acid column (16 g, Mallinckrodt, 100 mesh), eluting successively with 80 ml portions of hexane: ether mixture at stepwise-increasing ratios, and finally with ethyl acetate. The chromatogram and the activity of each fraction are shown in Fig. 31. 4. The wing-raising

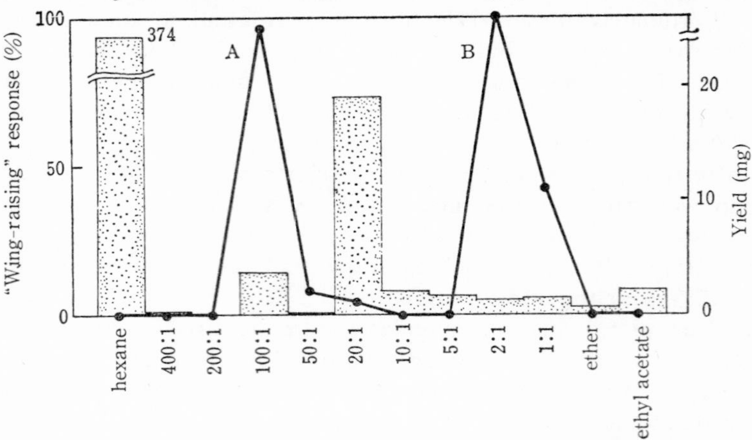

Fig. 31.4. Pheromonal activities of various eluates of the neutral fraction extracted from 1,600 female German cockroaches

pheromone activity was found in two distinct fractions, tentatively named Fractions A and B. Both could independently elicit wing-raising behavior in males.

The same number of male German cockroaches was extracted with hexane and fractionated by the same procedure. The neutral fraction (267 mg), which showed no activity, was chromatographed to separate two fractions corresponding to Fractions A and B of the females. No activity was found in the two fractions from the males.

[Experimental procedure 3] Isolation of the wing-raising pheromone[9)]

Two chemical factors responsible for male wing-raising in the sequential courtship behavior of the German cockroach were found in the female cuticular wax. Compound A (40 mg) was isolated from Fraction A obtained by extraction of 37,000 females, but compound B could not be isolated in this case, since it was present in such a small amount.

The German cockroach females were segregated from males after the adult ecdysis and reared for 1–2 weeks at about 28°C until sexual maturation. The mature females (1,000–3,000) in one batch were anesthetized with carbon dioxide and rinsed 3 times with hexane (a total of 1 ml per insect). The combined rinse of 224,000 females was passed through a short column packed with a mixture of Celite-545 (20 g) and anhydrous sodium sulfate (70 g), and concentrated under reduced pressure to give a waxy material (67 g).

This material was fractionated by column chromatography on silica gel (500 g, Wakogel C-200, 65 × 350 mm) with the following solvent systems to yield 18 fractions of 1 l each: Nos. 1 and 2, hexane alone; Nos. 3 and 4, hexane:benzene (2:1); Nos. 5–10, benzene; Nos. 11 and 12, benzene: ethyl acetate (30:1); Nos. 13 and 14, benzene:ethyl acetate (10: 1), Nos. 15 and 16, benzene:ethyl acetate (2:1), Nos. 17 and 18, ehyl acetate. Of these fractions, only fraction 7 was active on bioassay. This fraction was again subjected to chromatographic purification on a silicic acid column to give an active crystalline mass which was recrystallized twice from ethanol to yield compound A (239 mg) as colorless fine needles, mp 45–46°C.

As shown in Fig. 31.4, the active component(s) in Fraction B appeared to be more polar than compound A. However, fractions more polar than that containing compound A (No. 7) did not show activity. This unexpected problem had not been encountered in previous runs with a small number of females and was thought to be due to some chemical factor(s) accompanying the active component(s) on the initial column chromatography and masking its activity. A preliminary trial on a small quantity of each fraction using a combination of tlc separation and bioassay revealed that fractions 15 and 16 corresponded to Fraction B described previously, and contained the active component(s). The isolation of the active component from these fractions was carried out with the aid of bioassay according to the procedures shown in Table 31.2. First, the inactive oil (1.25 g) obtained by concentration of fraction 15 was chromatographed twice (20 g and 16 g) on columns of silicic acid, eluting with hexane containing ether in stepwise-increasing ratios. The first chromatography gave an active oil (220 mg) in the hexane:ether (10:1 to 5:1) fraction, which was subjected to a second chromatography to give an active oil (33 mg), tentatively named No. 15A, in the hexane:ether (10:1) fraction.

Concentration of fraction 16 gave an inactive oil (480 mg). An ethereal solution of this was shaken successively with dilute hydrochloric acid and sodium hydoroxide solution to separate the neutral fraction (345 mg) from the inactive acidic and basic fractions. The neutral fraction was chromatographed again on a silicic acid column (20 g) and fractionated into 10-ml fractions by eluting with hexane containing ether in stepwise-increasing ratios. Thirty fractions eluted with hexane:ehter (5:1) showed strong activity. Concentration of these fractions gave a crystalline mass (3.8 mg), from the middle fractions, and an oil (31 mg) from the others: these were tentatively named No. 16A and No. 16B, respectively.

Fractions 15A and 16B were combined and rechromatographed on a silicic acid column (7.5 g), using benzene containing ethyl acetate in stepwise-increasing ratios. Benzene:ethyl acetate (50:1) gave a highly active material (4.2 mg), which was combined with No. 16A and subjected to preparative tlc (20 × 20 cm, 0.75 mm thick, silica gel GF, E. Merck) developed with benzene:ethyl acetate (5:1). The activity was found in the band around *Rf* 0.4, and this was scraped off and extracted with ethyl acetate to give an active crystalline mass (4.6 mg). This mass was sufficiently pure for general spectrometric analyses and chemical conversion to derivatives for structural elucidation, and fruther purification was carried out by recrystallization from a small amount (less than 1 ml) of hexane to give fine needles of compound B (1.7 mg), mp 42–43° C. On gas-liquid chromatography, this compound showed a single peak at *Rf* 17.5 min. Under the same conditions, compound A and authentic hexatriacontane appeared at *Rf* 5.4 and 13.2 min, respectively.

The activities of the isolated compounds A and B are shown in Fig. 31.5; compound B is much more active than compound A and elicited the wing-raising response from all males tested at a concentration of 3 μg/ml in carbon tetrachloride. This figure also shows activities of compounds A and B synthesized after elucidation of their structures. It is interesting that the synthetic compounds showed the same level of activity as the iso-

TABLE 31.2. Isolation of the wing-raising pheromone from female German cockroaches.

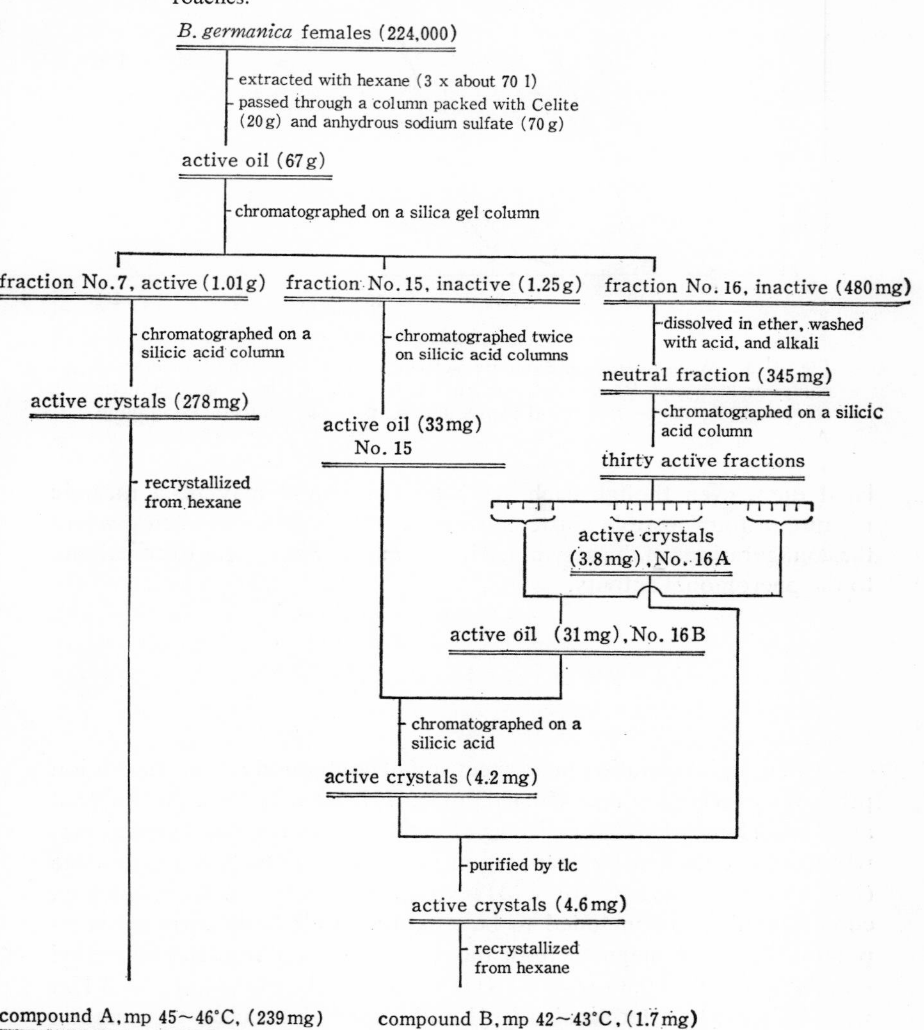
B. germanica females (224,000)
extracted with hexane (3 x about 70 l)
passed through a column packed with Celite (20 g) and anhydrous sodium sulfate (70 g)
active oil (67 g)
chromatographed on a silica gel column
fraction No. 7, active (1.01 g)
fraction No. 15, inactive (1.25 g)
fraction No. 16, inactive (480 mg)
chromatographed on a silicic acid column
chromatographed twice on silicic acid columns
dissolved in ether, washed with acid, and alkali
neutral fraction (345 mg)
chromatographed on a silicic acid column
active crystals (278 mg)
active oil (33 mg) No. 15
thirty active fractions
recrystallized from hexane
active crystals (3.8 mg), No. 16A
active oil (31 mg), No. 16B
chromatographed on a silicic acid
active crystals (4.2 mg)
purified by tlc
active crystals (4.6 mg)
recrystallized from hexane
compound A, mp 45~46°C, (239 mg)
compound B, mp 42~43°C, (1.7 mg)

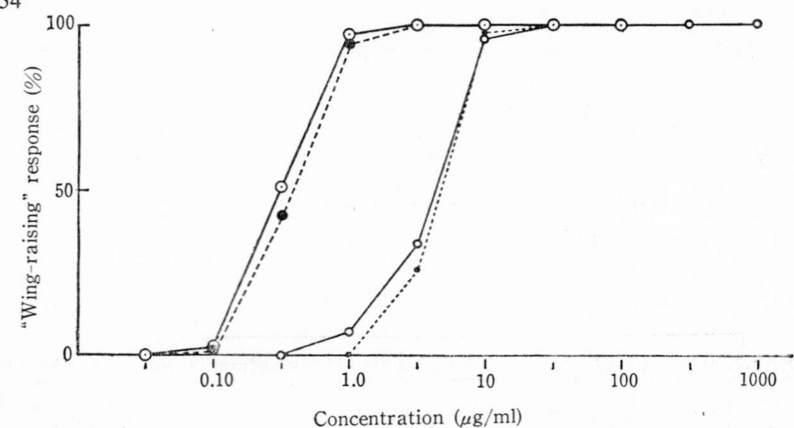

Fig. 31.5. Wing-raising pheromone activities of natural and synthesized compounds A and B. ⊙——⊙, natural compound B; ●·······●, synthetic compound B; ○——○, natural compound A; ●·······●, synthetic compound A.

lated ones, even though each synthetic compound may be a racemic mixture of four possible diastereoisomers. This is an unusual case where the configurations of the assymmetric carbons do not appear to contribute to the pheromonal activity.

31.3 Structure Elucidation[6,8,9]

In the high-resolution mass spectrum of compound A a molecular ion peak was observed at *m/z* 450.4803 (450.4797 for $C_{31}H_{62}O$). An infrared absorption band at 1,710 cm^{-1} suggested the presence of a carbonyl group, which was confirmed by preparation of the 2,4-dinitrophenylhydrazone (M^+ 630.5066 [630.5080 for $C_{37}H_{66}N_4O_4$], mp 55–56° C). Consequently, compound A was concluded to be a saturated aliphatic carbonyl compound. The proton magnetic resonance (pmr) spectrum showed four methyl signals: a triplet (δ 0.84, $J = 6.5$ Hz) and a doublet (δ 0.82, $J = 6.5$ Hz) were assigned to a terminal methyl and a methyl attached to a carbon atom of the long methylene chain, respectively; a singlet (δ 2.10) and a doublet (δ 1.04, $J = 6.8$ Hz) were assigned to –CO–CH_3 and –CH (CH_3)–CO–, respectively. The last methyl signal was coupled with the methine sextet (δ 2.49, $J = 6.8$ Hz). These pmr spectral data suggest that com-

$$X\text{–}\overset{29}{C}H_2\text{–}(CH_2)_{17}\text{–}\overset{11}{C}H(CH_3)\text{–}(CH_2)_7\text{–}\overset{3}{C}H(CH_3)\text{–}COCH_3$$

1 a : X = H, b : X = HO

pound A has the structural formula **1a**, in which the position of a mehtyl branch is uncertain. This formula was also supported by an intense fragment ion peak at *m/z* 72.0540 in the high-resolution mass spectrum, which corresponds to the fragment CH_3–CH = C^+(OH)–CH_3 (72.0574 for C_4H_8O) formed by McLafferty rearrangement of compound A.

The proposed structure **1a** was consistent with the ^{13}C-nuclear magnetic resonance (cmr) spectrum of compound A, in which the following signals were observed; a terminal methyl (δ 14.13), a branched methyl (δ 19.74), a carbonyl (δ 212.59) attached to both a methyl (δ 27.92) and a methine (δ 47.26) carrying a methyl (δ 16.19), as well as another methine and 24 methylenes. These assignments were based upon empirical chemical shifts for alkanes and carbonyl compounds,[10)] and the results of off-resonance decoupling experiments.

To determine the position of the methyl group on the long alkyl chain, compound A was reduced with hydrazine hydrate in concentrated alkali solution according to the Wolff-Kishner method to give the corresponding alkane (mp 27° C). The mass spectrum of this alkane showed two distinct ion peaks at *m/z* 183 ($C_{13}H_{27}$) and 281 ($C_{20}H_{41}$) as shown in Fig. 31.6. The high-resolution mass spectrum of compound A itself showed a significant ion peak at *m/z* 197.1910 (197.1916 for $C_{13}H_{25}O$). These data provide conclusive evidence that compound A is 3,11-dimethyl–2-nonacosanone **(1a)**.

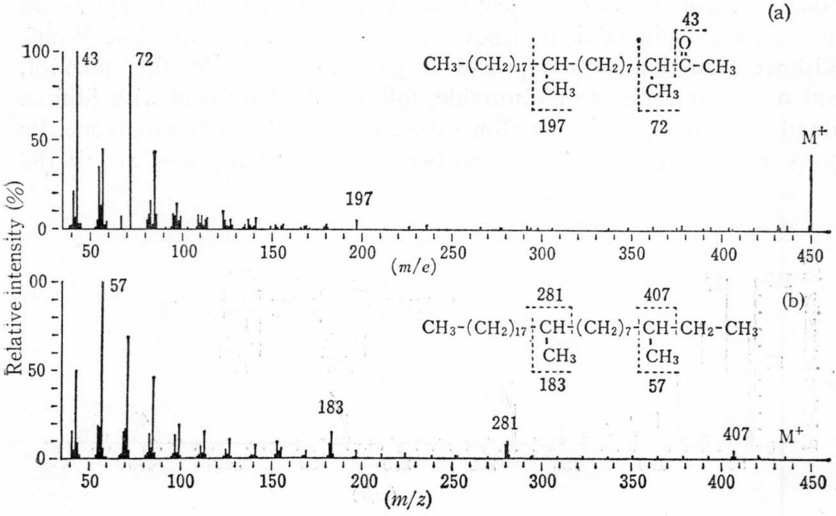

Fig. 31.6. Mass spectra of compound A (a) and the corresponding hydrocarbon (b).

The isolated compounds A and B are chiral compounds showing small positive optical rotations: $[\alpha]_D^{22} + 5.1°$ ($c = 3.54$ in hexane) and $[\alpha]_D^{18} + 7.1°$ ($c = 0.35$ in hexane), respectively. The S-configuration of the C-3 chiral center of compound A was indicated by ord and pmr studies using a chiral shift reagent.[11] Furthermore, Mori *et al.* successfully synthesized four diastereoisomers of compound A, of which the 3S,11S-isomer was confirmed to be identical with the naturally occurring compound A by comparison of their infrared spectra in the crystalline state and also by mixed melting point determination.[12]

The natural compound B exhibited a Cotton effect ($[\phi]_{305}^{20} + 620°$, $c = 0.24$ in ethanol) similar in both sign and amplitude to that of natural compound A. This suggests that natural compound B may also possess the 3S,11S configuration.

That compound B may be a compound having a hydroxymethyl group in place of the terminal methyl in compound A was indicated by the spectral data. The molecular ion peak at m/z 466 suggested the molecular formula $C_{31}H_{62}O_2$, and absorption bands at 3635 and 1715 cm^{-1} in the infrared spectrum indicated the presence of hydroxyl and carbonyl groups in the molecule. The hydroxyl group was confirmed by the formation of a monotrimethylisilylated-derivative (M^+; m/z 538). The pmr spectrum showed a signal for $-O-CH_2-CH_2-$ at δ 3.56 (2H, triplet, $J = 6.5$ Hz) coupling with the adjacent methylene at δ 1.5, indicating the hydroxyl to be primary, and other signals were the same as those observed in compound A. Conclusive identification of the position of the methyl group in the long chain of methylenes was obtained as follows. The Wolff-Kishner reduction of compound B gave the corresponding alkanol, which was converted to its bromide, followed by treatment with lithium aluminum tetradeuteride to afford the corresponding deutero-alkane. Its mass spectrum (Fig. 31.7) showed two prominent ion peaks, at m/z 183

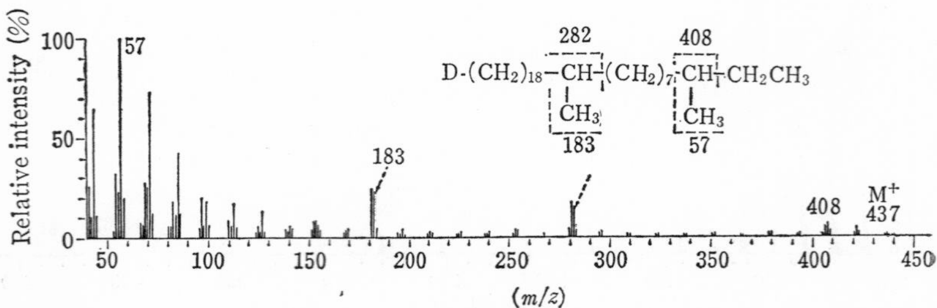

Fig. 31.7. Mass spectrum of the deutero-hydrocarbon derived from compound B.

and 282, of which only the latter showed a one mass unit shift from the corresponding ion peak at *m/z* 281 in the mass spectrum of the alkane derived from compound A (see Fig. 31.6), indicating compound B to be 29-hydroxy–3,11-dimethyl–2-nonacosanone (**1b**).

References

1) C. Kitamura, S. Takahashi, *Kontyu*, **41**, 383 (1973).
2) C. J. Persoons *et al.*, *Tetr. Lett.*, **1976**, 2055.
3) L. M. Roth, E. R. Willis, *Am. Midland Naturalist*, **47**, 65 (1952).
4) S. Ishii, *Appl. Ent. Zool.*, **7**, 226 (1972).
5) A. Butenandt *et al.*, *Z. Naturforsch. B*, **14**, 283 (1959).
6) R. Nishida, H. Fukami, S. Ishii, *Appl. Ent. Zool.*, **10**, 10 (1975).
7) T. Sato, R. Nishida, Y. Kuwahara, H. Fukami, S. Ishii, *Agr. Biol. Chem.*, **40**, 391 (1976).
8) R. Nishida, H. Fukami, S. Ishii, *Experientia*, **30**, 978 (1974).
9) R. Nishida, T. Sato, Y. Kuwahara, H. Fukami, S. Ishii, *J. Chem. Ecol.*, **2**, 449 (1976).
10) G. C. Levy, G. L. Nelson, *Carbon-13 Nuclear Magnetic Resonance for Organic Chemists*, Wiley-Interscience (1972).
11) R. Nishida, Y. Kuwahara, H. Fukami, S. Ishii, *J. Chem. Ecol.*, **5**, 287 (1979).
12) K. Mori, J. Suguro, S. Masuda, *Tetr. Lett.*, **1978**, 3447.

CHAPTER 32

Isolation and Bioassay of Ecdysones from Plants

Insects undergo dramatic changes during their growth and development from larvae directly or indirectly (through pupae) into reproducing imagines.[1-10] The processes are divided by cyclical moltings among which the larval-pupal transformation and the pupal-adult transformation are designated as metamorphosis. These significant transformations have fascinated biologists since olden times. The first contribution to this problem was reported by Kopeć (1917) who proved by ligation of larvae, extirpation of the brains and amputation of the nerve code of a certain insect that secretion from the brain of a hormonal substance is essential for moltings and metamorphosis. Since then, many biological contributions have appeared, and the mechanism of the moltings and metamorphosis can now be accounted for uniformly in terms of the endocrinological controls. Thus, the moltings and metamorphosis that occur during the growth and development of insectd are governed by the brain hormone, the prothoracic gland hormone and the corpus allatum hormone. The neurosecretory cells of the brain secrete a substance which triggers the prothoracic glands. This substance is called the brain hormone or the activation hormone, and is considered to be a protein of high molecular weight. When the prothoracic glands respond to this stimulus, they release the prothoracic gland hormone (the molting hormone or the metamorphosis hormone). This hormone, in the presence of the corpus allatum hormone, causes molting with the retention of the larval character, and in the absence of the corpus allatum hormone, elicits metamorphosis. The corpus allatum hormone alone has no effect on molting, but cooperates with the prothoracic gland hormone to bring about molting, though not metamorphosis. It is named the juvenile hormone because it causes insects to retain the juvenile structures. This hormone also possesses activity to mature the ovaries of female imagines. The juvenile hormone was first isolated from male imagos of the Cecropia moth (*Hyalophora cecropia*) and was shown to be sesquiterpenoids.

Pioneering work on the isolation of the prothoracic gland hormone was carried out by Becker (1941), who obtained considerably active fractions from the blowfly (*Calliphora erythrocephala*), but failed to isolate the pure hormone. This work was continued by Butenandt and Karlson who, after preliminary examinations on the pupae of the blowfly and later, on the pupae of the silkworm (*Bombyx mori*), carried out the extraction and fractionation of the male pupae (0.5 ton) of the silkworm. To monitor the fractionation, the *Calliphora* test devised by Fraenkel was improved and was an extremely useful tool. Finally, the prothoracic gland hormone was isolated in the crystalline state (25 mg) and termed ecdysone because it induces ecdysis. This hormone was later isolated from the same material in an improved yield using a modified procedure (Table 32.1). Afterwards Karlson showed in *Bombyx mori* the existence of a second hormone which was designated as β-ecdysone, and thus the name of the previously found ecdysone was changed to α-ecdysone. β-Ecdysone was later isolated from the pupae of the silkworm in a crystalline form and was also called 20-hydroxyecdysone and ecdysterone. Meanwhile a similar hormone was obtained from a kind of sea crayfish, *Jasus lalandei*, and termed crustecdysone. All these analogs were later proved to be identical. Eight kinds of molting hormones have so far been discovered in arthropods.

TABLE 32.1. Isolation of ecdysone from silkworm pupae.

Bombyx mori pupae (1,000 kg)
| extracted with 75% methanol and condensed
methanol solution (600 l)
| extracted with butanol
butanol solution
| washed with alkali and acid
neutral butanol extract
| partitioned with water and light petroleum
water-soluble fraction (232 g, 50 CU/mg)
| chromatographed over alumina
active fraction (35 g)
| subjected to counter-current distribution
active fraction (14.6 g, 2,500 CU/mg)
| chromatographed over alumina
active fraction (2.9 g)
| subjected to counter-current distribution
crude crystals (0.25 g, 100,000 CU/mg)
| crystallized from water
ecdysone

The structure elucidation of ecdysone carried out by Karlson *et al.* met with difficulty due to the small amount? of material available, and the inadequacy of available physico-chemical techniques. After ten years of chemical investigations, the structure (including stereochemistry) was established as **2** by single crystal X-ray analysis by Huber and Hoppe, and this was later substantiated by its synthesis. The establishment of the stereostructure of ecdysone as a reference compound was to contribute to the structure elucidation of its analogs later found in nature.

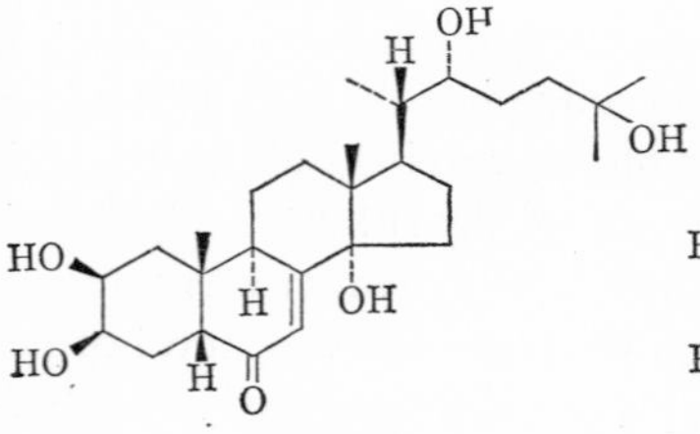

ecdysone **1** 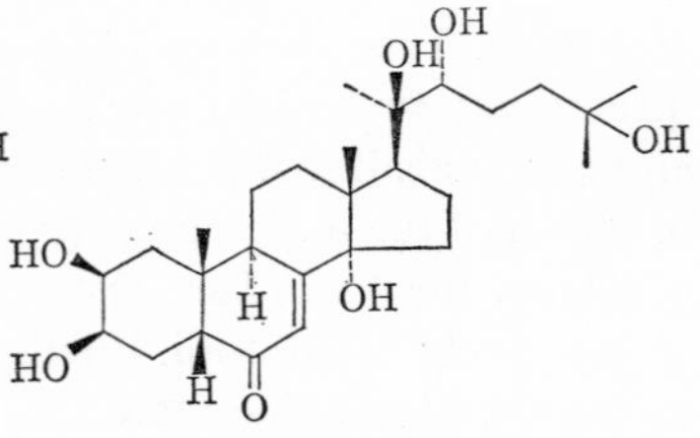   ecdysterone **2**

32.1 Bioassay of Ecdysones[1,11)]

As will be described later, extensive screening tests on a number of plant sources were performed immediately after the discovery of the ecdysones from the plant kingdom. For this purpose, not only the assay procedures already devised but also modified methods to deal with multiple samples were utilized. As a result, the molting hormone activity was found in numerous plants. In some cases, extraction was performed in order to isolate ecdysones. At the early stage of the investigations, exhaustive bioassay of the hormone activity in all the fractions obtained was very useful for following the ecdysones. As the properties of the ecdysones began to be clarified, however, bioassay of all the fractions was no longer necessary, and purification is now attained by assaying some fractions at key points.

Since several methods are known for assaying the ecdysones, a method should be selected which is suitable for the particular purpose and conditions in each case.

Bioassay of Ecdysones Using Flies

The oldest and most commonly used assay method for ecdysones is the *Calliphora* test utilizing the isolated abdomens of the blowfly (*C. erythrocephala*), which can be replaced by similar flies such as the flesh fly (*Sarcophaga peregrina*) and the house fly (*Musca domestica*). In this procedure, mature larvae of the fly are first ligated on the anterior third and isolated abdomens are prepared from the larvae which have pupated only in front of the ligature after a certain period, then a test solution is injected. After a certain period, the segments are scored for pupation. In the *Calliphora* test, the amount of the molting hormone which provokes 50–70% pupation of an isolated abdomen 24 hr after the injection, is defined as 1 *Calliphora* unit. In one system, complete pupation, incomplete pupation and weak pupation are scored as 1.0, 0.75 and 0.5, respectively, and the activity of a sample is judged from the overall percentage of the total scores divided by the number of surviving animals. This assay procedure is suitable for comparing the levels of activity of a number of samples. Although the *Musca* test is 3–4 times as sensitive as the *Calliphora* test, more skill is required because the body of the insect is smaller. In order to treat a large number of samples, a modified method was devised in which the *Musca* larvae are introduced by means of a thin-walled tube into small holes burned into rubber bands. This method of ligating larvae is reported to be much quicker than using thread.

The most important problem in this assay procedure is how to obtain efficiently isolated abdomens which can be used for testing. This ultimately depends on the timing of the ligation of larvae. This problem was examined in detail with *Sarcophaga peregrina* and, as a result, the optimum timing of ligation was determined in relation to the physiology of the larvae.

This assay procedure employing flies is sensitive (Table 32.2), essentially quantitative, and shows good reproducibility, and many plant materials have been assayed by this method for screening and for fractionation of their extracts.

TABLE 32.2. Molting hormone activities of ecdysones assayed in different species of insects.

Substance	ED_{50}(μg/individual)				
	Calliphora larvae (injection)	*Sarcophaga* larvae (injection)	*Musca* larvae (injection)	*Chilo* larvae (dipping)	*Samia* pupae (injection)
Ecdysone	0.018	0.035	0.005	—	2.5
Ecdysterone	0.017	0.018	0.005	0.63	5.0
Inokosterone	—	0.064	0.05	2.8	5.0
Ponasterone A	—	0.011	0.015	0.8	1.3
Ponasterone B	—	0.023	0.038	—	1.3
Ponasterone C	—	0.025	0.75	—	2.5
Cyasterone	—	0.016	—	0.35	0.13

[Experimental procedure 1] Bioassay of ecdysones using *Sarcophaga peregrina*

The imagines of the flesh fly are raised on 1% sucrose solution and, if necessary, on powdered milk. Eggs are laid on pig's liver in Petri dishes. After 4–5 days, the larvae become mature and creep out of the vessels. The larvae are collected, washed with water and stored in sealed containers with water. After 1–2 days (if necessary, 3–4 days), the larvae are placed on dried sand, anesthetized with ether after approximately 7 hr, and ligated on the anterior third. The front parts of the animals, which are pupated only in front of the ligature after 24 hr, are cut off and the isolated abdomens are injected with a test solution. After 24 hr, the larval segments are scored for pupation. The temperature is maintained at 25–26° C throughout the procedure. Kneaded yeast may be utilized instead of pig's liver but, in this case, more time is required for maturation, and matured larvae are less vivid and can be stored for a shorter time.

Bioassay of Ecdysones Using the Rice-Stem Borer Moth[6)]

Although the assay procedure using flies is an excellent method, it requires some time for the administration of sample solutions, and when testing plant extracts the test animals are sometimes killed by coexisting toxic substances. On the other hand, topical applications of sample solutions on insects are sometimes performed to determine the potency of insecticides. Thus, various conditions were tested to determine whether the ecdysones could act on insects through the skin, and it was found that topical applications of the ecdysones in organic solvents to larvae of the rice-stem borer moth (*Chilo suppressalis*) induced molting. On the basis of this observation, a unique *Chilo* dipping test was devised, in which larvae of *C. suppressalis* are reared on the seedlings of rice, ligated at the metathoraces 3 days before pupation, kept at 28° C for 24 hr, and simply dipped for 10 sec into a methanolic solution of a test material. After removal of the solvent by evaporation and maintenance at 28° C, the activity is evaluated in terms of pupation of the abdomens within 48 hr. Although this method is much less sensitive (Table 32.2) and may not be quantitative, it is easier to perform as compared with the injection methods. It may thus be useful to screen rapidly a wide range of plants if facilities for rearing many insects are available. This method was applied to test the extracts of numerous plant materials for molting hormone activity.

Bioassay of Ecdysones Using the Silkworm

Isolated abdomens of larvae and pupae, and brainless pupae of the silkworm have been used for biological examinations of the function of the prothoracic glands. They can also be applied for assaying ecdysones. Isolated abdomens prepared by ligation of unpigmented pupae are injected with a test solution. Activity is signaled by the observation of adult development of the abdomens in a certain period following the injection. In this method, the results are clear-cut and accurate because the judgement relates to an unambiguous phenomenon, imaginal development. However, there are some disadvantages, as follows: (1) the test objects are not always available throughout the year, (2) it requires a relatively long time and (3) its sensitivity is low (Table 32.2). Although artificially diapausing pupae which are deprived of their brains immediately after pupation can be used for the assay, the prothoracic glands can be activated by the prothoracic gland hormone itself, and consequently it is not clear that the function and activity of the molting hormone itself is determined by this procedure. In the case of isolation of the ecdysones from *Podocarpus* plants, brainless pupae (*Samia cynthia*) were employed for assay.

Bioassay of Ecdysones Using the Fruit Fly

As a rapid and sensitive method for testing the effect of ecdysone, a procedure using the chromosomes in cells of the salivary glands of dipteral insects was proposed. Thus, final instar larvae (at day 11 or 12) of the fruit fly (*Chironomus tetans*) are injected with ecdysone. After 15–30 min, certain parts of the chromosomes appear to be swelling; such swellings are called puffs. Judgement is performed by the microscopic observation of puffs at 1 hr after the injection. This method is thought to be less reliable and more difficult to carry out than the *Calliphora* test.

32.2 Isolation of Ecdysones from Plants

Unlike the isolation of the ecdysones from animal sources, which was achieved as a result of careful planning, the isolation of ecdysones from plant sources was accomplished fortuitously. Thus, during the course of a chemical study on the leaves of *Podocarpus nakaii*, a traditional medicine for cancer in Formosa, Nakanishi *et al.* isolated three constituents, determined the structure of the main constituent in comparison with that of ecdysone, recognized strong molting hormone activity, and named the compounds ponasterone A, B and C. Around that time, Takemoto *et al.* investigated the active principles of the roots of *Achyranthes fauriei* (said to have diuretic activity) and obtained two constituents, of which one was identified as ecdysterone **2** and the other was assumed to be a positional isomer designated as inokosterone; both possessed potent molting hormone activity. These were the first ecdysones to be recognized in the plant kingdom. These discoveries raised the possibility that other plant sources might contain similar substances and extensive screening tests were carried out on a vast number of plant materials. It was found that substances having the hormone activity are widely distributed in plant sources.

Among them, the probability of finding active substances is higher in the pteridophytes and gymnosperms and the occurrence is irregular in the angiosperms. Isolation of active principles was attempted from plants which showed the hormone activity, and the structures of over 40 ecdysones have so far been elucidated.

In the plant meterials from which ecdysones were isolated in the early work, the ecdysone contents were quite high, so that their isolation was fairly easy. However, when the search was expanded, lower ecdysone contents were sometimes found, and purification was sometimes difficult. The ecdysones are polar substances which are difficult to separate efficiently, and further analogs frequently coexist in a given plant material, making separation more difficult. For the extraction of ecdysones, polar solvents such as methanol, aqueous methanol, ethanol and aqueous ethanol were commonly utilized. If necessary, the solution can be diluted with water to remove precipitates. The extract can be effectively concentrated by partition. Thus, the non-polar fraction can be eliminated by extraction with a non-polar solvent or by partition between water and a non-polar solvent. The resulting polar fraction was extracted with aqueous ethyl acetate, butanol and tetrahydrofuran, yielding an active fraction which could be concentrated further by counter-current distribution. Further fractionation is usually performed by chromatography on active carbon, neutral alumina, silica gel and silicic acid. Reverse phase chromatography using Amberlite XAD-2 resin is also useful. When the quantity of the active fraction becomes small enough, separation by means of thin layer chromatography is also a powerful tool. If adequate separation cannot be achieved by the above methods, the mixture is acetylated. The resulting mixture of acetates can usually be separated by chromatography and each acetate hydrolyzed to generate the corresponding ecdysone. High-speed liquid chromatography (recently developed) is also a valuable technique. The most effective procedure for the extraction, purification and isolation of ecdysones varies from case to case, so that the isolation must be accomplished by a combination of a number of methods, as required. In any event, bioassay for molting hormone activity is extremely useful in the fractionation of the ecdysones.

The ecdysones of plant origin so far known are steroids possessing the 5β(H)-7-en-6-one system, with $2\beta,3\beta,14\alpha$-hydroxyl and usually the $20R,22R$-hydroxyl groups. Variation of the structure can involve further hydroxyls, an ether linkage and a lactone moiety. In the structural study of these ecdysones, since the chemical and physico-chemical properties of reference substances, such as eceysone **1** and ecdysterone **2**, are well known, comparison of the results of simple reactions and of the mass, uv, ir, nmr and ord spectral data of ecdysones and their acetates with those of the reference substances has permitted rather facile structure determination.

Isolation of the Ponasterones from *Podocarpus Nakaii* Leaves[4)]

Based on the experience obtained during the extraction and purification of the ponasterones from *P. nakaii*, the general fractionation procedure for ecdysones shown in Table 32.3 has been recommended from the viewpoint of ease of operation and good yield. The path A or B in Table 32.3 may be selected depending on whether a precipitate forms or not when the methanol solution is diluted with water to a 20% methanolic solution. In the extraction and fractionation of the ecdysones from *P. nakaii* and *P. macrophyllus*, path A was adopted. In the isolation of the ponasterones

TABLE 32.3. Isolation of the ponasterones from *Podocarpus nakaii* leaves.

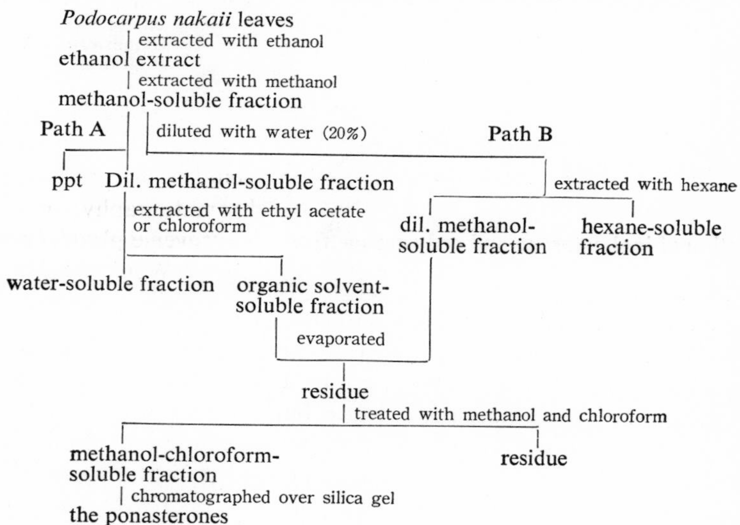

ponasterone A **3**

inokosterone **4**

from *P. nakaii,* the route shown in Table 32.4 was devised and reported to be preferrable (both easy and efficient).

TABLE 32.4. Isolation of the ponasterones from *Podocarpus nakaii* leaves.

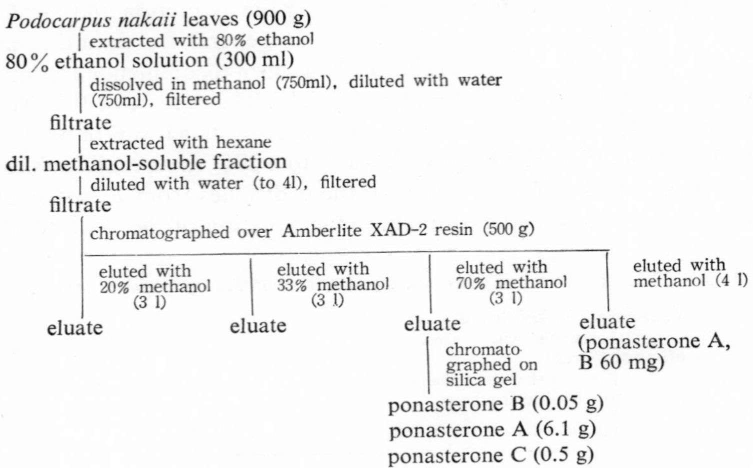

Isolation of inokosterone and ecdysterone from *Achyranthes fauriei* roots[5)]

In examining the constituents of *Achyranthes fauriei* roots, the plant material was extracted with methanol to give an extract, which was in turn extracted with water, and the water-soluble portion thus obtained was continuously extracted with ethyl acetate. Extraction of plant materials with methanol is a common procedure but extraction of the resulting methanolic extract with water and continuous extraction of this water-soluble portion with ethyl acetate are rarely used, except for special purposes. In the case of fractionation of the extract of *Achyranthes fauriei,* ecdysones were concentrated effectively by these procedures. Thus, alumina chromatography of the above ethyl acetate-soluble portion resulted in the facile isolation of a mixture of inokosterone and ecdysterone. The mixture was thought to be a homogenous substance at an early stage because the properties of the two compounds are quite similar. However, acetylation afforded two derivatives which could easily be spearated. After this, complete separation of both ecdysones was attempted. In the first procedure, the mixture was roughly separated by crystallization from water, because inokosterone is less soluble in water than ecdysterone, and both were separately crystallized from methanol-hexane to give the pure ecdysones ("Experimental procedure 2"). This procedure, however, requires much labor and time, and is not very efficient, though loss of the material can be avoided. In the second procedure, the mixture was separated by an acetylation-chromatographic separation-hydrolysis ("Experimental procedure 3"). Both compounds can be separated in fairly good yield by this technique if the acetates are hydrolyzed under mild conditions.

[Experimental procedure 2] Isolation of inokosterone and ecdysterone from *Achyranthes fauriei* roots[14]

Achyranthes fauriei roots (15 kg) were extracted with methanol (60 l each) 3 times. The methanol solution was condensed to 2 l and the deposited crystals of potassium nitrate were filtered off. Water was added to the filtrate and after extraction with ether, the wtaer-soluble portion was continuously extracted with ethyl acetate for 24 hr. The ethyl acetate-soluble portion was dissolved in a small amount of ethanol and adsorbed on an alumina column. Fractions eluted with ethyl acetate-ethanol (4:1) crystallized (yield 5 g). The crude crystals (5.0 g) were extracted with hot water and the less soluble portion was crystallized from ethanol-hexane to give inokosterone **4** as colorless needles (1.5 g). The more soluble portion of the crude crystals was repeatedly crystallized from ethanol-hexane, yielding ecdysterone **2** as colorless needles (0.5 g).

[Experimental procedure 3] Separation of a mixture of inokosterone and ecdysterone[14]

A mixture (0.5 g) of inokosterone and ecdysterone in pyridine (5 ml) was treated with acetic anhydride (10 ml) at room temperature for 24 hr. The precipitate obtained after addition of water was dissolved in a small quantity of ethyl acetate, applied to a column of alumina (50 g) and eluted with light petroleum-ethyl acetate (1:1). Fractions eluted first were crystallized from dil. ethanol to give inokosterone tetraacetate as colorless needles (0.15 g). Fractions eluted next were crystallized from ethyl acetatehexane, furnising ecdysterone triacetate as colorless needles (0.1 g).

Each acetate (100 mg) was heated on a steam bath for 1 hr with 10% sodium hydroxide solution (1 ml) in ethanol (10 ml). After cooling, water (10 ml) was added and the solution was charged on a column of Amberlite IR-120 resin (10 ml), which was washed with 50% methanol. The eluate and washings were combined and evaporated to dryness. Chromatography of the residue on alumina (10 g), elution with ethyl acetate-ethanol (4:1) and crystallization from ethanol-hexane gave inokosterone or ecdysterone as colorless needles, yield 30 mg or 25 mg, respectively.

Isolation of Cyasterone from *Cyathula capitata* Roots[5]

After the isolation of inokosterone and ecdysterone from *Achyranthes fauriei* roots, it was considered that the related *Cyathula capitata* might contain similar substances. Thus, extraction of the plant material was performed. In this case, however, the properties of the ecdysones were fairly well known, so that the fractionation was carried out most efficiently. The methanol extract was thus extracted with ethyl acetate and the ehtyl acetate-soluble portion was chromatographed to afford cyasterone **5**. Although nine kinds of ecdysones have so far been discovered from *C. capitata,* cyasterone is the main component and the quantities of the other ecdysones are very small. In fact, the extract of *C. capitata* gave only one peak for ecdysones on liquid chromatography using Amberlite XAD-2 resin. For the isolation of the minor ecdysones, silica gel chromatography was carried out repeatedly, but when complete separation was not achieved by this method, mixtures were separated by the acethylation-chromatographic separation-hydrolysis method.

[Experimental procedure 4] Isolation of cyasterone from *Cyathula capitata* roots

Cyathula capitata roots (15 kg) were extracted with methanol (25 l each) 5 times under reflux for 7 hr (each extraction) to give an extract (7.6 kg), which was in turn continuously extracted with ethyl acetate, yielding the ethyl acetate-soluble portion (170 g). Chromatography on alumina (750 g), eluting with ethyl acetate, and crystallization from methanol furnished cyasterone **5** as colorless needles (4.0 g).

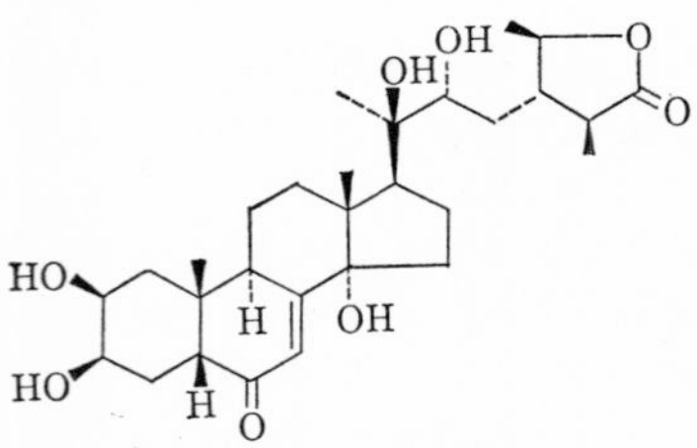

cyasterone **5**

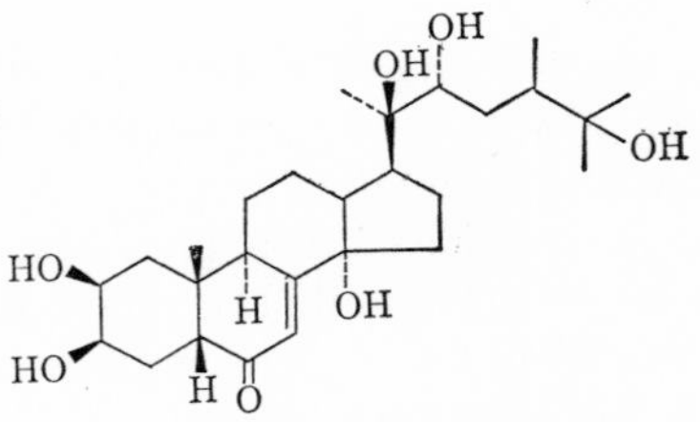

makisterone A **6**

Isolation of the Makisterones from *Podocarpus macrophyllus* Leaves

After the ponasterones had been isolated from *Podocarpus nakaii,* it was considered that analogs might be present in other *Podocarpus* plants. *P. macrophylls* leaves (1.2 tons) were then extracted and fractionated as shown in Table 32.5. Chromatography of the ethyl acetate-soluble portion on active carbon and alumina furnished ponasterone A **3** (120 g). Since the fractions eluted after ponasterone A contained several kinds of ecdysones, they were combined and subjected to chromatography on Amberlite XAD-2 resin to afford makisterone A **6** (yield 0.001 %). It was found that makiste-

TABLE 32.5. Isolation of the makisterones from *Podocarpus macrophyllus* leaves.

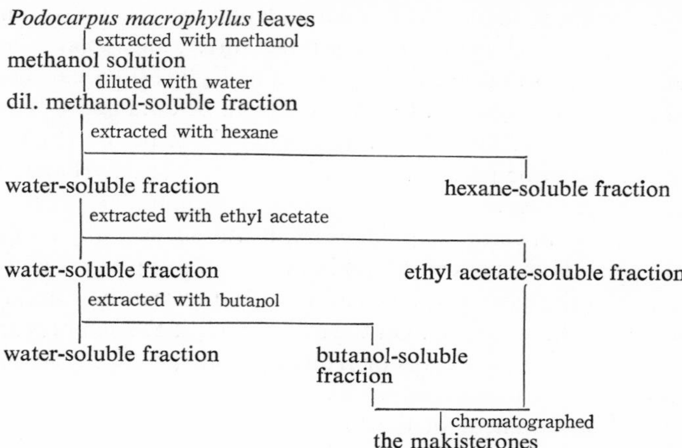

rone A was contained in the butanol-soluble portion in a fair quantity, so the mother liquor was repeatedly chromatographed to give two kinds of crude crystals, from which makisterones B, C and D were isolated by the acetylation-chromatographic separation-hydrolysis method (yield 0.0001 % each).

32.3 Isolation of the Ecdysones by Automatic Liquid Chromatography[6]

Chromatography of polyhydroxysteroids such as the ecdysones is generally performed with various adsorbents, such as alumina and silica gel, as the stationary phase. A serious disadvantage of chromatography with these adsorbents is that the column cannot be used repeatedly because of changes in the activity of the adsorbents. After examination of adsorbents and operating conditions, Hori found that the ecdysones could be effectively separated by an automatic liquid chromatographic method using an Amberlite XAD-2 resin column and eluting with various concentrations of dil. ethanol (20%–70%). The eluates were continuously monitored spectroscopically, and reproducible chromatograms could be obtained. Ecdysones could be estimated qualitatively and quantitatively from the elution times and the peak areas, respectively. When chromatograms were recorded at three wavelengths, 230, 250 and 300 nm, ecdysones could be distinguished from other substances (e.g. flavonoids) from the ratios of the absorbances. Since this method is a kind of reverse-phases absorption chromatography, the less polar alcohol becomes a solvent with stronger eluting power than the more polar water, and polar substrates are eluted earlier, so that coexisting sugars and glycosides can be easily separated from the ecdysones. Components can be obtained with ease by simply evaporating off the solvent. The column can be used repeatedly, and the method can be performed on a preparative scale. Therefore, the technique has been employed not only for estimation but also for preparative purposes, and in certain special cases, it was successfully applied for assaying the metabolites of ecdysones in insects. If samples contain minor impurities, as in the cases of plant roots the extract can be directly subjected to the chromatography. On the other hand, if the samples are very impure, as in the cases of leaves or fruits of plants, the extracts must be cleaned up by appropriate treatments.

With the development of the high-pressure liquid chromatographic technique, ecdysones can now be efficiently separated. Since the application of this technique is described elsewhere, only the separation of the two epimers of inokosterone is presented here (see below).

[Experimental procedure 5] Isolation of inokosterone and ecdysterone from *Achyranthes fauriei* roots[16)]

Extraction of *Achyranthes fauriei* roots (170 g) with methanol gave the methanol extract (14.3 g). The extract (2 g) was applied to an Amberlite XAD-2 resin column (36 × 1500 mm, 200–400 mesh, previously washed with various solvents), which was eluted first with 20% ethanol and then with aq. ethanol at a concentration which was continuously increased from 20% to 70%. Ecdysterones (5.6 mg) was obtained from fractions eluted earlier and inokosterone (4.8 mg) was obtained from those eluted later.

Separation of the Two Epimers of Inokosterone[17,18)]

Inokosterone is a type of ecdysone first isolated from *Achyranthes fauriei* roots, and its structure was assumed to be **4** (excluding stereochemistry). The configurations from C-1 to C-22 were later proved to be the same as those of the common ecdysones, but all the fragments derived from the side-chain showed no optical activity, so that inokosterone was concluded to be a mixture of the C-25 epimers. Separation of the epimers, however, could not be achieved. Recently ^{13}C -nmr spectroscopy and the use of a lanthanide shift reagent with the 2,3,22-triacetate and 2,3,22,26-tetraacetate of inokosterone established that inokosterone was a 1:2 mixture of the C-25 epimers. Based on the optical activity ($[\alpha]_{300}$ + 27.7°) of α-methylglutaric acid derived from inokosterone by metaperiodate oxidation followed by chromic acid oxidation, it was concluded that it was a

1:2 mixture of C-25 *R* and *S* epimers. In fact, inokosterone obtained from *Achyranthes fauriei* roots was later separated by means of high-pressure liquid chromatography and the ratio of inokosterones A and B was found to be 1:1.6. Analysis of inokosterone obtained from *Woodwardia orientalis* leaves by high-pressure liquid chromatography indicated that the ratio was 1:2.8, suggesting that the ratio varies depending upon the source.

[Experimental procedure 6] Separation of the epimers of inokosterone by high-pressure liquid chromatography[18]

Inokosterone (0.1% methanolic solution, 0.25 ml) was charged on a Permaphase ODS column (16.7 × 500 mm), which was eluted with aq. methanol at a concentration increasing continuously from 0% at a rate of 2% per min. Concentration of the appropriate eluates and crystallization from ethanol-ethyl acetate yielded inokosterones A and B as crystals.

References

1) S. Ishii, *Konchu no Seiri-kassei-busshitsu* (in Japanese), Nankodo (1969).
2) V. B. Wigglesworth (translated by T. Ito, M. Kobayashi), *Konchu Horumon* (in Japanese), Nankodo (1971).
3) P. Karlson, *Naturwissenschaften*, **53**, 445 (1966).
4) K. Nakanishi, M. Koreeda, *Kagaku no Ryoiki* (in Japanese), **22**, 597 (1968).
5) T. Takemoto, H. Hikino, *ibid.*, (in Japanese), **22**, 603 (1968).
6) T. Matsuoka, S. Imai, M. Sakai, M. Kamata, *Takeda Kenkyusho Nempo* (in Japanese), **28**, 221 (1969).
7) H. Hikino, Y. Hikino, *Progr. Chem. Org. Nat. Prod.*, **28**, 256 (1970).
8) D.H.S. Horn, *Naturally Occurring Insecticides* (ed. M. Jacobson *et al.*), p. 333, Marcel Dekker (1971).
9) K. Nakanishi, *Pure Appl. Chem.*, **25** 167 (1971).
10) H. Hikino, T. Takemoto, *Invertebrate Endocrinology and Hormonal Heterophylly* (ed. W. Burdette), p. 185 (1974).
11) K. Hasegawa, *Kagaku no Ryoiki* (in Japanese), **22**, 612 (1968).
12) T. Ohtaki, R. D. Milkman, C. M. Williams, *Proc. Natl. Acad. Sci. U.S.A.*, **58**, 981 1967).
13) D. A. Schooley, G. Weiss, K. Nakanishi, *Steroids*, **19**, 377 (1972).
14) T. Takemoto, S. Ogawa, N. Nishimoto, *Yakugaku Zasshi* (in Japanese), **87**, 1463, 1469 (1967).
15) H. Hikino, Y. Hikino, K. Nomoto, T. Takemoto, *Tetrahedron*, **24**, 4895 (1968).
16) M. Hori, *Steroids*, **14**, 33 (1969).
17) H. Hikino, M. Mohri, Y. Hikino, S. Arihara, T. Takemoto, H. Mori, K. Shibata, *Tetrahedron*, **32**, 3015 (1976).
18) S. Ogawa, A. Yoshida, R. Kato, *Chem. Pharm. Bull.*, **25**, 904 (1977).

CHAPTER 33

Biosynthesis of Ecdysones

The insect molting hormone, α-ecdysone, was first isolated in crystalline form from the silkworm by Butenant and Karlson in 1954, and the structure was determined in 1965. Soon after that, the second molting hormone, β-ecdysone (ecdysterone) was isolated and the structure was elucidated as 20-hydroxy–α-ecdysone. However, the amounts of the hormones isolated were too small for biological studies. Since 1966, when ecdysterone, inokosterone, makisterone and other phytoecdysones were independently isloated from plant sources by Takemoto and Nakanishi,[1] extensive biological studies of these hormones including biogenetic studies, have been carried out by many groups.

Pioneer work demonstrating the role of cholesterol (**3**) as a precursor of ecdysone was done by Karlson using *Calliphora erythrocephala*, though the incorporation was low (0.0001 %).[2] Recently, higher incorporations of labeled cholesterol into α-ecdysone (0.015%) and β-ecdysone (0.018% or 0.035%) have been reported in *in vivo* experiments,[3] though the site of ecdysone synthesis was not identified.

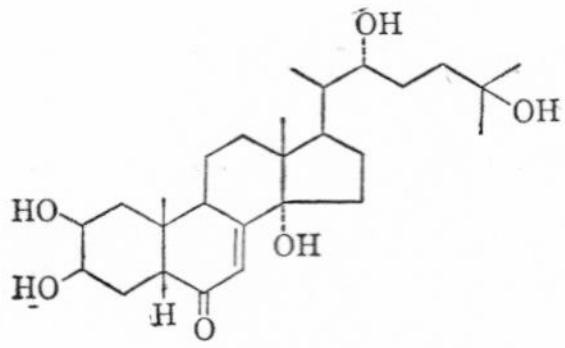

α–Ecdysone **1** 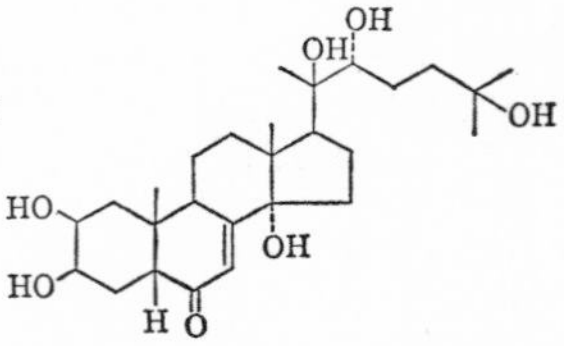 β–Ecdysone (Ecdysterone) **2**

In 1940, Fukuda[4] proposed that the prothoracic gland (PG) secreted molting hormone. However, there was no direct evidence to establish this. Endocrinological and chemical studies[5] during 1966 to 1970 on the insect molting hormone led to considerable controversy regarding the site of ecdysone synthesis. For instance, it was reported that cholesterol is converted into α-ecdysone (0.0008%) and β-ecdysone (0.00083%) in the isolated abdomen of larvade of the silkworm, and that the isolated ring glands do not convert cholesterol into α- or β-ecdysone.

Some investigators have shown that isolated PG produces the molting hormone *in vitro*.[6] However, the amounts of secreted hormone have been always too small to provide definitive proof. It was necessary to carry out organ culture of isolated PG, extraction of hormones from the culture system, and analysis by both bioassay and chemical studies. This approach, beginning with a fundamental improvement of the culture system, was adopted by Chino, Otaki and Sakurai in 1971. On the other hand, the author investigated the microanalysis of ecdysone by glc, using a technique similary to that used by Horning *et al.* for the profile analysis of human urinary steroids. In collaboration with Miyazaki's group, a microanalysis of insect molting hormones by gc was developed.

In this chapter, the method of microanalysis of the hormones is described first, followed by applications of this method to studies on the biosynthesis.

33.1 Separation and Identification of Ecdysones

Extraction of Zooecdysones

We encountered much greater difficulties in the extration and clean-up procedure for zooecdysones than for phytoecdysones due to the

Table 33.1. Counter-current distribution coefficients and *Rf* values (tlc) of zooecdysones (After J. N. Kaplanis *et al.*[8]).

Ecdysones	*K*†1	*Rf*†2
α-Ecdysone	3.54	0.23
Makisterone A	1.27	0.20
β-Ecdysone	0.52	0.15
3-Epi-β-ecdysone	0.52	0.17
26-Hydroxy-α-ecdysone	0.39	0.08
20,26-Dihydroxy-α-ecdysone	0.06	0.05

†1 Cyclohexane-*n*-butanol—water (5:5:10).
†2 Silica gel G; chloroform-ethanol (4:1).

presence of very small amounts of ecdysones with large amounts of lipids which were coextracted. Horn[7] used a countercurrent method for the purification of 2–deoxy-β-ecdysone and β-ecdysone from crayfish. Kaplanis[8] applied alumina and silica gel chromatography for the separation of egg ecdysone. The author[9] found that extraction with tetrahydofuran (THF) and subsequent absorption of ecdysone on Carplex are effective for the removal of lipids from ecdysones. Purification of the ecdysones was carried out by tlc. The distribution coefficients in countercurrent distribution and the *Rf* values on tlc are listed in Table 33.1. Examples of the extraction precdure for zooecdysones are described bleow.

[Experimental procedure 1] Extraction of ecdysones from crayfish[7]

Frozen crayfish (1 ton) were extracted with EtOH, then the EtOH was evaporated off *in vacuo*. The residue was extracted with hexane-isopropanol (1:3). The lipid was removed by pet. ether extraction. Defatted aq. extract was treated with sat. $(NH_4)_2SO_4$ and extracted twice with hexane-isopropanol (1:3). The combined organic layers were concentrated *in vacuo*. Counter-current extraction was carried out using n-BuOH–H_2O. The upper layer was concentrated and again subjected to counter-current extraction using $CHCl_3$–MeOH–H_2O (1:2:1). The MeOH layer was concentrated and extracted with $CHCl_3$–EtOH-aq. $KHCO_3$. The chloroform layer was concentrated and the 2-deoxy–β-ecdysone fraction was obtained by reverse-phase partition chromatography using n-BuOH–H_2O. The β-ecdysone fraction was chromatographed on CM-Sephadex and silicic acid columns to obtain 2.3 mg of β-ecdysone.

[Experimental procedure 2] Extraction and Purification of Zooecdysones[9]

Ten pupae were put into a 100-ml beaker, and 20 ml of acetone and dry ice were added. The beaker was covered with aluminium foil. The pupae were dried completely under reduced pressure at below 50°C. Next, 5 μg of cyasterone and 15 grams of sea sand were added to the residue, and the whole was ground to a fine powder in a mortar. The powder was put into a filter paper thimble and extracted with 100 ml of tetrahdrofuran (THF) in a modified Soxhlet apparatus for 24 hr. This apparatus was designed in such a way that a glass container for the filter paper thimble is heated by the solvent vapor. The THF extract was concentrated to half the original volume and 3 g of Carplex No. 80 (silicic acid) was added. The THF was then completely eliminated with a rotary evaporator. The residue was put into a filter paper thimble and extracted in the Soxhlet apparatus successively for 1 hr each with 59 ml of each of the following solvents: n-hexane, benzene, ether, and THF. The THF extract was passed through 3 g of silica gel (Merck, silica gel 60 for chromatography) in a column prepared in a glass filter (15 AG-3). The column was washed with THF. The eluate and washing were combined and THF was removed *in vacuo*. The residue was purified by preparative tlc (silica gel, 20 × 20 cm plate, thickness 500 μ) using $CHCl_3$–EtOH (2:1) as the developing solvent system. A wide band containing compounds with R_f values from 0.45 to 0.9 was extracted with THF using an ultrasonic generator (overall recovery yield, 75%), the THF was evaporated off, the residue was dissolved in 100 μl of trimethylsilylimidazole (TSIM), and the solution was heated at 100° C for 30 min. This solution was analyzed by mass fragmentography.

Separation of Ecdysones by HPLC

Ecdysone analogs can be effectively separated by hplc. Hori[10] first used hplc for the separation of ecdysones. He separated phytoecdysones by reverse phase hplc using Amberlite XAD-2. This method was applied for research on ecdysone metabolism by Moriyama.[11] Later, Poragel PN and Corasil II were used for ecdysone separation by hplc.[12] However, after testing several column packings, we found thzt Zorbax SIL was preferable. An example of separations was shown in Fig. 33.1.

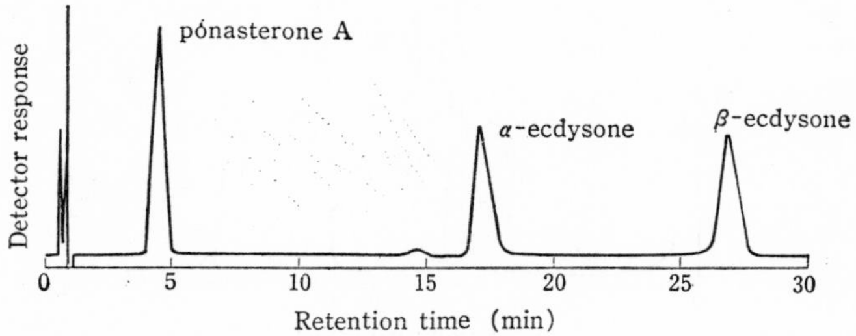

Fig. 33.1. Separation of ecdysones by hplc. Zorbax SIL column (25 cm × 2.1 mm i.d.), 50K, 7% ethanol–CH_2Cl_2.

Separation of Ecdysones by GLC[13]

The microdetermination of the insect molting hormones in biological systems has been carried out by bioassays with *Calliphora vincia, Musca domestica, Chilo suppressalis,* and other insects. However, a more sensitive and more specific determination of ecdysones was necessary for studies on the excretion, metabolism, and mode of action of the molting hormones. After extensive studies on the derivatization of ecdysones for glc analysis, we found that ecdysones can be analyzed as their trimethylilyl ethers (TNSi) or trimethylsilylether heptafluorobutyrates (TMSi-HFB), which can be detected at the subnanogram level by the FID or ECD detector.[9)]

The hydroxy groups at the 2, 3 and 14 positions of the steroidal skeleton can easily be trimethylsilylated with trimethylsilyl imidazole (TSIM) at room temperature for 30 min. The hydroxy groups at C-22 and C-25 of the side chain were also silylated at room temperature. However, the hydroxy group at C-20 is the most hindered, and is not silylated at room temperauture, though silylation can be effected at 100° C for 1 hr.

Thus phytoecdysones can be separated by glc either as the partially or completely silylated derivatives.[13)] In both cases, a single sharp peak is obtained. However, the completely silylated derivative is preferable be cause it shows higher intensity and minimum tailing. Separations of a mixture of phytoecdysones by glc using FID detection are shown in Fig. 33.2. The relative retention times are listed in Table 33.2.

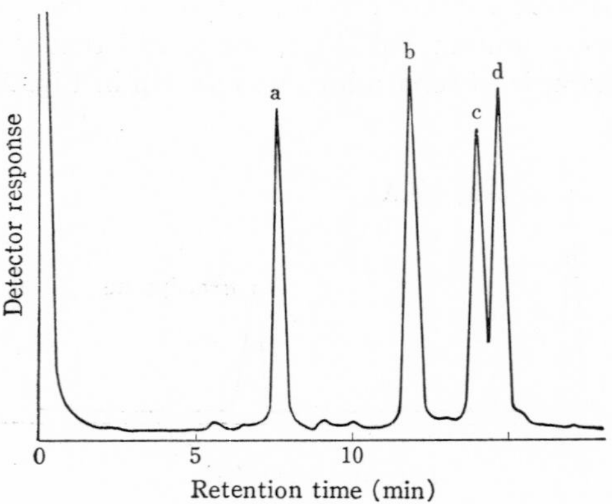

Fig. 33.2. Separation of a mixture of ponasterone A penta-TMS (a), ecdysterone hexa-TMS (b), inokosterone hexa-TMS (c) and Makisterone A hexa-TMS (d); 1.5% OV-101, 275°C, FID.

Heptafluorobutyryl (HFB) derivatives have helpful glc properties and can be used with either FID or ECD detection systems. Attempts to effect direct formation of HFB derivatieves of phytoecdysones with heptafluorobutyryl imidazole (HFBI) or heptafluorobutyryl anhydride (HFBAn) appear to result in incomplete reaction, as indicated by the presence of several peaks on glc. Satisfactory HFB derivative formation was achieved by an exchange reaction between TMS and HFB moieties. Thus, the cholesterol TMS derivative was obtained by dissolution in TSIM. After 5 min, HFBI was added to this solution and the whole was heated at 100° C

TABLE 33.2. Relative retention times of ecdysone and phytoecdysones.

Steroid	TMS derivatives		HFB derivatives
	Complete TMS	Partial TMS	
2β,3β,14α-Trihydroxy-5β-cholest-7-en-6-one	1.37 (tri)		0.61
α-Ecdysone **1**	2.62 (penta)		1.07
Ponasterone A	2.33 (penta)	2.23 (tetra)	1.00
Ecdysterone **2**	3.61 (hexa)	3.12 (penta)	1.86
Inokosterone	4.24 (hexa)	3.73 (penta)	1.49
Makisterone A	4.45 (hexa)	4.07 (penta)	2.27
Makisterone B	4.28 (hexa)	3.62 (penta)	1.52
Cyasterone	10.09 (penta)	9.54 (tetra)	4.94
Cholesteryl butyrate	1.00 (3.3 min)		1.00 (6.1 min)
Column packing	1.5% OV-101		1.5% OV-101
Column temp.	275°		260°
N_2 flow rate	40 ml/min		40 ml/min

for 30 min. Analysis of this solution by glc showed that TMS was exchanged completely for HFB by this procedure.

This reaction was applied to phytoecdysone analysis. For this purpose, 0.1 mg of ecdysone was dissolved in 20 μl of TSIM and after standing for 30 min at room temperature, the same amount of HFBI and 2 μl of HFBA were added and the solution was heated at 50° C for 2 hr (standing at room temperature overnight gave the same results). After dilution with methyl ethyl ketone, the solution was analyzed by glc using electron capture detection. Mass sepactrometric analysis indicated the structures 2,22-di-HFB–3, 14, 25-tri-TMSi for the α-ecdysone derivative and 2-HFB–3, 14, 20, 22, 25-penta-TMSi for the β-ecdysone derivative.[9)]

Analysis by GC-MS

Development of a glc separation procedure for ecdysones made it possible to apply gc-ms to ecdysone analysis, not only for structure determination of unknown ecdysones, but also for ultramicroanalysis of ecdysones by selected ion monitoring (mass fragmentography).

For more precise and specific determination of zooecdysones, the mass fragmentographic technique was applied to the completely silylated derivatives, which are more stable than partial TMS or HFB-TMS derivatives.

In the mass spectra of TMS derivatives of phytoecdysones, compounds having OTMS groups at C-20 and C-22 exhibit a base peak of *m/e* 561 (**4**). Therefore, this key fragment ion was chosen for the analysis of β-ecdysone. In the case of α-ecdysone, which does not give the peak at *m*/*e* 561, the strong peak *m*/*e* 564 (**5**) was selected. As an internal standard for this estimation, cyastarone, which gives the same fragment ion of *m*/*e* 561 was used.[9]

m/*e* = 561 **4** *m*/*e* = 564 **5**

The mass fragmentogram of the ecdysone fraction of ten 2-day-old pupae purified by the procedure described in "Experimental procedure 2", focusing at *m*/*e* 561 and *m*/*e* 564, is shown in Fig. 33.3. The amounts of α-ecdysone and ecdysterone were calculated to be 53 ng and 78 ng per pupa, respectively.[9]

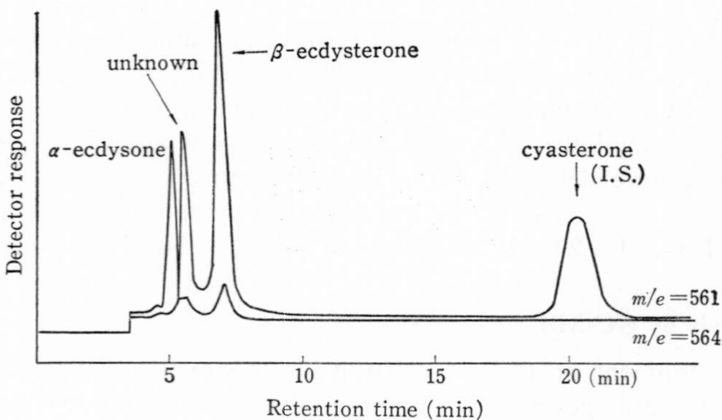

Fig. 33.3. Mass fragmentogram of ecdysones from 2-day-old silkworm pupae. The column was a 100 cm × 4 mm i.d. glass coil with 1.5% OV-101 on Chromosorb W HP, 80–100 mesh (Applied Science Lab.). The column temperature was 270° C; the ionization current was 60 μA; the voltages were 20 eV and 70 eV; the accelerating voltages were 1.75 KV and 3.5 KV; the ion source temperature was 290° C.

Radioimmunoassay of Ecdysones

Radioimmunoassay (RIA) is also a powerful technique for the microdetermination of steroid hormones. RIA was first applied for the detection of β-ecdysone by Borst and O'Conner.[14] β-Ecdysone was treated with aminooxyacetic acid to give the oxime acetic acid, **(6)** which was converted to the mixed anhydride and treated with bovine serum albumin. The protein-ecdysone conjugate was dialyzed and lyophilized. White rabbits were injected with this conjugate. The specificity of the antibodies obtained was examined by the RIA competition method. The inhibition of [^{3}H] β-ecdysone binding by rabbit antiserum was measured in the presence of unlabeled steroids.

α-Ecdysone and β-ecdysone can be detected by this method at a level of 200 pg or 25 times better sensitivity compared with bioassay. This method was used to detect α-ecdysone from a culture of PG of tobacco hornworm, *Manduca sexta*,[15] and from the ovary of adult mosquito, *Aedes agypti*.[16]

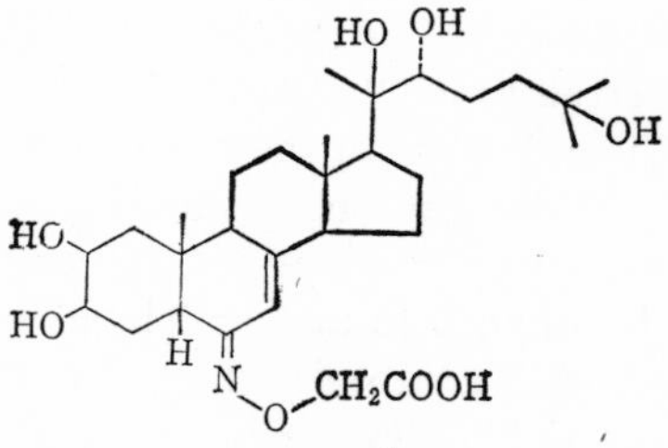

HO OH
OH
HO
HO
H
N
O
CH_2COOH

6

33.2 Biosynthesis of Ecdysones

Organ Culture of Prothoracic Gland

The organ culture of prothoracic glands of the silkoworm was carried out by the following procedure.[17] The prothoracic glands were dissected from last instar silkworm larvae 1 day after spinning and subjected to organ culture, in which one pair of glands was cultivated in 0.04 to 0.06 ml of culture medium for 5 to 7 days at 25° C in a hanging drop. After cultivation, the hormone was extracted from the glands or the culture medium and subjected to bioasssy with *Sarcophaga* test abdomens to determine the hormonal activity. The activity was calculated from the standard curve and puparium index for β-ecdysone and expressed in nanograms of β-ecdysone equivalent.

We cultivated the glands in synthetic media such as Grace's or Wyatt's.

The amount of hormone produced was very small (less than 5 ng per pair of glands), thus confirming the very poor yield of the hormone reported by others. We then tried culturing in insect hemolymph, since it was assumed that hemolymph would be the most physiologically satisfactory medium and that the cholesterol associated with hemolymph lipoproteins might be utilized for ecdysone biosynthesis. We used hemolymph from diapausing pupae of the Cynthia silkworm, after removing the hemocytes. However, on exposure to air the hemolymph darkened, because of the tyrosine-tyrosinase reaction. It may be possible to use reagents such as phenylthiourea to prevent such darkening, but such chemicals are poisonous or physiologically deleterious. Therefore, we applied the hemolymph to a Sephadex column to separate the protein fraction from the lowmolecular substances, including tyrosine. The protein fraction from a Sephadex G-25 column equilibrated with saline was collected and diluted with an equal volume of Wyatt's insect culture medium (omitting tyrosine), filtered with a Millipore filter, and then used as the culture medium.

The medium, referred to as "hemolymph medium," contained about 40% of the original hemolymph on the basis of the amount of protein, and did not blacken even after prolonged standing in air. We made another change in the culture system: an excess of oxygen (0.5 unit partial pressure) was continuously supplied to the small incubator in which the organ culture was performed, since anatomical fact that the prothoracic glands are associated with a highly developed tracheal system, and we thus considered that their function might require an ample supply of oxygen.

This improved culture system, in which the prothoracic glands were cultivated in hemolymph medium under a continuous supply of excess oxygen, led to the first successful result. A typical set of data (Table 33.3) reveals that, before cultivation, the hormonal activity is extremely low in the glands and completely in the culture medium, whereas, after

TABLE 33.3. Amounts of molting hormone found in glands and media before and after cultivation (expressed in nanograms of β-ecdysone equivalent per pair of glands). Culturing conditions are described in the text. (Source: ref. 17. Reproduced by kind permission of the American Association for the Advancement of Science, U.S.A.)

Medium or organ	Gland cultures (No.)	Amount of hormone (ng)
Before cultivation		
Prothoracic gland	40	2.5
Hemolymph medium (1 ml)	—	0.0
After cultivation		
Prothoracic gland	25	0.0
Hemolymph medium	25	120.0

cultivation, a considerable amount of activity is found in the medium and none in the glands themselves. This result indicates that as soon as the hormone synthesized by the glands it is secreted.

Table 33.4 shows the efficiency of various culture media; in comparison with glands cultured in hemolymph medium, those in synthetic media produce only a small amount of hormone. The reason why hemolymph medium is so effective in supporting hormone synthesis by the glands while synthetic media lack such capacity may at least partly be explained as follow. Since cholesterol is a precursor of ecdysone, the cholesterol present in the hemolymph, which is known to be associated with two major lipoproteins, lipoproteins I and II[18], may be utilized as a substrate, whereas the synthetic media contain no such available cholesterol. To test this assumption, we pruified lipoproteins I and II from hemolymph, and prepared culture media consisting of lipoprotein I or II and Wyatt's medium. These lipoprotein media appear to be more efficient than hemolymph on the basis of the protein amount (Table 33.4), suggesting that the above assumption is correct.

TABLE 33.4. Effect of culture medium on the synthesis of molting hormone by prothoracic glands. The amount of hormone found in culture medium after cultivation is expressed in nanograms of β-ecdysone equivalent per pair of glands. The amounts of proteins contained in one culture medium were 1.3 mg for hemolymph, 0.54 mg for lipoprotein I, and 0.50 mg for lipoprotein II.
(Source: ref. 17. Reproduced by kind permission of the American Association for the Advancement of Science, U.S.A.)

Medium	Gland cultures (No.)	Amount of hormone (ng)
Grace's	14	4.8
Wyatt's	24	17.7
Hemolymph	25	120.0
Lipoprotein I	21	148.0
Lipoprotein II	16	108.0

A continuous supply of excess oxygen was also beneficial. Glands cultured in air (0.2 oxygen partial pressure) produced only 25 ng (per pair of glands) of hormone equivalent to β-ecdysone, while the production in 0.5 oxygen partial pressure was approximately four times greater (97 ng).

Extraction and Identification of α-Ecdysone[17]

In order to determine the chemical nature of the molting hormone formed in our culture system, the hormonal fraction was extracted from the culture medium after cultivating 40 paris of glands. One half of the fraction was directly subjected to bioassay to determine the hormonal activity, and the other half was applied to a tlc plate (precoated silica gel plate, Merck) which was developed with a mixture of chloroform and methanol (2:1). Pure β-ecdysone was run at the same time as a marker. After developing the plate, the gel was eluted with tetrahydrofuran from several regions including the solvent front and the origin, and the extracts were bioassayed. The results revealed that all the hormonal activity was located in a distinct region corresponding to β-ecdysone ($Rf = 0.53$). It is well known that α-and β-ecdysone show very close and sometimes overlapping spots on tlc, so further evidence on the chemical nature of the hormone was obtained by high pressure liquid chromatography. Hormonal fraction extracted from culture medium equivalent to 200 paris of glands was first applied to tlc. A portion of the fraction eluted from the region corresponding to ecdysone was subjected to bioassay to determine the hormonal activity, and the remaining portion was subjected to liquid chromatography. The fractions were collected individually and bioassayed. A typical chromatogram (Fig. 33.4 a) shows eight or more peaks, indicating that the extract still contains many different substances. Fraction 8 was identical with α-ecdysone, not β-ecdysone, with respect to retention time. On the other hand, the results of the bioassay revealed that only fraction 8 had activity, which was equal to the total activity of the material applied. The fraction 8 material collected from another run of liquid chromatograppy was applied to a gas chromatography-mass spectrometry system as the trimethylsilyl (TMS) derivative. Ions of m/e 425, 474, and 564 which are characteristic of α-ecdysone TMS derivative[9] were monitored. As shown in Fig. 33.4b, the retention time of each of these three peaks exactly coincided with that (5.0 min) of authentic α-ecdysone TMS derivative, and the relative intensities of these three peaks were consistent with those of the authentic specimen.[9]

Our experimental results, together with existing knowledge on the insect molting hormone, lead to the conclusion that the prothoracic gland is indeed the site of α-ecsysone synthesis. Another research team working independently has provided chemical evidence that the isolated prothoracic glands of the tobacco hornworm, *Manduca sexta*[19] and cockroach, *Leucophaea maderae,*[20] produce α-ecdysone; this result is consistent with our conclusion.

3β-Hydroxy-5α-Cholestan-6-One, a Possible Precursor of Ecdysones

It is generally considered that cholesterol is a precursor in ecdysone biosynthesis. To substantiate the role of cholesterol in ecdysone synthesis, prothoracic glands (day 9, fifth instar) were cultured in Wyatt's insect

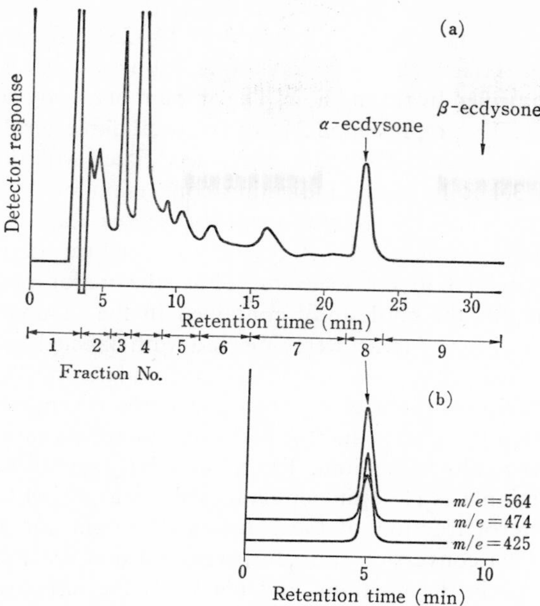

Fig. 33.4. (a) Liquid chromatogram of the fraction eluted from a tlc plate (see the text). High-pressure liquid chromatography: DuPont 830 liquid chromatograph; Zorbax SIL column (25 cm × 2.1 mm i.d.); mobile phase, CH_2Cl_2 and CH_3OH (9:1); flow rate, 0.3 ml/min; column pressure, 1000 psi; column temperature, 20°C; detector, ultraviolet photometer at 254 nm. (b) Mass fragmentogram of TMS derivative of fraction 8. Instrument, LKB 9000s MID PM 9060S; column, 1.5 % OV-101 on Chromosorb WHP 80 to 100 mesh 1 m × 4 mm i.d.); column temperature, 270°C; ionization current, 60 μa; ionization voltage, 20 eV; accelerating voltage, 3.5 kV; ion source temperature, 290°C. (Source: ref. 17. Reproduced by kind permission of the American Association for the Advancement of Science, U.S.A.)

culture medium in the presence and absence of cholesterol. The glands were maintained in culture for 5 days before α-ecdysone was extracted from the medium and bioassayed by Ohtaki's method. The results indicated that inclusion of cholesterol, emulsified with Tween 80, in the culture medium enhanced ecdysone production four- to fivefold to levels similar to those obtained in pure hemolymph. However, when the glands were cultured in medium to which only the emulsifying agent Tween 80 was added, a similar enhancement was observed.

These data suggest that the major precursors for ecdysone synthesis are contained within the gland, and no extraneous supply is required during the 5-day incubation period. The role of Tween 80 in activating ecdysone synthesis is not understood.

The extent of cholesterol incorporation from the medium into α-ecdysone was investigated by incubating 30 paris of prothoracic glands in Wyatt's medium containing [4-^{14}C] cholesterol (4.125×10^7 cpm, 380 nmole) emulsified with Tween 80. After 5 days, α-ecdysone was extracted from the medium and purified by tlc and hplc. The amount of α-ecdysone recovered was 7.6 nmole. This represented 2% of the total cholesterol (380 nmole) available in the medium. The radioactivity associated with the α-ecdysone fraction was 3.4×10^3 cpm, indicating conversion of 0.0082% of the the labeled cholesterol into α-ecdysone. From these data it can be calculated that only 0.4% of the total α-ecdysone produced was derived from the cholesterol provided in the medium. Thus, the major precursor of α-ecdysone was not the cholesterol contained in the incubation medium.

When last instar larvae were injected with labeled cholesterol, an unknown labeled sterol was found to accumulate in the prothoracic glands. Incubation of such glands resulted in the disappearance of this compound from the glands and the concomitant appearance of labeled α-ecdysone in the culture medium. The radioactivity associated with the α-ecdysone fraction was 118 cpm per gland which was 4% of that present in the labeled sterol fraction of the glands (2960 cpm per gland) before incubation. This conversion ratio far exceeded that (0.0082%) obtained with cholesterol in the medium, and identified the unknown sterol as a possible precursor of α-ecdysone.

The unknown sterol was purified and chemically identified by a combination of chromatographic and mass-spectrometric procedures. Prothoracic glands (400) were removed from larvae (day 9, fifth instar) and extracted with chloroform and methanol (2:1), and the extract was applied to tlc plates (the developing solvent was benzene and acetone, 2:1). The fraction corresponding to the unknown sterol was eluted and subjected to hplc together with the labeled marker; it was then collected and analyzed by gc-ms.

By direct comparision with a synthetic smaple, the unknown sterol was found to be 3β-hydroxy–5α-cholestan-6-one (7). The possibility that the 6-one is a precursor was further investigated by conversion of 3β-hydroxy-5α-[4-^{14}C] cholestan-6-one into α-ecdysone *in vitro*. Forty pairs of prothoracic glands were incubated for 5 days in 1 ml of Wyatt's medium containing the labeled 6-one (2.14×10^7 cpm, 175 nmole) emulsified with Tween 80; the sterol was prepared from [4-^{14}C] cholesterol. After pruification of α-ecdysone from the medium, the radioactivity associated with the α-ecdsyone fraction was 3.00×10^3 cpm which indicated conversion of 0.014%. The incorporation ratio was only a little higher than that of [4-^{14}C] cholesterol (0.0082%), but this may be due to dilution of the radioactive 6-one with the endogeneous sterol accumulated in the glands before cultivation.

The demonstration that 7-[^{3}H]-dehydrocholesterol may be incorpo-

cholesterol 3

7

OH

HO

H

O

rated into β-ecdysone, together with the observation that in some insects cholesterol is metabolized to 7-dehydrocholesterol, has led to the proposal that 7-dehydrocholesterol is an intermediate in ecdysone biosynthesis[21]. However, the significance of this pathway has not been clearly estabilshed. In the investigation reported by us the metabolism of labeled cholesterol into 3-βhydroxy-5α-cholestan-6-one was demonstrated *in vitro*, and the disappearance of this compound from incubated prothoracic glands was correlated with the appearance of α-ecdysone. Thus it can be concluded that the 6-one is a possile precursor of α-ecdysone biosynthesis in prothoracic glands.

The isolation of 2-deoxy-α-ecdysone[22] and 2-deoxy-β-ecdysone[23] from ovaries and eggs of the silkworm suggests the following biosynthetic pathway for ecdysone.

33.3 Sterol Metabolism in Insects[24]

Since insects lack the capacity for *de novo* sterol biosynthesis, they in general require a dietary or exogenous source of sterol for normal growth, development and reproduction. In phytophagous insects, exogenous phytosterols such as sitosterol (**8**), stigmasterol and campesterol undergo side chain dealkylation at the C-24 position to afford cholesterol (**3**), which serves as a precursor for ecdysone and is an important component of the cell membrane.

The observations that fucosterol eposide (**10**) yields desmosterol (**11**) upon boron trifluoride etherate treatment and that [3α-^{3}H]-fucosterol epoxide is converted to cholesterol in the silkworm led us to propose that in insects, fucosterol epoxide is a key intermediate in the conversion of sitosterol (**8**) into choleserol (**3**). The data subsequently reported by us and another group supported the following biogenetic pathway.

We have demonstrated that 24, 28-iminofucosterol (**14**), a structural analog of fucosterol epoxide, and cholesta-5,23,-24-trien-3β-ol (allene-II) (**13**) can strongly disrupt the normal growth and development of the silkworm. It was also found that stigmata-5,24(28), 28-trien-3β-ol (allene-I) (**12**) is a highly specific inhibitor for the steps involving fucosterol, that is, for the conversion of sitosterol to fucosterol and/or of fucosterol to the epoxide. It is interesting to note that silkworms reared on allene-I alone or in combination with sitosterol or fucosterol survived in the second instar for more than 20 days without reaching the third instar.

We have also investigated the stereochemistry of the fucosterol epoxide using *in vivo* and *in vitro* systems. In nutritional experiments,

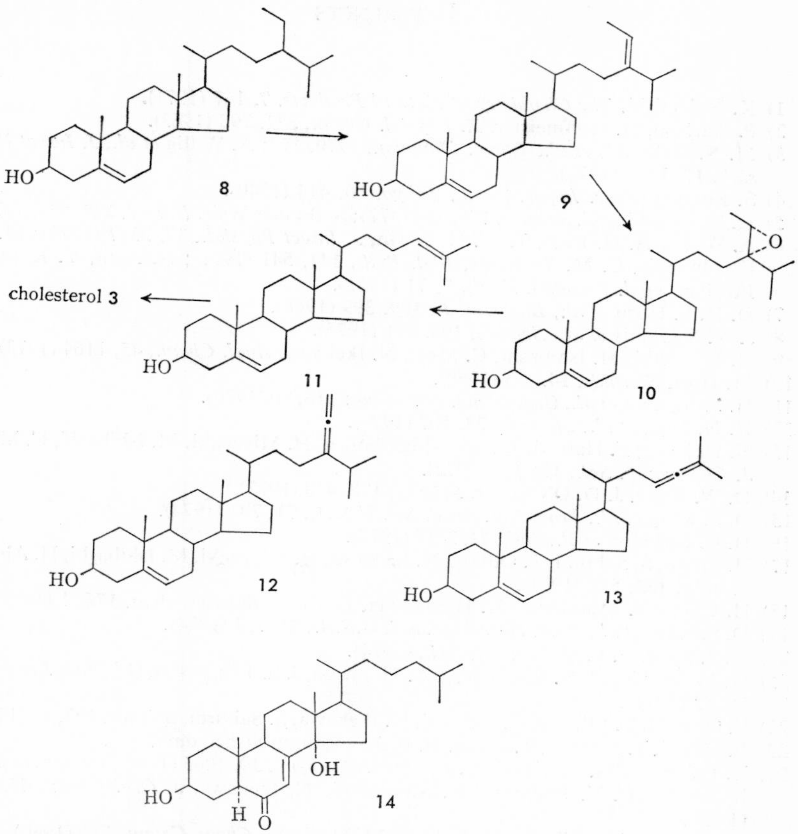

HO 8 HO 9 cholesterol 3 HO 11 HO 10 HO 12 HO 13 OH HO H O 14

isofucosterol epoxides were unable to support growth and development of silkworm larvae. Incubation of [3α-[3] H]-(24*R*, 28*R*)- and (24*S*, 28*S*)-fucosterol epoxides with a cell-free preparation from silkworm guts resulted in effective conversion into desmosterol and cholesterol; the (24*R*, 28*R*)-epoxide was a slightly better substrate. Fucosterol incubation yielded (24*R*, 28*R*)-epoxide and the (24*S*, 28*S*)-isomer in approximately equal amounts, while slightly preferential formation of the (24*R*, 28*R*)-isomer was observed in *in vivo* experiments. These results indicate that the stereospecificity in both the formation of the epoxide from fucosterol and its conversion to demosterol is low.

Recently, we isolated both isomers of (24*R*, 28*R*)-and (24*S*, 28*S*)-fucosterol epoxide from silkworm larvae, thus clearly confirming the epoxide pathway from sitosterol to cholesterol in insects.[25]

REFERENCES

1) K. Nakanishi, *The Chemistry of Natural Products,* **7**, 167 (1971).
2) P. Karlson, H. Hoffmeister, *Z. Physiol. Chem.*, **331**, 298 (1963).
3) M. N. Galbraith *et al.*, *Chem. Commun.*, **1970**, 179; A. Willig *et al.*, *J. Insect Physiol.*, **17**, 2317 (1971).
4) S. Fukuda, *Proc. Imper. Acad., Tokyo*, **16**, 414 (1940).
5) K. Nakanishi *et al. Science*, **176**, 51 (1972); S. Bonner Weir, *Nature*, **228**, 580 (1970).
6) A. Willig, H. H. Rees, T. W. Goodwin, *J. Insect Physiol.*, **17**, 2317 (1970); M. P. Kambysellis, C. M. Williams, *Biol. Bull.*, **141**, 541 (1971); N. Agui, Y. Kimura, Fukaya, *Appl. Entomol. Zool.*, **7**, 71 (1972).
7) D. H. S. Horn *et al.*, *Biochem. J.*, **109**, 399 (1968).
8) J. N. Kaplanis *et al.*, *Science*, **190**, 681 (1975).
9) H. Miyazaki, M. Ishibashi, C. Mori, N. Ikekawa, *Anal. Chem.*, **45**, 1164 (1973).
10) M. Hori, *Steroids*, **14**, 33 (1969).
11) H. Moriyama *et al.*, *Gen. Comp. Endocrinol.*, **15**, 80 (1970).
12) H. N. Nigg *et al.*, *Steroids*, **23**, 507 (1974).
13) N. Ikekawa, F. Hattori, J. Rubio-Lightbourn, H. Miyazaki, M. Ishibashi, C. Mori, *J. Chromatog. Sci.*, **10**, 233 (1972).
14) D. W. Borst, J. D. O'Conner, *Science*, **178**, 418 (1972).
15) D. S. King *et al.*, *Proc. Nat. Acad. Sci. U.S.A.*, **71**, 793 (1974).
16) H. H. Hagedon *et al.*, *ibid.*, **72**, 3255 (1975).
17) H. Chino, S. Sakurai, T. Ohtaki, N. Ikekawa, H. Miyazaki, M. Ishibashi, H. Abuki, *Science*, **183**, 529 (1974).
18) H. Chino, S. Murakami, K. Harashima, *Biochem. Biophys. Acta*, **176**, 1 (1969).
19) D. S. King *et al.*, *Proc. Nat. Acad. Sci. U.S.A.*, **71**, 793 (1974).
20) D. S. King, E. P. Marks, *Life Sci.*, (1974).
21) W. E. Robbins, J. N. Kaplanis, J. A. Thomson, J. S. Thompson, *Adv. Res. Entomol.*, **16**, 53 (1971).
22) E. Ohnishi, T. Mizuno, F. Chatani, N. Ikekawa, S. Sakurai, *Science,* **197**, 66 (1977).
23) E. Onishi, T. Mizuno, T. Ikeda, N. Ikekawa, unpublished data.
24) Y. Fujimoto, M. Morisaki, N. Ikekawa, *Biochem.*, **19**, 1065 (1980), references cited in this report. Y. Fujimoto, K. Murakami, N. Ikekawa, *J. Org. Chem.*, **45**, 566 (1980).
25) N. Ikekawa, Y. Fujimoto, A. Takasu, M. Morisaki, *Chem. Comm.*, 709 (1980).

CHAPTER 34

Strategy for Marine Toxins Research

The oceans cover 71 % of the earth's surface and include an abundance of fauna and flora: 500,000 animal species, accounting for 80 % of the earth's animal life, are estimated to live there. There are several thousand species of poisonous and venomous marine animals, but at present no more than 50 toxins are known whose molecular structures and pharmacological properties have been well elucidated. Although recent increasing awareness of ocean exploration has stimulated research on toxic marine organisms, our knowledge of the toxic substances involved lags far behind our knowledge of toxins elaborated by terrestrial organisms.

Research on a marine toxin, like other natural products research, involves specimen procurement, separation and purification, structure determination, synthesis, and identification of the biosynthetic route of the toxin. Research techniques for natural products are applicable to such steps as isolation and structure determination. However, the discovery of toxic species, procurement and handling of specimens, and bioassay are particularly time-consuming and important in the course of research on a marine toxin.

This chapter will be focused on these initial steps in the course of research on marine toxins and will deal with the general strategy for marine toxins research.

34.1 Discovery of Toxic Marine Organisms

To begin with, a toxic species must be selected, though in some cases there are obvious targets, e.g., an outbreak of food poisoning due to toxic fish or shellifish such as the Japanese ivory shell *Babylonia japanica*[1] and the northern blenny *Stichaeus grigorjewi,*[2] or large-scale fish mortality caused by a red-tide organism.[3]

Information from biologists or local people is often valuable. Some useful books[4–6] have been published, among which Halstead's monumental three–volume monograph "*Poisonous and Venomous Marine Animals of the World*" is most helpful.

Survey of Toxic Marine Organisms by Field Investigation

It is important to know whether an outbreak of food poisoning is caused by a naturally occurring toxin or bacterial contamination. Information can be obtained from the local health center, or an epidemiological survey carried out by interviewing the patients or their family members may be justified. Various marine toxins have been discovered from outbreaks of food poisoning, e.g., tetrodotoxin, saxitoxin, ciguatoxin, and surugatoxin.

Field surveys and local beliefs can also be helpful in seeking toxic species. The staff of the Laboratory of Marine Biochemistry at the University of Tokyo has performed field investigations on toxic marine organisms in the Amami and Ryukyu Islands since 1966, and found toxic crabs and other toxic marine animals as well as toxic fish. These investigations are described below.

[Experimental procedure 1] Field survey on toxic marine organisms[7]

Information on toxic marine organisms and food poisoning due to ingestion of marine products was obtained from organizations concerned with public health or fisheries, and from the fishermen's associations listed below in Okinawa, Miyako, and Ishigaki Islands, July 1966, and in Amami-Oshima and Tokunoshima Islands, November 1966. In cases of food poisoning, interviews were carried out with persons who had experienced poisoning and their family members or relatives, and the following details were obtained: causative species, part of the body eaten, cooking method, number of victims, and sysmptoms. Science teachers of local schools and staff members of the Oshima Branch of Kagoshima Prefectural Fisheries Experimental Station were also asked to collect information. Organizations visited were as follows.

the Ryukyu Islands

1) Department of Fisheries of the Ryukyu Government
2) Department of Public Health of the Ryukyu Government
3) Ryukyu Fisheries Research Laboratory
4) Economics Section of Miyako Regional Bureau
5) Economics Section of Yaeyama Regional Bureau
6) Nago Health Center
7) Miyako Health Center
8) Yaeyama Health Center
9) Union of Ryukyu Fishermen's Association
10) Miyako Branch of the Union of Ryukyu Fishermen's Association
11) Yaeyama Branch of the Union of Ryukyu Fishermen's Association
12) Motobu Fishermen's Association
13) Haneji Fishermen's Association
14) Ishikawa Fishermen's Association
15) Yonabaru Fishermen's Association
16) Katsuren Fishermen's Association
17) Naha District Fishermen's Association
18) Itoman Fishermen's Association
19) Hirara Fishermen's Association
20) Ishigaki Fishermen's Association

the Kyushu and Amami Islands

1) Faculty of Fisheries of the University of Kagoshima
2) Fisheries Section of Kagoshima Prefectural Office
3) Environmental Sanitation Section of Kagoshima Prefectural Office
4) Fisheries Section of Oshima Branch Office
5) Oshima Branch of the Kagoshima Prefectural Fisheries Expermental Station
6) Tokunoshima Town Office
7) Yoron Town Office
8) Yoron Health Center
9) Naze Health Center
10) Tokunoshima Health Center
11) Naze Fishermen's Association
12) Setouchi Fishermen's Association
13) Tokunoshima Fishermen's Association
14) Yamakawa Fishermen's Association

Questionnaires were sent out in November and December of 1966. The questionnaire adopted was called Questionnaire A, and had previously been used for the ciguatera survey in the South Pacific by a group from the University of Hawaii. It included questions on toxic fish, their habitats, outbreaks of poisoning, symptoms, local remedies, and toxic organisms other than fish. In the Ryukyus, the questionnaire was distributed to 21 fishermen's associations selected at the suggestion of the Department of Fisheries of the Ryukyu Government, and replies were received from 13 associations. In the Amami Islands it was distributed through the Fisheries Section of Oshima Branch Office and the Oshima Branch of Kagoshima Prefectural Fisheries Experimental Station.

Survey of Toxic Species by Screening Tests

It is desirable to test as many species as possible when screening for toxic species, and since the toxicity of marine organisms varies greatly in individuals, seasons and regions, many specimens collected from various places over a long period should be tested. We must consider the variation of toxicity in the parts of the body: for example in puffer fish[8] and toxic crabs.[9] Freshly collected or frozen specimens should be assayed, since some marine organisms produce toxic substances by the action of bacteria (e.g. histamine in scombroid fish[10]) and some marine toxins are deactivated by enzymic action.

A simple screening method is clearly desirable. We routinely prepared test solutions by the hot 70% ethanol method[11] (Table 34.1) and examined the toxicity towards the mouse *Mus masculers* and the killifish *Oryzias latipes,* as well as the hemolytic activity against rabbit red blood cells. Most toxins other than proteinaceous toxins[6] can be detected by this method, which is described below.

TABLE 34.1. Preparation of test solutions for bioassay.

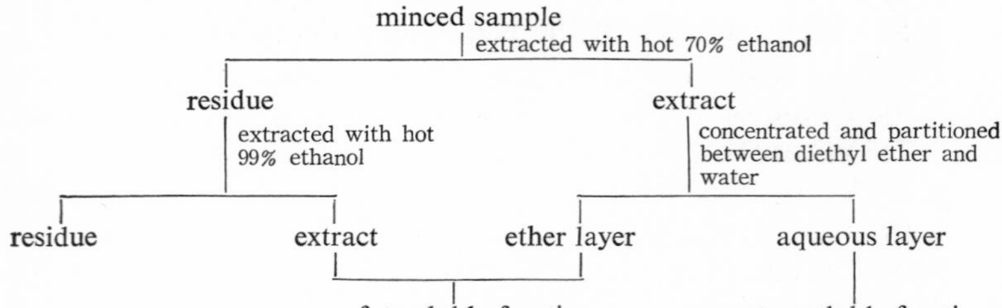

[Experimental procedure 2] Screening method for toxicity[11]

A minced specimen is extracted with 3 volumes of 70% ethanol for 30 min in a boiling water bath. After cooling, the mixture is filtered and the residue extracted in the same manner. The residue is then extracted with 3 volumes of 99% ethanol in the same way. The 70% ethanol extract is concentrated and partitioned between diethyl ether and water. The ether layer is combined with the ether-soluble material of the 99% ethanol extract and is designated as the fat-soluble fraction. The aqueous layer is freed from ether and made up to a given volume. Each fraction thus obtained is injected intraperitoneally into mice weighing 18–22 g at a dose equivalent to 2 g of fresh specimen. The fat-soluble fraction is administered to mice after emulsification with 1% Tween 60 saline solution. The water-soluble fraction sometimes kills mice due to the presence of a large amount of inorganic salts. This can be recognized from the symptoms; inorganic salts should be suspected when mice are killed rapidly. In this case, the sample can be tested after dilution to an appropriate concentration. Ichtyyotoxicity is examined by placing 5 killifish in 100 ml of test solution. The fish are observed for 24 hr. The fat-soluble fraction is dissolved in 0.5 ml of ethanol and then suspended in 100 ml of distilled water. Hemolytic activity is tested at a concentration of 1 mg per ml, as described later.

34.2 Procurement and Handling of Specimens

Except in the cases of sessile of inactive organisms, marine animals are rarely caught by researchers. Though scientists often collect specimens by scuba diving in the United States and European countries, it is usually more convenient to ask a professional collector or a fisherman to procure a particular specimen. In some cases there are legal restrictions, and a local fisheries experimental station, a fishermen's association, marine laboratory of local college may be able to help. As already mentioned, specimens should be collected from various places over a long period.

Toxic specimens are kept frozen, unless extraction is done upon collection. If no dry-ice is available, specimens can be cooled with a salt-ice mixture or ice and then frozen as soon as possible. The frozen samples are usually transported in dry-ice to our laboratory by air and kept in a freezer.

Some marine organisms lose their toxicity in a short time during frozen storage or while defrosting due to enzymic action. This can be prevented by heating in the field before storage. For instance, activity of the skin toxin of the gobies *Gobiodon* spp.[12)] was successfully maintained by preheating in ethanol before storage at low temperature.

Preservation of specimens in ethanol (palytoxin[13)]) or in formalin (tetrodotoxin[8)]) is sometimes effective when the properties of the toxin are known. However, this is not applicable to the crab toxin,[14)] identical with saxitoxin, which is labile in alkaline media, since the pH of crustacean tissues is about 9 soon after death. Sun-drying of specimens, which is used for research on echinoderm saponins,[6,15] is not desirable in many cases. In conclusion, a suitable preservation method must be selected on the basis of the properties of the specimen and its toxin.

Bioassay

It should be emphasized that it is most important in marine toxins research to establish a simple and sensitive bioassay method. Though the development of simple chemical assays for marine toxins has been attempted by many researchers, it is very difficult to determine trace amounts of toxin by a chemical assay without interference from contaminants. A bioassay, which is in many cases sensitive and specific, is usually far superior to chemical assay.

Information obtained from interviews with local people or from biologists can be helpful in establishing a bioassay. Such knowledge has been satisfactorily applied to research on the crab toxin.[9] On the basis of information from interviews that a person had died within several hours after ingestion of *miso* soup containing poisonous crabs, the toxicity of the crab was determined by injecting the aqueous extract into mice. Bioassay methods which have been used for the major marine toxins are listed in Table 34.2.

Mice, which are not only readily available, but are also easily handled, are most widely used. Lethality is usually adopted as the index, except in a few cases, such as mydriasis for surugatoxin.[1,6] and dermatitis for the toxin in the blue-green alga *Microcoleus lyngbyaceus*.[16] The last two methods were devised on the basis of the symptoms of the victims. A mouse assay is highly desirable for fast-acting toxins: tetrodotoxin[8] and saxitoxin[3] can kill mice weighing 20 g by injectin at a dose of 0.2 μg, thus the mouse assay is as sensitive as a colorimetric assay of a chemical. The dose-death time relationship not only allows us to estimate the amounts of a toxin with considerable accuracy, but may also indicate its nature, since the dose-death time curve is toxin-specific (Fig. 34.1). As described below, the crab toxin was postulated to be quite similar to saxitoxin, based on dose-death time curves obtained from an aqueous extract of the crab.[17]

[Experimental procedure 3] Preparation of dose-death time curve for the crab toxin

A minced specimen was extracted with 5 volumes of water for 30 min at 90°C. The mixture was centrifuged after cooling and the residue was extracted in the same manner. The combined extract was made up to a

TABLE 34.2. Bioassays used for the major marine toxins

Marine toxin	Source	Bioassay	References
Tetrodotoxin	puffer fish, goby (*Gobius criniger*)	lethality in mice	8, 27
Saxitoxin	bivalves, xanthid crabs, dinoflagellates (*Gonyaulax* spp.)	lethality in mice	3,4,6,14
Surugatoxin	Japanese ivory shell (*Babylonia japonica*)	mydriasis in mice	1,6
Ciguatoxin	ciguateric fish	lethality in mice	4, 6, 11, 19
Dinogunellin	northern blenny (*Stichaeus grigorjewi*), cabezon (*Scorpaenichthys marmoratus*)	lethality in in mice	2, 6
Aplysiatoxin	sea hare (*Stylocheilus longicauda*)	lethality in mice	6, 20
Nereistoxin	annelid (*Lumbrinereis brevicirra*)	ichthyotoxicity in killifish	6, 21
Palytoxin	zoanthids (*Palythoa* spp.)	lethality in mice	6, 13
Holothurins	sea cucumbers	hemolytic activity	4, 6, 15
Asterosaponins	starfish	hemolytic activity	4, 6, 15
Pahutoxin	boxfish (*Ostracion lentiginosus*)	ichthyotoxicity, hemolytic activity	6, 22
Grammistins	grammistid fish	hemolytic activity	6, 23, 25
Glenodinine	dinoflagellate (*Peridinium polonicum*)	ichthyotoxicity in killifish	6, 24
Lyngbyatoxin A	blue-green alga (*Microcoleus lyngbyaceus*)	dermatitis in mice	6, 16, 29

given volume. A 0.5 ml aliquot of the extract was injected intraperitoneally into mice weighing 20 g. The original solution was then diluted to a concentration that killed mice within 15–30 min. The dose-death time curve was obtained by plotting the mean death time of three mice against the concentration of the extract. An amount of toxin which kills mice in 15–30 min was defined as one mouse unit (MU). Total amounts of the toxin can be calculated from a concentration equivalent to 1 MU.

The above technique, however, is not applicable to a toxin that kills mice very slowly. In such a case, a minimum lethal dose may be calculated by injecting test solutions of various concentrations prepared from an original solution. Several mice should be used at each dilution level. An estimation of LD_{50} is preferable, since only a semiquantitative value is obtained by this method. However, it requires many mice, so that an LD_{50} value is usually obtained only for the purified toxin. An approximate to-

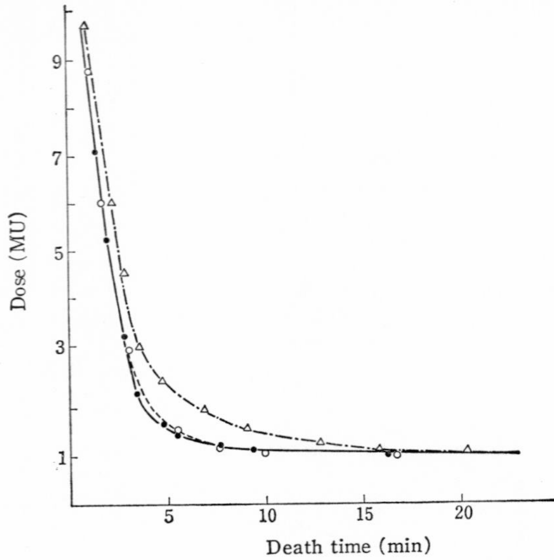

Fig. 34.1. Dose-death time relationship for the crab toxin. ●——●, crab toxin; ○······○, saxitoxin; △- • -△, tetrodotoxin.

xicity may be obtained in order to monitor the toxicity at each purification step. In the case of ciguatoxin research, one mouse unit was defined as the amount of toxin required to kill a mouse within 2 hr. Generally, the amount of toxin that is required to kill a 20 g mouse is defined as 1 MU.

Other test animals can be used if the mouse is too insensitive to a toxin. The cat *Felis catus* is very sensitive to a variety of toxins and is not only used for the detection of toxicity in many toxic species, but also for the screening of ciguateric fish.[11] However, it is harder to obtain and maintain them. The rat *Rattus norvegicus* is another popular test animal. Cats and rats are often used to investigate the oral toxicity of a toxin involved in food poisoning, since oral feeding of mice is quite troublesome.

Hemolytic or ichthyotoxic activity is a valid assay for toxins with surface-activity. By following the hemolytic activity, a small amount of toxin can be determined quantiatively. A red blood cell suspension prepared from blood of rabbits, humans, or fish is used for assay in this case. Our routine method will be described below. The killifish is most popular for ichthyotoxicity tests. Other small tropical fish or marine fish are also empolyed. It should be noted that the pH of the test medium can greatly affect ichthyotoxicity, as in the case of glenodinine.[24]

Chicks, frogs, crabs, shrimps, earthworms, insects, and others have also been used as test animals.

[Experimental procedure 4] Determination of Hemolytic Activity[25]

Preparation of 2% rabbit red blood cell suspension: One ml of freshly collected rabbit blood was washed three times with a mixture of 0.85% NaCl and 0.1 M phosphate buffer (pH 6.8) (4:1), and made up to 50 ml with 0.85% NaCl.

Determination of hemolytic activity First, 1 ml of test solution (1 mg of sample in 1 ml of 0.85% NaCl for screening test) was mixed with 1 ml of the blood cell suspension and the mixture was incubated for 30 min at 37° C. The reaction mixture was centrifuged after addition of 2 ml of 0.85% NaCl. The absorbance of the supernatant at 542 nm was recorded, and the hemolytic curve obtained by plotting absorbance against sample concentration (Fig. 34.2). A curve was also prepared using a standard saponin (Merck Erg B6). The amount of the hemolysin that produces 50% hemo-

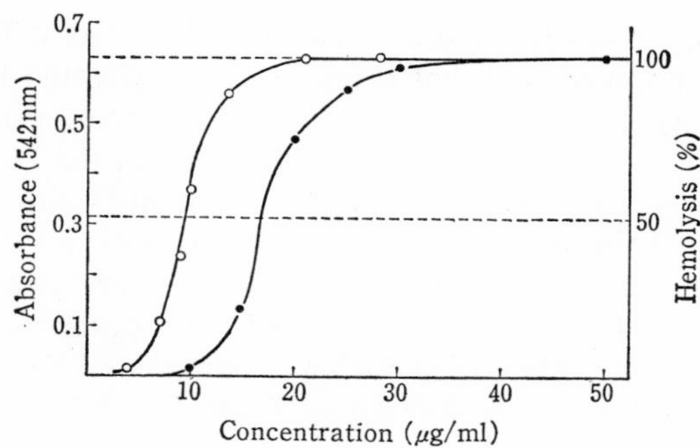

Fig. 34.2. Hemolytic activity of the *Ulva* hemolysin. ○, *Ulva* hemolysin; ●, standard saponin.

lysis was read off from the curve, and the hemolytic activity was expressed as SU per mg taking the activity of 1 mg of standard saponin as one saponin unit (SU). For example, if standard saponin evokes 50% hemolysis at a concentration of 17μg per ml while the toxin does so at 10 μg per ml, the hemolytic activity of the toxin is 1.7 SU per mg. A fat-soluble toxin can be initially dissolved in 0.2 ml of ethanol, then 0.2 ml of 1.7% NaCl and 0.6 ml of 0.85% NaCl added to the solution.

34.4 Separation and Purification

A great variety of marine toxins has been isolated:[3–6,26] proteins, peptides, lipids, pigments, glycosides, heterocyclics, and sulfur compounds.

Therefore, a suitable technique must be used, considering the chemical properties such as solubility, stability to acid or base, permeability, etc. A method as mild as possible should be employed, since many marine toxins are labile.[6] For instance, Sephadex G-10 was successfully used for the isolation of tetrodotoxin from the goby,[27] and Sephadexes or Bio-Gels are often quite useful for the purification of marine toxins.

Finally, the isolation of tetrodotoxin from the goby will be presented as an example of the application of these principles to real research.

Isolation of Tetrodotoxin from the Goby *Gobius criniger*[18,27]

The goby *G. criniger* attains a total length of 15 cm and is found in shallow bays or the mouth of rivers in tropical and subtropical regions. In 1969, Hashimoto and Noguchi[18] were informed by a biologist of the University of the Ryukyus that the goby had been used for killing field rats in Iriomote Island. They visited the island and confirmed that the fish had been applied to rice fields to terminate rats. Villagers at Tekebu, Amami-Oshima Island, also said that chickens and pigs could be killed in a short time by ingestion of the goby. A literature survey also showed that humans and ducks had died after eating the goby in the Philippines and Taiwan. Specimens of the goby collected in Amami-Oshima Island, Iriomote Island and Taiwan were all identified as *G. criniger*.

Specimens caught at Amami-Oshima Island were examined for toxicity by the method described earlier. Only the water-soluble fraction showed high toxicity. Mice died within 2 min when injected with high doses, and within 30 min at around the minimum lethal dose. This confirmed the presence of a fast-acting toxin in goby.

Since the symptoms of mice were very similar to those caused by tetr dotoxin, the dose-death time curve was obtained by the method adopted for tetrodotoxin (Fig. 34.3). The curve resembled that for tetrodotoxin, so the toxicity of specimens collected from various places was examined according to the method described in *"Shokuhin Eisei Kensa Shishin"*[28] (A Guide to Tests for Food Hygiene). The most striking feature of the toxicity pattern was the marked regional variation. All specimens from Tekebu in the north-east of Amami-Oshima Island were highly toxic, regardless of season, while all fish from Shinokawa on the west coast of the same island were nontoxic. Similar narrow regionality of toxicity was also recognized in the Ryukyus.

Since only small individual variations of toxicity were observed in a given area specimens collected in toxic areas were pooled for extraction of toxin.

The goby toxin was postulated to be closely related to tetrodotoxin rather than saxitoxin, based on the following properties.

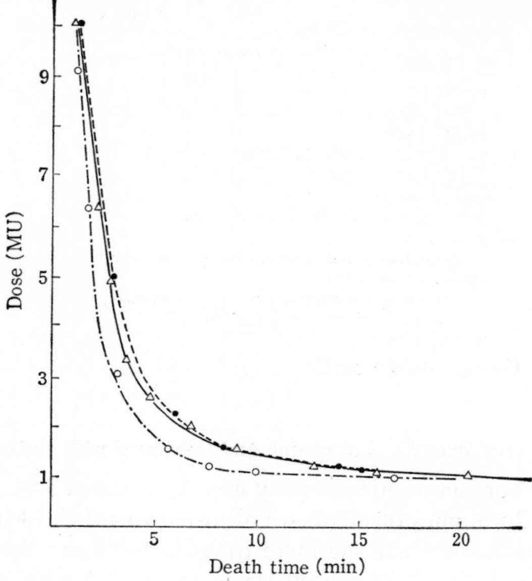

Fig. 34.3. Dose-death time relationship for the goby toxin. ●······●, Goby toxin; △——△, tetrodotoxin; ○- • -○, saxitoxin.

(1) The dose-death time relationship.

(2) The puffer fish *Chelonodon patoca* and the goby appeared to be completely normal when a dose of either tetrodotoxin or goby toxin 200 to 300 times the minimum lethal dose for other animals was injected. In contrast, the control damselfish *Abudefduf sexfasciatus* was killed within 7–15 min at the minimum lethal dose for mice.

(3) The principle was readily permeable through a cellophane membrane.

(4) It was heat-labile in either alkali- or HCl-containing media, but stable in acetic acid-containing media.

(5) It was soluble in water and methanol, but hardly soluble in ethanol; it was not extractable from aqueous solution with 1–butanol.

Since the toxin appeared to be quite similar to tetrodotoxin, the method using activated charcoal, Amberlite IRC-50, and silicic acid (Mallinckrodt) that had been applied for the isolation of tetrodotoxin[8)] by the groups of Tsuda, Hirata and Mosher was attempted, without success. After several attempts, Noguchi and Hashimoto[27)] devised a method using Sephadex G–10 and obtained a crystalline toxin by the following procedures.

TABLE 34.3. Isolation of tetrodotoxin from the goby *Gobius criniger*

minced goby (25 kg)
| extracted with methanol containing acetic acid (pH 5.0)
extract
| treated with activated charcoal
crude toxin
| Amberlite IRC-50 (NH_4^+) column chromatography
toxic fractions
| treated with activated charcoal
toxic fractions
| Sephadex G-10 column chromatography
strongly toxic fractions
| crystallized from a mixture of methanol, acetic acid, and diethyl ether
tetrodotoxin
(30 mg, MLD 9 μg/kg)

[Experimental procedure 5] Extraction and treatment with activated charcoal

A 1.2 kg portion of minced goby was extracted at 5 °C for 30 min by stirring with 2 volumes of methanol adjusted to pH 5.0 with 10% acetic acid. The extraction was done three times in total and the concentrated extract was centrifuged after adjustment to pH 8.5 with 2 N ammonia. A mixture of activated charcoal and Hyflo Super Cel (1:1) (3–4 g per 1 g of solid) was added to the supernatant, and the charcoal was washed with 10 volumes of water. The toxin was eluted with 1% acetic acid in 10% methanol. Yield, 3.8 g; toxicity, 1.7 mg/kg.

[Experimental procedure 6] Amberlite IRC-50 column chromatography and treatment with activated charcoal

Amberlite IRC-50 (NH_4^+ form) was added batchwise to the aqueous solution (10 ml) of the crude toxin in an amount just sufficient to adsorb all the toxin. The resin was packed into a small column and washed with 1–3 volumes of water. The column was eluted with 1.5 volumes of 10% acetic acid, followed by water. The eluate was collected with a fraction collector.

The toxic fractions were concentrated and adjusted to pH 8.5 with 1N ammonia. Activated charcoal (40–50g per 1 g of solid) was added and the charcoal was washed with 5 volumes of water on a Buchner funnel. The toxin was eluted with 1% acetic acid in 20% ethanol. Yield, 506 mg; toxicity, 250 μg/kg.

[Experimental procedure 7] Sephadex G-10 chromatography

The toxin preparation was dissolved in a small volume of water and applied to a column of Sephadex G–10 (2.6 × 100 cm). The column was washed with 1 l of water and then with 1% acetic acid to elute the toxin. Each fraction was tested for toxicity by mouse assay, and the strongly toxic fractions were combined. Yield, 83 mg; toxicity, 50 μg/kg.

[Experimental procedure 8] Crystalization

The toxin was dissolved in methanol containing acetic acid. Diethyl ether was added until the solution become cloudy. The solution was allowed to stand overnight in a refrigerator to give needles. After crystallization twice in the same manner, the toxin was obtained as prisms. A total of 30 mg of crystals was isolated from 25 kg of frozen goby.

The crystalline goby toxin was identical with tetrodotoxin in every respect; toxicity, melting point, combustion data, infrared and nmr spectra, and mobilities on thin layer chromatography.

References

1) Y. Hashimoto, K. Miyazawa, H. Kamiya, M. Shibota, *Bull. Japan. Soc. Sci. Fish.*, **33**, 661 (1967).
2) M. Hatano, Y. Hashimoto, *Toxicon*, **12**, 231 (1974).
3) Y. Shimizu, *Marine Natural Products–Chemical and Biological Perspectives* (ed. P. J. Scheuer), vol. 1, p. 1, Academic Press (1978).
4) B. W. Halstead, *Poisonous and Venomous Marine Animals of the World*, vols. 1–3, U. S. Government Printing Office (1965, 1967, 1970).
5) M. H. Baslow, *Marine Pharmacology*, Williams & Wilkins (1969).
6) Y. Hashimoto, *Marine Toxins and Other Bioactive Marine Metabolites*, Japan Scientific Societies Press (1979).
7) Y.,Hashimoto, S. Konosu, T. Yasumoto, H. Kamiya, *Bull. Japan. Soc. Sci. Fish.*, **35**, 316 (1969).
8) K. Tsuda, *Kagaku No Ryoiki Zokan* (in Japanese) No. 80, 9 (1967).
9) Y. Hashimoto, S. Konosu, T. Yasumoto, A. Inoue, T. Noguchi, *Toxicon*, **5**, 85 (1967).
10) T. Kawabata, *Shin Shokuhin Eiseigaku* (in Japanese), Dobunshoin (1972).
11) Y. Hashimoto, T. Yasumoto, H. Kamiya, T. Yoshida, *Bull. Japan. Soc. Sci. Fish.*, **35**, 327 (1969).
12) Y. Hashimoto, K. Shiomi, K. Aida, *Toxicon*, **12**, 523 (1974).
13) R. E. Moore, P. J. Scheuer, *Science*, **172**, 495 (1971).
14) T. Noguchi, S. Konosu, Y. Hashimoto, *Toxicon*, **7**, 325 (1969).
15) P. J. Scheuer, *Naturwissenschaften*, **58**, 547 (1971).
16) Y. Hashimoto, H. Kamiya, K. Yamazato, K. Nozawa, *Animal, Plant, and Microbial Toxins* (ed. A. Ohsaka *et al.*), vol. 1, p. 333, Plenum (1976).
17) S. Konosu, A. Inoue, T. Noguchi, Y. Hashimoto, *Toxicon*, **6**, 113 (1968).
18) Y. Hashimoto, T. Noguchi, *ibid.*, **9**, 79 (1971).
19) P. J. Scheuer, W. Takahashi, J. Tsutsumi, T. Yoshida, *Science*, **155**, 1267 (1967).
20) Y. Kato, P. J. Scheuer, *Pure Appl. Chem.*, **41**, 1 (1975).
21) T. Okaichi, Y. Hashimoto, *Bull. Japan. Soc. Sci. Fish.*, **28**, 930 (1962).
22) D. B. Boylan, P. J. Scheuer, *Science*, **155**, 52 (1967).
23) J. E. Randall, K. Aida, T. Hibiya, N. Mitsuura, H. Kamiya, Y. Hashimoto, *Publ. Seto Mar. Biol. Lab.*, **19**, 157 (1971).

24) Y. Hashimoto, T. Okaichi, LeDung Dang, T. Noguchi, *Bull. Japan. Soc. Sci. Fish.*, **34**, 528 (1968).
25) Y. Hashimoto, Y. Oshima, *Toxicon*, **10**, 279 (1972).
26) P. J. Scheuer, *Chemistry of Marine Natural Products,* Academic Press (1973).
27) T. Noguchi, Y. Hashimoto, *Toxicon*, **11**, 305 (1973).
28) Ministry of Health and Welfare, *Eisei Kensa Shishin* III, *Shokuhin Eisei Kensa Shishin* (*1*), *Shusan Shokuhin Eisei Kensaho* (in Japanese), p. 45, Kyodoishoshuppan (1959).
29) J. H. Cardellina II, F.-J. Marner, R. E. Moore, *Science*, **204**, 193 (1979).

CHAPTER 35

Extraction and Isolation of Surugatoxin

In September 1965, food poisoning occurred as a result of ingestion of a carnivorous gastropod, the Japanese ivory shell, *Babylonia japonica,* taken from Suruga Bay. The patients complained of visual disorders, mydriasis, abdominal distension, dryness of the mouth and constipation. The causative toxin was found to be present only in the mid-gut gland of gastropods collected in a very limited area of the bay (Fig. 35.1).

In 1968, when we started this study,[1,2] little work had been done on marine natural products, particularly on marine toxins, because natural product chemists were mainly interested in the constituents of organisms living on the land. A few reports dealing with tetrodotoxin and saxitoxin had appeared. This was partly due to difficulty in collecting specific marine organisms. In addition, methods for isolating marine natural products had not been established, and natural product chemists were not familiar with ocean and marine organisms.

Photographs of *Babylonia japonica* are shown in Fig. 35.2.

35.1 Extraction, Isolation and Bioassay

Collection of the Shells

An important and difficult problem in studies of marine natural products, especially marine toxins, is to obtain large amounts of homogeneous raw materials. Fortunately, several hundred kg of the shells could be easily

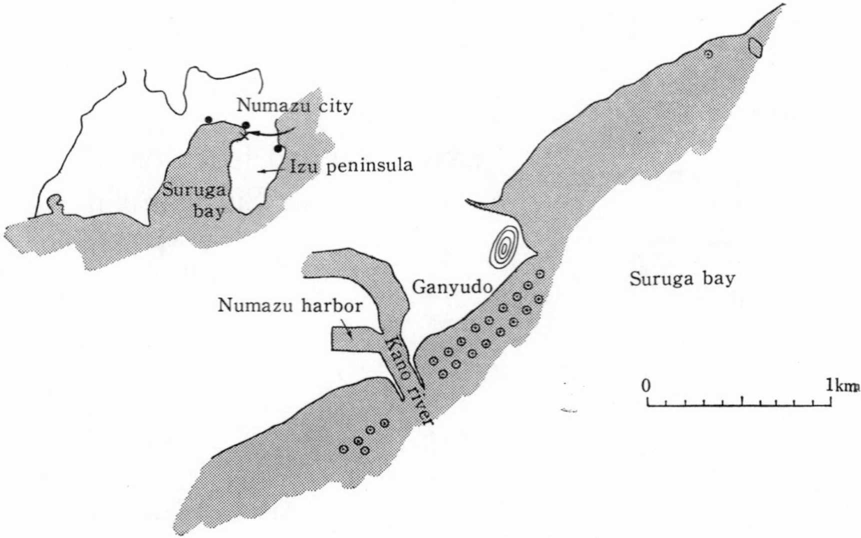

Fig. 35.1. Location of toxic Japanese ivory shells.

Fig. 35.2. (a) Japanese ivory shells (the right-hand one is toxic). (b) Mid-gut gland.

obtained for the study because the general public had been prohibited from collecting shells in the toxic area after the outbreak of poisoning. The shells were collected by fishermen using the special tool shown in Fig. 35.3.

Bioassay

It is extremely important in studies on the isolation of biologically active principles in natural products to select a simple and appropriate bioassy. Hashimoto reported that Pulewka's method, developed for the assay of atropine activity on mydriasis of mice, was useful for the detection of the toxin. The author adopted this method with a slight modification, as described below.

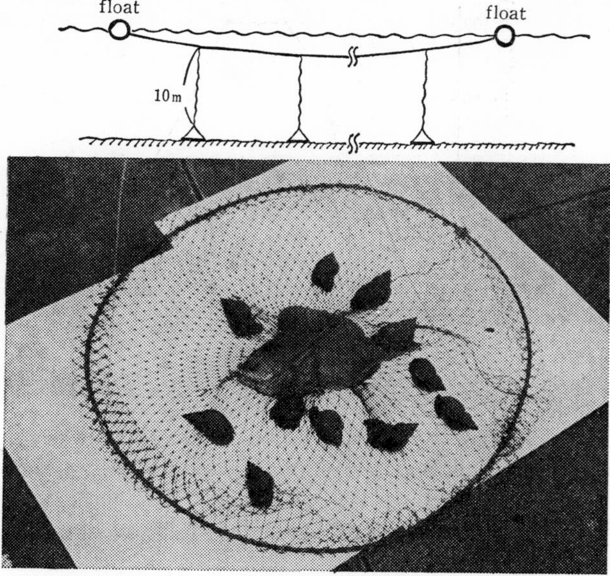

Fig. 35.3. The use of a special tool for collecting Japanese ivory shells.

[Experimental procedure 1] Bioassay

After weighing, the sample was diluted to an appropriate concentration with distilled water and 0.5 ml of the solution was administered subcutaneously into mice on the back. Mice having a pupil diameter of less than 0.5 mm were found to be suitable for the test. Forty minutes after the injection, the diameter of the pupil was measured under a microscope (16x). Mydriatic activity was scored as follows: the fully opened pupil was 100%, while no change of diameter of the pupil was taken as 0%.

Extraction and Isolation Procedures

Hashimoto and a group at the Shizuoka Prefectural Institute of Public Health[4–8] reported the following properties of the toxin after preliminary examinations: the toxin was located only in the mid-gut gland; it was extremely unstable to heat, mineral acid or alkali; it was readily soluble in water, but insoluble in organic solvents.

The author therefore adopted the following conditions for isolation of the toxin: the isolation procedure was carried out at as low a temperature as possible; the procedure was performed under nearly neutral conditions; light was excluded, since the crude extract was found to be sensitive to light.

Since neutral conditions were to be used, applicable siolation procedures were limited to the following: column chromatography using a neutral separator such as active charcoal or cellite; distribution chroma-

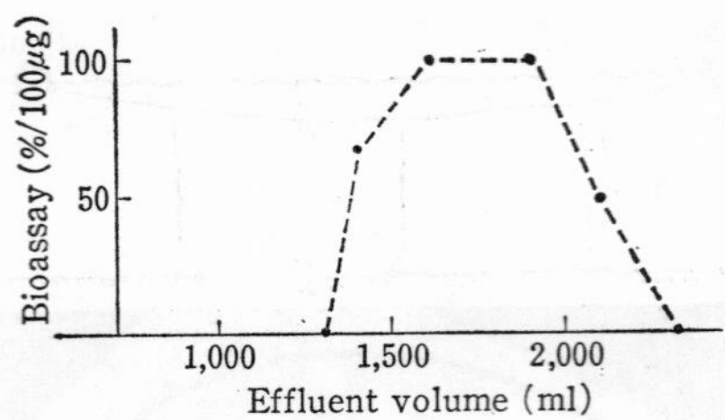

Fig. 35.4. Gel filtration of the AcOH extract on Sephadex G-25 (fine)

440 ~ 860 ml	dark brownish syrup	I_1
861 ~ 1260	brownish syrup⟶taurine	I_2
1261 ~ 1360	brownish syrup⟶taurine	I_2
1361 ~ 1540	yellow syrup	I_3
1541 ~ 1900	yellow powder	I_4

tography using cellulose powder; gel filtration with a molecular sieve; counter current distribution; ion exchange chromatography with a weakly acidic resin, since the toxin was fairly stable to acetic acid. After preliminary examination, gel filtration with Sephadex G-25 was found to be effective as a first step for isolation of the toxin.

The result of Sephadex G-25 (fine) gel filtration of the crude extract is shown in Fig. 35.4. The active fraction [I_4] evoked 80–100% mydriasis in mice at a minimum dose of 100 μg per mouse.

It was considered that relatively high molecular components were excluded from the crude extract by gel filtration through Sephadex G-25. The next step in the purification of the toxin was to employ an CM-Sephadex C-25 ion exchange chromatography, since the toxin was found to be fairly stable to acetic acid. As shown in Fig. 35.5, the active fraction I_4 was divided into two parts, II_2 and II_4, and both fractions evoked 100% my-

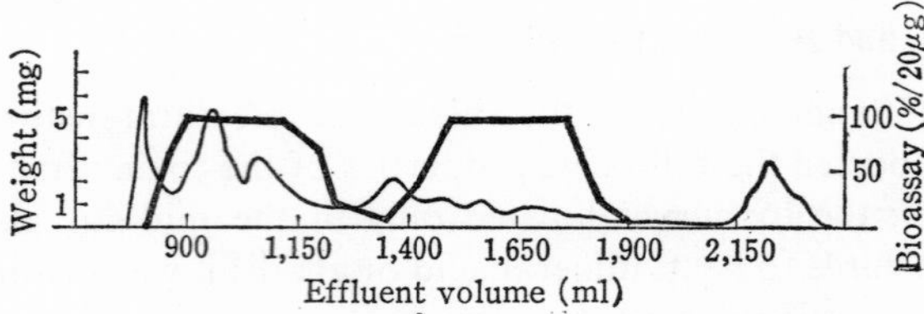

Fig. 35.5. Ion-exchange chromatography of I_4 on CM-Sephadex C-25.

750 ~ 830 ml	uracil II_1
831 ~ 1180	II_2 (900 ~ 1100)
1181 ~ 1430	II_3
1431 ~ 1680	II_4 (1500 ~ 1680)
1681 ~ 2170	
2171 ~	hypoxanthine II_5

driasis in mice at a minimum dose of 20 μg per mouse. II_2 was pale greenish-blue, having strong fluorecence, while II_4 was a light brown powder when isolated, but rapidly changed to a dark reddish substance. II_2 was obtained in 0.15% yield from the mid-gut gland. It was observed that non-active nucleic acid bases contaminating I_4 were effectively removed by the chromatography.

Fraction II_2 was finally purified on a Sephadex G-15 column, and the most active fraction, III_3, which evoked mydriasis at a minimum dose of 10 μg per mouse, was obtained as shown in Fig. 35.6. There was good accord between the elution peak and the peak of biological activity. III_3 was a pale yellow glass, and its activity was as strong as that of atropine sulfate.

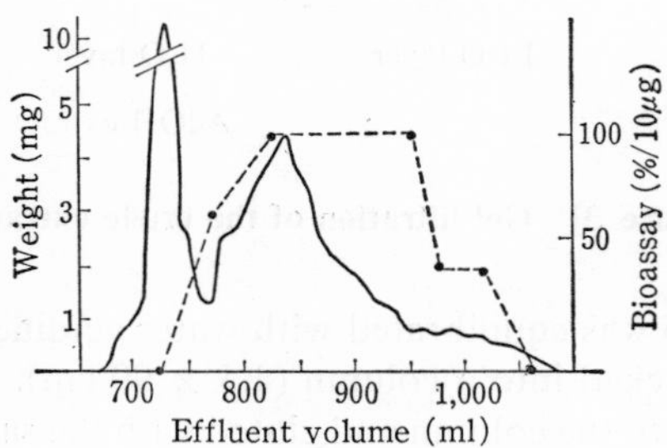

Fig. 35.6. Gel filtration of II_2 on Sephadex G-15

650 ~ 770 ml ——	inosin III_1
771 ~ 800	III_2
801 ~ 950	III_3(820 ~ 950)
951 ~	III_4

This active fraction, III_3, showed a single spot on thin layer chromatography, electrophoresis and paper chromatography. Colorless prisms were obtained from III_3 on standing for several months in the cold. This material was called surugatoxin (10 mg from 10 kg of the ivory shells).

[Experimental procedure 2] Preparation of crude extract

After cracking 100 kg of shells of the carnivorous gastropod, *Babylonia japonica,* captured at the Ganyudo area of Suruga Bay from August to September, 10 kg of dark gray mid-gut gland was obtained and frozen at -30° C. One kg of the mid-gut gland was ground throughly with a mixer and extracted twice with 5 l of 1% acetic acid each time. After centrifugation, the upper solution was mixed with ten volumes of acetone and the precipitate formed was again removed by centrifugation. The supernatant was concentrated to a brownish syrup under reduced pressure at 30° C, after the removal of ether-soluble fatty materials (Table 35.1).

TABLE 35.1. Procedure for extraction of toxin from *Babylonia japonica*

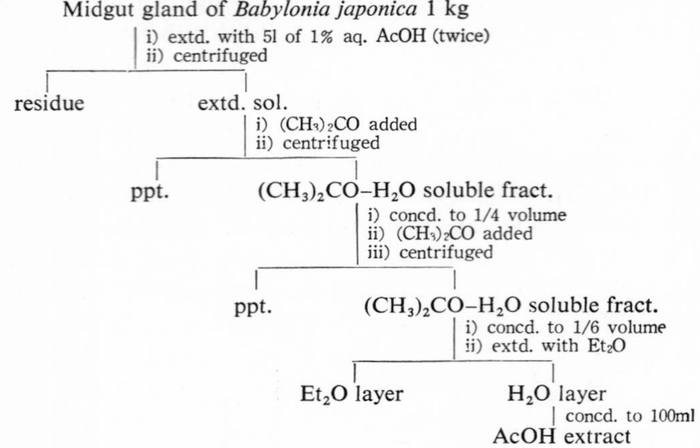

[Experimental procedure 3] Gel filtration of the crude extract through Sephadex G-25

Sephadex G-25 was equilibrated with water acidified with acetic acid to pH 3.9, then packed into a column (4.5×90 cm). The crude extract (20 ml) was applied to the column and eluted with the same aqueous acetic acid solution. After removal of the solvent under reduced pressure below 30° C, each fraction was tested for mydriatic activity by the method described above. The results are shown in Fig. 35.4. A yellow powder [I_4] which evoked 80–100% mydriasis in mice at a dose of 100 μg per mouse was obtained in 0.4% yield.

[Experimental procedure 4] Ion exchange of I_4 on CM-Sephadex C-25

Fraction I_4 (800 mg) was suspended in 40 ml of an aqueous acetic acid solution at pH 3.9 and the suspension was filtered. The filtrate was applied to a CM-Sephadex C-25 column equilibrated with aqueous acetic acid solution at pH 3.9 and developed with the same solvent. After removal of the solvent, the eluted fraction was assayed by the standard method. The results are shown in Fig. 35.5. Two active fractions, II_2 and II_4, were obtained. The mydriatic activities of II_2 and II_4 were both 100% at a dose of 20 μg per mouse.

[Experimental procedure 5] Gel filtration of II_2 on Sephadex G-15

II_2 was purified by gel filtration through Sephadex G-15 equilibrated with the solvent described above. The results are shown in Fig. 35.6. The most active fraction was eluted at 820–850 ml effluent volume, and the fraction evoked 100% mydriasis at a minimum dose of 10 μg per mouse.

35.2 Properties and Structure Determination of Surugatoxin

Surugatoxin obtained as described above (mp >300° C) was slightly soluble in water (1 mg/ > 15 ml) and insoluble in organic solvents. The spectral properties were as follows: uv λ_{max} (H_2O or 0.1 N HCl) 276 mn (ε 14,000), λ_{max} (0.1 N NaOH) 279 mn (ε 17,000); ir (KBr disc) 3200, 1720 (sh), 1695, 1640 cm^{-1}. X-ray analysis confirmed the molecular formula $C_{25}H_{26}N_5O_{13}Br + n\,H_2O$. The compound was substantially decomposed under the following conditions: in H_2O at 100° for 1hr; in 1N HCl at 20° for 24 hr; in 0.1 N NaOH at 20° for 24hr. It caused 100% mydriasis in mice at a dose of 10 μg per mouse; the minimum dose evoking mydriasis was 0.5 μg/g.

Degradation of surugatoxin in water, 1 N HCl or 2 N NH_4OH by heating at 100° C in a sealed tube followed by trimethylsilylation yielded a single product, which was identical with trimethylsilylated myo-inositol on gc (20 ± 3% (w/w) yield based on surugatoxin). Myo-inositol was also identified by gel filtration of the degradate on Sephadex G-10.

More vigorous degradation of surugatoxin (in 2 N NaOH at 100° C for 10 hr in a sealed tube) followed by trimethylsilylation yielded lumazine (6% w/w based on surugatoxin) as one of the products.

The unit formula $C_{25}H_{26}N_5O_{13}Br{:}7H_2O$ and the molecular structure (including the absolute configuration of surugatoxin) were finally determined by X-ray analysis of single crystals (Fig. 35.7).[9,10]

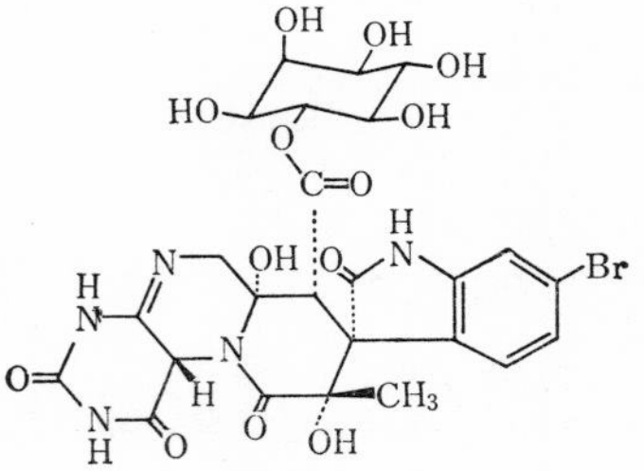

Fig. 35.7. Structure of surugatoxin.

[Experimental procedure 6] Degradation of surugatoxin

Surugatoxin (5 mg) was dissolved in 2 ml of 1 N HCl or 2 N NH_4 OH, sealed in a tube and heated on a water bath for 5 hr. After neutralization, the solvent was evaporated off to give the degradate [A]. Gel filtration of [A] on a Sephadex G-10 column was accomplished with water acidified with acetic acid to pH 3.9 as an elution solvent. Myo-inositol was eluted at 10 ml effluent volume (yield 0.9 mg).

[Experimental procedure 7] Identification of myo-inositol by gas chromatography

One mg of dried [A] or myo-inositol was dissolved in 1 ml of dry pyridine and a mixture of 0.2 ml of hexamethyldisilazane (HMDS) and 0.1 ml of trimethylchlorosilane (TMCS) was added. After vigorous stirring for 30 sec, the reaction mixture was allowed to stand at room temperature for 30 min. After removal of the solvent the residue was extracted with dry $CHCl_3$ and the $CHCl_3$ layer was analyzed. The gc conditions (Shimadzu GC 3AF machine, 1 kg/cm^2 N_2 carrier gas) were as follows: SE-30 column (0.3 × 150 cm) at 152° (R_t 16′ 12″) or 183° (R_t 4′ 29″). A 15% Reoplex 450 column (0.3 × 150 cm) was also used at 155° (R_t 11′ 42″) or 183° (R_t 8′ 54″).

[Experimental procedure 8] Identification of lumazine by gas chromatography

Surugatoxin (3 mg) was dissolved in 1 ml of 2 N NaOH and the solution was heated at 100° C for 10 hr in a sealed tube. After neutralization with dil. HCl the solvent was evaporated off under reduced pressure. The residue was dissolved in 0.5 ml of dry pyridine and a mixture of 0.4 ml of HMDS and 0.2 ml of TMCS was added. The mixture was heated at 150° C for 1.5 hr. After cooling, the mixture was diluted with dry pyridine and analyzed. The gc conditions (Shimadzu GC 3AF machine, 0.95 kg/cm^2 N_2 carrier gas) were as follows: OV-1 column (0.3 × 150 cm) at 150° (R_t 4′ 36″).

References

1) T. Kosuge, H. Zenda, A. Ochiai, N. Masaki, M. Noguchi, S. Kimura, H. Narita, *Tetr. Lett.*, **1972**, 2545.
2) H. Zenda, *Farumashia*, **7**, 646 (1972).
3) Y. Hashimoto, K. Miyazawa, H. Kamiya, M. Shibota, *Bull. Jap. Soc. Sci. Fish.*, **11**, 661 (1967).
4) M. Shibota, Y. Hashimoto, *ibid.*, **36**, 115 (1970).
5) M. Shibota, Y. Hashimoto, *ibid.*, **37**, 936 (1971).
6) S. Kimura, S. Sugiyama, *Nippon Koshu-eisei shi*, **14**, 1161 (1967).
7) K. Kimura, S. Sugiyama, H. Narita, *Annual Reports of Shizuoka Prefectural Institute of Public Health,* **16**, 163 (1972).
8) H. Narita, S. Kimura, *ibid.*, **16**, 173 (1972).
9) T. Yamamoto, T. Kosuge, H. Zenda, H. Ohba, *Bull. Jap. Soc. Sci. Fish.*, **42,** 1405 (1976).
10) E. Hayashi, S. Yamada, *Brit. J. Pharmac. Chemother.*, **53**, 207 (1975).

CHAPTER 36

The Extraction and Isolation of Tetrodotoxin from *Fugu* Ovaries

The first step in the investigation of *Fugu* tetrodotoxin was its isolation in pure crystalline form. However, this proved exceedingly difficult, and 40 years passed from the first report on *Fugu* toxin in 1909 by Tawara[1] until its successful isolation by Yokoo[2] (1950), Tsuda and Kawamura[3] (1952), Arakawa[4] (1956) and Hirata[5] (1957), independently. The ovaries used in these extractions were obtained from *Fugu rubripes rubripes ("torafugu"), Fugu vermicularis porphyreus ("mafugu"), Fugu pardalis ("higanfugu"),* and *Fugu stictonotos ("gomafugu").*

The isolated crystalline toxin (tetrodotoxin) obtained was identical from all sources. Structure determination was independently achieved by Woodward[6] (Harvard University), Hirata and Goto[7] (Nagoya University), and the Tsuda group[8] (Tokyo University-Sankyo Central Research Laboratories). All the research groups proposed the same structure at the IUPAC Symposium on the Chemistry of Natural Products held in Kyoto in April 1964. At this congress, we suggested that the toxin existed as a dimer. However, later X-ray crystallographic analysis of the toxin by the Woodward group[9] unambiguously determined the structure, showing it not to be a dimer. Tetrodotoxin possesses a unique orthoester structure, and almost all the carbons in this molecule have asymmetric substitution. In addition to these properties, this toxin is insoluble in organic solvents. These properties led us to believe that a total synthesis would be extremely difficult. However, these difficulties were successfully overcome in an elegant synthesis by Kishi.[10] Thus after 65 years of research since Tawara's first report on *Fugu* toxin, the problems involved in crystallization, structure determination and total synthesis of tetrodotoxin have been overcome. Although *Fugu* tetrodotoxin was assumed to be unique to this species, Mosher[11] found the same toxin not only in *Taricha torosa (Taricha, Triturus, Notophthalmus, Cynops)* but also in *Atelopus*[12] living in South America. In addition to tetrodotoxin, *Atelopus* contains chiliquitoxin,[13] which is structurally similar to *Fugu* tetrodotoxin. Hashimoto[14] also found a toxin identical to *Fugu* tetrodotoxin in "*tsumugihaze*" *(Gobius criniger)*. This suggests the possibility of a relatively wide distribution in nature, which might have biosynthetic implications.

36.1 Extraction, Isolation and Crystallization of Tetrodotoxin from *Fugu* Ovaries

Tawara published the first paper on *Fugu* toxin in the Japan Pharmaceutical Bulletin in 1909. According to this paper, after removal of salt adhering to the ovaries by washing with water the ovaries were mashed into a paste and soaked in hot water for 3 hr. After filtration, the filtrate was made acidic with AcOH and concentrated. The precipitate was filtered off. Lead acetate was added to the filtrate, and the newly formed precipitate was again filtered off. The addition of aq. NH_4OH to the filtrate gave a lead precipitate containing the toxin. The lead was removed by H_2S treatment and the filtrate was concentrated *in vacuo*. The residue was again dissolved in alcohol, and the insoluble fraction was filtered off. Ether was added to the filtrate, and the resulting precipitate dried. In this way, Tawara obtained the crude toxin (LD_{50}, 4.1 mg/kg) and named it tetrodotoxin after Tetraodontidae *("mafugu-ka")*. Based on the toxicity of pure tetrodotoxin (LD_{50} 9 μg/kg), this crude toxin is estimated to have contained only 0.2% toxin. After attempting Tawara's method Tsuda and Kawamura concluded that it was not suitable for the isolation of tetrodotoxin. Although repeating this procedure slightly increased the toxicity, the loss of the toxin was very great. They then turned to other methods (see "Experimental procedure 2").

In 1950, Yokoo obtained a crude toxin (LD_{50}, 800 μg/kg) from Tawara's toxin (LD_{50}, 10 mg/kg) by successive treatment with phosphotungstic acid, mercury picrate, phenylhydrazine, mercury picrate and picrolonic acid. Although this crude toxin (LD_{50}, 800 μg/kg) did not crystallize, alumina column chromatography was found to be effective, yielding a pure form of the toxin which would crystallize. Alumina column chromatography was then applied directly to obtain the pure crystalline toxin from the crude toxin $(LD)_{50}$, 10 mg/kg). Yokoo named the crystalline toxin obtained in this way spheroidine. Later, spheroidine was confirmed to be identical with tetrodotoxin isolated by Tsuda and Kawamura.

[Experimental procedure 1] The first crystallization of *Fugu* toxin by Yokoo (aluminum column chromatography)

The crude toxin (LD_{50}, 10 mg/kg) was prepared by Tawara's method. The ovaries (60 kg) of *Fugu rubripens rubripens ("torafugu")* were treated with 0.5% aq. HCHO, and the mixture was allowed to stand for one week. The supernatant was concentrated *in vacuo*. After 4 repetitions of the same procedure, 1.7 kg of the hygroscopic crude toxin (LD_{50}, 40 mg/kg) was obtained. The crude toxin (500 g) in 6 l of H_2O was treated with 300 g of lead acetate at 70° C with mechanical stirring. The resulting precipitate was filtered off and washed twice with 700 ml of hot water. The filtrate was combined. and the concentrated solution containing 700 g of lead acetate and 2.2 l of 25% NH_4OH was added with vigorous stirring at room temperature. The newly formed precipitate was collected by filtration and washed 3 times with 700 ml of 1.5% aq. NH_4OH. AcOH (250 ml) was added dropwise to the stirred precipitate containing the toxin in 3.5 l of H_2O, causing the solution to become clear. Enough H_2S was then bubbled through the solution to precipitate all the lead. The precipitate was filtered off, and the residue washed with 800 ml of hot water on a steam bath. The combined filtrate was concentrated *in vacuo* to yield a syrupy residue. The crude toxin obtained in this way had a toxicity (LD_{50}, 10 mg/kg) 4 times that of the original crude toxin (LD_{50}, 40 mg/kg); yield, 85 g.

The first attempt at crystallization iovolved successive treatments with phosphotungstic acid, picric acid, mercury picrate, phenylhydrazine, mercury picrate and picrolonic acid. This method provided crude toxin (LD_{50}, 800 μg/kg). In order to obtain crystalline toxin, 1.5 g of this fraction was subjected to alumina column chromatography. This chromatography produced a fraction 10 times as active as the original crude toxin; yield, 49 mg (LD_{50}, 90 μg/kg). This toxin (190 mg) was dissolved in 0.6 ml of H_2O and allowed to stand for one night to afford crystalline tetrodotoxin. Next, 2.4 ml of methanol was added and the mixture was allowed to stand for an additional one night, then the solvent was removed by decantation. The crystalline residue was dissolved in aq. AcOH, and methanol and, if necessary, ether were added to provide the pure tetrodotoxin. In this way, 13 mg of crystalline tetrodotoxin (LD_{50}, 10 μg/kg) was obtained. The improved method consists of directly subjecting the crude toxin (LO_{50} 10 mg/kg) to alumina column chromatography. Alumina (Merck), 250 mg, was packed in a glass column (5 × 35 cm) and washed with 250 ml of H_2O. The column was charged with 400 ml of H_2O containing 20 g of the crude toxin, and eluted with 400 ml of H_2O. The toxin adsorbed on the alumina was eluted with 3 l of hot water under a partial vacuum. The fraction containing the toxin was made acidic with aq. AcOH and concentrated *in vacuo* over a hot water bath to afford an oily residue. This was dissolved in 2 ml of H_2O, and 20 ml of methanol and 80 ml of ether were added to yield 1.4 g of the crude toxin (700 μg/kg) which could be isolated by decantation. This crude toxin was again dissolved in 2 ml of H_2O and made alkaline with aq. NH_4OH. The resulting precipitate was washed 5 times with 1 ml of H_2O, twice with 1 ml of methanol, and was again dissolved in 0.5 ml of H_2O containing a small amount of AcOH. The insoluble material was filtered off. After adding 30 ml of ether, the mixture was allowed to stand overnight to yield crystalline tetrodotoxin. One more recrystallization in a similar way afforded 16 mg of crystalline tetrodotixin (LD_{50}, 13–14 μg/kg).

At that time Tsuda and Kawamura were also investigating the crystallization of *Fugu* toxin. In 1952, they also succeeded in the isolation of crystalline tetrodotoxin (LD_{50}, 10 μg/kg) from Tawara's crude toxin (LD_{50} 6 mg/kg) partition column chromatography using a mixture of potato starch and celite and absorption chromatography using a mixture of charcoal and celite, followed by paper column chromatography in which 500 circular sheets of filter paper 5.5 cm in diameter were piled up to form a column (see "Experimental procedure 2"). However, this procedure did not provide sufficient tetrodotoxin for structure determination. This method was then improved gradually by monitoring the toxicity using the mouse *(Mus musculus)*. Finally, it was found that column chromatography using charcoal was effective for large-scale production (See "Experimental procedure 3"). By this method 400 kg of ovaries gave 3.8 g of crystalline tetrodotoxin. This method can be extended to quantities of 1,000 kg, and 10 g of the crystalline toxin can be obtained without difficulty. The ovaries were collected through a dealer at the fish market in Shimonoseki.

[Experimental procedure 2] Tsuda and Kawamura's method (Chromatography using circular filter paper)

Tawara's toxin (Ld_{50}, 6 mg/kg) obtained by treatment of the crude extracts from *Fugu* ovaries with lead acetate—qa. NH_4OH and H_2S, was subjected to partition column chromatography using a mixture of potato starch and celite, yielding a crude toxin (LD_{50}, 400 μg/kg). The toxicity of this toxin was increased to LD_{50} 200 μg/kg through absorption chromatography using charcoal and celite. However, crystalline tetrodotoxin could not be obtained even after this step. Next, chromatography using circular sheets of filter paper 5.5 cm in diameter (Toyo Roshi No. 2) was attempted. Five hundred sheets of filter tpaer were stacked to form a column. Those from the 50th to the 70th were impregnated with an aq. solution of 500 mg of tetrodotoxin possessing a toxicity of 200 μg/kg. The pile was then developed by the ascending method with 90% phenol-water. The sheets at *Rf* 0.0–0.1 were taken out, phenol was removed by washing with ether, and the sheets were extracted with cold water. The aqueous extract was concentrated *in vacuo,* washed again with ether, and dried under reduced pressure. In this way 200 mg of toxin (LD_{50}, 120 μg/kg) was obtained. After washing with alcohol, this toxin was dissolved in 0.5 ml of water containing 1 drop of AcOH and the insoluble fraction was filtered off. Next, 5 ml of ether was added and the mixture was allowed to stand overnight, yielding crystalline toxin, By repeating the same procedure, 10 mg of the crystalline toxin (LD_{50}, 10 μg/kg) was obtained.

[Experimental procedure 3][15] Tsuda and Kawamura's method for large-scale production (charcoal chromatography)

Fugu ovaries were purchased in winter through a fish dealer at a fish market in Shimonoseki. These ovaries were taken not only from *Fugu rubripes rubripes ("torafugu"),* but also from other species such as *Fugu vermicularis porphyreus ("mafugu")* and *Fugu pardalis ("higanfugu").* These ovaries were packed in 50 kg lots and kept in 3% HCHO to prevent decary.

Extraction was carried out as follows: enough water was added to 400 kg of the overies in 3% HCHO bring the total volume up to 600 l. After standing overnight, the ovaries were mechanically mashed and shifted to the reactor. Under mechanical stirring, the temperature in the reactor was raised to 90° C, and then quickly cooled to 30° C. In order to facilitate the separation, 15 kg of Hyflo Supercel was added and the mixture was centrifuged. H_2O (400 l) was then added to the residue. The mixture was mechanically stirred for 1 hr, and again centrifuged. The combined filtrate was concentrated to 40 l *in vacuo* below 50° C. After standing for 3 days the concentrated extract separated into an oily layer which contained no toxin, and an aqueous layer. Hyflo Supercel (2 kg) was added to 30 l of the aqueous layer and the mixture was filtered. The filtrate was concentrated *in vacuo* to yield a brown, sticky oil (15 kg).

The second step was the removal of formaldehyde. The above extract (15 kg) was stirred vigorously with 15 l of MeOH and 15 l of ether. The organic layer which separated on standing was removed. The newly formed precipitate was dissolved in 12 l of H_2O and concentrated below 50° C *in vacuo,* then 15 l of MeOH and 2 kg of Hyflo Supercel were added with mechanical stirring. After filtration, the residue was washed with 8 l of 70% aq. MeOH. The combined filtrate was concentrated below 55° C *in vacuo.* In this way, 11 kg of a brown oil was obtained. Based on the toxicity of this oil, it was estimated to contain more than 5 g of the toxin. Although crystalline tetrodotoxin is insoluble in H_2O, impure tetrodotoxin, which is not crystalline, is soluble in water. Prior to the next chromatography, the solution was adjusted to pH 4.2–5.0 by the addition of AcOH (about 50 ml) to prevent decomposition of the toxin.

Finally, charcoal chromatography and crystallization were carried out: 1.8 kg of MeOH extracts adjusted to pH 8–8.5 with aq. NH_4OH was charged on a glass column (17.5 cm × 90 cm) packed successively with 300 g of Hyflo Supercel and a mixture of charcoal (1.5 kg) and Hyflo Supercel (3 kg) in H_2O, and the column was eluted with aq. NH_4OH (pH 7.4–8.0). The first fraction was 50 l. Elution was continued with AcOH-MeOH-H_2O (0.8:10:100 by volume), and the fraction more basic than pH 6.6 was collected (about 14–16l). Finally, the main fraction (pH 6.6–4.0) was collected (about 16 l). By repeating this procedure 6 times, 11 kg of MeOH extract from the second step afforded 100 l of the main fraction, which was then condensed to 10 l below 50° C *in vacuo*. By bubbling H_2S into this solution, traces of metal were removed. The filtrate was concentrated to 700 ml *in vacuo*. After standing overnight, the resulting precipitate was filtered off. The brown, sticky filtrate was again concentrated *in vacuo*, yielding 85 g of the crude extract. The extract was dissolved in 1.7 l of MeOH, and the insoluble preicpitate was again filtered off. Next, 300 ml of aq.NH_4OH was added and the mixture was allowed to stand overnight at 0° C. The crude tetrodotoxin precipitated as crystals. The precipitate was collected by filtration, washed with H_2O and again dissolved in 20 ml of 5% aq. AcOH. The insoluble precipitate was filtered off. MeOH (150 ml) and ether (120 ml) were added to the filtrate. On standing, 4.5 g of the crude toxin was obtained. The crude toxin was again dissolved in 20 ml of 5 % aq. AcOH, and aq. NH_4OH was added under ice-water cooling to afford the crystalline tetrodotoxin. In this way, 3.8 g of the crystalline tetrodotoxin was obtained (LD_{50}, 10–13 μg/kg). By this method, 10 g of tetrodotoxin could be obtained from 1,000 kg of *Fugu* ovaries. This method of isolating tetrodotoxin gave quantities sufficient for structure determination.

Later, Arakawa (1956) and Hirata and Goto (1957) also indepdendently succeeded in the isolation of the crystalline tetrodotoxin. For large-scale production, Hirata and Goto developed a method using Amberlite IRC-50 and absorption by charcoal.

[Experimental procedure 4] The method of Hirata and Goto

Chopped fresh ovaries (100 kg) of *Spheroides rubripes* were soaked in water (100 l) with occasional stirring and, after 2 days, the supernatant was collected by decantation. The residue was extracted repeatedly with water and the combined aqueous extract (200 l) was boiled for a very short time, causing a large quantity of protein to coagulate. The slurry was filtered and the filtrate poured onto a column (8 cm × 2 m) packed with Amberlite IRC-50 resin (ammonium type, 8 l). The column was washed with water and eluted with 19% acetic acid (13 l) and water (10 l), successively. The first 3 l of the eluates had no toxicity towards mice and was discarded. The next fraction (10 l) was strongly toxic and was applied to the subsequent steps. The last fraction (10 l) contained a large amount of acetic acid and hence was used for the preparation of the 10% acetic acid solution required subsequently.

The toxic fraction was brought to pH 8–9 by the addition of NH_4OH, then Norit A (100 g) was added, and the mixture was shaken for 2 hr. The charcoal was collected by filtration and washed with water. This procedure was repeated 3 times and the combined charcoal was extracted with 20% EtOH containing 0.5% acetic acid (250 ml × 3). After concentration approx. 50 ml, the combined extracts were brought to pH 9 by the addition of NH_4OH, resulting in the separation of white precipitates which were collected by filtration, dissolved in *dil.* acetic acid and reprecipitated by the addition of NH_4OH, to give crude tetrodotoxin (1–2 g).

In 1958, Murtha also succeeded in isolating tetrodotoxin from *Fugu pardalis ("higanfugu").*

Recrystallization using the usual organic solvents was not applicable, because the toxin is insoluble in organic solvents. We found that reprecipitation is an effective method of purification. The toxin was dissolved in 5% aq. AcOH, and 5% NH_4OH was added to reprecipitate the crystalline tetrodotoxin. Hirata and Goto purified tetrodotoxin via the picrate, which could be recrystallized from hot water.

[Experimental procedure 5][16] Purification via the picrate

Crude tetrodotoxin (3.2 g) and picric acid (2.3 g) were dissolved in boiling water (29 ml) and the solution was filtered while hot. One cooling the filtrate, a crystalline precipitate speared, which, on recrystallization (3 times) from hot water, gave tetrodotoxin picrate as yellow needless (4.8 g). It darkened above 200° without melting. An analytical sample was dried at 80–100° for 20 hr *in vacuo*.

A solution of the picrate (4.7 g) in hot water was adjusted to pH 9 with NH_4OH and, after cooling, the precipitated solid was filtered off and washed with water. It was dissolved in a minimum amount of dil. acetic acid and precipitated by addition of NH_4OH; the yield of pure tetrodotoxin was 2.6 g.

36.2 Seasonal Variation of Toxin Content by Season Toxicity of Various Organs

The only species of *Fugu* possessing toxin are tetraodontidae *("mafugu -ka")* such as *F. vermicularis porphyreus ("Mafugu")*, *F. rubripes rubripes ("torafugu")*, *Lagocephalus lunaris ("sabafugu")*, and *F. vermicularis vermicularis ("shosaifugu")*. These fish are eaten in Japan after the organs containing the toxin have been removed. Tani[17] investigated the distribution of toxin in organs, the variation in toxicity levels among individuals and the variation of toxicity from season to season for 21 species living in the seas surrounding Japan. According to his research, the toxin was mainly present in the ovaries, liver, skin and intestines. The muscle tissue contains no toxin. There are no differences between the male and female fish as regards distribution of the toxin, except that the ovaries *("mako")* are toxic whereas the testicles *("shirako")* are not. Generally, the amount of toxin increases from December to April (winter) and decreases after the breeding season (May). However, according to recent research by Abe on 114 *F. vermicularis porphyreus ("mafugu")*, the skin and the liver showed the highest content of toxin from July to September (summer), and the lowest content from November to March (winter). More research may be required to reach a final conclusion.

36.3 Structure Determination

The toxin is insoluble in organic solvents and H_2O; however, it will dissolve in aq. AcOH and aq. mineral acid. It darkens at temperatures over 200° C, but does not decompose even at temperatures as high as 300° C Spectral properties: uv $[\alpha]_D^{52}$ -8.64 ($c = 8.55$ in aq. AcOH); ir ν_{max}^{Nujol} 3350, 3240, 1670, 1612, 1075 (cm^{-1}); characteristic nmr signals (CD_3COOD-D_2O) 2.2 1H (d) ($J = 10$ cps) 4a-H, 5.7 1H (d) ($J = 10$ cps) 4-H. At an early stage of the structure determination, the formula $C_{12}H_{19}O_9N_3$ was proposed, but the correct formula is $C_{11}H_{17}O_8N_3$. The use of ms was not very helpful for structure determination.

Degradation with Alkali or HI-Phosphine

The toxin is extremely insoluble in organic solvents, severely limiting the reactions available for structure determination. Although many attempts to obtain derivatives of the toxin were made, all resulted in failure. However, treatment of the toxin with 5% aq. KOH at 90–100° C yielded yellow crystals of $C_9H_9O_2N_3$ (uv 278 nm); this was nmaed the C_9-base. Oxalic acid was also obtained in this degradation reaction.

The C_9-base was synthetically confirmed to be 2-amino-6-hydroxymethyl-8-hydroxyquinazoline **1**. The toxin was heated under reflux in HI-phosphine. The crude product was oxidized with potassium ferricyanide, $K_3[Fe(CN)_6]$, to yield yellow crystals of $C_9H_9N_3$, which was also synthetically confirmed to be 2-amino-6-methyl quinazoline **2**. Hirata and Goto

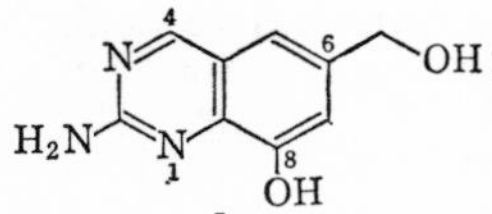

1

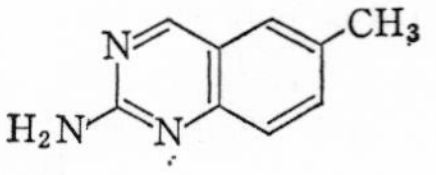

2

obtained 2-amino-6-hydroxyquinazoline by decomposition with sulfuric acid, and Woodward obtained the C_9-base by alkali degradation.

Tetrodonic Acid

Although the structure of the C_9-base could be easily determined, no information concerning the 8 oxygens which the toxin possesses could be obtained from this base. In the extraction process of the toxin, prolonged heating in water greatly reduced the yield. Thus, the toxin was heated under reflux in water. As described before, the toxin is insoluble in all organic solvents except for acids. However, the toxin slowly dissolved in water under vigorous reflux. After 20 hr, a clear solution resulted, which gave white crystals upon concentration *in vacuo*. These crystals could be recrystallized from hot water, and were named tetrodonic acid, $C_{11}H_{17}O_8N_3 \cdot H_2O$, after tetrodotoxin. Hirata and Goto used the name tetrodoic acid. Unlike tetrodotoxin, this acid is not toxic (LD_{50}, $\gg$ 30,000 μg i.v.). This acid afforded the C_9-base upon alkali degradation, like tetrodotoxin. The 6-methyl compound (**2**) was also obtained with HI-phosphine, and HCHO with $NaIO_4$ (1 mol). These observations suggest that tetrodonic acid is similar in structure to tetrodotoxin.

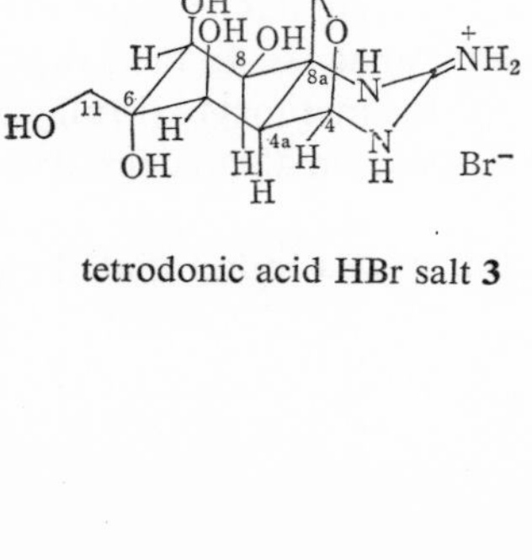

tetrodonic acid HBr salt **3**

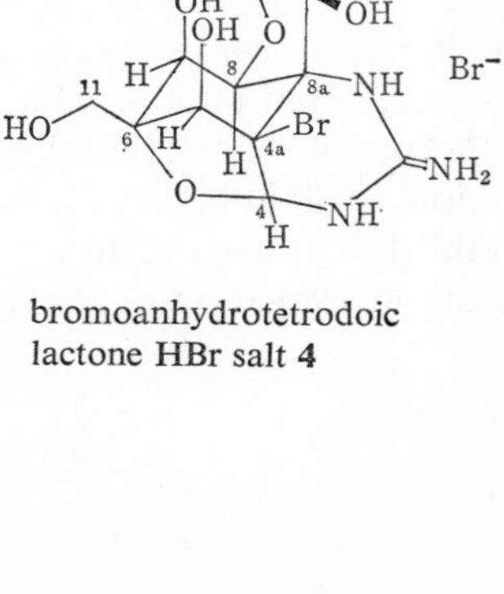

bromoanhydrotetrodoic lactone HBr salt **4**

The structure of tetrodonic acid was determined by X-ray crystallographic analysis of its HBr salt (**3**). Hirata and Nitta also succeeded in structure determination of the bromoanhydrotetrodoic lactone HBr salt (**4**), obtained in several steps from tetrodotoxin, by X-ray crystallographic analysis. Comparison of these two compounds showed differences in the configuration at C-9, which suggests that epimerization occurred in one of the compounds during the derivation from tetrodotoxin. As tetrodonic acid deuterated at C-9 was obtained by refluxing the toxin in D_2O (from nmr data), the toxin was presumed to be a derivative of **5**.

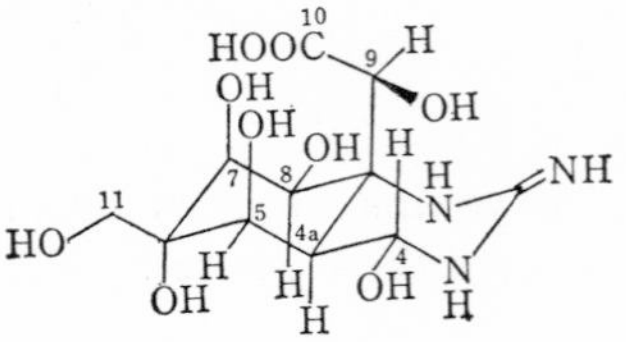

5

Anhydrotetrodotoxin and Tetrodotoxin

Treatment of the HCl salt of the toxin with pyridine-Ac_2O at 0–5° C for 2 days afforded the polyacetate, which on standing in MeOH yielded partially deacetylated diacetylanhydrotetrodotoxin $C_{11}H_{13}O_7N_3\,(CH_3CO)_2$ as crystals. Anhydrotetrodotoxin $C_{11}H_{15}O_7N_3$ (Hirata's anhydroepitetrodotoxin) was obtained by treatment of the diacetate with aq. NH_4OH. These crystalline compounds do not show distinctive melting points of the type usually observed in organic compounds. The anhydro compound was also obtained in the following way. After heating the toxin in formic acid at 82–85° C for 2 hr, excess formic acid was removed *in vacuo* to afford the polyformate. On standing in MeOH–ether, monoformylanhydrotetrodotoxin HCOOH salt $C_{11}H_{14}O_7N_3 \cdot HCO \cdot HCOOH$ was obtained as crystals, which afforded anhydrotetrodotoxin on treatment with aq. NH_4OH.

Tetrodotoxin and anhydrotetrodotoxin afforded methoxytetrodotoxin $C_{11}H_{16}O_7N_3 \cdot OMe$ on treatment with MeOH-dry HCl, and deoxytetrodotoxin $C_{11}H_{17}O_7N_3$ on catalytic reduction with Pt–H_2 over a long period of time. These derivatives were considered to be similar in structure to tetrodotoxin, based on nmr data and elemental analysis. In fact, the diacetate, the monoformate, the anhydrotoxin and the methoxytoxin all yielded tetrodonic acid in boiling water and the C_9-base on alkali degradation, though the deoxytoxin did not.

In the investigation of the anhydrotoxin, it was found that the anhydrotoxin is convertible to tetrodotoxin in 5% HCl at room temperature in good yield, which was very important information. The molecular weight of the monoformate HCOOH salt was determined to be 390 by X-ray crystallography. This value is compatible with the formula $C_{11}H_{14}O_7N_3 \cdot HCO \cdot HCOOH$ (monoformate HCOOH salt) deduced from elemental analysis. X-ray crystallographic analysis of the diacetate HI salt also indicated that the toxin consisted of 11 carbons. Therefore, the molecular formula $C_{12}H_{19}O_9N_3$ proposed in the early stage of research was revised to $C_{11}H_{17}O_8N_3$.

The conversion of the toxin to the anhydrotoxin by the removal of 1 mol of H_2O and of the anhydrotoxin to the toxin by the addition of 1 mol of H_2O suggest that the toxin possesses an appropriate functional group. The ir spectra of the toxin and the anhydrotoxin show no signal indicative of carboxylic acid or lactone. This suggests that the carboxyl function exists as the orthoester bound with an internal OH group.

The diacetate HI salt, which was obtained as crystals by the addition of ether and MeOH to the diacetate in a small amount of 40% HI, was subjected to X-ray crystallographic analysis. Thus, the structure of the diacetate, the anhydrotoxin and tetrodotoxin were unambigously determined.

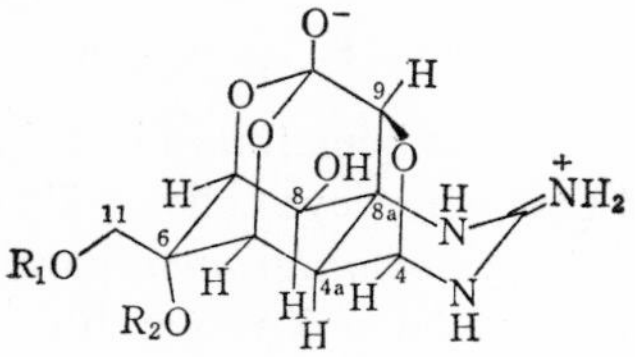

6,11-diacetylanhydro-tetrodotoxin ($R_1 = R_2 = Ac$)
11-monoformyl-anhydrotetrodotoxin
($R_1 = HCO$, $R_2 = H$)
anhydrotetrodotoxin
($R_1 = R_2 = H$)

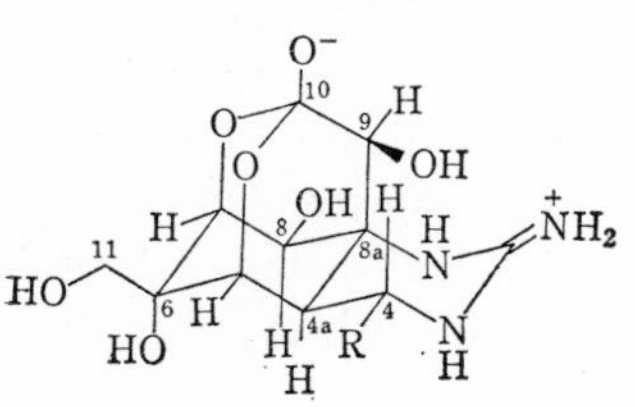

tetrodotoxin ($R = OH$)
methoxytetrodotoxin
($R = OCH_3$)
deoxytetroxdtoxin ($R = H$)

36.4 The Toxicities of Various Toxins

Comparison of the toxicities of tetrodotixin derivatives and other natural toxins.

Compound	LD_{50}(μg/kg)
Tetrodotoxin	8.22–
Anhydrotetrodotoxin	4,140
6,11-Diacetylanhydrotetrodotoxin	>5,000
11-Monoformylanhydrotetrodotoxin	~3,000
Methoxytetrodotoxin	341
Deoxytetrodotoxin[18]	84.5
Tetrodonic acid	>30,000
Batrachotoxin[18]	2.0
Saxitoxin[19]	5.0–10.0
Palytoxin[20]	0.6
Strychnine	500

Palytoxin is the most powerful known toxin among the natural toxins. The second strongest is batrachotoxin. Tetrodotoxin and saxitoxin are almost equal in toxicity. The toxocity of toxin derivatives decreases in the order deoxytetrodotoxin > methoxytetrodotoxin > anhydrotetrodotoxin ≫ tetrodonic acid. Tetrodonic acid is not toxic.

36.5 Distribution of Tetrodotoxin in Nature

Tetrodotoxin, named after tetraodontidae, has been found in animals other than *Fugu*. In 1930, Twitty reported that when eye cells of *Taricha torosa* were transplanted to the embryo of *Ambystoma tigrinum,* paralysis was observed for a while. In 1960, Mosher[11] undertook research on the paralytic component of *Taricha torosa,* and isolated 200 mg of toxin from the eggs (100 kg) of *Taricha torosa.* This toxin was named tarichatoxin. However, this toxin was later confirmed to be identical with tetrodotixin. Consequently, the distribution of tetrodotoxin was extended to 2 species of *Taricha,* 1 species of *Notophthalums,* 2 species of *Cynops* and 4 species of *Triturus*. Tetrodotoxin in *Taricha torosa* was mainly observed in the skin, eggs, ovaries, muscle and blood.

In 1971, based on stories that a certain fish near Iriomote island (Okinawa) and the Amami islands is poisonous, Hashimoto[14] found that *"tsumugihaze" (Tobius crinizes)* possesses a powerful toxin. By successive treatment of 25 kg of MeOH extracts from *"tsumugihaze"* with charcoal, Amberlite IRC-50, charcoal and Sephadex G-10, 30 mg of the toxin was isolated and confirmed to be identical with tetrodotoxin by direct comparison with a standard sample. At present, the only *"haze"* knlown to possess the toxin is *"tsumugihaze" (Gobius criniger)*.

Identification of extremely samll amounts of the toxin is apt to be extremely difficult. However, no abnormalities are observed when *"tsumugihaze"* toxin is injected into *Fugu* or vice versa, suggesting that the toxins are identical. Actually, this approach was aoplied to the identification of tarichatoxin with tetrodotoxin.

In 1975, Mosher and Kim[12,13] isolated tetrodotoxin from the skin of frogs such as *Atelopus* var. *ambulatorius, A. various varius* and *A. chiriquiensis* living in Costa Rica. Chiriquitoxin, which is presumed to be structurally smilar to tetrodotoxin, was obtained from *Atelopus chiriquiensis* accompanied by tetrodotoxin. This toxin has almost the same toxicity as tetrodotoxin. Its molecular weight is 73 units more than that of tetrodotoxin (mw 319) and its nmr spectrum is quite similar to that of tetrodotoxin. However, its structure has not yet been determined. Although the existence of tetrodotoxin is limited to *Atelopus,* a frog native to South America seems to possess another toxin, batrachotoxin, which is more toxic than tetrodotoxin. In addition, an octopus *(Octopus maculosus)* about 10 cm in length living in the waters off Australia possesses a toxin. Bites from this octopus have been known to cause paralysis for a short time, and in some cases death. This toxin was named maculotoxin. Although its structure has not yet been clarified, pharmaceutical research by Freeman[21] showed this toxin to be more similar to tetrodotoxin than to saxitoxin.

The author would like to dedicate this chapter to the late Dr. Kawamura, who participated in research on tetrodotoxin from its early stages.

References

1) R. Tawara, *J. Pharm. Soc. Japan* (in Japanese), **29**, 587 (1909).
2) A. Yokoo, *Bull. Chem. Soc. Japan* (in Japanese), **71**, 590 (1950).
3) K. Tsuda, M. Kawamura, *J. Pharm. Soc. Japan*, **72**, 187, 771 (1952).
4) H. Arakawa, *Bull. Chem. Soc. Japan* (in Japanese), **77**, 1295 (1956).
5) Y. Hirata, H. Kakizawa, Y. Okumura, presented at the 10th congress of Japan Chemical Society(1957).
6) R. B. Woodward, *Pure Appl. Chem.*, **9**, 49 (1964).
7) T. Goto, K. Takahashi, Y. Kishi, Y. Hirata, *Tetrahedron*, **21**, 2059 (1965).
8) K. Tsuda, C. Tamura, R. Tachikawa, K. Sakai, O. Amakasu, M. Kawamura, S. Ikuma, *Chem. Pharm. Bull.*, **12**, 634, 1357 (1964).
9) R. B. Woodward, J. Z. Gougoutas, *J. Am. Chem. Soc.*, **86**, 5030 (1964).
10) Y. Kishi *et al.*, *ibid.*, **94**, 9219 (1972).
11) H. S. Mosher *et al.*, *Science*, **134**, 474 (1964); **144**, 1100 (1964); **140**, 295 (1963).
12) Y. H. Kim, G. B. Brown, H. S. Mosher, F. A. Fuhrman, *ibid.*, **189**, 151 (1975).
13) H. S. Mosher, Y. H. Kim, F. A. Fuhrman, presented at the 34th Congress of the Japanese Chemical Society (1976).

14) T. Noguchi, Y. Hashimoto, *Toxicon*, **11**, 305 (1973).
15) K. Tsuda, M. Kawamura, unpublished data.
16) T. Goto, K. Takahashi, Y. Kishi, Y. Hirata, *Bull. Chem. Soc. Japan* (in Japanese), **85**, 508 (1964).
17) I. Tani, *Research on the toxicity of Fugu in Japanese waters* (*Nihon san fugu no chudokugaku teki kenkyu*) Teikoku Tosho (1945).
18) T. Kokuyama, J. W. Daly, B. Witkopf, *J. Am. Chem. Soc.*, **91**, 3931 (1969).
19) H. A. Rapport *et al.*, *ibid.*, **93**, 7344 (1971); E. J. Schantz *et al.*, *ibid.*, **97**, 1238 (1975)
20) R. E. Moore, P. J. Scheuer, *Science* **172**, 495 (1971).
21) S. E. Freeman, R. J. Turner, *Toxicol. Appl. Pharm.*, **16**, 681 (1970).

Reviews on *Fugu*

i) K. Tsuda, *Kagaku no Ryoiki* (zokan go 80) (in Japanese), p. 9, Nankodo (1967).
ii) A. Yokoo, *Zikken Kagaku Koza* (*in Japanese*) p. 381, Maruzen (1958).

CHAPTER 37

Bioluminescent Substances

When *Cypridina* was ground in cold water, the aqueous solution gave a pale blue light, which gradually faded. A solution extracted from *Cypridina* with hot water did not give light, but when these two extracts were mixed, the mixture gave light again.

This phenomenon can be explained as follows: the cold water extract contained both a luminescent compound and an enzyme which reacted with each other to give light emission; the former was consumed, and finally the enzyme was left. On the other hand, treatment with hot water destroyed the enzyme, and the remaining extract contained only the luminescent substrate. The extracts with hot and cold water, when mixed, gave light by reaction between the enzyme and the luminescent compound.

Similar bioluminescent systems have been found in more than 15 species of bioluminescent organisms, such as the firefly, luminescent bacteria, etc. Such an enzyme and luminescent compound are called in general terms luciferase and luciferin, respectively, and the reaction is called a "luciferin–luciferase reaction" (L–L reaction).[1)] Generally, each bioluminescent organism has its own luciferin and luciferase. Incidentally, there are many organisms having quite different bioluminescent systems, as will be described later.

In order to obtain a complete bioluminescence system, both the luciferin and luciferase must be isolated. Luciferases can be obtained by ordinary extraction techniques for enzymes. In this chapter, therefore, isolation techniques for luciferins are mainly dealt with.

In general, very small amounts of luciferin are present in an individual organism. The material is labile to oxidation by oxygen, and readily decomposes during extraction. Extraction thus requires special precautions for protecting the luciferin from oxidation; it should be performed under hydrogen, argon or carbon dioxide, and/or at low temperature. Oxygen dissolved in the solvent used should be removed by flushing with nitrogen or, if possible, argon before use. Photo-labile compounds should be protected from light and processed as quickly as possible. It is most important to try to obtain raw material with the highest possible concentration of luciferin. Amounts of the luminescent component in an individual organism fluctuate very greatly with the season, the method of collection, and the method of storage. As many bioluminescent organisms produce light emission in response to external stimulation, one should take great care to minimize the consumption of the luminescent compound during collection. Because the luciferins exist in extremely small amounts, a very sensitive detection method is required to monitor their extraction. Instruments for detecting luminescence due to the L-L reaction or chemiluminescence are suitable, or in other cases, fluorescence may be suitable for detecting active species.

37.1 Isolation of *Cypridina* Luciferin

The crustacean *Cypridina hilgendorfii* is a rice grain-shaped species, 2–3 mm in diameter, whose L-L reaction was first found by Harvey (1917) at Shimoda, Japan. This animal does not give *in vivo* bioluminescence but ejects the luciferin and luciferase from the body when stimulated. Organisms completely dried in the sun and stored in a desiccator give light when ground with water. The luminescence system is stable almost infinitely in the absence of water. This suggests that the luciferin and luciferase are stored in separate organs of the body. Many workers have tried to isolate *Cypridina* luciferin from the dried organisms. However, the luciferin **1** was not isolated in crystalline form until Shimomura *et al.* succeeded in crystallizing it in 1957.[2] In 1960, Haneda *et al.* extracted the luciferin directly in good yield from frozen *Cypridina* in dry ice without drying it under the sun. (This method, i.e., freezing with dry ice and storing at −20° or −80° C, is generally useful for isolating luciferins from bioluminescent organisms.)

The poor yields of luciferin from dried *Cypridina* may be due to decomposition during drying in the sun.

[Experimental procedure 1] Isolation and Crystallization of Luciferin 1

Completely dried *Cypridina* (500 g) was pulverized and packed in a

Soxhlet-type extraction apparatus fitted with a dropping funnel and a gas inlet tube, and capable of being evacuated. It was defatted by refluxing with benzene under reduced pressure (ca. 80 mm Hg) for 50 hr. The benzene extract was discarded, wet *Cypridina* was dried *in vacuo* (ca. 12 hr), and then the apparatus was filled with purified hydrogen gas obtained by passing ordinary hydrogen successively through a soda-lime tube, over red-hot metallic copper packed in a silica tube, and through a washing bottle containing conc. H_2SO_4. The extraction was carried out with absolute methanol (1.2 l) under reduced pressure (ca. 80 mm Hg of H_2). After 50 hr, MeOH was evaporated off *in vacuo,* the apparatus was filled again with H_2, and 0.5 N HCl (850 ml) was added to the residual paste. After heating in a boiling water bath for 10 min under an H_2 atmosphere, the yellow solution was cooled to *ca.* 0° C, transferred into a separating funnel, and extracted with ether. The ethereal extract was discarded, the acidic aqueous solution was extracted several times with *n*-BuOH (total volume *ca.* 350 ml), and the BuOH extract was purified by a modification of Anderson's benzoylation method as follows. The butanol extract was benzoylated with benzoyl chloride (3 ml of benzoyl chloride per 100 ml of butanol solution) at 0° C; the orange color faded gradually as the benzoylation of **1** proceeded. After 10 min, the reaction mixture was washed with an equal volume of distilled water and diluted with 8 to 10 volumes of water. The resulting aqueous butanol solution was extracted several times with ether. After concentration of the combined ether extracts *in vacuo,* the apparatus was filled with H_2 and 0.5 N HCl (700 ml) was added to the residual solution. The benzoyl luciferin was then hydrolyzed by heating in a boiling water bath for 30 min under an H_2 atmosphere. After cooling, the acid solution was extracted with ether to remove impurities, and then with BuOH to extract luciferin.

This modified Anderson's benzoylation method was repeated with the BuOH extract. The resulting butanol solution of purified luciferin was neutralized with a saturated $NaHCO_3$ solution and concentrated *in vacuo* to give a reddish residue.

The residue was dissolved in a small amount of MeOH (2 to 3 ml), adsorbed on a chromatography column of cellulose powder, and developed with EtOAc:EtOH:H_2O (5:2:3). Fractions corresponding to a yellow band having yellow flourescence contained luciferin, and were evaporated to dryness *in vacuo*. The residue was extracted with a small amount of MeOH and the extract was evaporated to dryness. A "purified" luciferin was obtained as an orange powder.

The "purified" luciferin was dissolved in dilute MeOH, acidified with conc. HCl, and allowed to stand overnight in a desiccator filled with H_2. Orange-red crystals of luciferin precipitated in a yield of about 3 mg.

The luminescence activity induced by luciferase was 37,000 times that of the starting material, dried *Cypridina hilgendorfii*. Although the absorption spectra (uv/vis) of the purified and crystalline luciferin were nearly identical, the hydrolysis of purified luciferin gave more than 20 amino acids, whereas that of the crystalline material gave only 3 amino acids. It is clear that the crystallization removed a substantial amount of impurities.

Because of the presence of a guanidine group, the luciferin was crystallized in a large excess of HCl, which decreased the solubility of the resulting hydrochloride. It was later found that HBr was better for crystallization. Luciferin hydrochloride is orange-red, but the hydrobromide is, unexpectedly, pale brown.

Kishi *et al.*[3] developed a new method for the preparation of crystalline luciferin. *Cypridina* is frozen with dry ice just after collection and stored under the same conditions. This sample is ground with dry ice in MeOH and filtered. The filtrate is concentrated under reduced pressure and the residual solution is dissolved in dil. HCl. Luciferin is extracted from the solution with butanol after defatting with benzene, and is purified by chromatography as described above. An alumina column can be used instead of the cellulose powder column. Argon is more suitable for protecting the luciferin in the flask and column than H_2 gas, since argon is heavier than air and therefore can isolate the compound from oxygen even in open systems. This method is generally superior, and the yield of luciferin is markedly improved: ca. 20 mg of crystalline luciferin can be obtained from 1 kg of wet *Cypridina.*

The overall scheme of *Cypridina* bioluminescence is as follows

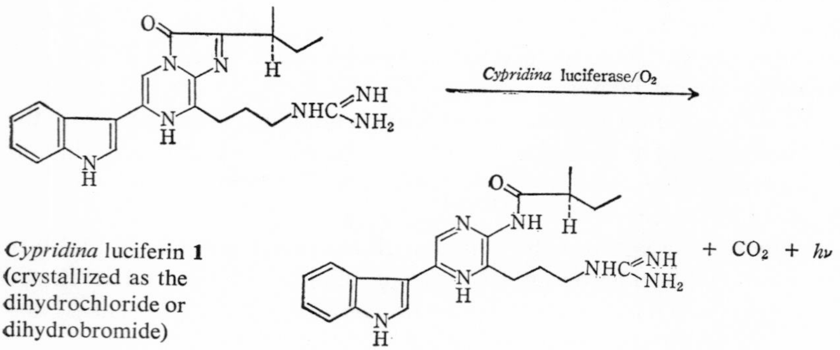

Cypridina luciferase/O_2 →

+ CO_2 + $h\nu$

Cypridina luciferin **1** (crystallized as the dihydrochloride or dihydrobromide)

Cypridina oxyluciferin **2** (crystallized as the dihydrochloride)

37.2 Isolation of Firefly Luciferin

Firefly luciferin **3** was first crystallized from a North American firefly *(Photinus pyralis)* by Bitler and McElroy (9 mg from 15,000 individuals).[4] Kishi *et al.* isolated **3** by a similar method from *genji-botaru,"* a Japanese firefly *(Luciola cruciata)*.[5]

[Experimental procedure 2] Isolation of firefly luciferin 3

Fireflies *(Luciola cruciata)* were frozen with dry ice, and their abdomens were cut off and defatted by grinding in cold acetone followed by filtration. The acetone-defatted and pulverized abdomens (233 g) of *Luciola cruciata* (12,000 individuals) were extracted with hot water. The aqueous extract was acidified with HCl and extracted several times with ethyl acetate. The combined extracts were concentrated and the residue was chromatographed on cellulose powder, eluting with EtOAc-EtOH-H_2O (5:2:3) to yield a green-fluorescent fraction and a blue-fluorescent fraction. The former fraction was found to contain two fluorescent substances. The mixture was further chromatographed on a DEAE-cellulose column, with aqueous NaCl gradient elution. The eluted major fluorescent fraction was acidified with HCl and extracted with ethyl acetate. The ethyl acetate extract was concentrated and the residue crystallized from MeOH to give crystalline luciferin (5.5 mg).

The blue-fluorescent substance was new pterin, luciopterin **7**. The reaction scheme of firefly bioluminescence is as follows.

firefly luciferase
ATP+Mg^{2+}+O_2

+ CO_2 + $h\nu$

firefly luciferin **3** firefly oxyluciferin **4**

37.3 Isolation of Firefly Oxyluciferin (an Emitter)

It has been suggested that the emitter of firefly bioluminescence has the structure **4** on the basis of investigations of the chemiluminescence mechanism.[6] White *et al.* reported that attempted isolation of **4** from a spent solution of luciferin yielded only degradation products of unknown structure.[6] Suzuki *et al.* succeeded in synthesizing pure **4** in spite of its instability.[7] Subsequently, they attempted to isolate **4** from fireflies using the synthetic **4** as an authentic sample for comparison.

[Experimental procedure 3] Confirmation of production of 4 during firefly *in vivo* bioluminescenece

Japanese fireflies *("genji-botaru; Luciola cruciata)*, killed by dipping in liquid N_2 and stored in a deep-freezer (−20° C), were used in the following experiments.

Fireflies (7 females) were allowed to stand for 15 min at room temperature (*ca.* 22° C, light produced) and their lanterns were cut off. After grinding in MeOH cooled by the addition of dry ice, the slurry was filtered and the filtrate was evaporated down *in vacuo* (10^{-5} mm Hg). The residue was chromatographed on Avicel tlc plates (Table 37.1).

TABLE 37.1. *Rf* Values of products of firefly bioluminescence on tlc.†[1]

Solvent	Oxyluciferin 4	Luciferin 3	Dehydro-luciferin 6	Luciopterin 7	Unknown
MeOH–H_2O (1:1)	0.12 B ++†[2]	0.80 YG +	0.60 B ±	0.78 V +++	0.26 B + 0.42 B +
95% EtOH–1 N NH_4OAc (7:3)	0.86 YG +†[2]	0.64 YG ++	0.23 YG +	0.35 V ++	0.96 ±
n-BuOH–HOAc–H_2O	0.75 B +†[2]	0.81 Y ++	0.58 Y ±	0.16 Y ++	0.21 Y + 0.38 B ± 0.45 B ±

†[1] Color of fluorescence is indicated as follows: B, blue; Y, yellow; YG, yellowish green; G, green; V, violet. Symbols (+, ±, etc.) indicate the intensity of fluorescence. Avicel tlc (0.25 mm), detected by irradiation with a fluorescent lamp (366 nm).

†[2] This spot was not detected in fireflies before bioluminescence.

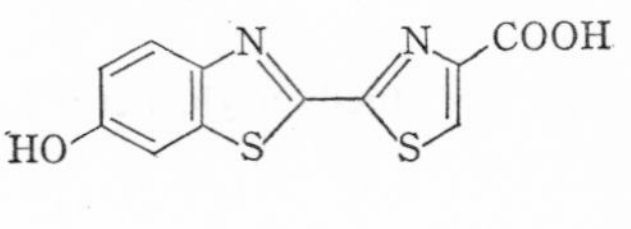

firefly dehydroluciferin **6**

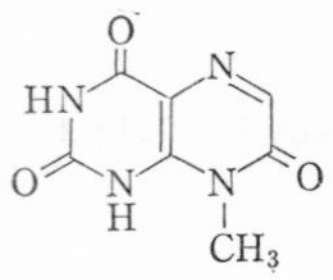

luciopterin **7**

The above experiments were also carried out using fireflies not allowed to stand at room temperature (tlc data, Table 37.1).

The tlc was performed in the dark. Fluorescent spots could be detected by uv irradiation (366 nm), but are photodecomposed by this treatment.

Since the extract was contaminated by a large quantity of fat, only a very small amount of the sample could be applied to the tlc plate; 7 fireflies were the maximum for one tlc plate (20 × 20 cm). In spite of these restrictions, the production of oxyluciferin **4** during *in vivo* bioluminescence was confirmed. Further characterization of **4** was carried out by acetylation followed by isolation of the photostable diacetate **5**.

[Experimental procedure 4] Isolation of a derivative (5) of the product of firefly *in vivo* bioluminescence

Fireflies (100 individuals) were allowed to stand for 15 min at room temperature, then their lanterns were ground in Ac_2O (10 ml) and pyridine (2 ml) under external cooling with dry ice, and the mixture was allowed to stand for 15 min at room temp. The solvents were evaporated off *in vacuo,* the residue was extracted with ether and the extracts were chromatographed on Avicel plates (10 × 20 cm, 1.0 mm thick, 7 plates, MeOH:H_2O (1:1)). The fraction at *Rf* 0.70 was eluted with ether and rechromatographed on silica gel plates (20 × 20cm, 2.2 mm thick, 3 plates, ether:CH_2Cl_2 (10:1)). After repeated chromatographic separations (3 times) on silica gel the spot at *Rf* 0.84 was eluted with ether to give oxyluciferin diacetate **5**. The diacetate was sufficiently stable for isolation and measurements of its uv and ms spectra. The above experiment was repeated using fireflies which had not been allowed to stand at room temperature.

37.4 Bioluminescent Component(s) in Aequorea Photoprotein

A jellyfish, *Aequorea aequorea,* gives light from the edge of the bowl, but does not utilize an L-L reaction system. In 1962 Shimomura *et al.* isolated a protein named aequorin, which gives light simply on addition of a trace of Ca^{++} ions and does not require molecular O_2.[8] The photoprotein must have at least a fluorescent chromophore for the production of luminescence but attempted isolation of such a chromophore was not successful.

However, when a spent solution of the protein after luminescence was treated with HCl, a low molecular fluorescent compound separated out. The fluorescent compound was identified as **9** and named coelenteramide.[9]

The main part of this structure is identical with that of *Cypridina* oxyluciferin, suggesting that their structural changes during bioluminescence are similar. These findings led to the suggestion that the original chromophore in aequorin before luminescence may have the structure **8**.

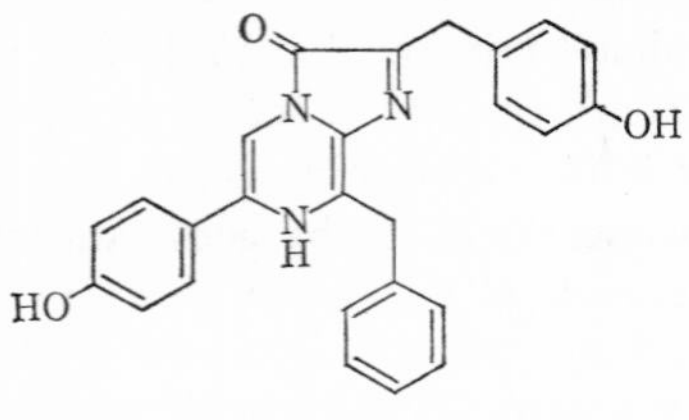

coelenteramide **9**

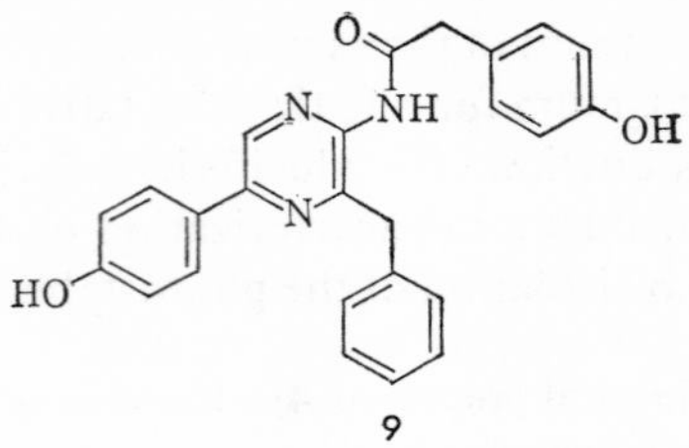

We synthesized compound **8**, which was incubated by Shimomura *et al.* with the spent protein; aequorin was reconstituted in the solution.

37.5 Isolation of *Watasenia* Oxyluciferin (Squid)

Watasenia scintillans is a species of small luminous squid (5 cm trunk) found in Toyama Bay, Japan. A number of tiny lanterns are scattered all over the body, but three large ones 1 mm in diameter are located at the tip of each ventral pair of arms, and can be seen as black spots.

Attempted extractions of the luminescent substance to obtain a photoprotein or L-L reaction system were reported by many workers without success. This may be partly because the squid dies easily, and the luminescence is lost after death. Accordingly, there have been few chemical studies on its luminescence, even though this interesting luminous squid is a single species, which can be obtained in very large quentities.

Consideration of the bioluminescence mechanisms of other luminescent systems suggested that the luminescent product might be fluorescent. Hence, even though the luminous system could not be isolated, the structure of the system might be elucidated by isolating the fluorescent substance, determining its structure, and assuming that the fluorescent substance is a product of the bioluminescence. From this point of view, therefore, Goto *et al.* investigated fluorescent substances in the lanterns of the *Watasenia* squid.[10)]

[Experimental procedure 5] Isolation of *Watasenia* Oxyluciferin 10

Fresh arms of *Watasenia* were frozen and stored in dry ice. The light organs (165 g) of *Watasenia* (ca. 10,000 individuals) were collected using forceps. Portions of 20 g of them were ground with silica sand in a mortar, defatted with ether (30 ml), and extracted with MeOH (40 ml). The residue was extracted with MeOH (40 ml) under reflux (30 min). The combined methanolic extracts were dried by adding BuOH in order to facilitate evaporation. Residual oil was subjected to Avicel tlc once with *n*-BuOH:AcOH:H_2O (3:1:2) and twice with *n*-BuOH:AcOH:H_2O (4:2:1) to give three fluorescent compounds, A (4 mg), B (trace), and C (11 mg).

Compound A has the structure **10**, which is very similar to *Cypridina* oxyluciferin **2** and the *Aequorea* coelenteramide **9** and it appeared to be the product of *Watasenia* bioluminescence. It was therefore named *Watasenia* oxyluciferin.

Compound B is the deacylated product (**11**) of A. The structure of C has not been elucidated.

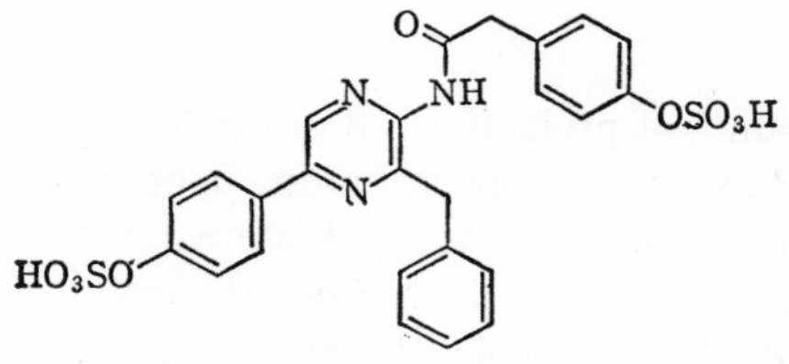

compd. A
Watasenia oxyluciferin **10**

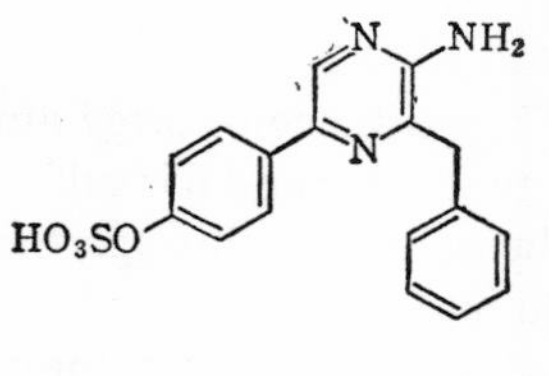

compd. B
Watasenia etioluciferin **11**

37.6 Isolation of *Watasenia* Preluciferin

As described in section 37.5, *Watasenia* oxyluciferin **10**, which appeared to be the product of bioluminescence, was isolated by the investigation of fluorescent substances. However, it was not clear whether *Watasenia* bioluminescence belongs to an L-L or a photoprotein system.

Consideration of the structure **10** and the general mechanism of bioluminescent systems suggested that *Watasenia* luciferin (or the chromophore in the photoprotein) has the structure **12** or a related structure. If such a compound exists in *Watasenia* bodies, it could chemiluminesce in an aprotic polar solvent such as DMSO (addition of a strong base might give a better result). In fact, Inoue *et al.* investigated the chemiluminescence in DMSO of parts of lyophilized *Watasenia*, and found that its liver luminesces strongly, while almost no luminescence was observed in other parts.[11)]

[Experimental procedure 6] Isolation of *Watasenia* preluciferin 8

Lyophilized *Watasenia* livers (200 g) were defatted thoroughly with CH_2Cl_2 and then MeOH:CH_2Cl_2 (1:15). The residue was extracted several times with MeOH and the yellowish-green extracts were dried *in vacuo*. The residue was subjected to a silica gel column chromatography, developing with MeOH:CH_2Cl_2 (1:3). Yellowish-green fluorescent fractions were collected, dried, and then purified by silica gel tlc, developing with MeOH: CH_2Cl_2 (1:7). A yellow band at *Rf* 0.69 was eluted, and gave yellow crystals (43 mg). Repeated tlc and recrystallizations from MeOH gave orange-yellow rods, 15 mg (mp 175–178° C). This compound has the structure **8**, which is not a sulfate, as expected, but a substance assumed to be a precursor (preluciferin).

37.7 Isolation of *Watasenia* Luciferin

The existence of large amounts of preluciferin **8** in the liver suggests either that compund **8** is sulfated to give the true bioluminescent compound after being passed through the blood vessels into the lanterns, or that it is carried into the lanterns after sulfation in the liver. Since the sulfate could not be found in the liver, Inoue *et al.* attempted to extract the sulfate from the lanterns on the ventral pair of arms.[12] The methanolic extracts gave oxyluciferin **10** as described above, but no luciferin **12**. However, the residue gave preluciferin on extraction after being acidified with HCl. This can be explained if luciferin **12** combines with some insoluble material in the residue which is released by acid, and is hydrolyzed at the same time to give preluciferin **8**. It is likely therefore that luciferin **12** exists in an insoluble form in the lantern. We therefore used a base to isolate the luciferin, since an acid might hydrolyze the sulfuric ester.

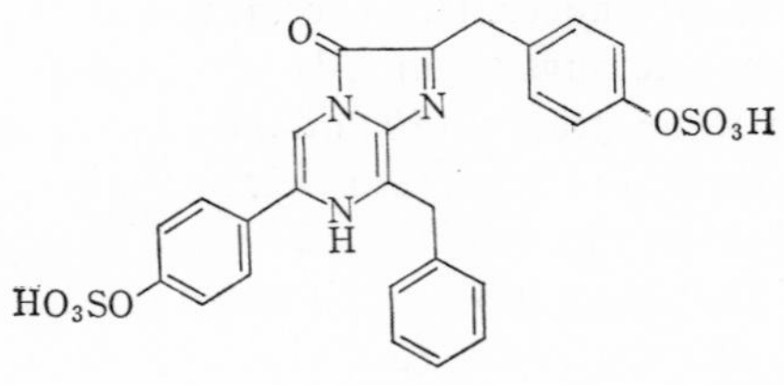

Watasenia luciferin **12**
(Na salt)

[Experimental procedure 7] Isolation of *Watasenia* luciferin 12

Lyophilized arm photophores (2.0 g, ca. 2,500 individuals) were washed with ether in a mortar, and the residue was extracted with oxygen-free MeOH. The extracts contained oxyluciferin **10**. The residue was stirred with MeOH (15 ml) containing 3% NaOMe at room temperature with complete exclusion of oxygen. The extract was neutralized with acetic acid and concentrated *in vacuo*. The residue was chromatographed on a silica gel column, eluting with oxygen-free MeOH:CH_2Cl_2 (1:3), to give a yellow fraction which provided 6.7 mg of a compound. It was further purified by tlc to give a yellow compound (ca. 40 μg), the structure of which was identified by comparison of its *Rf* values on tlc, mobility on electrophoresis, and uv spectrum with those of synthetic **12**. Since chemiluminescent compounds like this are very sensitive to oxygen in general, especially under basic conditions, they must be handled under rigorously oxygen-free conditions.

37.8 Isolation of *Cavernularia* Oxyluciferin (an Emitter)

The sea cactus, *Cavernularia obesa habereri,* belongs to *Coelenterata,* like *Aequorea* and *Renilla,* and has both L-L and photoprotein systems. The bioluminescent substances of the two systems are thought to be similar.

Sea cactus contracts to ca. 15 cm in length during the day and stretches to ca. 50 cm at night. On stimulation, it luminesces brightly. Shimomura *et al.* extracted the emitter as follows.[13]

[Experimental procedure 8] Isolation of the emitter 9 from the sea cactus

Three specimens of *Cavernularia* in a contracted state were cut into slices 3 mm thick, and stimulated to produce bright luminescence by putting them into 100 ml of 0.8 M KCl containing 0.01 M $CaCl_2$. The slices in this solution were repeatedly pressed and squeezed until the luminescence finally died away, and the whole preparation was mixed with 500 ml of MeOH and then filtered. The filtrate was evaporated down to 100 ml of an aqueous solution, then extracted with ether, and the ether layer was re-extracted with 0.025 N NaOH. The aqueous alkaline layer, after washing **4** times with ether, was neutralized with dry ice and extracted with ether. The ether extract containing a blue fluorescent compound was further purified by two successive silica gel tlc procedures, developing with water-saturated ether. The purified compound was identified as coelenteramide **9** from its behavior on tlc, and its uv and mass spectra. The yield was estimated to be ca. 15 μg by uv spectrometry.

References

1) T. Goto, *Bioluminescence* (in Japanese), Kyoritsu Publ. Co. (1975).
2) L. Shimomura, T. Goto, Y. Hirata, *Bull. Chem. Soc. Japan*, **30**, 929 (1957).
3) Y. Kishi, T. Goto, Y. Hirata, O. Shimomura, F. H. Johnson, *Tetr. Lett.*, **1966**, 3427.
4) B. Bitler, W. D. McElroy, *Arch. Biochem. Biophys.*, **72**, 358 (1957).
5) Y. Kishi, S. Matsuura, S. Inoue, O. Shimomura, T. Goto, *Tetr. Lett.*, **1968**, 2847.
6) P. J. Plant, E. H. White, W. D. McElroy, *Biochem. Biophys. Res. Commun.*, **31**, 98 (1968); E. H. White, E. Rapaport, T. A. Hopkins, H. H. Seliger, *J. Am. Chem. Soc.*, **91**, 2178 (1969).
7) N. Suzuki, T. Goto, *Tetr. Lett.*, **1971**, 2021; *Tetrahedron*, **28**, 4075 (1972).
8) O. Shimomura, F. H. Johnson, Y. Saiga, *J. Cell. Comp. Physiol.*, **59**, 223 (1962).
9) O. Shimomura, F. H. Johnson, *Tetr. Lett.*, **1973**, 2963.
10) T. Goto, H. Iio, S. Inoue, H. Kakoi, *ibid.*, **1974**, 2321.
11) S. Inoue, S. Sugiura, H. Kakoi, K. Hashizume, T. Goto, H. Iio, *Chem. Lett.*, **1975**, 141.
12) S. Inoue, H. Kakoi, T. Goto, *Tetr. Lett.*, **1976**, 2971.
13) O. Shimomura, S. Inoue, T. Goto, *Chem. Lett.*, **1975**, 247.

Reviews

i) E. N. Harvey, *Bioluminescence*, Academic Press (1952).

ii) F. H. Johnson, M. Florkin, E. Stots (eds.), *Comprehensive Biochemistry*, p. 79, Elsevier (1967).

iii) F. McCapra, *Accounts Chem. Res.*, **9**, 201 (1976) [chemical mechanisms in bioluminescence].

iv) T. Goto, *Seibutsu Butsuri* (in Japanese), **12**, 69 (1972) [chemiluminescence and bioluminescence].

v) T. Goto, *Kagaku* (in Japanese), **43**, 673 (1973) [bioluminescence of coelenterates].

vi) T. Goto, *Kagaku-no-Ryoiki* (in Japanese), **30**, 569 (1976) [structure determination of luciferins].

vii) T. Goto, in F. Korte (ed.), *Methodicum Chimicum*, pp. 282–287 (1977) [luciferins and luciferases].

CHAPTER 38

Isolation of Cardiotonic Constituents from Japanese Toads

Toads belong to the *Bufo* family of the Anura order of Amphibia and are found throughout the world. The skin secretions of local toads such as *Bufo bufo gargaizans* CANTORON and *Bufo melanostrictus* SCHNEIDER, known as *"Ch'an Su"* in China and as *"Senso"* in Japan, have been employed for centuries as galenical cardiac or diuretic preparations.

Since the first report by Pelletier in 1817, extensive chemical studies on cardiotonic constituents in toad venom have been carried out by many investigators, such as Wieland, Mayer, Kondo, and Kotake and their co-workers, because the material is readily available. It has been demonstrated that toad venom contains cardiotonic steroids and indoleamines, together with peptides, amino acids, fatty acids, polysaccharides, and sterols as minor components.

The cardiotonic steroids having the bufadienolide structure, characterized by an α-pyrone ring at the C-17 position, are divided into two groups, bufogenin and its 3-suberoylarginine ester, bufotoxin. Almost all studies on the cardiotonic steroids in toad venom have been focused on unconjugated bufogenins which are obtainable with relative ease from the natural source. In contrast, the isolation and characterization of conjugated bufadienolides have been less well studied, because of the greater technical difficulty due to their high polarity. Only vulgarobufotoxin **1** and marinobufotoxin **2** have so far been separated in the pure state from the tropical toad, *Bufo marinus*. We collected 1800 Japanese toads, *Bufo vulgaris formosus* BOULENGER, which had gathered for spawning at ponds and swamps in the northeastern district of Japan in early spring. These toads

bufogenin

suberoylarginine

vulgarobufotoxin **1** marinobufotoxin **2**

were sacrificed by freezing in dry ice-acetone. The skins were immediately flayed off and extracted with ethanol (100 l) at room temperature for 3–4 weeks. The ethanolic extract was concentrated and partitioned in an ether-water system and then in an ethyl acetate-water system, yielding the genin and conjugate fractions, and shown in Table 38.1.

TABLE 38.1. Separation of genin and conjugate fractions.

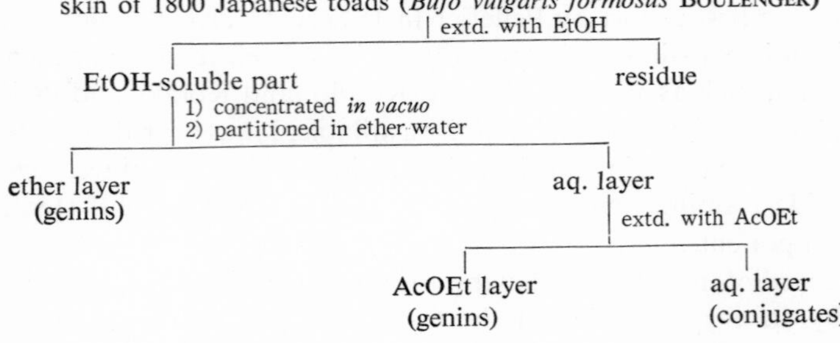
skin of 1800 Japanese toads (*Bufo vulgaris formosus* BOULENGER)
extd. with EtOH
EtOH-soluble part
residue
1) concentrated *in vacuo*
2) partitioned in ether-water
ether layer (genins)
aq. layer
extd. with AcOEt
AcOEt layer (genins)
aq. layer (conjugates)

38.1 ISOLATION OF BUFOTOXINS AND THEIR HOMOLOGS

As a preliminary experiment for the present study, we synthesized several cardiotonic steroid conjugates and their analogs[1,2] and confirmed that these compounds are extractable quantitatively with Amberlite XAD-2 or -4 resin. Based upon these data, the separation of conjugated steroids derived from the skin of toad was carried out as shown in Table

TABLE 38.2. Separation of conjugated bufogenins.

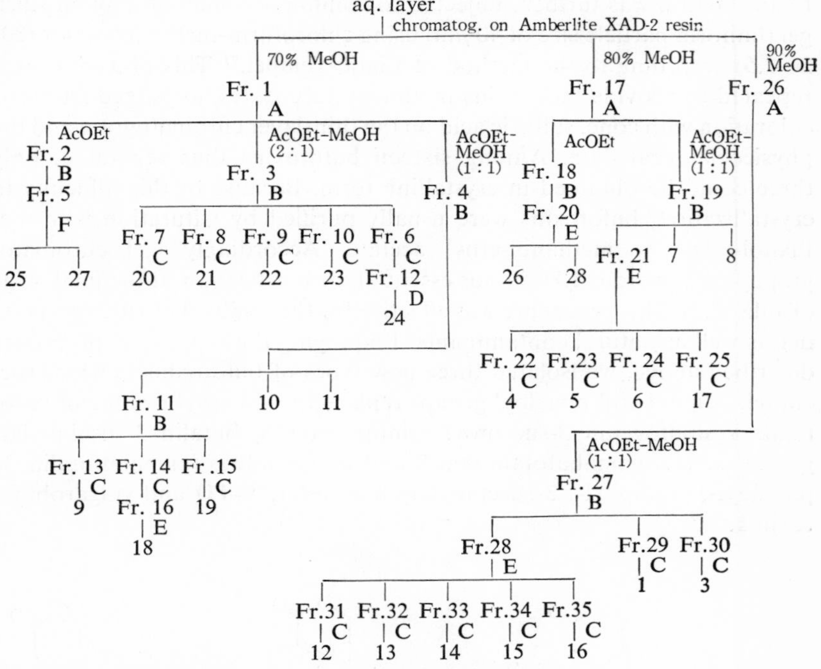

A: chromatog. on silica gel eluting with AcOEt–MeOH.
B: chromatog. on silica gel eluting with $CHCl_3$–MeOH–H_2O (80:20:2.5.).
C: chromatog. on Sephadex LH-20 eluting with MeOH.
D: chromatog. on Sephadex LH-20 eluting with H_2O.
E: high-performance liquid chromatography.
F: preparative tlc on silica gel with $CHCl_3$–MeOH–AcOH (90:5:0.2).

38.2. The aqueous layer was diluted with 2–3 volumes of water and percolated through a column of Amberlite XAD-2 resin. After thorough washing with water, the conjugated steroid fraction was eluted successively with 60%, 70%, 80% and 90% methanol. Indoleamines, bufochromes (pteridine derivatives) and amino acids were obtained by elution with 60% methanol, bufogenin 3-sulfates with 70% methanol and bufogenin hemidicarboxylates with 70–90% methanol. This porous polystyrene resin proved to be suitable not only for the extraction of polar compounds from the aqueous phase but also for reversed phase chromatography, and simplified the subsequent purification procedure. The hemidicarboxylates and sulfates of bufogenins and bufotoxins were subjected to dry column chromatography on silica gel, eluting in a stepwise manner with ethyl acetate, ethyl acetate–methanol (2:1) and ethyl acetate–methanol (1:1). The bufotoxin fraction was further subjected to column chromatography on silica gel (uniform particle size of 40 μm) using chloroform-methanol-water (80:20:2.5) according to the method of Linde-Tempel.[3] This procedure was repeated to provide bufotoxins in almost pure states, as judged from the coloration with conc. sulfuric acid on the thin-layer chromatogram and the physicochemical data. Among sixteen bufotoxins thus separated, only three could be obtained in crystalline form. Because of this difficulty in crystallization, bufotoxins were usually purified by trituration with methanol-ether or methanol-ethyl acetate. Accordingly, gel chromatography on Sephadex LH-20 was essential for purification at the final stage (Table 38.2). This procedure was effective for the removal of silica gel powder as well as natural contaminants. Employing the separation procedure described above, we isolated three new types of bufotoxins in which succinoyl, adipoyl and pimeloyl groups replace the suberoyl residue of bufotoxin, resibufogenin 3-succinoylarginine ester **3**, bufalin 3-succinoylarginine ester **7**, gamabufotalitoxin **8** and its homologs (gamabufotalin 3–pimeloyl-, 3-adipoyl-, 3-succinoylarginine ester, **9–11**) and vulgarobufotoxin **1**.[4–7]

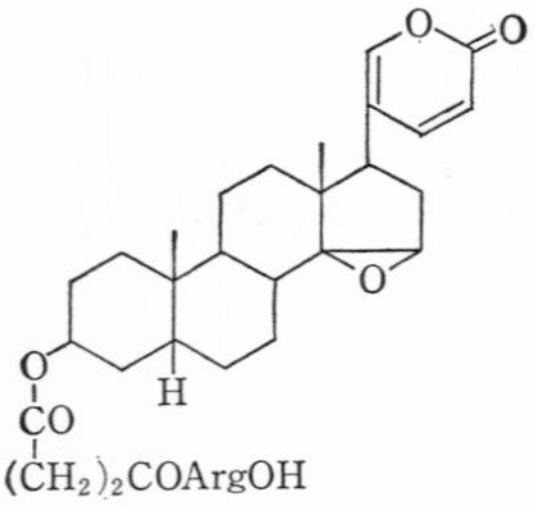

$(CH_2)_2COArgOH$

resibufogenin 3-succinoylarginine ester 3

$(CH_2)_nCOArgOH$

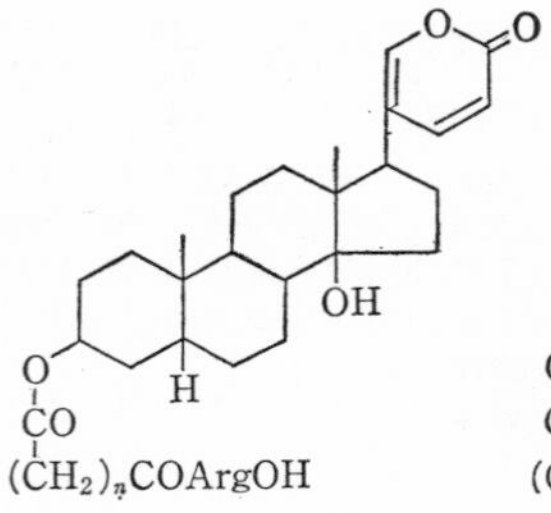

4 ($n = 6$)
5 ($n = 5$)
6 ($n = 4$)
7 ($n = 2$)

bufalitoxin and its homologs

$(CH_2)_nCOArgOH$

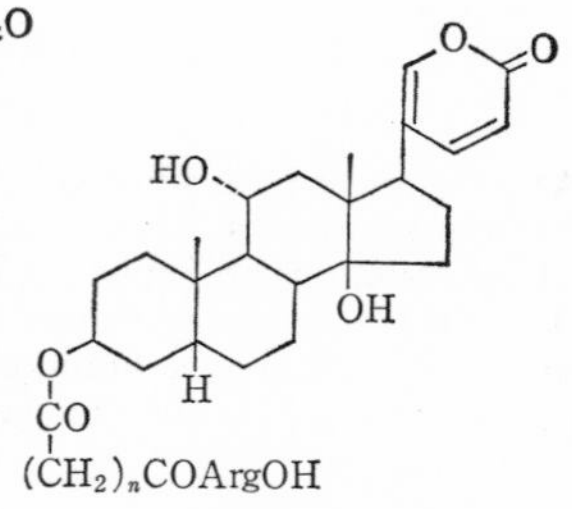

8 ($n = 6$)
9 ($n = 5$)
10 ($n = 4$)
11 ($n = 2$)

gamabufotalitoxin and its homologs

These separation procedures were somewhat tedious and time-consuming, but the use of high-performance liquid chromatography overcame this difficulty. The toad venom was extracted with chloroform-methanol (1:1) and the extract was divided into three fractions (Fractions I-III) by chromatography on silica gel with choloroform-methanol-water (80:20:2.5). Each fraction was subjected to reversed phase chromatography on a μ-Bondapak C_{18} column (1/4″ × 1′) (Waters Assoc., Inc., Milford) using methanol-water or tetrahydrofuran-water as the mobile phase. Twenty

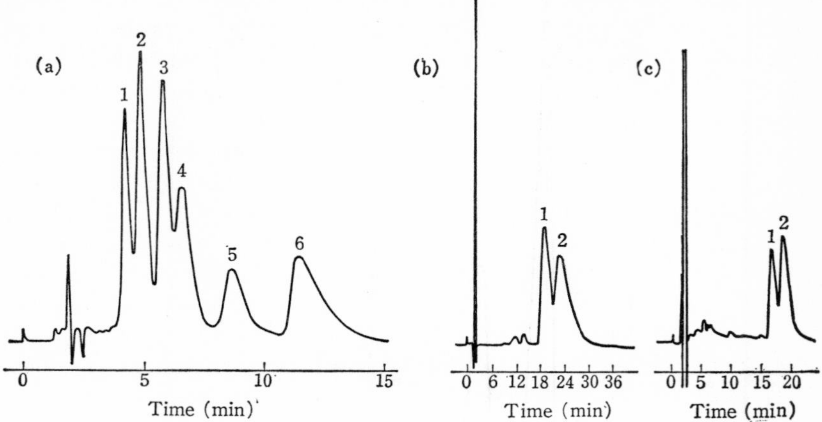

Fig. 38.1. Chromatograms of bufotoxin homologs in the venom of Japanese toads (Fraction I).[8] (a) Conditions: μ-Bondapak C_{18} column, methanol-water (2:1), 1.5 ml/min, 254 nm. 1, Cinobufotalitoxin, marinobufotoxin; 2, cinobufagin 3-succinoylarginine ester; 3, vulgarobufotoxin; 4, cinobufagin 3-adipoylarginine ester, resibufogenin 3-adipoylarginine ester; 5, cinobufagin 3-pimeloylarginine ester, resibufogenin 3-pimeloylarginine ester; 6, cinobufotoxin, resibufotoxin. (b) Conditions: μ-Bondapak C_{18} column, tetrahydrofuran-water (2:5), 2 ml/min, 280 nm. 1, marinobufotoxin; 2, cinobufotalitoxin. (c) Conditions: μ-Bondapak C_{18} column, tetrahydrofuran-water (1:2), 1.5 ml/min, 280 nm. 1, resibufotoxin; 2, cinobufotoxin.

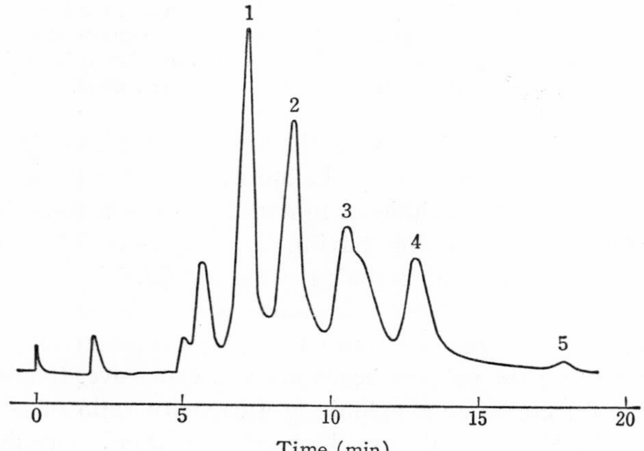

Fig. 38.2. Chromatogram of bufotoxin homologs in the venom of Japanese toads (Fraction II).[8] Conditions: μ-Bondapak C_{18} column, methanol-water (2:1), 1 ml/min, 280 nm. 1, Arenobufotoxin; 2, resibufogenin 3-succinoylarginine ester; 3, bufalin 3-adipoylarginine ester; 4, bufalin 3-pimeloylarginine ester; 5, bufalitoxin.

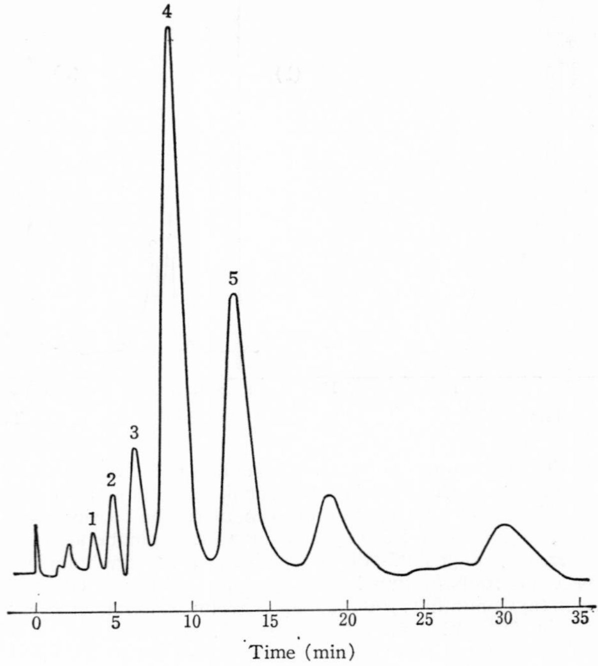

Fig. 38.3. Chromatogram of bufotoxin homologs in the venom of Japanese toads (Fraction III).[8] Conditions: μ-Bondapak C_{18} column, methanol-water (5:4), 1.5 ml/min, 280 nm. 1, Gamabufotalin 3-succinoylarginine ester; 2, gamabufotalin 3-adipoylarginine ester; 3, gamabufotalin 3-pimeloylarginine ester; 4, gamabufotalitoxin; 5, bufalin 3-succinoylarginine ester.

bufotoxins and their homologs were distinctly resolved, as illustrated in Figs. 38.1 through 38.3.[8] Almost all the bufotoxins could be obtained on a preparative scale by this technique. In particular, it was suitable for the separation of bufotoxin homologs having various dicarboxylic acid moieties. After several hours, four homologs were obtained in amounts, (approximately 20 mg each) sufficient for structural elucidation.

Careful clean-up is required prior to high-performance liquid chromatography because the column beads are still expensive. In addition, it is not suitable for compounds exhibiting a capacity ratio of > 10 on a preparative scale. However, the development of new columns should increase the scope of this technique for the separation of biologically active substances from natural sources. For instance, the isolation of cardenolides by gel chromatography has recently been reported.[9] A suitable liquid chromatography system might well give superior results.

The combination of chromatography on Amberlite XAD-2 resin and silica gel, gel filtration and liquid chromatography afforded cinobufotalitoxin **16**, arenobufotoxin **17**, cinobufotoxin **12** and its homologs **13**–**15**, and bufalitoxin **4** and its homologs **5**,**6**.[10] In 1970, Meyer and his co-

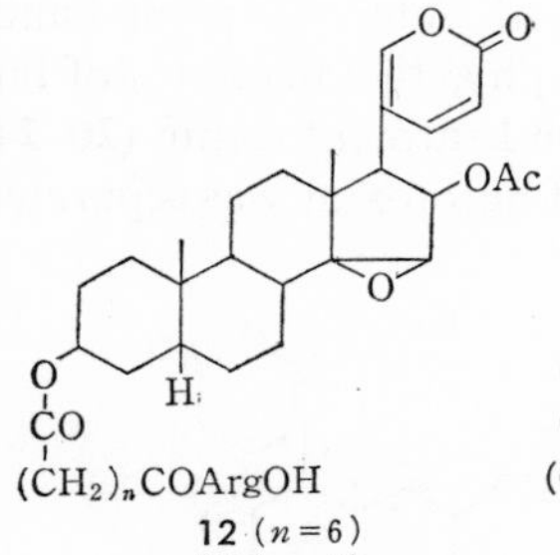

OAc O H O CO $(CH_2)_n$COArgOH
12 ($n=6$)
13 ($n=5$)
14 ($n=4$)
15 ($n=2$)
cinobufotoxin and its homologs

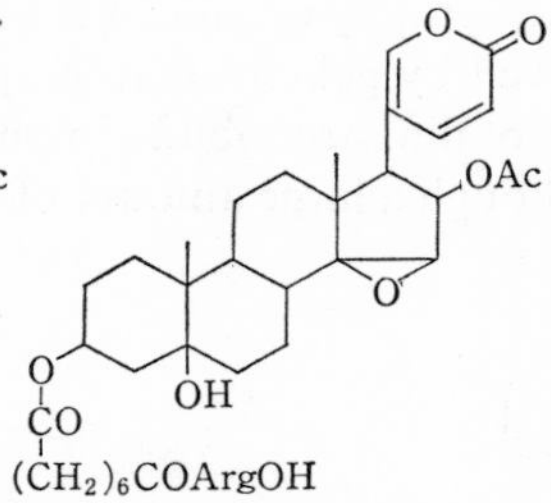

OAc O OH O CO $(CH_2)_6$COArgOH
cinobufotalitoxin 16

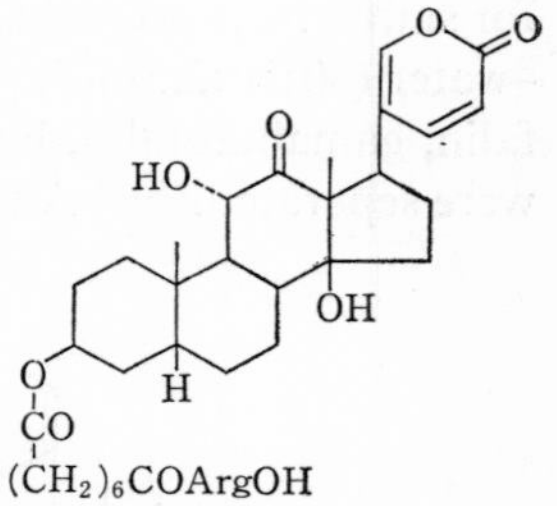

HO O OH H O CO $(CH_2)_6$COArgOH
arenobufotoxin 17

workers demonstrated the presence of cardenolides and their 3-hemisuberates in "*Ch'an Su.*"[11–13] We also sought to isolate cardiotonic steroids, employing Kedde's reaction characteristic for the butenolide ring as a marker, and identified conjugated cardenolides named cardenobufotoxin, **18** and **19**, in living animals.[14]

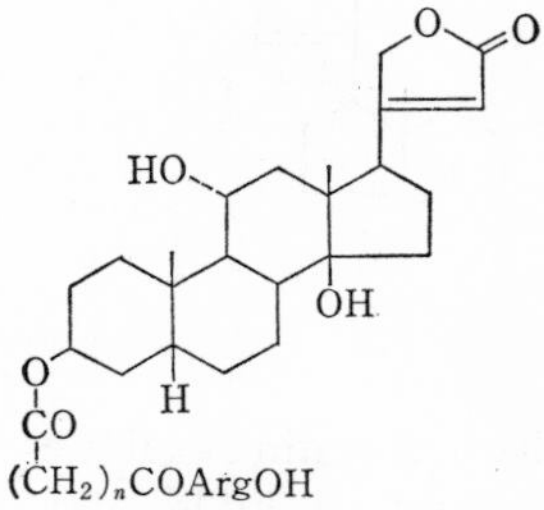

HO OH H O CO $(CH_2)_n$COArgOH
18 ($n=6$)
19 ($n=5$)
cardenobufotoxin

38.2 Isolation of Bufogenin 3-Sulfates

Careful observation of chromatographic behavior led us to find, unexpectedly, bufogenin 3-sulfates in the skin of toads. The eluate obtained with 70% methanol on Amberlite XAD-2 chromatography was further subjected to column chromatography on silica gel using ethyl acetate-methanol (2:1) as a solvent. A new conjugate fraction was eluted before the bufotoxin fraction. These compounds gave lower *Rf* values than bufotoxins on tlc with chloroform–methanol–water (80:20:2.5), were negative to Sakaguchi's reagent and positive to the barium-rhodizonate reagent for sulfate. Upon chromatography on silica gel with chloroform–methanol–water (80:20:2.5) followed by gel chromatography, the 3-sulfates of bufalin, gamabufotalin, bufotalin, arenobufagin and sarmentogenin (**20–24**) were separated.[15,16] Although a large amount of bufotoxins was separated

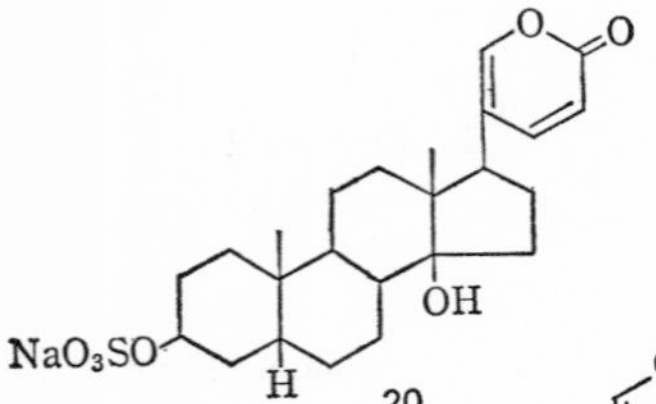

20 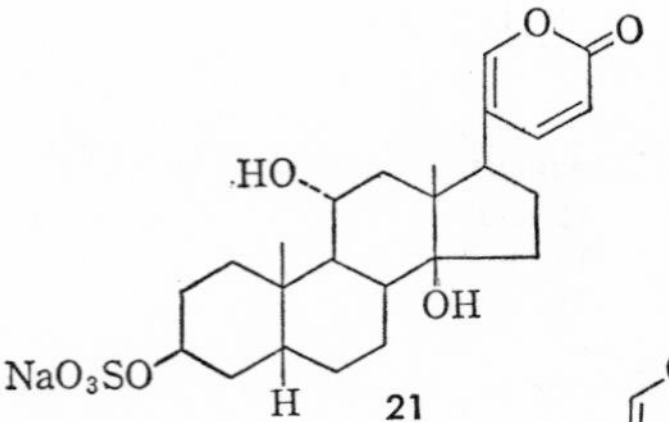 **21** 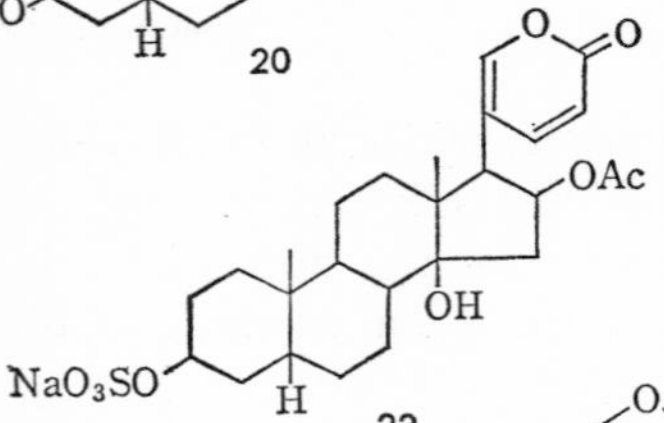 **22** 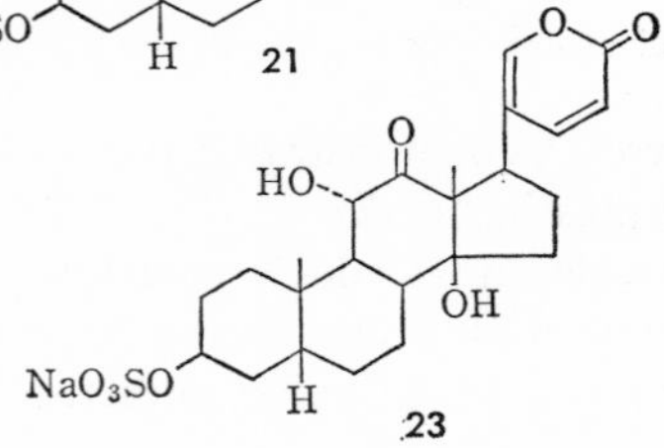 **23** 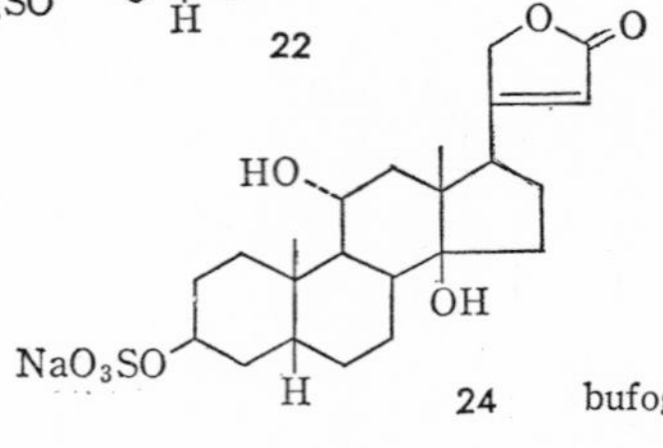 **24** bufogenin 3-sulfates

from the skin of toads, the bufogenin 3–sulfates were obtained in yields of only a few milligrams. The water-soluble sulfates can be readily separated by liquid chromatography (Fig. 38.4),[8] but desalting with an ionization suppressor in the mobile phase should be considered for preparative liquid chromatography.

38.3 Isolation of Bufogenin 3-Hemidicarboxylates

Kamano *et al.*[17] and Höriger *et al.*[12] reported the isolation of bufogenin 3-hemisuberates from *"Ch'an Su"*. In their studies, column chromatography on silica gel and preparative tlc with methylene chloride-me-

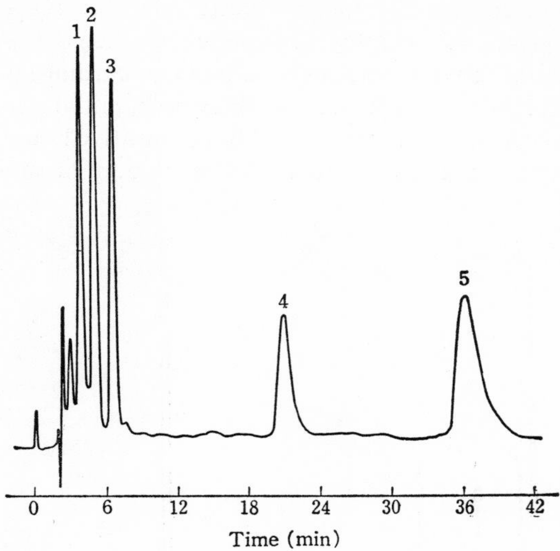

Fig. 38.4. Chromatogram of bufogenin 3-sulfates.[8] Conditions: μ-Bondapak C_{18} column, methanol-0.03 M $NH_4H_2PO_4$ (2:3), 1.5 ml/min, 254 nm. 1, Sarmentogenin 3-sulfate; 2, gamabufotalin 3-sulfate; 3, arenobufagin 3-sulfate; 4, bufotalin 3-sulfate; 5, bufalin 3-sulfate.

HO, OH, H, O, CO, $(CH_2)_6COOH$ **25**

HO, O, OH, H, O, CO, $(CH_2)_6COOH$ **26**

HO, OH, H, O, CO, $(CH_2)_6COOH$ **27**

OH, O, H, O, CO, $(CH_2)_2COOH$ **28**

bufogenin 3-hemidicarboxylates

thanol (19:1) and chloroform–isopropanol–acetic acid (94:5:1) as developing solvents were principally employed. It should be noted that a developing solvent containing acetic acid is not suitable for cardiotonic steroids which have an acid-labile oxido ring. A crude sample, initially purified by chromatography on silica gel, was subjected to preparative liquid chromatography using μ-Bondapak C_{18} (Fig. 38.5).[8] The amount of sample to be

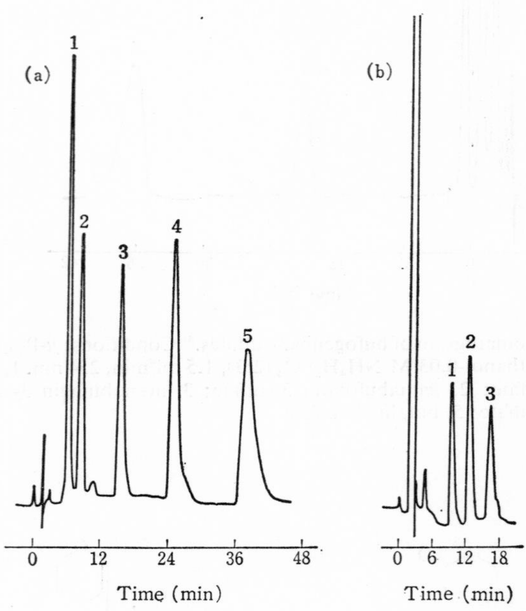

Fig. 38.5. Chromatograms of bufogenin dicarboxylic acid half esters. (a) Conditions: μ-Bondapak C_{18} column, methanol-0.02 M $NH_4H_2PO_4$ (2:1), 2 ml/min, 254 nm. 1, Desacetylcinobufagin 3-hemisuccinate, sarmentogenin 3-hemisuberate, gamabufotalin 3-hemisuberate; 2, arenobufagin 3-hemisuberate; 3, bufotalin 3-hemisuberate; 4, bufalin 3-hemisuberate; 5, cinobufagin 3-hemisuberate. (b) Conditions: μ-Bondapak C_{18} column, tetrahydrofuran-0.02 M $NH_4H_2PO_4$ (2:3), 1.5 ml/min, 254 nm. 1, Desacetylcinobufagin 3-hemisuccinate; 2, sarmentogenin 3-hemisuberate; 3, gamabufotalin 3-hemisuberate.

applied is dependent upon the composition and concentration of the buffer solution. Even though satisfactory resolution is obtained with a small quantity of specimen, leading or tailing often occurs in separation on a preparative scale. By means of liquid chromatography we were able to isolate the 3-hemisuberates of gamabufotalin, arenobufagin and sarmentogenin (**25–27**) and desacetylcinobufagin 3–hemisuccinate **28** from the skin of toads.

38.4 Isolation of Bufogenins

Numerous papers dealing with the isolation of bufogenins from toad venom and *"Ch'an Su"* have appeared.[18–20] We separated resibufogenin **29**, cinobufagin **30**, bufalin **32**, arenobufagin **33** and gamabufotalin **34**

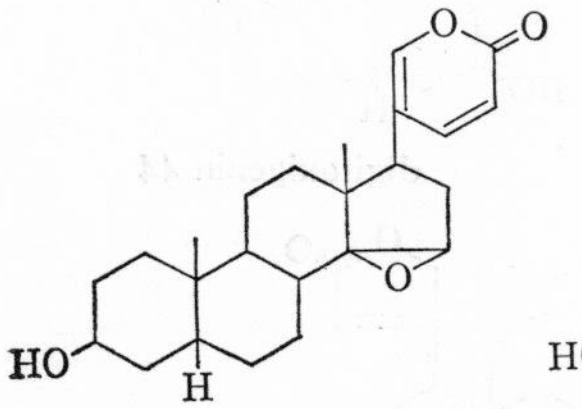

resibufogenin **29**

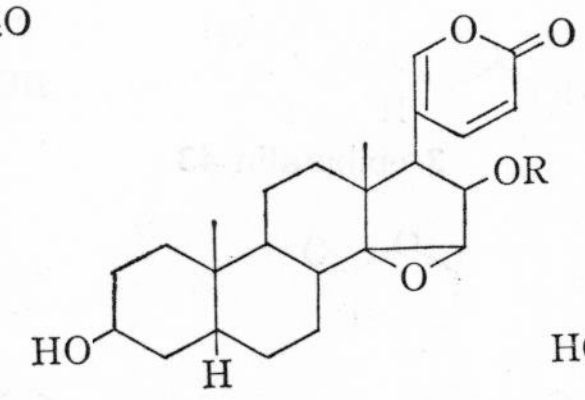

cinobufagin **30** (R=Ac)
desacetylcinobufagin **31** (R=H)

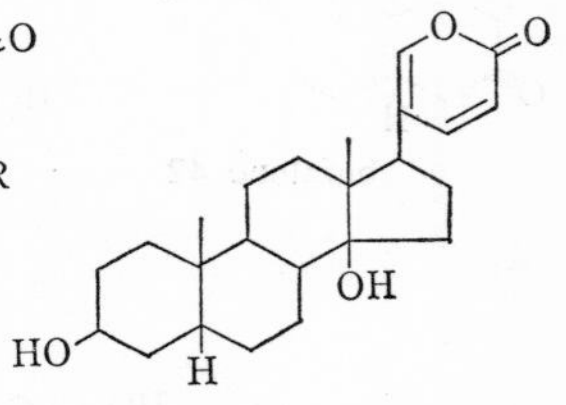

bufalin **32**

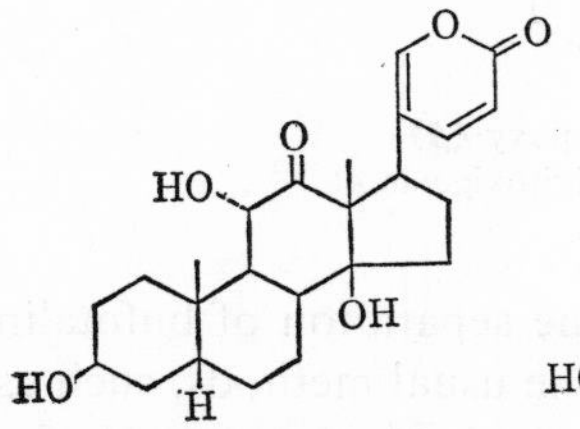

arenobufagin **33**

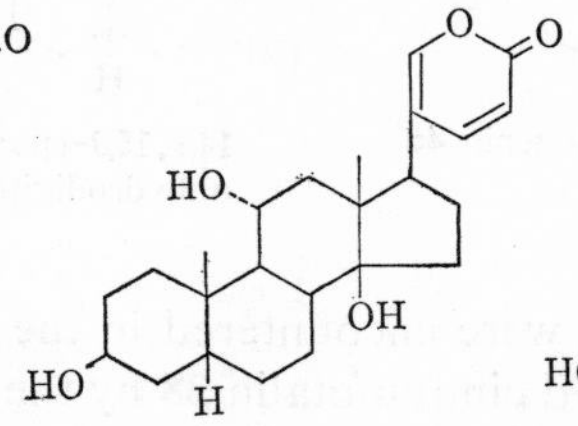

gamabufotalin **34**

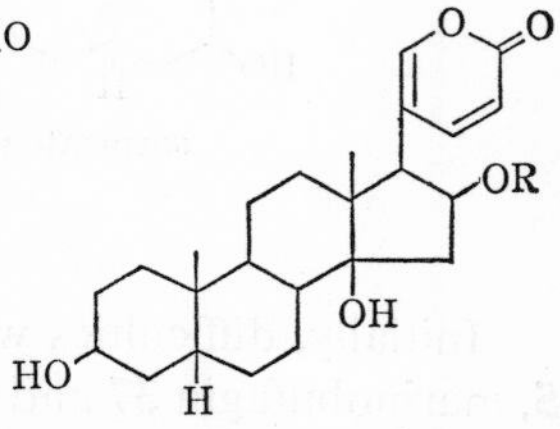

bufotalin **35** (R=Ac)
desacetylbufotalin **36** (R=H)

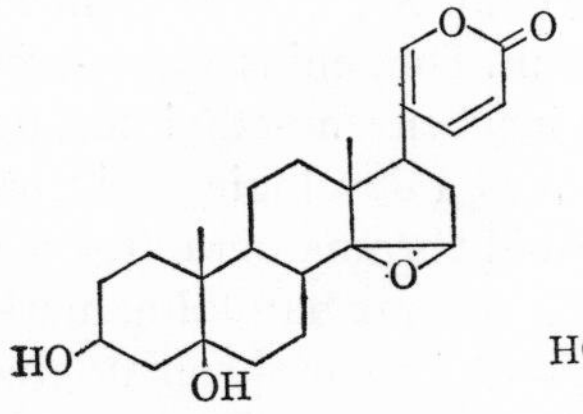

marinobufagin **37**

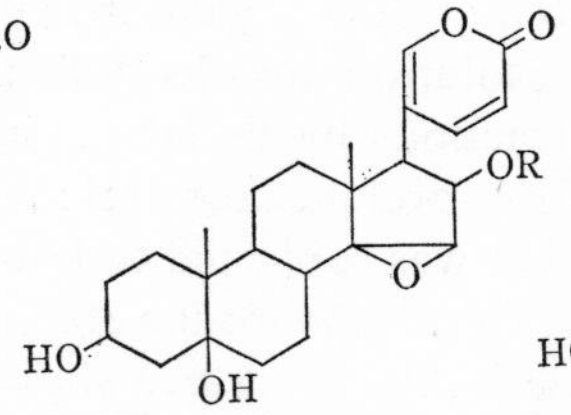

cinobufotalin **38** (R=Ac)
desacetylcinobufotalin **39** (R=H)

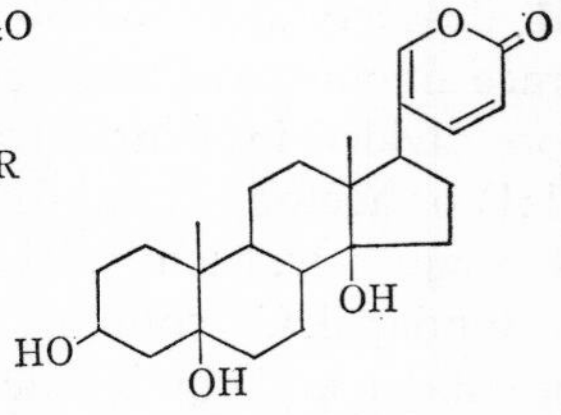

telocinobufagin **40**

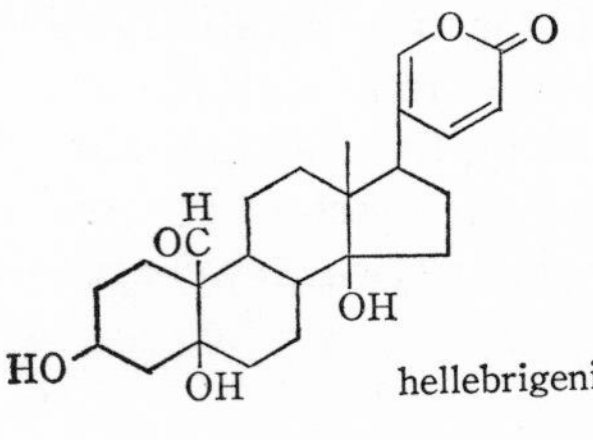

hellebrigenin **41**

from toad venom. In addition, three Kedde-positive cardenolides, digitoxigenin **44**, sarmentogenin **45** and 14β,15β-epoxy-《β》-anhydrodigitoxigenin **46**, were first isolated from the skin of Japanese toads.[14)]

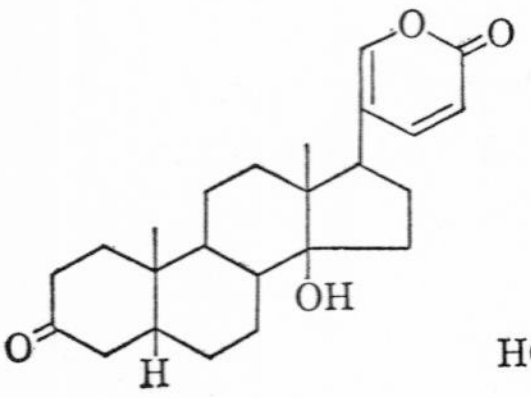

bufalone 42

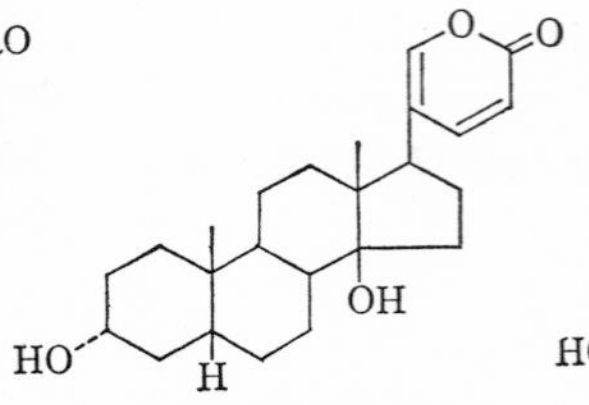

3-epibufalin 43

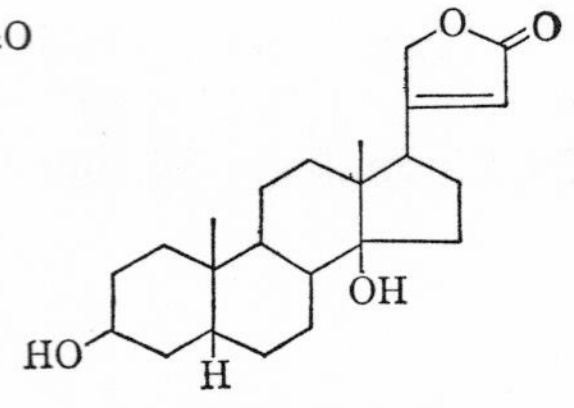

digitoxigenin 44

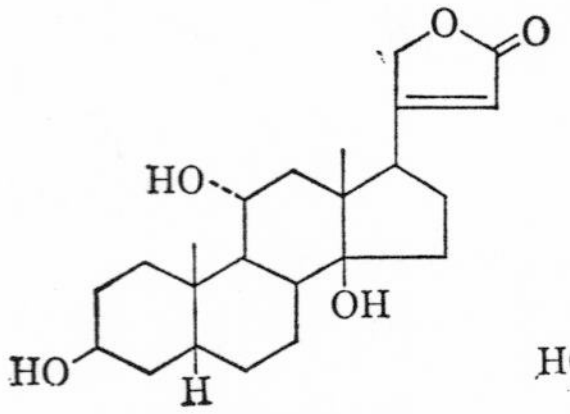

sarmentogenin 45

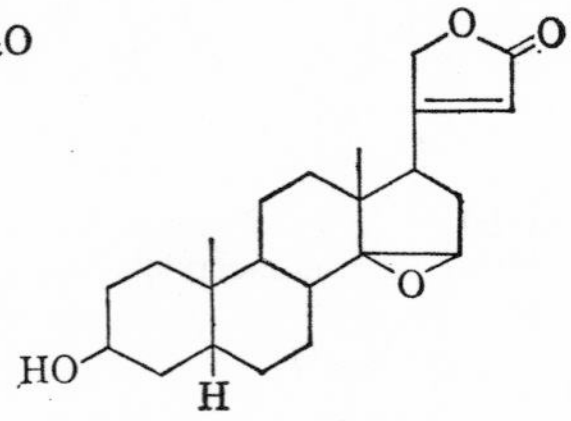

14β,15β-epoxy-〈β〉-anhydrodigitoxigenin 46

Initially, difficulties were encountered in the separation of bufotalin **35**, marinobufagin **37** and cinobufotalin **38** by the usual methods, such as column chromatography and tlc. More polar compounds than gamabufotalin **34**, namely telocinobufagin **40**, desacetylcinobufagin **31**, hellebrigenin **41**, desacetylbufotalin **36** and desacetylcinobufotalin **39**, were present in trace amounts and their resolution was also difficult. The genins were therefore divided into three fractions by tlc employing benzene-ethyl acetate (1:1) or acetone-chloroform-cyclohexane (1:1:1) as a developing solvent. Subsequently, each fraction was subjected to liquid chromatography on a μ-Bondapak C_{18} column (Figs. 38.6 through 38.8). Among bufotalin, marinobufagin and cinobufotalin, the latter two showed similar chromatographic behavior, appearing as a single peak when methanol–water (5:4) was employed as a mobile phase. The use of tetrahydrofuran–water (1:2) was found to be suitable for the separation and was applicable on a preparative scale. On chromatography with methanol-water (5:4) as a mobile phase on the same column, the fraction more polar than gamabufotalin was resolved with a separation factor of $>$ 1.[21)] However, the existance of bufalone **42** and 3-epibufalin **3**, which were reported by Iseli *et al.*,[20)] could not be confirmed.

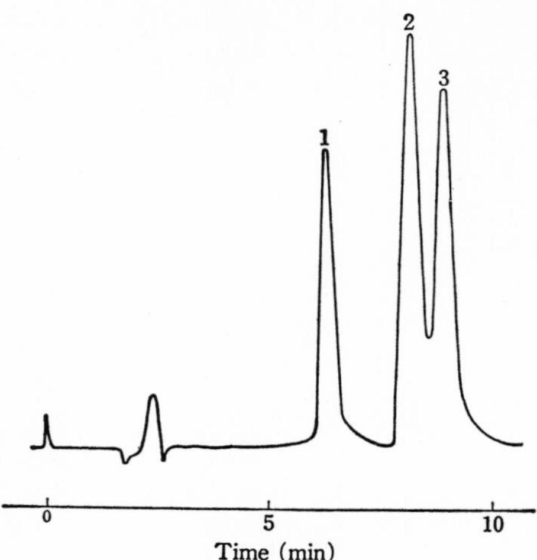

Fig. 38.6. Chromatogram of bufogenins in the venom of Japanese toads (Fraction I).[21] Conditions: μ-Bondapak C_{18} column, acetonitrile-water (1:1), 1.5 ml/min, 280 nm. 1, Bufalin; 2, cinobufagin; 3, resibufogenin.

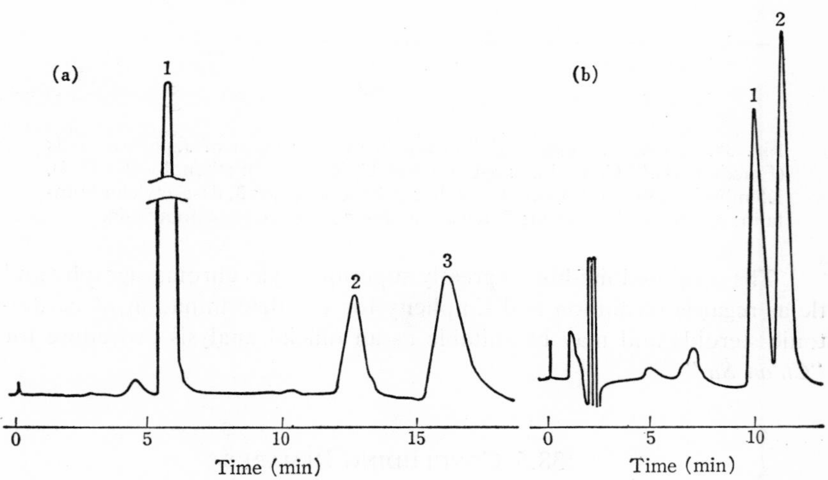

Fig. 38.7. Chromatograms of bufogenins in the venom of Japanese toads (Fraction II).[21] (a) Conditions: μ-Bondapak C_{18} column, methanol-water (5:4), 1.5 ml/min, 280 nm. 1, Arenobufagin; 2, bufotalin; 3, marinobufagin, cinobufotalin. (b) Conditions: μ-Bondapak C_{18} column, tetrahydrofuran-water (1:2), 1.5 ml/min, 280 nm. 1, marinobufagin; 2, cinobufotalin.

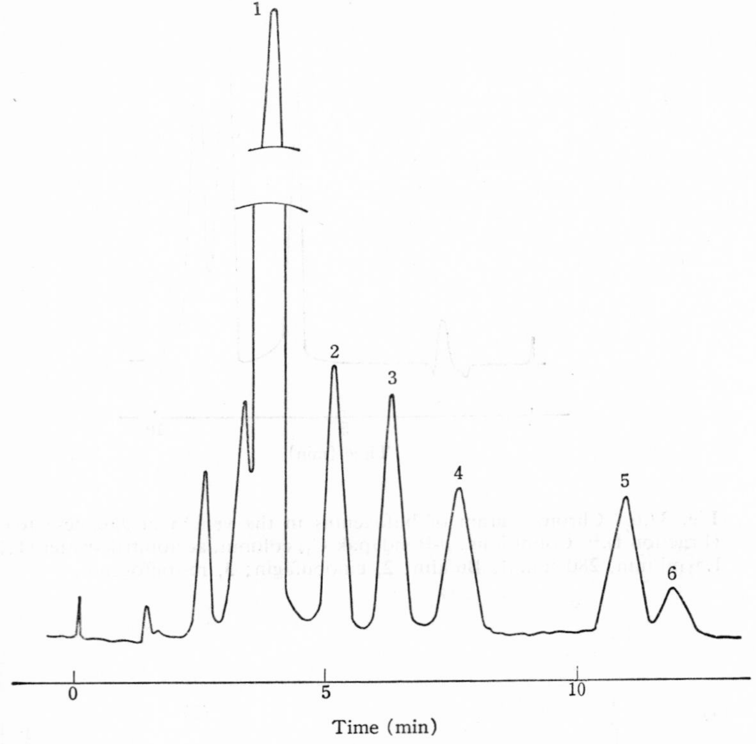

Fig. 38.8. Chromatogram of bufogenins in the venom of Japanese toads (Fraction III).[21] Conditions: μ-Bondapak C_{18} column, methanol-water (5:4), 1.5 ml/min, 280 nm. 1, Gamabufotalin; 2, hellebrigenin; 3, desacetylcinobufotalin; 4, desacetylbufotalin; 5, telocinobufagin; 6, desacetylcinobufagin.

The proposed method is greatly superior to gas chromatography and tlc as regards resolution and simplicity for the determination of cardiotonic steroids and may be suitable as an official analysis procedure for *"Ch'an Su."*

38.5 Concluding Remarks

Even though chemical studies on cardiotonic steroids in toads have continued over a period of centuries, we detected a variety of novel compounds, in particular conjugated bufadienolides and cardenolides, in the Japanese toad by employing modern separation techniques.

It seems very likely that chemistry of natural products will deal increasingly with trace amounts of biologically active substances in the future. The isolation and characterization of new cardiotonic steroids in toad venom provides an interesting example for improvement of the methodology.

References

1) K. Shimada, Y. Fujii, T. Nambara, *Chem. Ind.*, **1972**, 258.
2) K. Shimada, Y. Fujii, T. Nambara, *Chem. Pharm. Bull.*, **21**, 2183 (1973).
3) H. O. Linde-Tempel, *Helv. Chim. Acta*, **53**, 2188 (1970).
4) K. Shimada, Y. Fujii, E. Mitsuishi, T. Nambara, *Chem. Ind.*, **1974**, 342.
5) K. Shimada, Y. Fujii, E. Mitsuishi, T. Nambara, *Tetr. Lett.*, **1974**, 467.
6) K. Shimada, Y. Fujii, T. Nambara, *Chem. Ind.*, **1974**, 963.
7) K. Shimada, Y. Fujii, Y. Niizaki, T. Nambara, *Tetr. Lett.*, **1975**, 653.
8) K. Shimada, M. Hasegawa, K. Hasebe, Y. Fujii, T. Nambara, *J. Chromatogr.*, **124**, 79 (1976).
9) G. Züllich, K. H. Damm, B. P. Lisboa, *ibid.*, **115**, 117 (1975).
10) K. Shimada, Y. Sato, Y. Fujii, T. Nambara, *Chem. Pharm. Bull.*, **24**, 118 (1976).
11) N. Höriger, H. H. A. Linde, K. Meyer, *Helv. Chim. Acta*, **53**, 1503 (1970).
12) N. Hröiger, D. Živanov, H. H. A. Linde, K. Meyer, *ibid.*, **53**, 1993 (1970).
13) N. Höriger, D. Živanov, H. H. A. Linde, K. Meyer, *ibid.*, **53**, 2051 (1970).
14) Y. Fujii, K. Shimada, Y. Niizaki, T. Nambara, *Tetr. Lett.*, **1975**, 3017.
15) K. Shimada, Y. Fujii, T. Nambara, *ibid.*, **1974**, 2767.
16) Y. Fujii, K. Shimada, T. Nambara, *Chem. Ind.*, **1976**, 614.
17) Y. Kamano, H. Yamamoto, Y. Tanaka, M. Komatsu, *Tetr. Lett.*, **1968**, 5673.
18) M. Komatsu, Y. Kamano, M. Suzuki, *Bunseki Kagaku* (in Japanese), **14**, 1049 (1965).
19) M. Komatsu, S. Okano, *ibid.*, **15**, 1115 (1966).
20) E. Iseli, M. Kotake, Ek. Weiss, T. Reichstein, *Helv. Chim. Acta,* **48**, 1093 (1965).
21) K. Shimada, M. Hasegawa, K. Hasebe, Y. Fujii, T. Nambara, *Chem. Pharm. Bull.*, **24**, 2995 (1976).

CHAPTER 39

Extraction and Other Experiments with Bovine Rhodopsin

39.1 Introduction

The vertebrate eye

The inner layer of the vertebrate eye (Fig. 39.1) is called the retina, and is a light-sensitive layer containing optic cells. The peripheral area of the retina consists mainly of the long rod cells responsible for black-and-white vision, whereas the fovea region contains the cone cells responsible for color recognition. The rod and cone cells are linked to the optic nerves, which penetrate the retina and connect to the brain; the point of penetration contains no optic cells and hence is called the blind spot. The rod cells not only far outnumber the cone cells but also are much more sensitive; therefore as dusk sets in we first lose the sensation of color.

The outer segment of the cone cells consists of folded plasma membrane and contains the color visual pigment iodopsin. However, iodopsin is difficult to collect and is far less well studied than rhodopsin, which is present in the rod outer segment (R.O.S.).

In the case of the most intensively investigated rhodpsin, i.e., bovine rhodopsin, the R.O.S. contains ca. 1,500 hollow pancake-like discs, 50 μm thick. The pancake crust is a typical biological membrane consisting of 40% phosphatidylethanolamine, phosphatidylcholine, phosphatidylserine other phospholipids, fatty acid, and cholesterol, 40% protein (85% of which is rhodopsin), and 20% carbohydrate and other components. The dumbbell-shaped mass in the middle-bottom drawing of Fig. 39.1 which

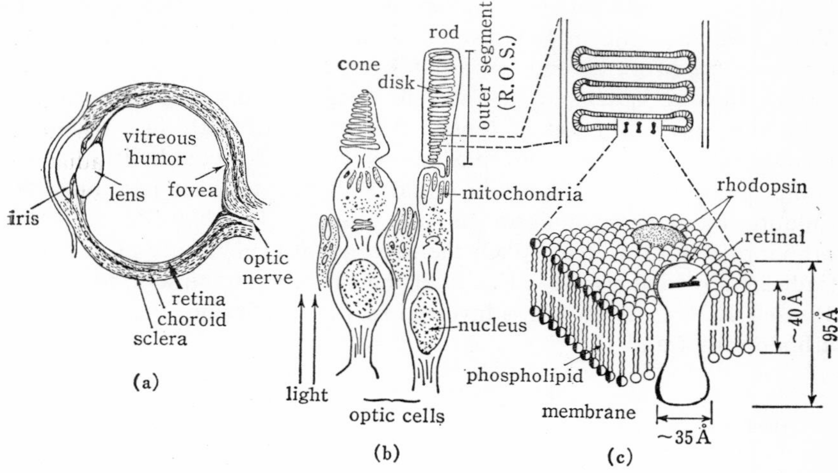

Fig. 39.1. Structure of the vertebrate eye.

penetrates the disk membrane is the visual pigment rhodopsin; the black bar represents the chromophore retinal which lies approximately at right angles to the incident light so that it can absorb the incident light with high efficiency.

Rhodopsin is a hydrophobic protein with an estimated molecular weight of 40,000; it contains saccharide, lipid and steriod moieties which play crucial roles in its function. So far, it has been impossible to obtain rhodopsin in a crystalline form. Until recently, it was thought that the mechanism leading to visual transduction was an outflow of Ca^{++} ions from the disk with a resultant blocking of the Na^{+} permeability and subsequent build-up of electric potential, but the process is now thought to be far more complex and the details are obscure.[1)]

Photobleaching of Rhodopsin

Bovine rhodopsin (which contains 11-*cis*-retinal as the chromophore, λ_{max} 500 nm) is stable at room temperature if kept in the dark, but undergoes a complex series of transformations initiated by light and eventually gives the liberated chromophore as all-*trans*-retinal together with the apoprotein, opsin. This process is called bleaching because the initial orange-red mass is discolored into a pale-yellow mass. The *trans* chromophore is considered to be isomerized back to the 11-*cis* form by the action of an isomerase, which regenerates rhodopsin; however, the details are not known. The early intermediate bathorhodopsin (called "batho" because of its red-shifted maximum at 543 nm) was first observed by Yoshizawa and Kito in 1958[2] upon irradiation of rhodopsin at liq. N_2 temperature. More recently,[3] it has been shown that it is a true intermediate at room temperature as well, being formed within 6 psec. Since light only travels 0.1 nm in 1 psec, this is an extremely fast reaction. Hypsorhodopsin is a new intermediate which was detected by irradiating bathorhodopsin at > 540 nm in liquid He[4] but its exact relation with rhodopsin and bathorhodopsin is still obscure. The half-life of bathorhodopsin is 10 nsec at 30° C; the subsequent changes are all thermal (indicated by solid arrows in Fig. 39.2), the half-life of lumirhodopsin being 18 μsec at 26° C. A proton is involved in the equilibrium between metarhodopsins I and II.[5]

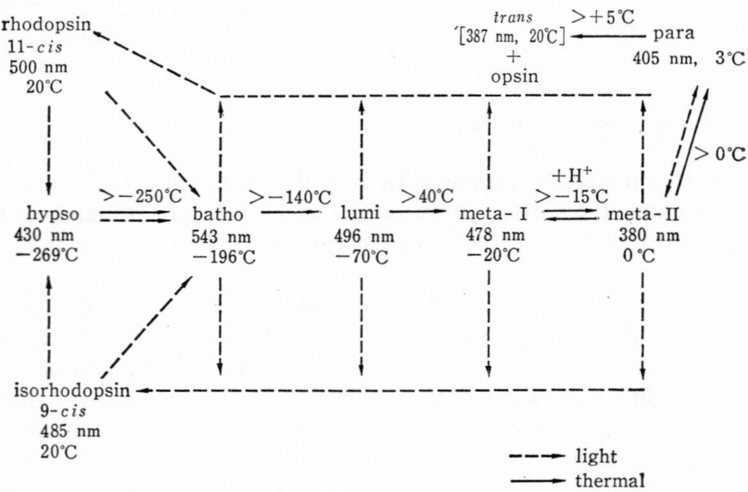

Fig. 39.2. Photobleaching and thermal bleaching of cattle rhodopsin.

Provided that the appropriate temperatures shown in Fig. 39.2 are maintained, it is possible to measure the absorption and circular dichroism (CD) spectra of the various rhodopsins. However, there is still no satisfactory explanation for these shifts. For example, several hypotheses have been put forward to explain the 543 nm maximum of bathorhodopsin.[6] Photoirradiation of batho and other intermediate rhodopsins (indicated by broken arrows) leads to the presence of isorhodopsin (containing 9-*cis*-retinal as the chromophore) in the equilibrium mixture.[7]

Fig. 39.2 summarizes the current status of our knowledge. Many details are unclear. What is meta-II or para (meta-III) ?[8] How does pararhodopsin yield opsin and *trans*-retinal? The changes up to meta-II, which occur at room temperature in the msec time range are directly involved in visual transduction. Subsequent changes are much slower and are probably related to the regeneration of rhodopsin rather than to transduction.

The Binding of Rhodopsin and Retinal

The pioneering studies of Wald and co-workers[8] proved that the rhodopsin chromophore consists of 11-*cis*-retinal. The chromophore in turn is bound to the terminal amino group of a lysine residue via a protonated Schiff base (SBH^+) (Fig. 39.3). Rhodopsin is not reduced by $NaBH_4$, but upon irradiation in an $NaBH_4$ solution the Schiff linkage

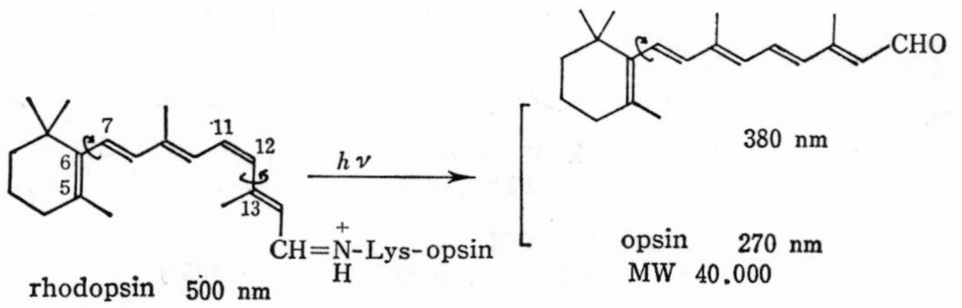

Fig. 39.3. The binding between rhodopsin and retinal.
The arrows around the C-6/C-7 and C-12/C-13 bonds denote that they are not planar.

eventually yields retinyllysine after hydrolysis.[9] When rhodopsin is treated with $NaBH_4$ at 60° C in the dark, however, it gives a product in which retinal is bound to the amino terminal of phosphatidylethanolamine.[10] This suggested that retinal was bound to the lipid and led to considerable confusion. However, it is now clear that retinal is bound to lysine, because retinyllysine was obtained when rhodopsin was treated in the dark with $NaB(CN)_3H$, which penetrates the lipid layer,[11] and because a peak corresponding to $C = NH^+$ stretch was observed at 1655–1660 cm^{-1} in the resonance laser Raman spectrum of rhodopsin.[12] The hypothesis advanced by Lugtenberg[6] that rhodopsin contains a simple Schiff base link rather than a protonated one, and that bathorhodopsin has a retro-retinal structure resulting from proton translocation is in conflict with these experimental results; a conclusive result could be obtained by means of studies with [C-5–CD_3] retinal.

The λ_{max} of a protonated Schiff base (SBH^+) formed between retinal and butylamine is only at 440 nm and not at 500 nm. What is this shift due to? Theoretical calculations show that if the counteranion of the

NH+ group is separated from NH+ in a vacuum by some distance, a value exceeding 500 nm can be attained. However, there is no experimental evidence for this.

The Conformation of 11-*cis*-Retinal in Solution

We will first consider the conformation of free 11-*cis*-retinal in solution. Two rotational isomers, "s-*trans*" and "s-*cis*" are conceivable around the C-6/C-7 and C-12/C-13 single bonds due to steric hindrance. X-ray crystallographic measurements have shown that crystalline 11-*cis*-retinal adopts conformations **2** (Fig. 39.4), i.e., 6-s-*cis* (ϕ_{6-7} 40°)/12-s-*cis* (ϕ_{12-13} 39°);[14] *trans*-retinal also favors a 6-s-*cis* conformer **1** with ϕ_{6-7}

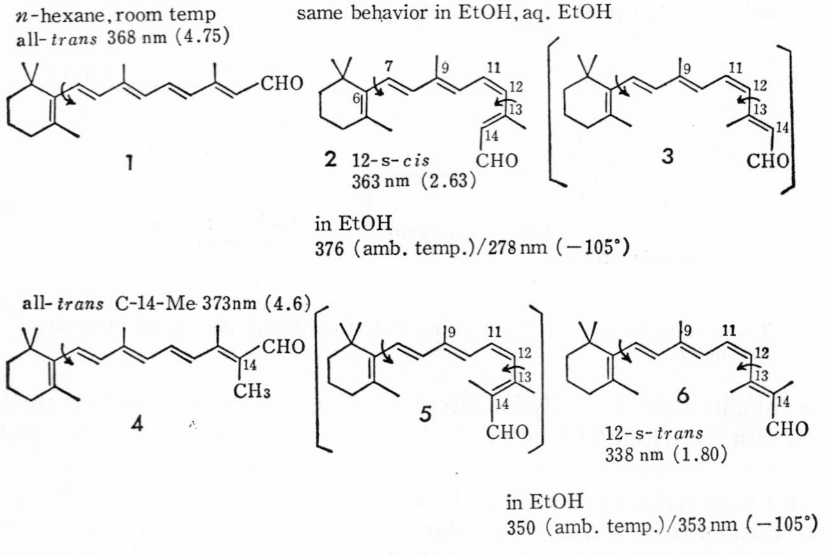

n-hexane, room temp
all-*trans* 368 nm (4.75)
same behavior in EtOH, aq. EtOH

1 CHO

2 12-s-*cis* CHO
363 nm (2.63)

3 CHO

in EtOH
376 (amb. temp.)/278 nm (−105°)

all-*trans* C-14-Me 373nm (4.6)

4 CHO CH3

5 CHO

6 12-s-*trans* CHO
338 nm (1.80)

in EtOH
350 (amb. temp.)/353 nm (−105°)

Fig. 39.4. Conformation of 11-*cis*-retinal in solution. Arrows denote non-planarity.

59°.[15] If the 11-*cis* isomer were to adopt the 12-s-*trans* conformer **3**, theoretically it can be predicted that 11-*cis*-retinal should absorb 30 nm towards the blue in comparison with the all-*trans* isomer; however, as shown in Fig. 39.4 the 11-cis isomer absorbs only 5 nm towards the blue (363 nm *vs.* 368 nm, in hexane), and hence it was predicted that in solution, the 11-*cis*-retinal is 12-s-cisoid.[16]* Nuclear magnetic resonance (NMR)

* The prediction that the λ_{max} of **2** is red-shifted in comparison with that of **3** can be understood qualitatively in terms of the fact that the absorptions of homoannular *cis*-dienes are red-shifted from those of the corresponding hetero-annular *trans*-dienes.

measurements in acetone revealed nuclear Overhauser enhancements between 10-H and 13-Me as well as 14-H; this indicates that there is an equilibrium between **2** and **3**.[16,17]

In order to investigate this point, 14-methylretinal containing an extra methyl at C-14 was prepared.[18] As shown in Fig. 39.5 the retinals were synthesized by reactions empolyed in carotenoid synthesis, such as the Emmons, Witting and Reformatsky reactions;[19] a synthesis employing $TiCl_4$ has also been reported.[20]

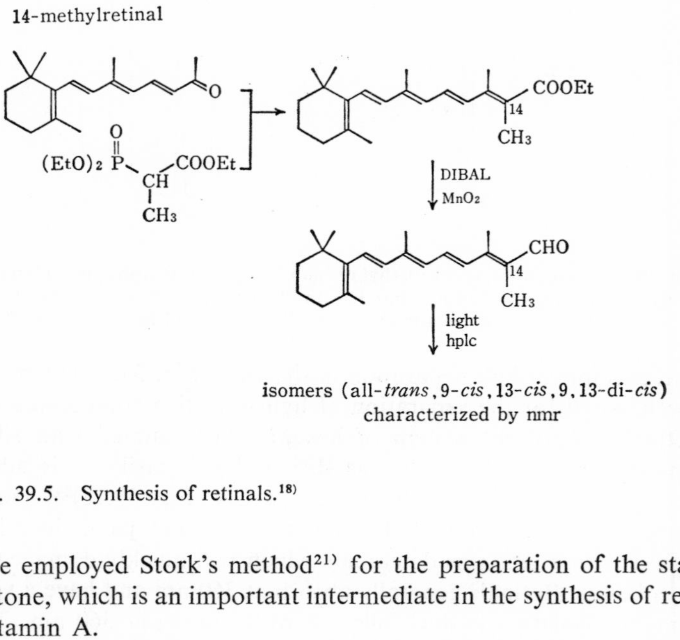

14-methylretinal

$(EtO)_2P(O)CH(CH_3)COOEt$

COOEt 14 CH_3

DIBAL MnO_2

CHO 14 CH_3

light hplc

isomers (all-*trans*, 9-*cis*, 13-*cis*, 9, 13-di-*cis*) characterized by nmr

Fig. 39.5. Synthesis of retinals.[18]

We employed Stork's method[21] for the preparation of the starting C_{18}-ketone, which is an important intermediate in the synthesis of retinals and vitamin A.

One of the objectives of our stuides is to clearify the steric environment of the opsin binding site. One method to achieve this is to synthesize retinals which have predictable steric structures, to bind them to opsin and to study the spectroscopic and other properties of rhodopsins thus formed. In general, the most stable all-*trans* form is synthesized, then this is photoisomerized to give the various double bond isomers which are separated by hplc, and the separated geometrie isomers are identified by high-field (200 MHz or higher) NMR (see Fig. 39.10). Since retinals are readily ozidized and tend to isomerize at room temperature, the pure isomers were stored under argon in a sealed tube in the cold (−60° C) and dark.

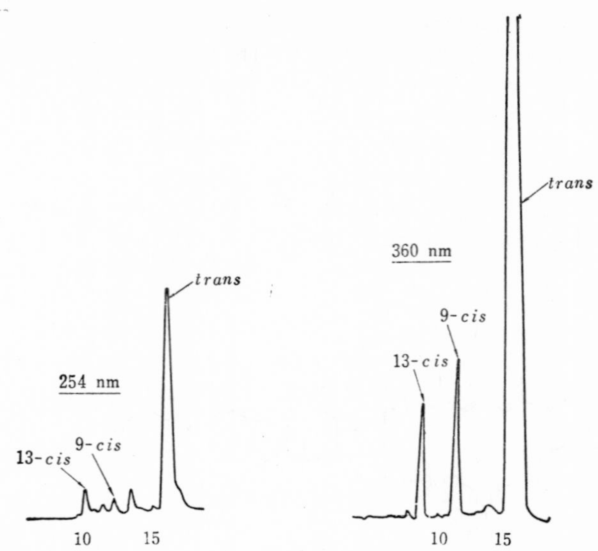

Fig. 39.6. Separation of 14-methyl retinals by hplc illustrating the differences between detections at 254 nm and 360 nm. [μ-Porasil (2 ft $\times$ 2 ft analytical); 2 ml/min; 1% Et_2O in hexane (Waters 201/Schoeffel).] See also Fig. 39.18.

An example of hplc separation is shown in Fig. 39.6. The method is very well suited for the separation of light-sensitive compounds such as the retinals. A solvent system of hexane-ether coupled with silica gel packing materilas such as Corasil II[18)] and μ-Porasil[22,23)] is adequate. However, a reverse phase column, e.g., μ-CN (Fig. 39.18)[24)] should be used when separating retinals beleached from rhodopsin, since in such cases the solution contains detergents which will quickly damage normal phase columns. Retinals generally absorb at 360 nm and have a valley at 254 nm; the isolation is thus followed by scanning at 360 nm. The advantage of scanning at 360 nm is demonstrated in Fig. 39.6. The same amounts of retinals were injected in both cases, but when the isolation is monitored at 254 nm, it is seen that the non-retinal contaminant peaks are relatively strong. A major drawback in studies with model retinals is the difficulty in securing a sufficient quantity. In the case shown in Fig. 39.6, it would take 4 days to collect 2 mg of each of the 14-methylretinals even with a semi-prep 3/8 inch column. As yet there is no efficient method to collect 50 mg lots of pure isomers.* Furthermore, since photoisomerization

* The Waters prep LC-500 can be used for separating 1 mg to several g quantities, provided proper care is taken. However, HPLC is still the ultimate process for final purification.

gives a complex mixture, it is better to separately prepare the desired geometric isomer. In the case of 14-methylretinal (Fig. 39.5), the desired 11-*cis* isomer was synthesized from an appropriate intermediate since the photoisomerization yield was unsatisfactory. The final stages of syntheses, separations, and characterizations of model retinals must be carried out under red light and require experience and patience. It is not unusual for the synthesis of a particular retinal to take a year or longer.

The 11-*cis*-14-methylretinal cannot adopt a 12-s-*cis* conformation **5** (Fig. 39.4), and its λ_{max} was indeed 35 nm blue-shifted relative to the all-*trans* isomer **4**, as predicted.[18,23] The same tendency was seen in aqueous ethanol. The maximum of **6** is at 338 nm, while that of **2** is at 363 nm; the difference of 25 nm is too large to be attributed to the effect of one methyl group and therefore must be due to a difference in conformation. Namely, in solution, the preponderant conformer of 11-*cis*-retinal is 12-s-*cis* **2** whereas that of 11-*cis*-14-methylretinal is 12-s-*trans*.

The Conformation of 11-*cis*-Retinal in Rhodopsin

The synthetic 11-*cis*-methylretinal fortunately binds with cattle opsin[18,23] to yield a rhodopsin absorbing at 502, 340 and 280 nm (in 2% 2% Ammonyx LO*, 67 mM phosphate buffer, pH 8.0); these maxima are

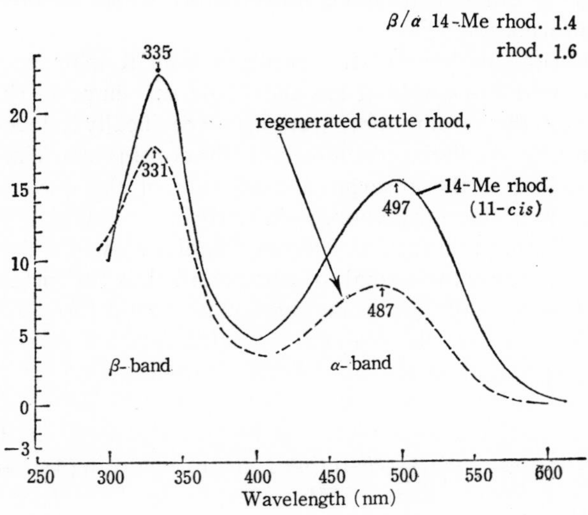

Fig. 39.7. Circular dichroism (CD) spectra of rhodopsin and 14-methyl-rhodopsin in 2% Triton X-100 at room temperature.

* A nonionic micellar detergent consisting of a 2:1 mixture of *N*-oxido-lauryl-dimethylamine and *N*-oxidotetradecyldimethylamine.

similar to those of genuine rhodopsin at 498, 350 and 280 nm.[23] The origin of the two Cotton effects in the CD (Fig. 39.7) is still unknown, but it seems reasonable to consider them to be due to two causes: (a) coupling[25] between the retinal chromophore and an aromatic amino acid within the binding site and (b) a chiraly distorted retinal chromophore. In any event, the CD undoubtedly reflects the rhodopsin conformation. The CD spectra measured in Ammonyx LO[23] or in 2% Triton X-100[18]*1 demonstrate that 14-methylrhodopisn and natural bovine rhodopsin have very similar conformaions. As mentioned earlier (Fig. 39.4), the 14-methyl group prevents 11-*cis*-retinal from adopting the 12-s-*cis* conformation **5** and confines it to the 11-s-*trans* conformation **6**, and thus we conclude that the 11-*cis*-retinal moiety in natural rhodopsin also has a 12-s-transoid conformation.*2

Since rhodopsins are insoluble in water, various detergents are employed for spectroscopic measurements. However, it is important that comparisons be made in the same detergent solutions, as it has been shown that the CD curves[26] and photobleaching products[22,26,27] are dependent on the detergent. A single photon induces the simultaneous isomerization of two double bonds when rhodopsin is excited with a 300 nsec pulse laser in the nonionic detergent sodium desoxycholate; this can be interpreted as being due to different retinal conformations, which in turn reflect the protein conformation.[26,28]

Subsequent experiments with resonance laser Raman spectroscopy[29] also suggest that 11-*cis*-retinal has the 12-s-*trans* shape in rhodopsin.[12] In this method, the chromophore (retinal) is specifically induced to scatter incident light by irradiating the molecule (rhodopsin) with a laser having a wavelength close to the absorption maximum of the chromophore. In the present case, since the apoprotein absorbs at 280 nm whereas the retinal chromophore absorbs at 500 nm, irradiation with a 568.2 nm Kr^+ laser will induce only the retinal to scatter, and thus this procedure gives the Raman spectrum of the chromophore in its bound form. Since rhodopsin is photosensitive, the Raman measurements are carried out with the aid of a special device so that the same molucule is not hit twice by light.[12]

The Raman spectroscopy results are depicted in Fig. 39.8. The bottom run is that for crystalline 11-*cis*-retinal, which, according to X-ray results, has a 12-s-cisoid conformation with a twist angle of 39°. For the sake of convenience, we will assign the 1018 cm^{-1} peak to the C–CH_3 stretch of

*1 A nonionic micellar detergent (polyoxyethylene ($n = 9 - 10$) octylphenol).

*2 The 11-*cis*-retinal cannot adopt a planar 12-s-*cis* or *trans* conformation because of steric hindrance. The *cis* and *trans* designations simply mean that the 12-s-bond is inclined towards a cisoid or transoid shape.

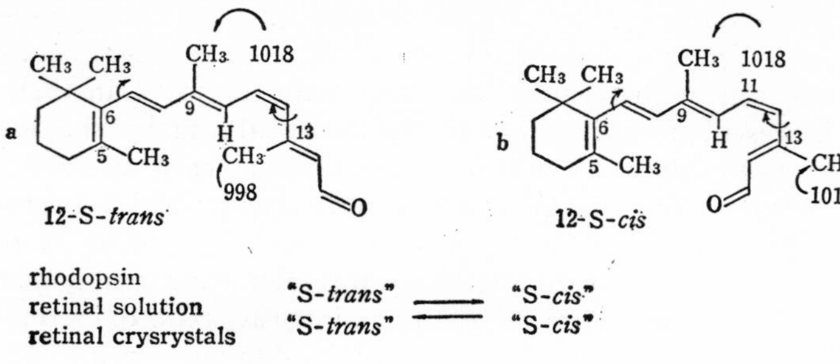

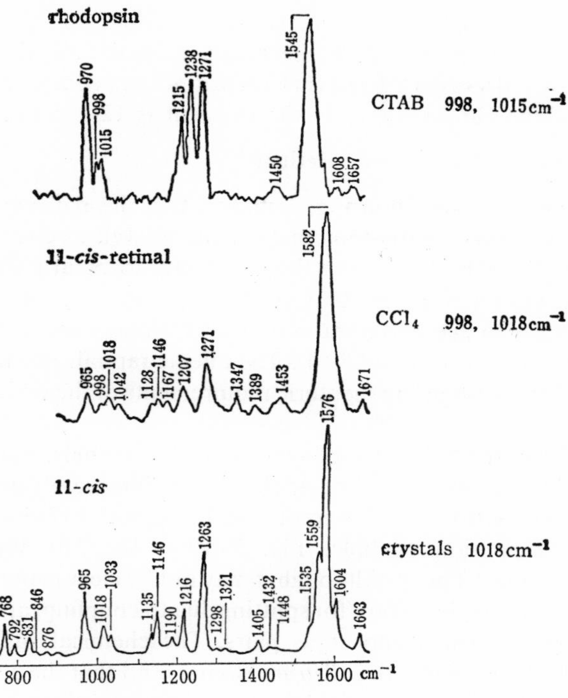

Fig. 39.8. Resonance laser Raman spectra.

methyls which are "outside", namely to 9-Me in the 12-s-*trans* and the 13-Me in the 12-s-*cis* forms; the 965 cm^{-1} band is attributed to a C–H bending mode[30] and can be disregarded for the present purposes. In CCl_4, a weak peak is present at 997 cm^{-1} in addition to the 1018 cm^{-1} peak. This can be interpreted as follows: a portion of 11-*cis*-retinal exists in the 12-s-*trans* conformer and therefore a band appears at 998 cm^{-1}, which is assigned to a C–CH_3 stretch of the "inside" methyl group of 11-*cis*-12-s-*trans*-retinal. In rhodopsin, as measured in CTAB*, it is interesting to note that the 1015 and 998 cm^{-1} bands are present in an intensity ratio of ca. 1:1. If we assume that the intensities of the "outside" 9-Me and "inside" 13-Me bands are approximately equal, these Raman results, *i.e.*, the 1:1 ratio, suggest that the 11-*cis* retinal moiety in rhodopsin adopts exclusively the 12-s-transoid form. Retinals containing 9-CD_3, 13-CD_3 and 9, 13-$(CD_3)_2$ are being prepared to check this deduction. If the assignments of the 1018 and 998 cm^{-1} bands are correct, we can conclude that the conformation of the chromophore in the pigment is 12-s-transoid.

Rhodopsins Derived from Allenic Retinals[31]

The allenic retinals shown in Fig. 39.9 have two chiral centers at C-5 and C-7 and hence exist as diastereomeric pairs as well as enantiomers. In addition, the allenic bond requires the side chiain to be at right angles to the cyclohexene ring, a feature that makes the general shape of the allenic retinals similar to that of natural retinals, which are distorted around the C-6/C-7 bond (Figs. 39.3 and 39.4). The allenic retinals are therefore unique systems for investigating the steric requirements of the ring-binding site in opsin.

Fig. 39.9 shows the hplc trace of syntetic allenic retinals. The central 9-*cis* peak is split into two; the same applies to the intense all-*trans* peak when the detector sensitivity is lowered. The 13-*cis* and 9,13-di-*cis* peaks also split upon repetition of hplc. Fig. 39.10 is the 270 MHz nmr spectrum of the strongest peak. Although it may not be clear in the reduced figure, all signals are single, except for splitting due to coupling, apart from the asterisked peaks which appear as pairs. The chemical shift and *J* values show that the isomer is the all-*trans* isomer, and that the asterisked peaks are due to the diastereomeric differences. Namely, these peaks are assignable to 1-Me, 9-Me and 8-H; it is clear from the diastereomeric structures shown in Fig. 39.10 that it is these protons which are in different environments and therefore absorb at differet chemical shifts. All other isomers separated by hplc were structurally identified and shown to be diastereomeric mixtures.

---

* A cationic micellar detergent (cetyltrimethylammonium bromide).

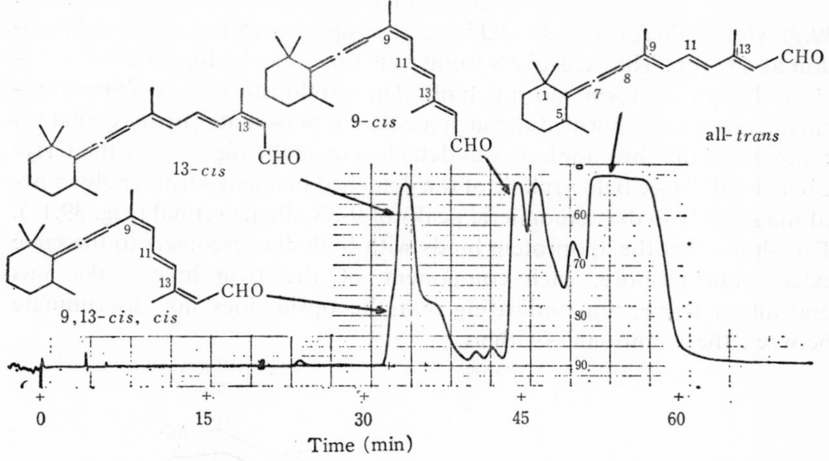

Fig. 39.9. Separation of 5-hydro-6,7-allenic retinals by hplc. [μ-Porasil (1 ft $\times$ 0.25 in; $\times$ 2); 2 ml/min; 1% Et_2O in hexane

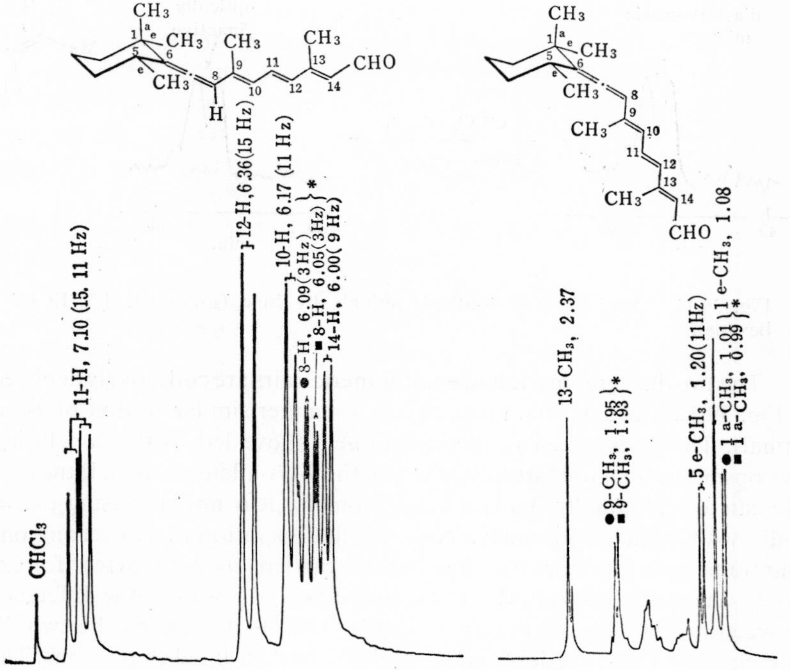

Fig. 39.10. Nuclear magnetic resonance spectra (270MHz, in $CDCl_3$) of diastereomeric all-*trans*-5-hydro-6,7-allenic retinals.

Incubation of the 9-*cis* diastereomeric mixture (central peak in Fig. 39.9) with opsin fortunately yielded a rhodopsin with the longest absorption at 455 nm. The 9,13-di-*cis* isomer also gives a rhodopsin but the 13-*cis* and all-*trans* forms do not bind. The rhodopsin formed from 9-*cis*-allenic retinal was purified through a calcium phosphate (hydroxyapatite) column and the chromophore was detached from the pigment by the CH_2–Cl_2 method[24] (see below); hplc of the detached pigment surprisingly showed the same twin diastereomieric peaks of 9-*cis* allenic retinal (Fig. 39.11). This shows that the apoprotein binds with both diastereomers to the same extent. Furthermore, each component of the twin hplc peaks was optically inactive, thus indicating that the opsin does not discriminate between the enantiomers either.

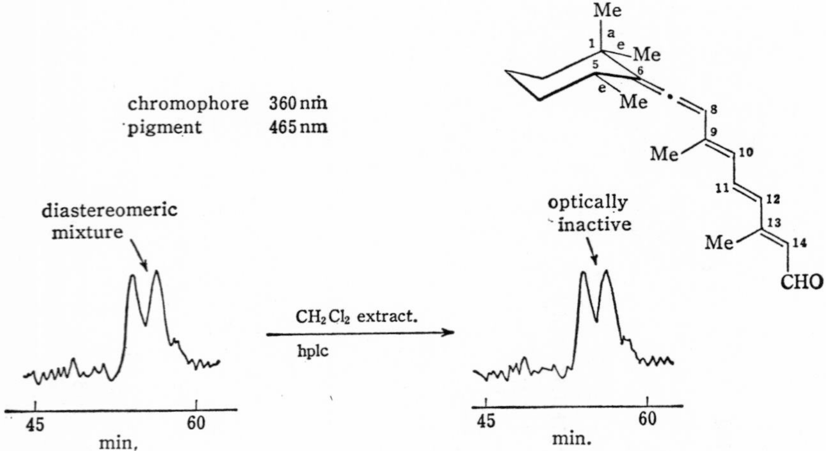

Fig. 39.11. Analysis of 6-*cis*-allenic retinals by hplc (μ-Porasil; 1% Et O/hexane).

The two diastereomeric and enantiomeric pairs are collectively depicted in Fig. 39.12 (the side chain is drawn in a manner similar to that of 9-*cis*-retinal). The unexpected results mentioned above led to the conclusion that opsin binds all four stereoisomers of the 9-*cis*-allenic retinal. If the protein can accept equally the two diastereomers, it is not surprising that it binds with both enantiomers. Namely, the operation of converting one enantiomer into its mirror image can be carried out by moving, for example, the dotted 1-axial Me to the dotted 5-axial position; the difference between the two mirror images is smaller than that between the two diastereomers, which are depicted as solid and broken-lined chair forms. The experiments with allenic retinals show that, unlike most other proteins, the the structural stereospecificity of the opsin molecule is quite unrestrictive

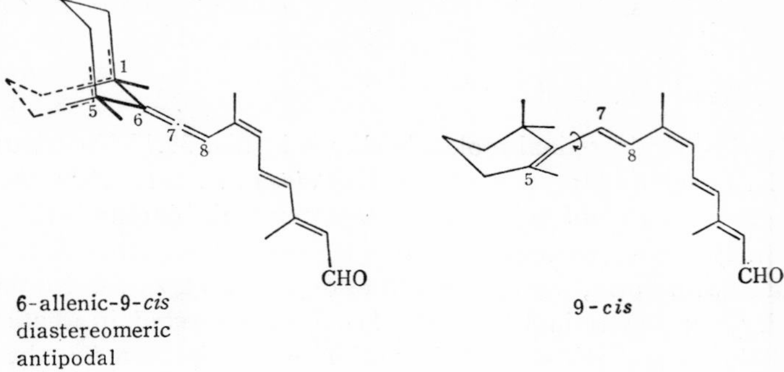

Fig. 39.12. 6,7-Allenic-9-*cis*-retinals and 9-*cis*-retinal. The two chair-form cyclohexanes indicate the two diastereomers. The operation of transferring 1-ax-Me (in the dotted cyclohexane structure) to 5-ax-Me leads to the enantiomer.

as far as the "β-ionone binding site" is concerned.[31,32] Insertion of an extra CH_2 group between the tips of the two chairs (or in the opened jaw of the crocodile) gives an adamantylallenic retinal. We have recently

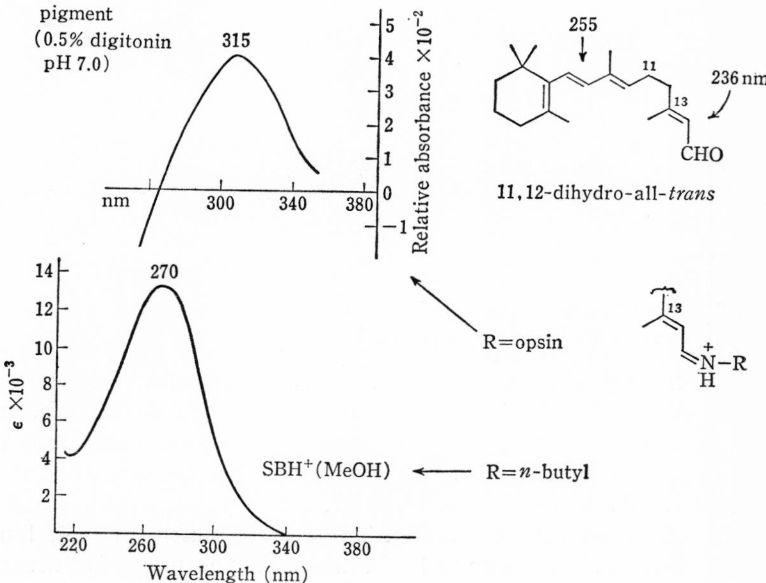

Fig. 39.13. The uv spectra of 11,12-dihydroretinal and 11,12-dihydrorhodopsin.

synthesized this molecule to further check the properties of the binding site;[33] binding studies have not yet been performed.

A Nonbleachable Rhodopsin[34]

The 11-dihydroretinal molecule shown in Fig. 39.13 has a saturated C-11/C-12 bond and therefore has neither *cis* nor *trans* forms; moreoever, the single bond should be flexible enough to permit binding of this retinal to opsin. 11-Dihydroretinals were synthesized and the all-*trans* form was incubated with opsin, whereupon a rhodopsin with a λ_{max} of 315 nm was formed. Since this retinal lacks the 11-ene, as expected, irradiation with white light did not change the absorption of the pigment, *i.e.*, the 11,12-dihydrorhodopsin is not bleached by room light. If the 315 nm peak is indeed due to the protonated Schiff base linkage of the 233 nm enal chromophore, the large red-shift is similar to that encountered in the natural series, *i.e.*, 500 nm for rhodopsin and 376 nm for 11-*cis*-retinal. Further studies on this large bathochromic shift may cast further light on the enigmatic red-shift which has been one of the central problems in vision science. Further studies with dihydroretinals and dihydrorhodopsins are in progress (see Addendum Refs. C,D,E,H).

39.2 Extraction of Bovine Rhodopsin and Other Experiments

Bovine retinae*1 can either be purchased*2 in lots of 50 which are prepared from dark-adapted cattle before slaughtering or can be collected from slaughterhouses. Next, an experiment with 100 rod outer segments is described (adapted after Papermaster and Dreyer[35]).

[Experimental procedure 1] Isolation of rod outer segments (R.O.S.)
The following solutions were prepared.

1) 10 mM Tris acetate buffer, pH 7.4 (1,600 ml).
2) Homogenate solution (350 ml). Sucrose 34% (w/v), 2 mM magnesium acetate, 65 mM sodium chloride dissolved in 5 mM Tris acetate buffer, pH 7.
3) 67 mM phosphate buffer, pH 7.
4) Sucrose gradient solution. Tris acetate buffer, pH 7.4, 1 mM solution is made up to 1 mM magnesium acetate, then sucrose is added as follows (w/v %):

---

*1 Not for culinary consumption!
*2 Hormal Corporation, Austin, Minn., U.S.A.

No. 1 gradient, 35% (50 ml);
No. 2 gradient, 26% (90 ml);
No. 3 gradient, 24.5% (50 ml).

All operations were carriedt at 0–4° C under dim red light (see Table 39.1).

Steps 1 and 2: Retinae (100) and 200 ml of homogenate solution were placed in a 250 ml Erlenmeyer flask with a stopper and shaken

TABLE 39.1. Isolation of rod outer segments (steps 1–12 correspond to those described in the text)

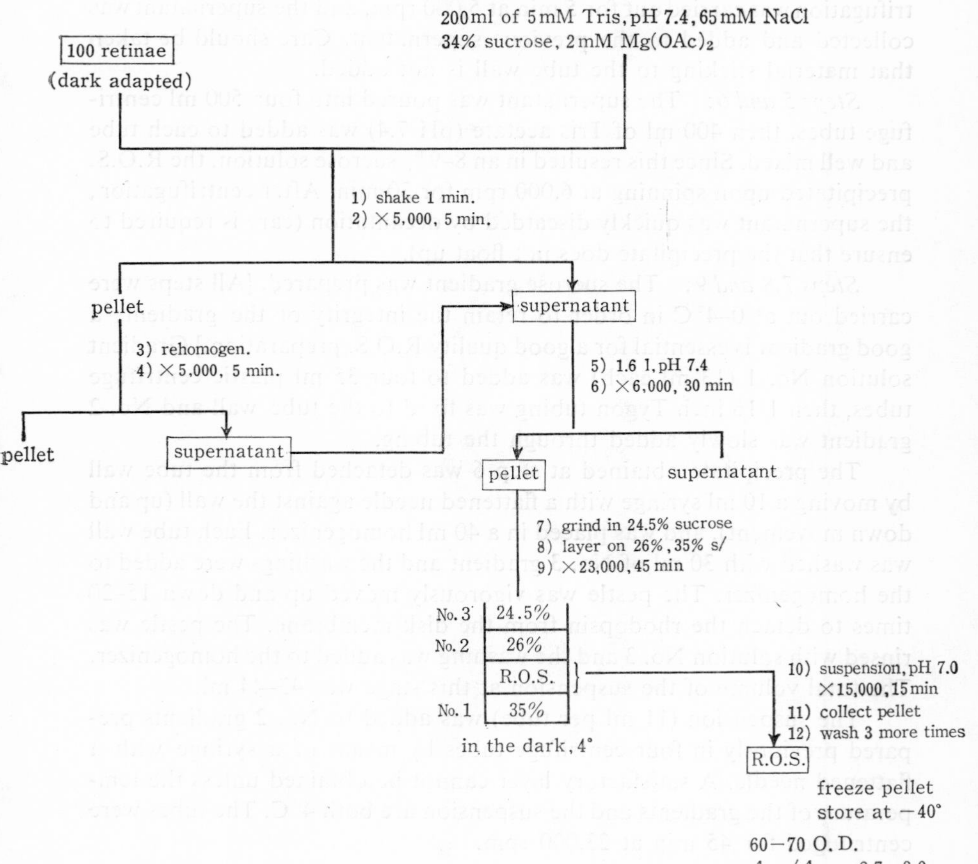

vigorously for 1 hr. [This process cleaves the narrow portion (Fig. 39.1) connecting the R.O.S. to the rod inner segment. If shaken too vigorusly, the cell fragments become too fine and render subsequent purification difficult. The shaking is accompanied by foaming.] The homogenate was placed in 6 centrifuge tubes of 40 ml, and centrifuged for 5 min at 5,000 rpm. The supernatant was collected. Some R.O.S. adhered to the tube walls but since this was contaminated, it was added to the precipitate (pellet).

Steps 3 and 4: To each centrifuge tube, 10–15 ml of homogenate solution was added. R.O.S. was transferred to the supernatant* by gently pressing the agar-like precipitate three times with a Teflon pestle. Centrifugation was carried out for 5 min at 5,000 rpm, and the supernatant was collected and added to the previous supernatant. Care should be taken that material sticking to the tube wall is not added.

Steps 5 and 6: The supernatant was poured into four 500 ml centrifuge tubes, then 400 ml of Tris acetate (pH 7.4) was added to each tube and well mixed. Since this resulted in an 8–9% sucrose solution, the R.O.S. precipitated upon spinning at 6,000 rpm for 30 min. After centrifugation, the supernatant was quickly discatded by decantation (care is required to ensure that the precipitate does not float up).

Steps 7,8 and 9: The sucrose gradient was prepared. [All steps were carried out at 0–4° C in order to retain the integrity of the gradient; a good gradient is essential for a good quality R.O.S. preparation.] Gradient solution No. 1 (13 ml each) was added to four 35 ml plastic centrifuge tubes, then 1/16 inch Tygon tubing was fixed to the tube wall and No. 2 gradient was slowly added through the tubing.

The precipitate obtained at step 6 was detached from the tube wall by moving a 10 ml syringe with a flattened needle against the wall (up and down movement), and was placed in a 40 ml homogenizer. Each tube wall was washed with 30 ml of No. 3 gradient and the washings were added to the homogenizer. The pestle was vigorously moved up and down 15–20 times to detach the rhodopsin from the disk membrane. The pestle was rinsed with solution No. 3 and the washing was added to the homogenizer. The total volume of the suspension at this stage was 43–44 ml.

The suspension (11 ml per tube) was added to No. 2 gradients prepared previously in four centrifuge tubes by means of a syringe with a flattened needle. A satisfactory layer cannot be obtained unless the temperature of the gradients and the suspension are both 4° C. The tubes were centrifuged for 45 min at 23,000 rpm.

---

* R. O. S. floats when the sucrose concentration exceeds 34–35%, and sinks when it is less.

Steps 10, 11 and 12: The dark red band (since the experiment was carried out under red light the band looked black) between No. 1 and No. 2 gradients was transferred to a 40 ml centrifuge tube with the 10 ml syringe/flattened needle. Care must be taken to avoid contamination of the transferred preparation with impure rhodopsin present between No. 2 (26% layer) and No. 1 (24.5% layer) or with the layer containing nerve tissues. Some loss of pure rhodopsin must be accepted in order to avoid such contamination. The rhodopsin was divided into six 40 ml centriguge tubes. Each tube was treated with 67 mM phosphate buffer to decrease the sucrose concentration, and the solutions were mixed well and centifuged for 15 min at 15,000 rpm. The supernatant was discarded. This process was repeated three times. The final R.O.S. precipitate was added to 67 mM phosphate buffer and stored at -40° C.

The yield was 6–70 O.D.*1 per 100 retinae, and the purity, A_{280}/A_{500}, was 2.7–3.0.*2

[Experimental procedure 2] Purification of rhodopsins

The following is based on the procedures of Shichi,[36] Ebrey[37] and co-workers (see Table 39.2).

The following buffers were prepared from 1 N HCl and were adjusted to pH 7. ALO denotes Ammonyx LO; DTT stands for dithiothreitol (Cleland reagent), which was added to the buffer the previous day or on the day of usage. NaPi denotes sodium phosphate buffer.

1) Dissolving buffer: 3% ALO, 10 mM imidazole, 15 mg DTT (100 ml).
2) Starting buffer: 1% ALO, 10 mM imidazole, 150 mg DTT (1 liter).
3) No. 1 Washing buffer: 1% ALO, 40 mM NaPi, 75 mg DTT (500 ml).
4) No. 2 Washing buffer: 1% ALO, 150 mM NaPi, 75 mg DTT (500 ml).
5) Imidazole buffer: 10 mM imidazole, 15 mg of DTT (100 ml).

A calcium phosphate (HTP) column was prepared as follows. HTP (2 g) (Bio-Rad Co.) was added to 100 ml of starting buffer and the flask was shaken gently; stirring rods and stirrers should be avoided since the particles are destroyed. The solution was left to stand for 15–20 min, then the supernatant was decanted off. The starting buffer (100 ml) was added and the supernatant was again discarded. The same volume of starting buffer was then added to form a suspension.

A 25 cm x 1.5 cm column was half-filled with the starting buffer, then

*1 Optical density or absorbance. For natural bovine rhodopsin, this was measured at the longest-wavelength maximum, 500 nm.

*2 In Fig. 39.14 (rhodopsin absorption spectrum), the 280 nm band consists of absorptions from retinal and opsin whereas the 500 nm band is solely due to retinal absorption. The contaminants in rhodopsin preparations are exotic proteins which absorb at 280 nm. Experimentally, the purest rhodopsin preparation gives a value of *ca.* 1.8.

the HTP suspension was added and allowed to settle for at least 2 hr. The supernatant on the column bed[*1] was removed with a pipet, to leave a 4 mm depth of liquid. The column was washed with twice the bed volume of starting buffer (*ca.* 75 ml) with 1 m × 0.3 cm Tygon tubing. No air bubbles should be formed, and the bed should not be exposed to air.

Phosphate buffer containing 25 O.D. rhodopsin was centrifuged (15 min, 12,000 rpm) and the precipitate was washed several times with imidazole buffer to remove the phosphate solution. The rhodopsin was dissolved in the minimum amount of dissolving buffer (3% ALO), the O.D. was measured and the solution was applied to the column. The column was washed with *ca.* 30 ml of starting buffer[*2]; No, 1 washing buffer was added when the level of starting buffer fell to 4 mm above the bed. Aliquots of 3 ml were collected and monitored by absorption measurement. The unwanted lipids, proteins, and (in the case of regenerated rhodopsin) excess chromophore are eluted with this solution. The absorption spectra of the elutes were taken. When a 500 rm maximum began to appear, *i.e.*, when rhodopsin emerges, the column was eluted with No. 2 washing buffer (30–50 fractions were usually collected with No. 1 buffer). The 500 nm maximum disappeared and the chromatography was completed after about 35 fractions had been collected with the No. 2 washing buffer. Yield of rhodopsin, 60%; half of this was eluted with the No. 1 buffer, and the other half with the No. 2 buffer. The purest rhodopsin was eluted around the change-over point from No. 1 to No. 2 solution, the A_{280}/A_{500} ratio being approximately 1.8.

TABLE 39.2. The purification of rhodopsins (under dim red light at 0° C)

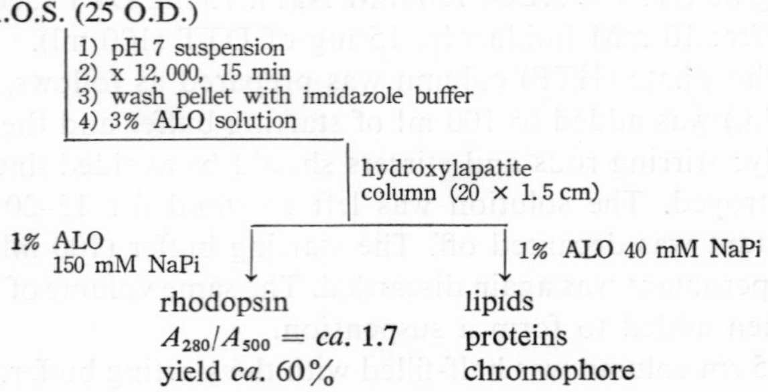

Fig. 39.14 shows the absorption spectrum of rhodopsin regenerated from 14-methyl–11-*cis*-retinal. The A_{280}/A_{500} vlaue is 1.8, so that this is a

[*1] The column portion containing the filling (or support) is called the bed.
[*2] Some retinals, e.g., 11-dihydroretinal, are eluted at this stage by excess detergent.

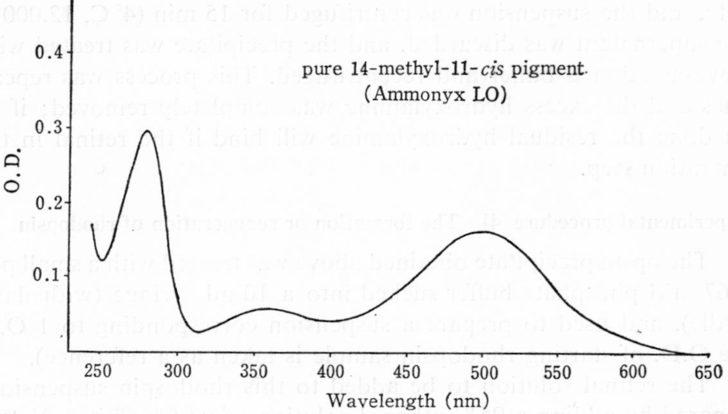

Fig. 39.14. Absorption spectrum of 14-methyl-11-*cis*-rhodopsin.

rhodopsin preparation of very high purity. The spectrum of purified natural rhodopsin has a similar shape.

[Experimental procedure 3] Photobleaching of rhodopsin*

Take the requisite O.D. amount of rhodopsin in the dark.

For example, if the rhodopsin concentration is 4.5 O.D./ml in 67 mM phosphate buffer suspension, and 18 O.D. is necessary, then take 4 ml of the suspension. To this is added 1 μl of pH 6.5 1 M NH_2OH hydrochloride solution per 1 O.D. rhodopsin, and shake the test tuble in front of a sun lamp. The dark red solution bleached to a pale-yellow color after a few minutes [i.e., the 11-*cis* chromophore isomerizes to all-*trans*, is relaeased from the opsin (the 500 nm absorption is shifted to 370 nm), and is bound in the form of an oxime by the NH_2OH (Table 39.3)].

Phosphate buffer (67mM) was added when the bleaching was com-

TABLE 39.3. Photobleaching of rhodopsins

e.g., 4 ml of 4.5 O.D./ml solution (18 O.D.)
↓ 1) 18μl of 1 M $NH_2OH \cdot HCl$, pH 6.5
2) $h\nu$ few minutes
bleached (opsin + all-*trans*-ret-oxime) pigment
↓ 3) suspension, pH 7, × 12,000, 15 min
4) collect pellet
5) repeat 5 more times
opsin

* It is only at this stage in experiments dealing with retinals and rhodopsins that one can operate safely in the light. In most other stages, experiments are carried out under dim red light at temperatures below 4° C, *i.e.* under conditions in a bar in winter with the heater off.

plete, and the suspension was centrifuged for 15 min (4° C, 12,000 rpm). The supernatant was discarded, and the precipitate was treated with the above-mentioned buffer and recentrifuged. This process was repeated 6 times and the excess hydroxylamine was completely removed; if this is not done the residual hydroxylamine will bind if the retinal in the regeneration step.

[Experimental procedure 4] The formation or regeneration of rhodopsin

The opsin precipitate obtained above was treated with a small portion of 67 mM phosphate buffer sucked into a 10 ml syringe (with flattened needle), and used to prepare a suspension corresponding to 1 O.D./ml (the O.D. of starting rhodopsin sample is taken as a reference).

The retinal solution to be added to this rhodospin suspension was prepared by adding a 95% ethanol solution of 0.05% Triton X-100 followed by 1 ml of 67 mM phosphate buffer to 1 mg of retinal. This solution was diluted 10-fold and 100–fold with methanol, and the retinal concentration was determined from the ε values. With most model retinals a satisfactory concetration for UV measurements is obtained by 100-fold dilution.

The opsin suspension prepared previously was divided into two 1 O.D. fractions and the remainder (Table 39.4). Nothing was added to (1); 2 O.D. of 9-*cis*-retinal was added to (2) as a standard; the test retinal was added to (3), the amount typically being 40–50 O.D. of test retinal to 10 O.D. opsin. The air in each test-tube was replaced with nitrogen or argon in the dark, then the tubes were stoppered and shaken gently for 3 hr in the dark. Unnatural retinals usually require incubation for more than 3

TABLE 39.4. Regeneration of rhodopsins

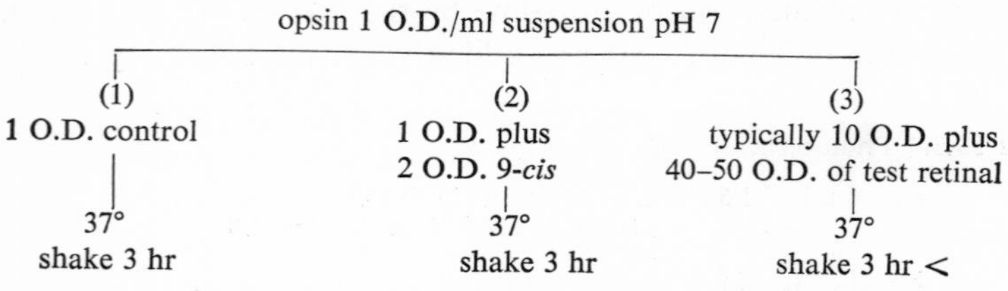

× 12, 000, 10 min

pellet

supernatant

1) 1 ml of 2% ALO in KPi, pH 7,650–250 nm abs.
2) bleach and remeasure
3) subtract curve 2 from 1

hr. 7-*cis*-Retinals form rhodopsins sluggishly and require a 150–fold longer incubation than 11-*cis*-retinal.[38)]

After incubation, 1 ml was taken from each tube and centrifuged for 10 min at 12,000 rpm in the dark at 4° C. The supernatant was removed with a pipet, 1 ml of 2% ALO potassium phosphate buffer added was to the precipitate and the absorption in the range of 250–650 nm was recorded. After measurement, the pigment was photobleached, the absorption remeasured and the difference curve taken. This is the absorption spectrum of rhodopsin. The control solution (1) is necessary to check whether the original rhodopsin is completely bleached or not, whereas solution (2) is required to ascertain the regeneration properties of the opsin preparation.

39.3 Detachment of Retinals from Rhodopsins and their Identification

Rhodopsins are not only unstable to light but also to temperature; at room temperature the rhodopsins gradually bleach even in the dark. On the other hand, it is extremely important to identify the retinal after detachment from rhodopsin without isomerizing the double bond. The first characterization of 11-*cis*-retinal as the rhodopsin chromophore by Wald and co-workers was carried out by bringing the pigment to room temperature and then identifying the detached chromophore. However, since most model retinals are more temperature-sensitive, retinals isolated by photobleaching or thermal bleaching probably will not reflect correctly the geometrical isomer(s) which had been bound to opsin. Furthermore, as in the case of bacteriorhodopsin (see below) the apoprotein may be bound to a mixture of retinals.

We have recently discovered a simple detachment/identification method which employs neither light nor heat and which can be carried out in the presence of detergents that usually damage expensive hplc columns.[39)]

[Experimental procedure 5] The detachment of retinal

A rhodopsin suspension was shaken with CH_2Cl_2 and centrifuged. The CH_2Cl_2 layer was collected, dried and subjected to reverse phase hplc. Presumably the rhodopsin is denatured during the shaking with CH_2Cl_2, so that the Schiff base linkage is exposed to water and undergoes spontaneous hydrolysis.

Initially the hplc separation was carried out with μ-Porasil columns, which were repidly damaged by the detergents and other polar substances extracted into the CH_2Cl_2 layer. However, since this was the only

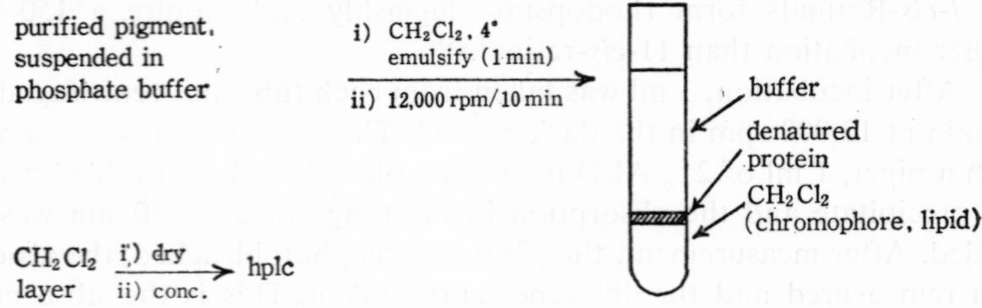

Fig. 39.15. Detachment of retinal using neither light nor heat.

mild method available to quantify retinals*1 in their original geometric isomeric forms, it was applicable to many problems. Some examples are shown in Fig. 39.16.[22] As expected, regenerated rhodopsin and isorhodopsin gave single peaks corresponding to 11-*cis*- and 9-*cis*-retinals, respectively. Although 9,13-di-*cis*-retinal was reported not to bind to opsin, our experiments with 14-methyl-retinal unexpectedly showed that the di-*cis*-isomer of this product did bind to opsin. Experiments with 9,13-di-*cis*-retinal then showed that this also bound to opsin.[18)*2] It is conceivable

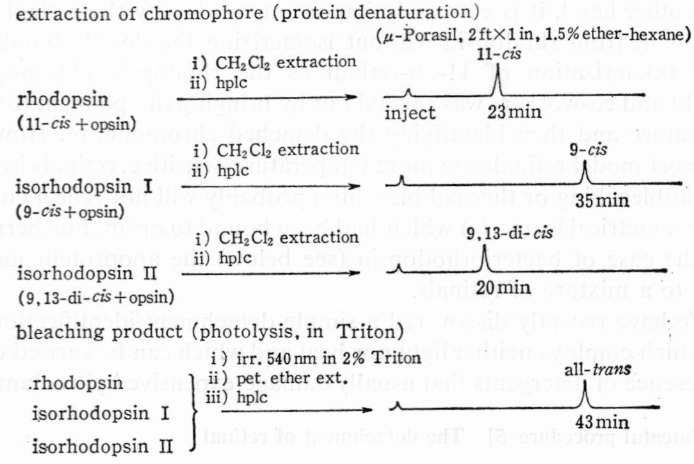

Fig. 39.16. The CH_2Cl_2/hplc method. Except for the usage of μ-Porasil, the results are similar to those of Fig. 39.18.[22]

---

*1 The retinals can be quantified by dividing the hplc peak areas by the appropriate ε values.

**2 We initially named the new rhodopsin isorhodopsin II. However, subsequent studies by us and others have led to a large number of artificial rhodopsins. It is more appropriate to identify the pigment by its specific retinal; thus, in the present case the pigment should be named 9,13-di-*cis*-rhodopsin.

that a pigment could have been formed from 9,13-di-*cis*-retinal as a result of its partial isomerization to 9-*cis*-retinal during incubation; however, this is clearly ruled out by the third run in Fig. 39.16. The CH_2Cl_2/hplc method has played an important role in the analysis of photochemical products.[22,26,27)]

Column Packing Materials

It is worthwhile to give some thought to column packing material at this stage (Fig. 39.17). The schematic drawing may not be accurate but is convenient to grasp the concept. The normal phase chromatography employs hexane/ether for the separation of retinals[18,40,41)] (Fig. 39.6). The polar aldehyde group is loosely bound to the OH and CN groups of silica gel, μ-CN and other particles; the weakly nucleophilic double bonds of various retinal isomers can also be regarded as being involved in weak bonding with the OH and CN groups. The affinity to the packing material will depend on the *cis/trans* isomerism, the weakly bonded retinals (e.g., the bent retinal depicted on the far left, Fig. 39.17) being eluted more readily. Since μ-Porasil and Corasil columns are damaged by polar solvents such as methanol and water, we checked whether columns such as μ-CN and C_{18}, which can be used either in the normal or reverse phase mode, could be employed for retinal separation.

We found that μ-CN columns, which are used for separating compounds lacking polar founctionalities, could not separate retinal isomers.

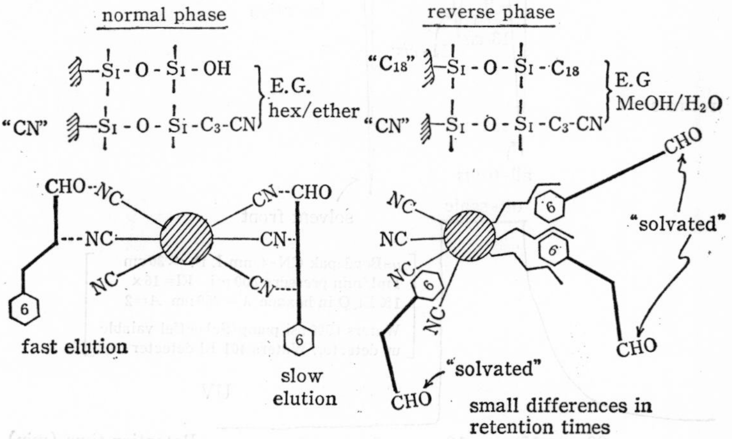

Fig. 39.17. Schematic drawing of retinal separation by normal and reverse phase hplc procedures.

This can be qualitatively explained by the right-hand drawing in Fig. 39.17, Thus, the hydrophobic cyclohexene ring is embedded hydrophobically in the particles so that the polar side chain is exposed. However, the aldehyde group is solvated in polar systems such as aqueous MeOH, and the extent of polarization is more or less independent of the *cis/trans* isomerism; furthermore, the hydrophobic bonding of the ring moiety to the particles is also unaffected by *cis/trans* isomerism. Hence separation of retinals by reverse phase hplc is not satisfactory. A combination of normal and reverse phase hplc with μ-CN was therefore investigated, and this gave satisfactory results.[39]

[Experimental procedure 6] Separation of a retinal mixture

A suspension of rhodopsin containing detergents and other polar impurities was emulsified by rapidly pumping in and out CH_2Cl_2 through a syringe at 0° C, and the emulsion was centriguged for 10 min at 10,000 rpm, at 4° C. The CH_2Cl_2 layer was dried over anhydrous sodium sulfate, and filtered. The solvent was removed *in vacuo*, then the residue was treated with 20 μl of MeOH and the solution was injected into the hplc machine. Usage of hexane/ether results in adsorption of the detergent on top of the column and elution of only the retinals. Combined detection has

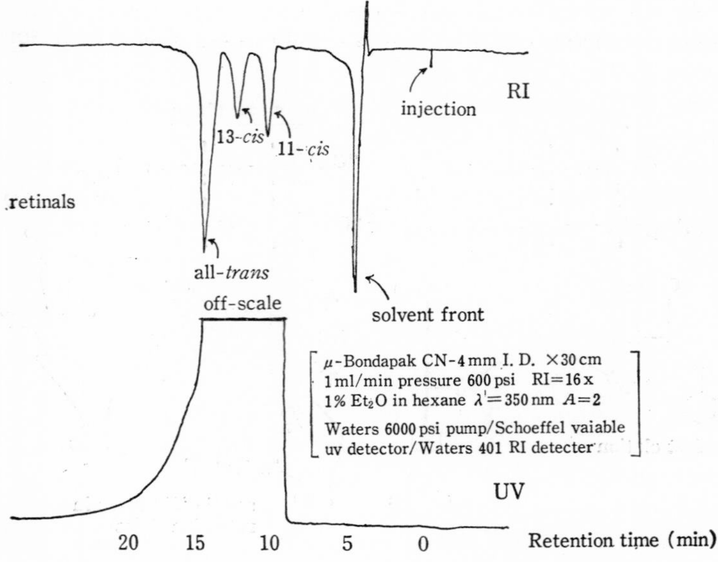

Fig. 39.18. Separation of retinals with a μ-CN column. The column is not damaged by membranes and detergents.

the advantage that the lower sensitivity of RI allows separate detection of isomers in cases when the UV absorbance goes off-scale (Fig. 39.18). The μ-CN column can be repeatedly used for retinal separation. The polar materials adsorbed can be eluted by occasional passage of a polar solvent. The column can be regenerated by passage of 100 ml each of chloroform, methanol and water, followed by washing with methanol and chloroform, and finally with a mixed hexane/ether system. The retention times of various retinals are not affected by column regeneration.

This method is very simple and can be applied to the analysis of various rhodopsins of obscure retinal composition.

Chromophore of Bacteriorhodopsin

Finally the application of the above procedures to bacteriorhodopsin will be briefly described. Although this is not a visual pigment, bacteriorhodopsin[42)] has attracted great interest in recent years because it is contained in the purple membrane of a halophilic bacterium, *Halobacterium halobium*, which is one of the simplest membrane systems involved in converting light into chemical energy; it is thus a good model for understanding membranes. The bacteriorhodopsin has two cycles, involving the light-adapted (bR^{LA}_{570}, λ_{max} at 570 nm) and dark-adapted (bR^{DA}_{560}, λ_{max} 560 nm) forms. As in the case of rhodopsins, the cycles have several intermediates, which in turn are involved in the proton pumping required for converting solar energy into chemical ATP synthetic energy. The results of analysis by the CH_2Cl_2 method have shown[43)] that bR^{LA}_{570} contains all-*trans*, that bR^{DA}_{560} contains a 1:1 mixture of all-*trans* and 13-*cis*, and that an important intermediate designated M_{412} contains 13-*cis*-retinal as the chromophore. The implications of these findings are not clear at this stage, but they show that the *cis-trans* isomerization of the chromophore plays a crucial role in the proton pumping.

39.4 Concluding Remarks

We initiated our studies on visual pigments in 1973 with the guidance and help of Drs. Tom Ebrey and Barry Honig, who were both in the Biology Department of Columbia University. Our "team" consisted of only one graduate student, Wan Kit Chan, who single-handedly started the synthesis, purification, etc., of model retinals. The number of workers in this area has since expanded so that it is now one of our major research topics. Although it has taken us years to become accustomed to the handling of visual pigments, the investigation of visual pigments and the related bacteriorhodopsin is most challenging and intriguing becase of its totally interdisciplinary nature, encompassing the fields of biology, physics, chemistry, molecular biology, electrophysiology, biochemistry, etc. We regard this as a new branch of natural products chemistry and hope to contribute to its clarification via a basically organic chemical approach. I am grateful to my colleageues whose works are quoted in the references.

References

1) H. Saibil, M. Chabre, D. Worcester, *Nature*, **262**, 266 (1976).
2) T. Yoshizawa, Y. Kito, *ibid.*, **201**, 340 (1958).
3) G. E. Busch, M. L. Applebury, A. A. Lamola, P. M. Rentzepis, *Proc. Natl. Acad. Sci. U.S.A.*, **69**, 2802 (1972).
4) T. Yoshizawa, S. Horiuchi, *Biochemistry and Physiology of Visual Pigments* (ed. H. Langer), p. 69, Springer Verlag (1973).
5) J. Rapp, J. Wiesenfeld, E. W. Abrahamson, *Biochem. Biophys. Acta*, **201**, 119 (1970).
6) A. Warshel, *Nature*, **260**, 679 (1976); M. R. Fransen, W. C. M. M. Luyten, J. van Thuijl, J. Lugtenberg, P. A. A. Jansen, P. J. G. M. van Breugel, F. J. M. Daemen, *ibid.*, **260**, 726 (1976); K. van der Meer, J. J. C. Mulder, J. Lugtenberg, *Photochem. Photobiol.*, **23**, 363 (1976).
7) G. Wald, *The Nobel Lectures*, 1967, The Nobel Foundation (1968).
8) R. Hubbard, G. Wald, *J. Gen. Physiol.*, **36**, 269 (1952); G. Wald, *Nature*, **219**, 800 (1968).
9) D. Bounds, *ibid.*, **216**, 1178 (1967); M. Akhtar, P. T. Blosse, P. B. Dewhurse, *Chem. Commun.*, **1967**, 13.
10) R. P. Poincelot, E. W. Abrahamson, *Biochemistry*, **9**, 1820 (1970).
11) R. S. Fager, P. Sejnowski, E. W. Abrahamson, *Biochem. Biophys. Res. Commun.*, **47**, 1244 (1972).
12) A. Lewis, R. S. Fager, E. W. Abrahamson, *J. Raman Spectry.*, **1**, 465 (1973); A. R. Oseroff, R. H. Callender, *Biochemistry*, **13**, 4243 (1974); R. Mathies, A. R. Oseroff, L. Stryer, *Proc. Natl. Acad. Sci. U.S.A.*, **73**, 1 (1976); R. H. Callender, A. Doukas, R. Crouch, K. Nakanishi, *Biochemistry* **15**, 1621 (1976).
13) M. Suzuki, T. Komatsu, H. Kitajima, *J. Phys. Soc. Japan*, **37**, 177 (1974); T. Komatsu, H. Suzuki, *ibid.*, **40**, 1725 (1976).
14) R. Gilardi, I. L. Karle, J. Karle, W. Sperling, *Nature*, **232**, 187 (1971).
15) T. Hamanaka, T. Mitsui, T. Ashida, M. Kakudo, *Acta Crystallogr.* B**28**, 214 (1972).
16) B. Honig, M. Karplus, *Nature*, **229**, 558 (1971); B. Honig, A. Warshel, M. Karplus, *Acc. Chem. Res.*, **8**, 92 (1975).
17) R. Rowan, A. Warshel, B. D. Sykes, M. Karplus, *Biochemistry*, **13**, 970 (1974).
18) W. K. Chan, K. Nakanishi, T. G. Ebrey, B. Honig, *J. Am. Chem. Soc.*, **96**, 3642 (1974).
19) H. Mayer, O. Isler, *Carotenoids* (ed. L. Isler), p. 325, Birkhaüser Verlag (1971).
20) T. Mukaiyama, A. Ishida, *Chem. Lett.*, **1975**, 1201.
21) G. Stork, G. A. Kraus, *J. Am. Chem. Soc.*, **98**, 2351 (1976).
22) R. Crouch, V. Purvin, K. Nakanishi, T. Ebrey, *Proc. Natl. Acad. Sci. U.S.A.*, **72**, 1538 (1975).
23) T. Ebrey, R. Govindjee, B. Honig, E. Pollock, W. K. Chan, R. Crouch, A. Yudd, K. Nakanishi, *Biochemistry*, **14**, 3933 (1975).
24) F. G. Pilkiewicz, M. J. Pettei, A. P. Yudd, K. Nakanishi, *Exptl. Eye Res.*, **24**, 421 (1977).

25) N. Harada, K. Nakanishi, *Acc. Chem. Res.*, 5, 257 (1972); N. Harada, S. L. Chen, K. Nakanishi, *J. Am. Chem. Soc.*, **97**, 5345 (1975).
26) W. H. Waddell, A. P. Yudd, K. Nakanishi, *ibid.*, **98**, 238 (1976).
27) W. H. Waddell, R. Crouch, K. Nakanishi, N. J. Turro, *ibid.*, **98**, 4189 (1976).
28) K. Nakanishi, *Pure Appl. Chem.*, **49**, 333 (1977).
29) M. Tsuboi, *Gendai Kagaku*, **47**, 40 (1975).
30) A. Warshel, M. Karplus, *J. Am. Chem. Soc.*, **96**, 5677 (1974).
31) K. Nakanishi, A. P. Yudd, R. K. Crouch, G. L. Olson, H.-C. Cheung, R. Govindjee, T. G. Ebrey, D. J. Patel, *ibid.*, **98**, 236 (1976).
32) H. Matsumoto, T. Yoshizawa, *Nature*, **258**, 523 (1975).
33) V. Balogh-Nair, K. Nakanishi, unpublished. See Addendum Ref. G.
34) M. Gawinowicz, J. Sabol, V. Balogh-Nair, K. Nakanishi, *J. Am. Chem. Soc.*, **99**, 7720 (1977).
35) D. S. Papermaster, W. J. Dreyer, *Biochemistry*, **13**, 2438 (1974).
36) H. Shichi, M. Lewis, F. Irreverre, A. Stone, *J. Biol. Chem.*, **244**, 529 (1969).
37) T. Ebrey, *Vision Research*, **11**, 1007 (1971).
38) W. J. DeGrip, R. S. H. Lin, V. Ramamurthy, A. Asato, *Nature*, **262**, 416 (1976).
39) F. G. Pilkiewicz, M. J. Pettei, A. P. Yudd, K. Nakanishi, *Exptl. Eye Res.*, 421 (1977).
40) A. E. Asato, R. S. H. Liu, *J. Am. Chem. Soc.*, **97**, 4128 (1975).
41) J. P. Rotmans, A. Kropf, *Vision Research*, **15**, 1301 (1975).
42) H. Shichi, *Kagaku to Seibutsu*, **12**, 724 (1974); W. Stoeckinius, *Scientific American*, **234**, 38 (1976); D. Oesthehalt, *Angew. Chem. Int. Ed.*, **15**, 19 (1976); M. Sumper, H. Reitmeier, D. Oesterhalt, *ibid.*, **15**, 187 (1976).
43) M. J. Pettei, A. P. Yudd, K. Nakanishi, W. Stoeckinius, *Biochemistry*, **16**, 1955 (1977).

On visual pigments

a) T. Yoshizawa, F. Tokunaga, *Kagaku*, **43**, 610 (1973).
b) G. Wald, *Science*, **162**, 230 (1968).
c) T. Tomita, *Quart. Rev. Biophys.*, **3**, 179 (1970).
d) T. G. Ebrey, B. Honig, *ibid.*, **8**, 2 (1975).
e) W. A. Hagins, *Ann. Rev. Biophys. Bioenz.*, **1**, 131 (1972).
f) B. Honig, T. G. Ebrey, *ibid.*, **3**, 151 (1974).
g) H. Langer (ed.), *Biochemistry and Physiology of Visual Pigments*, Springer-Verlag (1973).
h) *Acc. Chem. Res.* (Special Issue on the Chemistry of Vision), 8 (1975). This special issue may be purchased for $3.00 from: Journal Dept., Am. Chem. Soc., 1155 Sixteenth St., N. W., Washington, D. C. 20036, U.S.A.).
i) O. Isler (ed.), *Carotenoids*, Birkhaüser Verlag (1971).

ADDENDUM (JULY, 1980)

This article is a direct translation from the 1976 Japanese version and has not been brought up-to-date. Subsequent publications from our laboratory are listed below.

A) A Convenient Synthesis of Stereochemically Pure Retinoids. The Synthesis of 10,14-Dimethyl Retinals. S. P. Tanis, R. H. Brown, K. Nakanishi, *Tetrahedron Lett.*, **869** (1978).
B) Double Point Charge Model for Visual Pigments; Evidence from Dihydro-

rhodopsins. K. Nakanishi, V. Balogh-Nair, M. A. Gawinowicz, M. Arnaboldi, M. Motto, B. Honig, *Photochem. Photobiol.*, **29**, 657 (1979).

C) Hydroretinals and Hydrorhodopsins. M. Arnaboldi, M. G. Motto, K. Tsujimoto, V. Balgoh-Nair, K. Nakanishi, *J. Am. Chem. Soc.*, **101**, 7082–7084 (1979).

D) An External Point-Charge Model for Wavelength Regulation in Visual Pigments. B. Honig, U. Dinur, K. Nakanishi, V. Balogh-Nair, M. A. Gawinowicz, M. Arnaboldi, and M. G. Motto, *ibid.*, **101**, 7084–7086 (1979).

E) Through-Space Electrostatic Effects in Electronic Spectra. Experimental Evidence for the External Point-Charge Model of Visual Pigments. M. Sheves, K. Nakanishi, B. Honig, *ibid.*, **101**, 7086–7088 (1979).

F) A Versatile Synthesis of Retinoids via Condensation of the Side-Chain to Cyclic Ketones. F. Derguini, V. Balogh-Nair, K. Nakanishi, *Chem. Commun.*, 977 (1979).

G) Adamantyl Allenic Rhodpsin. Leniency of the Ring Binding Site in Bovine Opsin. R. A. Blatchly, J. D. Carriker, V. Balogh-Nair, K. Nakanishi, *J. Am. Chem. Soc.*, **102**, 2495 (1980).

H) Non-bleachable Rhodopsins retaining the Full Natural Chromophire. H. Akita, S. P. Tanis, M. Adams, V. Balogh-Nair, K. Nakanishi, *J. Am. Chem. Soc.*, **102,** 6370. (1980)

I) An External Point-charge Model for Bacteriorhodopsin to Account for its Purple Color. K. Nakanishi, V. Balogh-Nair, M. Arnaboldi, K. Tsujimoto, B. Hong, *J. Am. Chem. Soc.,* **102,** 7947 (1980).

J) Opsin Shifts in Bovine Rhodopsin and Bacteriorhodopsin. Comparison of Two External Point-Charge Models. M. G. Motto, M. Sheves, K. Tsujimoto, V. Balogh-Nair, K. Nakanishi, *J. Am. Chem. Soc.,* **102,** 7945 (1980).

K) Incorporation of 11,12-Dihydroretinal into the Retinae of Vitamin A-Deprived Rats. R. Crouch, S. Katz, K. Nakanishi, M. A. Gawinowicz, V. Balogh-Nair, *Photochem. Photobiol*, **33,** 91 (1981).

Index of Bioassay

Index of Experimental Procedures

Plant and Animal Index

A

B

C

D

E

F

G

H

I

J

L

M

N

O

P

R

S

T

U

V

W

Y,Z

Subject Index

C

D

M

N

O

P

Q, R

S